Geophysical Monograph Series

Geophysical Monograph Series

232 **Microstructural Geochronology: Planetary Records Down to Atom Scale** *Desmond Moser, Fernando Corfu, James Darling, Steven Reddy, and Kimberly Tait (Eds.)*
233 **Global Flood Hazard: Applications in Modeling, Mapping, and Forecasting** *Guy Schumann, Paul D. Bates, Giuseppe T. Aronica, and Heiko Apel (Eds.)*
234 **Pre-Earthquake Processes: A Multidisciplinary Approach to Earthquake Prediction Studies** *Dimitar Ouzounov, Sergey Pulinets, Katsumi Hattori, and Patrick Taylor (Eds.)*
235 **Electric Currents in Geospace and Beyond** *Andreas Keiling, Octav Marghitu, and Michael Wheatland (Eds.)*
236 **Quantifying Uncertainty in Subsurface Systems** *Celine Scheidt, Lewis Li, and Jef Caers (Eds.)*
237 **Petroleum Engineering** *Moshood Sanni (Ed.)*
238 **Geological Carbon Storage: Subsurface Seals and Caprock Integrity** *Stephanie Vialle, Jonathan Ajo-Franklin, and J. William Carey (Eds.)*
239 **Lithospheric Discontinuities** *Huaiyu Yuan and Barbara Romanowicz (Eds.)*
240 **Chemostratigraphy Across Major Chronological Eras** *Alcides N.Sial, Claudio Gaucher, Muthuvairavasamy Ramkumar, and Valderez Pinto Ferreira (Eds.)*
241 **Mathematical Geoenergy: Discovery, Depletion, and Renewal** *Paul Pukite, Dennis Coyne, and Daniel Challou (Eds.)*
242 **Ore Deposits: Origin, Exploration, and Exploitation** *Sophie Decree and Laurence Robb (Eds.)*
243 **Kuroshio Current: Physical, Biogeochemical, and Ecosystem Dynamics** *Takeyoshi Nagai, Hiroaki Saito, Koji Suzuki, and Motomitsu Takahashi (Eds.)*
244 **Geomagnetically Induced Currents from the Sun to the Power Grid** *Jennifer L. Gannon, Andrei Swidinsky, and Zhonghua Xu (Eds.)*
245 **Shale: Subsurface Science and Engineering** *Thomas Dewers, Jason Heath, and Marcelo Sánchez (Eds.)*
246 **Submarine Landslides: Subaqueous Mass Transport Deposits From Outcrops to Seismic Profiles** *Kei Ogata, Andrea Festa, and Gian Andrea Pini (Eds.)*
247 **Iceland: Tectonics, Volcanics, and Glacial Features** *Tamie J. Jovanelly*
248 **Dayside Magnetosphere Interactions** *Qiugang Zong, Philippe Escoubet, David Sibeck, Guan Le, and Hui Zhang (Eds.)*
249 **Carbon in Earth's Interior** *Craig E. Manning, Jung-Fu Lin, and Wendy L. Mao (Eds.)*
250 **Nitrogen Overload: Environmental Degradation, Ramifications, and Economic Costs** *Brian G. Katz*
251 **Biogeochemical Cycles: Ecological Drivers and Environmental Impact** *Katerina Dontsova, Zsuzsanna Balogh-Brunstad, and Gaël Le Roux (Eds.)*
252 **Seismoelectric Exploration: Theory, Experiments, and Applications** *Niels Grobbe, André Revil, Zhenya Zhu, and Evert Slob (Eds.)*
253 **El Niño Southern Oscillation in a Changing Climate** *Michael J. McPhaden, Agus Santoso, and Wenju Cai (Eds.)*
254 **Dynamic Magma Evolution** *Francesco Vetere (Ed.)*
255 **Large Igneous Provinces: A Driver of Global Environmental and Biotic Changes** *Richard. E. Ernst, Alexander J. Dickson, and Andrey Bekker (Eds.)*
256 **Coastal Ecosystems in Transition: A Comparative Analysis of the Northern Adriatic and Chesapeake Bay** *Thomas C. Malone, Alenka Malej, and Jadran Faganeli (Eds.)*
257 **Hydrogeology, Chemical Weathering, and Soil Formation** *Allen Hunt, Markus Egli, and Boris Faybishenko (Eds.)*
258 **Solar Physics and Solar Wind** *Nour E. Raouafi and Angelos Vourlidas (Eds.)*
259 **Magnetospheres in the Solar System** *Romain Maggiolo, Nicolas André, Hiroshi Hasegawa, and Daniel T. Welling (Eds.)*
260 **Ionosphere Dynamics and Applications** *Chaosong Huang and Gang Lu (Eds.)*
261 **Upper Atmosphere Dynamics and Energetics** *Wenbin Wang and Yongliang Zhang (Eds.)*
262 **Space Weather Effects and Applications** *Anthea J. Coster, Philip J. Erickson, and Louis J. Lanzerotti (Eds.)*
263 **Mantle Convection and Surface Expressions** *Hauke Marquardt, Maxim Ballmer, Sanne Cottaar, and Jasper Konter (Eds.)*
264 **Crustal Magmatic System Evolution: Anatomy, Architecture, and Physico-Chemical Processes** *Matteo Masotta, Christoph Beier, and Silvio Mollo (Eds.)*
265 **Global Drought and Flood: Observation, Modeling, and Prediction** *Huan Wu, Dennis P. Lettenmaier, Qiuhong Tang, and Philip J. Ward (Eds.)*
266 **Magma Redox Geochemistry** *Roberto Moretti and Daniel R. Neuville (Eds.)*
267 **Wetland Carbon and Environmental Management** *Ken W. Krauss, Zhiliang Zhu, and Camille L. Stagg (Eds.)*
268 **Distributed Acoustic Sensing in Geophysics: Methods and Applications** *Yingping Li, Martin Karrenbach, and Jonathan B. Ajo-Franklin (Eds.)*
269 **Congo Basin Hydrology, Climate, and Biogeochemistry: A Foundation for the Future (English version)** *Raphael M. Tshimanga, Guy D. Moukandi N'kaya, and Douglas Alsdorf (Eds.)*
269 **Hydrologie, climat et biogéochimie du bassin du Congo: une base pour l'avenir (version française)** *Raphael M. Tshimanga, Guy D. Moukandi N'kaya, et Douglas Alsdorf (Éditeurs)*
270 **Muography: Exploring Earth's Subsurface with Elementary Particles** *László Oláh, Hiroyuki K. M. Tanaka, and Dezso˝ Varga (Eds.)*
271 **Remote Sensing of Water-Related Hazards** *Ke Zhang, Yang Hong, and Amir AghaKouchak (Eds.)*
272 **Geophysical Monitoring for Geologic Carbon Storage** *Lianjie Huang (Ed.)*
273 **Isotopic Constraints on Earth System Processes** *Kenneth W. W. Sims, Kate Maher, and Daniel P. Schrag (Eds.)*
274 **Earth Observation Applications and Global Policy Frameworks** *Argyro Kavvada, Douglas Cripe, and Lawrence Friedl (Eds.)*
275 **Threats to Springs in a Changing World: Science and Policies for Protection** *Matthew J. Currell and Brian G. Katz (Eds.)*
276 **Core-Mantle Co-Evolution: An Interdisciplinary Approach** *Takashi Nakagawa, Taku Tsuchiya, Madhusoodhan Satish-Kumar, and George Helffrich (Eds.)*
277 **Compressional Tectonics: Plate Convergence to Mountain Building (Tectonic Processes: A Global View, Volume 1)** *Elizabeth J. Catlos and Ibrahim Çemen (Eds.)*
278 **Extensional Tectonics: Continental Breakup to Formation of Oceanic Basins (Tectonic Processes: A Global View, Volume 2)** *Ibrahim Çemen and Elizabeth J. Catlos (Eds.)*
279 **Strike-Slip Tectonics: Oceanic Transform Faults to Continental Plate Boundaries (Tectonic Processes: A Global View, Volume 3)** *Ibrahim Çemen and Elizabeth J. Catlos (Eds.)*
280 **Landscape Fire, Smoke, and Health: Linking Biomass Burning Emissions to Human Well-Being** *Tatiana V. Loboda, Nancy H. F. French, and Robin C. Puett (Eds.)*
281 **Clouds and Their Climatic Impacts: Radiation, Circulation, and Precipitation** *Sylvia C. Sullivan and Corinna Hoose (Eds.)*

Geophysical Monograph 281

Clouds and Their Climatic Impacts

Radiation, Circulation, and Precipitation

Sylvia C. Sullivan
Corinna Hoose
Editors

This Work is a co-publication of the American Geophysical Union and John Wiley and Sons, Inc.

AGU
ADVANCING EARTH
AND SPACE SCIENCES

WILEY

Published under the aegis of the AGU Publications Committee

Matthew Giampoala, Vice President, Publications
Carol Frost, Chair, Publications Committee
For details about the American Geophysical Union visit us at www.agu.org.

The right of Sylvia C. Sullivan and Corinna Hoose to be identified as the editors of this work has been asserted in accordance with law.

Registered Office
John Wiley & Sons, Inc., 111 River Street, Hoboken, NJ 07030, USA

Editorial Office
111 River Street, Hoboken, NJ 07030, USA

For details of our global editorial offices, customer services, and more information about Wiley products visit us at www.wiley.com.

Wiley also publishes its books in a variety of electronic formats and by print-on-demand. Some content that appears in standard print versions of this book may not be available in other formats.

Library of Congress Cataloging-in-Publication Data

Names: Sullivan, Sylvia C., editor. | Hoose, Corinna, editor.
Title: Clouds and their climatic impacts : radiation, circulation, and precipitation / editors Sylvia C. Sullivan, Corinna Hoose.
Description: Hoboken, NJ : Wiley-American Geophysical Union, 2024. | Series: Geophysical monograph series | Includes index.
Identifiers: LCCN 2023027576 (print) | LCCN 2023027577 (ebook) | ISBN 9781119700319 (cloth) | ISBN 9781119700333 (adobe pdf) | ISBN 9781119700340 (epub)
Subjects: LCSH: Clouds. | Climatic changes.
Classification: LCC QC921 .C567 2024 (print) | LCC QC921 (ebook) | DDC 551.57/6–dc23/eng/20230815
LC record available at https://lccn.loc.gov/2023027576
LC ebook record available at https://lccn.loc.gov/2023027577

Cover image: Monsoon clouds over the Bay of Bengal from the International Space Station. Credit: NASA
Cover design: Wiley

Set in 9.5/12pt TimesNewRomanMTStd by Straive, Chennai, India

Dedicated in memory of our colleague, mentee, and friend, Marco Paukert.

CONTENTS

Part III Clouds and Circulation

Part IV Clouds and Precipitation

Part V Outlook

LIST OF CONTRIBUTORS

Andrew S. Ackerman
NASA Goddard Institute for Space Studies
New York, NY, USA

Alessandro Battaglia
Department of Environment, Land and Infrastructure Engineering
Polytechnic University of Turin
Turin, Italy
and
Department of Physics and Astronomy
University of Leicester
Leicester, UK

Gilles Bellon
Department of Physics
The University of Auckland
Auckland, New Zealand
and
Centre National de Recherches Météorologiques
Université de Toulouse, Météo France, CNRS
Toulouse, France

Tom Beucler
Institute of Earth Surface Dynamics
University of Lausanne
Lausanne, Switzerland
and
Department of Earth System Science
University of California Irvine
Irvine, CA, USA
and
Department of Earth and Environmental Engineering
Columbia University
New York, NY, USA

Scott Braun
NASA Goddard Space Flight Center
Greenbelt, MD, USA

Florent Brient
Laboratoire de Météorologie Dynamique
Institut Pierre Simon Laplace
Sorbonne University, CNRS
Paris, France

Grégory V. Cesana
Center for Climate Systems Research
Columbia University
New York, NY, USA
and
NASA Goddard Institute for Space Studies
New York, NY, USA

Sudip Chakraborty
Institute for Harnessing Data and Model Revolution in the Polar Regions
Department of Information Systems
University of Maryland, Baltimore County
Baltimore, MD, USA
and
NASA Jet Propulsion Laboratory
California Institute of Technology
Pasadena, CA, USA

David Delene
Department of Atmospheric Sciences
University of North Dakota
Grand Forks, ND, USA

Bastiaan van Diedenhoven
SRON Netherlands Institute for Space Research
Leiden, The Netherlands
and
Center for Climate System Research
Columbia University
New York, NY, USA

Tra Dinh
Department of Physics
The University of Auckland
Auckland, New Zealand

Wenhao Dong
Cooperative Programs for the Advancement of Earth System Science
University Corporation for Atmospheric Research
Boulder, CO, USA
and
NOAA Geophysical Fluid Dynamics Laboratory
Princeton, NJ, USA
and
Department of Earth System Science
Ministry of Education Key Laboratory for Earth System Modeling
Institute for Global Change Studies
Tsinghua University
Beijing, China

Imme Ebert-Uphoff
Cooperative Institute for Research in the Atmosphere
Colorado State University
Fort Collins, CO, USA
and
Department of Electrical and Computer Engineering
Colorado State University
Fort Collins, CO, USA

Zhe Feng
Atmospheric Sciences and Global Change Division
Pacific Northwest National Laboratory
Richland, WA, USA

Michelle E. Frazer
Department of Geosciences
Pennsylvania State University
University Park, PA, USA
and
Program in Atmospheric and Oceanic Sciences
Princeton University
Princeton, NJ, USA

Blaž Gasparini
Department of Meteorology and Geophysics
University of Vienna
Vienna, Austria

Pierre Gentine
Department of Earth and Environmental Engineering, and Columbia Climate School
Columbia University
New York, NY, USA

Franziska Glassmeier
Department of Geoscience and Remote Sensing
Delft University of Technology
Delft, The Netherlands

Hamish Gordon
Department of Chemical Engineering, and Center for Atmospheric Particle Studies
Carnegie Mellon University
Pittsburgh, PA, USA

Jan O. Haerter
Complexity and Climate
Leibniz Centre for Tropical Marine Research
Bremen, Germany
and
Constructor University
Bremen, Germany
and
Niels Bohr Institute
Copenhagen University
Copenhagen, Denmark

David S. Henderson
Space Science and Engineering Center
University of Wisconsin–Madison
Madison, WI, USA

Fabian Hoffmann
Meteorological Institute
Faculty for Physics
Ludwig Maximilian University of Munich
Munich, Germany

Corinna Hoose
Institute of Meteorology and Climate Research
Karlsruhe Institute of Technology
Karlsruhe, Germany

Emma Järvinen
Institute of Meteorology and Climate Research
Karlsruhe Institute of Technology
Karlsruhe, Germany

Olivier Jourdan
Laboratoire de Météorologie Physique
Université Clermont Auvergne/OPGC/CNRS
Clermont-Ferrand, France

Seiji Kato
NASA Langley Research Center
Hampton, VA, USA

Maki Kikuchi
Earth Observation Research Center
Japan Aerospace Exploration Agency
Ibaraki, Japan

Yanluan Lin
Department of Earth System Science
Ministry of Education Key Laboratory for Earth System Modeling
Institute for Global Change Studies
Tsinghua University
Beijing, China

Guosheng Liu
Department of Earth, Ocean and Atmospheric Science
Florida State University
Tallahassee, FL, USA

Simone Lolli
Institute of Methodologies for Environmental Analysis
National Research Council of Italy
Tito, Italy
and
CommSensLab
Department of Signal Theory and Communications
Universitat Politècnica de Catalunya
Barcelona, Spain

Nicholas J. Lutsko
Scripps Institution of Oceanography
University of California San Diego
La Jolla, CA, USA

Nathan Magee
Department of Physics
The College of New Jersey
Ewing, NJ, USA

Daniel T. McCoy
Department of Atmospheric Science
University of Wyoming
Laramie, WY, USA

Caroline Muller
Information and System Sciences
Institute of Science and Technology Austria
Klosterneuburg, Austria
and
Laboratoire de Météorologie Dynamique
Paris Sciences Lettres/Institut Pierre Simon Laplace
Paris, France

Johannes Mülmenstädt
Department of Atmospheric Sciences and Global Change
Pacific Northwest National Laboratory
Richland, WA, USA

Timothy A. Myers
Cooperative Institute for Research in Environmental Sciences
University of Colorado
Boulder, CO, USA
and
Physical Science Laboratory
National Oceanic and Atmospheric Administration
Boulder, CO, USA
and
Lawrence Livermore National Laboratory
Livermore, CA, USA

Steven Neshyba
Department of Chemistry
University of Puget Sound
Tacoma, WA, USA

Hossein Parishani
Ansys, Inc.
Irvine, CA, USA
and
Department of Earth System Science
University of California Irvine
Irvine, CA, USA

Michael Pritchard
Department of Earth System Science
University of California Irvine
Irvine, CA, USA

Stephan Rasp
Google Research
San Francisco, CA, USA
and
ClimateAi
San Francisco, CA, USA
and
Meteorological Institute
Faculty for Physics
Ludwig Maximilian University of Munich
Munich, Germany

Kathleen A. Schiro
Department of Environmental Sciences
University of Virginia
Charlottesville, VA, USA

Martin Schnaiter
Institute of Meteorology and Climate Research
Karlsruhe Institute of Technology
Karlsruhe, Germany

Ryan C. Scott
NASA Langley Research Center
Hampton, VA, USA

Steven C. Sherwood
Climate Change Research Centre and ARC Centre of Excellence for Climate Extremes
University of New South Wales
Sydney, Australia

Georgia Sotiropoulou
Laboratory of Atmospheric Processes and their Impacts
School of Architecture, Civil & Environmental Engineering
Ecole Polytechnique Fédérale de Lausanne
Lausanne, Switzerland
and
Department of Meteorology, and
Bolin Center for Climate Research
Stockholm University
Stockholm, Sweden
and
currently at the Department of Physics
National and Kapodistrian University of Athens
Athens, Greece

Hui Su
Department of Civil and Environmental Engineering
The Hong Kong University of Science and Technology
Hong Kong SAR
and
Department of Atmospheric and Oceanic Sciences
University of California Los Angeles
Los Angeles, CA, USA

Sylvia C. Sullivan
Department of Chemical & Environmental Engineering, and
Department of Hydrology & Atmospheric Science
University of Arizona
Tucson, AZ, USA
and
Institute of Meteorology and Climate Research
Karlsruhe Institute of Technology
Karlsruhe, Germany

Kentaroh Suzuki
Atmosphere and Ocean Research Institute
The University of Tokyo
Chiba, Japan

Ivy Tan
Department of Atmospheric and Oceanic Sciences
McGill University
Montreal, Canada
and
formerly at
NASA Goddard Space Flight Center
Greenbelt, MD, USA
and affiliated with
University of Maryland Baltimore County
Baltimore, MD, USA

Patrick C. Taylor
Climate Science Branch
NASA Langley Research Center
Hampton, VA, USA

Christopher R. Terai
Atmospheric, Earth, and Energy Division
Lawrence Livermore National Laboratory
Livermore, CA, USA

Thibault Vaillant de Guélis
Science Systems and Applications, Inc.
Hampton, VA, USA
and
NASA Langley Research Center
Hampton, VA, USA

Raphaela Vogel
Meteorological Institute
Department of Earth System Sciences
University of Hamburg
Hamburg, Germany
and
Laboratoire de Météorologie Dynamique
Institut Pierre Simon Laplace
Sorbonne University, CNRS
Paris, France

Fritz Waitz
Institute of Meteorology and Climate Research
Karlsruhe Institute of Technology
Karlsruhe, Germany

Manfred Wendisch
Leipzig Institute for Meteorology
University of Leipzig
Leipzig, Germany

Guanglang Xu
Institute of Meteorology and Climate Research
Karlsruhe Institute of Technology
Karlsruhe, Germany

Lauren Zamora
Earth System Science Interdisciplinary Center
University of Maryland
College Park, MD, USA
and
NASA Goddard Space Flight Center
Greenbelt, MD, USA

Mark D. Zelinka
Atmospheric, Earth, and Energy Division
Lawrence Livermore National Laboratory
Livermore, CA, USA

Ming Zhao
NOAA Geophysical Fluid Dynamics Laboratory
Princeton, NJ, USA

PREFACE

Understanding and adjusting to climate under anthropogenic influence is now one of our greatest challenges. In our various emission-dependent projections of climate, clouds are–and have been for quite some time–the primary source of uncertainty. In 1979, Jule Charney and coauthors assembled a report for the National Research Council explaining how much surface temperature should increase in response to a doubling of atmospheric carbon dioxide concentrations, a value called the equilibrium climate sensitivity. In their analysis, they described the net effect of clouds as "an extremely difficult question to answer" (National Academy of Sciences, 1979). Thirty-five years later, the fifth Assessment Report of the Intergovernmental Panel on Climate Change (IPCC) stated that "clouds and aerosols continue to contribute the largest uncertainty to estimates and interpretations of the Earth's changing energy budget" (Boucher et al., 2013). In the past five years, the World Climate Research Programme has declared the coupling of clouds to circulation and their influence on the equilibrium climate sensitivity a Grand Challenge, meaning a transformative and high-priority topic of research.

The intractability of the net climate impact of clouds is first a problem of scale. Clouds evolve rapidly relative to other components of the Earth system. High-frequency observation or model output is necessary to capture their life cycle. Clouds also exist at spatial scales smaller than the mesh of a typical global climate model and thus require parameterization. Even in cloud-resolving models, the microphysics of cloud droplets and ice crystals must be parameterized. Second, the climate impacts of clouds are tied to cloud type and location. Clouds come in a variety of flavors: ice, liquid, or mixed phase; geometrically or optically thin or deep; and isolated or organized in larger fields. Climate impacts depend on which of these types exists and where it exists, be it in the tropics or in the Arctic, over land or over ocean. Finally, the climate impacts of clouds are multifaceted. If we understand the Earth as a control system, a large range of cloud feedbacks can act to amplify or diminish direct warming. Clouds are embedded in the large-scale atmospheric circulation, but they also determine it, changing temperature gradients by means of their absorption or remission of radiation. And clouds are the source of precipitation, perhaps the climate variable of greatest relevance to building socioeconomic resilience.

Despite these challenges, we are making progress. The Technical Summary of the sixth IPCC Assessment Report notes that uncertainties in the equilibrium climate sensitivity, as well as in the transient climate response, have been lessened in the new generation of climate models, in part thanks to "a 50% reduction in the uncertainty range of cloud feedbacks" and an accounting for aerosol-cloud interactions (Arias et al., 2021). Multi-decade satellite observations, global coverage of ground-based instrument networks, and model resolutions sufficient to resolve atmospheric convection are all fueling this progress. Indeed, a strength of cloud and climate research is the variety of techniques brought to bear on different questions. The momentum and technical diversity of this field are primary motivators for this monograph.

The research attention that has been devoted to clouds and climate is not only impressive, but also intimidating. In the introductory chapter of this volume, we discuss the exponential growth in publications on clouds and climate over the past five decades. Since 1970, the literature on clouds and climate has doubled every 8 years, a rate twice that for scientific literature in general. More than 600 articles were published per year throughout the 2010s, and by 2019, unique authors studying clouds and climate had increased by a factor of 70 relative to 1989. To extract meaning from this overwhelming body of work, review efforts like this monograph are increasingly necessary. Many existing books address a subset of cloud types or climate impacts, and a few recent ones tackle the whole range of these types and impacts. What makes this volume unique is lead authorship by early career scientists. Those who will push the field forward over the coming decades have assessed the state of knowledge in their specialties and posed what they believe will be the most pressing questions for the years ahead.

We have structured this book with the diversity of cloud types and impacts in mind. We begin with three chapters discussing liquid droplet and ice crystal formation on aerosol particles and the feedbacks associated with absorption and emission by these cloud hydrometeors. Five chapters present radiatively influential cloud types: Arctic mixed-phase, extratropical, tropical low, and tropical deep organized. Three chapters are devoted to cloud-circulation coupling at the global, meso, and individual cloud scales; and an additional three chapters discuss precipitation – its efficiency, extremes, and measurement by satellites. We close with an outlook on the role of machine learning for understanding cloud physics and dynamic and their importance to climate.

Our goal is to provide a "lay of the land" for new graduate students in this diverse and rapidly evolving field of research. For those gaining independence as researchers, our hope is to direct their time and energy toward high-impact problems. And for those even farther along their career trajectory, we wish to facilitate collaborations across subfields and highlight the foci of younger researchers.

The dedication of our lead authors has made this project possible. Thanks go to AGU Publications Director, Dr. Jenny Lunn, the AGU Books Editorial Board, and the editorial staff at Wiley – Dr. Rituparna Bose, Lesley Fenske, Noel McGlinchey, Vaishali Rajasekar, Sangaprabha Mohan, and Layla Harden – for their efforts in preparing the volume. We would also like to acknowledge the many reviewers, whose time and expertise have ensured the quality of all chapters. And finally, to our readers, thank you for your interest in and support of this field.

Sylvia C. Sullivan
Department of Chemical and Environmental Engineering
University of Arizona, USA
Institute of Meteorology and Climate Research
Karlsruhe Institute of Technology, Germany
Corinna Hoose
Institute of Meteorology and Climate Research
Karlsruhe Institute of Technology, Germany

REFERENCES

Arias, P. A., Bellouin, N., Coppola, E., et al. (2021). Technical Summary. *In Climate Change 2021: The Physical Science Basis. Contribution of Working Group I to the Sixth Assessment Report of the Intergovernmental Panel on Climate Change* (pp. 35–141). Cambridge University Press, Cambridge, UK.

Boucher, O., Randall, D., Artaxo, P., et al. (2013). *Clouds and aerosols. In Climate Change 2013: The Physical Science Basis. Contribution of Working Group I to the Fifth Assessment Report of the Intergovernmental Panel on Climate Change* (pp. 571–657). Cambridge University Press, Cambridge, UK.

Charney, J. G., Arakawa, A., Baker, D. J., et al. (1979). *Carbon dioxide and climate: A scientific assessment*. National Academy of Sciences, Woods Hole, MA.

1

Science of Cloud and Climate Science: An Analysis of the Literature Over the Past 50 Years

Sylvia C. Sullivan[1,2] and Corinna Hoose[2]

ABSTRACT

Clouds pose a particularly difficult challenge within Earth's climate system. They are relatively small in spatiotemporal scale but still have a strong influence on radiative fluxes, global circulation, and precipitation patterns. Increasing research attention has been devoted to them over the past 50 years, and we give a summary of the resulting body of scientific literature in this introductory chapter. Articles on clouds and climate are doubling every 8 years, a rate about twice that of scientific publications generally. This expanding number of publications correlates with more citations, but citation rates have also slowed in the most recent decade, despite a growing number of atmospheric science students. We show some basic "science of science" (SciSci) analyses of the clouds and climate literature, such as authorship networks or abstract text mining for techniques, and suggest that further SciSci analyses may help us to process the proliferation of results on clouds and climate and optimize how we do research in the crucial years ahead.

1.1. RESEARCH ON CLOUDS AND CLIMATE

Clouds have a multifaceted impact on the climate. They affect the terrestrial radiative balance by reflecting visible and ultraviolet radiation from the Sun and absorbing infrared radiation from the Earth surface. They are also coupled to the large-scale circulation. Cloud formation is determined by where circulation patterns bring moisture aloft, while the reflection and absorption of radiation by clouds also feeds back on circulation. Finally, precipitation is generated from clouds. The intensity, frequency, and duration of precipitation are determined by cloud structure and dynamics.

Climatic impacts of clouds are also multiscale, involving a huge range of processes, from new particle formation at the nanometer scale up to atmospheric wave propagation over thousands of kilometers. Simulating or measuring these 15 orders of magnitude challenges our computational and observational tools. And with phenomena as diverse as turbulence and nucleation, and forms as varied as cumulonimbus towers and stratocumulus decks, clouds challenge our ability to simplify or generalize.

This combination of impact and complexity has piqued the interest of ever more researchers and funding agencies and given rise to an imposing body of scientific literature. As in other scientific fields, this rapid growth in publication has motivated systematic review and meta-analysis, in this case through entities like the Intergovernmental Panel on Climate Change (IPCC) or the Climate Model Intercomparison Projects (CMIP). The fifth assessment report of the IPCC contained a chapter dedicated to clouds and climate (Boucher et al., 2013), and the most recent report devotes sections to cloud feedbacks and

[1]*Department of Chemical & Environmental Engineering, and Department of Hydrology & Atmospheric Science, University of Arizona, Tucson, AZ, USA*

[2]*Institute of Meteorology & Climate Research, Karlsruhe Institute of Technology, Karlsruhe, Germany*

Clouds and Their Climatic Impacts: Radiation, Circulation, and Precipitation, Geophysical Monograph 281, First Edition.
Edited by Sylvia C. Sullivan and Corinna Hoose.

DOI: 10.1002/9781119700357.ch01

water cycle changes with warming (Douville et al., 2021; Forster et al., 2021). An endorsed model intercomparison project focused exclusively on cloud feedbacks (CFMIP) exists within the experimental design of the most recent CMIP (Webb et al., 2017).

This monograph represents one of such increasingly important community-wide review efforts. With primarily early-career lead authors, each chapter provides a review of the existing scientific literature and open questions in a given subfield. The volume is intended to act as a resource for graduate students, both to orient in their new subfield and gain fluency in related ones. For young scientists, it may direct their energy toward high-impact questions or help them formalize their future research program, and for more established scientists, it may generate ideas for collaboration.

Chapters have been organized around the climatic impacts of clouds on radiation, circulation, and precipitation described in Figure 1.1. Chapters 2 through 4 focus on the formation of liquid droplets and ice crystals on particulates and how this cloud formation can affect radiative fluxes, both at the surface and in the atmosphere. Chapters 5 through 9 review research on the most climatically relevant cloud types including Arctic mixed-phase, extratropical, tropical marine low, and tropical organized deep. Cloud-circulation coupling at the global, meso, and micro scales are covered in Chapters 10 through 12. The final part discusses precipitation efficiency, extremes, and measurements (Chapters 13–15).

1.1.1. Science of Science for Clouds and Climate

As an imposing body of scientific literature might already suggest, clouds and climate research at the institute or university level has tended to answer questions by performing new simulations or analyses, rather than by synthesizing output from existing studies. There could be many possible reasons for this focus on generating new output, for example, limited code and data documentation and accessibility in the past or cost and difficulty of storing and analyzing large volumes of data. In any case, systematic review at the community level is fueled by this very large number of individual studies. In this introductory chapter, we posit that analyzing how these individual studies are produced in a kind of "science of science" (SciSci) is another meta-exercise that would be useful for the clouds and climate community.

SciSci is an emerging area of study that combines scientometrics and the sociology of science to understand and optimize how science is done, from the emergence of new paradigms to the career path of students. As Fortunato et al. (2018) note in their review, SciSci has been driven in recent years by increasingly quantitative and accessible data on publications in databases like Scopus or Web of Science and by collaboration of natural, computational, and social scientists. Although some SciSci findings are domain- or culture-specific, a number of generalizable statements can be made. For example, networks of scientific concepts, tools, and authors tend to densify over time, indicating risk aversion and the tendency to select questions and collaborations conservatively (Foster et al., 2015; Fortunato et al., 2018). Such densification can be dangerous, as a small subset of authors cite one another and reinforce established hypotheses in an echo chamber effect.

SciSci analyses have not yet been done for clouds and climate research to our knowledge. Two decades ago, Geerts (1999) aggregated data on atmospheric science

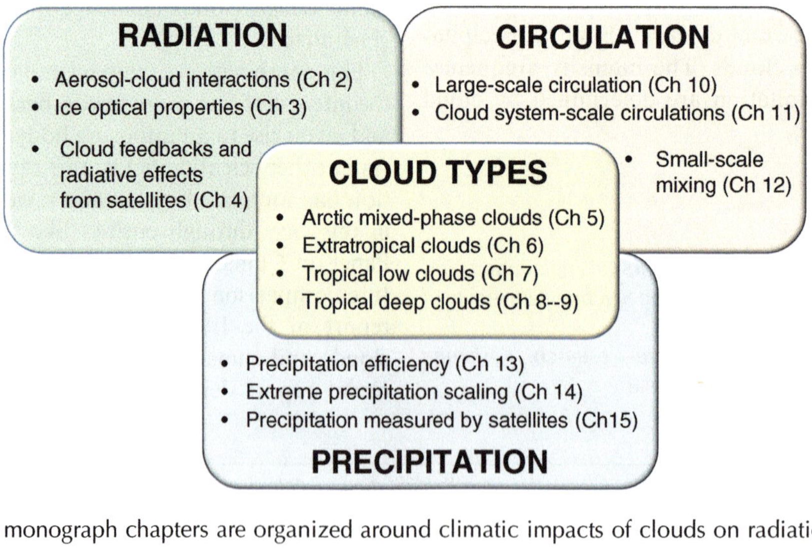

Figure 1.1 The monograph chapters are organized around climatic impacts of clouds on radiation, circulation, and precipitation.

publications and found an increasing number of journals, pages per journal, and words per page in articles published between 1965 and 1995. But we have not found studies building upon this one or studies focused exclusively on climatic impacts of clouds. As a field with burgeoning student interest and the pressure to inform climate policy reliably and rapidly, clouds and climate research could benefit from such meta-study. To catalyze these efforts, we perform preliminary SciSci analyses in this introductory chapter.

1.1.2. Publication Data and Methods

We draw our publication data from the Scopus database of Elsevier, which contains abstracts and citations of peer-reviewed academic literature. A set of Boolean keywords define the Scopus query for publications on clouds and climate, as well as those filtered by theme (Table 1.1). For example, for all publications in the field, we require that the title, abstract, or keywords contain both the term "clouds" and the term "climate" (TITLE-ABS-KEY("clouds" AND "climate")). We also require that the document be an English language journal article past the review stage (LIMIT TO(PUBSTAGE("final")) AND LIMIT-TO(DOCTYPE,"ar") AND LIMIT-TO (LANGUAGE,"English")) and that the journal domain be Earth or environmental sciences (LIMIT-TO (SUBJAREA("EART") OR LIMIT-TO(SUBJAREA ("ENVI")). We find a non-negligible number of publications on Mars for some queries and additionally omit these (NOT TITLE-ABS-KEY("Mars")).

Scopus queries are also used to classify techniques by searching all clouds and climate abstracts for keywords. For example, we classify abstracts that contain the strings "model," "parameterization," "simulation," "GCM," "trajector," or "radiative transfer" as modeling work. Abstracts can be classified as employing multiple techniques if they satisfy multiple queries. The anonymous Author ID field of Scopus queries is used with the Python networkx package to generate authorship networks. Finally, the titles associated with some queries are used to generate word clouds, using the Python WordCloud package. WordCloud takes the title text and fills space with individual words, sized by their frequency. A set of default "stop words" is omitted (listed in https://github.com/amueller/wordcloud/blob/master/wordcloud/stopwords) to which we add the following words: *cloud, climate, using, change, comparison, evaluation, effect, global, study, effects, atmospheric, changes, response, based, part, characteristic,* and *influence*. By eliminating these words, concrete article themes and techniques emerge more clearly from the title text.

The queries in Table 1.1 are denoted **Clouds and Climate** for all publications on clouds and climate; **Impacts** for publications decomposed into cloud impacts on radiation, circulation, and precipitation corresponding to the monograph sections; **Chapters** for publications decomposed according to the monograph chapters; and **Techniques** for identification of the techniques used in the abstracts of the clouds and climate publications. With the queries in Table 1.1, advanced searches can be reproduced at www.scopus.com/search/form.uri?display=advanced. Publication data for the figures have been downloaded on 3 September 2020 and are shown only through the end of 2019. The only non-Scopus data used concerns doctorates granted and funding awarded for the United States, which we take from the National Center for Science and Engineering Statistics (NCSES) Survey Data at https://ncsesdata.nsf.gov/home. For US doctorate degrees granted in geosciences, we look at the Survey of Earned Doctorates at https://ncsesdata.nsf.gov/builder/sed with Doctorate Recipients as our measure and Academic Discipline: Geosciences, atmospheric sciences, and ocean sciences as our dimension. For US annual funding, we look at the Survey of Federal Funds for Research and Development at https://ncsesdata.nsf.gov/builder/ffs with Research Obligations as our measure and Fields of Study: Atmospheric Science as our dimension.

1.2. PUBLICATIONS ON CLOUDS AND CLIMATE

We quantify the rapid growth of interest in clouds and climate with the output of the Scopus **Clouds and Climate** query. We find that the number of articles published on clouds and climate has been doubling every 8 years since 1970 (Fig. 1.2a). While an average of 27 articles per year were published in the 1980s, this rate had increased more than tenfold by the 2000s. In the 2010s, an average of 660 articles per year were published. By comparison, Price (1963) and Fortunato et al. (2018) find a 15-year doubling period for scientific articles more generally, while Milojević (2015) cites 16- and 19-year doubling periods for physics and biomedical articles, respectively.

We could also consider how this growth compares to that of other subfields in climate science. Searching instead for publications on *ocean and climate* or *air pollution and climate*, these have doubling times of 7.9 and 8.5 years, respectively, comparable to that of clouds and climate (not shown). Publications on the *biosphere and climate* are growing somewhat more gradually with a doubling time of 8.9 years, while publications on the *land surface*, the *cryosphere*, or the *carbon cycle* have stronger recent growth relative to cloud research, doubling in only 6.6 years, 5.1 years, and 6.0 years, respectively. These doubling rates show that publication growth for climate research is about twice as fast as that for general scientific research.

Table 1.1 Boolean syntax for Scopus database queries. We filter for English-language journal articles ("ar") in the finalized publication stage ("final"), classified as Earth and planetary sciences ("EART") or environmental sciences ("ENVI"). Filters other than TITLE-ABS-KEY are held constant for all searches and denoted *other filters*.

Queries	
Clouds and Climate	TITLE-ABS-KEY("clouds" AND "climate") AND NOT TITLE-ABS-KEY("Mars") AND LIMIT-TO(PUBSTAGE("final")) AND LIMIT-TO(DOCTYPE,"ar") AND LIMIT-TO(SUBJAREA,"EART") OR LIMIT-TO(SUBJAREA,"ENVI") AND LIMIT-TO(LANGUAGE,"English")
Impacts + *other filters*	TITLE-ABS-KEY("clouds" AND "climate" AND "circulation")
	TITLE-ABS-KEY("clouds" AND "climate" AND "radiati*")
	TITLE-ABS-KEY("clouds" AND "climate" AND "precipitation")
Chapters + *other filters*	TITLE-ABS-KEY("equilibrium climate sensitivity" OR "cloud feedback*")
	TITLE-ABS-KEY(("radiative transfer" AND "climate" AND "cloud") OR "cloud radiative effect*")
	TITLE-ABS-KEY("cloud classification")
	TITLE-ABS-KEY("cloud microphysics" OR ("cloud" AND "microphysics") OR "microphysics parameterization")
	TITLE-ABS-KEY("aerosol indirect effect" OR "aerosol-cloud interaction")
	TITLE-ABS-KEY("atmospher*" AND ("dynamical core" OR "primitive equations"))
	TITLE-ABS-KEY("radiative-convective equilibrium" OR "convective organization" OR "convective aggregation" OR "organized convection")
	TITLE-ABS-KEY("cloud-circulation coupling" OR ("cloud" AND "large-scale circulation"))
	TITLE-ABS-KEY("cloud" AND "field campaign")
	TITLE-ABS-KEY("cloud" AND "ground-based measurement")
	TITLE-ABS-KEY(("cloud" AND "machine learning") OR ("cloud" AND "causal inference"))
Techniques	ABS("in-situ" OR "flight" OR "campaign" OR "aircraft" OR "rocket" OR "drone")
	ABS("model" OR "parameterization" OR "simulation" OR "GCM" OR "trajector" OR "radiative transfer")
	ABS("reanalys" OR "emission")
	ABS("satellite" OR "CERES" OR "TRMM" OR "ISCCP" OR "remote sens" OR "retrieval" OR "imager" OR "CALIPSO" OR "CloudSat" OR "MODIS" OR "mission")
	ABS("ground-based" OR "station" OR "meteorological observator" OR "rain gauge" OR "site" OR "SHEBA" OR "flux tower")
	ABS("laboratory" OR "chamber" OR "chemical characteriz")
	ABS("observation")

Not only is publication number on clouds and climate rapidly rising, the number of unique journals has also increased over time, doubling roughly every 11 years (Fig. 1.2b). In the 1980s, with the focus more often on meteorology, the *Journal of Atmospheric Sciences* published the most articles. By the 1990s, spurred by the release of the Charney Report and the launch of early geostationary satellites, interest broadened to the role of clouds in climate with more publications now in the *Journal of Climate* and *Journal of Geophysical Research*. Although only initiated in 2001, *Atmospheric Chemistry and Physics* (*ACP*) already published the most articles by the 2010s. As an open-access publisher, *ACP* output has also driven the large increase in open-access publication percentage over the past four decades from less than 10% in the 1980s to more than 50% in the late 2010s (Fig. 1.2c).

1.2.1. Citation and Readability

Does this expanding body of literature correspond to higher readership and, hence, citation? For the five

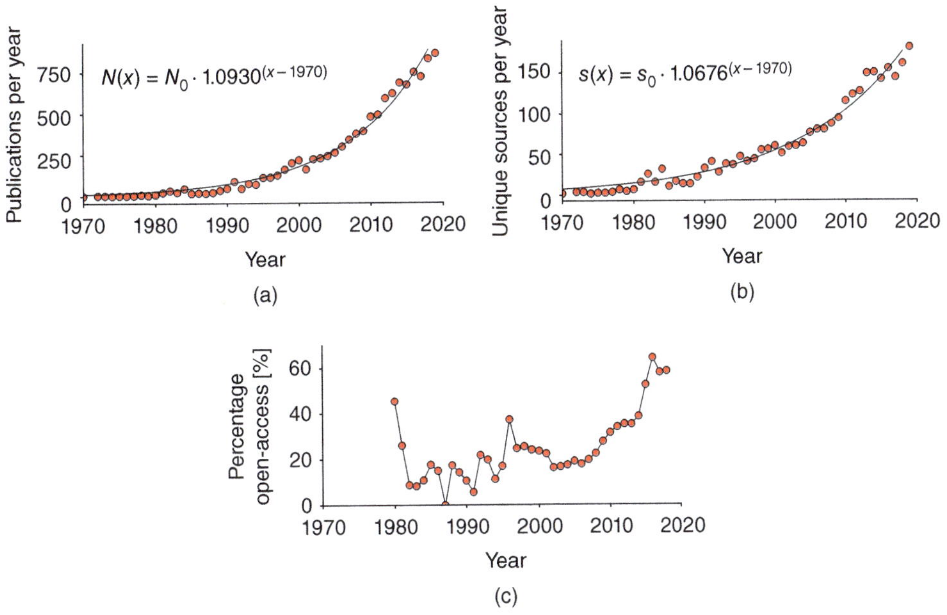

Figure 1.2 Atmospheric science publications, particularly open-access ones, and their unique sources have all increased dramatically over the past 50 years. Total number of publications per year from 1970 through 2019 with the **Clouds and Climate** query in Table 1.1 and their exponential fit (panel a). Number of unique journals for these publications per year from 1970 through 2019 and their exponential fit (panel b). Percentage of publications that have been open-access over time (panel c).

most prolific journals publishing on clouds and climate, higher annual publication counts do correlate with higher annual citation counts from 1970 through 2019, but not always very strongly (Fig. 1.3). In spite of – or perhaps because of – its open-access policy, high publication rates in *ACP* correlate least strongly with high citation rates. From linear fits to these publication-citation scatter plots, annual citation count per journal increases by 33 for each additional article published. This annual citation increase per publication is largest for the *Journal of Atmospheric Sciences* at almost 60.

Without filtering for journal, cumulative distributions show that cloud and climate citation peaks for articles written between 1995 and 2005 (Fig. 1.4a,b). We may expect decreasing citation rates for more recent years, as the time since publication shortens, but the decreasing citation rates prior to 1995 indicate that older literature has been cited less than more recent literature. In the late 1990s, almost 20% of articles have more than a hundred citations, and only 10% are cited five times or fewer, counter to the highly skewed nature of most citation distributions in other fields in which many articles are never cited and a very small number are very highly cited (Price, 1965).

To further quantify citation trends, we employ the Kullback-Leibler divergence:

$$D_{KL} = (P \| Q) = \sum_{x_i} P(x_i) \log_2 \left(\frac{P(x_i)}{Q(x_i)} \right) \qquad (1.1)$$

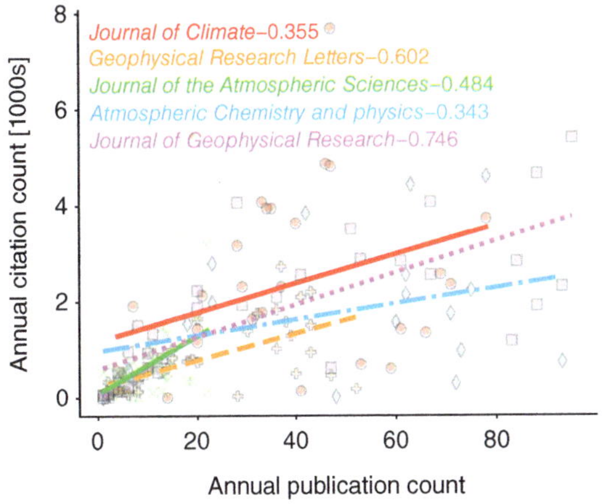

Figure 1.3 Publication and citation number are positively correlated, but not always strongly. Annual publication counts from 1970 through 2019 scattered against annual citation counts over the same period for the five most prolific atmospheric science publications; *Journal of Climate* and *Journal of the Atmospheric Sciences* are shown as solid traces, *Geophysical Research Letters* as a dashed trace, *Atmospheric Chemistry and Physics* as a dotted-dashed trace, and *Journal of Geophysical Research* as a dotted trace. Correlation coefficients are given next to the publication name in the legend.

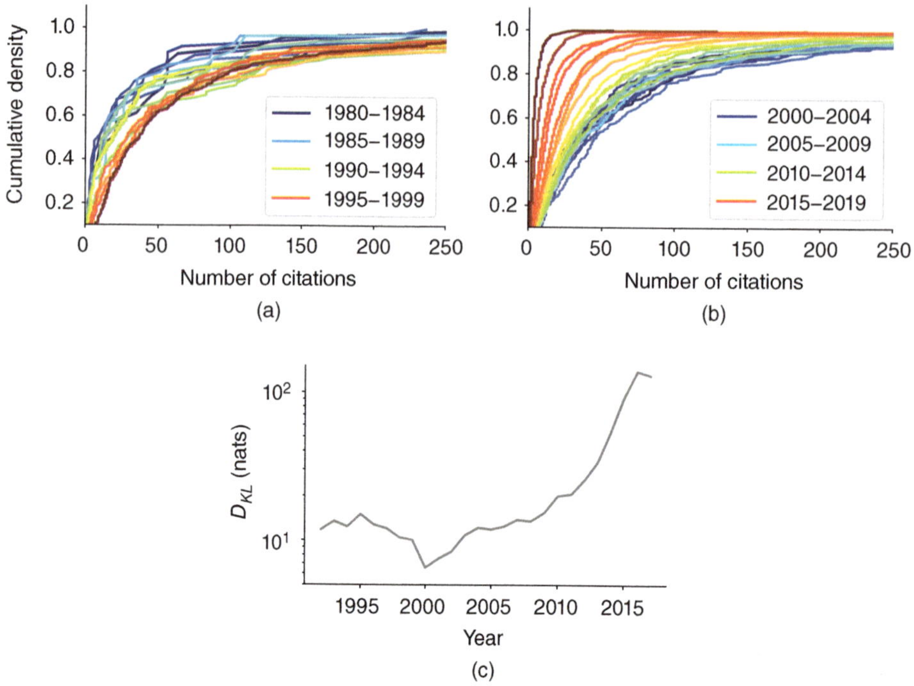

Figure 1.4 Citation rates in clouds and climate research have slowed relative to publication rate in the most recent decade. Total number of citations per article per year from 1980 to 1999 (panel a) and from 1999 to 2019 (panel b). Time series of the Kullback-Leibler divergence between the cumulative density of citations in a given year and the proceeding year (panel c).

where P and Q are two probability distribution functions of annual citation numbers, x_i. The larger the value of D_{KL}, the more different the distributions P and Q are. Stated more formally, the larger the value of D_{KL}, the more information (in nats when D_{KL} is evaluated with the natural logarithm) would be lost in replacing P by Q. D_{KL} is not symmetric, and we always take the preceding year as the reference distribution (Q). Calculating a 5-year running mean of this pairwise D_{KL} yields relatively stable values of between 10 and 15 until 2007; thereafter, D_{KL} increases monotonically up to a most recent value of 140 (Fig. 1.4c). These increasing D_{KL} values indicate that citation growth in recent years is not maintaining pace with publication growth.

Decreased readability could contribute to this seeming drop in readership in recent years. The assessment of atmospheric science literature in Geerts (1999) found an increasing number of journals, pages per journal, and words per page in articles published between 1965 and 1995. While we reproduce this increasing trend in article length, page number per article, as a crude metric of readability, has been growing far less rapidly than publication or journal numbers, at an average of only a page per decade since 1970 (Fig. 1.5a). Geerts (1999) also proposed that lagging US federal funding and the plateauing number of PhD students could slow these publication rates in coming years. We find instead that, while US funding has converged or even dropped in recent years, the number of US students in geosciences has continued to grow (Fig. 1.5b). Equivalent data are not as readily available for other regions, however, and these US trends may not be representative globally.

1.3. THE ROLE OF CLOUDS IN RADIATION, CIRCULATION, AND PRECIPITATION

To understand how research effort has been devoted across the climatic impacts of clouds, we next decompose the publication trends from section 1.2, using the **Impacts** queries from Table 1.1. Publications concerning radiation have been most numerous in the past three decades (Fig. 1.6a), averaging approximately 50 per year in the 1990s, more than twice that many in the 2000s, and more than five times that many in the 2010s. In word clouds of the titles from these publications on clouds and radiation (section 1.1.2), studies of aerosol and using simulations are dominant, and the most commonly studied regions are the Arctic and the tropics (Fig. 1.6b). Feedback quantification and the surface energy budget also emerge as common topics.

Publications concerning precipitation have increased most rapidly among the three impacts, doubling every 7 years since 1970 (Fig. 1.6a). In the associated title word

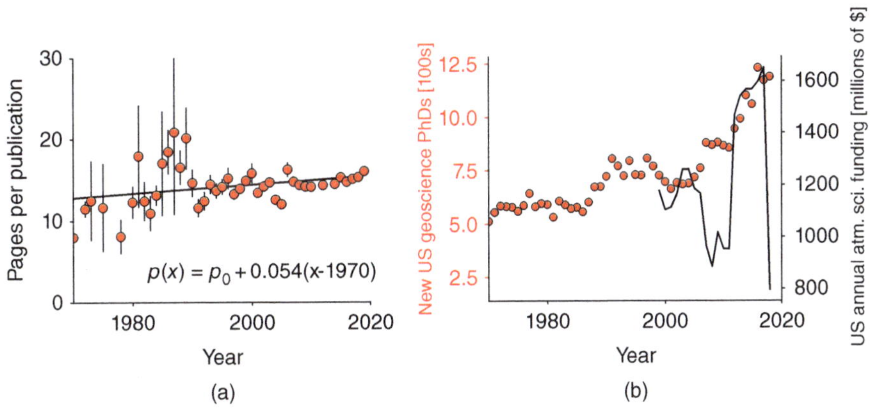

Figure 1.5 Articles have become only slightly less readable with page number as a metric, while potential readership has doubled since 2000 with graduating doctorates as a metric. Average number of pages per publication per year including standard error and a linear fit in black (panel a). US doctoral degrees awarded in geosciences, including both atmospheric and oceanic sciences, and US federal funding awarded to atmospheric sciences projects (panel b) with data for the last panel come from the US National Center for Science and Engineering Statistics Survey of Earned Doctorates and Survey of Federal Funds for Research and Development (see section 1.1.2).

clouds, model simulation and satellite observation appear about equally, and the Arctic and the tropics emerge again as important areas of study, as does China (Fig. 1.6c). Influence of aerosol and convection or convective parameterization also appear prominently. Perhaps because of the computational demand associated with scale separation, growth has been slowest for publications on clouds and circulation with a doubling time of 10 years. Even more than cloud-radiation studies, these are heavily simulation-oriented (Fig. 1.6d). Aerosol, feedbacks, and surface interactions are again frequent topics, but less so than for clouds and radiation.

We can further decompose these trends into topics covered in monograph chapters (Fig. 1.7, **Chapters** queries in Table 1.1). Cloud microphysics is the most active subtheme with almost 400 publications in 2019 alone (Fig. 1.7 inset). Its growth was particularly rapid in the early to mid-1990s, increasing sevenfold over 6 years. Precipitation measurement has also been a very active area of research since the mid-1990s with 160 publications in 2019 alone. Slow and steady progress over the past three decades has characterized the optimization of dynamical cores and extension of ground-based measurements. Research output on cloud-circulation coupling has also been more gradual, again possibly because of the computational challenge to simulate all involved scales. Most dramatic is recent growth in machine learning work. While almost no studies used machine learning in 2012, there were 200 publications in 2019 alone (see also Chapter 16: Machine Learning for Clouds and Climate).

1.4. METHODOLOGY IN CLOUDS AND CLIMATE

1.4.1. Techniques

The word clouds in Figure 1.6b–d reveal some of the most common techniques used in clouds and climate research. We can organize these techniques around pillars of modeling, in situ or ground-based measurement, and satellite retrieval, and we next search the clouds and climate abstract text in the **Techniques** queries of Table 1.1 (section 1.1.2).

With this classifications of abstracts, almost 70% mention some kind of modeling (Fig. 1.8a). Coarse-resolution global climate models (GCMs) still dominate these modeling abstracts (45.8%), although high-resolution or cloud-resolving models account for another 10% each (Fig. 1.8b). Radiative transfer calculations make up another 18%, while less than 10% of abstracts are attributed to each of large-eddy simulation, trajectory analyses, and idealized or parcel simulations.

But modeling alone characterizes only 22% of abstracts; it is more often used in combination with satellite, flight, reanalysis, or ground-based data (Fig. 1.8c). Similarly, almost 40% of studies use remote sensing products, but only 6% use remote sensing products exclusively. Ground-based measurements appear in 25% of abstracts, and flight measurements, given their more limited duration, appear in 8%. Laboratory studies are the most infrequent at only 3%, although these abstracts are also hardest to classify with their varied methods and keywords. While models and satellite data provide good spatiotemporal coverage and robust statistics for studies, we conclude that clouds and climate research is fundamentally driven by a synergy of different techniques.

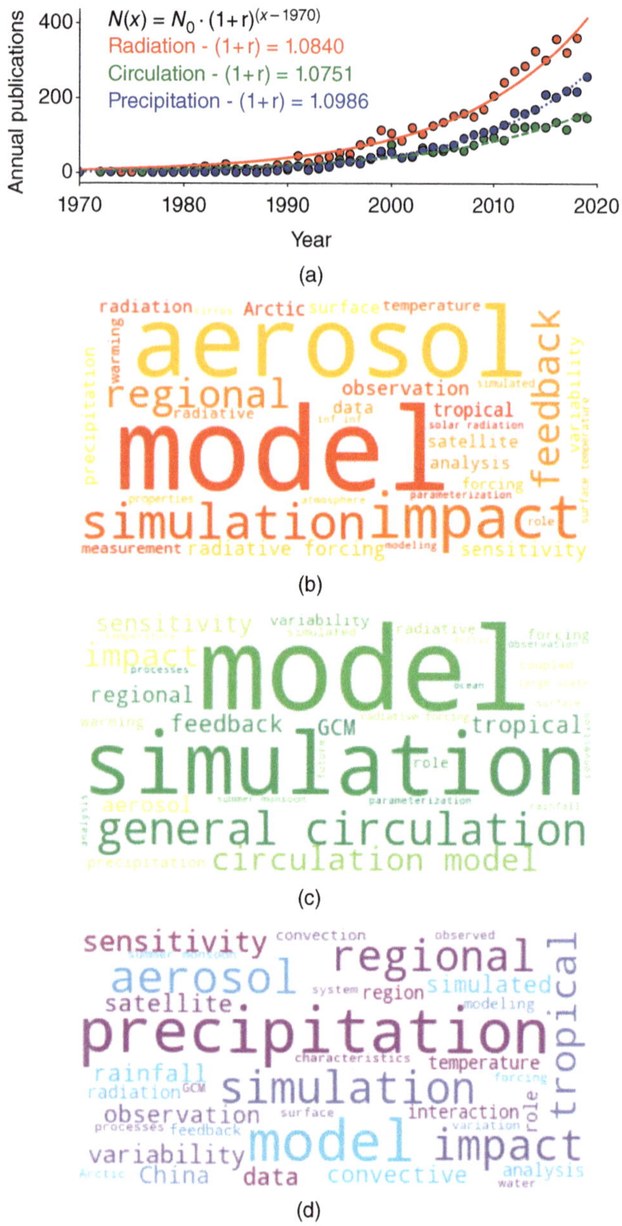

Figure 1.6 While publications on clouds and radiation have been most numerous, those on clouds and precipitation have increased most rapidly. Publication trends decomposed into cloud impacts on radiation, circulation, and precipitation, as well as their exponential fits, based on the Scopus **Impacts** queries (panel a). Word clouds generated from the article titles for cloud impacts on radiation (panel b), circulation (panel c), and precipitation (panel d). Font size is in proportion to the frequency of the word.

1.4.2. Authorship

Along with publication trends and techniques, we can also examine the personal aspect of how clouds and climate research is done. Do certain authors always work with the same coauthors? Do two or three "author clusters" characterize certain subfields? We use anonymous Author IDs from the **Impacts** and **Chapters** searches to generate author networks. Circular networks visualize the increase in authorship complexity over time dramatically; in such networks, nodes represent authors, edges represent coauthorship, and node color indicates degree, or coauthor number in our case.

We give an illustrative example of coauthorship evolution in the circular network of Figure 1.9, taken from the **Impacts** query on clouds and radiation. In 1989, this network contained only 21 nodes, or authors, with an average degree, or coauthor number, of 1.52. A decade later in 1999, the same network contained 294 authors with an average coauthor number of 3.74. Author number increases by more than an order of magnitude for the other **Impacts** queries in the decade between 1989 and 1999 also (not shown). Authorship in the most recent decade is too complex to present in this layout, but by 2019, author numbers involved in clouds and climate research had increased more than seventyfold since 1989.

We can describe networks not only with their degree, but also with their density. Network density refers to the degree normalized by the maximum possible number of edges, or the absolute coauthor number relative to its possible maximum value. Looking again at Figure 1.9, while absolute coauthorship increased, relative coauthorship, as quantified by density, decreased sixfold between 1989 and 1999 for work on clouds and radiation. For work on clouds and precipitation, the drop in relative coauthorship is even more dramatic at thirtyfold between 1989 and 1999. In the two decades from 1999 to 2019, relative coauthorship dropped another two- to threefold. The takeaway is that while those in clouds and climate research are collaborating more in absolute terms, they are also directly involved in less and less of the total body of research produced by the community.

Increases in absolute coauthorship may suggest increased objectivity in our results, but as noted above, densification of author networks can actually have an opposite echo chamber effect in which author clusters reinforce their own hypotheses. This kind of clustering cannot be seen in a circular network, so we present authorship networks for the **Chapters** queries also in a spring layout (Fig. 1.10). A spring layout is generated by a force-directed algorithm in which edges are interpreted as springs governed by Hooke's law, and nodes are arranged to minimize energy in the network.

These force-directed graphs for authorship tend to have a few studies with big groups of authors and many with smaller groups (Fig. 1.10). Ideally, there would be some overlap and not too much distance between the "hub" studies with large coauthorship and the "rim" studies with limited coauthorship. Work on cloud-radiative effects

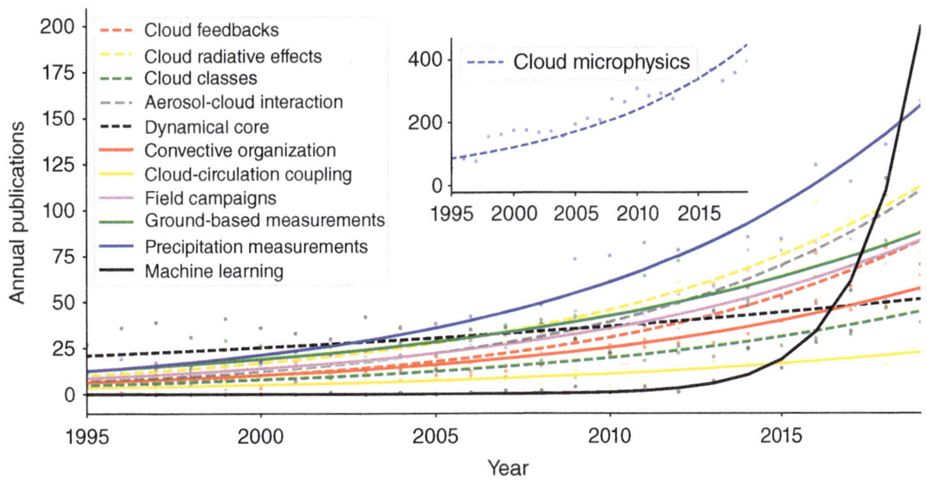

Figure 1.7 Recent growth in machine learning publication has been dramatic. Cloud microphysics and precipitation measurement are also very active areas of research. Publication trends decomposed into chapter topics, as well as their exponential fits, based on the Scopus **Chapters** queries. The inset y-axis for microphysics publications is two times larger than the primary y-axis.

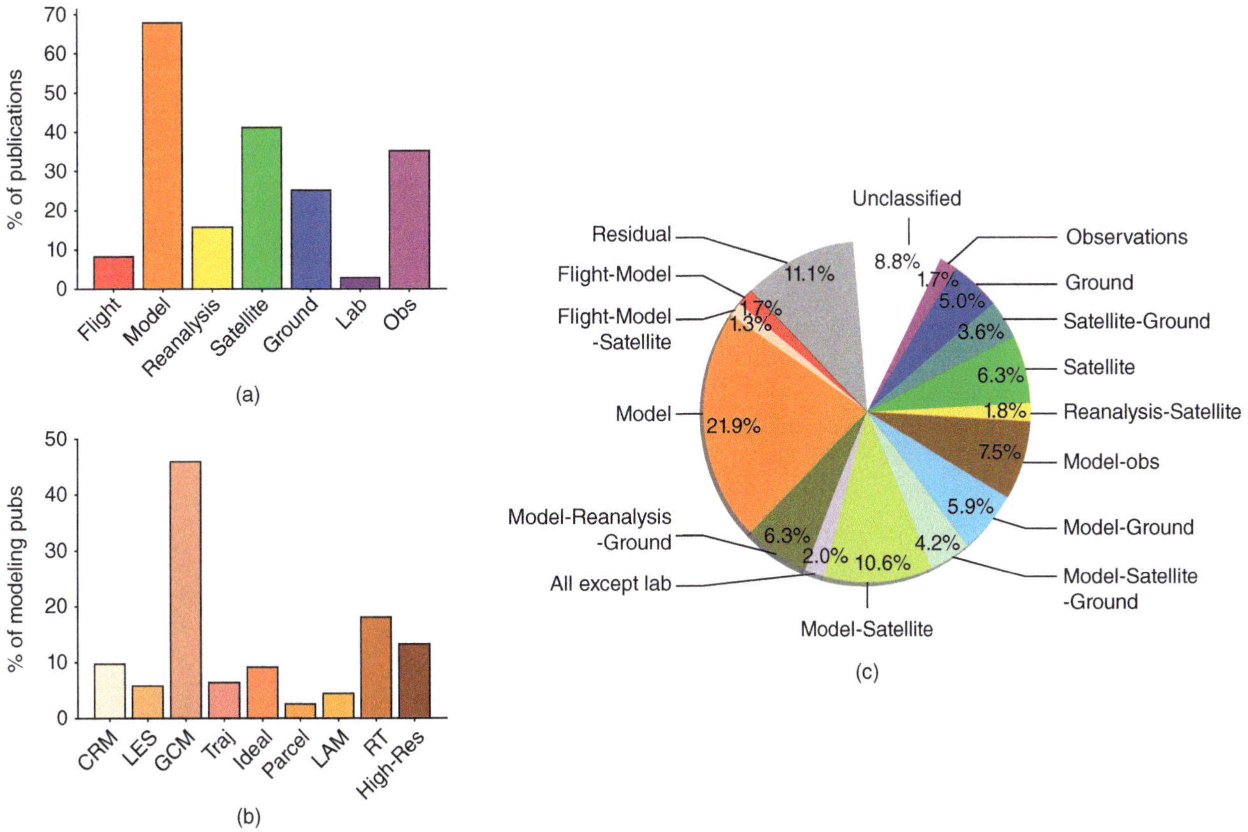

Figure 1.8 About 40% of studies use a combination of modeling and/or satellite data, but a great diversity of techniques are used to study clouds and climate. Percentage of techniques mentioned in **Clouds and Climate** abstracts, identified in the **Techniques** query (panel a). **Obs** designates unspecified observations, and studies on emissions inventories are grouped into the **Reanalysis** category. Additional filtering of modeling publications into model type (panel b). The following acronyms are used to label: Cloud-resolving model (CRM), global climate model (GCM), trajectories (Traj), idealized or RCE setups (Ideal), limited-area or regional modeling (LAM), radiative transfer calculations (RT), and high-resolution simulation (High-Res). Overall distribution of techniques or technique combinations (panel c). The generic Observations (Obs) classification is removed if another observational classification, for example, satellite or ground-based was made.

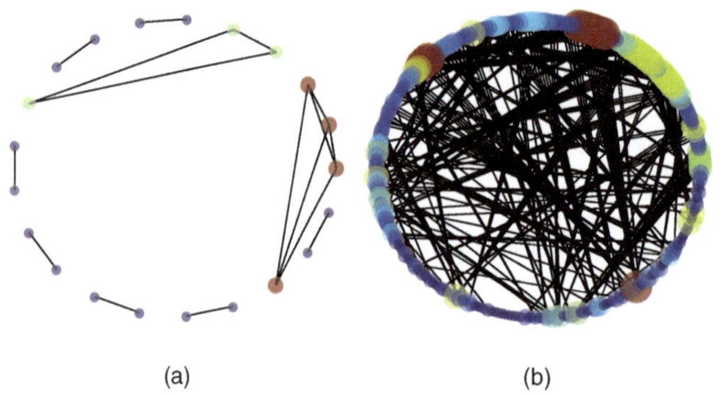

Figure 1.9 Over a decade, author number increased by a factor of 14 and the degree of coauthorship more than doubled. Evolution of author networks between 1989 (panel a) and 1999 (panel b) for the themes of clouds, climate, and radiation. The higher the degree of the nodes, the larger and darker (more red in the color version) they are.

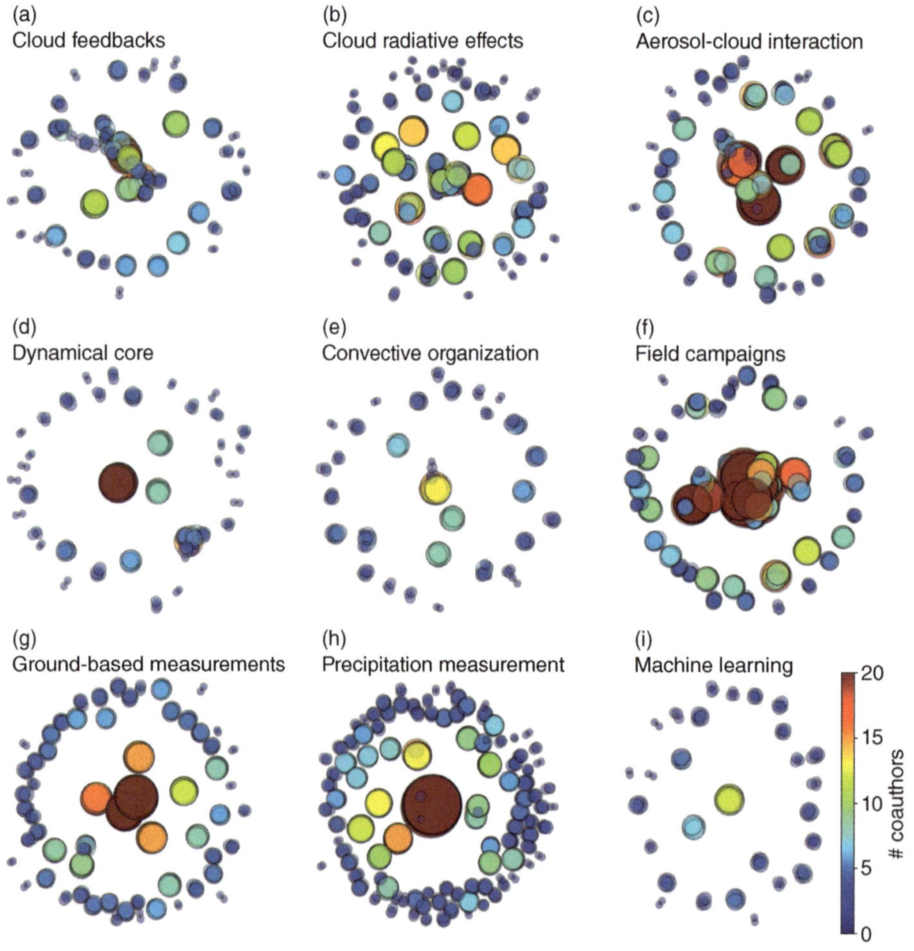

Figure 1.10 Work across cloud and climate research is typically characterized by a "hub" of studies with large coauthorship and a "rim" of smaller collaborations. Sample author networks for 2015 for several of the chapter themes listed in Table 1.1.

exhibits such a structure in which collaborations of different size are fairly well mixed, as does work on aerosol-cloud interaction and precipitation measurement, although to a lesser degree. In contrast, work on dynamical cores is characterized by a few studies with large authorship and many smaller scale efforts that do not overlap at all; however, such separation may be due to model-specific development in this case. We also see the cooperative nature of field campaigns with many more "hub" studies than the other subfields, as well as the nascence of machine learning techniques with no large collaborations yet. Such visualizations give us an idea of where author clusters are becoming isolated and where additional community-level work or cross-institutional collaboration could be beneficial.

1.5. SUMMARY AND OUTLOOK

In this introductory chapter, we have used basic SciSci analyses to analyze metatrends in clouds and climate research. We have also motivated the layout of the overall monograph into impacts of clouds on radiation, circulation, and precipitation. While the largest number of studies exists on clouds and radiation, recent growth in precipitation studies has been most rapid. Output on precipitation measurement is outpaced only by that on cloud microphysics and the recent boom in machine learning. Cloud-circulation coupling remains the least explored impact, given the large range of involved spatiotemporal scales; feasibility of large-domain storm-resolving simulations should bring new insight to this area. We also present three important takeaways from our analyses:

1. **As publications and interest in clouds and climate accelerate, the need to organize and reproduce existing work is becoming more important.** Publications on clouds and climate are doubling every 8 years with almost two articles published per day over the last decade. The number of different journals that publish this work is also increasing with a doubling time of 11 years. At the same time, cumulative distributions and distance metrics of citation indicate that readership in recent years may not be keeping pace with this burgeoning literature. As author networks expand, we are less likely to have been involved in or even heard of the efforts from others in our subfield. In the face of these trends, we need to make special effort at the individual or team level to process this information, for example using news aggregators, automatic search alerts, or journal clubs (Landhuis, 2016). Regular meetings of smaller subfield communities are also crucial to promote idea exchange and prevent too much "author clustering."

2. **A synergy of techniques are fueling this growth, and we should continue to take advantage of complementary tools.** Only a third of abstracts on clouds and climate research can be classified as using exclusively satellite data, ground-based measurement, *or* modeling. Instead, it is the combination of simulation and observation that drives the majority of research. We find that almost half of simulation work is still done with coarse-resolution GCMs, although high-resolution models account for another quarter. Increasing computational power means that cloud- and storm-resolving models with kilometer- or hectometer-scale resolutions will become increasingly important in the next few years (Stevens et al., 2020). A new generation of satellite measurements will also become available over the next years, for example, with the upcoming launch of EarthCARE or the outcomes of the NASA Aerosol, Cloud, Convection, and Precipitation assessment (Gettelman et al., 2021; Illingworth et al., 2015). We emphasize the role that a synergy of models and measurements have played over the past five decades and its continued importance going forward.

3. **Small-scale cloud processes constitute the highest publication numbers, suggesting their intractability. Orienting the recent surge in machine learning toward these problems could afford progress.** Publications in microphysics, precipitation measurement, and aerosol-cloud interaction have been the most numerous over the past 10 years. These research areas represent the smallest scales involved in cloud formation and evolution: droplet activation and coalescence and ice nucleation and aggregation. In the past 5 years, work in machine learning on clouds and climate has also exploded from 7 publications in 2012 to 198 in 2019. How machine learning techniques can assist in the persistent small-scale problems of cloud physics is worthy of exploration. Such efforts are already underway, for example, in using Bayesian inference for rain microphysics or neural networks to represent aerosol effective radii and refractive indices (e.g., Llerena et al., 2018; Morrison et al., 2019). Neural networks have also been used to represent aspects of raindrop formation by droplet collision-coalescence and could be an even more promising route for ice microphysics, characterized by closed-form solutions or property tables rather than the coupled differential equations involved in the stochastic collection equation (e.g., Gettelman et al., 2020; Seifert & Rasp, 2020).

While these SciSci analyses have been straightforward, they still give insight into which topics and tools are receiving the most attention. We suggest that, with its synthesis of scientometrics and sociology, SciSci can also complement traditional review to help us understand discrepancies and biases in how clouds and climate research is done. For example, publication data mining through tools like PaperHunter (http:// paperhunter.net), Connected Papers (https://www.connectedpapers.com), or ScholarSight (http://scholarsight.org) could be used to pinpoint contradictory studies or quantify the range in uncertain parameters across many studies. With more comprehensive citation trees or text mining for technique identification, we would lose track of past and originating ideas less readily, avoid repeating certain studies unnecessarily, or reproduce other studies more intentionally. Such meta-level analyses would allow us to answer questions like, Are our hypotheses biased by authorship clusters? *or* Are we providing sufficient funding to the intersection of subfields? Given the socioeconomic costs of climate change, clouds and climate research is done in a uniquely urgent environment, and we posit that SciSci could help to optimize how this work is done in the decisive years ahead.

ACKNOWLEDGMENTS

S.C. Sullivan acknowledges funding from DFG Project 02.056750161 (TropiC) and from the German Ministry of Education and Research (BMBF) and FONA: Research for Sustainable Development (www.fona.de) under grant 01LK1509A. C. Hoose acknowledges funding from the European Research Council (ERC) under the European Union's Horizon 2020 research and innovation programme under grant agreement No 714062 (ERC Starting Grant "C2Phase"). Both authors thank two anonymous reviewers for their feedback that has improved the manuscript. S.C. Sullivan thanks Aiko Voigt for feedback on an early version of this work and Philipp Gasch for discussions about this topic.

AVAILABILITY STATEMENT

Information about databases and search terms are given in Methods and Table 1.1, and the results of these searches are also directly available at https://doi.org/10.5281/zenodo.5524758. Codes to reproduce figures are provided at https://github.com/sylviasullivan/CloudLit.

REFERENCES

Boucher, O., Randall, D., Artaxo, P., Bretherton, C., Feingold, G., Forster, P., et al. (2013). Clouds and aerosols. In: *Climate Change 2013: The Physical Science Basis. Contribution of Working Group I to the Fifth Assessment Report of the Intergovernmental Panel on Climate Change*. Cambridge, UK: Cambridge University Press.

Douville, H., Raghavan, K., Renwick, J., Allan, R. P., Arias, P. A., Barlow, M., et al. (2021). Water cycle changes. In *Climate Change 2021: The Physical Science Basis. Contribution of Working Group I to the Sixth Assessment Report of the Intergovernmental Panel on Climate Change*. Cambridge, UK: Cambridge University Press.

Forster, P., Storelvmo, T., Armour, K., Collins, W., Dufresne, J.-L., Frame, D., et al. (2021). The Earth's energy budget, climate feedbacks, and climate sensitivity. In *Climate Change 2021: The Physical Science Basis. Contribution of Working Group I to the Sixth Assessment Report of the Intergovernmental Panel on Climate Change*. Cambridge, UK: Cambridge University Press.

Fortunato, S., Bergstrom, C. T., Börner, K., Evans, J. A., Helbing, D., Milojević, S., et al (2018). Science of science. *Science*, *359* (6379), eaao0185. doi: 10.1126/science.aao0185

Foster, J. G., Rzhetsky, A., & Evans, J. A. (2015). Tradition and innovation in scientists' research strategies. *American Sociological Review*, *80* (5), 875–908. doi: 10.1177/0003122415601618

Geerts, B. (1999). Trends in atmospheric science journals: A reader's perspective. *Bulletin of the American Meteorological Society*, *80*, 639–652.

Gettelman, A., Carmichael, G. R., Feingold, G., Silva, A. M. D., & van den Heever, S. C. (2021). Confronting future models with future satellite observations of clouds and aerosols. *Bulletin of the American Meteorological Society*, *102* (8), E1557–E1562. doi: 10.1175/BAMS-D-21-0029.1

Gettelman, A., Chen, J., & Gagne, D. J. (2020). Machine learning rain formation and impacts on clouds, precipitation, and radiation. In *EGU General Assembly Conference Abstracts*, 4636.

Illingworth, A. J., Barker, H. W., Beljaars, A., Ceccaldi, M., Chepfer, H., Clerbaux, N., et al. (2015). The EarthCARE satellite: The next step forward in global measurements of clouds, aerosols, precipitation, and radiation. *Bulletin of the American Meteorological Society*, *96* (8), 1311–1332.

Landhuis, E. (2016). Scientific literature: Information overload. *Nature*, *535*, 457–458.

Llerena, C., Müller, D., Adams, R., Davey, N., & Sun, Y. (2018). Estimation of microphysical parameters of atmospheric pollution using machine learning. In *Lecture Notes in Computer Science Book Series* (LNTCS, vol. 11139). (pp. 579–588). International Conference on Artificial Neural Networks (ICANN), Springer.

Milojević, S. (2015). Quantifying the cognitive extent of science. *Journal of Information*, (*9*), 962–973. doi: 110.1016/j.joi.2015.10.005

Morrison, H., van Lier-Walqui, M., Kumjian, M. R., & Prat, O. P. (2019). A Bayesian approach for statistical-physical bulk parameterization microphysics. Part I: Scheme description. *Journal of Climate*, *77*, 1019–1041. doi: 10.1175/JCLI-D-19-0070.1

Price, D. J. S. (1963). *Little science, big science*. New York, NY: Columbia University Press.

Price, D. J. S. (1965). Networks of scientific papers. *Science*, *149* (3638), 510–515. doi: 10.1126/science.149.3683.510

Seifert, A., & Rasp, S. (2020). Potential and limitations of machine learning for modeling warm-rain cloud microphysical processes. *Journal of Advances in Modeling Earth Systems*, *12* (12), e2020MS002301. doi: 10.1029/2020MS002301

Stevens, B., Acquistaoace, C., Hansen, A., Heinze, R., Klinger, C., Klocke, D., et al. (2020). The added value of large-eddy and storm-resolving models for simulating clouds and precipitation. *Journal of the Meteorological Society of Japan*, *98*, 395–435. doi: 10.2151/jmsj.2020-021

Webb, M. J., Andrews, T., Bodas-Salcedo, A., Bony, S., Bretherton, C. S., Chadwick, R., et al. (2017). The cloud feedback model intercomparison project (CFMIP) contribution to CMIP6. *Geoscientific Model Development*, *10* (1), 359–384.

Part I
Clouds and Radiation

2

An Overview of Aerosol-Cloud Interactions

Hamish Gordon[1], Franziska Glassmeier[2], and Daniel T. McCoy[3]

ABSTRACT

Aerosol-cloud interactions refer to the group of atmospheric processes by which aerosols influence cloud properties, and sometimes also processes by which clouds affect aerosols. The effect of these atmospheric processes on Earth's radiative balance is potentially large, but uncertain. When combined with uncertainties in aerosol concentrations that result from emissions and aerosol processes, the uncertainty in aerosol-cloud interactions dominates the overall uncertainty in our knowledge of radiative forcing of Earth's climate. Aerosols affect clouds primarily by changing the number of cloud condensation and ice nuclei, "indirect effects," and sometimes also the temperature of the cloud, "semi-direct effects." Changes in cloud processes in response to aerosol-cloud interactions may cause significant adjustments to cloud macrophysical properties such as coverage and condensate amount. Aerosol-cloud interaction research focuses on understanding the atmospheric processes at work, mainly by analyzing observation data, performing laboratory experiments, and building models to simulate how aerosols influence clouds. In this review, we outline the relevant atmospheric science and highlight some promising techniques that have been applied recently to better understand aerosol-cloud interactions and their implications for radiative balance, such as Gaussian process emulation. This chapter is intended to provide background to subsequent chapters in this series of monographs and as an introduction for graduate students to current research in the field of aerosol-cloud interactions.

2.1. INTRODUCTION AND MOTIVATION

Aerosols are of fundamental importance to clouds. Without aerosols, relative humidities must reach 420% before condensation of water vapor will occur on ions, the next suitable surface in the atmosphere (Mason, 1960; Wilson, 1900). Such high saturation ratios would be unlikely to occur near Earth's surface, as water would instead condense as dew, and so without aerosols, fogs and boundary-layer clouds would presumably be rare or absent. Conversely, in a world without clouds, atmospheric aerosols would also look very different, presumably existing in substantially higher concentrations.

2.1.1. The Importance of Aerosol-Cloud Interactions for Climate

The relevant question for climate, however, is how much difference to clouds is made by perturbations to a natural background level of aerosols. Changing the number of aerosols capable of acting as nuclei for water vapor to condense upon (cloud condensation nuclei, or CCN) leads to a change to the cloud droplet concentration, which in turn changes the reflectivity or albedo of

[1]*Department of Chemical Engineering and Center for Atmospheric Particle Studies, Carnegie Mellon University, Pittsburgh, PA, USA*

[2]*Department of Geoscience and Remote Sensing, Delft University of Technology, Delft, The Netherlands*

[3]*Department of Atmospheric Science, University of Wyoming, Laramie, WY, USA*

Clouds and Their Climatic Impacts: Radiation, Circulation, and Precipitation, Geophysical Monograph 281, First Edition.
Edited by Sylvia C. Sullivan and Corinna Hoose.

DOI: 10.1002/9781119700357.ch02

clouds. When kept separate from any change to the cloud condensate (liquid or ice mixing ratio) resulting from changing aerosol concentrations, this effect on cloud albedo is referred to as the Twomey effect (Twomey, 1974, 1977). In addition to the albedo, perturbations to cloud droplet concentrations also perturb microscopic processes inside clouds. These include collision-coalescence leading to the formation of warm rain (the Albrecht effect (Albrecht, 1989)), the condensation of supersaturated water vapor, the rate of sedimentation of cloud droplets, and the time needed for any given cloud droplet to evaporate. Changing the number of aerosols capable of acting as ice nuclei, the surfaces on which water can freeze, similarly influences the radiative properties of clouds and the probability of precipitation. Together, all of these processes by which aerosols affect cloud albedo and microphysical properties are known as indirect effects of aerosols on Earth's radiative balance. This distinguishes these aerosol effects from the "direct" effects of aerosols on radiation due to scattering and absorption, and from the thermodynamical perturbations to nearby clouds that result from aerosol absorption, termed semi-direct effects by Hanse et al. (1997).

The most obvious consequence of aerosol-cloud interactions, hereafter abbreviated as ACIs, for long-term radiation balance is apparent when the natural background level of aerosols is considered to be the pre-industrial atmosphere, and the perturbation is the total anthropogenic change in aerosol since pre-industrial times. Anthropogenic aerosols have led to an increase in the amount of radiation reflected by the Earth that has offset between 20% and 50% of the warming due to increased carbon dioxide concentrations over the industrial period (Bellouin et al., 2020; Charlson et al., 1992; Twomey, et al., 1984). The majority of this aerosol effect is due to the "indirect forcing of climate" by aerosols via ACIs, rather than due to the direct radiative effects of the aerosols themselves. The radiative forcing of Earth's climate due to the strengthening of the Twomey effect over time is usually abbreviated as RF_{aci}. When changes in cloud microphysical properties driven by aerosol, leading to *adjustments* in cloud macrophysical properties (e.g., area, thickness), are included, the resulting radiative forcing is termed an "effective" radiative forcing (ERF_{aci}). The large uncertainty in ERF_{aci}, which was estimated by Bellouin et al. (2020) to lie between -1.73 and $-0.27\ \mathrm{W m^{-2}}$ (16–84% range), is the main motivation for the sustained efforts by the atmospheric community to better understand ACIs. Hansen et al. (2011) are more blunt, remarking that "Aerosol uncertainty is the principal barrier to quantitative understanding of on-going climate change". Thus, ACIs are critical to a treatment of the role of clouds in the climate system.

ACIs have numerous other more subtle implications for climate, as they are involved in several important feedback loops (Gettelman et al., 2016) and effects on atmospheric circulation (e.g., Mann et al., 2018; Wilcox et al., 2020). It is likely that changes in cloud cover and thickness in response to a warming atmosphere will be affected by changing aerosol concentrations as the climate warms, and, in turn, ERF_{aci} will be strongly affected by changing clouds in a future climate.

2.1.2. Outline and Aims of this Review

ACIs have been the subject of numerous recent review articles (Bellouin et al., 2020; Fan et al., 2016; Kreidenweis et al., 2019; Mülmenstädt et al., 2018; Rosenfeld et al., 2014; Seinfeld et al., 2016; Storelvmo, 2017; Tao et al., 2012) and we are aware of others in preparation. This contribution is selective rather than comprehensive and is intended to complement these alternative perspectives. We provide a brief introductory overview intended to provide context for later chapters in this monograph, followed by a characterization of some emerging trends in current research. We begin with a brief overview of the ACIs of interest in different meteorological regimes. We then discuss some key warm-cloud processes in detail: activation of aerosols, rain formation processes, and albedo. We follow with an overview of some new methodological approaches to disentangling ACIs. This is followed by discussion of cold clouds, and finally field experiments and satellite observations. To keep our review to a reasonable length, we do not discuss how clouds affect aerosols, although we do acknowledge that such effects may also have feedbacks to, and implications for, Earth's climate, as discussed by Carslaw et al. (2010) and Tegen and Schepanski (2018).

2.2. HOW AEROSOLS AFFECT DIFFERENT CLOUD TYPES

ACIs vary widely depending on the meteorological regime in which they occur. Some of the different regimes are illustrated in Figure 2.1.

Warm clouds are those which contain no ice and in which the temperature does not extend below 0°C; we note that supercooled liquid clouds are not warm. In warm non-raining low clouds (stratus and stratocumulus) and fogs found in the subtropics and midlatitudes, the size of the droplets for a given liquid water content (LWC) depends on the droplet number concentration. This dependence in turn leads to a significant and relatively well-understood effect of increasing droplet concentration on cloud albedo (RF_{aci}). The other possible ACIs of interest are evaporation-entrainment and

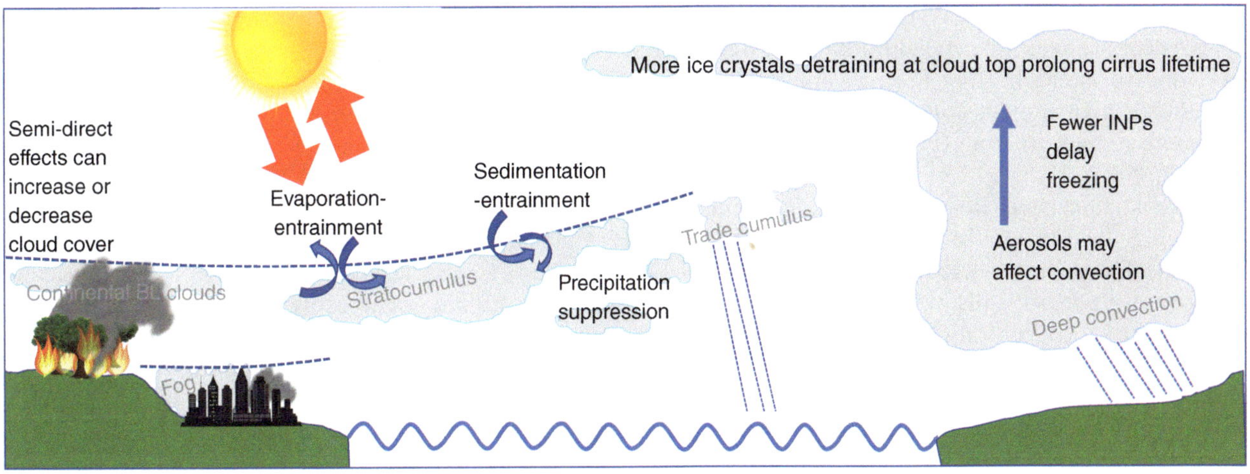

Figure 2.1 Schematic depicting some of the possible aerosol effects on clouds in the tropics and subtropics. Dashed lines denote temperature inversions.

sedimentation-entrainment feedbacks (Hill et al., 2009), which are also discussed in Chapter 12: Small-Scale Mixing of Clouds with Their Environment. Addressing sedimentation-entrainment feedbacks first, a higher droplet concentration leads to a slower sedimentation rate of droplets. Ackerman et al. (2004) showed that slower sedimentation will increase the entrainment of dry air into clouds, leading to more droplets evaporating, reducing the cloud albedo and water content. Bretherton et al. (2007) found that this increased entrainment was the result of more cloud water being concentrated at the top of the cloud when sedimentation is slower. As a result, evaporation rates at the cloud top are increased in proportion to the increased LWC, and entrainment is promoted by evaporative cooling. Furthermore, smaller droplets evaporate more quickly, promoting more turbulence and also increasing entrainment of dry air: the evaporation-entrainment feedback (Homann & Feingold, 2019; Small et al., 2009; Wang et al., 2003).

In warm raining clouds with low aerosol concentrations, a further ACI may become important: the Albrecht effect (Albrecht, 1989). This aerosol and precipitation regime is found frequently in the remote marine atmosphere. Recent research suggests the Albrecht effect is most important in very clean clouds, perhaps even only in clouds where the droplet concentration is below around 30 cm^{-3}, although this threshold is very uncertain. When aerosol concentrations are this low, droplets are large, and so fewer collision-coalescence events must take place to grow a cloud droplet into a drizzle or rain drop. Conversely, drizzle is suppressed in polluted conditions. Evidence stems from large eddy simulations and correlations in observations (e.g., Dagan al., 2015; Gryspeerdt, Goren, et al., 2019; Khairoutdinov & Kogan, 2000; Mann et al., 2014). The same mechanism may be active in slowing down precipitation processes at much higher droplet concentrations, for example in orographic clouds (Givati & Rosenfeld, 2004). However, some droplets nucleated on very large soluble aerosols, known as giant cloud condensation nuclei, grow efficiently into raindrops (Johnson, 1982) due to their high solute concentrations (Jensen & Nugent, 2017). Increasing concentrations of these large aerosols should, up to a point, hasten precipitation onset and reduce cloud liquid water paths (LWP) (Posselt & Lohmann, 2008).

Tropical trade cumulus, the topic of Chapter 7: Tropical Marine Low Clouds, and also congestus and deep convective clouds tend to have higher vertical velocities (greater than 1.0 ms^{-1}). In these clouds, concentrations of large aerosols are generally low (below 200 cm^{-3} in the accumulation mode), as these aerosols are often depleted by rainfall. Therefore, when compared to other cloud types, supersaturations may reach higher values, and smaller aerosols may be activated as droplets. Several authors have explored possible links between aerosol concentrations and the magnitude and altitude of latent heat release during condensation and freezing in clouds, and, by extension, if aerosols can influence how much condensation occurs in liquid and how much in ice phases of deep clouds. The effect of aerosols on deep clouds is discussed further in section 2.7. In brief, in more polluted clouds, in theory the larger surface area of aerosols should lead to more condensation at lower altitudes than in cleaner clouds. The lower altitude of the condensation and resultant latent heat release in more polluted clouds can affect updraft speeds but there is a lack of consensus in the community on the strength of these latent heating effects (Fan et al., 2007; Grabowski & Morrison, 2016, 2020).

In mixed-phase clouds, as found in cyclones and tropical deep convection, in addition to invigoration effects due to latent heating (Fan et al., 2018; Khain et al., 2004), high droplet concentrations may suppress or promote droplet freezing (Pruppacher & Klett, 2010; Yin et al., 2005). Delayed formation of warm rain may enhance cold rain later (Rosenfeld et al., 2008), as well as changing the radiative properties of the cloud top. If droplets freeze homogeneously, more droplets lead directly to more ice crystals, while if they freeze heterogeneously the ice concentration will depend more on the concentration of ice nuclei, as well as on poorly understood ice multiplication processes. Larger ice crystals have substantially higher sedimentation rates, and thus aerosols also affect the lifetime of cirrus clouds (Kärcher & Lohmann, 2003) and the mixed-phase clouds found at high latitudes.

2.3. AEROSOL ACTIVATION

With the exception of the semi-direct effect, all ACIs depend on the influence of aerosols on cloud droplet concentration. The major source of liquid droplet concentration in all warm clouds, and probably most cold clouds, is aerosol activation, and the source rate is determined by the cooling rate (usually the result of upward air motion), by the aerosol concentration and properties, and by the temperature. Sinks are droplet evaporation, droplet freezing, warm rain formation, and riming. Bellouin et al. (2020) estimate that cloud droplet concentrations increased by between 5% and 17% from 1850 to the period 2005–2015.

The activation of an aerosol particle to a cloud droplet occurs when the supersaturation of water vapor exceeds the critical supersaturation for that particle S_c. The supersaturation is generally the result of an air parcel cooling as it rises. S_c is given by

$$S_c = \left(\frac{4A^3}{27\kappa r_d^3} \right)^{1/2} \tag{2.1}$$

where A is proportional to the ratio of surface tension to temperature, κ is the volume-weighted hygroscopicity, and r_d is the dry radius of the aerosol (Ghan et al., 2011). Thus, for a given aerosol particle, the critical supersaturation depends principally on its size (Dusek et al., 2006) but also on hygroscopicity (Köhler, 1936; Petters & Kreidenweis, 2007) and surface tension (Prisle et al., 2012; Shulman et al., 1996). Larger, more hygroscopic particles with lower surface tension activate at lower supersaturations than smaller, less hygroscopic particles with higher surface tensions, and surface-bulk partitioning of molecules, especially surfactants, is also important in determining surface tension and hygroscopicity (Petters & Kreidenweis, 2013; Sullivan et al., 2009). The dependence of activation on hygroscopicity, sulfuric acid's key role in new aerosol formation, and large anthropogenic emissions of sulfur compounds together explain why the radiative forcing of climate by ACIs is more strongly dependent on sulfate than on any other aerosol species (Boucher & Lohmann, 1995; Charlson et al., 1992).

Calculating whether the critical supersaturation is exceeded requires accounting for the number of aerosol particles present, as aerosols take up water hygroscopically before they activate. If cloud droplets are already present they should also be accounted for. Well above cloud base in warm clouds (where activation is less likely), the cloud droplets, not the aerosols, likely dominate the sink of water vapor, and the supersaturation S is given by

$$S = \frac{w}{2\pi G} \frac{dq_l}{dz} \left(\frac{1}{N_d^2 q_l} \right)^{1/3} \tag{2.2}$$

where w is the vertical air velocity, G is a growth factor whose most important dependence is on temperature, q_l is LWC, z is height, and N_d is droplet concentration (Yang et al., 2019). A similar equation (Korolev & Mazin, 2003; Politovich & Cooper, 1988; Squires, 1952) does not require knowledge of $\frac{dq_l}{dz}$, but relies on the assumption of quasi steady state between the source of supersaturation, the updraft, and the sink of water vapor to droplets. This may be a good approximation even at the base of the cloud, provided turbulent fluctuations in vertical velocity are correctly accounted for (Prabhakaran et al., 2020). However, more complicated parameterizations accounting for the sink of water vapor to aerosols are usually used at cloud base (Ghan et al., 2011; Twomey, 1959); the most commonly used is that of Abdul-Razzak and Ghan (2000) which adopts the framework of Köhler (1936). More detailed models, often with bin-resolved cloud microphysics, are able to treat activation without the 'saturation adjustment' approximation that is typically used in weather and climate models (Khain et al., 2000). This approximation is the assumption that water vapor condenses on droplets or ice crystals sufficiently quickly that supersaturation is reduced to zero at the end of each model time step, at which point water vapor and condensate are in equilibrium. Detailed models without this assumption must treat supersaturation "prognostically," which means they must keep track of how it evolves from one time step to the next. These models are especially needed when activation within existing clouds is important (Fridlind et al., 2004), and potentially also fogs (Schwenkel & Maronga, 2019; Thouron et al., 2012). In fogs, supersaturation may be generated by radiative cooling, not by updraft velocities. Recently, however, approaches have been developed that will enable activation in fog or activation above cloud base to be approximated to some extent in larger scale weather and climate models (Gordon et al., 2020; Poku et al., 2021; Wang et al., 2013; Yang et al., 2015).

Both aerosol size distributions (Kulkarni & Wang, 2006; Liu et al., 1974; Wang & Flagan, 1989) and CCN concentrations at different vapor supersaturations (Chuang et al., 2000; Hudson & Squires, 1976) can be measured accurately by in situ instrumentation aboard aircraft, as can cloud droplet size distributions (Baumgardner et al., 2001; Knollenberg, 1976) and updraft speeds (Axford, 1968; Wood et al., 1997). However, a perfect closure study of aerosol activation in the real atmosphere is elusive, mainly because droplets are not generally nucleated in the same place or at the same time as they are detected. Activation can be better observed in chamber experiments such as the Michigan Technological University Pi chamber (Prabhakaran et al., 2020; Shaw et al., 2020). This technology for cloud physics studies has stood the test of time (Aitken, 1881) and still offers great promise for ever more quantitative studies of the critical interactions between turbulence and microphysics.

2.4. WARM CLOUD ALBEDO

In the preceding section, we described the activation of aerosol into cloud droplets. Here, we describe how the ability of aerosol to control cloud droplet number concentration (N_d) allows aerosol to directly affect cloud albedo (α). Alteration in cloud albedo by aerosol through changing the number concentration of cloud droplets, separately to any changes to the cloud condensate amount, is known as the first indirect effect or Twomey effect (Twomey, 1977).

Based on Lorenz-Mie theory, the optical depth (τ) of a cloud is the extinction cross-section of a collection of liquid droplets with geometric radii (r) and size distribution of cloud droplets ($n_d(r)$)

$$\tau = \int_{z=0}^{h} \int_{0}^{\infty} Q_e \pi r^2 n_d(r) dr dz \tag{2.3}$$

where Q_e is the efficiency of extinction, z is vertical distance and h is the geometric cloud height, and $Q_e \approx 2$ for liquid cloud droplets. The effective radius (r_e) proportional to scattering area, is

$$r_e = \frac{\int_0^\infty r \cdot \pi r^2 n_d(r) dr}{\int_0^\infty \pi r^2 n_d(r) dr} \tag{2.4}$$

and LWC within some volume through the cloud is

$$LWC = \frac{4\pi}{3} \rho_{liq} \int_0^\infty r^3 n_d(r) dr \tag{2.5}$$

Dividing equation (2.5) by the integrand of the integral with respect to z in equation (2.3), assuming $Q_e \sim 2$, rearranging, and integrating with respect to height yields in terms of τ

$$\tau = \int_{z=0}^{h} \frac{3}{2\rho_{liq}} LWC(z)/r_e(z) dz \tag{2.6}$$

Following the original presentation of S. Twomey (1977), we wish to represent this in terms of the number concentration of cloud droplets N_d. This is useful because N_d depends mainly on the number concentration of CCN, and unlike r_e is to a first approximation independent of LWC. N_d is the LWC divided by droplet mass

$$N_d(z) = LWC(z) \cdot \left(\rho_{liq} \frac{4\pi r(z)^3}{3} \right)^{-1} \tag{2.7}$$

To calculate N_d from satellite data, a constant factor $k = r^3/r_e^3$ is inserted to allow the effective radius rather than the true radius to be used in this equation. An important feature of real clouds that must also be taken into account when measuring N_d from space is the degree to which they are subadiabatic (f_{ad}). We may represent the LWC of a cloud as a function of adiabaticity and cloud vertical extent

$$LWC(z) = f_{ad} \Gamma_{ad} z \tag{2.8}$$

where Γ_{ad} is the adiabatic rate of increase of LWC with height, a function of pressure and temperature. Using equations (2.8) and (2.7) in equation (2.6) we may write cloud optical depth as a function of N_d, cloud geometric depth and adiabaticity

$$\tau = A \cdot (f_{ad} \Gamma_{ad})^{2/3} N_d^{1/3} h^{5/3} \tag{2.9}$$

where A is a constant equal to $(243\pi/250\rho_{liq}^2)^{1/3}$. Thus, cloud optical depth is a weak function of N_d. This may also be written in terms of LWP instead of cloud thickness using the integrated version of equation (2.8)

$$\tau = N_d^{1/3} \cdot LWP^{5/6} \cdot A \cdot \left(\frac{32}{f_{ad} \Gamma_{ad}} \right)^{1/6} \tag{2.10}$$

Increasing N_d increases cloud optical depth and yields the first indirect effect or aerosol-cloud radiative effect. The dependence of cloud albedo, or reflectance, on N_d is further muted: assuming conservative scattering allows us to approximate (Lacis & Hansen, 1974)

$$\alpha \sim \frac{\tau}{\tau + \eta} \tag{2.11}$$

where η is a constant that varies depending on the value of the asymmetry parameter (usually given symbol g) and on assumptions in the derivation of the equation. The asymmetry parameter is the intensity-weighted average of the cosine of the scattering angle. For a typical g of 0.85, η ranges between 7 (Petty, 2006) and 13.3 (Tornow et al., 2020). An interesting feature of the

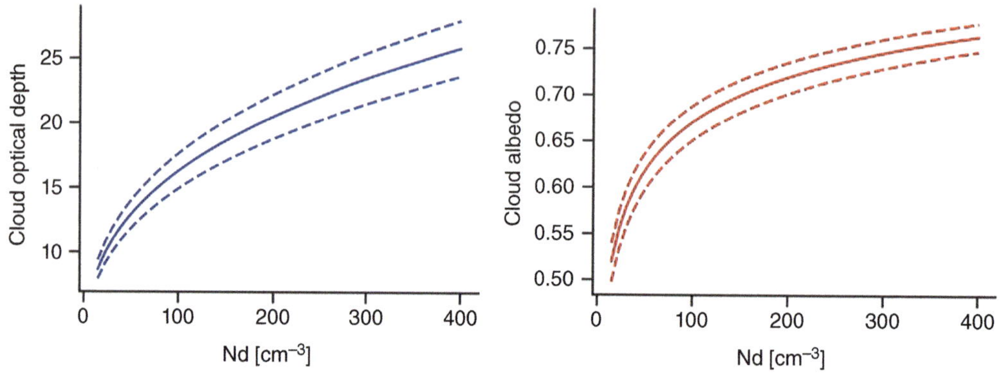

Figure 2.2 Cloud optical depth (τ) and albedo (α) as a function of N_d. Solid lines show a cloud with $LWP = 100\,\mathrm{g\,m^{-3}}$ and dashed lines show $LWP = 90\,\mathrm{g\,m^{-3}}$ and $110\,\mathrm{g\,m^{-3}}$. Cloud albedo is derived following the approximation in Lacis and Hansen (1974).

1/3 power to which N_d is raised and the relationship between α and τ is that at high N_d the effects of increasing N_d yield diminished returns. One consequence of this is the importance of the pre-industrial state in determining ERF_{aci} (Carslaw et al., 2013). The dependence of τ on N_d and α on τ is shown in Figure 2.2.

Aerosol-induced changes in cloud albedo translate into changes to the radiative balance at the top of the atmosphere R,

$$dR = d(R_{\mathrm{SW}}^{\mathrm{in}}[A_{\mathrm{cld}}c + (1-c)A_{\mathrm{clr}}] - R_{\mathrm{LW}}^{\mathrm{out}}) \tag{2.12}$$

where $R_{\mathrm{SW}}^{\mathrm{in}}$ denotes incoming shortwave radiation, $R_{\mathrm{LW}}^{\mathrm{out}}$ outgoing longwave radiation, c cloud fraction, and $A_{\mathrm{clr/cld}}$ the clear- and cloud-sky albedo at the top of the atmosphere. The longwave term in equation (2.12) can be neglected in warm clouds unless aerosol changes in cloud top temperature are of interest or the clouds are thin (e.g., Garrett et al., 1 February 2002). Furthermore, changes in $R_{\mathrm{SW}}^{\mathrm{in}}$ are usually also negligible, at least on sub-decadal timescales. Diamond et al. (2020) provide a recipe to relate changes in radiative balance to the albedo of a scene with partial cloud cover, accounting for the transmissivity and albedo of the atmosphere above clouds. Over ocean in unpolluted situations, when both the surface albedo and the albedo of the atmosphere above the clouds are small and the change in the cloud fraction is not too large, it may be possible to approximate equation (2.12) as

$$dR \propto d(\alpha \cdot c) \tag{2.13}$$

where α is the cloud albedo. However, over land or sea ice, in polluted conditions, and likely also for high or deep clouds (because the approximation relies on the above-cloud albedo being similar to the clear-sky albedo), equation (2.13) would not be valid.

2.5. APPROACHES TO DETERMINING SUSCEPTIBILITY

Aerosol effects on the radiative properties of clouds can be quantified by relating a change in scene albedo $dR = d(\alpha \cdot c)$ (equation (2.13)) to the perturbation in cloud droplet concentration dN_d that caused the change in scene albedo,

$$\frac{\mathrm{d}\ln R}{\mathrm{d}\ln N_d} = \frac{\mathrm{d}\ln \alpha}{\mathrm{d}\ln N_d} + \frac{\mathrm{d}\ln c}{\mathrm{d}\ln N_d}, \tag{2.14}$$

where the logarithmic derivative $\mathrm{d}\ln R/\mathrm{d}\ln N_d = (\mathrm{d}R/R)/(\mathrm{d}N_d/N_d)$ is introduced for convenience and will broadly be referred to as *susceptibility*, following Platnick and Twomey (1994). The perturbation in cloud droplet number concentration is assumed to be the result of a change in aerosol number concentration dN_a, not the result of precipitation or thermodynamic changes in the cloud. Its numerical value corresponds to the percentage change in R that is obtained for a 1% change in N_d. Expressing cloud albedo as a nested function of N_d,

$$\alpha = \alpha\{\tau[N_d, \mathrm{LWP}(N_d)]\}, \tag{2.15}$$

where $\alpha(\tau)$ is given by equation (2.11) and $\tau(N_d, \mathrm{LWP})$ by equation (2.10). Applying the chain rule provides the following simplification for warm clouds:

$$\begin{aligned}\frac{\mathrm{d}\ln \alpha}{\mathrm{d}\ln N_d} &= \frac{\partial \ln \alpha}{\partial \ln \tau}\frac{\mathrm{d}\ln \tau}{\mathrm{d}\ln N_d}\\ &= \frac{\partial \ln \alpha}{\partial \ln \tau}\left(\frac{\partial \ln \tau}{\partial \ln N_d} + \frac{\partial \ln \tau}{\partial \ln \mathrm{LWP}}\frac{\mathrm{d}\ln \mathrm{LWP}}{\mathrm{d}\ln N_d}\right)\\ &= (1-\alpha)\left(\frac{1}{3} + \frac{5}{6}\frac{\mathrm{d}\ln \mathrm{LWP}}{\mathrm{d}\ln N_d}\right)\end{aligned} \tag{2.16}$$

Note that partial derivatives (symbol: ∂) quantify changes that arise with respect to one specific variable, keeping all other variables at fixed values. Partial logarithmic derivatives of a power law like equation (2.10) correspond to its exponents. They are thus independent of any constant pre-factors. Derivatives indicated by "d" comprise total changes, including nested dependencies. Such dependencies can encapsulate very complex relationships, for which no theoretical expressions are available; the LWP response to droplet number perturbations, $\mathrm{d}\ln \mathrm{LWP}/\mathrm{d}\ln N_\mathrm{d}$, for example, depends on the full complexity of cloud dynamics.

Summarizing equations (2.14) and (2.16) yields the overall susceptibility of a partially cloudy scene (Bellouin et al., 2020),

$$\frac{\mathrm{d}\ln R}{\mathrm{d}\ln N_d} = \frac{1-\alpha}{3} + \frac{5(1-\alpha)}{6}\frac{\mathrm{d}\ln \mathrm{LWP}}{\mathrm{d}\ln N_d} + \frac{\mathrm{d}\ln c}{\mathrm{d}\ln N_d} \tag{2.17}$$

$$= \frac{\partial \ln R}{\partial \ln N_d} + \frac{\partial \ln R}{\partial \ln \mathrm{LWP}}\frac{\mathrm{d}\ln \mathrm{LWP}}{\mathrm{d}\ln N_d} + \frac{\partial \ln R}{\partial \ln c}\frac{\mathrm{d}\ln c}{\mathrm{d}\ln N_d} \tag{2.18}$$

where the first equality, like equation (2.16), strictly applies only to warm clouds. The second equality assumes that it is more generally suitable to quantify variability in R based on a functional relationship R=f(N_d, LWP, c). We can identify the first term on the right-hand side of equation (2.18) as a quantification of the Twomey effect, that is, the change of R with N_d, when keeping LWP (and c) constant. The second and third terms correspond to the response of LWP and c to changes in droplet number, respectively. These terms are usually referred to as LWP and cloud fraction *adjustments*.

Traditionally, logarithmic derivatives are empirically quantified as linear regression slopes in log-log space. Total derivatives correspond to one-dimensional linear regression (e.g., Quaas et al., 2009). For partial derivatives, other dependencies must be kept constant through stratification of the dataset (e.g., McComiskey et al., 2009), or controlled for by applying multi-linear regression (Fig. 2.3b).

The pragmatic linear regression approach faces methodological and conceptual difficulties. First of all, the linear regression approach only quantifies correlation but does not guarantee causality, that is, it is not clear if observed variability in LWP is indeed caused by variability in N_d (Gryspeerdt, Goren, et al., 2019; McCoy, Field, et al., 2020). As an example of a correlation between LWP and N_d that is not caused by cloud processes, consider a dataset that either samples under moist and clean oceanic conditions or under dry and polluted continental conditions. A second problem related to the regression approach is that the variability exhibited by a dataset varies with the *scale of data aggregation* (Bender et al., 2019; McComiskey & Feingold, 2012; Possner et al., 2016; Schutgens et al., 2016). Here, the scale of data aggregation refers to how data are averaged at a scale greater than the *process scale* at which ACI occurs before analysis is performed (McComiskey & Feingold, 2012).

A linear regression between the logarithms of quantities further assumes a power law relationship with constant exponents. It is thus not taken into account that such a relationship may only be a valid approximation under certain conditions. As an example, equation (2.10) is only valid if a broadening of the droplet size distribution from collision-coalescence can be neglected (Feingold et al., 1997). We consider LWP adjustments as another example of this challenge: Compare LWP adjustments in a raining and a non-raining warm cloud. For the raining state, we expect $\mathrm{d}\ln \mathrm{LWP}/\mathrm{d}\ln N_d > 0$ from rain suppression. For the non-raining state, droplet-size effects on entrainment dominate and $\mathrm{d}\ln \mathrm{LWP}/\mathrm{d}\ln N_d < 0$. In addition to this *state dependence*, which is tied to the internal state of a cloud field, the LWP adjustment may also depend on external cloud controlling factors, for example, on above-cloud moisture (Chen et al., 2014), which control the entrainment response to droplet size. We denote this as *regime dependence*. Both *state and regime dependence* of susceptibility can be accounted for by a stratification of the dataset prior to quantifying the aerosol-cloud relationship, in our example based on the occurrence of rain, and based on above-cloud moisture (Gryspeerdt, Goren, et al., 2019). Biases from imbalanced sampling within a specific state and regime, for example, in the form of sparsely sampled very high N_d in the non-precipitating regime for high above-cloud RH, can be addressed through joint histograms (e.g., Gryspeerdt, Goren, et al., 2019) or, equivalently, through the use of bin-average values (Glassmeier et al., 2021; Rosenfeld et al., 2019). Finally, the response dLWP to $\mathrm{d}N_d$ may not be instantaneous but evolve over time so that susceptibility is *time-dependent* on the time that has evolved since the aerosol conditions of a cloud system were perturbed (Glassmeier et al., 2021). The next section will discuss recent approaches to address these challenges.

2.6. NEW METHODOLOGICAL APPROACHES

The proliferation of new, increasingly sophisticated observations and new simulations with increasingly complex models and/or increasingly high grid resolution has led to an explosion in the availability of data that contains information on ACIs. To make sense of the data, new statistical approaches drawing on increasingly popular machine learning algorithms can be employed. We highlight here only a small sample of the techniques available, chosen according to our own familiarity with them.

2.6.1. Gaussian-Process Emulation to Address State Dependence

If the albedo susceptibility is not dependent on the state of the cloud field, in particular the absolute values of N_d, LWP, and c, the scene albedo R varies with cloud properties according to a power law, or log-log linear function,

$$R = i \cdot N_d^{\alpha} \mathrm{LWP}^{\beta} c^{\gamma} \Leftrightarrow \ln R = \ln i + \alpha \ln N_d + \beta \ln \mathrm{LWP} + \gamma \ln c, \qquad (2.19)$$

where i, α, β, γ are constants that can be determined from multi-linear regression. Figure 2.3b illustrates a two-dimensional version of equation (2.19).

For a state-dependent susceptibility, scene albedo $R = f(N_d, \mathrm{LWP}, c)$ is given by a more complicated function f. Figure 2.3a illustrates such a functional relationship for our two-dimensional example. A power law approximation is possible for limited regions of the N_d-LWP state space, in particular for fully overcast states with $c \approx 1$. Beyond this region, the power law approximation cannot capture the full complexity of the R-state space relationship as indicated by the missing curvature of scene albedo isolines in Figure 2.3b. While a power law relationship can be determined from (log-log) linear regression, mapping-out relationships like the one illustrated in Figure 2.3a requires surrogate models, or emulators, for how R, or the complex function f, respectively, varies across state space. *Gaussian-process emulation* provides such a description. We refer the reader to Rasmussen and Williams (2006) and Lee et al. (2011) for a detailed explanation. The scene-albedo surface in Figure 2.3a corresponds to the mean of a two-dimensional Gaussian process (a functional analog of a multivariate Gaussian distribution), whose mean and covariance functions have been constrained by a set of training data points. This effectively corresponds to an interpolation between the known training data points with thoroughly quantified uncertainty (the latter is not illustrated in Fig. 2.3). Relationships obtained in this manner can then be differentiated to obtain state-dependent partial derivatives as illustrated in Figure 2.3c,d.

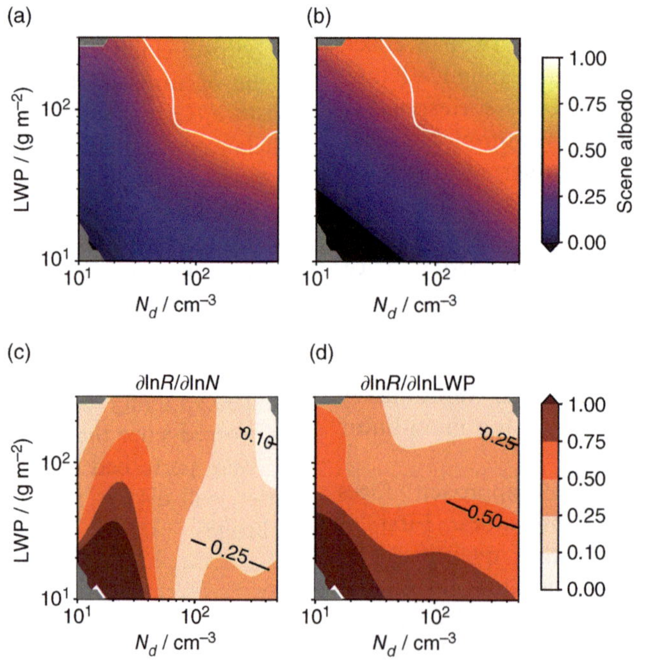

Figure 2.3 Scene albedo R' as a function of cloud droplet number concentration, N_d, and liquid-water path, LWP, as obtained from (a) Gaussian process emulation representing the general function $R' = f'(N_d, \mathrm{LWP}) = f[N_d, \mathrm{LWP}, c(N_d, \mathrm{LWP})]$ and (b) bi-linear regression with an underlying model $R' = \ln i' + \alpha' \ln N_d + \beta' \ln \mathrm{LWP}$ (cf. equation (2.19)). The white contour indicates where cloud fraction approaches unity, $c \approx 1$, such that fully overcast conditions are expected in the upper right corner of the N_d-LWP space. Comparing (a) and (b) in this region illustrates that the bi-linear model (b) is an appropriate approximation of the more complex relationship (a) under such conditions. (c, d) State-dependent values for partial derivatives (color contours) as derived from (a) and according to equation (2.17) (black line contours, only shown in the non-raining regime where they apply). Plots are adapted from Glassmeier et al. (2019) and apply to stratocumulus as modeled with large eddy simulations under the specific boundary conditions outlined in the reference.

2.6.2. Tendency Emulation to Address Time Dependence of Adjustments

Uncertainty in adjustments constitute an important uncertainty in overall ERF_{aci} (Bellouin et al., 2020). Forcing from adjustments in LWP was estimated to be between 0 and 0.56 Wm^{-2} and from adjustments in cloud cover to be between −1.14 and 0 Wm^{-2} (Bellouin et al., 2020) based on observational constraints. Thus, methods to leverage observations to constrain adjustments are impactful. LWP adjustments, $\text{adj} = \mathrm{d}\ln \text{LWP}/\mathrm{d}\ln N_d$, as well as cloud-fraction adjustments, $\mathrm{d}\ln c/\mathrm{d}\ln N_d$, aim to quantify the effect of perturbing certain cloud processes, like precipitation formation and entrainment, on the state of the cloud as captured by LWP and c. (A more direct, but also less feasible, way of doing so would be to quantify aerosol effects on process rates; see Mülmenstädt et al. (2020) for an interesting approach along these lines). These perturbations to process rates over time manifest in perturbations to the cloud state. In particular, they are absent at the moment of perturbation, $\text{adj}(t = 0) = 0$, where t indicates time since the perturbation. For stratocumulus cloud fraction, Christensen et al. (2020) show that initially strong aerosol effects arising from aerosol-enhanced cloud formation decrease over time. They specifically observe that differences in cloud fraction become less pronounced over a timescale of days. Also for stratocumulus, Glassmeier et al. (2021) have shown that the value of time-dependent adjustments, $\text{adj} = \text{adj}(t)$, evolves toward a steady state value $\text{adj}(\infty)$ according to

$$\text{adj}(t) = \text{adj}(\infty)\left[1 - \exp\left(-\frac{t}{\tau_{adj}}\right)\right], \tag{2.20}$$

where the equilibration timescale of the adjustment,

$$\tau_{\text{adj}} = \tau\left[1 - d \cdot \text{adj}(\infty)\right], \tag{2.21}$$

scales with the steady state value and $\tau = 9.6$ h and $d = 1.6$ are specific to the non-raining Sc studied. Glassmeier et al. (2021) determine the value of $\text{adj}(\infty)$ from short-term LWP tendencies, or LWP time series, respectively. Emulation of such tendencies $v = \mathrm{d}\ln \text{LWP}/\mathrm{d}t$ allows the authors to represent the LWP evolution of an Sc deck as a flow field $v(N_d, \text{LWP})$ in the space of Sc states spanned by N_d and LWP. Figure 2.4 illustrates such a flow field and its relationship to $\text{adj}(\infty)$.

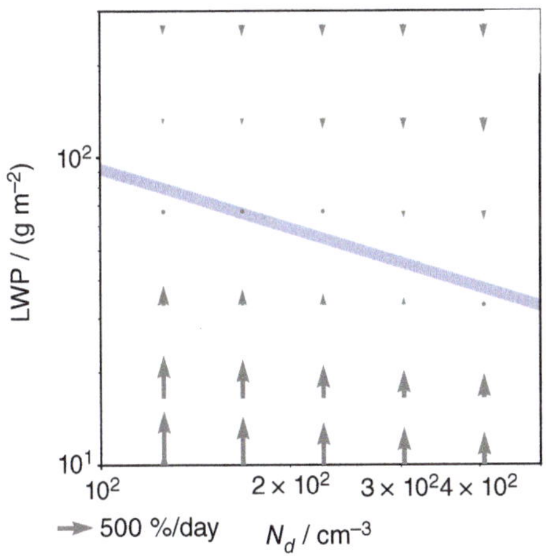

Figure 2.4 Liquid water path tendencies $v = \mathrm{d}\ln \text{LWP}/\mathrm{d}t$ as function of N_d and LWP. The LWP steady state, characterized by $v = 0$, is indicated by the blue line with negative slope (gray in monochrome version). Its slope corresponds to the steady state LWP adjustment $\text{adj}(\infty)$ defined in the text. See Glassmeier et al. (2021) for details.

2.6.3. Causality of LWP Adjustments

Examining covariances between aerosol, cloud microphysics (e.g., N_d), and cloud macrophysical quantities is inherently causally ambiguous. Covariances between these quantities can result from external factors as well as from real ACI (Stevens & Feingold, 2009), in contrast to the steady state adjustments discussed in section 2.6.2. In the case illustrated in Figure 2.5, N_d may affect LWC through adjustments (the strength of this effect is characterized as $d\ln LWP/d\ln N_d$, as discussed in section 2.5). Any change in liquid content may in turn affect precipitation processes in clouds that are precipitating or near to precipitating. However, precipitation strongly affects N_d through coalescence scavenging. Wood et al. (2012) estimate that precipitation scavenging reduces N_d by a factor of 2–3 over remote oceans. Precipitation also reduces concentrations of CCN by removing both interstitial and activated aerosol through wet scavenging. Air

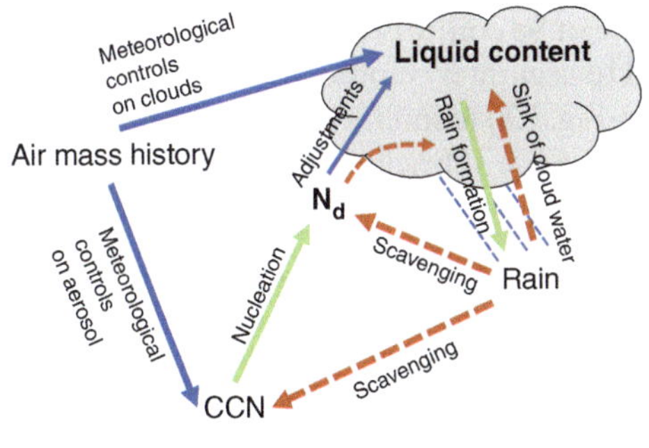

Figure 2.5 A schematic of causal links between cloud, aerosol, precipitation, and meteorology in warm clouds (Gryspeerdt, Goren, et al., 2019; McCoy, Field, et al., 2020; Michibata & Suzuki, 2020). Green lines (gray in monochrome version) indicate a positive relationship. Red (dashed in monochrome version) denotes a negative relationship. Blue (black in monochrome version) is ambiguous.

mass history affects both aerosol and clouds, inducing further potentially spurious covariance between N_d and cloud macrophysical properties and complicating the constraint of $d\ln LWP/d\ln N_d$ from observations (Mauger & Norris, 2007). As a specific example of the covariability discussed in section 2.5, the present-day configuration of sources and sinks of aerosol, with strong continental aerosol sources and strengthening precipitation away from continents, yields a spurious negative $d\ln LWP/d\ln N_d$ in important cloudy regions such as the stratocumulus-cumulus transition off the Chilean coast (Engstrom et al., 2015; Gryspeerdt, Goren, et al., 2019; Mauger & Norris, 2007; McCoy, Field et al., 2020; Wood et al., 2012).

Several methods have been proposed to cut this causal Gordian Knot. These must rely on either modeling (where causality can be controlled) or examining case studies where causality is mostly unambiguous (discussed below in section 2.6.4). An example of the empirical approach is given by Gryspeerdt, Goren, et al. (2019), who use variability observed by Moderate Resolution Imaging Spectroradiometer (MODIS) stratified according to volcanic emissions of aerosol to isolate a causal signal. Some examples of the modeling-based approach are given by D. T. McCoy et al. (2018) and McCoy, Field, et al. (2020), who utilize global simulations to study extratropical weather regimes where the effects of precipitation on aerosol and N_d are disabled to isolate the effect $N_d \rightarrow LWP$ and $N_d \rightarrow c$. The methodology assumes that models are able to replicate the effects of precipitation scavenging on N_d and CCN in a realistic manner and that nonlinear feedbacks between adjustments and precipitation are negligible (e.g., precipitation is primarily determined by the state of the atmosphere, not by adjustments).

In short, observational inference must either use a case study where causality is less opaque, or use a model where causality can be controlled (but where the model must be assumed to accurately reproduce the non-ACI parts of the system in Fig. 2.5) to infer $d\ln LWP/d\ln N_d$. In the following section, we will briefly discuss the use of case studies to study ACIs.

2.6.4. Causality Inference from Transient Events

As discussed in the preceding section, causal ambiguity greatly hampers our ability to cleanly infer the strength of ACIs from observations. Examination of case studies where changes in aerosol are disentangled from meteorology and precipitation has been used in many studies to constrain ACIs. These so-called *"opportunistic experiments"* in aerosol forcing offer an opportunity to assign causality to ACI inferred from observations (Christensen et al., 2021).

The canonical example of an opportunistic experiment is the bright lines left in cloud cover by ships (so-called ship tracks), which have been observed since the first satellite images of Earth (Conover, 1966). Ship tracks are a common feature of oceanic clouds in the anthropocene and many studies have examined their behavior in an attempt to isolate the effect of aerosol on clouds (Ackerman, Toon, Taylor, et al., 2000; Diamond et al., 2020; Goren & Rosenfeld, 2014; Gryspeerdt, Smith et al., 2019; Hobbs et al., 2000; Toll et al., 2017, 2019; Thornton et al., 2017) (Web of Science suggests >70 papers since 1990), although recent work shows that estimates of the radiative effect of ACI and LWP adjustments in Sc from ship tracks are up to 200% too large when compared to the climatological effect (Glassmeier et al., 2021).

While less visually striking, many other opportunistic experiments have been discovered and used to constrain ACI (M. Christensen et al., 2022). These include volcanic emissions of sulfur dioxide from strong emitters such as Kīlauea (Hawaii) (Fig. 2.6), Holuhraun (Iceland), and the Vanuatu archipelago (Carn et al., 2017; McCoy & Hartmann, 2015; Mace & Abernathy, 2016; McCoy et al., 2018; Malavelle et al., 2019; Yuan et al., 2011) as well as smaller sources such as the volcanoes on the Aleutian and Sandwich islands (Gassó, 2008). Volcanoes have the capability to substantially affect boundary layer aerosol and by extension cloud over duration of the eruption. However, this relatively short time period means that meteorological variability may make changes in cloud properties difficult to interpret (McCoy et al., 2018). Steady changes in aerosol sources in step with economic activity and emissions controls, most notably in East Asia and North America, give a longer timescale

Figure 2.6 MODIS Aqua imagery of the 2008 eruption of Kīlauea, Hawaii. MODIS thermal anomalies are shown in red (impossible to see by eye in monochrome version). The OMI boundary layer SO_2 retrieval is overlaid in orange (large gray pixels in monochrome version).

during which the effects of aerosol and meteorological variability can be more cleanly separated (Bennartz, 2007; McCoy et al., 2018). Some examples of emissions controls that have had changes in cloud microphysics associated with them include the implementation in the United States of the 2009 Clean Air Interstate Rule and 2010 Acid Rain Program. In the People's Republic of China, similar controls include the twelfth 5-year Plan and preparations for the 2009 Beijing Olympic Games (McCoy et al., 2018). However, the slower nature of these changes introduces the confounding factors of internal variability and climate change (Manaster et al., 2017; Norris et al., 2016; Zelinka et al., 2020). The cycle of aerosol emissions between weekdays and weekends has also been suggested as a path toward assigning causality in ACI (Quaas et al., 2009). More generally, the contrast between polluted and pristine regions (Hamilton et al., 2014) may be considered to aggregate anthropogenic activity into a single opportunistic experiment (Koren et al., 2014; McCoy, McCoy, et al., 2020).

2.6.5. Ensemble Approaches to Uncertainty Quantification and Reduction in GCMs

General circulation models (GCMs) must represent ACIs to model Earth's climate and are also frequently used to study ACIs on large scales. Uncertainties in GCMs are associated with their initial conditions and internal variability, their "structural" representation of, or omission of, key processes, and their uncertain input datasets and parameters. The uncertainty associated with a subset of the uncertain input parameters, the "parametric uncertainty," can be quantified using perturbed parameter ensembles. In this approach, a large number of simulations are run with parameters varied, or inputs scaled, over plausible uncertainty ranges determined by the modeling team or other experts.

Some GCM ensembles aimed at ACI studies use designs where one parameter (or set of parameters) is varied at a time, and the effect of that parameter is determined directly (e.g., Lohmann & Ferrachat, 2010; Liu et al., 2018). Others vary many parameters at once and infer the importance of each parameter from subsequent analysis. This procedure allows interactions between parameters to be studied. It is feasible to vary around 50 parameters in today's GCMs in this way if around 200 GCM simulations of a year are integrated. To formally quantify the parametric uncertainty, tools such as Gaussian process emulation (see section 2.6.1) are then used to interpolate over the 50-dimensional parameter space and the range of model output variables such as radiative forcing in the emulated simulations is a measure of the parametric uncertainty in that output variable. Such techniques have been applied to understand parametric uncertainty in several climate models (e.g., Sexton et al., 2012; Qian et al., 2018). Focusing on ACIs, radiative forcing from the Twomey effect in a chemical transport model was calculated by Carslaw et al. (2013) and similar techniques were applied to a GCM ensemble described by Yoshioka et al. (2019) with a wider range of possible aerosol indirect forcing mechanisms by Regayre et al. (2018). These authors found that parametric uncertainty in the aerosol forcing is dominated by uncertainties in aerosol emissions and in the parameterization of clouds and their radiative effects. Studies of ACIs in a convective cloud model by Johnson et al. (2015) highlighted the high sensitivity of a convective cloud to aerosol concentrations and to the representation of cloud microphysics.

More recent work is focused on using observations to constrain the parameter space by eliminating implausible model variants, and thereby reducing uncertainty in the model output radiative forcing estimate (Johnson et al., 2020; Regayre et al., 2019; Watson-Parris et al., 2020). At the time of writing in 2021, these authors were able to reduce the uncertainty in their (ERF_{aci}) estimate by only 7%, though the direct aerosol effective radiative forcing uncertainty was reduced by 33%. Reducing the uncertainty is intrinsically difficult because not all parameters can be easily constrained and because each value of radiative forcing in the plausible range can be achieved with many different combinations of inputs, that is, a situation of *equifinality* (Lee et al., 2016). However, there is scope to reduce the uncertainty further by refining the statistical techniques and by using a more diverse set of complementary observations.

2.6.6. Regime Classification

The multiscaled nature of ACI means that patterns of space and time covariability cannot be restricted to a single length scale. One example is the structure of open and closed cellular convection in clouds (Fig. 2.7). This bi-stable morphology resembles Bernard-Rayleigh convection and occurs across the globe, but was not described until the first images of clouds from space revealed their structure (Agee, 1984). It is hypothesized that suppression of precipitation by aerosols may inhibit the transition from closed to open cells (Rosenfeld et al., 2006). This has been examined in satellite (Christensen et al., 2020; Goren & Rosenfeld, 2012, 2015) and aircraft (Abel et al., 2020) observations and in high-resolution simulations (Grosvenor et al., 2017; Possner et al., 2018). While structures such as open and closed cells present a simulation challenge at all but the highest resolutions (Kazil et al., 2011), machine learning presents an objective pathway toward classifying these convective regimes (Muhlbauer et al., 2014; McCoy, Wood, et al., 2017;

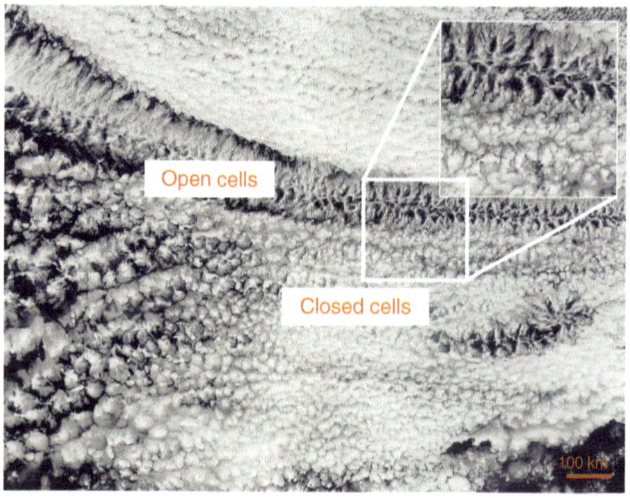

Figure 2.7 MODIS Terra imagery of open and closed cellular convection over the Pacific, 1 February 2016.

Wood & Hartmann, 2006). Recent analysis based on training a convolutional neural network provides a global examination of the properties of pockets of open cells, although it assigns a relatively slight radiative impact to closing all such pockets in the present-day (Watson-Parris et al., 2021).

Similar regime classification techniques exist for larger atmospheric structures, such as midlatitude cyclones. Approaches exist to identify cyclones and fronts (Field & Wood, 2007; Graf et al., 2017; Naud et al., 2010). Once the cyclones are isolated, the covariability of aerosol and clouds within them can be examined (Grandey, Stier, & Wagner, 2013; Grandey, Stier, Grainger, et al., 2013; Naud et al., 2017). Regime classification in large datasets of systems where variability in meteorological state and precipitation is substantial provides an opportunity to understand the causal linkages in the aerosol-cloud-precipitation system (McCoy et al., 2018). More complex regime classification techniques that characterize more of the frontal structure (Spensberger & Sprenger, 2018) are likely to yield additional insight into ACI occurring in this regime.

More broadly, k-means clustering offers a regime classification approach that can be applied to both observations and models with minimal assumptions (Williams & Webb, 2009). Application of this technique by Gryspeerdt and Stier (2012) demonstrated that the observationally inferred strength of ACIs (the sensitivity of N_d to aerosol optical depth, AOD) is greatest in the stratiform cloud regime, and this sensitivity was much more difficult to detect in cirrus and deep convective clouds. Further machine learning approaches to understanding clouds and climate are discussed in Chapter 16: Machine Learning for Clouds and Climate.

2.7. AEROSOL EFFECTS ON ICE AND MIXED-PHASE CLOUDS

Our discussion so far has focused on warm clouds; introducing the ice phase and generalizing to other cloud types lead to much additional complexity and associated uncertainty. Because warm clouds are shallow and their longwave emission temperatures are comparable to that of the surface, longwave radiative effects can usually be neglected in equation (2.12). For cirrus (Ci), in contrast, the longwave effect even dominates the shortwave effect (L'Ecuyer et al., 2019). To quantify aerosol effects on cold clouds, it is therefore necessary to extend equation (2.18) such that it accounts for perturbations to concentrations of ice nucleating particles (INP), the number of ice crystals N_i and ice water path. To date, a comprehensive quantification of the effects of aerosols on the radiative effects of cold clouds is missing, with available estimates largely limited to global modeling studies (Gettelman et al., 2012; Gryspeerdt et al., 2020; Heyn et al., 2017; Jensen & Toon, 1992; Lohmann et al., 2008; Penner et al., 2009). We therefore restrict our discussion to a qualitative overview of aerosol effects on cold clouds, as summarized in Table 2.1, but see also Chapter 4: Ice Particle Properties and Cirrus Emissivity. We do not discuss the details of the ice nucleation process or the role of INP chemical composition in it, and instead refer the reader to recent dedicated reviews by Murray et al. (2012) or Kanji et al. (2017).

We distinguish the formation of in situ cirrus from that of liquid-origin, or anvil cirrus. In situ Ci can form either *homogeneously*, that is, through homogeneous freezing of solution droplets, or *heterogeneously*, that is through deposition freezing onto the surfaces of INP. Both mechanisms for Ci formation can coexist and cloud response depends on their relative importance (e.g., Barahona et al., 2010; Gettelman et al., 2012; Zhu & Penner, 2020). Aerosol effects on Ci are strongest if the aerosol perturbation leads to a significant shift from homogeneous to heterogeneous ice nucleation because this means replacing the freezing of many solution droplets by ice nucleation on much fewer INP. The result is the formation of drastically fewer, but much larger ice crystals (Kärcher & Lohmann, 2003). The effect is enhanced by faster cloud dissipation through faster sedimentation of the larger crystals. A deliberate shift of homogeneous Ci formation to heterogeneous formation and faster sedimentation by seeding Ci with efficient INP has been discussed as a possible form of solar radiation management (Gasparini & Lohmann, 2016; Lohmann & Gasparini, 2017; Mitchell & Finnegan, 2009; Penner et al., 2015; Storelvmo et al., 2014). A third pathway of Ci formation is through the homogeneous freezing of liquid water droplets (in contrast to the solution

Table 2.1 Cloud response to aerosol perturbations by cloud and aerosol type. Cloud response is separated into the initial albedo response and adjustments. Processes targeted by suggestions for solar radiation management are highlighted by bold font, processes targeted by weather modification attempts in italics. Note that INP are considered non-soluble. See main text for details.

Cloud type	Aerosol type	Initial albedo response	Adjustments
Warm	CCN	**Higher albedo**	Entrainment-evaporation enhancement, **rain suppression/increased lifetime**
Homogeneous Ci	CCN	Slightly increased albedo	
Homogeneous Ci	INP	Lower albedo	**Faster sedimentation/reduced lifetime**
Heterogeneous Ci	INP	Slightly increased albedo	
Anvil Ci	CCN	Higher albedo	Slower sedimentation/increased lifetime
Mixed phase	INP	Glaciation/reduced albedo	*Faster precipitation formation/reduced lifetime/reduced orographic spill-over*
Mixed phase	CCN	Higher albedo	Evaporation/WBF enhancement, riming suppression
Deep convective	INP	Glaciation/reduced albedo due to glaciation	*Hail suppression*
Deep convective	CCN	Higher albedo	Decreased supersaturation/updraft invigoration, domain-wide adaptation of convective activity

droplets discussed earlier) in the anvil regions of deep convective clouds, which occurs at temperatures <−38°C. The number of cloud droplets available for homogeneous freezing depends on the number of CCN such that more but smaller droplets lead to more but smaller crystals (analogous to the Twomey effect) and anvils that persist longer due to reduced sedimentation velocities (Fan et al., 2013; Gryspeerdt, Mülmenstädt, et al., 2018; Morrison & Grabowski, 2011; Wang et al., 2020; Zhao et al., 2019). Increasing INP concentrations in a cloud that already freezes heterogeneously can also increase ice crystal number concentration (Zhou & Penner, 2014). Likewise, increasing the abundance of CCN/solution droplets in a homogeneous Ci may increase ice crystal number concentration (Gettelman et al., 2012; Gryspeerdt, Mülmenstädt, et al., 2018; Kärcher & Lohmann, 2003). Overall, however, the coupling of solution droplets and INP to Ci crystal number is much weaker in comparison to meteorological controls like updraft and temperature than for CCN and cloud droplet number (Gryspeerdt, Mülmenstädt, et al., 2018; Kärcher & Lohmann, 2002, 2003).

Ice crystals in mixed-phase clouds result from the freezing of cloud droplets that contain INP. For such crystals, the Wegener-Bergeron-Findeisen (WBF) (Bergeron, 1935; Findeisen, 1938; Storelvmo & Tan, 2015; Wegener, 1911) process leads to rapid growth of ice crystals at the expense of cloud water, that is, *glaciation* of the cloud. This means that ice crystal size is not controlled by the number of INP. The most important effect of perturbations to the INP concentration is therefore perturbations to the glaciation state, rather than changes in ice crystal size.

Glaciation frequently results in the formation of precipitation. *Glaciogenic cloud seeding*, that is, the deliberate introduction of efficient INP, usually silver iodide (Langmuir, 1950; Vonnegut, 1958), has thus been attempted as a means to shift orographic precipitation pattern from the leeward to the windward side of mountain ridges. While a chain of cloud processes observed during the SNOWIE field campaign (Tessendorf et al., 2019) has been documented (French et al., 2018), a statistically significant documentation of the efficiency of glaciogenic seeding remains challenging (Flossmann et al., 2019). Glaciogenic cloud seeding is also attempted to inhibit hail formation by favoring the ice-pathway of precipitation, that is, ice-ice collection and rimed snow, over the mixed-phase pathway leading to hail. A comprehensive series of experiments in Switzerland concluded the absence of a statistically significant effect of a randomized seeding strategy on hail kinetic energy (Federer et al., 1986), while a more recent study reports a significant effect for a moderately increased sample size and considering variability in seeding strength rather than randomized seeding (Dessens et al., 2006).

Pathways in mixed-phase precipitation formation can also be perturbed by an increase in CCN. Smaller cloud droplets reduce the efficiency of riming but increase evaporation, which in turn enhances the efficiency of the WBF process. Both sensitivities are notably weaker than that of autoconversion to droplet size (Glassmeier &

Lohmann, 2016). This is only one example of a tendency of adjustments in mixed-phase clouds to compensate each other, in line with the notion of clouds as *buffered* systems (Stevens & Feingold, 2009). Furthermore, the total amount of precipitation is subject to external constraints, which enforce a re-distribution of precipitation formation pathways in response to aerosol, rather than an overall change. For the example of stratiform orographic precipitation, such a constraint arises from the total condensation allowed for by orographic lifting (Glassmeier & Lohmann, 2018).

The effect of additional CCN on deep convective clouds has led to much debate. Considering individual updrafts, Rosenfeld et al. (2008) suggested an "ice-phase" pathway where smaller droplets would be more prone to cross the freezing level, which would result in convective invigoration through additional latent heat of freezing, and also precipitation formation and sedimentation. A recent theoretical study by Igel and van den Heever (2021) emphasizes the point made by Rosenfeld et al. (2008) and Grabowski and Morrison (2020) that the latent heat of freezing is approximately balanced, but slightly exceeded, by the energy needed to loft the condensate, so invigoration can actually only occur when the sedimentation of precipitation reduces the weight of the air parcel, making it more buoyant. Igel and van den Heever (2021) and Grabowski and Morrison (2016, 2020) agree that as a result the overall invigoration effect is often very weak.

Perhaps more effective is the "liquid-phase" pathway for convective invigoration described by Fan et al. (2007), Koren et al. (2014), Grabowski and Morrison (2016) and Fan et al. (2018). In this case, a convective invigoration is the result of additional latent heat of condensation due to a reduced in-cloud supersaturation under high-aerosol conditions. The invigoration, which is manifested in increased updraft speeds, can result in the "secondary" activation of aerosols well above the cloud base in comparison to unperturbed clouds. These aerosols were either too small to activate at cloud base or were entrained into the cloud from its edges. As these aerosols act to reduce the supersaturation, they may also feed back to affect the altitude at which the latent heat is released. However, isolating these complex effects of aerosols on deep convective clouds in observations is challenging, perhaps even impossible (Fan et al., 2018; Grabowski, 2018; Miltenberger et al., 2018).

When considering the embedding of convection in its environment beyond invigoration effects on individual cloud parcels (Morrison & Grabowski, 2013), the "embedding pathway" suggests that invigoration effects are likely restricted to localized areas rather than spread over entire regions (Abbott & Cronin, 2021; Morrison & Grabowski, 2013). Increased aerosol loading is also observed to lead to more, rather than stronger, convection (Abbott & Cronin, 2021; Blossey et al., 2018; Grabowski & Morrison, 2020). Grabowski and Morrison (2020) have associated this with mesoscale circulations, which they also suggest as a likely explanation for lightning enhancements in shipping lanes (Thornton et al., 2017). In a recent study, Abbott and Cronin (2021) moreover describe a "humidity-entrainment" mechanism for invigoration, which attributes invigoration to increased humidity caused by aerosol-enhanced detrainment and evaporation. As well as questioning the relevance of original "ice-phase" pathway for individual updrafts (Grabowski & Morrison, 2020), recent research has also suggested that invigoration effects are likely restricted to small areas rather than spread over entire regions (Abbott & Cronin, 2021).

2.8. SEMI-DIRECT EFFECTS

Semi-direct effects of aerosols on clouds were first so described by Hansen et al. (1997) and are associated with the effect of absorbing aerosols on local temperatures. The magnitude of the effects can be very large (tens of Wm^{-2}) locally, but both regional and global effects are poorly understood. The lack of understanding stems from two key sources: first, the effects depend strongly on the location of aerosols relative to clouds (Johnson et al., 2004), and second, calculation of semi-direct radiative effects requires an atmospheric model, and the simulated effects depend both on (frequently large) model biases in cloud cover and optical thickness and on internal variability.

Just as cloud droplet number concentrations are critical to understanding indirect effects, heating rates in the vicinity of clouds are critical to understanding semi-direct effects. Instantaneous heating rates generally range from 0.1 to 3 $Kday^{-1}$ (Gordon et al., 2018; Johnson, 2005; Panicker et al., 2014), which of course decrease substantially when annual or monthly means are considered (Allen et al., 2019). Local temperature changes resulting from the heating rates can reach up to a few Kelvin.

The effect of absorbing aerosols on cloud cover can be either positive or negative. Absorbing aerosols inside clouds can lead to clouds evaporating, while absorbing aerosols above temperature inversions can strengthen the temperature inversion by heating the air. This second effect can lead to additional cloud cover as the entrainment of dry air from above the boundary layer into clouds is reduced if the inversion is stronger (Gordon et al., 2018; Johnson et al., 2004; Koch & Del Genio, 2010). However, for above-cloud aerosols to have a strong effect, the aerosol layer must be very close to the cloud layer (Herbert et al., 2020). On the other hand, absorbing aerosols below clouds can also enhance convection and therefore cloud cover in some cloud regimes (Feingold et al., 2005; McFarquhar & Wang, 2006).

Semi-direct effects are likely important in cloudy regions where concentrations of absorbing aerosols, mainly black carbon but also dust, are high. Regional studies have focused on three key areas, among others. In the Amazon rainforest, fires are frequent in the dry season, and different modeling studies find semi-direct effects of different signs (Archer-Nicholls et al., 2016; Feingold et al., 2005; Koren et al., 2004; Liu et al., 2020), and magnitudes of order 5 Wm^{-2} when averaged over a day or a few days. In the southeast Atlantic stratocumulus deck, which is in the outflow of the world's largest source of biomass burning aerosol in southern Africa, semi-direct effects could exceed 10 Wm^{-2} locally, but the sign is again uncertain (Gordon et al., 2018; Lu et al., 2018; Mallet et al., 2019; Yamaguchi et al., 2015). Similarly, large effects may prevail in very polluted and cloudy areas of South and East Asia (Menon et al., 2002) or nearby oceans (Ackerman, Toon, Stevens, et al., 2000). However, the semi-direct effects vary between and even within models, depending, for example, on whether convection is parameterized or explicit (Archer-Nicholls et al., 2016). The constraint that overlying aerosols must be close to cloud layers to exert negative semi-direct effects places challenging requirements on both models and observing instruments such as CALIOP, which are prone to errors of a few hundred meters in precisely locating the height of aerosol layers (Das et al., 2020; Rajapakshe et al., 2017; Shinozuka et al., 2020).

Global modeling studies are similarly diverse in their prediction of semi-direct effects, which is not surprising given that they often cannot even agree on direct effects of absorbing aerosol (Bond et al., 2013; Stier et al., 2013). Global modeling studies have frequently investigated semi-direct effects in the context of aerosol effects on atmospheric dynamics (Mahajan et al., 2012); in this context, they are a subset of "rapid adjustments." Recent Precipitation Driver and Response Model Intercomparison Project (PDRMIP) simulations with CMIP5 models found negative semi-direct effects of black carbon largely cancel positive direct effects (Stjern et al., 2017). However, one noteworthy study (Allen et al., 2019) attempts to reduce model biases via observational constraints and finds semi-direct effects are globally positive, in contrast to the CMIP5 models. These global model intercomparisons also highlight the importance of semi-direct effects on high clouds in climate models, which may be a promising avenue for future work with more detailed models. More generally, the lack of consensus even on how semi-direct effects influence low clouds invites additional modeling and observational studies at a range of scales.

2.9. FIELD EXPERIMENTS

For decades, field campaigns have been conducted to better understand ACIs. Usually, these involve either aircraft or surface measurements or both, although tethered or untethered balloons, drones or airships are also promising as the slower pace permits more detailed sampling of individual clouds. The scale and scope of ACI-focused field campaigns has steadily increased since early observations of CCN by Woodcock (01 Oct. 1953) and others, who collected the aerosols on glass slides mounted on their aircraft, or sometimes stretched spider's webs (Twomey, 1954). Here, we discuss only a selection of the most recent field campaigns dedicated to ACIs for which major scientific outcomes are known, with apologies for the inevitable omissions.

The important role of CCN in determining the properties of low clouds at low latitude was investigated in campaigns in the southeast Atlantic between 2016 and 2018 (Zuidema et al., 2016). In the ORACLES (Redemann et al., 2021), CLARIFY (Haywood et al., 2021), LASIC (Zuidema et al., 2018), and AEROCLO-sa (Formenti et al., 2019) campaigns, the effect of smoke plumes from Africa on the important stratocumulus-to-cumulus transition in the region is potentially critical to their radiative effects (Yamaguchi et al., 2015; Zhou et al., 2017). A highlight of these campaigns for ACIs was the demonstration that entrainment of free tropospheric air masses into the boundary layer can be suppressed by pockets of open cells (Abel et al., 2020).

Low-level clouds at low latitudes have also been targeted by the EUREC^{4}A, and ATOMIC campaigns in the Barbados trade cumulus region (Bony et al., 2017; Stevens et al., 2021) were more focused on clouds than aerosols, but their extremely large-scale (four aircraft, four research vessels, two unmanned aerial vehicles [UAVs], and a surface station) and broad-scope meant ACIs were also investigated. We anticipate insights into the role of Saharan dust in suppressing precipitation in trade cumulus clouds (Stevens et al., 2021).

At the crossover point from low-latitudes to midlatitudes, the ACE-ENA campaign from the Azores in 2017 and 2018 (Zheng et al., 2021) studied both marine stratocumulus and clouds associated with extra-tropical cyclones. The current ACTIVATE campaign is also likely to yield new insights into low-level clouds, also in the North Atlantic, but further north and west in a similar region to that sampled by the NAAMES campaign (Behrenfeld et al., 2019). Concurrently, a series of field experiments (CAPRICORN, SOCRATES, MARCUS, and MICRE) probed low clouds in the Southern Ocean (McFarquhar et al., 2020). So far, a common theme of all of these campaigns has been how clouds can create favorable environments for the formation of new aerosol particles (McCoy et al., 2021; Sanchez et al., 2018; Zheng et al., 2021), which may subsequently grow to act as CCN. The proposed mechanism in the subtropics (at the Azores) involves new particle formation in clean regions of the upper boundary layer, while in the Southern Ocean and north-western North Atlantic it appears to happen at higher altitude, facilitated by upward transport of precursors, perhaps in midlatitude cyclones.

The Southern Ocean campaigns also made extensive measurements of INP concentrations and found them to be much lower than observed earlier by Bigg (1973). INP concentrations are hypothesized to be an important contributor to climate model biases in the region (Murray et al., 2021; Vergara-Temprado et al., 2018). The deficiency of supercooled liquid in the Southern Ocean is now addressed in the latest climate model simulations informed by these field campaigns (Gettelman et al., 2020). Likewise in the northern latitudes, ongoing analysis of the MOSAIC campaign data from the RV Polarstern, which drifted in Arctic ice over the 2019–2020 winter season, is likely to yield additional insights into the role of ACIs in the rapidly changing Arctic climate.

Deep clouds and their interactions with aerosols have also been studied extensively recently, though aircrafts do not usually sample convective cores so in situ measurements are more limited. The synergistic Go-Amazon and ACRIDICON-CHUVA campaigns in the Amazon elucidated the aerosol-cloud life cycle in the Amazon rainforest (Andreae et al., 2018; Wang & Zhang, 2016). As in the more recent ATom campaigns (Williamson et al., 2019), intense new particle formation in the upper troposphere was observed and hypothesized to supply CCN to boundary-layer clouds. The significance of upper-tropospheric aerosol production processes for clouds was appreciated earlier by Clarke et al. (1998) and others but it can now be studied in unprecedented detail, and is one aim of the CAFÉ-Brazil mission of 2022. Complementary measurements of ACIs in deep clouds have been collected more recently by the $CAMP^2EX$ mission in fall 2019, which has studied cloud effects on aerosol in air masses near the Maritime Continent (Hilario et al., 2021). Early data suggest wet scavenging is a dominant process shaping aerosol size distributions in the free troposphere.

Aircraft measurements yielding detailed in situ information are complemented by short- and long-term remote sensing from the surface as well as from space (section 2.10). Sometimes, clouds can be sampled directly in this way: in fog, detailed in situ measurements can be obtained at the surface over long periods directly. Two recent notable field campaigns on aerosol-fog interactions are C-FOG in eastern Canada (Fernando et al., 2021) focused on coastal fog and WIFEX in the Indo-Gangetic Plain between 2015 and 2019 (Ghude et al., 2017). While a critical research question in all fog studies concerns the activation of aerosol at low supersaturation (e.g., Mazoyer et al., 2019), this is complicated in the severely polluted environment of the Indo-Gangetic Plain by high concentrations of absorbing aerosol (Safai et al., 2019) which could potentially lead to changes in supersaturation due to heating effects.

When clouds are higher, surface remote sensing measurements using radar, lidar, and microwave techniques are key to long-term studies of ACIs. The most prominent surface remote sensing measurements are those of the US Department of Energy's ARM program, now in its 28th year of operations (Mather & Voyles, 2013), which now has three permanent stations at the Southern Great Plains, North Slope of Alaska, and Graciosa Island sites as well as several mobile facilities which participated in, or were the focus of, some of the campaigns listed above, namely, LASIC, MOSAIC, and MARCUS. The increased precision and sophistication of cloud radars (Kollias et al., 2020), improved calculations of cloud microphysical properties from surface retrievals (e.g., Wu et al., 2020; Yang et al., 2019), and the introduction of tethered balloon measurements (Dexheimer et al., 2019) are likely the most exciting recent innovations for studies of ACIs.

2.10. NEW SATELLITE PRODUCTS

Many of the techniques for constraining ACI we describe in earlier sections rely on satellite retrievals. Satellite observations are attractive for several reasons: they can provide observations of remote, pristine regions that are critical in setting forcing from ACI (Carslaw et al., 2013; Hamilton et al., 2014; McCoy, McCoy et al., 2020), their large data volume allows detection of relatively small amounts of ACI-induced variability relative to meteorologically-induced variability (Bender et al., 2019); and they can allow long-term observations of regions that reveal trends and responses to transient events (Bennartz et al., 2011; Li et al., 2018; McCoy et al., 2018; Malavelle et al., 2019). We will briefly survey common satellite observations used in evaluating ACI.

Constraint of the first indirect effect in liquid clouds by satellite requires knowledge of aerosol that is relevant to CCN and the cloud microphysical state (Nakajima et al., 2001; Quaas et al., 2020). While passive near-IR and visible retrievals by satellites are sensitive to cloud droplet radius(r_e), the more physically informative variable describing ACI in warm clouds is N_d (Wood, 2012) (see section 2.4). This quantity can be calculated with some assumptions using retrievals of τ and r_e (Bennartz & Rausch, 2017; Grosvenor et al., 2018; Grosvenor & Wood, 2014; Nakajima et al., 2001) (section 2.4). However, these calculations require that clouds be plane-parallel, which may potentially lead to substantial regional and seasonal biases (Grosvenor & Wood, 2014; Zhang et al., 2016, 2019). Similar retrievals are possible utilizing microwave LWP (Bennartz, 2007), polarimeter droplet radius (Alexandrov et al., 2012; Hasekamp et al., 2019), and lidar observations of optical depth (Hu et al., 2007; Li et al., 2018). Each of these retrievals has assumptions that are made in the retrieval process and thus strengths and weaknesses relative to other retrievals (Grosvenor et al., 2018). Observations of aerosol properties from space are possible and may be coupled with observations of N_d to yield an estimate of the first indirect effect (Gryspeerdt et al., 2017; Nakajima et al., 2001; Quaas et al., 2020). However, the use of vertically integrated aerosol can be problematic when aerosol and clouds are not vertically collocated. Furthermore, when aerosol and cloud are vertically collocated, aerosol swelling by cloud substantially complicates the relationship of AOD to aerosol number concentration (Christensen et al., 2017; Twohy et al., 2009), and this and other sources of bias tend to be most important in clean regimes where clouds are most sensitive to aerosols (Ma et al., 2018). As an alternative, studies have utilized aerosol reanalysis that attempt to account for near-cloud aerosol swelling (Bellouin et al., 2013; McCoy, Bender, et al., 2017) or have screened aerosol retrievals for nearby cloud (Christensen et al., 2017).

Constraints on aerosol-cloud adjustments in liquid clouds require knowledge of cloud macrophysical state and microphysical state (Albrecht, 1989; Gryspeerdt, Goren, et al., 2019). As discussed above, N_d is the most relevant variable relating to microphysical state. However, inference of the sensitivity of cloud macrophysical state to N_d is complicated due to non-distinct causality (see sections 2.6.3 and 2.6.4). The most common satellite observations of warm cloud properties considered are in-cloud LWP (Malavelle et al., 2019; Gryspeerdt, Goren, et al., 2019), cloud areal extent (Albrecht, 1989; Gryspeerdt et al., 2016), and area-mean LWP (McCoy et al., 2018; McCoy, Field, 2020). This list is by no means conclusive and other quantities less directly related to shortwave radiative forcing have been examined, such as cloud top height (Mace & Abernathy, 2016). These properties may be derived from a multitude of different satellite instruments, although they are most commonly examined in the context of passive spectroradiometers such as the MODIS (Platnick et al., 2003). However, it has been argued that passive microwave observations of liquid water path (Elsaesser et al., 2017) share less information with N_d calculated from τ and r_e, are more directly comparable to GCMs, and are less sensitive to cloud heterogeneity (McCoy, Field, et al., 2020). On the other hand, the fact that microwave liquid water path observations do not share shortwave scattering information with the calculation of N_d also means that they are less directly related to the shortwave radiative properties, such as cloud fraction and τ, which are primarily derived from visible wavelength light.

Satellite-based inferences of the aerosol effect on cold clouds are much more difficult to make than inferences of effects on warm clouds, and the effect of ice-phase microphysics on ERF_{aci} is thought to be small relative to liquid (Bellouin et al., 2020). However, lidar (e.g., CALIOP) and combined lidar and radar (e.g., Cloudsat) observations of cloud phase and ice crystal number concentration show covariation with dust, indicative of glaciation, as well as potential signatures of INP (Gryspeerdt, Sourdeval, et al., 2018; Sourdeval et al., 2015, 2016 2018; Hu et al., 2010; Tan et al., 2014). Observations of ice water content and path are still highly uncertain, but can be made using both passive and active observations (Jiang et al., 2012).

2.11. OUTLOOKS

Radiative forcing by ACIs has been considered quantitatively by Wigley (1989) and then by the Second IPCC Assessment Report, where it was assessed to be between −1.5 and 0 Wm^{-2}. Top-down constraints such as the temperature differences between hemispheres, the observed temperature record, as well as sulfur flux estimates (Schwartz, 1988) were all employed. The radiative forcing due to the first indirect effect is referred to as RF_{aci} and the forcing including "rapid adjustments," which include cloud adjustments to aerosols or "second indirect effects" is referred to as effective radiative forcing or ERF_{aci} (Boucher et al., 2014). In 2020, the ERF_{aci} from 1850 to the present-day was estimated by Bellouin et al. (2020) as between −1.73 and −0.27 Wm^{-2} (16–84% range), based mainly on global modeling and satellite-derived estimates. When top-down considerations are included, the overall aerosol forcing range becomes −2.0 to −0.35 Wm^{-2}, this time a 5–95% confidence interval including aerosol-radiation interactions. A subset of CMIP6 models yield a similar estimate of −1.41 to −0.21 Wm^{-2} (Smith et al., 2020). To obtain the uncertainty range quoted here, we have simply doubled the standard deviation of the different models calculated by Smith et al. (2020) to approximate the 5–95% confidence interval. This uncertainty from the multi-model intercomparison is slightly smaller than the Bellouin et al. (2020) estimate, but may grow as more models participate. It also does not account for the large uncertainties in emissions inventories, which are prescribed for the intercomparison and are therefore identical between models, nor does it account for correlation between models (Knutti et al., 2010). Some approaches yield much the same results, but some do not. For example, McCoy, McCoy, et al. (2020) obtain an estimate for the RF_{aci} of −1.6 to −0.6 Wm^{-2} by contrasting the Northern and Southern Hemispheres, but the estimate obtained by Johnson et al. (2020) from an uncertainty quantification exercise on a single GCM of −2.88 to −1.28 Wm^{-2} (5 − 95%) is rather larger and the central value is significantly more negative. The uncertainty thus remains high, but should decrease as anthropogenic emissions are reduced and the central value of the present-day forcing becomes less negative. As this central value decreases, global temperature rise will accelerate (Andreae et al., 2005), although perhaps not substantially (Shindell & Smith, 2019). Either way, the feasibility of a controlled utilization of ACIs in geoengineering approaches to temporarily moderate temperature rise may need to be explored (National Academies of Sciences & Medicine, 2021). Here, "feasibility" has to be understood in the broadest sense, ranging from the level of cloud physics via our modeling capabilities of unintended side effects to the level of governance and ethics.

Solar radiation management by stratospheric aerosol injection is associated with risks to the ozone layer or to long-term biogeochemical cycles. Marine cloud brightening (Latham, 2002; Latham et al., 2012), in contrast, is based on an easily reversible intervention, whose exploration has, moreover, large overlap with improving our fundamental understanding of ACI in stratocumulus. While small-scale marine cloud brightening experiments would offer a laboratory for ACI studies (Wood & Ackerman, 2013; Wood et al., 2017), an improved understanding of ACI directly informs the feasibility and limitations of marine cloud brightening strategies (Glassmeier et al., 2021). The main problems associated with marine cloud brightening from a geoengineering perspective are that it is not clear whether the technique would be sufficiently effective in enhancing cloud albedo, and it might, at least temporarily, disturb tropospheric circulation. However, we expect to see significant new work in the coming decade.

There are some other potential avenues for breaking the deadlock in ACI research to reduce the forcing uncertainty. New records of pre-industrial aerosols may come to light, for example, in ice cores (Carslaw et al., 2017) or less obvious sources such as historical artworks (Gryspeerdt, 2019; Zerefos et al., 2014). Global cloud resolving models (Stevens et al., 2019) are likely critical to understanding both forcing and feedbacks (Schneider et al., 2017; Terai et al., 2020). The incorporation of fully prognostic aerosols in such models is also under way either by testing aerosols and ACIs in various kinds of multiscale models (Gordon et al., 2020; Wang et al., 2011; Zhang et al., 2012), or by running global models with aerosols at progressively higher resolution. To our knowledge, the highest spatial grid resolution of a global model with prognostic aerosols is 3.5 km for a 2-week integration (Sato et al., 2016).

New machine learning approaches to better understand the proliferation of data that we already have are likely to play a major role in reducing forcing uncertainty, but require caution as results from machine learning algorithms are only as good as the data or models that are used to train and test them. For example, parametric uncertainty quantification says nothing about uncertainties or biases that cannot be easily parameterized, so can only ever predict part of the full uncertainty. Happily, it is a complementary part of the uncertainty to that previously explored by comparing different models. Machine learning can also be embedded in models directly to allow some processes to be treated more efficiently or with more complexity (e.g., Brenowitz & Bretherton, 2018; Krasnopolsky et al., 2005). While these approaches are challenging, due, for example, to the need to satisfy conservation laws and other basic requirements of physical models, constraints to overcome these problems are rapidly being developed (e.g., Beucler et al., 2021), paving the way toward more widespread adoption of promising machine learning algorithms.

As global temperatures climb, we foresee two further topics gaining new importance. The first is aerosol effects on cloud feedbacks. Gettelman et al. (2016) showed that different aerosol concentrations in the same model can lead to different cloud feedbacks as well as different radiative forcing and demonstrated the use of an "aerosol kernel," inspired by the radiative kernels of Soden et al. (2008), to disentangle forcing and feedbacks. Mathematically and conceptually, this kernel has some similarities to susceptibility. The second topic is the role of ACIs in weather forecasting and environmental engineering research. While both weather and climate modification techniques are rightly controversial, the interest in seeding of precipitation in clouds (French et al., 2018) is likely to grow as the planet warms and many formerly fertile areas are likely to become arid and barren. It may thus become crucial to revisit questions of the theoretical and practical feasibility of weather modification. Similarly, marine cloud brightening (Latham, 2002) may be seen as a potential way to leverage ACIs to influence regional climate. Semi-direct effects affect weather in heavily polluted regions (Ding et al., 2013; Gautam et al., 2007; Ma et al., 2020) and numerical weather prediction models are increasingly catching up with climate models in adopting more sophisticated aerosol schemes (Benedetti & Vitart, 2018). Widespread inclusion of ACIs in weather forecasting models is still on the horizon, but becomes more feasible every year.

Thus, despite the likelihood that global aerosol loadings decrease in the coming decades, ACIs will remain important. Continued efforts to understand them better will likely be repaid many times over with substantially improved capabilities for environmental prediction at all spatial and temporal scales.

ACKNOWLEDGMENTS

We are grateful to Leighton Regayre for discussions. Hamish Gordon acknowledges the support of Carnegie Mellon University and the NASA Roses program under grant number 80NSSC19K0949. Daniel McCoy acknowledges the support of the University of Wyoming. Franziska Glassmeier acknowledges support from The Branco Weiss Fellowship Society in Science, administered by the ETH Zürich, and from a Veni grant of the Dutch Research Council (NWO).

REFERENCES

Abbott, T. H., & Cronin, T. W. (2021). Aerosol invigoration of atmospheric convection through increases in humidity. *Science, 371*(6524), 83–85. doi: 10.1126/science.abc5181

Abdul-Razzak, H., & Ghan, S. J. (2000, Mar). A parameterization of aerosol activation: 2. Multiple aerosol types. *Journal of Geophysical Research: Atmospheres, 105*(D5), 6837–6844. doi: 10.1029/1999JD901161

Abel, S. J., Barrett, P. A., Zuidema, P., Zhang, J., Christensen, M., Peers, F., et al. (2020). Open cells exhibit weaker entrainment of free-tropospheric biomass burning aerosol into the south-east Atlantic boundary layer. *Atmospheric Chemistry and Physics, 20*(7), 4059–4084. doi: 10.5194/acp-20-4059-2020

Ackerman, A. S., Kirkpatrick, M. P., Stevens, D. E., & Toon, O. B. (2004, Dec). The impact of humidity above stratiform clouds on indirect aerosol climate forcing. *Nature, 432*(7020), 1014–1017. doi: 10.1038/nature03174

Ackerman, A. S., Toon, O. B., Stevens, D. E., Heymsfield, A. J., Ramanathan, V., & Welton, E. J. (2000). Reduction of tropical cloudiness by soot. *Science, 288*(5468), 1042–1047. doi: 10.1126/science.288.5468.1042

Ackerman, A. S., Toon, O. B., Taylor, J. P., Johnson, D. W., Hobbs, P. V., & Ferek, R. J. (2000, Aug). Effects of aerosols on cloud albedo: Evaluation of Twomey's parameterization of cloud susceptibility using measurements of ship tracks. *Journal of the Atmospheric Sciences, 57*(16), 2684–2695. doi: 10.1175/1520-0469(2000)057⟨2684: EOAOCA⟩2.0.CO;2

Agee, E. M. (1984). Observations from space and thermal convection: An historical perspective. *Bulletin of the American Meteorological Society, 65*(9), 938–949. doi: 10.1175/1520-0477(1984)065⟨0938:ofsatc⟩2.0.co;2

Aitken, J. (1881). *Dust, fogs, and clouds* (Vol. 23) (No. 591). Nature Publishing Group, 384–385. doi: 10.1038/023384a0

Albrecht, B. A. (1989, Sep). Aerosols, cloud microphysics, and fractional cloudiness. *Science, 245*(4923), 1227–1230. doi: 10.1126/science.245.4923.1227

Alexandrov, M. D., Cairns, B., Emde, C., Ackerman, A. S., & van Dieden-hoven, B. (2012, Oct). Accuracy assessments of cloud droplet size retrievals from polarized reflectance measurements by the research scanning polarimeter. *Remote Sensing of Environment, 125*, 92–111. doi: 10.1016/j.rse.2012.07.012

Allen, R. J., Amiri-Farahani, A., Lamarque, J.-F., Smith, C., Shindell, D., Hassan, T., & Chung, C. E. (2019). Observationally constrained aerosol–cloud semi-direct effects. *NPJ Climate and Atmospheric Science, 2*(1). doi: 10.1038/s41612-019-0073-9

Andreae, M. O., Afchine, A., Albrecht, R., Holanda, B. A., Artaxo, P., Barbosa, H. M. J., et al. (2018, Jan). Aerosol characteristics and particle production in the upper troposphere over the Amazon Basin. *Atmospheric Chemistry and Physics, 18*(2), 921–961. doi: 10.5194/acp-18-921-2018

Andreae, M. O., Jones, C. D., & Cox, P. M. (2005, June). Strong present-day aerosol cooling implies a hot future. *Nature, 435*(7046), 1187–1190. doi: 10.1038/nature03671

Archer-Nicholls, S., Lowe, D., Schultz, D. M., & McFiggans, G. (2016). Aerosol– radiation-cloud interactions in a regional coupled model: The effects of convective parameterisation and resolution. *Atmospheric Chemistry and Physics, 16*(9), 5573–5594. doi: 10.5194/acp-16-5573-2016

Axford, D. N. (1968, Aug). On the accuracy of wind measurements using an inertial platform in an aircraft, and an example of a measurement of the vertical mesostructure of the atmosphere. *Journal of Applied Meteorology, 7*(4), 645–666. doi: 10.1175/1520-0450(1968)007

Barahona, D., Rodriguez, J., & Nenes, A. (2010). Sensitivity of the global distribution of cirrus ice crystal concentration to heterogeneous freezing. *Journal of Geophysical Research: Atmospheres, 115*(D23). doi: https://doi.org/10.1029/2010JD014273

Baumgardner, D., Jonsson, H., Dawson, W., O'Connor, D., & Newton, R. (2001, Oct). The cloud, aerosol and precipitation spectrometer: A new instrument for cloud investigations. *Atmospheric Research, 59–60*, 251–264. doi: 10.1016/S0169-8095(01)00119-3

Behrenfeld, M. J., Moore, R. H., Hostetler, C. A., Graff, J., Gaube, P., Russell, L. M., et al. (2019). The North Atlantic Aerosol and Marine Ecosystem Study (NAAMES): Science motive and mission overview. *Frontiers in Marine Science, 6*, 122. doi: 10.3389/fmars.2019.00122

Bellouin, N., Mann, G. W., Woodhouse, M. T., Johnson, C., Carslaw, K. S., & Dalvi, M. (2013, Mar). Impact of the modal aerosol scheme GLOMAP-mode on aerosol forcing in the Hadley Centre Global Environmental Model. *Atmospheric Chemistry and Physics, 13*(6), 3027–3044. doi: 10.5194/acp-13-3027-2013

Bellouin, N., Quaas, J., Gryspeerdt, E., Kinne, S., Stier, P., Watson-Parris, D., Stevens, B. (2020, Mar). Bounding global aerosol radiative forcing of climate change. *Reviews of Geophysics, 58*(1), e2019RG000660. doi: 10.1029/2019RG000660

Bender, F.-M., Frey, L., McCoy, D. T., Grosvenor, D. P., & Mohrmann, J. K. (2019). Assessment of aerosol-cloud–radiation correlations in satellite observations, climate models and reanalysis. *Climate Dynamics, 52*(7–8), 4371–4392.

Benedetti, A., & Vitart, F. (2018, Oct). Can the direct effect of aerosols improve subseasonal predictability? *Monthly*

Weather Review, *146*(10), 3481–3498. doi: 10.1175/MWR-D-17-0282.1

Bennartz, R. (2007, Jan). Global assessment of marine boundary layer cloud droplet number concentration from satellite. *Journal of Geophysical Research*, *112*(D2), D02201. doi: 10.1029/2006JD007547

Bennartz, R., Fan, J., Rausch, J., Leung, L. R., & Heidinger, A. K. (2011). Pollution from China increases cloud droplet number, suppresses rain over the East China Sea. *Geophysical Research Letters*, *38*(9). doi: 10.1029/2011gl047235

Bennartz, R., & Rausch, J. (2017, Aug). Global and regional estimates of warm cloud droplet number concentration based on 13 years of AQUA-MODIS observations. *Atmospheric Chemistry and Physics*, *17*(16), 9815–9836. doi: 10.5194/acp-17-9815-2017

Bergeron, T. (1935). On the physics of cloud and precipitation. In *Procés-verbaux des séances de l'Association de Météorologie (Cinquiéme assemblée générale, Lisbonne)* (Vol. II: Mémoires, pp. 156–178). Imprimerie Paul Dupont.

Beucler, T., Pritchard, M., Rasp, S., Ott, J., Baldi, P., & Gentine, P. (2021, Mar). Enforcing analytic constraints in neural networks emulating physical systems. *Physical Review Letters*, *126*, 098302. doi: 10.1103/PhysRevLett.126.098302

Bigg, E. K. (1973, Sept). Ice nucleus concentrations in remote areas. *Journal of Atmospheric Sciences*, *30*(6), 1153–1157. doi: 10.1175/ 1520-0469(1973)030⟨1153:INCIRA⟩2.0.CO;2

Blossey, P. N., Bretherton, C. S., Thornton, J. A., & Virts, K. S. (2018). Locally enhanced aerosols over a shipping lane produce convective invigoration but weak overall indirect effects in cloud-resolving simulations. *Geophysical Research Letters*, *45*(9305–9313).

Bond, T. C., Doherty, S. J., Fahey, D. W., Forster, P. M., Berntsen, T., Deangelo, B. J., et al. (2013, Jun). Bounding the role of black carbon in the climate system: A scientific assessment. *Journal of Geophysical Research Atmospheres*, *118*(11), 5380–5552. doi: 10.1002/jgrd.50171

Bony, S., Stevens, B., Ament, F., Bigorre, S., Chazette, P., Crewell, S., et al. (2017, Nov). EUREC^{4}A: A field campaign to elucidate the couplings between clouds, convection and circulation. *Surveys in Geophysics*, *38*(6), 1529–1568. doi: 10.1007/s10712-017-9428-0

Boucher, O., & Lohmann, U. (1995). The sulfate-CCN-cloud albedo effect. *Tellus B*, *47*(3), 281–300. doi: https://doi.org/10.1034/j.1600-0889.47.issue3.1.x

Boucher, O., Randall, D., Artaxo, P., Bretherton, C., Feingold, G., Forster, P., et al. (2014). Clouds and aerosols. In T. Stocker et al. (Eds.), *Climate Change 2013 – The Physical Science Basis: Working Group I Contribution to the Fifth Assessment Report of the Intergovernmental Panel on Climate Change* (p. 571–658). Cambridge University Press. doi: 10.1017/CBO9781107415324.016

Brenowitz, N. D., & Bretherton, C. S. (2018, Jun). Prognostic validation of a neural network unified physics parameterization. *Geophysical Research Letters*, *45*(12), 6289–6298. doi: 10.1029/2018GL078510

Bretherton, C. S., Blossey, P. N., & Uchida, J. (2007). Cloud droplet sedimentation, entrainment efficiency, and subtropical stratocumulus albedo. *Geophysical Research Letters*, *34*(3). doi: 10.1029/2006GL027648

Carn, S. A., Fioletov, V. E., McLinden, C. A., Li, C., & Krotkov, N. A. (2017). A decade of global volcanic SO_2 emissions measured from space. *Scientific Reports*, *7*, 44095. doi: 10.1038/srep44095https://www.nature.com/articles/srep44095#supplementary-information

Carslaw, K. S., Boucher, O., Spracklen, D. V., Mann, G. W., Rae, J. G. L., Woodward, S., & Kulmala, M. (2010, Feb). A review of natural aerosol interactions and feedbacks within the Earth system. *Atmospheric Chemistry and Physics*, *10*(4), 1701–1737. doi: 10.5194/acp-10-1701-2010

Carslaw, K. S., Gordon, H., Hamilton, D. S., Johnson, J. S., Regayre, L. A., Yoshioka, M., & Pringle, K. J. (2017, Mar). Aerosols in the pre-industrial atmosphere. *Current Climate Change Reports*, *3*(1), 1–15. doi: 10.1007/ s40641-017-0061-2

Carslaw, K. S., Lee, L. A., Reddington, C. L., Mann, G. W., & Pringle, K. J. (2013, Dec). The magnitude and sources of uncertainty in global aerosol. *Faraday Discussions*, *165*, 495–512. doi: 10.1039/c3fd00043e

Charlson, R. J., Schwartz, S. E., Hales, J. M., Cess, R. D., Coakley, J. A., Hansen, J. E., & Hofmann, D. J. (1992, Jan). Climate forcing by anthropogenic aerosols. *Science*, *255*(5043), 423–430. doi: 10.1126/science.255.5043.423

Chen, Y.-C., Christensen, M. W., Stephens, G. L., & Seinfeld, J. H. (2014). Satellite-based estimate of global aerosol-cloud radiative forcing by marine warm clouds. *Nature Geoscience*, *7*, 643–646.

Christensen, M., Gettelman, A., Cermak, J., Dagan, G., Diamond, M., Douglas, A., et al. (2021). Opportunistic experiments to constrain aerosol effective radiative forcing. *Atmospheric Chemistry and Physics, 2021*, 1–60. doi: 10.5194/acp-2021-559

Christensen, M. W., Jones, W. K., & Stier, P. (2020, Jul). Aerosols enhance cloud lifetime and brightness along the stratus-to-cumulus transition. *Proceedings of the National Academy of Sciences*, *117*(30), 17591–17598.

Christensen, M. W., Neubauer, D., Poulsen, C. A., Thomas, G. E., McGarragh, G. R., Povey, A. C., et al. (2017). Unveiling aerosol–cloud interactions. Part 1: Cloud contamination in satellite products enhances the aerosol indirect forcing estimate. *Atmospheric Chemistry and Physics*, *17*(21), 13151–13164. doi: 10.5194/acp-17-13151-2017

Chuang, P. Y., Nenes, A., Smith, J. N., Flagan, R. C., & Seinfeld, J. H. (2000, Aug). Design of a CCN instrument for airborne measurement. *Journal of Atmospheric and Oceanic Technology*, *17*(8), 1005–1019. doi: 10.1175/1520-0426(2000)017⟨1005:DOACIF⟩2.0.CO;2

Clarke, A. D., Varner, J. L., Eisele, F., Mauldin, R. L., Tanner, D., & Litchy, M. (1998, Jul). Particle production in the remote marine atmosphere: Cloud outflow and subsidence during ACE 1. *Journal of Geophysical Research: Atmospheres*, *103*(D13), 16397–16409. doi: 10.1029/97JD02987

Conover, J. H. (1966, Nov). Anomalous cloud lines. *Journal of Atmospheric Sciences*, *23*(6), 778–785. doi: 10.1175/1520-0469(1966)023⟨0778:ACL⟩2.0.CO;2

Dagan, G., Koren, I., & Altaratz, O. (2015). Aerosol effects on the timing of warm rain processes. *Geophysical Research Letters*, *42*(11), 4590–4598. doi: https://doi.org/10.1002/2015GL063839

Das, S., Colarco, P. R., & Harshvardhan, H. (2020, Mar). The influence of elevated smoke layers on stratocumulus clouds over the SE Atlantic in the NASA Goddard Earth Observing System (GEOS) model. *Journal of Geophysical Research: Atmospheres*, *125*(6). doi: 10.1029/2019JD031209

Dessens, J., Berthet, C., & Sanchez, J. L. (2006). A sensitivity test for hail prevention assessment with hailpad measurements. *Journal Weather Modification*, *38*, 44–50.

Dexheimer, D., Airey, M., Roesler, E., Longbottom, C., Nicoll, K., Kneifel, S., et al. (2019). Evaluation of arm tethered-balloon system instrumentation for supercooled liquid water and distributed temperature sensing in mixed-phase arctic clouds. *Atmospheric Measurement Techniques*, *12*(12), 6845–6864. doi: 10.5194/amt-12-6845-2019

Diamond, M. S., Director, H. M., Eastman, R., Possner, A., & Wood, R. (2020). Substantial cloud brightening from shipping in subtropical low clouds. *AGU Advances*, *1*(1), e2019AV000111.

Ding, A. J., Fu, C. B., Yang, X. Q., Sun, J. N., Petäjä, T., Kerminen, V.-M., et al. (2013). Intense atmospheric pollution modifies weather: A case of mixed biomass burning with fossil fuel combustion pollution in eastern China. *Atmospheric Chemistry and Physics*, *13*(20), 10545–10554. doi: 10.5194/acp-13-10545-2013

Dusek, U., Frank, G. P., Hildebrandt, L., Curtius, J., Schneider, J., Walter, S., et al. (2006, Jun). Size matters more than chemistry for cloudnucleating ability of aerosol particles. *Science*, *312*(5778), 1375–1378. doi: 10.1126/science.1125261

Elsaesser, G. S., O'Dell, C. W., Lebsock, M. D., Bennartz, R., Greenwald, T. J., & Wentz, F. J. (2017). The Multi-Sensor Advanced Climatology of Liquid Water Path (MAC-LWP). *Journal of Climate*, *30*(24), 10193–10210. doi: 10.1175/jcli-d-16-0902.1

Engstrom, A., Bender, F. A. M., Charlson, R. J., & Wood, R. (2015). Geographically coherent patterns of albedo enhancement and suppression associated with aerosol sources and sinks. *Tellus B: Chemical and Physical Meteorology*, *67*, 26442. DOI: https://doi.org/10.3402/tellusb.v67.26442.

Fan, J., Leung, L. R., Rosenfeld, D., Chen, Q., Li, Z., Zhang, J., & Yan, H. (2013, Nov). Microphysical effects determine macrophysical response for aerosol impacts on deep convective clouds. *Proceedings of the National Academy of Sciences of the United States of America*, *110*(48), E4581–E4590. doi: 10.1073/pnas.1316830110

Fan, J., Rosenfeld, D., Zhang, Y., Giangrande, S. E., Li, Z., Machado, L. A. T., et al. (2018, Jan). Substantial convection and precipitation enhancements by ultrafine aerosol particles. *Science*, *359*(6374), 411–418. doi: 10.1126/science.aan8461

Fan, J., Wang, Y., Rosenfeld, D., & Liu, X. (2016, Nov). Review of aerosol–cloud interactions: Mechanisms, significance, and challenges. *Journal of the Atmospheric Sciences*, *73*(11), 4221–4252. doi: 10.1175/JAS-D-16-0037.1

Fan, J., Zhang, R., Li, G., & Tao, W.-K. (2007). Effects of aerosols and relative humidity on cumulus clouds. *Journal of Geophysical Research: Atmospheres*, *112*(D14). doi: https://doi.org/10.1029/2006JD008136

Federer, B., Waldvogel, A., Schmid, W., Schiesser, H. H., Hampel, F., Schweingrube, M., et al. (1986). Main results of Grossversuch IV. *Journal of Applied Meteorology*, *25*, 917–957.

Feingold, G., Boers, R., Stevens, B., & Cotton, W. R. (1997). A modeling study of the effect of drizzle on cloud optical depth and susceptibility. *Journal of Geophysical Research*, *102*(D12), 13527–13534.

Feingold, G., Jiang, H., & Harrington, J. Y. (2005, Jan). On smoke suppression of clouds in Amazonia. *Geophysical Research Letters*, *32*(2), L02804. doi: 10.1029/ 2004GL021369

Fernando, H. J. S., Gultepe, I., Dorman, C., Pardyjak, E., Wang, Q., Hoch, S. W., et al. (2021, Feb). C-FOG: Life of coastal fog. *Bulletin of the American Meteorological Society*, *102*(2), E244–E272. doi: 10.1175/BAMS-D-19-0070.1

Field, P. R., & Wood, R. (2007, Jan). Precipitation and cloud structure in midlatitude cyclones. *Journal of Climate*, *20*(2), 233–254. doi: 10.1175/jcli3998.1

Findeisen, W. (1938). Die kolloidmeteorologischen Vorgänge bei der Niederschlagsbildung (colloidal meteorological processes in the formation of precipitation). *Meteorologische Zeitschrift*, *55*, 121–133.

Flossmann, A. I., Manton, M., Abshaev, A., Bruintjes, R., Murakami, M., Prabhakaran, T., & Yao, Z. (2019, Aug). Review of advances in precipitation enhancement research. *Bulletin of the American Meteorological Society*, *100*(8), 1465–1480. doi: 10.1175/BAMS-D-18-0160.1

Formenti, P., D'Anna, B., Flamant, C., Mallet, M., Piketh, S. J., Schepanski, K., et al. (2019, Jul). The aerosols, radiation and clouds in southern Africa field campaign in Namibia: Overview, illustrative observations, and way forward. *Bulletin of the American Meteorological Society*, *100*(7), 1277–1298. doi: 10.1175/BAMS-D-17-0278.1

French, J. R., Friedrich, K., Tessendorf, S. A., Rauber, R. M., Geerts, B., Rasmussen, R. M., et al. (2018, Jan). Precipitation formation from orographic cloud seeding. *Proceedings of the National Academy of Sciences*, *115*(6), 1168–1173.

Fridlind, A. M., Ackerman, A. S., Jensen, E. J., Heymsfield, A. J., Poellot, M. R., Stevens, D. E., et al. (2004, Apr). Evidence for the predominance of mid-tropospheric aerosols as subtropical anvil cloud nuclei. *Science*, *304*(5671), 718–722. doi: 10.1126/science.1094947

Garrett, T. J., Radke, L. F., & Hobbs, P. V. (2002, Feb). Aerosol effects on cloud emissivity and surface longwave heating in the Arctic. *Journal of the Atmospheric Sciences*, *59*(3), 769–778. doi: 10.1175/1520-0469(2002)059⟨0769:AEOCEA⟩2.0.CO;2

Gasparini, B., & Lohmann, U. (2016). Why cirrus cloud seeding cannot substantially cool the planet. *Journal of Geophysical Research: Atmospheres*, *121*(9), 4877–4893. doi: https://doi.org/10.1002/2015JD024666

Gassó, S. (2008). Satellite observations of the impact of weak volcanic activity on marine clouds. *Journal of Geophysical Research: Atmospheres*, *113*(D14). doi: 10.1029/2007JD009106

Gautam, R., Hsu, N. C., Kafatos, M., & Tsay, S.-C. (2007). Influences of winter haze on fog/low cloud over the Indo-Gangetic plains. *Journal of Geophysical Research: Atmospheres*, *112*(D5). doi: https://doi.org/10.1029/2005JD007036

Gettelman, A., Bardeen, C. G., McCluskey, C. S., Järvinen, E., Stith, J., Bretherton, C., et al. (2020). Simulating observations of Southern Ocean clouds and implications for climate. *Journal of Geophysical Research: Atmospheres*, *125*(21), e2020JD032619. doi: https://doi.org/10.1029/2020JD032619

Gettelman, A., Lin, L., Medeiros, B., & Olson, J. (2016, Sep). Climate feedback variance and the interaction of aerosol forcing and feedbacks. *Journal of Climate*, *29*(18), 6659–6675. doi: 10.1175/JCLI-D-16-0151.1

Gettelman, A., Liu, X., Barahona, D., Lohmann, U., & Chen, C. (2012). Climate impacts of ice nucleation. *Journal of Geophysical Research Atmospheres*, *117*(20), 1–14. doi: 10.1029/2012JD017950

Ghan, S. J., Abdul-Razzak, H., Nenes, A., Ming, Y., Liu, X., Ovchinnikov, M., et al. (2011, Apr). Droplet nucleation: Physically-based parameterizations and comparative evaluation. *Journal of Advances in Modeling Earth Systems*, *3*(4). doi: 10.1029/2011ms000074

Ghude, S. D., Bhat, G., Prabhakaran, T., Jenamani, R., Chate, D., Safai, P., et al. (2017). Winter fog experiment over the Indo-Gangetic plains of India. *Current Science*, 767–784.

Givati, A., & Rosenfeld, D. (2004, Jul). Quantifying precipitation suppression due to air pollution. *Journal of Applied Meteorology*, *43*(7), 1038–1056. doi: 10.1175/1520-0450(2004)043⟨1038:QPSDTA⟩2.0.CO;2

Glassmeier, F., Hoffmann, F., Johnson, J. S., Yamaguchi, T., Carslaw, K. S., & Feingold, G. (2019). An emulator approach to statocumulus susceptibility. *Atmospheric Chemistry and Physics*, *19*, 10191–10203.

Glassmeier, F., Hoffmann, F., Johnson, J. S., Yamaguchi, T., Carslaw, K. S., & Feingold, G. (2021). Aerosol-cloud-climate cooling overestimated by ship-track data. *Science*, *371*, 485–489.

Glassmeier, F., & Lohmann, U. (2016). Constraining precipitation susceptibility of warm, iceand mixed-phase clouds with microphysical equations. *Journal of the Atmospheric Sciences*, *73*(12), 5003–5023.

Glassmeier, F., & Lohmann, U. (2018). Precipitation susceptibility and aerosol buffering of warm and mixed-phase orographic clouds in idealized simulations. *Journal of the Atmospheric Sciences*, *75*(4), 1173–1194.

Gordon, H., Field, P. R., Abel, S. J., Barrett, P., Bower, K., Crawford, I., et al. (2020, Sep). Development of aerosol activation in the double-moment Unified Model and evaluation with CLARIFY measurements. *Atmospheric Chemistry and Physics*, *20*(18), 10997–11024. doi: 10.5194/acp-20-10997-2020

Gordon, H., Field, P. R., Abel, S. J., Dalvi, M., Grosvenor, D. P., Hill, A. A., et al. (2018, Oct). Large simulated radiative effects of smoke in the south-east Atlantic. *Atmospheric Chemistry and Physics*, *18*(20), 15261–15289. doi: 10.5194/acp-18-15261-2018

Goren, T., & Rosenfeld, D. (2012). Satellite observations of ship emission induced transitions from broken to closed cell marine stratocumulus over large areas. *Journal of Geophysical Research: Atmospheres*, *117*(D17).

Goren, T., & Rosenfeld, D. (2014, Mar). Decomposing aerosol cloud radiative effects into cloud cover, liquid water path and Twomey components in marine stratocumulus. *Atmospheric Research*, *138*, 378–393. doi: 10.1016/j.atmosres.2013.12.008

Goren, T., & Rosenfeld, D. (2015). Extensive closed cell marine stratocumulus downwind of Europe–A large aerosol cloud mediated radiative effect or forcing? *Journal of Geophysical Research: Atmospheres*, *120*(12), 6098–6116.

Grabowski, W. W. (2018, Oct). Can the impact of aerosols on deep convection be isolated from meteorological effects in atmospheric observations? *Journal of the Atmospheric Sciences*, *75*(10), 3347–3363. doi: 10.1175/JAS-D-18-0105.1

Grabowski, W. W., & Morrison, H. (2016, Sep). Untangling microphysical impacts on deep convection applying a novel modeling methodology. Part II: Double-moment microphysics. *Journal of the Atmospheric Sciences*, *73*(99), 3749–3770. Retrieved from http://journals.ametsoc.org/doi/10.1175/JAS-D-15-0367.1. doi: 10.1175/JAS-D-15-0367.1

Grabowski, W. W., & Morrison, H. (2020). Do ultrafine cloud condensation nuclei invigorate deep convection? *Journal of the Atmospheric Sciences*, *77*(7), 2567–2583. doi: 10.1175/JAS-D-20-0012.1

Graf, M. A., Wernli, H., & Sprenger, M. (2017). Objective classification of extratropical cyclogenesis. *Quarterly Journal of the Royal Meteorological Society*, *143*(703), 1047–1061.

Grandey, B. S., Stier, P., Grainger, R. G., & Wagner, T. M. (2013). The contribution of the strength and structure of extratropical cyclones to observed cloudaerosol relationships. *Atmospheric Chemistry and Physics*, *13*(21), 10689–10701. doi: 10.5194/acp-13-10689-2013

Grandey, B. S., Stier, P., & Wagner, T. M. (2013). Investigating relationships between aerosol optical depth and cloud fraction using satellite, aerosol reanalysis and general circulation model data. *Atmospheric Chemistry and Physics*, *13*(6), 3177–3184. doi: 10.5194/acp-13-3177-2013

Grosvenor, D. P., Field, P. R., Hill, A. A., & Shipway, B. J. (2017). The relative importance of macrophysical and cloud albedo changes for aerosol-induced radiative effects in closed-cell stratocumulus: Insight from the modelling of a case study. *Atmospheric Chemistry and Physics*, *17*(8), 5155–5183.

Grosvenor, D. P., Sourdeval, O., Zuidema, P., Ackerman, A., Alexandrov, M. D., Bennartz, R., et al. (2018, Jun). Remote sensing of droplet number concentration in warm clouds: A review of the current state of knowledge and perspectives. *Reviews of Geophysics*, *56*(2), 409–453. doi: 10.1029/2017RG000593

Grosvenor, D. P., & Wood, R. (2014). The effect of solar zenith angle on MODIS cloud optical and microphysical retrievals within marine liquid water clouds. *Atmospheric Chemistry & Physics*, *14*(14), 7291–7321. doi: 10.5194/acp-14-7291-2014

Gryspeerdt, E. (2019). Ruskin and meteorology. In S. F. Cooper & R. Johns (Eds.), *Ruskin, turner and the storm cloud*. Paul Holberton Publishing.

Gryspeerdt, E., Goren, T., Sourdeval, O., Quaas, J., Mülmenstädt, J., Dipu, S., et al. (2019, Apr). Constraining the aerosol influence on cloud liquid water path. *Atmospheric Chemistry and Physics*, *19*(8), 5331–5347. doi: 10.5194/acp-19-5331-2019

Gryspeerdt, E., Mülmenstädt, J., Gettelman, A., Malavelle, F. F., Morrison, H., Neubauer, D., et al. (2018). Ice crystal number concentration estimates from lidar-radar satellite remote sensing Part 2: Controls on the ice crystal number concentration. *Atmospheric Chemistry and Physics*, *18*, 14351–14370.

Gryspeerdt, E., Mülmenstädt, J., Gettelman, A., Malavelle, F. F., Morrison, H., Neubauer, D., et al. (2020). Surprising similarities in model and observational aerosol radiative forcing estimates. *Atmospheric Chemistry and Physics*, *20*(1), 613–623. doi: 10.5194/acp-20-613-2020

Gryspeerdt, E., Quaas, J., & Bellouin, N. (2016). Constraining the aerosol influence on cloud fraction. *Journal of Geophysical Research: Atmospheres*, *121*(7). doi: 10.1002/2015JD023744

Gryspeerdt, E., Quaas, J., Ferrachat, S., Gettelman, A., Ghan, S., Lohmann, U., et al. (2017, May). Constraining the instantaneous aerosol influence on cloud albedo. *Proceedings of the National Academy of Sciences*, *114*(19), 4899–4904. doi: 10.1073/pnas.1617765114

Gryspeerdt, E., Smith, T. W. P., O'Keeffe, E., Christensen, M. W., & Goldsworth, F. W. (2019, Nov). The impact of ship emission controls recorded by cloud properties. *Geophysical Research Letters*, *46*, 12,547–12,555. https://doi.org/10.1029/2019GL084700

Gryspeerdt, E., Sourdeval, O., Quaas, J., Delanoë, J., Krämer, M., & Kühne, P. (2018, Oct). Ice crystal number concentration estimates from lidarradar satellite remote sensing – Part 2: Controls on the ice crystal number concentration. *Atmospheric Chemistry and Physics*, *18*(19), 14351–14370. Retrieved from https://acp.copernicus.org/articles/18/14351/2018/ doi: 10.5194/acp-18-14351-2018

Gryspeerdt, E., & Stier, P. (2012). Regime-based analysis of aerosol-cloud interactions. *Geophysical Research Letters*, *39*(21). doi: https://doi.org/10.1029/2012GL053221

Hamilton, D. S., Lee, L. A., Pringle, K. J., Reddington, C. L., Spracklen, D. V., & Carslaw, K. S. (2014). Occurrence of pristine aerosol environments on a polluted planet. *Proceedings of the National Academy of Sciences*, *111*(52), 18466–18471. doi: 10.1073/pnas.1415440111

Hansen, J., Sato, M., Kharecha, P., & von Schuckmann, K. (2011). Earth's energy imbalance and implications. *Atmospheric Chemistry and Physics*, *11*(24), 13421–13449. doi: 10.5194/acp-11-13421-2011

Hansen, J., Sato, M., & Ruedy, R. (1997, Mar). Radiative forcing and climate response. *Journal of Geophysical Research Atmospheres*, *102*(D6), 6831–6864. doi: 10.1029/96JD03436

Hasekamp, O. P., Gryspeerdt, E., & Quaas, J. (2019, Nov). Analysis of po-larimetric satellite measurements suggests stronger cooling due to aerosol-cloud interactions. *Nature Communications*, *10*(1), 5405.

Haywood, J. M., Abel, S. J., Barrett, P. A., Bellouin, N., Blyth, A., Bower, K. N., et al. (2021). The cloud–aerosol–radiation interaction and forcing: Year 2017 (CLARIFY-2017) measurement campaign. *Atmospheric Chemistry and Physics*, *21*(2), 1049–1084. doi: 10.5194/acp-21-1049-2021

Herbert, R. J., Bellouin, N., Highwood, E. J., & Hill, A. A. (2020, Feb). Diurnal cycle of the semi-direct effect from a persistent absorbing aerosol layer over marine stratocumulus in large-eddy simulations. *Atmospheric Chemistry and Physics*, *20*(3), 1317–1340. doi: 10.5194/acp-20-1317-2020

Heyn, I., Block, K., Mülmenstädt, J., Gryspeerdt, E., Kühne, P., Salzmann, M., & Quaas, J. (2017). Assessment of simulated aerosol effective radiative forcings in the terrestrial spectrum. *Geophysical Research Letters*, *44*(2), 1001–1007. doi: https://doi.org/10.1002/2016GL071975

Hilario, M. R. A., Crosbie, E., Shook, M., Reid, J. S., Cambaliza, M. O. L., Simpas, J. B. B., et al. (2021). Measurement report: Long-range transport patterns into the tropical northwest Pacific during the CAMP^{2}Ex aircraft campaign: Chemical composition, size distributions, and the impact of convection. *Atmospheric Chemistry and Physics*, *21*(5), 3777–3802. doi: 10.5194/acp-21-3777-2021

Hill, A. A., Feingold, G., & Jiang, H. (2009, May). The influence of entrainment and mixing assumption on aerosol-cloud interactions in marine stratocumulus. *Journal of the Atmospheric Sciences*, *66*(5), 1450–1464. doi: 10.1175/2008JAS2909.1

Hobbs, P. V., Garrett, T. J., Ferek, R. J., Strader, S. R., Hegg, D. A., Frick, G. M., et al. (2000, Aug). Emissions from ships with respect to their effects on clouds. *Journal of the Atmospheric Sciences*, *57*(16), 2570–2590. doi: 10.1175/1520-0469(2000)057⟨2570: EFSWRT⟩2.0.CO;2

Hoffmann, F., & Feingold, G. (2019). Entrainment and mixing in stratocumulus: Effects of a new explicit subgrid-scale scheme for large-eddy simulations with particle-based microphysics. *Journal of the Atmospheric Sciences*, *76*(7), 1955–1973.

Hu, Y., Vaughan, M., McClain, C., Behrenfeld, M., Maring, H., Anderson, D., et al. (2007). Global statistics of liquid water content and effective number concentration of water clouds over ocean derived from combined CALIPSO and MODIS measurements. *Atmospheric Chemistry and Physics*, *7*(12), 3353–3359. doi: 10.5194/acp-7-3353-2007

Hu, Y. X., Rodier, S., Xu, K. M., Sun, W. B., Huang, J. P., Lin, B., et al. (2010, Oct). Occurrence, liquid water content, and fraction of supercooled water clouds from combined CALIOP/IIR/MODIS measurements. *Journal of Geophysical Research-Atmospheres*, *115*, 13. doi: 10.1029/2009jd012384

Hudson, J. G., & Squires, P. (1976, Jul). Improved continuous flow diffusion cloud chamber. *Journal of Applied Meteorology*, *15*(7), 776–782. doi: 10.1175/1520-0450(1976)015⟨0776:AICFDC⟩2.0.CO;2

Igel, A. L., & van den Heever, S. C. (2021). Invigoration or enervation of convective clouds by aerosols? *Geophysical Research Letters*, *48*(16), e2021GL093804. doi: https://doi.org/10.1029/2021GL093804

Jensen, E. J., & Toon, O. B. (1992). The potential effects of volcanic aerosols on cirrus cloud microphysics. *Geophysical Research Letters*, *19*(17), 1759–1762. doi: https://doi.org/10.1029/92GL01936

Jensen, J. B., & Nugent, A. D. (2017, Mar). Condensational growth of drops formed on giant sea-salt aerosol particles. *Journal of the Atmospheric Sciences*, *74*(3), 679–697. doi: 10.1175/JAS-D-15-0370.1

Jiang, J. H., Su, H., Zhai, C., Perun, V. S., Del Genio, A., Nazarenko, L. S., et al. (2012). Evaluation of cloud and water vapor simulations in CMIP5 climate models using NASA "A-Train" satellite observations. *Journal of Geophysical Research: Atmospheres*, *117*(D14), D14105. doi: 10.1029/2011JD017237

Johnson, B. T. (2005, Jul). Large-eddy simulations of the semidirect aerosol effect in shallow cumulus regimes. *Journal of Geophysical Research: Atmospheres*, *110*(D14). doi: 10.1029/2004JD005601

Johnson, B. T., Shine, K. P., & Forster, P. M. (2004, Apr). The semi-direct aerosol effect: Impact of absorbing aerosols on marine stratocumulus. *Quarterly Journal of the Royal Meteorological Society*, *130*(599 PART B), 1407–1422. doi: 10.1256/ qj.03.61

Johnson, D. B. (1982, Feb). The role of giant and ultra-giant aerosol particles in warm rain initiation. *Journal of Atmospheric Sciences*, *39*(2), 448–460. doi: 10.1175/1520-0469(1982)039⟨0448:TROGAU⟩2.0.CO;2

Johnson, J. S., Cui, Z., Lee, L. A., Gosling, J. P., Blyth, A. M., & Carslaw, K. S. (2015). Evaluating uncertainty in convective cloud microphysics using statistical emulation. *Journal of Advances in Modeling Earth Systems*, *7*(1), 162–187. doi: https://doi.org/10.1002/2014MS000383

Johnson, J. S., Regayre, L. A., Yoshioka, M., Pringle, K. J., Turnock, S. T., Browse, J., et al. (2020). Robust observational constraint of uncertain aerosol processes and emissions in a climate model and the effect on aerosol radiative forcing. *Atmospheric Chemistry and Physics*, *20*(15), 9491–9524. doi: 10.5194/acp-20-9491-2020

Kanji, Z. A., Ladino, L. A., Wex, H., Boose, Y., Burkert-Kohn, M., Cziczo, D. J., & Krämer, M. (2017, Jan). Overview of ice nucleating particles. *Meteorological Monographs*, *58*, 1.1–1.33. doi: 10.1175/ AMSMONOGRAPHS-D-16-0006.1

Kärcher, B. & Lohmann, U. (2002). A parameterization of cirrus cloud formation: Homogeneous freezing including effects of aerosol size. *Journal of Geophysical Research: Atmospheres*, *107*(D23), AAC 9-1–AAC 9-10. doi: https://doi.org/10.1029/2001JD001429

Kärcher, B. & Lohmann, U. (2003). A parameterization of cirrus cloud formation: Heterogeneous freezing. *Journal of Geophysical Research: Atmospheres*, *108*(D14). doi: https://doi.org/10.1029/2002JD003220

Kazil, J., Wang, H., Feingold, G., Clarke, A. D., Snider, J. R., & Bandy, A. R. (2011, Aug). Modeling chemical and aerosol processes in the transition from closed to open cells during VOCALS-REx. *Atmospheric Chemistry and Physics*, *11*(15), 7491–7514. doi: 10.5194/acp-11-7491-2011

Khain, A., Ovtchinnikov, M., Pinsky, M., Pokrovsky, A., & Krugliak, H. (2000, Dec). Notes on the state-of-the-art numerical modeling of cloud microphysics. *Atmospheric Research*, *55*(3–4), 159–224. doi: 10.1016/S0169-8095(00)00064-8

Khain, A., Pokrovsky, A., Pinsky, M., Seifert, A., & Phillips, V. (2004 Dec). Simulation of effects of atmospheric aerosols on deep turbulent convective clouds using a spectral microphysics mixed-phase cumulus cloud model. Part I: Model description and possible applications. *Journal of the Atmospheric Sciences*, *61*(24), 2963–2982. doi: 10.1175/JAS-3350.1

Khairoutdinov, M., & Kogan, Y. (2000, Jan). A new cloud physics parameterization in a large-eddy simulation model of marine stratocumulus. *Monthly Weather Review*, *128*(1), 229–243. doi: 10.1175/1520-0493(2000)128⟨0229:ANCPPI⟩2.0.CO;2

Knollenberg, R. G. (1976). *Three New Instruments for Cloud Physics Measurements: The 2-D Spectrometer, the Forward Scattering Spectrometer Probe, and the Active Scattering Aerosol Spectrometer*. In *Preprints, International Conference on Cloud Physics* (pp. 554–561). American Meteorological Society.

Knutti, R., Furrer, R., Tebaldi, C., Cermak, J., & Meehl, G. A. (2010, May). Challenges in combining projections from multiple climate models. *Journal of Climate*, *23*(10), 2739–2758. doi: 10.1175/2009JCLI3361.1

Koch, D., & Del Genio, A. D. (2010, Aug). Black carbon semi-direct effects on cloud cover: Review and synthesis. *Atmospheric Chemistry and Physics*, *10*(16), 7685–7696. doi: 10.5194/acp-10-7685-2010

Köhler, H. (1936, Jan). The nucleus in and the growth of hygroscopic droplets. *Transactions of the Faraday Society*, *32*, 1152–1161. doi: 10.1039/ TF9363201152

Kollias, P., Bharadwaj, N., Clothiaux, E. E., Lamer, K., Oue, M., Hardin, J., et al. (2020, May). The arm radar network: At the leading edge of cloud and precipitation observations. *Bulletin of the American Meteorological Society*, *101*(5), E588–E607. doi: 10.1175/BAMS-D-18-0288.1

Koren, I., Dagan, G., & Altaratz, O. (2014). From aerosol-limited to invigoration of warm convective clouds. *Science*, *344*(6188), 1143–1146. doi: 10.1126/science.1252595

Koren, I., Kaufman, Y. J., Remer, L. A., & Martins, J. V. (2004). Measurement of the effect of Amazon smoke on inhibition of cloud formation. *Science*, *303*(5662), 1342–1345. doi: 10.1126/science.1089424

Korolev, A. V., & Mazin, I. P. (2003). Supersaturation of water vapor in clouds. *Journal of the Atmospheric Sciences*, *60*(24), 2957–2974. doi: 10.1175/1520-0469(2003)060⟨2957:SOWVIC⟩2.0.CO;2

Krasnopolsky, V. M., Fox-Rabinovitz, M. S., & Chalikov, D. V. (2005, May). New approach to calculation of atmospheric model physics: Accurate and fast neural network emulation of longwave radiation in a climate model. *Monthly Weather Review*, *133*(5), 1370–1383. doi: 10.1175/MWR2923.1

Kreidenweis, S. M., Petters, M., & Lohmann, U. (2019, Jan). 100 years of progress in cloud physics, aerosols, and aerosol chemistry research. *Meteorological Monographs*, *59*, 11.1–11.72. doi: 10.1175/amsmonographs-d-18-0024.1

Kulkarni, P., & Wang, J. (2006, Oct). New fast integrated mobility spectrometer for real-time measurement of aerosol size distribution-I: Concept and theory. *Journal of Aerosol Science*, *37*(10), 1303–1325. doi: 10.1016/j.jaerosci.2006.01.005

Lacis, A. A., & Hansen, J. E. (1974). A parameterization for the absorption of solar radiation in the Earth's atmosphere. *Journal of the Atmospheric Sciences*, *31*, 118–133.

Langmuir, I. (1950). Control of precipitation from cumulus clouds by various seeding techniques. *Science*, *112*(2898), 35–41.

Latham, J. (2002). Amelioration of global warming by controlled enhancement of the albedo and longevity of low-level maritime clouds. *Atmospheric Science Letters*, *3*(24), 52–58. doi: https://doi.org/10.1006/asle.2002.0099

Latham, J., Bower, K., Choularton, T., Coe, H., Connolly, P., Cooper, G., et al. (2012). Marine cloud brightening. *Philosophical Transactions of the Royal Society A: Mathematical, Physical and Engineering Sciences*, *370*(1974), 4217–4262. doi: 10.1098/rsta.2012.0086

L'Ecuyer, T. S., Hang, Y., Matus, A. V., & Wang, Z. (2019). Reassessing the effect of cloud type on Earth's energy balance in the age of active spaceborne observations. Part I: Top of atmosphere and surface. *Journal of Climate*, *32*(19), 6197–6217.

Lee, L. A., Carslaw, K. S., Pringle, K. J., Mann, G. W., & Spracklen, D. V. (2011). Emulation of a complex global aerosol model to quantify sensitivity to uncertain parameters. *Atmospheric Chemistry and Physics*, *11*(23), 12253–12273.

Lee, L. A., Reddington, C. L., & Carslaw, K. S. (2016, May). On the relationship between aerosol model uncertainty and radiative forcing uncertainty. *Proceedings of the National Academy of Sciences*, *113*(21), 5820–5827. doi: 10.1073/pnas.1507050113

Li, J., Jian, B., Huang, J., Hu, Y., Zhao, C., Kawamoto, K., et al. (2018, Aug). Long-term variation of cloud droplet number concentrations from space-based Lidar. *Remote Sensing of Environment*, *213*, 144–161. doi: https://doi.org/10.1016/j.rse.2018.05.011

Liu, B. Y., Whitby, K. T., & Pui, D. Y. (1974). A portable electrical analyzer for size distribution measurement of submicron aerosols. *Journal of the Air Pollution Control Association*, *24*(11), 1067–1072. doi: 10.1080/ 00022470.1974.10470016

Liu, L., Cheng, Y., Wang, S., Wei, C., Pöhlker, M. L., Pöhlker, C., et al. (2020). Impact of biomass burning aerosols on radiation, clouds, and precipitation over the Amazon: Relative importance of aerosol–cloud and aerosol–radiation interactions. *Atmospheric Chemistry and Physics*, *20*(21), 13283–13301. doi: 10.5194/acp-20-13283-2020

Liu, Y., Zhang, K., Qian, Y., Wang, Y., Zou, Y., Song, Y., et al. (2018). Investigation of short-term effective radiative forcing of fire aerosols over North America using nudged hindcast ensembles. *Atmospheric Chemistry and Physics*, *18*(1), 31–47. doi: 10.5194/acp-18-31-2018

Lohmann, U., & Ferrachat, S. (2010). Impact of parametric uncertainties on the present-day climate and on the anthropogenic aerosol effect. *Atmospheric Chemistry and Physics*, *10*(23), 11373–11383. doi: 10.5194/acp-10-11373-2010

Lohmann, U., & Gasparini, B. (2017). A cirrus cloud climate dial? *Science*, *357*(6348).

Lohmann, U., Spichtinger, P., Jess, S., Peter, T., & Smit, H. (2008, Oct). Cirrus cloud formation and ice supersaturated regions in a global climate model. *Environmental Research Letters*, *3*(4), 045022. doi: 10.1088/1748-9326/3/4/045022

Lu, Z., Liu, X., Zhang, Z., Zhao, C., Meyer, K., Rajapakshe, C., et al. (2018). Biomass smoke from Southern Africa can significantly enhance the brightness of stratocumulus over the Southeastern Atlantic ocean. *Proceedings of the National Academy of Sciences*, *115*(12), 2924–2929. doi: 10.1073/pnas.1713703115

Ma, P.-L., Rasch, P. J., Chepfer, H., Winker, D. M., & Ghan, S. J. (2018, Jul). Observational constraint on cloud susceptibility weakened by aerosol retrieval limitations. *Nature Communications*, *9*, 2640. doi: 10.1038/s41467-018-05028-4

Ma, Y., Ye, J., Xin, J., Zhang, W., Vil-Guerau de Arellano, J., Wang, S., et al. (2020). The stove, dome, and umbrella effects of atmospheric aerosol on the development of the planetary boundary layer in hazy regions. *Geophysical Research Letters*, *47*(13), e2020GL087373. doi: https://doi.org/10.1029/2020GL087373

Mace, G. G., & Abernathy, A. C. (2016). Observational evidence for aerosol invigoration in shallow cumulus downstream of Mount Kilauea. *Geophysical Research Letters*, *43*(6), 2981–2988. doi: 10.1002/2016GL067830

Mahajan, S., Evans, K. J., Truesdale, J. E., Hack, J. J., & Lamarque, J.-F. (2012, Dec). Interannual tropospheric aerosol variability in the late twentieth century and its impact on tropical Atlantic and west African climate by direct and semidirect effects. *Journal of Climate*, *25*(23), 8031–8056. doi: 10.1175/JCLI-D-12-00029.1

Malavelle, F. F., Haywood, J. M., Mercado, L. M., Folberth, G. A., Bellouin, N., Sitch, S., & Artaxo, P. (2019, Jan). Studying the impact of biomass burning aerosol radiative and climate effects on the Amazon rainforest productivity with an Earth system model. *Atmospheric Chemistry and Physics*, *19*(2), 1301–1326. doi: 10.5194/acp-19-1301-2019

Mallet, M., Nabat, P., Zuidema, P., Redemann, J., Sayer, A. M., Stengel, M., et al. (2019, Apr). Simulation of the transport, vertical distribution, optical properties and radiative impact of smoke aerosols with the ALADIN regional climate model during the ORACLES-2016 and LASIC experiments. *Atmospheric Chemistry and Physics*, *19*(7), 4963–4990. doi: 10.5194/acp-19-4963-2019

Manaster, A., O'Dell, C. W., & Elsaesser, G. (2017). Evaluation of cloud liquid water path trends using a multidecadal record of passive microwave observations. *Journal of Climate*, *30*(15), 5871–5884. doi: 10.1175/jcli-d-16-0399.1

Mann, J. A. L., Christine Chiu, J., Hogan, R. J., O'Connor, E. J., L'Ecuyer, T. S., Stein, T. H. M., & Jefferson, A. (2014). Aerosol impacts on drizzle properties in warm clouds from ARM Mobile Facility maritime and continental deployments. *Journal of Geophysical Research: Atmospheres*, *119*(7), 4136–4148. doi: https://doi.org/10.1002/2013JD021339

Mann, M. E., Rahmstorf, S., Kornhuber, K., Steinman, B. A., Miller, S. K., Petri, S., & Coumou, D. (2018). Projected changes in persistent extreme summer weather events: The role of quasi-resonant amplification. *Science Advances*, *4*(10). doi: 10.1126/sciadv.aat3272

Mason, B. J. (1960, Jan). Nucleation of water aerosols. *Discussions of the Faraday Society*, *30*, 20–38. doi: 10.1039/DF9603000020

Mather, J. H., & Voyles, J. W. (2013, Mar). The ARM climate research facility: A review of structure and capabilities. *Bulletin of the American Meteorological Society*, *94*(3), 377–392. doi: 10.1175/BAMS-D-11-00218.1

Mauger, G. S., & Norris, J. R. (2007). Meteorological bias in satellite estimates of aerosol-cloud relationships. *Geophysical Research Letters*, *34*(16). doi: 10.1029/ 2007gl029952

Mazoyer, M., Burnet, F., Denjean, C., Roberts, G. C., Haeffelin, M., Dupont, J.-C., & Elias, T. (2019). Experimental study of the aerosol impact on fog microphysics. *Atmospheric Chemistry and Physics*, *19*(7), 4323–4344. doi: 10.5194/acp-19-4323-2019

McComiskey, A., & Feingold, G. (2012). The scale problem in quantifying aerosol cloud indirect effects. *Atmospheric Chemistry and Physics*, *12*, 1031–1049.

McComiskey, A., Feingold, G., Frisch, A. S., Turner, D. D., Miller, M. A., Chiu, J. C., et al. (2009). An assessment of aerosol-cloud interaction in marine stratus clouds based on surface remote sensing. *Journal of Geophysical Research, 114*(D09203).

McCoy, D. T., Bender, F. A.-M., Mohrmann, J. K. C., Hartmann, D. L., Wood, R., & Grosvenor, D. P. (2017, Feb). The global aerosol-cloud first indirect effect estimated using MODIS, MERRA, and AeroCom. *Journal of Geophysical Research: Atmospheres, 122*(3), 1779–1796. doi: 10.1002/2016JD026141

McCoy, D. T., Field, P., Gordon, H., Elsaesser, G. S., & Grosvenor, D. P. (2020). Untangling causality in midlatitude aerosolcloud adjustments. *Atmospheric Chemistry and Physics, 20*(7), 4085–4103. doi: 10.5194/acp-20-4085-2020

McCoy, D. T., Field, P. R., Schmidt, A., Grosvenor, D. P., Bender, F. A. M., Ship-way, B. J., et al. (2018, Apr). Aerosol midlatitude cyclone indirect effects in observations and high-resolution simulations. *Atmospheric Chemistry and Physics, 18*(8), 5821–5846. doi: 10.5194/acp-18-5821-2018

McCoy, D. T., & Hartmann, D. L. (2015). Observations of a substantial cloud-aerosol indirect effect during the 2014-2015 Bardabunga-Veidivotn fissure eruption in Iceland. *Geophysical Research Letters, 42*(23), 10,409–10,414. doi: 10.1002/2015gl067070

McCoy, I. L., Bretherton, C. S., Wood, R., Twohy, C. H., Gettelman, A., Bardeen, C. G., & Toohey, D. W. (2021). Influences of recent particle formation on southern ocean aerosol variability and low cloud properties. *Journal of Geo-physical Research: Atmospheres, 126*(8), e2020JD033529. doi: https://doi.org/10.1029/2020JD033529

McCoy, I. L., McCoy, D. T., Wood, R., Regayre, L., Watson-Parris, D., Grosvenor, D. P., et al. (2020). The hemispheric contrast in cloud microphysical properties constrains aerosol forcing. *Proceedings of the National Academy of Sciences, 117*(32), 18998–19006. doi: 10.1073/pnas.1922502117

McCoy, I. L., Wood, R., & Fletcher, J. K. (2017). Identifying meteorological controls on open and closed mesoscale cellular convection associated with marine cold air outbreaks. *Journal of Geophysical Research: Atmospheres, 122*(21), 11,678–11,702. doi: 10.1002/2017JD027031

McFarquhar, G. M., Bretherton, C., Marchand, R., Protat, A., DeMott, P. J., Alexander, S. P., et al. (2020, Nov). Observations of clouds, aerosols, precipitation, and surface radiation over the southern ocean: An overview of capricorn, marcus, micre and socrates. *Bulletin of the American Meteorological Society*, 1–92. doi: 10.1175/BAMS-D-20-0132.1

McFarquhar, G. M., & Wang, H. (2006, Apr). Effects of aerosols on trade wind cumuli over the Indian Ocean: Model simulations. *Quarterly Journal of the Royal Meteorological Society, 132*(616), 821–843. doi: 10.1256/qj.04.179

Menon, S., Hansen, J., Nazarenko, L., & Luo, Y. (2002). Climate effects of black carbon aerosols in China and India. *Science, 297*(5590), 2250–2253. doi: 10.1126/ science.1075159

Michibata, T., & Suzuki, K. (2020, Jun). Reconciling compensating errors between precipitation constraints and the energy budget in a climate model. *Geophysical Research Letters, 47*(12), e2020GL088340. doi: 10.1029/2020GL088340

Miltenberger, A. K., Field, P. R., Hill, A. A., Shipway, B. J., & Wilkinson, J. M. (2018, Jul). Aerosolcloud interactions in mixed-phase convective clouds. Part 2: Meteorological ensemble. *Atmospheric Chemistry and Physics, 18*(14), 10593–10613. doi: 10.5194/acp-18-10593-2018

Mitchell, D. L., & Finnegan, W. (2009). Modification of cirrus clouds to reduce global warming. *Environmental Research Letters, 4*, 045102.

Morrison, H., & Grabowski, W. W. (2011). Cloud-system resolving model simulations of aerosol indirect effects on tropical deep convection and its thermodynamic environment. *Atmospheric Chemistry and Physics, 11*, 10503–10523.

Morrison, H., & Grabowski, W. W. (2013). Response of tropical deep convection to localized heating perturbations: Implications for aerosol-induced convective invigoration. *Journal of the Atmospheric Sciences. 70*(11), 3533–3555.

Muhlbauer, A., McCoy, I. L., & Wood, R. (2014). Climatology of stratocumulus cloud morphologies: Microphysical properties and radiative effects. *Atmospheric Chemistry and Physics, 14*(13), 6695–6716. doi: 10.5194/acp-14-6695-2014

Mülmenstädt, J., Nam, C., Salzman, M., Kretzschmar, J., L'Ecuyer, T. S., Lohmann, U., et al. (2020). Reducing the aerosol forcing uncertainty using observational constraints on warm rain processes. *Science Advances, 6*(eaaz6433).

Mülmenstädt, J. & Feingold, G. (2018). The radiative forcing of aerosol–cloud interactions in liquid clouds: Wrestling and embracing uncertainty. *Current Climate Change Reports, 4*(1), 23–40. (This article is part of the Topical Collection on Aerosols and Climate) doi: 10.1007/s40641-018-0089-y

Murray, B. J., Carslaw, K. S., & Field, P. R. (2021). Opinion: Cloud-phase climate feedback and the importance of ice-nucleating particles. *Atmospheric Chemistry and Physics, 21*(2), 665–679. doi: 10.5194/acp-21-665-2021

Murray, B. J., O'Sullivan, D., Atkinson, J. D., & Webb, M. E. (2012). Ice nucleation by particles immersed in supercooled cloud droplets. *Chemical Society Reviews, 41*, 6519–6554. doi: 10.1039/ C2CS35200A

Nakajima, T., Higurashi, A., Kawamoto, K., & Penner, J. E. (2001). A possible correlation between satellite-derived cloud and aerosol microphysical parameters. *Geophysical Research Letters, 28*(7), 1171–1174. doi: 10.1029/2000GL012186

National Academies of Sciences, Engineering and Medicine. (2021). *Reflecting sunlight: Recommendations for solar geoengineering research and research governance.* Washington, DC: The National Academies Press. doi: 10.17226/25762

Naud, C. M., Del Genio, A. D., Bauer, M., & Kovari, W. (2010, Jun). Cloud vertical distribution across warm and cold fronts in CloudSat-CALIPSO data and a general circulation model. *Journal of Climate, 23*(12), 3397–3415. doi: 10.1175/2010jcli3282.1

Naud, C. M., Posselt, D. J., & van den Heever, S. C. (2017). Observed covariations of aerosol optical depth and cloud cover in extratropical cyclones. *Journal of Geophysical Research: Atmospheres, 122*(19), 10,338–10,356. doi: 10.1002/ 2017JD027240

Norris, J. R., Allen, R. J., Evan, A. T., Zelinka, M. D., O'Dell, C. W., & Klein, S. A. (2016). Evidence for climate change in the satellite cloud record. *Nature, 536*(7614), 72–75. doi: 10.1038/nature18273

Panicker, A., Pandithurai, G., Safai, P., Dipu, S., Prabha, T., & Konwar, M. (2014). Observations of black carbon induced semi direct effect over North-East India. *Atmospheric Environment*, *98*, 685–692. doi: https://doi.org/10.1016/j.atmosenv.2014.09.034

Penner, J. E., Chen, Y., Wang, M., & Liu, X. (2009). Possible influence of anthropogenic aerosols on cirrus clouds and anthropogenic forcing. *Atmospheric Chemistry and Physics*, *9*(3), 879–896. doi: 10.5194/acp-9-879-2009

Penner, J. E., Zhou, C., & Liu, X. (2015). Can cirrus cloud seeding be used for geoengineering? *Geophysical Research Letters*, *42*, 8775–8782.

Petters, M. D., & Kreidenweis, S. M. (2007, Apr). A single parameter representation of hygroscopic growth and cloud condensation nucleus activity. *Atmospheric Chemistry and Physics*, *7*(8), 1961–1971. doi: 10.5194/acp-7-1961-2007

Petters, M. D., & Kreidenweis, S. M. (2013, Jan). A single parameter representation of hygroscopic growth and cloud condensation nucleus activity. Part 3: Including surfactant partitioning. *Atmospheric Chemistry and Physics*, *13*(2), 1081–1091. doi: 10.5194/acp-13-1081-2013

Petty, G. (2006). *A first course in atmospheric radiation*. Sundog Publishing.

Platnick, S., King, M. D., Ackerman, S. A., Menzel, W. P., Baum, B. A., Riedi, J. C., & Frey, R. A. (2003, Feb). The MODIS cloud products: Algorithms and examples from Terra. *IEEE Transactions on Geoscience and Remote Sensing*, *41*(2), 459–473. doi: 10.1109/tgrs.2002.808301

Platnick, S., & Twomey, S. (1994). Determining the susceptibility of cloud albedo to changes in droplet concentration with the advanced very high resolution radiometer. *Journal of Applied Meteorology*, *33*, 334–347.

Poku, C., Ross, A. N., Hill, A. A., Blyth, A. M., & Shipway, B. (2021). Is a more physical representation of aerosol activation needed for simulations of fog? *Atmospheric Chemistry and Physics*, *21*(9), 7271–7292. doi: 10.5194/acp-21-7271-2021

Politovich, M. K., & Cooper, W. A. (1988). Variability of the supersaturation in cumulus clouds. *Journal of the Atmospheric Sciences*, *45*(11), 1651–1664. doi: 10.1175/1520-0469(1988)045⟨1651:VOTSIC⟩2.0.CO;2

Posselt, R., & Lohmann, U. (2008). Influence of Giant CCN on warm rain processes in the ECHAM5 GCM. *Atmospheric Chemistry and Physics*, *8*(14), 3769–3788. doi: 10.5194/acp-8-3769-2008

Possner, A., Wang, H., Wood, R., Caldeira, K., & Ackerman, T. P. (2018). The efficacy of aerosolcloud radiative perturbations from near-surface emissions in deep open-cell stratocumuli. *Atmospheric Chemistry and Physics*, *18*(23), 17475–17488.

Possner, A., Zubler, E., Lohmann, U., & Schär, C. (2016, May). The resolution dependence of cloud effects and ship-induced aerosol-cloud interactions in marine stratocumulus. *Journal of Geophysical Research: Atmospheres*, *121*(9), 4810–4829. doi: 10.1002/2015JD024685

Prabhakaran, P., Shawon, A. S. M., Kinney, G., Thomas, S., Cantrell, W., & Shaw, R. A. (2020, Jul). The role of turbulent fluctuations in aerosol activation and cloud formation. *Proceedings of the National Academy of Sciences of the United States of America*, *117*(29), 16831–16838. doi: 10.1073/pnas.2006426117

Prisle, N. L., Asmi, A., Topping, D., Partanen, A.-I., Romakkaniemi, S., Dal Maso, M., et al. (2012, Mar). Surfactant effects in global simulations of cloud droplet activation. *Geophysical Research Letters*, *39*(5). doi: 10.1029/2011GL050467

Pruppacher, H., & Klett, J. (2010). Diffusion growth and evaporation of water drops and snow crystals. In *Microphysics of clouds and precipitation* (pp. 502–567). Dordrecht: Springer Netherlands. doi: 10.1007/978-0-306-48100-0 13

Qian, Y., Wan, H., Yang, B., Golaz, J.-C., Harrop, B., Hou, Z., et al. (2018, Dec). Parametric sensitivity and uncertainty quantification in the version 1 of E3SM atmosphere model based on short perturbed parameter ensemble simulations. *Journal of Geophysical Research: Atmospheres*, *123*(23), 13,046–13,073. doi: 10.1029/2018JD028927

Quaas, J., Arola, A., Cairns, B., Christensen, M., Deneke, H., Ekman, A. M. L., et al. (2020, Dec). Constraining the Twomey effect from satellite observations: Issues and perspectives. *Atmospheric Chemistry and Physics*, *20*(23), 15079–15099. doi: 10.5194/acp-20-15079-2020

Quaas, J., Ming, Y., Menon, S., Takemura, T., Wang, M., Penner, J. E., et al. (2009). Aerosol indirect effects general circulation model intercomparison and evaluation with satellite data. *Atmospheric Chemistry and Physics*, *9*(22), 8697–8717.

Rajapakshe, C., Zhang, Z., Yorks, J. E., Yu, H., Tan, Q., Meyer, K., et al. (2017, Jun). Seasonally transported aerosol layers over southeast Atlantic are closer to underlying clouds than previously reported. *Geophysical Research Letters*, *44*(11), 5818–5825. doi: 10.1002/2017GL073559

Rasmussen, C. E., & Williams, C. K. I. (2006). *Gaussian processes for machine learning*. London: MIT press.

Redemann, J., Wood, R., Zuidema, P., Doherty, S. J., Luna, B., LeBlanc, S. E., et al. (2021). An overview of the ORACLES (observations of aerosols above clouds and their interactions) project: Aerosol–cloud–radiation interactions in the southeast Atlantic basin. *Atmospheric Chemistry and Physics*, *21*(3), 1507–1563. doi: 10.5194/acp-21-1507-2021

Regayre, L., Schmale, J., Johnson, J., Tatzelt, C., Baccarini, A., Henning, S., et al. (2019). The value of remote marine aerosol measurements for constraining radiative forcing uncertainty. *Atmospheric Chemistry and Physics*, 1–11. doi: 10.5194/acp-2019-1085

Regayre, L. A., Johnson, J. S., Yoshioka, M., Pringle, K. J., Sexton, D. M., Booth, B. B., et al. (2018, Jul). Aerosol and physical atmosphere model parameters are both important sources of uncertainty in aerosol ERF. *Atmospheric Chemistry and Physics*, *18*(13), 9975–10006. doi: 10.5194/acp-18-9975-2018

Rosenfeld, D., Andreae, M. O., Asmi, A., Chin, M., de Leeuw, G., Donovan, D. P., et al. (2014, Dec). Global observations of aerosol-cloud-precipitation-climate interactions. *Reviews of Geophysics*, *52*(4), 750–808. doi: 10.1002/2013RG000441

Rosenfeld, D., Kaufman, Y., & Koren, I. (2006). Switching cloud cover and dynamical regimes from open to closed Benard cells in response to the suppression of precipitation by aerosols. *Atmospheric Chemistry and Physics*, *6*(9), 2503–2511.

Rosenfeld, D., Lohmann, U., Raga, G. B., O'Dowd, C. D., Kulmala, M., Fuzzi, S., et al. (2008, Sep). Flood or drought:

How do aerosols affect precipitation? *Science*, *321*(5894), 1309–1313. doi: 10.1126/science.1160606

Rosenfeld, D., Zhu, Y., Wang, M., Zheng, Y., Goren, T., & Yu, S. (2019, Feb). Aerosol-driven droplet concentrations dominate coverage and water of oceanic low-level clouds. *Science*, *363*(6427). doi: 10.1126/science.aav0566

Safai, P. D., Ghude, S., Pithani, P., Varpe, S., Kulkarni, R., Todekar, K., et al. (2019). Two-way relationship between aerosols and fog: A case study at IGI airport, New Delhi. *Aerosol and Air Quality Research*, *19*(1), 71–79. doi: 10.4209/aaqr.2017.11.0542

Sanchez, K. J., Chen, C.-L., Russell, L. M., Betha, R., Liu, J., Price, D. J., et al. (2018). Substantial seasonal contribution of observed biogenic sulfate particles to cloud condensation nuclei. *Scientific Reports*, *8*(1), 1–14.

Sato, Y., Miura, H., Yashiro, H., Goto, D., Takemura, T., Tomita, H., & Nakajima, T. (2016, May). Unrealistically pristine air in the Arctic produced by current global scale models. *Scientific Reports*, *6*, 26561. doi: 10.1038/srep26561

Schneider, T., Teixeira, J., Bretherton, C. S., Brient, F., Pressel, K. G., Schär, C., & Siebesma, A. P. (2017, Jan). Climate goals and computing the future of clouds. *Nature Climate Change*, *7*(1), 3–5. doi: 10.1038/nclimate3190

Schutgens, N. A. J., Gryspeerdt, E., Weigum, N., Tsyro, S., Goto, D., Schulz, M., & Stier, P. (2016, May). Will a perfect model agree with perfect observations? The impact of spatial sampling. *Atmospheric Chemistry and Physics*, *16*(10), 6335–6353. doi: 10.5194/acp-16-6335-2016

Schwartz, S. E. (1988, Dec). Are global cloud albedo and climate controlled by marine phytoplankton? *Nature*, *336*(6198), 441–445. doi: 10.1038/336441a0

Schwenkel, J., & Maronga, B. (2019, May). Large-eddy simulation of radiation fog with comprehensive two-moment bulk microphysics: Impact of different aerosol activation and condensation parameterizations. *Atmospheric Chemistry and Physics*, *19*(10), 7165–7181. doi: 10.5194/acp-19-7165-2019

Seinfeld, J. H., Bretherton, C., Carslaw, K. S., Coe, H., DeMott, P. J., Dunlea, E. J., et al. (2016, May). Improving our fundamental understanding of the role of aerosol-cloud interactions in the climate system. *Proceedings of the National Academy of Sciences of the United States of America*, *113*(21), 5781–5790. doi: 10.1073/pnas.1514043113

Sexton, D. M. H., Murphy, J. M., Collins, M., & Webb, M. J. (2012, Jun). Multivariate probabilistic projections using imperfect climate models. Part I: Outline of methodology. *Climate Dynamics*, *38*(11–12), 2513–2542. doi: 10.1007/s00382-011-1208-9

Shaw, R. A., Cantrell, W., Chen, S., Chuang, P., Donahue, N., Feingold, G., et al. (2020, Jul). Cloud–aerosol–turbulence interactions: Science priorities and concepts for a large-scale laboratory facility. *Bulletin of the American Meteorological Society*, *101*(7), E1026–E1035. doi: 10.1175/BAMS-D-20-0009.1

Shindell, D., & Smith, C. J. (2019, Sep). Climate and air-quality benefits of a realistic phase-out of fossil fuels. *Nature*, *573*(7774), 408–411. doi: 10.1038/s41586-019-1554-z

Shinozuka, Y., Saide, P. E., Ferrada, G. A., Burton, S. P., Ferrare, R., Doherty, S. J., Zuidema, P. (2020). Modeling the smoky troposphere of the southeast Atlantic: A comparison to ORACLES airborne observations from September of 2016. *Atmospheric Chemistry and Physics*, *20*(19), 11491–11526. doi: 10.5194/acp-20-11491-2020

Shulman, M. L., Jacobson, M. C., Carlson, R. J., Synovec, R. E., & Young, T. E. (1996, Feb). Dissolution behavior and surface tension effects of organic compounds in nucleating cloud droplets. *Geophysical Research Letters*, *23*(3), 277–280. doi: 10.1029/95GL03810

Small, J. D., Chuang, P. Y., Feingold, G., & Jiang, H. (2009). Can aerosol decrease cloud lifetime? *Geophysical Research Letters*, *36*(L16806).

Smith, C., Kramer, R., Myhre, G., Alterskjær, K., Collins, W., Sima, A., et al. (2020). Effective radiative forcing and adjustments in CMIP6 models. *Atmospheric Chemistry and Physics*, 1–37. doi: 10.5194/acp-2019-1212

Soden, B. J., Held, I. M., Colman, R., Shell, K. M., Kiehl, J. T., & Shields, C. A. (2008, Jan). Quantifying climate feedbacks using radiative kernels. *Journal of Climate*, *21*(14), 3504. doi: 10.1175/2007JCLI2110.1

Sourdeval, O., C.-Labonnote, L., Baran, A. J., Mülmenstädt, J., & Brogniez, G. (2016, Oct). A methodology for simultaneous retrieval of ice and liquid water cloud properties. Part 2: Near-global retrievals and evaluation against A-Train products. *Quarterly Journal of the Royal Meteorological Society*, *142*(701), 3063–3081. doi: 10.1002/qj.2889

Sourdeval, O., C-Labonnote, L., Baran, A. J., & Brogniez, G. (2015). A methodology for simultaneous retrieval of ice and liquid water cloud properties. Part I: Information content and case study. *Quarterly Journal of the Royal Meteorological Society*, *141*(688), 870–882.

Sourdeval, O., Gryspeerdt, E., Krämer, M., Goren, T., Delanoë, J., Afchine, A., et al. (2018, Oct). Ice crystal number concentration estimates from lidar–radar satellite remote sensing. Part 1: Method and evaluation. *Atmospheric Chemistry and Physics*, *18*(19), 14327–14350. doi: 10.5194/acp-18-14327-2018

Spensberger, C., & Sprenger, M. (2018, Jan). Beyond cold and warm: An objective classification for maritime midlatitude fronts. *Quarterly Journal of the Royal Meteorological Society*, *144*(710), 261–277. doi: 10.1002/ qj.3199

Squires, P. (1952, Mar). The growth of cloud drops by condensation. I: General characteristics. *Australian Journal of Scientific Research A Physical Sciences*, *5*, 59. doi: 10.1071/PH520059

Stevens, B., Bony, S., Farrell, D., Ament, F., Blyth, A., Fairall, C., et al. (2021). EUREC^{4}A. *Earth System Science Data*, *13*(8), 4067–4119. Retrieved from https://essd.copernicus.org/articles/13/4067/2021/ doi: 10.5194/essd-13-4067-2021

Stevens, B., & Feingold, G. (2009, Oct). Untangling aerosol effects on clouds and precipitation in a buffered system. *Nature*, *461*(7264), 607–613. doi: 10.1038/nature08281

Stevens, B., Satoh, M., Auger, L., Biercamp, J., Bretherton, C. S., Chen, X., et al. (2019, Sep). DYAMOND: The DYnamics of the Atmospheric general circulation Modeled On Non-hydrostatic Domains. *Progress in Earth and Planetary Science*, *6*(1), 61. doi: 10.1186/s40645-019-0304-z

Stier, P., Schutgens, N. A. J., Bellouin, N., Bian, H., Boucher, O., Chin, M., et al. (2013, Mar). Host model uncertainties in aerosol radiative forcing estimates: Results from the AeroCom

Prescribed intercomparison study. *Atmospheric Chemistry and Physics*, *13*(6), 3245–3270. doi: 10.5194/acp-13-3245-2013

Stjern, C. W., Samset, B. H., Myhre, G., Forster, P. M., Hodnebrog, Ø., Andrews, T., et al. (2017). Rapid adjustments cause weak surface temperature response to increased black carbon concentrations. *Journal of Geophysical Research: Atmospheres*, *122*(21), 11,462–11,481. doi: https://doi.org/10.1002/2017JD027326

Storelvmo, T. (2017, Aug). Aerosol effects on climate via mixed-phase and ice clouds. *Annual Review of Earth and Planetary Sciences*, *45*(1), 199–222. doi: 10.1146/annurev-earth-060115-012240

Storelvmo, T., Boos, W. R., & Herger, N. (2014). Cirrus cloud seeding: A climate engineering mechanism with reduced side effects? *Philosophical Transactions of the Royal Society A: Mathematical, Physical and Engineering Sciences*, *372*(2031), 20140116. doi: 10.1098/rsta.2014.0116

Storelvmo, T., & Tan, I. (2015). The Wegener-Bergeron-Findeisen process? Its discovery and vital importance for weather and climate. *Meteorologische Zeitschrift*, *24*(4), 455–461. doi: 10.1127/metz/2015/0626

Sullivan, R. C., Moore, M. J. K., Petters, M. D., Kreidenweis, S. M., Roberts, G. C., & Prather, K. A. (2009, May). Effect of chemical mixing state on the hygroscopicity and cloud nucleation properties of calcium mineral dust particles. *Atmospheric Chemistry and Physics*, *9*(10), 3303–3316. doi: 10.5194/acp-9-3303-2009

Tan, I., Storelvmo, T., & Choi, Y.-S. (2014). Spaceborne lidar observations of the ice-nucleating potential of dust, polluted dust, and smoke aerosols in mixedphase clouds. *Journal of Geophysical Research: Atmospheres*, *119*(11), 6653–6665. doi: 10.1002/2013JD021333

Tao, W.-K., Chen, J.-P., Li, Z., Wang, C., & Zhang, C. (2012). Impact of aerosols on convective clouds and precipitation. *Reviews of Geophysics*, *50*(2). doi: https://doi.org/10.1029/2011RG000369

Tegen, I., & Schepanski, K. (2018). Climate feedback on aerosol emission and atmospheric concentrations. *Current Climate Change Reports*, *4*(1), 1–10.

Terai, C. R., Pritchard, M. S., Blossey, P., & Bretherton, C. S. (2020). The impact of resolving subkilometer processes on aerosol-cloud interactions of low-level clouds in global model simulations. *Journal of Advances in Modeling Earth Systems*, *12*(11), e2020MS002274. doi: 10.1029/2020MS002274

Tessendorf, S. A., French, J. R., Friedrich, K., Geerts, B., Rauber, R. M., Rasmussen, R. M., et al. (2019, Jan). A transformational approach to winter orographic weather modification research: The SNOWIE project. *Bulletin of the American Meteorological Society*, *100*(1), 71–92. doi: 10.1175/BAMS-D-17-0152.1

Thornton, J. A., Virts, K. S., Holzworth, R. H., & Mitchell, T. P. (2017). Lightning enhancement over major oceanic shipping lanes. *Geophysical Research Letters*, *44*(17), 9102–9111. doi: 10.1002/2017GL074982

Thouron, O., Brenguier, J.-L., & Burnet, F. (2012, May). Supersaturation calculation in large eddy simulation models for prediction of the droplet number concentration. *Geoscientific Model Development*, *5*(3), 761–772. doi: 10.5194/gmd-5-761-2012

Toll, V., Christensen, M., Gassó, S., & Bellouin, N. (2017). Volcano and ship tracks indicate excessive aerosol-induced cloud water increases in a climate model. *Geophysical Research Letters*, *44*(24), 12,492–12,500. doi: 10.1002/2017GL075280

Toll, V., Christensen, M., Quaas, J., & Bellouin, N. (2019, Aug). Weak average liquid-cloud-water response to anthropogenic aerosols. *Nature*, *572*(7767), 51–55. doi: 10.1038/s41586-019-1423-9

Tornow, F., Domenech, C., Barker, H. W., Preusker, R., & Fischer, J. (2020, Jul). Using two-stream theory to capture fluctuations of satellite-perceived TOA SW radiances reflected from clouds over ocean. *Atmospheric Measurement Techniques*, *13*(7), 3909–3922. doi: 10.5194/amt-13-3909-2020

Twohy, C. H., Kreidenweis, S. M., Eidhammer, T., Browell, E. V., Heymsfield, A. J., Bansemer, A. R., et al. (2009, Jan). Saharan dust particles nucleate droplets in eastern Atlantic clouds. *Geophysical Research Letters*, *36*(1). doi: 10.1029/2008GL035846

Twomey, S. (1954). The composition of hygroscopic particles in the atmosphere. *Journal of Meteorological*, *11*, 334–338.

Twomey, S. (1959, May). The nuclei of natural cloud formation. Part II: The supersaturation in natural clouds and the variation of cloud droplet concentration. *Geofisica Pura e Applicata*, *43*(1), 243–249. doi: 10.1007/BF01993560

Twomey, S. (1974). Pollution and the planetary albedo. *Atmospheric Environment (1967)*, *8*(12), 1251–1256. doi: https://doi.org/10.1016/0004-6981(74)90004-3

Twomey, S. (1977, Jul). The influence of pollution on the shortwave albedo of clouds. *Journal of the Atmospheric Sciences*, *34*(7), 1149–1152. doi: 10.1175/1520-0469(1977)034⟨1149:TIOPOT⟩2.0.CO;2

Twomey, S. A., Piepgrass, M., & Wolfe, T. L. (1984, Nov). An assessment of the impact of pollution on global cloud albedo. *Tellus B*, *36B*(5), 356–366. doi: 10.1111/j.1600-0889.1984.tb00254.x

Vergara-Temprado, J., Miltenberger, A. K., Furtado, K., Grosvenor, D. P., Shipway, B. J., Hill, A. A., et al. (2018, Mar). Strong control of Southern Ocean cloud reflectivity by ice-nucleating particles. *Proceedings of the National Academy of Sciences of the United States of America*, *115*(11), 2687–2692. doi: 10.1073/pnas.1721627115

Vonnegut, B. (1958). Early work on silver iodide smokes for cloud seeding. *Final Report of the Advisory Committee on Weather Control*, *2*, 283–285.

Wang, H., Easter, R. C., Rasch, P. J., Wang, M., Liu, X., Ghan, S. J., et al. (2013, Jun). Sensitivity of remote aerosol distributions to representation of cloud–aerosol interactions in a global climate model. *Geoscientific Model Development*, *6*(3), 765–782. doi: 10.5194/gmd-6-765-2013

Wang, M., Ghan, S., Ovchinnikov, M., Liu, X., Easter, R., Kassianov, E., Morrison, H. (2011). Aerosol indirect effects in a multi-scale aerosol-climate model PNNL-MMF. *Atmospheric Chemistry and Physics*, *11*(11), 5431–5455. doi: 10.5194/acp-11-5431-2011

Wang, S., Wang, Q., & Feingold, G. (2003, Jan). Turbulence, condensation, and liquid water transport in numerically simulated nonprecipitating stratocumulus clouds. *Journal of the Atmospheric Sciences*, *60*(2), 262–278. doi: 10.1175/1520-0469(2003)060⟨0262:TCALWT⟩2.0.CO;2

Wang, S. C., & Flagan, R. C. (1989, Jan). Scanning electrical mobility spectrometer. *Journal of Aerosol Science*, *20*(8), 1485–1488. doi: 10.1016/0021-8502(89) 90868-9

Wang, Y., Su, H., Jiang, J. H., Xu, F., & Yung, Y. L. (2020). Impact of cloud ice particle size uncertainty in a climate model and implications for future satellite missions. *Journal of Geophysical Research: Atmospheres*, *125*(6), e2019JD032119. doi: https://doi.org/10.1029/2019JD032119

Wang, Y., & Zhang, G. J. (2016, Dec). Global climate impacts of stochastic deep convection parameterization in the NCAR-CAM5. *Journal of Advances in Modeling Earth Systems*, *8*(4), 1641–1656. doi: 10.1002/2016MS000756

Watson-Parris, D., Bellouin, N., Deaconu, L. T., Schutgens, N. A. J., Yoshioka, M., Regayre, L. A., et al. (2020, May). Constraining uncertainty in aerosol direct forcing. *Geophysical Research Letters*, *47*(9), e2020GL087141. doi: 10.1029/2020GL087141

Watson-Parris, D., Sutherland, S. A., Christensen, M. W., Eastman, R., & Stier, P. (2021, Feb). A large-scale analysis of pockets of open cells and their radiative impact. *Geophysical Research Letters*, *48*(6), e2020GL092213. doi: 10.1029/2020GL092213

Wegener, A. (1911). *Thermodynamik der atmosphäre*. Leipzig: Verlag von Johann Ambrosius Barth.

Wigley, T. M. L. (1989, Jun). Possible climate change due to SO_2-derived cloud condensation nuclei. *Nature*, *339*(6223), 365–367. doi: 10.1038/339365a0

Wilcox, L. J., Liu, Z., Samset, B. H., Hawkins, E., Lund, M. T., Nordling, K., et al. (2020). Accelerated increases in global and Asian summer monsoon precipitation from future aerosol reductions. *Atmospheric Chemistry and Physics*, *20*(20), 11955–11977. doi: 10.5194/acp-20-11955-2020

Williams, K. D., & Webb, M. J. (2009, Jul). A quantitative performance assessment of cloud regimes in climate models. *Climate Dynamics*, *33*(1), 141–157. doi: 10.1007/s00382-008-0443-1

Williamson, C. J., Kupc, A., Axisa, D., Bilsback, K. R., Bui, T. P., Campuzano-Jost, P., et al. (2019, Oct). A large source of cloud condensation nuclei from new particle formation in the tropics. *Nature*, *574*, 399–403. doi: 10.1038/s41586-019-1638-9

Wilson, C. (1900, Jan). IX. On the comparative efficiency as condensation nuclei of positively and negatively charged ions. *Philosophical Transactions of the Royal Society of London. Series A, Containing Papers of a Mathematical or Physical Character*, *193*, 289–308. doi: 10.1098/rsta.1900.0009

Wood, R. (2012). Stratocumulus clouds. *Monthly Weather Review*, *140*(8), 2373–2423. doi: 10.1175/MWR-D-11-00121.1

Wood, R., Ackerman, T., Rasch, P., & Wanser, K. (2017, Jul). Could geo-engineering research help answer one of the biggest questions in climate science? *Earth's Future*, *5*(7), 659–663. doi: 10.1002/2017EF000601

Wood, R., & Ackerman, T. P. (2013, Dec). Defining success and limits of field experiments to test geoengineering by marine cloud brightening. *Climatic Change*, *121*(3), 459–472. doi: 10.1007/s10584-013-0932-z

Wood, R., & Hartmann, D. L. (2006, May). Spatial variability of liquid water path in marine low cloud: The importance of mesoscale cellular convection. *Journal of Climate*, *19*(9), 1748–1764. doi: 10.1175/jcli3702.1

Wood, R., Leon, D., Lebsock, M., Snider, J., & Clarke, A. D. (2012, Oct). Precipitation driving of droplet concentration variability in marine low clouds. *Journal of Geophysical Research: Atmospheres*, *117*(D19). doi: 10.1029/2012jd018305

Wood, R., Stromberg, I. M., Jonas, P. R., & Mill, C. S. (1997, Aug). Analysis of an air motion system on a light aircraft for boundary layer research. *Journal of Atmospheric and Oceanic Technology*, *14*(4), 960–968. doi: 10.1175/1520-0426(1997)014⟨0960:AOAAMS⟩2.0.CO;2

Woodcock, A. H. (1953, Oct). Salt nuclei in marine air as a function of altitude and wind force. *Journal of Atmospheric Sciences*, *10*(5), 362–371. doi: 10.1175/1520-0469(1953)010⟨0366:SNIMAA⟩2.0.CO;2

Wu, P., Dong, X., Xi, B., Tian, J., & Ward, D. M. (2020, May). Profiles of MBL cloud and drizzle microphysical properties retrieved from ground-based observations and validated by aircraft in-situ measurements over the Azores. *Journal of Geophysical Research: Atmospheres*, *125*(9). doi: 10.1029/2019JD032205

Yamaguchi, T., Feingold, G., Kazil, J., & McComiskey, A. (2015). Stratocumulus to cumulus transition in the presence of elevated smoke layers. *Geophysical Research Letters*, *42*(23), 10,478–10,485. doi: https://doi.org/10.1002/2015GL066544

Yang, F., McGraw, R., Luke, E. P., Zhang, D., Kollias, P., & Vogelmann, A. M. (2019, Nov). A new approach to estimate supersaturation fluctuations in stratocumulus cloud using ground-based remote-sensing measurements. *Atmospheric Measurement Techniques*, *12*(11), 5817–5828. doi: 10.5194/amt-12-5817-2019

Yang, Q., Easter, R. C., Campuzano-Jost, P., Jimenez, J. L., Fast, J. D., Ghan, S. J., et al. (2015, Aug). Aerosol transport and wet scavenging in deep convective clouds: A case study and model evaluation using a multiple passive tracer analysis approach. *Journal of Geophysical Research: Atmospheres*, *120*(16), 8448–8468. doi: 10.1002/2015JD023647

Yin, Y., Carslaw, K. S., & Feingold, G. (2005). Vertical transport and processing of aerosols in a mixed-phase convective cloud and the feedback on cloud development. *Quarterly Journal of the Royal Meteorological Society*, *131*(605), 221–245. doi: https://doi.org/10.1256/qj.03.186

Yoshioka, M., Regayre, L. A., Pringle, K. J., Johnson, J. S., Mann, G. W., Par-tridge, D. G., et al. (2019). Ensembles of global climate model variants designed for the quantification and constraint of uncertainty in aerosols and their radiative forcing. *Journal of Advances in Modeling Earth Systems*, *11*(11), 3728–3754. doi: https://doi.org/10.1029/2019MS001628

Yuan, T., Remer, L. A., & Yu, H. (2011). Microphysical, macrophysical and radiative signatures of volcanic aerosols in trade wind cumulus observed by the A-Train. *Atmospheric Chemistry and Physics*, *11*(14), 7119–7132. doi: 10.5194/acp-11-7119-2011

Zelinka, M. D., Myers, T. A., McCoy, D. T., Po-Chedley, S., Caldwell, P. M., Ceppi, P., et al. (2020, Jan). Causes of higher climate sensitivity in CMIP6 models. *Geophysical Research Letters*, *47*(1), e2019GL085782. doi: 10.1029/2019GL085782

Zerefos, C. S., Tetsis, P., Kazantzidis, A., Amiridis, V., Zerefos, S. C., Luterbacher, J., et al. (2014). Further evidence of important environmental information content in red-to-green ratios as depicted in paintings by great masters. *Atmospheric Chemistry and Physics*, *14*(6), 2987–3015. doi: 10.5194/acp-14-2987-2014

Zhang, Y., Karamchandani, P., Glotfelty, T., Streets, D. G., Grell, G., Nenes, A., et al. (2012, Oct). Development and initial application of the global-through-urban weather research and forecasting model with chemistry (GU-WRF/Chem). *Journal of Geophysical Research (Atmospheres)*, *117*(D20), D20206. doi: 10.1029/2012JD017966

Zhang, Z., Song, H., Ma, P. L., Larson, V. E., Wang, M., Dong, X., & Wang, J. (2019). Subgrid variations of the cloud water and droplet number concentration over the tropical ocean: Satellite observations and implications for warm rain simulations in climate models. *Atmospheric Chemistry and Physics*, *19*(2), 1077–1096. doi: 10.5194/acp-19-1077-2019

Zhang, Z., Werner, F., Cho, H. M., Wind, G., Platnick, S., Ackerman, A. S., et al. (2016). A framework based on 2-D Taylor expansion for quantifying the impacts of subpixel reflectance variance and covariance on cloud optical thickness and effective radius retrievals based on the bispectral method. *Journal of Geophysical Research: Atmospheres*, *121*(12), 7007–7025. doi: 10.1002/2016JD024837

Zhao, B., Wang, Y., Gu, Y., Liou, K.-N., Jiang, J. H., Fan, J., et al. (2019). Ice nucleation by aerosols from anthropogenic pollution. *Nature Geoscience*, *12*(8), 602–607.

Zheng, G., Wang, Y., Wood, R., Jensen, M. P., Kuang, C., McCoy, I. L., et al. (2021). New particle formation in the remote marine boundary layer. *Nature Communications*, *12*(1). doi: 10.1038/s41467-020-20773-1

Zhou, C., & Penner, J. E. (2014). Aircraft soot indirect effect on large-scale cirrus clouds: Is the indirect forcing by aircraft soot positive or negative? *Journal of Geophysical Research: Atmospheres*, *119*(19), 11,303–11,320. doi: https://doi.org/10.1002/2014JD021914

Zhou, X., Ackerman, A. S., Fridlind, A. M., Wood, R., & Kollias, P. (2017). Impacts of solar-absorbing aerosol layers on the transition of stratocumulus to trade cumulus clouds. *Atmospheric Chemistry and Physics*, *17*(20), 12725–12742. doi: 10.5194/acp-17-12725-2017

Zhu, J., & Penner, J. E. (2020). Radiative forcing of anthropogenic aerosols on cirrus clouds using a hybrid ice nucleation scheme. *Atmospheric Chemistry and Physics*, *20*(13), 7801–7827. doi: 10.5194/acp-20-7801-2020

Zuidema, P., Redemann, J., Haywood, J., Wood, R., Piketh, S., Hipondoka, M., & Formenti, P. (2016, Jul). Smoke and clouds above the southeast Atlantic: Upcoming field campaigns probe absorbing aerosol's impact on climate. *Bulletin of the American Meteorological Society*, *97*(7), 1131–1135. doi: 10.1175/BAMS-D-15-00082.1

Zuidema, P., Sedlacek, A. J., Flynn, C., Springston, S., Delgadillo, R., Zhang, J., et al. (2018, May). The Ascension Island boundary layer in the remote southeast Atlantic is often smoky. *Geophysical Research Letters*, *45*(9), 4456–4465. doi: 10.1002/2017GL076926

3

Ice Crystal Complexity and Link to the Cirrus Cloud Radiative Effect

Emma Järvinen[1], Bastiaan van Diedenhoven[2], Nathan Magee[3], Steven Neshyba[4], Martin Schnaiter[1], Guanglang Xu[1], Olivier Jourdan[5], David Delene[6], Fritz Waitz[1], Simone Lolli[7], and Seiji Kato[8]

ABSTRACT

The aim of this chapter is to present emerging information on ice crystal morphology in the nano- and microscale from various sources, and explore its impact on ice crystal's single-scattering properties and ultimately on cirrus cloud radiative effect. Increasing amount of observations in the last decade have shown that cirrus ice crystals invariably contain some degree of facet roughness, hollowness, or other morphological complexities that make the crystal shape deviate from that of a pristine hexagonal shape. These morphological complexities are found to be present in a wide variety of environmental conditions. A set of physical and statistical complexity parameters is used to quantify ice crystal complexity depending on the application. Although laboratory studies have shown that complexity parameters tend to increase with increasing crystal growth rate, observations in natural cirrus indicate that complexity parameters only weakly vary at cirrus temperatures below 230 K. Morphological complexities affect the optical properties of ice crystals. Implementing the optical properties of complex ice crystals in climate models has shown that the radiative effect of crystal complexity is a cooling of 1–2 Wm^{-2} globally. Due to the prevalence of ice crystal morphological complexities and their influence on cirrus radiative properties, the radiative effect of crystal complexity should be taken into account in future climate models. However, work is still needed to transfer the increasing knowledge of the physical nature of ice crystal complexity into optical particle models and eventually into parameterizations used in models.

3.1. INTRODUCTION

Cirrus clouds are ubiquitous in Earth's atmosphere. At any given moment, they cover approximately 40–60% of Earth's surface and in the tropics the cirrus cloud coverage can reach up to 70% (e.g., Mace et al., 2009; Matus & L'Ecuyer, 2017; Sassen et al., 2008). Our understanding of cirrus clouds and their role in the climate system has improved greatly in the past decades thanks to new capabilities in modeling, improvements in remote sensing,

[1]*Institute of Meteorology and Climate Research, Karlsruhe Institute of Technology, Karlsruhe, Germany*

[2]*Center for Climate System Research, Columbia University, New York, NY, USA; now at SRON Netherlands Institute for Space Research, Leiden, The Netherlands*

[3]*Department of Physics, The College of New Jersey, Ewing, NJ, USA*

[4]*Department of Chemistry, University of Puget Sound, Tacoma, WA, USA*

[5]*Laboratoire de Météorologie Physique, Université Clermont Auvergne/OPGC/CNRS, Clermont-Ferrand, France*

[6]*Department of Atmospheric Sciences, University of North Dakota, Grand Forks, ND, USA*

[7]*Institute of Methodologies for Environmental Analysis, National Research Council of Italy, Tito, Italy; and CommSensLab, Department of Signal Theory and Communications, Universitat Politècnica de Catalunya, Barcelona, Spain*

[8]*NASA Langley Research Center, Hampton, VA, USA*

Clouds and Their Climatic Impacts: Radiation, Circulation, and Precipitation, Geophysical Monograph 281, First Edition.
Edited by Sylvia C. Sullivan and Corinna Hoose.

DOI: 10.1002/9781119700357.ch03

increased accumulation of in situ data (e.g., with aircraft and balloon-borne measurements), dedicated laboratory and cloud chamber measurements, and development of new instrumentation. While it is generally agreed that cirrus clouds are an important cloud type both from climate and weather perspectives, it is also acknowledged that large gaps exist in our knowledge related to cirrus clouds, leading to substantial uncertainties and possible systematic biases in our predictions of Earth's future climate (Boucher et al., 2013).

Partly, these uncertainties arise from inadequate understanding of microscale physics. Based on data gathered with airborne cloud instruments, it is well known that cirrus ice crystals cover a wide range of crystal sizes from a few micrometers for newly formed crystals to several millimeter sized precipitation particles (e.g., Krämer et al., 2020). It is also known that, unlike spherical liquid droplets, cirrus ice crystals manifest impressively myriad geometries (e.g., Baran, 2009; Lawson et al., 2019). The basic shape of ice crystals often resembles a sixfold radial symmetry that can be attributed to the hexagonal molecular structure of ice. Three facets are most commonly observed: the basal, the prismatic, and the pyramidal. Laboratory studies have shown that the growth rates of these facets are determined by temperature and relative humidity leading to different crystal habits, such as columnar crystals, plate-like crystals, and bullet rosettes (Bailey & Hallett, 2002, 2004, 2009). Variations in atmospheric conditions and interactions with other crystals and supercooled droplets lead to more complex crystal structures. One commonly used scheme to classify ice crystals contains 80 classes for snow and ice (Magono & Lee, 1966).

Recently, we have become aware that the morphological diversity of ice crystals extends beyond the crystal habits. Technological developments in the past decades have made it possible to inspect the sub-micron scale structure of ice surfaces using electron microscope technology (Magee et al., 2014, 2020; Neshyba et al., 2013; Pfalzgraff et al., 2010) and by inspecting high-resolution light scattering patterns of individual ice crystals (Schnaiter et al., 2016; Ulanowski et al., 2006, 2014). These studies have confirmed that ice crystal sub-micron scale structures are present in wide range of environmental conditions both in laboratory grown crystals (Magee et al., 2014; Neshyba et al., 2013; Schnaiter et al., 2016; Voigtländer et al., 2018) and in natural clouds (Järvinen et al., 2018a; Järvinen et al., 2018b; Magee et al., 2020; Ulanowski et al., 2014), and that there is substantial variation of crystal structure within habit classes, specifically in the number and aspect ratios of their components (Fridlind et al., 2016; Um & McFarquhar, 2007; Um et al., 2015).

Ice crystal morphology plays an important role in the interactions with electromagnetic radiation. For example, most notable to the human eye are optical phenomena known as halos and parhelia frequently seen around the Sun or Moon, records of which date back at least to Aristotle (Tape & Moilanen, 2006). Such phenomena are readily explained by pristine hexagonal facets: the 22° halo, for example, can be understood as resulting from the passage of light rays through two non-neighboring prismatic facets. Microscopic structures on the crystal's surface lead to randomization of the reflection angles on the crystal boundary and, thus, suppress the visibility of these optical phenomena. It was estimated that the contribution by pristine crystals to the total scattering cross-section needs to be at least 10–40% for halos to be observable, depending on crystal shape (Forster et al., 2020; van Diedenhoven, 2014).

On a global scale, ice crystal optical properties, together with cirrus cloud fraction, height and ice water path define the cirrus cloud radiative effect (CRE). According to our current understanding, the cirrus and ice cloud globally and annually averaged top-of-the-atmosphere (TOA) CRE is slightly warming, at about 5 Wm^{-2} (Gasparini & Lohmann, 2016; Hong et al., 2016). However, there is a significant source of uncertainty related to the magnitude and even sign of the cirrus and ice cloud CRE due to uncertainty in cloud macro- and microphysical properties. General circulation models (GCMs) aim to predict cloud macro- and microphysical properties and the cirrus CRE in current and future climate states. Most models include schemes to diagnose vertically resolved cloud fraction and ice water path, while recent advances in cloud microphysical parameterizations also allow models to predict ice number concentrations and related ice sizes (del Genio et al., 1996; Morrison & Gettelman, 2008; Zhao et al., 2017). While these modeled ice cloud properties constrain the cloud optical thickness, spectral absorption, and emissivity relatively well, they do not constrain the ice crystal scattering properties, which are largely determined by ice crystal morphological properties, such as crystal size (Zhang et al., 1999), shape (Hong et al., 2009; Mishchenko et al., 1995; Macke et al., 1998; Yang et al., 2007; Zhang et al., 1999; Zhao et al., 2018), and ice crystal morphological complexity (Järvinen et al., 2018a; Yi et al., 2013).

For a more accurate estimation of the radiative effects of cirrus and ice clouds, a better knowledge of ice crystal morphological and optical properties as a function of cloud type, location, height, and temperature is needed. Moreover, existing and emerging new information needs to be synthesized to gain a better understanding of cirrus properties relevant for climate modeling and weather forecasting. Recent review papers have provided an overview of cirrus observations and occurrence (Heymsfield et al., 2017), ice crystal sizes and cirrus ice water contents (IWCs) (Krämer et al., 2020), ice crystal habits (Lawson et al., 2019), optical properties (Yang et al., 2018), and developments in space-borne remote sensing capabilities (van Diedenhoven, 2018). However, a review of the emerging knowledge of ice crystal morphology at the nano- and microscale is still missing. Since it has been shown that ice crystal optical properties are sensitive to morphological complexity, such as surface roughness (e.g., Yang et al., 2013), it is important to summarize the work on ice crystal morphological complexity to facilitate consensus between laboratory, field, and theoretical work.

In this chapter, we give an overview of the current work on ice crystal morphological complexity and its effects on ice cloud optical properties. We report both laboratory and field observations of complex ice crystals and aim to unify the terminology used in the field. First, section 3.2 gives a qualitative overview of ice crystal nano- and microscale morphological features and introduces these in the context of size scale. We also present the different metrics used to quantify ice crystal morphological complexity in different applications. Section 3.3 summarizes how crystal complexity can be quantified using (1) optical microscopy, (2) high-resolution angular light scattering patterns, and (3) space-borne remote sensing methods and what we have learned of the link between the degree of crystal morphological complexity and environmental conditions using these methods. Section 3.4 discusses the effects of crystal complexity on the single-scattering properties of ice crystals. We give an overview of the methods used to describe the interactions between electromagnetic light and ice crystals, introduce the state-of-the-art optical models for complex ice crystals, and discuss in situ and remote sensing observations of ice crystal angular light scattering properties and ice cloud asymmetry parameter. Section 3.5 discusses how ice crystal optical properties are linked to the cirrus CRE, the magnitude of cirrus CRE as a function of cloud optical thickness and geographical location, and the uncertainty in CRE caused by ice crystal morphological complexity. Finally, we give a summary of the present knowledge in section 3.6 and give recommendations for future research directions in section 3.7.

3.2. ICE CRYSTAL MORPHOLOGICAL COMPLEXITY ACROSS SCALES

Here, we aim to give a comprehensive, albeit not exhaustive, overview of types of complexity found in natural ice crystals. One challenge in this regard concerns terminology: although most observed ice crystals are not pristine in nature, there is no uniform terminology or classification for ice crystal morphological complexity. Several types of morphological complexity are identified from laboratory and field measurements, including facet roughness, atypical facet angles, hollowness, inclusion of air bubbles and aerosol particles, rounded surfaces, conjoined prisms, aggregation of single crystals, riming, fracturing of ice crystals in ice-ice collisions, and sublimation as well as post-sublimation regrowth of micro facets.

Since identifying different morphological complexity types on a single crystal is not always possible, a useful way of addressing this challenge is to organize morphological complexity according to its spatial scale, as shown in Figure 3.1. Relevant scales are nanoscale, which includes most of the facet roughness features, and microscale that contains complexity types visible in optical microscopy. Since often morphological complexity lies between nano- and microscales, it has been found appropriate to use the term "mesoscopic complexity" (not to be confused with "mesoscale" commonly used to indicate a scale between micro and macro, e.g., in meteorological applications) to describe features from ~100 nm to a few tens of micrometers (Järvinen et al., 2018b; Magee et al., 2014; Neshyba et al., 2013; Schnaiter et al., 2016). The terms mesoscopic complexity and "mesoscopic roughness" have been used in conjunction with cryoscopic scanning electron microscopy (CSEM) observations (Magee et al., 2014; Neshyba et al., 2013) and high-resolution light scattering patterns (Järvinen et al., 2018a; Järvinen et al., 2018b). Schmitt et al. (2016) used the term "complexity" to describe the transition from single crystals to complex (aggregated) particles as seen in airborne cloud particle imagers. Here, we suggest to use the term "morphological complexity" (later referred to as "complexity") to describe any morphological deviation from that of a pristine, ideal hexagonal prism.

3.2.1. Types of Morphological Complexity

Figures 3.2 and 3.3 provide a glimpse of various morphological complexities that have been observed in atmospheric ice particles. When looking at laboratory or in situ images of natural ice particles (e.g., Fig. 3.3), it quickly becomes apparent that many distinct types of complexity occur, and several of these complexities are often seen overlaid on a single particle. Nearly all of these categories of complexity have been observed by multiple observing platforms and can be seen in a spectrum ranging from barely detectable to dramatically affecting the morphology of the entire particle or the geometry of the facet surfaces.

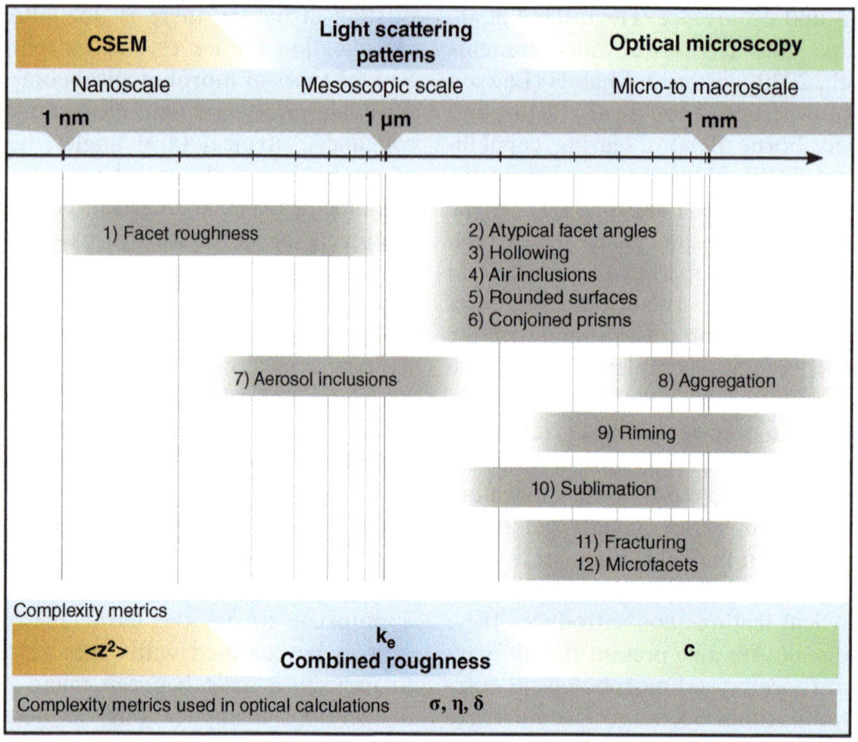

Figure 3.1 Overview of different types of ice crystal morphological complexity ranging scales from nano- to microscale and the methods to quantify crystal complexity in different scales: Cryoscopic scanning electron microscopy (CSEM), high-resolution light scattering patterns, and optical microscopy. The resulting complexity metrics can be divided into physical complexity metrics, which are the mean square of the local gradients of surface heights ($< Z^2 >$) that can be retrieved from CSEM, the optical complexity parameters k_e and combined roughness that can be retrieved from high-resolution light scattering patterns, and the complexity parameter C that can be inferred from Cloud Imaging Probes, and to statistical complexity metrics that are used in computations to simulate the optical properties of ice crystals. The degree of facet roughness in the statistical methods can be described by width of the surface slope distribution specified by σ, η, or δ. A detailed discussion of the different complexity metrics can be found in section 3.2.2.

Facet Roughening

Ice crystal facets are frequently observed to be roughened – deviating from a flat, smooth planar surface. The characteristics of these deviations vary with respect to symmetry, scale, and geometric organization relative to the crystal lattice. For example, regular linear ridging is often observed running across prism facets, parallel with the basal plane (e.g., crystals 1b and 1c in Fig. 3.2 and Fig. 3.3. in Pfalzgraff et al., 2010). On basal facets, roughening is typically observed to be less symmetrically organized than on prism facets, sometimes with seemingly random roughening (crystal 1a in Fig. 3.2), and occasionally with dendritic patterning. It is also worth noting that this facet roughening is sometimes apparent in optical microscopy, but often only resolved by high magnification electron microscopy. Some facets reveal roughening at micron-scale, but others require nanoscale observational resolution to detect roughening patterns. Notably, similar roughening patterns have been observed in both lab-grown ice crystals (Magee et al., 2014; Neshyba et al., 2013; Pfalzgraff et al., 2010) and captured cirrus particles (Magee et al., 2020). In addition, facet roughening has been observed during both growth and sublimation. Generally, growth mode roughening tends to appear more regular and symmetric relative to sublimation roughening.

Atypical Facet Angles

Atmospheric ice crystals also sometimes display crystal facet angles that diverge from a regular hexagonal polygons (Murray et al., 2015, and references therein). It is observed for ice crystals to present as "truncated triangles," with alternating longer and shorter widths of prism facets (crystals 2 and 7 in Fig. 3.2). Although apparently less common, ice particles with rectangular or rhombic faceting have been observed as well. The frequency of

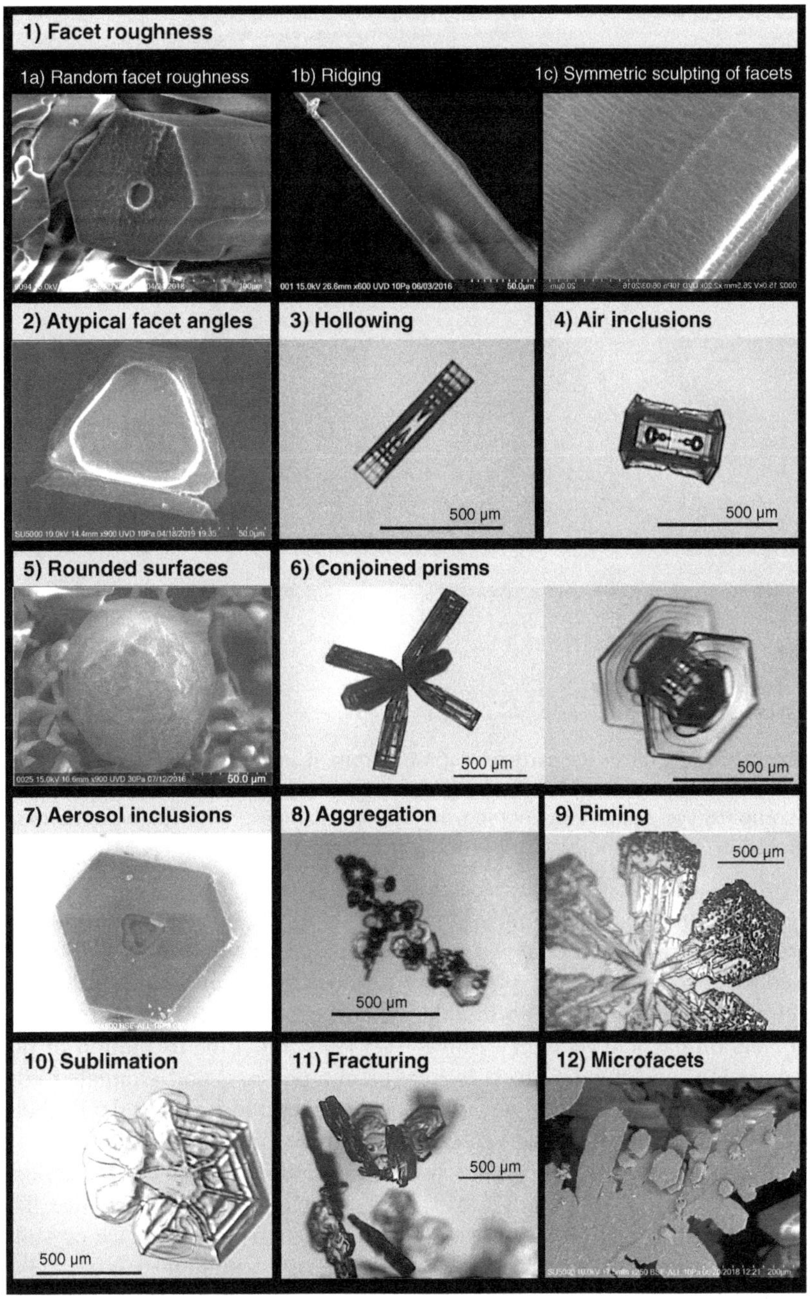

Figure 3.2 Examples of different types of morphological complexity seen on individual ice crystals. Example crystals 1, 2, 5, 7, and 12 were grown under CSEM and crystals 3, 4, 6, 8, 9, 10 and 11 were captured in situ in ice clouds.

these atypical particle symmetry under various atmospheric conditions is still uncertain, though in some samples, it may approach ~25% of crystals (Table 3.1).

Hollowing

Some degree of hollowing of ice crystals appears to be the rule rather than the exception in atmospheric ice particles. While pristine hexagonal plates and columns with no apparent hollowing are observed in the atmosphere, they have been reported infrequently. Schmitt & Heymsfield (2007) reported fractions up to 80% of hollow ends in bullet rosette and columnar crystals in midlatitude cirrus with hollow-fraction increasing with column and bullet length, and Walden et al. (2003) reported that

Figure 3.3 Approximately 500 cirrus ice particles captured from near the top of a synoptically forced cirrus cloud at ~11km altitude, and temperature −50°C. Particles from this single cloud top show many different habits and multiple types of complexity (e.g., facet roughening, atypical facet angles, hollowing, rounded surfaces, conjoined prisms, aerosol inclusion, sublimation, fracturing, and mircrofacets).

all bullet-shaped ice crystals and most columns were frequently hollow in precipitation at South Pole. As with all categories of observed ice particle complexity, hollowing is observed in a spectrum from being barely noticeable to having profound impacts on particle geometry. Some particles show slight depressions and pits on facets (e.g., example crystals 1a, 2, and 7 in Fig. 3.2), while other particles demonstrate sheath-like hollowing that extends through the length of the particle (example crystal 3 in Fig. 3.2). Recent work by Nelson & Swanson (2019) demonstrates images and mechanics of the early formation stages of hollowing and air inclusions on lab-grown ice crystals, which develop through thin layers of laterally protruding growth, sourced from a flux of surface-mobile molecules.

Air Inclusions and Air Pockets

Ice crystals with air inclusions or air pockets are thought to be common in cirrus clouds. The inclusion of air bubbles or pockets adds to the ice crystal complexity and can significantly affect ice crystal optical properties (Hong & Minnis, 2015; Macke, Mishchenko, et al., 1996; Tang et al., 2017; Xie et al., 2009). Modeling studies have suggested that both facet roughness and air inclusions are needed to explain polarimetric satellite observations (Tang et al., 2017). Air inclusions can either form when air dissolved in the liquid phase gets trapped in the heterogeneous freezing process (Maeno, 1967) or when a liquid water layer forms enveloping the ice crystal and is incorporated when the layer freezes. Nelson & Swanson (2019) studied the formation of pockets (corner pockets, planar pockets, and elongated edge pockets) in laboratory generated ice crystals. Pockets were formed when a sublimated ice crystal was regrown, where the process was largely driven by surface transport across the boundary of a face to the advancing edge. Therefore, occurrence of pockets in atmospheric ice crystals can give evidence of sublimation and regrowth cycles.

Occurrence of Rounded Surfaces

Rounded surfaces are also frequently found on atmospheric ice particles, originating from several apparent mechanisms. Ice particles that nucleate from small droplets will initially retain a nearly spherical droplet shape, although other surface roughness may still be layered on the rounded surface (see crystal 5 in Fig. 3.2). Freezing of small droplets to quasi-spherical shapes has been frequently reported in convective cirrus (e.g., Järvinen et al., 2016) and in ice fog situations (Schmitt et al., 2013). Of course, these drops will often then form the nucleus of a faceted single ice crystal or poly-crystalline particle that grows by vapor deposition ("droxtals"), but the transition from rounded to faceted may sometimes proceed slowly, incompletely, or not at all, depending on the local temperature and saturation regime (Gultepe et al., 2017). Another common observation of rounded surfaces occurs with highly sublimated ice particles (see crystal 10 in Fig. 3.2). Although perhaps less recognized, it is important to note that some sublimation conditions have been observed to lead to complex, sharp asperities on ice surfaces. Further research is necessary to distinguish between the circumstances that lead to sublimation rounding or sublimation roughening. An additional category of rounded ice surfaces has been observed to occur in actively growing single crystals, in a zone intermediate between faceted basal and pyramidal planes (e.g., Fig. 1 in Pfalzgraff et al., 2010).

Conjoined Prisms

Complex ice particles that appear as "conjoined prisms" are very frequently observed in natural cirrus and mixed-phase clouds. These conjoined prisms may sometimes be single crystals with highly variable aspect ratios (capped columns a.k.a. "TIE-fighters," see crystal 6b in Fig. 3.2), though most commonly, conjoined prisms appear as polycrystals growing from a common nucleation nexus. Bullet rosettes (see crystal 6a in Fig. 3.2) are the most well-recognized manifestation, though polycrystals with plate-dominated aspect ratio, and particles with mixed aspect ratio also appear regularly (Bailey & Hallett, 2009). There is some suggestion that the frequency of such mixed or planar polycrystals may be somewhat under-appreciated since they may sometimes be classified as "irregular" by automated categorization schemes (Stoelinga et al., 2007).

Aerosol Inclusions

The importance of understanding the role of ice-protect nucleating aerosol particles has long been recognized as a key issue in atmospheric ice clouds. Chapter 2: An overview of aerosol-cloud interactions provides a more complete description of ice crystal formation upon aerosol surfaces. However, relatively less attention has been paid to the scavenging of aerosol particles by ice crystals. Small aerosol particles (nm to μm scales) embedded or adhered to ice surfaces are typically not resolvable by airborne cloud particle imaging probes, but recent CSEM results suggest that adhered and included aerosols on ice crystals are common (Magee et al., 2020). Initial results suggest that the mean density of aerosol inclusions is highly variable, and likely dependent on air mass origins. Furthermore, as ice particles undergo a large degree of sublimation and erosion, adhered aerosols tend to aggregate toward the particle center, and gradually come to dominate the mass of the particle as the ice approaches complete sublimation. As with riming, such aerosol particles also often appear to be affected by particle aerodynamics, adhering in greatest density near particle corners (see crystal 7 in Fig. 3.2). Energy dispersive X-ray spectroscopy (EDX) promises to aid in measuring the chemical composition of the aerosol particles.

Aggregation

Most frequently associated with the formation of snowflakes in precipitating mixed-phase clouds, aggregation has been occasionally observed in mixed-phase and ice clouds. The occurrence of aggregated particles appears to decrease with decreasing temperature, particle number concentration, and particle size (Schmitt et al., 2016); however, several observers suggest that charged ice particles may aid in the formation of ice particle chains in cirrus clouds with significant electrification, for example, in convective anvils (see crystal 8 in Fig. 3.2 and Connolly et al., 2005; Stith et al., 2014).

Riming

Riming describes the accretion of small, supercooled liquid droplets on the surface of atmospheric ice particles (see crystal 9 in Fig. 3.2). It is most significant in mixed-phase precipitation-bearing clouds. However, the riming process may also play a significant role in amplifying complexity of ice particles within liquid-origin cirrus clouds and appears likely to contribute potential origin-points for new growth of microfacets.

Sublimation

Atmospheric ice crystals first nucleate, then subsequently grow and sublimate, sometimes cycling through multiple cycles of growth and ablation. It is therefore not surprising that observations of atmospheric ice crystals frequently evince marked indications of the sublimation process. Frequently, sublimation appears to result in smooth, rounded edges compared to the sharp edges of actively growing crystals (see crystal 10 in Fig. 3.2). However, exceptions to this rule have been observed in both directions. For example, on occasion, sublimation has been observed to yield a facet surface of apparently randomly sharpened ridges and peaks, and also to contribute to patterned roughening in the form of sculpted facet ridging. Conversely, actively growing ice crystals can also present smooth surfaces (see rounded surfaces above).

Fracturing

Ice-ice collisions, complementary to the formation of aggregates, lead to the breaking-up of individual ice crystals (e.g., Takahashi, 1993). Fracturing is most frequent on relatively large particles ($D \geq 1$ mm) such as dendrites and previously aggregated or rimed crystals. Further, fracturing can also take place during the sublimation process or be the result of cracking during the freezing of supercooled liquid droplets (e.g., Lauber et al., 2018).

Table 3.1 Summary of different complexity types, the typical temperature (T) range they form, expected ice saturation conditions, and estimation of their frequency.

Complexity type	Typical T range	Sub- or supersaturated conditions	Frequency	References
Facet roughening	Across all temperatures	Both	Always present	Pfalzgraff et al. (2010); Neshyba et al. (2013); Magee et al. (2014, 2020)
Atypical facet angles	Usually in T< −70°C but also observed in precipitation	Both	Frequent (up to 25%)	Murray et al. (2015, and references therein)
Hollowing	Across all temperatures	Both	Very frequent (up to 80%)	Schmitt & Heymsfield (2007); Nelson & Swanson (2019)
Air inclusions and air pockets	Across all temperatures	Both[a]	Frequent	Maeno (1967); Nelson & Swanson (2019)
Occurrence of rounded surfaces	Usually T< −50°C	Mostly sub-saturated	Frequent	Pfalzgraff et al. (2010); Schmitt et al. (2013); Järvinen et al. (2016)
Conjoined prisms	−20 to −50°C	Both	Very frequent	(Bailey & Hallett, 2009; Lawson et al., 2019)
Aerosol inclusions	Across all temperatures	Both	Uncertain	Magee et al. (2020)
Aggregation	Usually T> −40°C	Both	Frequent	Connolly et al. (2005); Schmitt et al. (2016)
Riming	T> −38°C (when liquid droplets are present)	Both	Frequent	Ono (1969); Hallett & Mossop (1974); Mosimann et al. (1993); Avila et al. (2009)
Sublimation	Across all temperatures	Sub saturated	Frequent	
Fracturing	Around −15°C	Both	Not so frequent	Takahashi (1993); Lauber et al. (2018)
Microfacets	Uncertain	Both[a]	Frequent	Neshyba et al. (2013); Magee et al. (2014, 2020)

[a] Potential indication of growth and sublimation cycles.

Microfacets

The growth of small new facets upon larger, preexisting ice crystal facets that can match the preexisting lattice structure of the underlying crystal appears to occur with some frequency in both mixed-phase and ice clouds. Rimed cloud droplets from mixed-phase clouds adhered onto ice particles seem to serve as a potential re-nucleation points (see crystal 12 in Fig. 3.2). Perhaps more pervasively, repeating cycles of growth and sublimation may lead to microfaceting, with each phase cycle contributing to increasing particle complexity (Voigtländer et al., 2018).

3.2.2. Complexity Metrics

Several metrics describing the degree of morphological complexity at various scales have been proposed and used in the literature. Such metrics can be divided into two categories, namely (1) physical complexity metrics derived from CSEM images, high-resolution light scattering patterns, or optical microscopy, and (2) optical complexity metrics used to simulate the optical effects of ice crystal complexity. Here, we give an overview of the two categories.

Physical Complexity Metrics

An extensive assembly of parameters describing surface roughness has been defined in industrial settings. The ISO Standard 25178, for example, defines parameters representing the distribution of heights above a reference plane, the periodicity, anisotropy, and curvature of features within that plane, void volumes, and many others. Different parameters have become established in various fields. For example, in optical engineering applications the root mean square roughness σ_{rms} is of particular interest and widely used for the quantification of diffuse light scattering from "smooth" optical grade surfaces. By contrast, ice surface morphology research has had a much more narrow focus, emphasizing the local gradients of surface height in the horizontal directions, Z_x and Z_y, and its square, $Z^2 = Z_x^2 + Z_y^2$. Two statistical properties of Z^2, moreover, have received the most focus in ice research: its mean, designated here as $< Z^2 >$, and its standard deviation, designated σ.

A few points about this focus are worth emphasizing here. First, the reason for the emphasis on Z_x and Z_y is the close connection between their distribution and the angles of incidence and refraction and reflection of light impinging on a textured, but otherwise planar, surface. This relationship may be obtained by Snell's law when the wavelength of light is much smaller than the spatial scale of surface texture. Second, the primary reason for limiting focus to statistical properties $< Z^2 >$ and σ is practical rather than fundamental: most available sources of information, for example, the optical scattering properties of surfaces, do not justify a more detailed description. Recent CSEM work is beginning to provide more detailed information, for example, about the kurtosis, skewness, and so on of these distributions, and such work has itself brought up fundamental questions, especially about the spatial scale of the roughness, that have yet to be resolved. A third point concerns the relationship between $< Z^2 >$ and σ on one hand, and model (analytical) distributions of the gradient. For example, while $< Z^2 >$ and σ may (and do) form the basis of parameterizations of such models (see, e.g., the dependence of Weibullian σ_{wb} and η, Butterfield et al. (2017)), they derive from the actual distribution of Z^2, and therefore stand independently of such models and their parameterizations.

In many applications, the degree of morphological complexity is measured indirectly, for example, by measuring the distribution of scattered light around a specular reflection (Lu et al., 2006). Such methods are used in industrial settings for probing the quality (or smoothness) of produced surfaces and, lately, have found their way into atmospheric sciences. The degree of complexity of three-dimensional (3-D) individual ice crystals can be quantified from high-resolution light scattering patterns measured using the Small Ice Detector Mark 3 (SID-3) instrument (Schnaiter et al., 2016; Ulanowski et al., 2014). Examples of SID-3 scattering patterns of individual ice crystals with different degrees of complexity can be seen in Section 3.3. The degree of complexity is related to the image texture of the measured light scattering patterns that are analyzed using a gray-level co-occurrence matrix (GLCM) method (Lu et al., 2006). GLCM is a frequency matrix describing how often pairs of gray levels occur in a texture image for pixels separated by a certain distance Δd and along a certain direction. Speckle pattern texture features can then be extracted from the GLCM. For analyzing ice crystal complexity, the relevant texture features are "combined roughness" (Ulanowski et al., 2014) and the "normalized energy feature parameter" (k_e) (Lu et al., 2006; Schnaiter et al., 2016). The combined roughness is weighted so that has bounds between 0 and 1 for most particles whereas k_e approximately varies between 4 and 6. Although both combined roughness and k_e have been shown to correlate with actual physical surface complexity (Schnaiter et al., 2016), there is no unequivocal link between these parameters and $< Z^2 >$ or σ. Also, it should be noted that this method is sensitive to all types of complexities with scales around the wavelength of the incident light (see Fig. 3.1) so discrimination between different types of complexities discussed in section 3.2.1 is not possible.

Whereas the previously introduced metrics describe crystal complexity at the nano- and mesoscopic scales, Schmitt & Heymsfield (2014) introduced a unitless complexity parameter C that describes crystal complexity in micro- and macroscale derived from ice crystal microscopic images observed by the cloud particle imaging (CPI) probe. The complexity parameter C is inferred from a combination of the area of the circle with the smallest area that covers the particle, the projected area of the particle, and the perimeter of the particle. Schmitt et al. (2016) showed that C generally increases with particle size and the number of monomers in an aggregate particle.

Optical Complexity Metrics

Complexity metrics in category (2) are used in combination with computational methods to simulate the optical properties of ice crystals (see more details on light scattering models in section 3.4.1). They can be divided into methods that generate the optical effect of complexity (a) by using the tilted-facet (TF) method, where the normal of the crystal facet surface is tilted randomly from its nominal orientation, (b) by generating complex, randomized morphologies, or (c) by simulating the bidirectional scattering distribution function (BSDF) around the specular reflected and the transmitted ray direction. Although method (c) is mainly used in the field of technical optics, we will show in section 3.4.2 how this knowledge could be transferred into the field of ice crystal light scattering.

The TF approach (a) is generally applicable to simulations of ice optical properties based on geometric optics. For example, Macke et al. (1996) used a set of random tilt angles distributed uniformly between $[0, \theta_t^{max}]$ to specify the degree of complexity caused by facet roughness or crystal distortion, where the degree of complexity is defined as

$$\delta = \theta_t^{max} / 90^\circ . \tag{3.1}$$

Yang & Liou (1998b) specified the degree of surface complexity caused by facet roughness using a first-order Gram-Charlier density function (i.e., Gaussian distribution) motivated by Cox & Munk (1954) who studied the optical effect of sea surface roughness. Here, the facets are tilted randomly according to the distribution

$$P(Z_x, Z_y) = \frac{1}{\sigma_g^2 \pi} \exp\left(-\frac{Z_x^2 + Z_y^2}{\sigma_g^2}\right), \quad (3.2)$$

where (as described above) Z_x and Z_y are local gradients of the surface height in the x and y directions and σ_g is a complexity parameter controlling the degree of roughness. Figure 3.4 shows three examples of hexagonal ice crystals with varying degree of complexity generated using equation (3.2).

Later, Shcherbakov et al. (2006) introduced a more general Weibull distribution to the TF approximation, namely

$$P(Z_x, Z_y) = \frac{1}{\sigma_{wb}^2 \pi} \left(\frac{Z_x^2 + Z_y^2}{\sigma_{wb}^2}\right)^{\eta-1} \exp\left[-\left(\frac{Z_x^2 + Z_y^2}{\sigma_{wb}^2}\right)^{\eta}\right], \quad (3.3)$$

where σ_{wb} is the complexity parameter and η is an additional shape parameter in the Weibull distribution. If $\eta = 1$, then equation (3.3) reduces to equation (3.2).

Simulations implementing the TF method have shown that ice crystal optical properties are largely equivalent regardless of used optical complexity metrics (Geogdzhayev & van Diedenhoven, 2016; Neshyba et al., 2013). Therefore, the TF method has been used to simulate the optical effects of complexity across multiple scales from facet roughness (Liu et al., 2013; Macke, Mueller, et al., 1996) to internal structures, such as air inclusions (Hong & Minnis, 2015; Macke, Mishchenko et al. 1996), to macroscale complexities, such as aggregation (Letu et al., 2016).

The approach (b) to simulate the optical effect of complexity is to generate complex, randomized morphologies. Examples of complex morphologies used for ice crystals are Koch fractals (Macke, Mueller, et al., 1996), Voronoi particle (Letu et al., 2016), Gaussian roughness on random spheres (Nousiainen & McFarquhar, 2004; Nousiainen & Muinonen, 2007; Um & McFarquhar, 2011), and particles prescribed by Chebyshev polynomial series (McFarquhar et al., 2002) and superspheroids

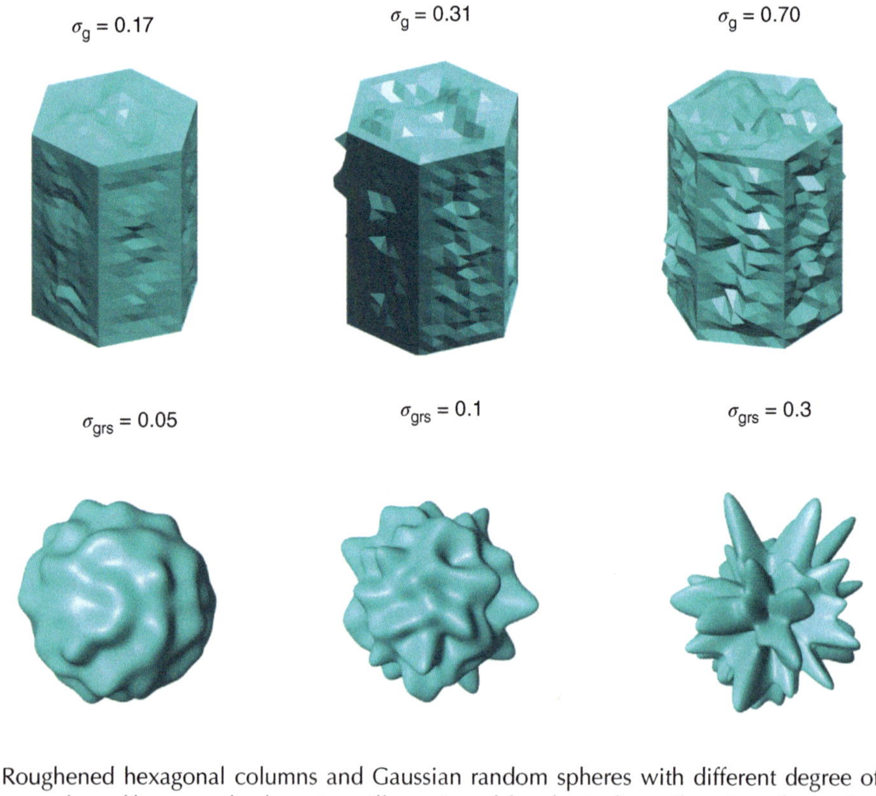

Figure 3.4 Roughened hexagonal columns and Gaussian random spheres with different degree of complexity. Note that the roughened hexagonal column is an illustration of the physical manifestation of complexity generated by the TF approach. In the actual ray-tracing calculation, the whole facet is tilted when encountered by the ray so that geometrical optics assumption is valid.

(Sun et al., 2022). These approaches all include parameters that alter the shapes, and which may be interpreted as complexity metrics. For example, the shape of the Gaussian random sphere geometry is described by a covariance function of radius as follows:

$$\mathbf{r}(\vartheta, \varphi) = \frac{a}{\sqrt{1+\sigma_{grs}^2}} \exp[s(\vartheta, \varphi)\hat{\mathbf{e}}_{\mathbf{r}}] \quad (3.4)$$

where a is the mean radius, σ_{grs} is the relative standard deviation of radius controlling the degree of complexity, s is the so-called logradius, and $\hat{\mathbf{e}}_{\mathbf{r}}$ is a unit vector in spherical coordinates. Examples of Gaussian random spheres with different degree of complexity are shown in Fig. 3.4. A downside of such approaches is that the morphologies are generally inconsistent with general hexagonal structure of ice and may not be considered realistic for natural ice crystals. Therefore, limited calculations have been performed for hexagonal prisms with randomized roughness structures added to their surfaces (Collier et al., 2016; Liu et al., 2013). A benefit of some of these approaches is that they are applicable to methods to calculate the optical properties for small size parameters (i.e., small ice crystals and/or at long wavelengths) for which the geometric optics, and thus the TF approach, cannot be applied (Kahnert et al., 2012; Nousiainen & Muinonen, 2007).

The approach (c) is used in technical optics to describe the residual light scattering of optical grade surfaces and is available in most optical design and engineering ray-tracing software. The broadening of the specular reflection and transmission peaks is described using, for example, the *ABg* model (Pfisterer, 2011). Based on scalar diffraction theory, these types of model are parameterized to describe the angular dependency of radiant power that is scattered out of the specular and transmitted rays. The *ABg* model is defined as

$$\mathrm{BSDF}(\beta - \beta_0) = \frac{A}{B + (\beta - \beta_0)^g}[sr^{-1}], \quad (3.5)$$

where BSDF, the bidirectional scattering distribution function, is the ratio between scattered and incident power, $\beta = sin(\theta_{scattered})$, $\beta_0 = sin(\theta_{specular})$, A/B the diffuse scattering fraction in the peak direction, and g the slope with which the power decreases away from the specular and transmitted beams. $B^{1/g}$ then defines the rollover angle where the BSDF starts to change from a constant (Lambertian) angular dependency to a power law decrease for larger angles. This angle is typically less than a degree. The *ABg* model is valid if the surface roughness is isotropic and the root mean square roughness parameter σ_{rms} is of the order of the incident wavelength or less (commonly only a few %). Therefore, the *ABg* model can be used to describe sub-micron scale facet roughness when performing measurements with visible light.

While the aforementioned optical complexity metrics (a–c) are introduced to simulate the optical effects of complexity, recently, Xu et al. (2022) introduced a complexity metric that can be derived from the angular scattering function due to the *reflection – refraction* of light rays, namely, the C_{p} parameter, defined as follows:

$$C_p = \left(\sum_{l=0}^{\infty} |\hat{c}_{GO,l}|\right)^{-1}, \quad (3.6)$$

where $\hat{c}_{GO,l}$ are the expansion coefficients of the angular scattering function using Legendre polynomials. It is found by Xu et al. (2022) that this parameter is strongly correlated with the distortion parameter δ introduced by Macke et al. (1996), and potentially with other complexity metrics as well because these complexity metrics (a–c) often lead to similar effects on phase function. A significant advantage of this approach is that the C_{p} parameter is retrievable by polar nephelometers, making the comparisons among models and measurements quite convenient.

Linking Physical and Optical Complexity Metrics

As our capabilities to observe and quantify physical complexities of ice crystals continue to grow, it is becoming increasingly important to establish "crossover" linkages between the aforementioned physical and optical complexity metrics. Apart from providing a more comprehensive picture of complexity, such linkages could be used in the service of broader objectives, for example, to clarify how cirrus CREs might respond to changes in environmental conditions and climate, or to improve the information obtained from remote sensing algorithms.

This goal remains an active area of research. Schnaiter et al. (2016) attempted to link the SID-3 complexity parameter k_e to the distortion parameter σ_{grs} describing the distortion in the Gaussian random sphere model introduced above (equation (3.4). They used ADDA/GDT-matrix approach to simulate SID-3 scattering patterns and analyzed these patterns using the GLCM. A clear correlation between k_e and σ_{grs} was found, where a k_e of 4.6 corresponded to a significant σ_{grs} of 0.3. k_e was saturated at σ_{grs} value of 0.5. Since the study was restricted to relatively small crystal sizes of 7.8 μm (due to computational limitations) and to a single non-hexagonal crystal shape (Gaussian random sphere), more work along these lines is required to reach a comprehensive crossover between k_e and optical parameterizations. In a similar vein, Neshyba et al. (2013) showed that the sensitivity of the 22° halo to ice crystal complexity depends crucially on the symmetry of prismatic-facet roughening: when that roughening is azimuthally anisotropic (occurs in rows), the halo is degraded to a far less extent than when the roughening is azumithally isotropic; both symmetries have been observed in real cirrus clouds, although the environmental conditions that control their occurrence are unknown.

3.3. OBSERVATIONS OF COMPLEX CRYSTALS

In this section, we summarize how crystal complexity has been quantified using (1) CSEM methods and optical microscopy, (2) high-resolution angular light scattering patterns, and (3) space-borne remote sensing observations. Methods (1) and (2) have been applied both for laboratory-grown and in situ ice crystals. We also discuss what we have learned so far about the link between ice crystal complexity and prevailing environmental conditions.

3.3.1. Microscopic Observations of Ice Morphology Associated with Complexity

Here, we summarize what is known about the morphologies of surface complexity at the nano- and mesoscopic scales. The methods underpinning these observations may be divided into field observations versus laboratory measurements on the one hand, and high-resolution optical microscopy versus CSEM on the other hand. We will briefly consider insights from these in various combinations.

Considering first the category of field optical microscopy (e.g., Walden et al., 2003), it is evident that complexity in naturally occurring ice crystals is strongly influenced by the underlying crystal symmetry. For example, both the surface texturing appearing in crystal 10 in Fig. 3.2 and the hollowing in crystal 3 in Fig. 3.2 clearly respect the sixfold crystal symmetry of the basal facet. (Indeed, it is tempting to speculate that hollowing of basal facets is an extreme form of symmetric basal surface texturing.) Crystal 3 in Fig. 3.2 also hints at another symmetry, appearing as lines running perpendicular to the long (crystallographic) c axis; however, it is not clear from this image whether this is a feature of the outer surface of the prismatic facets, or of the inner (hollowed) surface. Field-obtained optical images have also shown the existence of myriad other forms of complexity, for example, the conjoined prisms, riming, aggregation, and fracturing shown in Fig. 3.2.

A second category of investigation, laboratory studies using optical microscopy, offers the possibility of performing growth and ablation experiments through the manipulation of environmental variables. Such studies have revealed the occurrence of air inclusions, the compounding effects of multiple growth/ablation/regrowth cycles, and insights into the mechanisms by which crystal facets increase in area (Bacon & Swanson, 2000; Nelson & Swanson, 2019).

A third category of investigation is CSEM studies of laboratory-grown ice (Magee et al., 2014; Pfalzgraff et al., 2010). This approach is capable of providing greater surface detail than is possible with optical microscopy, albeit at the price of losing the ability to peer into a crystal, since CSEM is a surface-only technique. The aforementioned twofold symmetry appearing in crystal 3 in Fig. 3.2 is frequently observed in CSEM images of ice, where it is unambiguously attributable to texturing on the outside surface of prismatic facets (Pfalzgraff et al., 2010, see also examples in panels 1b and 1c). CSEM work has also shown that non-symmetrical (i.e., random) sculpting regularly occurs on facets of all types, sometimes in combination with symmetric roughening, and sometimes not (see crystals 1a–c in Fig. 3.2). The quantitative topographical capability of CSEM (which so far has not been worked out for optical microscopy of ice) has proven a key asset in this regard (Butterfield et al., 2017). Such work has revealed, for example, that the vertical relief of facet roughening, while typically not more than a few micrometers, tends to be larger for symmetric roughening than for non-symmetric roughening. Model representations invested with explicit, CSEM-derived surface topologies offer the additional advantage of permitting study of the light scattering properties in a more realistic way than Gaussian or Weibull (i.e., random-variable) methods permit (Neshyba et al., 2013) (described in more detail in section 3.4.1).

A fourth category is CSEM studies of ice captured in the field (Magee et al. (2020); see Fig. 3.3 for an example). This approach provides unique insights into the diversity of real-world ice crystal complexity (e.g., facet roughening, atypical facet angles, hollowing, rounded surfaces, conjoined prisms, and aerosol inclusion), and offers the possibility of guiding laboratory experiments seeking mechanistic origins of these forms of complexity.

3.3.2. Deriving Mesoscopic Complexity from High-Resolution Angular Light Scattering Patterns

Ice crystal mesoscopic complexity of small ice crystals in the sub-50 μm size range can be investigated by analyzing high-resolution angular light scattering patterns measured with the SID-3 instrument as discussed in section Physical Complexity Metrics. This method has been applied for laboratory-generated ice crystals and ice analoge particles (Schnaiter et al., 2016; Ulanowski et al., 2006; Voigtländer et al., 2018), and to investigate the mesoscopic complexity of cirrus ice crystals in airborne studies over midlatitudes (Järvinen et al., 2018b; Ulanowski et al., 2014) and the tropics (Järvinen et al., 2018a).

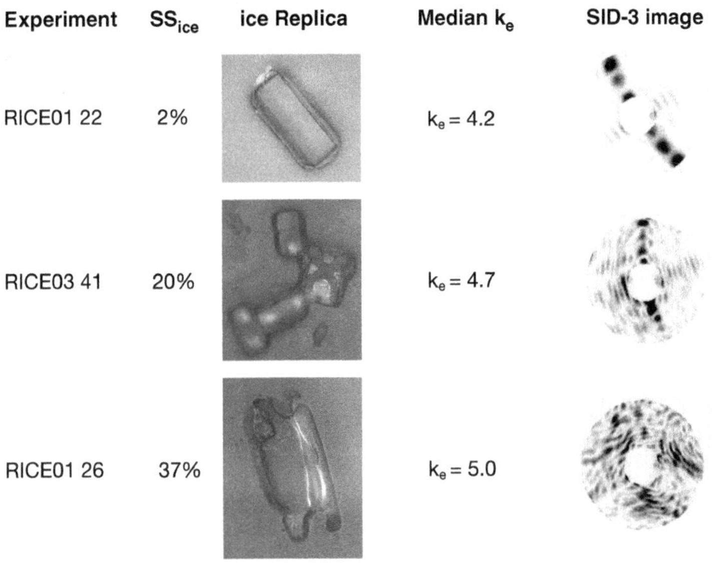

Figure 3.5 Examples of ice crystal replicas sampled during three crystal growth experiments performed at the AIDA cloud chamber. Ice crystals were grown in three different supersaturation conditions where the median ice supersaturation (SS_{ice}) was 2%, 20%, and 37%. The resulting ensemble median complexity parameters (k_e) were 4.2, 4.7, and 5.0, respectively. The SID-3 images show a representative scattering pattern for a single ice crystal from the same experiment having a complexity parameter corresponding to the median complexity parameter of the crystal ensemble. All ice crystals were below 100 μm.

Schnaiter et al. (2016) performed a series of laboratory experiments at the Aerosol Interactions and Dynamics in the Atmosphere (AIDA) cloud chamber where ice crystals were grown and sublimated at defined super- and subsaturated ice conditions at the cirrus temperature range between −40 and −60°C. The mesoscopic complexity of the ice crystals was quantified using the complexity parameter k_e as defined in section Physical Complexity Metrics. Figure 3.5 shows a visualization of ice crystal replicas from three of the experiments and a representative single-particle high-resolution angular light scattering patterns. The examples show that increase in ice supersaturation during growth is accompanied by visual increase in hollowing of the crystals. Note that conclusions about facet roughness cannot be made in the replicas. The single-particle high-resolution scattering patterns reflect this change so that increase of the number and decrease of the size of the speckles are seen in the images leading to an increase in k_e. In the AIDA experiments, a linear dependency was found between the growth ice supersaturation and the complexity parameter k_e for one temperature range in case of heterogeneously nucleated ice crystals. Figure 3.6 shows the fit to the laboratory measurements. It can be seen that only complex crystals ($k_e > 4.6$ or $\sigma_{grs} > 0.3$) were generated in experiments with starting temperature of 233 K, whereas at a colder starting temperature of 225 K ice crystals with complexity parameters below 4.6 were grown at ice supersaturations below 20%. Ice crystals with complexity parameter below 4.6 are considered as optically "pristine" so that the spatial light scattering is mainly defined by the habit and not crystal complexity, that is, showing a lack of speckles (see Fig. 3.5).

Similarly, Voigtländer et al. (2018) investigated how combined roughness parameter was affected by crystal growth conditions using the Ice Roughness Investigation System (IRIS) laboratory setup. They also found that, under laboratory conditions, complexity parameter correlates with the crystal growth conditions so that a higher saturation ratio with respect to ice led to higher complexity parameters. Furthermore, Voigtländer et al. (2018) found that repeated growth and sublimation cycles led to a memory effect and gradual "ratcheting up" of crystal complexity.

Despite the evidence from laboratory experiments showing that complexity parameters correlate with crystal growth conditions, no clear correlation between complexity parameters and simultaneously measured environmental variables has been found in natural cirrus clouds using the SID-3 method (Järvinen et al., 2018b; Ulanowski et al., 2014). This is also summarized in Fig. 3.6 that shows re-analysis of SID-3 data from two airborne campaigns from the tropics (ACRIDICON-CHUVA) and from the midlatitudes (ML-CIRRUS) containing both convective and in situ cirrus. The measured k_e values are grouped into four temperature bins and shown as a

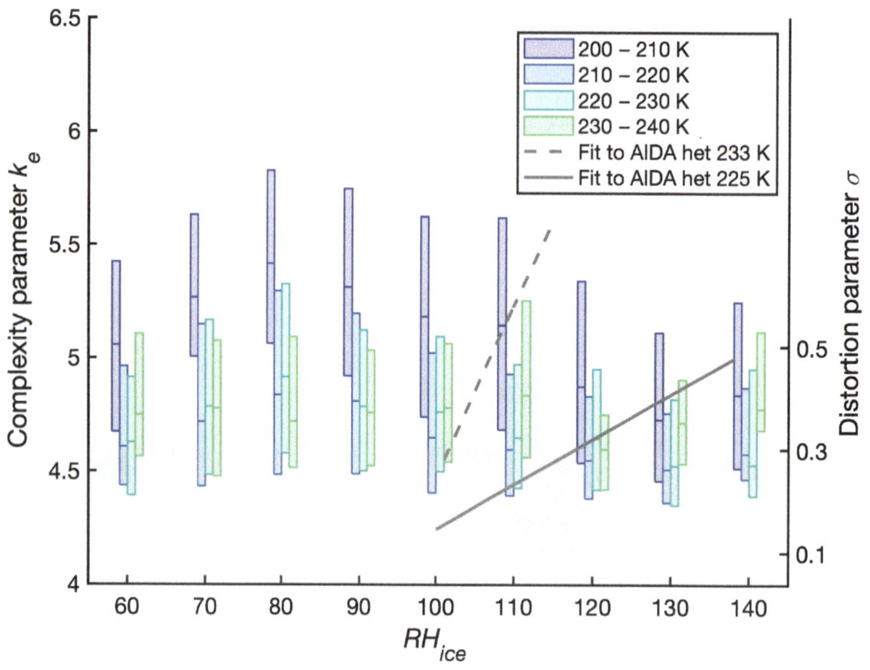

Figure 3.6 Statistical analysis of ice crystal complexity values retrieved from SID-3 measurements collected during the Midlatitude CIRRUS (ML-CIRRUS) and the ACRIDICON-CHUVA campaigns. The boxes represent the interquartile range for different temperature ranges and the line represents the median values. The gray solid and dashed lines represent the linear dependency that was observed in cloud chamber simulation experiments for heterogeneously nucleated ice crystals with initial chamber temperatures of 233 K (dashed line) and 223 K (solid line). The shaded gray area (below $\sigma \leq 0.3$) illustrates the range of k_e values where the ice crystals can be considered to be optically pristine. Source: Adapted from Schnaiter et al. (2016) and Järvinen et al. (2018a).

function of simultaneously measured relative humidity with respect to ice (RH_{ice}). The data were originally published in Järvinen et al. (2018a). Although no clear trend is seen in the median complexity parameters as a function of simultaneously measured RH_{ice}, some trend toward higher complexity values in the lowest temperature bin (200–210 K) in compared to higher temperatures is seen. Also, on average higher complexity values are found in subsaturated ($RH_{ice} < 100\%$) conditions compared to conditions where $RH_{ice} > 100\%$, which could indicate that sublimation can increase complexities as suggested by CSEM studies. In general, atmospheric measurements with the SID-3 instrument have shown a high degree of ice crystal complexity independent of the measurement location, cirrus type, or even cloud type (cirrus or mixed-phase cloud) (Järvinen et al., 2018a).

Understanding the origin of crystal complexity in natural ice clouds is a topic of ongoing research (see section 3.7). Since SID-3 observations are limited in geographical location and crystal size, other methods are needed to overcome these limitations, such as remote sensing techniques reviewed in the next section.

3.3.3. Remote Sensing of Ice Crystal Complexity

Multi-angle observations of solar reflectances of cloud tops contain information on ice crystal complexity since ice crystal single-scattering properties depend on crystal morphological properties (see section 3.4.1 for more details). As recently shown by Zhang et al. (2020), adding polarization measurements to multi-angle reflectance observations substantially increases the total degrees of freedom to infer ice properties. This is because effects of multiple scattering are minimized in the polarimetric signals leading to rather direct observations of the linear polarization element of the single-scattering phase matrix that depends on ice crystal's microphysical properties. An overview of past studies on ice crystal complexity and shape using multi-angle observations can be found in van Diedenhoven (2018) and a comprehensive study using Multi-angle Imaging SpectroRadiometer (MISR) data combined with Moderate Resolution Imaging Spectroradiometer (MODIS) observations was recently published by Wang et al. (2019). Here, we summarize recent findings of occurrence of complex ice crystals.

Wang et al. (2019) showed using complexity parameter σ_g (as defined by equation (3.2) in section 3.2.2)

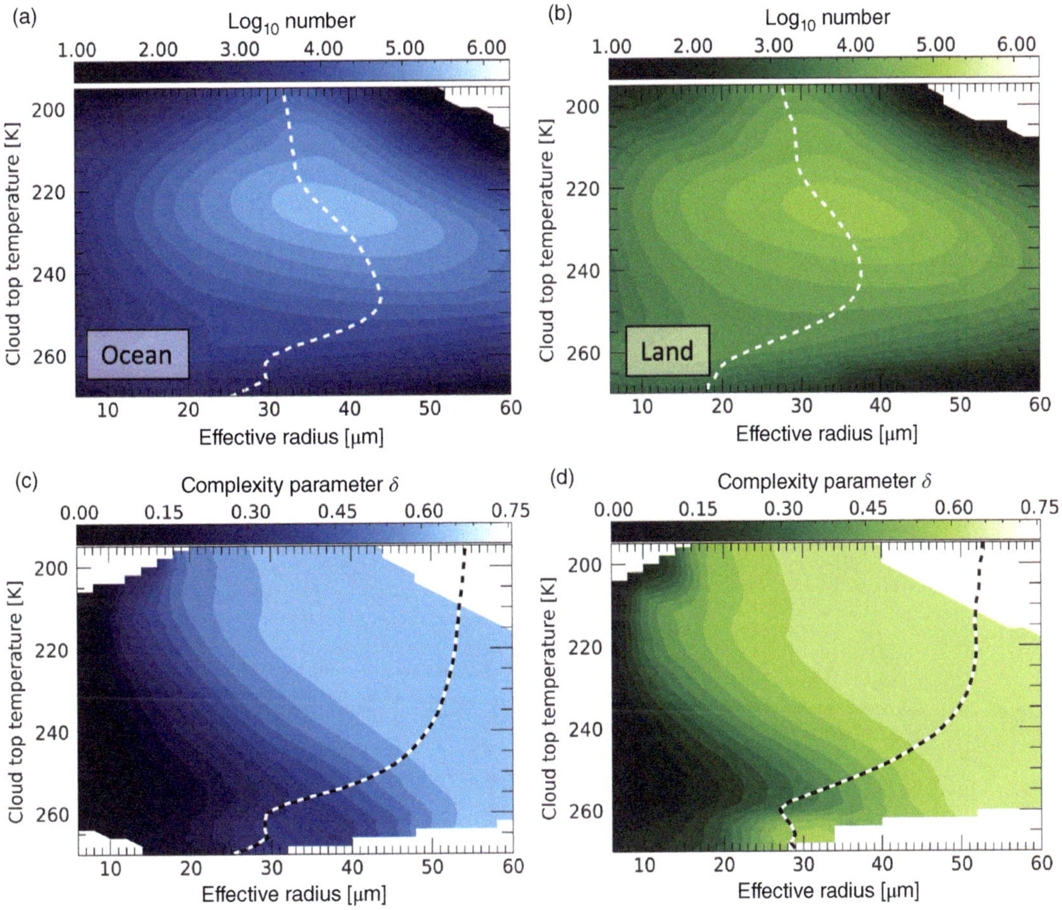

Figure 3.7 Observed number of thick ice clouds with a given combination of cloud top temperature and effective radius over ocean (a) and land (b) and complexity parameter δ as a function of cloud top temperature and effective radius over ocean (c) and land (d). The dashed lines in panels (a) and (b) show the mean effective radius, while the dashed lines in panels (c) and (d) show the mean complexity parameters indicated by the axis labels on top of the color/shade scale bar.

that ice crystal complexity varies weakly with cloud top temperature below 230 K. Furthermore, Wang et al. (2019) found that the occurrence of particles with low complexity is substantially larger for effective radii below 20 μm compared to larger crystals. Similar results were found by van Diedenhoven et al. (2020) who analyzed a year of combined observations of the MODIS spectrometer and POLarization and Directionality of the Earth's Reflectances (POLDER) polarimeter in the A-train satellite constellation. Their approach used hexagonal plates and columns with varying aspect ratios and levels of complexity as proxies for the components of more complex particles. Note that this study used complexity parameter δ (equations (3.1) of section 3.2.2) with a maximum value of 0.7. Only ice clouds with optical thickness above 5 were included. Multi-angle polarimetry allows a very robust filtering of liquid containing cloud tops, leading to a sampling of ice-topped clouds ranging up to the melting level (van Diedenhoven, Fridlind, et al., 2012). The shortwave infrared bands on MODIS also allow ice effective radius to be inferred. In total, nearly 85 million samples are included in their dataset.

Figure 3.7 shows the number of observed thick ice clouds with a given combination of cloud top temperature and effective radius over ocean and land, in addition to the mean effective radius as a function of cloud top temperature and the variation of complexity parameter δ with cloud top temperature and effective radius. In general, mean effective radii peak for cloud tops at approximately 245 K and are larger over ocean compared to over land. In agreement with the findings of Wang et al. (2019), the complexity parameter δ increases with effective radius, while its variation with cloud top temperature is relatively minimal above 230 K, where mean δ is close to its maximum value of 0.7. For warmer tops, δ generally

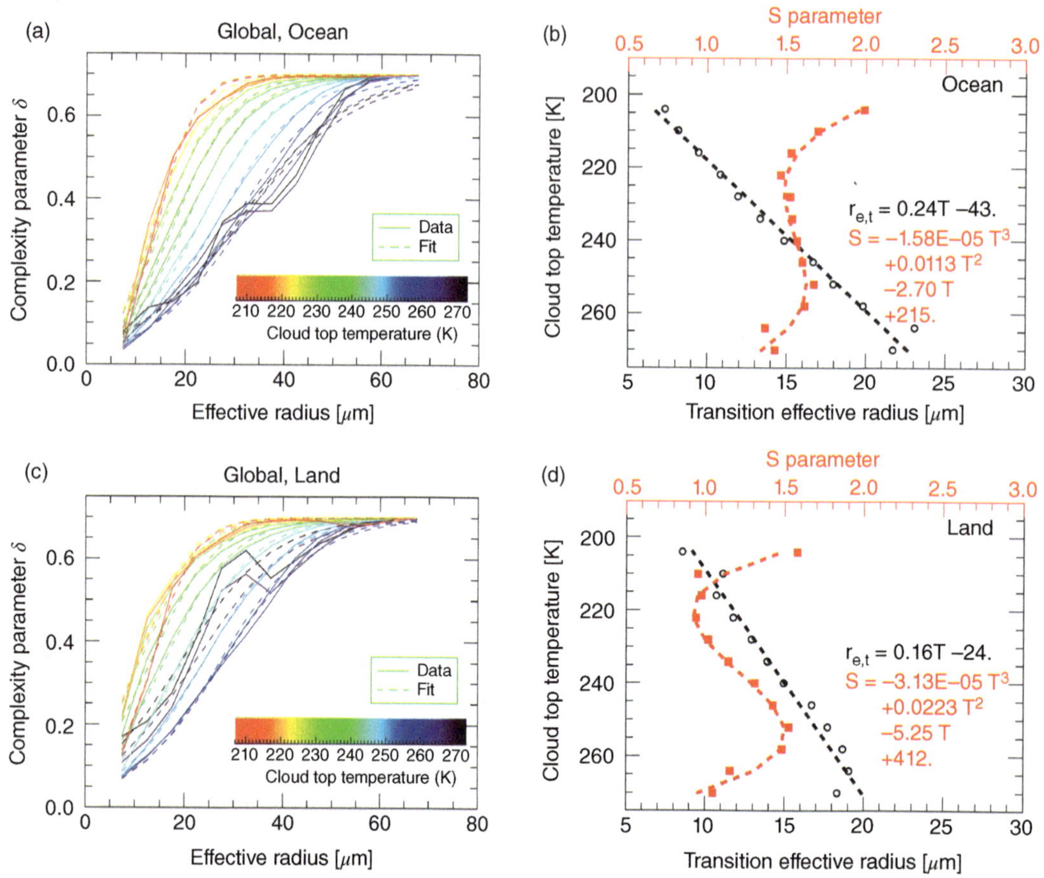

Figure 3.8 Mean complexity parameter δ as a function of effective radius for various cloud top temperature bins as observed (solid lines) over ocean (a) and land (c). Dashed lines show fits using equation (3.7). Panels (b) and (d) show fit parameters $r_{e,t}$ (black, circles) and S (gray/red, squares) from equation (3.7) as a function of cloud top temperature and corresponding polynomial fits.

decreases with cloud top temperature for a given effective radius, also leading to a decrease of mean δ with cloud top temperature. Interestingly, a minimum in δ occurs at around 260 K, which corresponds to the temperature where crystals with extreme aspect ratios are modeled and observed (Jensen et al., 2017; van Diedenhoven et al., 2020).

Similar vertical variation of the complexity parameter shown in Fig. 3.7 is observed independent of the season and latitude (van Diedenhoven et al., 2020) and has also been confirmed by aircraft remote sensing (van Diedenhoven, 2021) in subtropical convective clouds. Previous studies have also shown an apparent trend of decreasing complexity with increasing latitude (e.g., Chepfer et al., 2001; Cole et al., 2014; van Diedenhoven et al., 2020). However, this trend is largely explained by a positive trend of cloud top temperature with latitude. Furthermore, van Diedenhoven (2021) showed significantly higher values of δ for convective cloud tops over ocean that were subsaturated with respect to ice, compared to supersaturated conditions, which is consistent with the SID-3 observations shown in Fig. 3.6.

Figure 3.8 shows the mean complexity parameter δ as a function of effective radius for various cloud top height bins for clouds over ocean (a) and land (c). A systematic increase of complexity with effective radius is observed, which is consistent with the findings of Wang et al. (2019) discussed above. The rate of increase, however, decreases with increasing temperature. Interestingly, the functional dependence of the complexity parameter on effective radius and its variation with temperature is very similar to the dependence of crystal complexity parameter C (see section 3.2.2) as a function of crystal maximum dimension at various temperatures as derived from in situ observations of the CPI as described by Schmitt et al. (2016). Analogous to the analysis by Schmitt et al. (2016), the relation between complexity parameter δ and effective radius can be well described by

$$\delta(r_e) = 0.7\left(\frac{\tanh(r_e/r_t - S)}{2} + 0.5\right), \qquad (3.7)$$

where r_t represents the transition effective radius and S determines the slope. Fits of equation (3.7) to observations over ocean and land at various cloud top temperatures are shown in Figures 3.8a and 3.8c. The variation of δ with effective radius is generally well described by equation (3.7), except for the cloud tops warmer than 260 K. The cause of the different behavior for these very warm ice clouds is not known, although we note that the sampling is relatively sparse for these cloud tops. Consistent with the analysis of Schmitt et al. (2016), Figures 3.8b and 3.8d show that the transition size is a linear function of cloud top temperature, for which the least squares fit parameters are shown in the figure. Although Schmitt et al. (2016) remark that no significant trends were found in parameter S of their implementation of equation 3.7, an oscillation in S with cloud top temperature is seen in Figure 3.8d that can be well described by a third-order polynomial.

The transition effective radius may be interpreted as the size above which a transition to mostly complex particles occur. Figure 3.8 suggests that r_t approximately corresponds to the effective radius above which the complexity parameter δ is greater than about 0.2. A plausible explanation for the decrease of r_t with decreasing temperature is the increased lifetime of these ice crystals, which could be caused by slower growth and sublimation rates and slower sedimentation of small crystals that are more typical of colder temperatures. Woods et al. (2018) estimated that cirrus ice crystals in the tropical tropopause layer (TTL), with mean diameters typically below 30 μm, will fall at a rate of only 2 km per day, increasing the likelihood of these crystals to be exposed to greater variations of environmental conditions leading to increased complexity. Furthermore, at colder temperatures subsaturated conditions may be more prevalent leading to more complexity on average (van Diedenhoven, 2021, see also Fig. 3.6).

A limitation of the space-borne remote sensing approaches discussed above is that the observations represent ice crystal properties at the tops of thick ice clouds, which do not necessarily reflect the temperature dependence of crystal properties within deep clouds, or the properties of thinner cirrus clouds. Furthermore, since these remote sensing approaches are based on optical complexity metrics as discussed in section 3.2.2, which represent several physical crystal complexities across multiple scales, the actual nature of the crystal complexities cannot be deduced directly from these observations. It should be also noted that there is a tendency for the remote sensing techniques to yield the maximum-allowed complexity parameter for most observations (e.g., Hioki et al., 2016; van Diedenhoven et al., 2020). This is likely a consequence of the asymptotic behavior of simulated phase matrices at large complexity parameters and may suggest that the complexity parameterizations in these simulations are not sufficiently representative of highly complex crystals in natural clouds.

Besides observations of multi-angle solar reflectances, the polarization Light Detection and Ranging (lidar) technique is widely used to investigate the microphysical properties of cirrus clouds. The lidar depolarization ratio is used as an indicator of particle non-sphericity since the linear depolarization ratio of non-spherical scatterers can significantly differ from zero. Furthermore, it is believed that the magnitude of the linear depolarization ratio can offer some information on ice crystal morphology. Global CALIPSO-CALIOP statistics show a yearly averaged ice cloud depolarization ratio around 0.34–0.37, with a trend of lidar depolarization (1) increasing with increasing cloud altitude or (2) decreasing temperature and (3) decreasing with latitude (Baum et al., 2011; Okamoto et al., 2020; Sassen & Zhu, 2009; Sassen et al., 2012). Ray-tracing calculations have shown that different randomly oriented pristine ice crystal habits generate different backscattering depolarization ratios (Takano & Liou, 1995), whereas T-matrix calculations and laboratory studies have shown that the linear depolarization ratio has a strong size dependence for small size parameters ($x < 30$) (Mishchenko & Sassen, 1998; Schnaiter et al., 2012), which might explain the maximum depolarization values observed in contrail cirrus (Freudenthaler et al., 1996). In a similar manner, it has been suggested that crystal complexity can lead to changes in the linear depolarization ratio. Theoretical studies with distorted hexagonal columns have shown that linear depolarization ratio has a nonlinear dependency with the distortion parameter (Borovoi et al., 2015). Simulations with complex crystals tend to yield to a better agreement with annually averaged CALIOP data (Baum et al., 2011). Due to the complicated relationship between crystal morphology and the linear depolarization ratio as well as the limitations to simulate the scattering properties of complex ice crystals for the exact backward direction observed by lidars (Borovoi et al., 2014), the quantitative interpretation of depolarization ratios with respect to the complexity metrics discussed in section 3.2.2 remains uncertain. An additional limitation of lidar depolarization measurements is their sensitivity to oriented crystals. Even if a tilt to the lidar is applied, as is the case for CALIOP, the presence of a small fraction of oriented crystals can substantially reduce the depolarization ratio, further complicating the interpretation of the signal (Okamoto et al., 2020).

3.3.4. Synthesis of Ice Crystal Complexity Observations

The origin of ice crystal complexity in natural clouds is not yet well understood. However, some general trends are observed: (1) both in situ and remote sensing observations (Figs. 3.6 and 3.7) have shown that complexity parameters increase with decreasing temperature, and that (2) higher complexity parameters are observed in subsaturated compared to supersaturated conditions (Fig. 3.6 and van Diedenhoven (2021)). Furthermore, (3) remote sensing observations indicate that complexity parameters increase with increasing effective radius (Fig. 3.7). On the other hand, ice crystals tend to be smaller at lower temperatures so the complexity is, for a given radius, larger for lower temperatures. Laboratory studies have shown that growing ice crystals at higher temperatures would promote complexity through faster crystal growth rates. This can explain the observed increase in complexity parameter with increasing crystal effective radius but cannot explain the increased complexity at lower temperatures.

Increased complexity at lower temperatures might be promoted through sublimation and regrowth cycles in long-persisting cirrus layers near the TTL. Both CSEM and laboratory studies have shown that repeated growth and sublimation cycles can increase the complexity parameters. Sublimation and regrowth is also found to contribute to formation of air pockets in laboratory studies (see section 3.3.1). It is evident that more work is needed to understand the detailed morphological properties of the highest lying cold cirrus ice particles, especially since often they are the first condensed matter encountered by solar light.

3.4. LIGHT SCATTERING BY COMPLEX CRYSTALS

3.4.1. Computer Simulation of Angular Light Scattering by Ice Crystals

Computer simulations are an important tool for gaining fundamental knowledge of scattering and absorption properties of ice crystals, and much research effort has been invested toward this direction in recent decades. Due to their non-sphericity, Lorenz-Mie theory of light scattering by spherical particles can only serve as a first-order approximation to the problem and is not able to reproduce many of the angular scattering or polarization features of aspherical ice crystals. On the other hand, geometric-optics methods can explain much of the fascinating optical phenomena, such as the 22° and 46° halos associated with some cirrus clouds, which result from refractive rays through the prism angles of 60° and 90°. Conventional geometric-optics methods are applicable to faceted particles with a relatively large size parameter, defined as the ratio of particle size to wavelength.

$$x = 2\pi r/\lambda, \tag{3.8}$$

where r is the radius of the equal-volume-sphere. To compute the light scattering properties, one must generate different crystal shapes using computer techniques. Results for many shapes, such as quasi-spherical, column-type, plate-type, and bullet rosette (as shown in Fig. 3.9), have been presented in the literature (e.g., Macke, Mueller et al., 1996; Yang et al., 2018). The databases by Yang et al. (2013) and Bi & Yang (2017) are widely used and include optical properties for solar and infrared wavelengths for crystals of a large range of sizes and nine different shapes.

The scattering phase matrix (often denoted as $\mathbf{P}(\Theta)$, where Θ is the scattering angle) is a fundamental scattering property of ice crystals, which transforms incident Stokes parameters into scattered Stokes parameters. For an ensemble of randomly oriented ice particles, the scattering phase matrix generally has a block-diagonal form:

$$\begin{pmatrix} I \\ Q \\ U \\ V \end{pmatrix} = \frac{\Omega_{eff}}{4\pi} \begin{pmatrix} P_{11}(\Theta) & P_{12}(\Theta) & 0 & 0 \\ P_{21}(\Theta) & P_{22}(\Theta) & 0 & 0 \\ 0 & 0 & P_{33}(\Theta) & P_{34}(\Theta) \\ 0 & 0 & -P_{34}(\Theta) & P_{44}(\Theta) \end{pmatrix} \begin{pmatrix} I_0 \\ Q_0 \\ U_0 \\ V_0 \end{pmatrix}, \tag{3.9}$$

where $\Omega_{\mathrm{eff}} = \sigma_{\mathrm{sca}}/R^2$, σ_{sca} is the scattering cross-section, and R is the distance between the particle and the point of observation. The element $\mathbf{P_{11}}(\Theta)$, also called the *phase function*, describes the total intensity of the scattered light as a function of the scattering angle and, therefore, is relevant for radiative transfer calculations. For the two-stream radiative transfer approaches applied in most climate models, this element can be represented by the asymmetry parameter, g, defined as the cosine-weighted integral of $P_{11}(\Theta)$:

$$g = \frac{1}{2}\int_{-1}^{1} P_{11}(\Theta) cos(\Theta) dcos(\Theta) \tag{3.10}$$

In geometric-optics methods, one applies geometric ray-tracing to account for light ray refractions and (external or internal) reflections. Geometric ray-tracing applies Snell's law and the Fresnel formulas, and can be considered an asymptotic approach that becomes applicable when a locally smooth surface is much larger than the incident wavelength, that is, $x \gg 1$. The hit-and-miss Monte Carlo method was applied by, for example, Takano & Liou (1995), Macke, Mueller et al. (1996), and Yang & Liou (1998a), to trace the photons in complex ice crystals with polarization and absorption.

In addition to refraction and reflection, the contribution from diffraction in the far field, or Fraunhofer diffraction, must be determined in order to compute the phase function correctly. Conventionally, diffraction by ice crystals is computed according to Babinet's principle, which states that the diffraction pattern is determined by

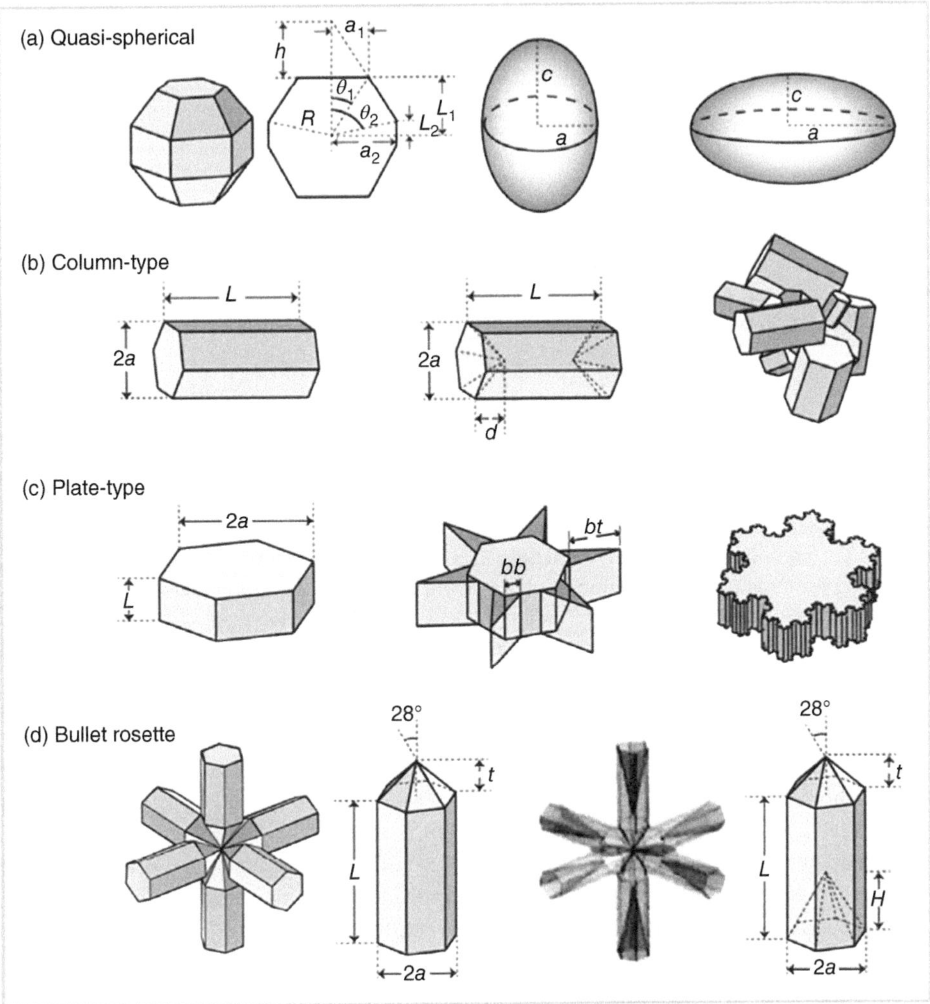

Figure 3.9 Computer-generated shapes for ice crystals. Source: Adapted from Liou & Yang (2016).

its geometric projection on a plane normal to the incident direction. This leads to a contribution to the phase function that is highly peaked in the forward scattering direction. The diffraction asymmetry parameter is generally approximately unity, but decreases for size parameters below about 100. In addition to the application of the Babinet's principle for diffraction computation, Hesse et al. (2012) extended the method to cases of oblique apertures, with good agreement with numerical exact results. The conventional geometric ray-tracing method assumes that the contributions of diffraction and Fresnel rays are equal, and the extinction efficiency is 2 (Bi et al., 2014). Such methods are only applicable to relatively large size parameters above 30–40 (Takano et al., 1992). To overcome these limitations, Yang & Liou (1996b) developed an improved method by mapping the near field to the far field based on the electromagnetic equivalence theorem. This mapping can be realized by either the surface integral equation or the volume integral equation.

The refraction and reflection calculations using geometric ray-tracing by definition do not depend on particle size for non-absorbing wavelengths (e.g., at visible wavelengths). Furthermore, as discussed above, the diffraction contribution is only sensitive to particle size for size parameters below about 100. Hence, for size parameters greater than about 100, the phase matrix and g derived by geometric optics calculations do not depend substantially on particle size for non-absorbing wavelengths, but only on crystal morphology (see Fig. 3.10). Note that morphology itself may depend on size, leading, in turn, to size-dependent scattering properties. At absorbing wavelengths, the refraction and reflection contributions are affected by light absorption, while the diffraction is not. Hence, the relative contribution of diffraction to the total phase function increases with increasing absorption (i.e., decreasing single-scattering albedo (SSA)), leading to an increase of g (Macke, Mueller, et al., 1996). As SSA decreases with particle

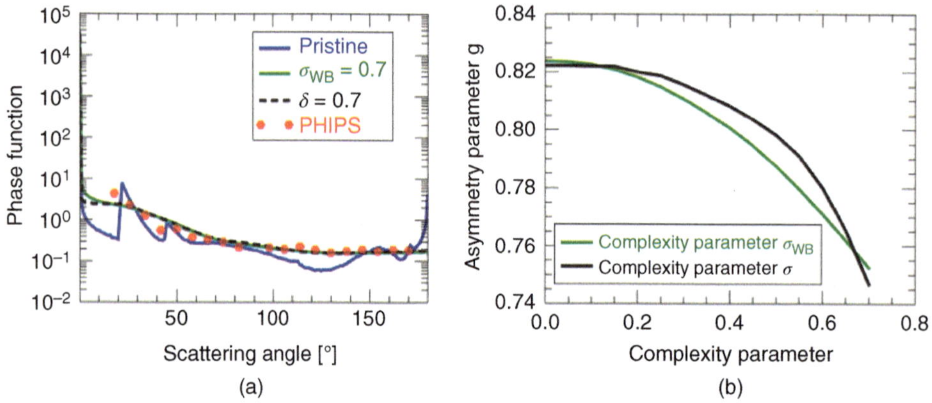

Figure 3.10 Panel (a) shows the phase functions of a hexagonal columnar prism with an aspect ratio of 0.5 (defined as the ratio of crystal basal plane width and prism length) without any complexity (dark gray/blue) and with complexity added using uniform facet tilt (equation (3.1) with $\delta = 0.7$ (black dashed line) and with complexity added using a Weibull distribution (equation (3.3) with $\sigma_{wb} = 0.7$ and $\eta = 0.75$ (gray/green). The gray/red hexagons represent observations of columnar crystals with aspect ratio restricted between 0.4 and 0.6 (defined as the ratio between the width perpendicular to maximum dimension and the maximum dimension) measured with the Particle Habit Imaging and Polar Scattering (PHIPS) probe. Panel (b) shows the decrease of asymmetry parameter with increasing parameter σ_{wb} (gray/green) and δ (black).

size, g increases with particle size at such absorbing wavelengths.

Geometric-optics methods are often applied to faceted particles with smooth surfaces. To account for ice crystal complexity, a TF approach is generally used to mimic the effect of facet roughness by changing the outgoing direction of a traced ray in the ray-tracing process, as discussed in section 3.2.2. This random facet tilt procedure can diminish the halos produced by pristine ice crystals, that is, the peaks in the scattering function at 22° and 46° scattering angle. A typical example is shown in Figure 3.10a, where the phase function of a pristine hexagonal column is compared to that of a column with complexity parameterized by using a uniform facet tilt (equation (3.1) with $\delta = 0.7$ and using a Weibull distribution for the facet tilt (equation (3.3) with $\sigma_{wb} = 0.7$ and $\eta = 0.75$. The two methods to parameterize the optical effects induced by crystal complexity lead to a similarly smooth phase functions in Figure 3.10a. Both parameterizations accurately represent the observed median angular scattering function of columnar crystals in the side and back-scattering angular range. However, there is a significant discrepancy for scattering angles < 50°, most likely because the complete randomization of the crystal surfaces in the model diminishes the halo features from less roughened facets that might still be present in the actual particle population. Furthermore, an increase in complexity generally decreases g of the scattering phase function (Macke, Mueller, et al., 1996), as demonstrated in Figure 3.10b.

In addition to crystal facet roughness, the TF approach has also often been used to simulate the optical effects of other types of crystal complexities, such as internal structures (Hong & Minnis, 2015; Macke, Mishchenko, et al., 1996) and macro-scale complexities (Letu et al., 2016). Here, it has to be emphasized that even though the TF method is widely used, it represents a rather non-physical approach to the light scattering properties of complex ice crystals, and it has not yet proven to be a good proxy for all the different types of complexities, with many having scales around the incident wavelength or smaller (i.e., outside the geometrical optics region) observed in real atmospheric ice crystals (see section 3.2.1).

In addition to geometric-optics methods, large efforts have been invested into the development of numerically accurate methods, such as the T-matrix method (Bi et al., 2013; Draft Johnson, 1988; Mishchenko & Travis, 1998; Waterman, 1971), the discrete-dipole approximation (DDA) (Draine & Flatau, 1994; Purcell & Pennypacker, 1973; Yurkin et al., 2007; Yurkin & Hoekstra, 2011), the finite-difference time domain (FDTD) (Sun et al., 1999; Taove & Umashankar, 1990; Yang & Liou, 1996a), and pseudo-spectral time domain (PSTD) methods (Liu, 1997; Liu et al., 2012). These methods can be used for accurate simulation of optical properties of ice crystals possessing a high degree of morphological complexity. In addition, they can be applied to examine the accuracy of the traditional geometric-optics treatments, for example, of diffraction (Mishchenko & Macke, 1998), and facet roughness (Collier et al., 2016; Riskilä et al., 2021). However, these methods are more suitable for simulations of crystals with relatively small size parameter, that is, for small crystals and/or long wavelengths, because (unlike geometric optics methods) computation times increase significantly with increasing size parameter.

3.4.2. In situ Observations of Ice Crystal Angular Light Scattering Functions and Ice Cloud Asymmetry Parameter

Direct measurement of the angular distribution of light scattered from ice crystals is important for developing optical particle models, such as those discussed in section 3.4.1, that are used to parameterize cloud optical properties in climate models. The ice crystal angular light scattering function can be measured in situ using nephelometers. There are several types of nephelometers (Barkey et al., 2012), all containing a collimated light source (usually a laser) and a (set of) detector(s), that can be either stationary or moving, measuring the scattered light intensity at desired angles from near-forward scattering to near-backscattering angles. Since it is technically difficult to measure the scattered light down to 0° and 180°, some assumptions for the fraction of scattered intensity outside the measurement range need to be made in order to derive g (Gerber et al., 2000; Xu et al., 2022).

Nephelometer measurements have been performed since the 1960s. The early measurements were conducted in the laboratory for artificially generated ice clouds (Dugin et al., 1977; Human, 1970; Rimmer & Saunders, 1997; Sassen & Liou, 1979; Sasaki et al., 1998). These studies reported poor agreement between experimental and theoretical angular scattering functions when assuming spherical particles or simple habits, such as infinitely long cylinders or pristine hexagonal particles. Since the 1990s, in situ measurements of the angular scattering function (and g) have been available from airborne polar nephelometers (e.g., Gayet et al., 1998; Garrett et al., 2001, 2003). In this section, we present angular scattering function measurements of a single oriented ice crystal and give an overview of in situ observations of g in different geographical locations.

Angular Light Scattering of Individual Oriented Ice Crystals

Attempts to experimentally characterize the angular scattering from single ice crystals having well-defined orientation have been quite limited. Yet, testing ray-tracing or other optical models against single oriented ice crystals would bring valuable information on how crystal complexity is affecting light scattering in detail. A first step in this direction was undertaken by Shcherbakov et al. (2006), who connected a polar nephelometer in tandem to a cloud particle imager. With this setup they measured angular light scattering functions and habits for individual ice crystals from diamond dust events at the South Pole. Although the two measurements were not performed on exactly the same individual crystals, the measurement setup ensured that the two instruments sampled the same crystal population in a statistical sense. They report highly variable angular light scattering functions with distinct peaks that are attributed to quasi-specular reflections from crystal facets of oriented ice crystals. Ray-tracing analysis with TF complexity metrics, for example, assuming Weibull distributed surface height gradients (equation 3.3), revealed that roughness parameters corresponding to deeply rough surfaces are necessary to reproduce the observed features in the angular scattering functions. While broad quasi-specular reflections could be reproduced by this approach, significant differences between modeled and measured angular scattering functions were observed for scattering angles between 125° and 160°. These discrepancies at backward angles were attributed by Shcherbakov et al. (2006) to shortcomings of the geometric-optics approximation and, in particular, the neglect of diffraction effects on individual elements of the surface roughness texture in the ray-tracing approach.

The Particle Habit and Polar Scattering (PHIPS) probe is an airborne single particle instrument that was developed to fill the knowledge gap in understanding the fundamental light scattering properties of individual atmospheric ice particles. Basically, PHIPS measures both the ice particle habit and its orientation, as well as the corresponding angular light scattering function for the *same* crystal at the *same* orientation. This is achieved by combining a stereo-microscopic imager with a polar nephelometer in a single instrument (Abdelmonem et al., 2016; Schnaiter et al., 2018). With PHIPS it is now possible to reconstruct the habit and orientation of observed individual ice crystals in order to test and refine optical models by comparing their light scattering results with corresponding measurements. In this way, existing optical models can be tested for their performance in reproducing ice crystal light scattering characteristics for the different morphological complexity types and ranges presented in section 3.2.

Figure 3.11a shows an example of a columnar ice crystal captured by PHIPS during the SOCRATES project over the Southern Ocean. The 3-D reconstruction of the crystal revealed a size of 40 μm and 86 μm for the a-axis and c-axis lengths, respectively. The optical engineering software FRED (Photon Engineering, Tucson, AZ) was used to perform ray-tracing calculations for this crystal in order to simulate the PHIPS light scattering measurement. Ray-tracing calculations were conducted for smooth facets as well as for roughened surfaces by activating either tilted-surface Weibull statistics or the ABg surface roughness diffraction model on individual facets of the crystal (see section 3.2.2 for details). While the hemispheric irradiation map of scattered light shows

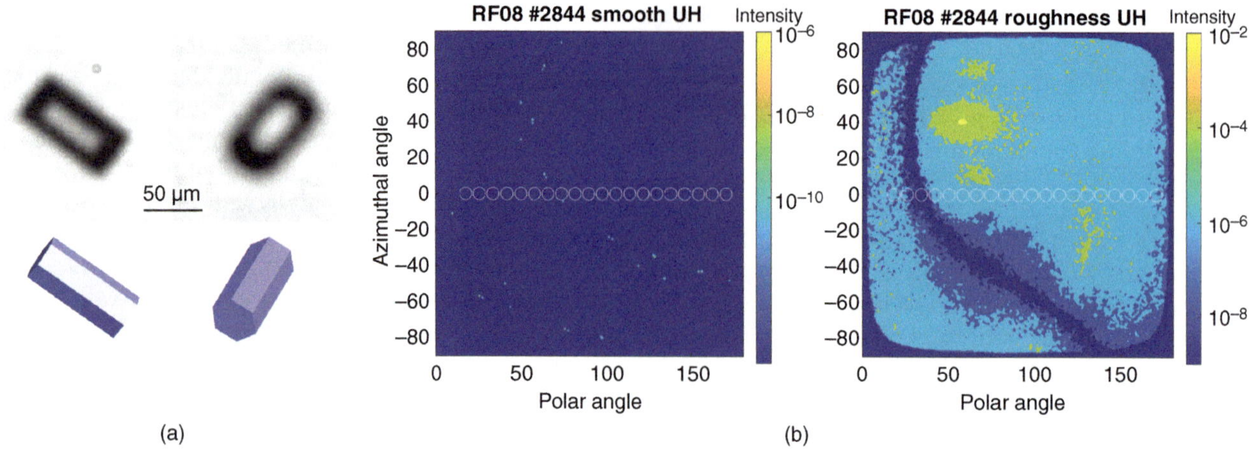

Figure 3.11 Columnar ice crystal (crystal number 2844) captured by PHIPS during the SOCRATES project over the Southern Ocean (a). The spatial light scattering of the single oriented crystal was simulated assuming smooth facets (smooth) and facets with added nanoscale surface roughness (roughness). Panel (b) shows the corresponding irradiation maps of light scattered in the upper hemisphere (UH), where the detectors are located. The color scale illustrates the scattering intensity so that lighter shade/color corresponds to a higher intensity and darker shade/color to a lower intensity. The white circles overlaid in the irradiation maps illustrate the solid angles of the nephelometer detectors. (a) Stereo-microscopic image of a columnar ice crystal and its 3-D reconstruction. (b) Hemispheric irradiation maps.

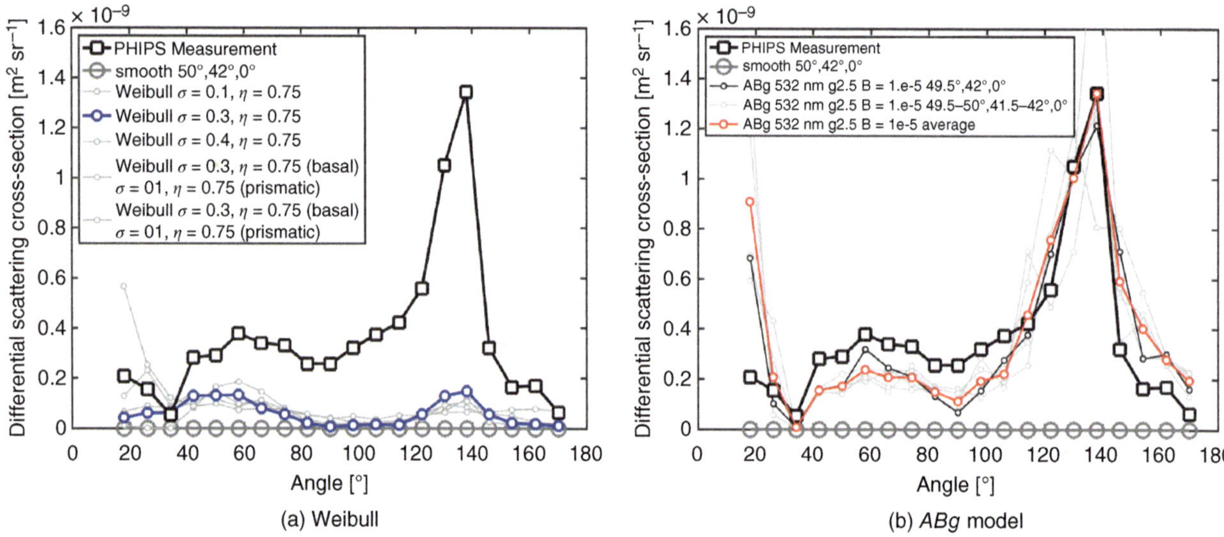

Figure 3.12 Measured and modeled differential scattering cross-sections of a columnar ice crystal displayed in Figure 3.11. Measurements are represented with black squares. Round symbols show calculations assuming rough surfaces where surface roughness was simulated using TF method and Weibull statistics (a) and the *ABg* model (b), respectively.

distinct ray spots in the case of smooth crystal facets, the scattered radiation is more evenly distributed across the hemisphere when the basal facets of the crystal were assumed to be rough (Fig. 3.11b).

The angular scattering function measured for this crystal by PHIPS, displayed in Figure 3.12, shows a distinct maximum in the 120° to 150° angular range, with a peak in the 138° channel, a less-marked and broad maximum in the 40° to 80° angular range, followed by a shallow minimum around 90°, a sharp minimum at 34°, and a significant non-zero scattering cross-section for the other channels. Interestingly, the ray-tracing result for the crystal with smooth facets does not predict scattered radiation in any of the PHIPS detection channels. Only if surface roughness is assumed for the basal facets of the crystal, do the ray-tracing calculations predict an angular light scattering profile that resembles the measurement. The measured angular scattering function of this crystal (at this specific orientation) can, therefore, be solely explained by surface roughness on the basal facets, and the two optical complexity metrics introduced in section 3.2.2 can be tested for this case. While the TF approach with Weibull statistics equation (3.3) follows the general trend of the measured scattering function, it shows significant deviations in feature position and shape and, in particular, the strength of predicted light scattering (Fig. 3.12a). Yet, the ABg roughness diffraction model equation (3.5) mimics the measurement quite well (Fig. 3.12b). This is true not only for the position and shape of the observed features but also for the strength of the predicted light scattering, which indicates that, in a case of this specific crystal, a physical (diffraction) based model gives better results than a TF approximation that ignores diffraction on the sub-wavelength scale features of the surface. It is clear that a final assessment of the performance of these two approaches (or any other approach to simulate the optical effect of crystal complexity) cannot be made based on one specific case. Rather, this example emphasizes the value of measuring and analyzing angular light scattering functions of individual oriented ice crystals.

Ice Cloud Asymmetry Parameter Derived from In situ Measurements

Here, we report observations of g from three different in situ instruments: the polar nephelometer (PN), the Cloud Integrating Nephelometer (CIN), and the PHIPS probe. All three instruments measure light scattering of cloud particles approximately in the size range from a few micrometers up to 1mm in diameter and in an angular range from near-forward direction to near-backward direction. Different methods are used to derive g from the measurements of scattered intensities. To derive g from PN (CIN) measurements, it is assumed that the fraction (f) of energy scattering into angles smaller than 15° (10°) is constant. For PN, it is assumed that $f = 0.56$ regardless of the cloud composition, whereas for CIN f varies between 0.52 and 0.57 (Gerber et al., 2000). It is estimated that the absolute error in g derived from PN measurements is ± 0.04, and the relative error for CIN is 3%. For PHIPS, g was retrieved by fitting an optical particle model from the database of Yang et al. (2013) to the measurements. The best fit was achieved using model particles composed of severely roughened column aggregates. Recently, Xu et al. (2022) introduced a method for deriving the asymmetry parameter from a polar nephelometer, without relying on specific light scattering models.

Table 3.2 summarizes the retrieved values for g and also values for simultaneously retrieved extinction coefficient, β. Values given for g and β are averages for one airborne measurement campaign, except in the case of PHIPS, where data are averaged over several campaigns. In general, g values below 0.79 are observed, which is less than expected for pristine hexagonal crystals (e.g., Fig. 3.10b). The apparently systematic difference in g between the different nephelometers can be partly explained through the uncertainty in the methods used to retrieve g, as discussed above. Additionally, in situ measurements cover a smaller spatial and temporal scale compared to, for example, satellite observations, and, therefore, the magnitude of g also depends on the sampled cloud conditions. Also, no inter-comparison studies have been reported between the different nephelometers, which would help to evaluate the differences.

Figure 3.13 shows g as a function of in-cloud temperature and RH_{ice} measured with the PN in three airborne campaigns. These campaigns covered different cirrus types and origins: in the CIRCLE campaign most of the sampled cirrus clouds were either of convective outflows or synoptic origin whereas in the interhemispheric differences in cirrus properties from anthropogenic emissions (INCA) campaigns mostly synoptic or wave cirrus with lower vertical velocities (from 1 to 35 cm s^{-1}) were measured. Figure 3.13 shows that g decreases slightly from 0.765 to 0.755 with decreasing temperature in the INCA campaigns, presumably as a result of change in ice crystal size and morphology. In the CIRCLE campaign, g increased toward colder temperatures, which might be explained by the presence of high concentrations of quasi spherical or sublimating small ice crystals near the cloud top of convective anvil cirrus, which have been shown in some occasions to lead to an increase in g (Gayet et al., 2012; Järvinen et al., 2016). Small "pristine" plates were also observed close to the cloud top of synoptic cirrus, resulting in reduced side scattering and enhanced forward scattering (i.e., higher g). Panel (b) in Fig. 3.13 shows the same set of data for INCA campaigns but now as a function of RH_{ice}. Only a slight increase in g with increasing RH is observed toward RH values around 130%.

The observed low g between 0.75 and 0.79 for clouds with temperature ranging from 235 K to 210 K support the findings discussed in previous section. Both theoretical (see section 3.4.1) and laboratory studies (Järvinen et al., 2016; Schnaiter et al., 2016; Ulanowski et al., 2006) have shown that g decreases with increasing crystal complexity. It has been proposed that crystal complexity is the main driver of ice crystals' angular scattering behavior (and thus g) in the solar wavelengths (Järvinen et al., 2018a; Schnaiter et al., 2016) so that the underlying ice crystal habit becomes less important in an optical sense. This would partly explain why

Table 3.2 Values of asymmetry parameter (g) and extinction coefficient (β) estimated from previous in situ measurements in ice clouds.

Cloud type Instrument	g	β (km^{-1})	References
ML cirrus (Cirrus 98′)	0.79 [0.78–0.79]	0.86 [0.5–1.5]	Jourdan et al. (2003)
PN (800 nm)			
NH cirrus (INCA 2000)	0.767 [0.76–0.77]	0.63 [0.29–1.3]	Gayet et al. (2004)
PN (804 nm)			
SH cirrus (INCA 2000)	0.767 [0.76–0.77]	0.49 [0.22–0.97]	Shcherbakov et al. (2005)
PN (800 nm)			
ML cirrus (CIRCLE-2 2007)	0.778 [0.77–0.78]	0.4 [0.2–1]	Mioche et al. (2010)
PN (804 nm)			
Polluted cirrus (CONCERT 2008,2011)	0.77 [0.76–0.79]	0.21 [0.05–0.6]	Chauvigné et al. (2018)
PN (804 nm)			
Several cloud types	0.75		Järvinen et al. (2018a)
PHIPS (532 nm)			
Arctic ice clouds	0.74±0.03	~0–30	Gerber et al. (2000)
CIN (635 nm)			Garrett et al. (2001)
Florida anvil	0.75±0.01		Garrett et al. (2003)
CIN (635 nm)			

Note: ML = Midlatitude, NH = Northern Hemisphere, and SH = Southern Hemisphere.

similar g is measured in different types of clouds with different origin (liquid-origin or in situ), although it is known that the ice crystals habits can be very different. However, the limitation of in situ measurements is their spatial restriction, whereas a more global picture of g can be gained through space-borne remote sensing observations.

3.4.3. Ice Cloud Asymmetry Parameter Inferred from Space-Borne Remote Sensing Observations

Cloud optical thickness and effective radius inferred from a bi-spectral method using visible and shortwave infrared reflectances are substantially sensitive to g of the assumed optical model in the retrieval algorithm. If the true optical thickness and asymmetry parameter at visible wavelengths are τ_{true} and g_{true}, and the assumed asymmetry parameter is g_{mod}, the relative bias in the retrieved optical thickness τ_{retr} can be approximated by

$$\tau_{retr} \approx \tau_{true}\frac{1-g_{true}}{1-g_{mod}}, \tag{3.11}$$

while the relative bias in retrieved effective radius is approximately the inverse of the optical thickness bias (Stephens et al., 2001; van Diedenhoven, Cairns, et al., 2012; van Diedenhoven, Cairns, et al., 2020). Hence, reducing uncertainties in the magnitude of g at visible wavelengths is crucial to reducing uncertainties in retrieved ice cloud optical thickness and effective radius. For a recent review of ice optical models used for such retrievals, we refer to Yang et al. (2018).

As discussed by Yang et al. (2018), equation (3.11) can be used to estimate g at the shortwave if a collocated retrieval of optical thickness and/or effective radius is available from another technique that is less sensitive to the ice particle scattering properties. One of the earliest examples of such an approach was presented by Stephens et al. (1990), who used a combination of airborne visible and thermal infrared (TIR) measurements over cirrus to conclude that g close to 0.7 is consistent with the data. More recently, Holz et al. (2016) used optical thickness inferred from passive thermal infrared (TIR) as a reference to evaluate the optical model for the latest MODIS ice cloud bi-spectral retrievals. That work showed that, for severely roughened aggregates of columns, a value of g of about 0.75, on average, agrees with the TIR retrievals. Furthermore, they showed this ice model is also consistent with CALIOP lidar observations. Note that both TIR and lidar retrievals of optical thickness are applicable to clouds with optical thickness less than about 3–5. Similar agreement between TIR and visible cloud optical thickness retrievals is obtained using a two habit model

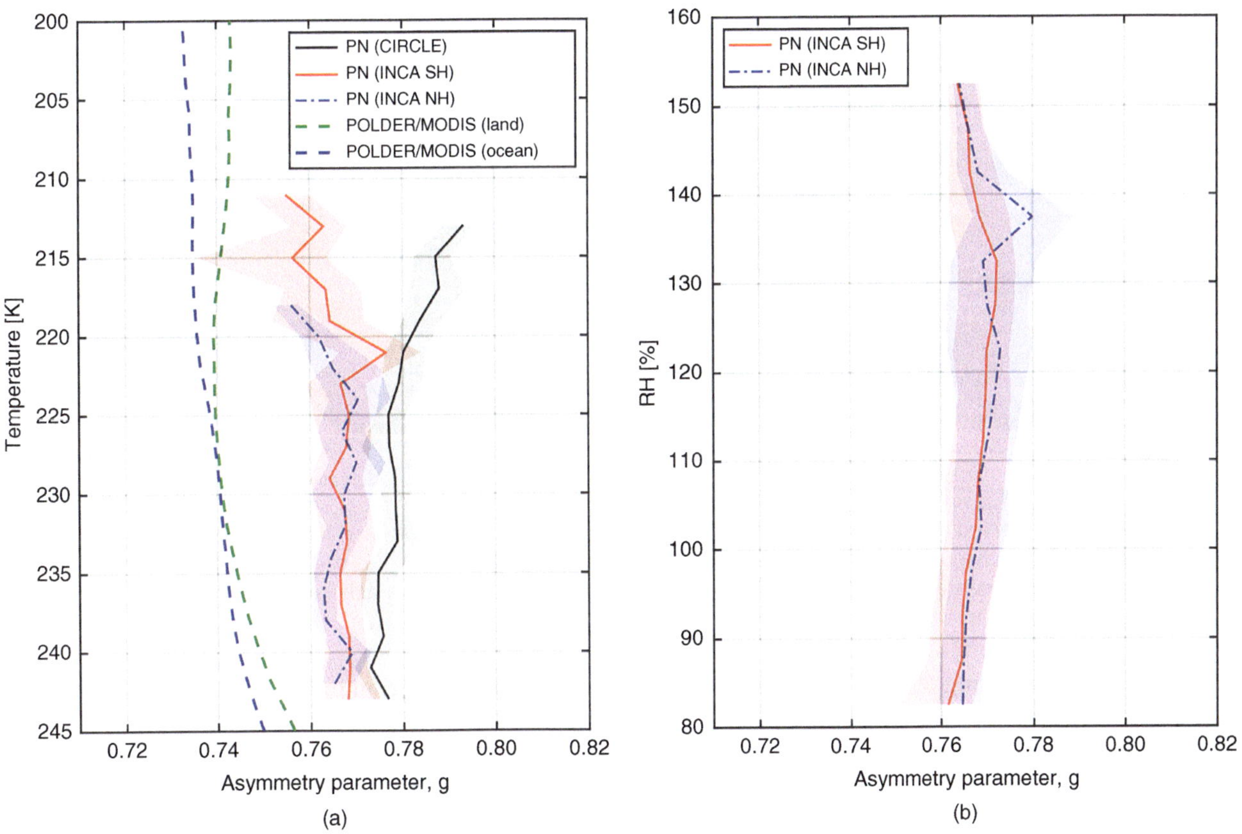

Figure 3.13 Asymmetry parameter (g) as a function of temperature (a) and relative humidity (RH) with respect to ice (b). Measurements were performed with the Polar Nephelometer (PN) probe in three airborne campaigns: CIRCLE (black solid line), INCA in Southern Hemisphere (SH, gray/red solid line), and INCA in Northern Hemisphere (NH, gray/blue dash-dotted line). RH data were not available for CIRCLE campaign. Curves represent median values and shaded areas the 25th and 75th percentile. The dashed lines in panel (a) illustrate the mean asymmetry parameter observations retrieved from POLDER and MODIS observations shown in Figure 3.14.

that also has a g near 0.75 (Liu et al., 2014). However, the assumption of isothermal cloud layers and the neglect of scattering effects may lead to biases in operational TIR retrievals (Saito et al., 2020), while lidar-derived optical thicknesses are affected by uncertainties in lidar ratios (Yorks et al., 2011).

Asymmetry parameters can also be estimated indirectly by inferring crystal morphology from lidar depolarization or multi-angular measurements, as discussed in section 3.3, and using g associated with that shape as the retrieved value. This approach applied to multi-angle polarimetry was evaluated with simulated measurements based on several smooth and roughened ice crystal habits by van Diedenhoven, Cairns, et al. (2012), who conclude that g can be retrieved within 5% for individual measurements.

Several studies showed that ice models with g near 0.75, such as the severely roughened aggregates of columns or the roughened two habit model, lead to a general good agreement with global satellite measurements (Liu et al., 2014; Letu et al., 2016; Yang et al., 2013; Yang et al., 2018). Similarly, the crystal shape and complexity parameter δ retrieved from global POLDER observations of thick ice clouds as discussed in section 3.3 and shown in Figure 3.7 can be used to infer g, of which mean values as a function of cloud top temperature and effective radius are shown in Figure 3.14. As discussed by van Diedenhoven et al. (2020), the globally averaged g agrees with the value of 0.75 inferred by other studies, but substantial variation is seen. Specifically, mean g is observed to increase with increasing cloud top temperature, especially for cloud top temperatures above about 240 K (Fig. 3.14). Furthermore, g is found to decrease with increasing ice effective radius. This variation of g is mainly affected by the variation of crystal complexity with cloud top temperature and effective radius, as shown in Figure 3.7. The value of mean g around 0.75 and its weak dependence on temperature for cloud tops colder than 240 K is generally consistent with in situ observations discussed in section 3.4.2. Furthermore, the

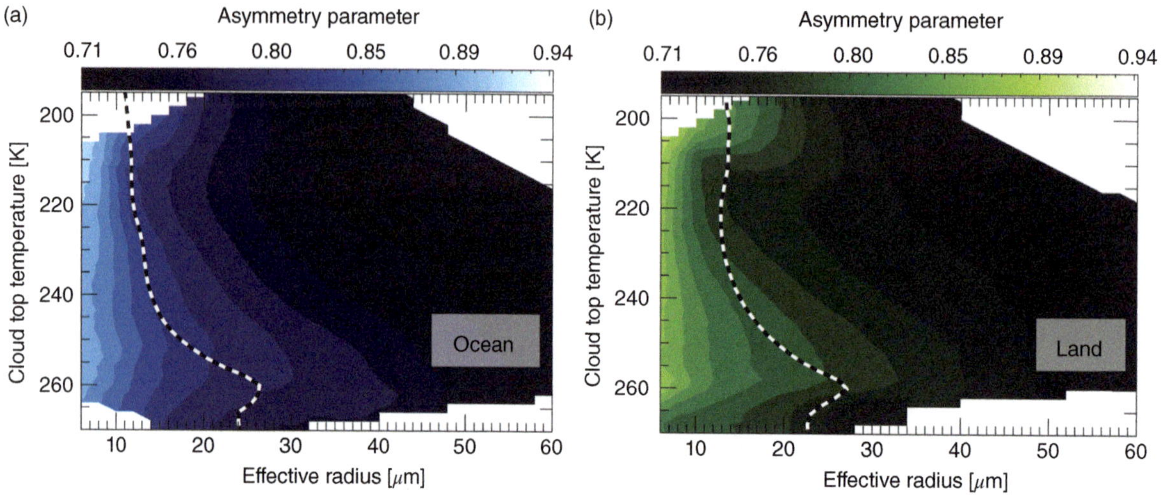

Figure 3.14 Observed global average of cloud top ice asymmetry parameters as a function of cloud top temperature and effective radius for thick ice clouds over ocean (a) and land (b). The dashed line shows the mean asymmetry parameters as a function of cloud top temperature indicated by the axis labels shown on top of the color/shade scale bar.

increase of g with increasing temperature above 240 K is consistent with the parameterization of Baran (2012).

3.5. THE IMPACT OF CRYSTAL COMPLEXITY ON ICE CLOUD RADIATIVE EFFECT

Ice cloud CRE, and radiation in general, is driven by three ice crystal single-scattering optical properties: the extinction cross-section (σ_{ext}), SSA, and g. For a given wavelength, σ_{ext} and SSA are mostly dependent on size, while g is also influenced by ice crystal complexity as discussed above. Other factors affecting CRE – independent from ice crystal morphological properties – are solar zenith angle, cloud height, surface temperature, and surface reflectivity. Because radiative transfer takes place in a volume, SSA and g need to be integrated over the size distribution of ice crystals weighted by extinction or scattering cross-sections (e.g., Twomey, 1977) to get the bulk optical properties. Similarly, the bulk extinction coefficient is computed by integrating σ_{ext} over the particle size distribution.

In radiative transfer, a frequently used quantity is the mass extinction coefficient, defined as the extinction coefficient (denoted here as β_{ext}) divided by the mass density of ice crystals (IWC). The resulting mass extinction coefficient, which we denote here as β_m, therefore has dimensions area/mass. Also of interest is the dimensionless total cloud optical thickness (or "optical depth"), obtained from these quantities as the profile-integrated product of IWC and β_m.

Ice cloud CRE at TOA is primarily dependent on the cloud optical thickness. Crystal complexity affects foremost g, which creates a source of uncertainty in the estimations of ice cloud CRE for a given cloud optical thickness and height both in climate models and remote sensing techniques. For example, in remote sensing, the scaled optical thickness (a product of optical thickness and $(1 - g)$, equation (3.11) determines the radiance used for retrievals. Furthermore, while climate models generally predict the vertical variation of IWC and sometimes crystal size, they generally do not constrain crystal morphology or complexity, which is therefore a source of uncertainty when relating the models' predictions to CRE. Previous estimates show that uncertainties in g need to be generally less than about 5% to determine the top of atmosphere fluxes in atmospheres overcast by ice cloud to within 5%, but these uncertainty limits depend on many factors (Vogelmann & Ackerman, 1995).

In this section, we first provide a general discussion on the estimation of ice cloud CRE, after which uncertainties in CRE related to ice crystal complexity are discussed. A more general discussion of cloud-radiative effects from a remote sensing perspective is provided in Chapter 4: Cloud-Radiation Interactions and Cloud-Climate Feedbacks.

3.5.1. Estimating Ice Cloud CRE

There are several different ways to estimate the radiative effect of ice clouds. This effect can be estimated (1) directly from observations of radiances or irradiances (Ramanathan et al., 1989), (2) using remotely retrieved or in situ observations of ice cloud optical properties (e.g., Campbell et al., 2016; Lolli, Campbell, et al., 2017; Campbell et al., 2021; Kato et al., 2008; L'Ecuyer et al., 2008), or (3) using modeled cloud properties (e.g., Soden et al., 2004). The first approach is based on direct observations of radiation. However, it is sometimes difficult to separate cloud effects from other radiative effects that might accompany a cloud, such as variations in water vapor and temperature (Loeb et al., 2018; Sohn et al., 2010). The second approach is subject to retrieval errors, although (fortunately) when cloud properties are derived from observed radiances at discrete wavelengths observed by satellite, retrieval errors are partially canceled in irradiance computations (Loeb et al., 2018). The third approach is used by all climate models. These estimates are affected by ice crystal morphology and complexity in numerous ways: not only via optical properties of individual ice particles, but also via the occurrence and lifetime of cirrus clouds (Gasparini et al., 2019; Hartmann et al., 2018).

In the first approach discussed above, the CRE is defined as the irradiance difference between all-sky conditions in which the cloud fraction can vary from 0 to 1 (hereinafter all-sky) $F_{all,y}^{TOA}$ and clear-sky conditions (i.e., no cloud is present, hereinafter clear-sky) $F_{clr,y}^{TOA}$,

$$CRE_y^{TOA} = F_{all,y}^{TOA}(x_{all}) - F_{clr,y}^{TOA}(x_{clr}), \quad (3.12)$$

where the subscript y indicates either SW or LW, x_{all} is atmospheric and surface variables under all-sky conditions (e.g., specific humidity, temperature profiles, and aerosols) and x_{clr} is those under clear-sky conditions. Generally, specific humidity is larger when clouds are present compared to clear-sky conditions. Similar CREs can be defined at the surface or net atmosphere (net TOA minus net surface irradiance). To a good approximation, TOA irradiance under all-sky conditions is a linear combination of cloudy-sky (i.e., overcast) irradiance and clear-sky irradiance weighed by their area fraction,

$$F_{all,y}^{TOA}(x_{all}) = fF_{cld,y}^{TOA}(x_{cld}) + (1-f)F_{clr,y}^{TOA}(x_{clr}), \quad (3.13)$$

where f is the cloud fraction and x_{cld} is the atmospheric state under overcast conditions. For approaches (2) and (3) discussed above, CRE can be computed by removing clouds while other meteorological conditions stay the same to isolate the effects of clouds on radiation. Therefore, the second term on the right side of equation (3.13) is equivalent to $F_{clr,y}^{TOA}(x_{all})$.

Under daytime conditions, ice clouds generally increase reflected SW irradiance at the TOA and reduce outgoing LW irradiance, relative to clear-sky conditions; these have opposing effects on the ice cloud CRE. The magnitude and sign of the ice cloud CRE therefore depend on numerous environmental conditions, including the underlying surface albedo, and physical characteristics of the cloud (e.g., altitude, temperature, particle size and morphology, and ice-water path/optical depth) (Campbell et al., 2016; Min et al., 2010). In contrast, in the absence of incoming solar radiation (i.e., at night), all clouds warm the planet by trapping the outgoing thermal radiation.

The Magnitude of CRE as a Function of Cloud Optical Thickness

The CRE defined by equation (3.12) is also used to derive the CRE observationally. For example, TOA irradiances are derived from Clouds and the Earth's Radiant Energy System (CERES) observations of broadband radiances. An overview of the CERES project is given by Wielicki et al. (1996). An empirical relationship is used to convert the radiance to irradiance (Loeb et al., 2005; Su et al., 2015). The relationship depends on scene type of the radiance measured in a CERES footprint. The scene type includes cloud fraction over the footprint, water phase, optical thickness, top height of clouds within the footprint, and surface type. CERES instruments are currently on NASA's Terra and Aqua, Suomi NPP, and NOAA-20 satellites. For Terra and Aqua satellites, radiances measured by MODIS are used to determine the scene type. The theoretical basis of how the irradiance is derived from radiance observations from CERES instruments is given by Kato et al. (2020).

Figure 3.15 shows TOA CRE derived from CERES observations over the tropics (15°N to 15°S) as a function of cloud optical thickness and cloud top height. As expected, cirrus TOA SW CRE and LW CRE appear with opposing signs, and magnitudes that increase with cloud optical thickness. Moreover, the magnitude of SW CRE for a given optical thickness is seen to increase with cloud top height; this is attributable to a reduction in absorption by underlying water vapor when clouds are at higher altitude. In addition, LW CRE for a given cloud optical thickness is seen to increase with cloud top height, attributable to an increased difference between the effective cloud temperature and the temperature of the underlying surface. Progressing from thin to thick, the figure shows that the daytime total (SW + LW) CRE by tropical cirrus is positive up to a threshold optical thickness of ~3 (reaching a maximum of 29 W m^{-2} prior to that point); clouds thicker than that threshold exert a negative CRE. For high cirrus, with cloud top height above 310 hPa, this threshold is larger, ~9. It is also evident from the figure that the magnitude of the total CRE for thick cirrus decreases with cloud top height because of a partial cancellation of SW and LW CREs. It is also noteworthy that the trends shown in Figure 3.15 are likely to vary depending on the temperature difference between the cloud and its underlying surface and on the albedo of the underlying surface (not shown).

Since the last decades, passive radiometric sensors from satellite platforms are widely used, thanks to their global coverage and long data records. Nevertheless, those sensors are not able to resolve thin sub-visible cirrus

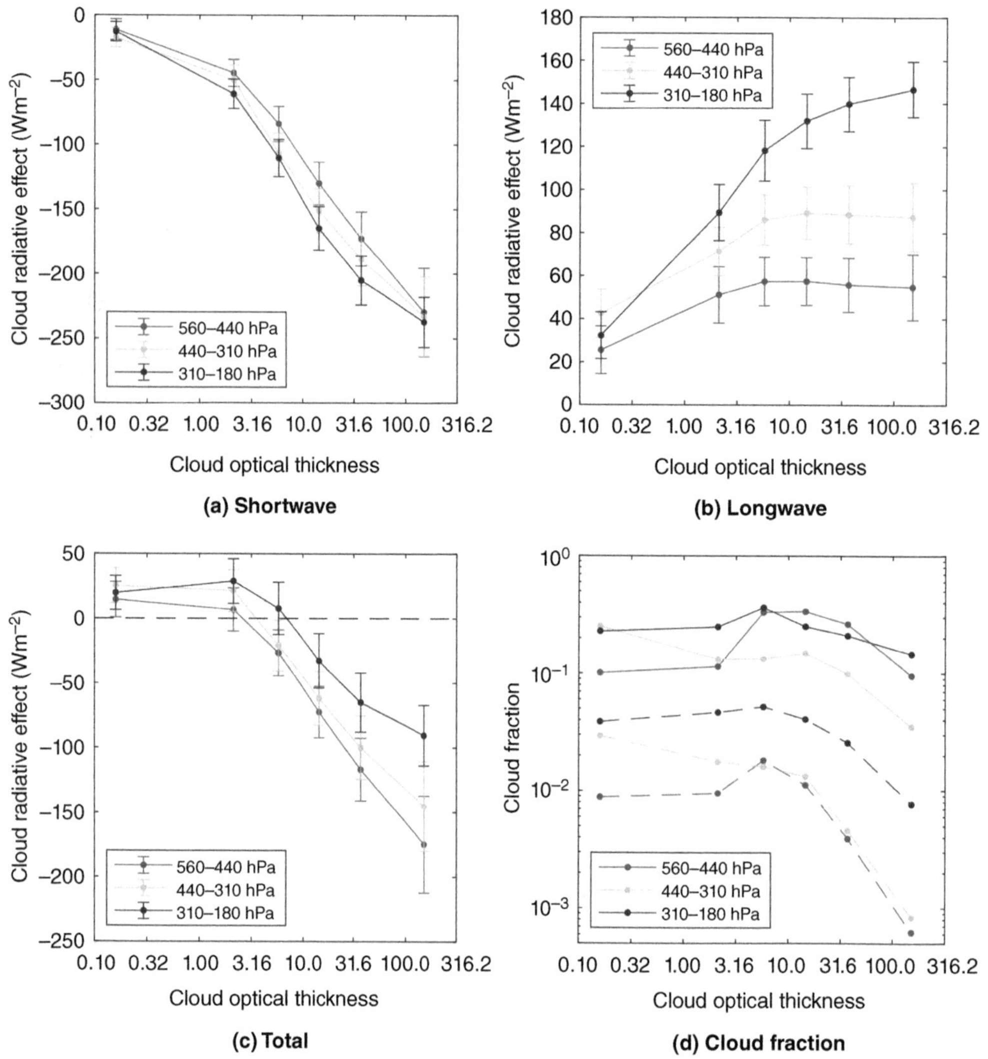

Figure 3.15 Shortwave (a), longwave (b), and total (shortwave + longwave) (c) cloud radiative effect of overcast clouds present over tropics (15°N to 15°S) at the top-of-atmosphere as a function of cloud optical thickness. The cloud top height range is indicated by the legend. Bars represent the standard deviation. Positive values indicate warming (i.e., downward positive). Panel (b) shows cloud fraction derived from MODIS. Solid and dashed lines are maximum and mean cloud fractions of monthly 1° × 1° gridded values. Cloud properties are derived from daytime observations from MODIS on Terra and Aqua taken from July 2002 through June 2005. Cloud radiative effects are derived from observations by CERES on Terra and Aqua using Edition 4A FluxByCloudTpe. Source: Adapted from Sun et al. (2022).

clouds with an optical depth lower than 0.3, as shown by a recent study (McHardy et al., 2021). On the other hand, the lidar technology shows a higher spatial and temporal resolution, topped with a unique spectral sensitivity. For this reason, lidar is a well-suited instrument for atmospheric profiling the optical, microphysical, and geometrical properties of translucent ice clouds as cirrus. Differently from the passive sensors, the ground-based lidars provide only punctual measurements. Furthermore, the lidar technology shows some issues in detecting cirrus clouds, as shown in Thorsen et al. (2013). In some degree, the discrepancies in cirrus cloud detection between active and passive remote sensing techniques can be due to the fact that lidar and passive sensors detect different sets of clouds. Cirrus CRE at TOA can be computed through equation (3.12), using the lidar-retrieved vertical profile of the cirrus cloud extinction coefficient as input in the radiative transfer model. Details on single-wavelength lidar equation inversion and its relative uncertainty can be found in Lolli, Di Girolamo, et al. (2017) and Lolli (2021). Other variables needed for the inversion, for example, the molecular extinction and backscattering coefficients, can

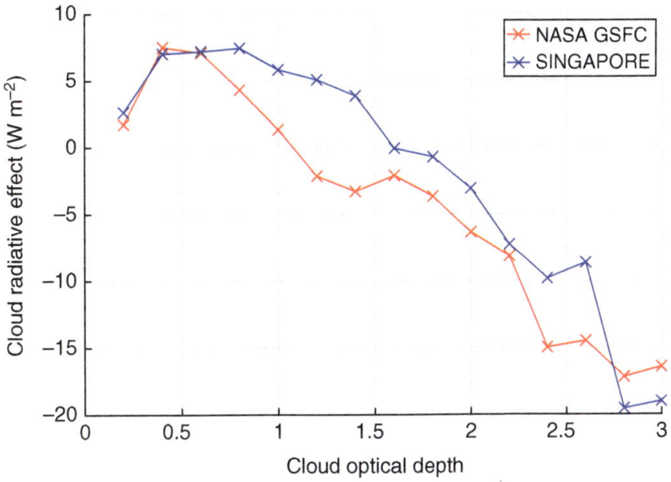

Figure 3.16 Instantaneous daytime net CRE of cirrus clouds as a function of cloud optical depth at Singapore and at NASA GSFC over 1 year of lidar data observations (2010 NASA GSFC and 2012 Singapore).

be obtained using ancillary data and theory, while for the cloud coefficients an assumption is needed to solve the lidar equation (Campbell et al., 2016).

A multi-year analysis of cirrus cloud CRE has been carried out for two ground-based lidar stations, permanent sites of the NASA Micropulse Lidar Network (MPLNET, Welton et al. (2001)). Figure 3.16 shows instantaneous daytime CRE as function of the optical depth from ground-based lidar observations at Singapore and at NASA Goddard Space Flight Center (GSFC). The instantaneous daytime CRE show similar trends with the diurnally averaged CRE values obtained from space-borne observations shown in Figure 3.15. It is interesting to highlight that for equatorial latitudes, there is a shift of the maximum CRE to higher cloud optical thickness with respect to midlatitudes (0.75 for Singapore, 0.45 for GSFC). During daytime, cirrus clouds are warming agents up to an optical depth of 1.6 in the tropics, versus 1.1 in midlatitudes. Overall, daytime clouds at the equator are more warming agents of the Earth-atmosphere system with respect to midlatitude clouds. This is clearly visible in Fig. 3.16, where the Singapore curve has overall higher values than the NASA GSFC curve. Partly, this difference can be explained by the differences in the solar zenith angle and surface temperatures.

3.5.2. Uncertainty in Cirrus CRE Estimations Caused by Crystal Complexity

The influence of ice crystal complexity on the global cirrus CRE has been investigated in studies of Yi et al. (2013) and Järvinen et al. (2018a). Yi et al. (2013) replaced the standard bulk optical properties (β_m, SSA, g) in the Community Atmosphere Model (CAM5.1) with those calculated using smooth and severely roughened ice crystals, assuming a mixture of various habits (Baum et al., 2011). Figure 3.17 shows the difference in the bulk optical

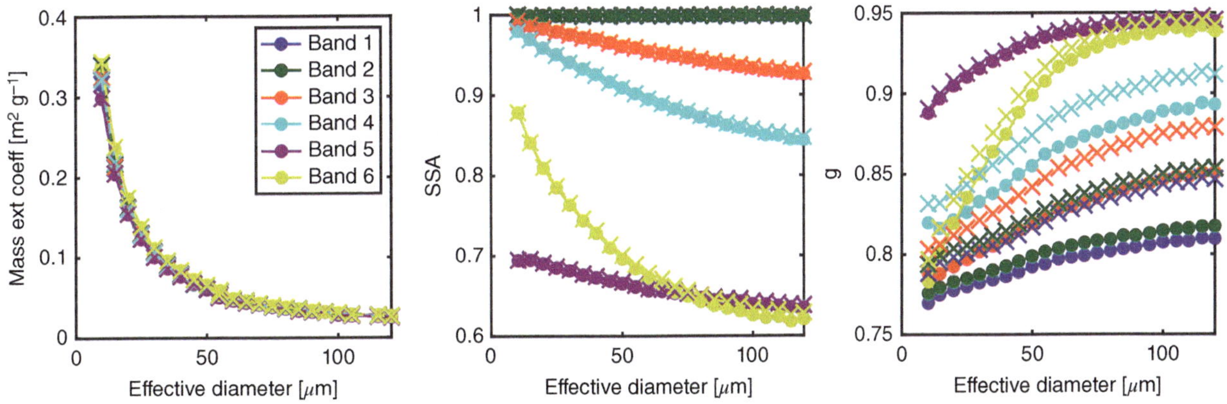

Figure 3.17 The bulk optical properties β_m, SSA, and g as a function of effective diameter for different SW bands for severely roughened crystals (dotted lines) and smooth crystals (crosses). Source: Adapted from Yi et al. (2013).

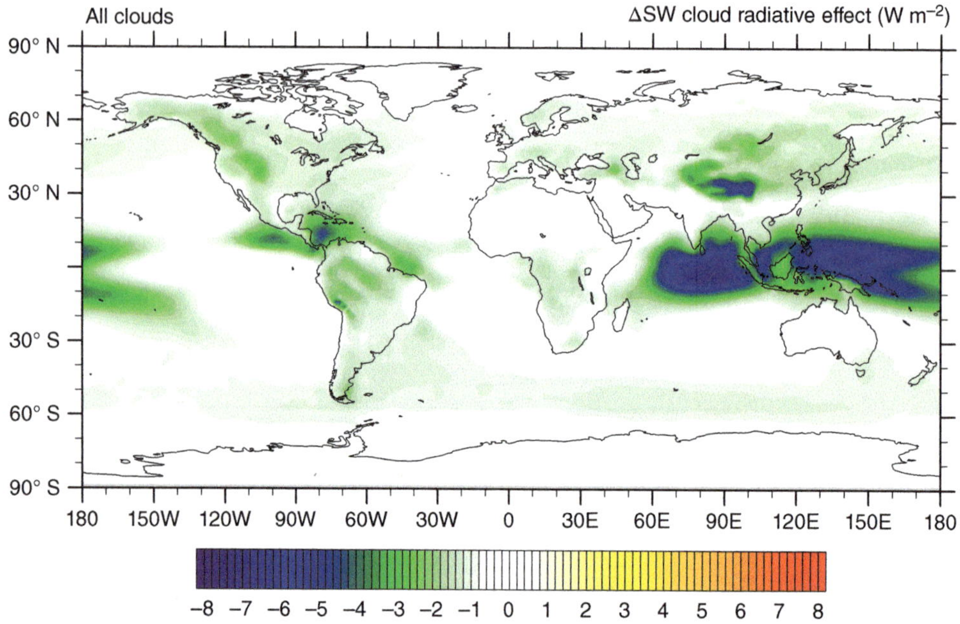

Figure 3.18 Difference in the global SW CRE after changing the current parameterization of g in the ECHAM-HAM model through a new parameterization calculated using severely roughened column aggregates (Yang et al., 2013). The severely roughened column aggregate model was found to best represent in situ measurement of ice cloud angular scattering function. Note that the difference is always negative or zero. Source: Järvinen et al., 2018a / European Geosciences Union / CC BY 4.0.

properties due to addition of facet roughness (using TF method). The primary effect of crystal complexity is seen in the decrease of g in the SW, whereas β_m and SSA are insensitive to crystal complexity. The largest difference in g is seen in the first four SW bands from 0.2 to 0.7 μm. When implementing the results in a radiative transfer model, Yi et al. (2013) found a significant global mean radiative effect of complexity (defined as the difference between simulations using smooth and severely roughened models) of $-1.46\,\mathrm{W\,m^{-2}}$. The net effect was mostly affected by the difference in the SW ($-1.83\,\mathrm{W\ m^{-2}}$), whereas the LW difference was more insignificant ($0.37\,\mathrm{W\ m^{-2}}$). Note that g of severely roughened crystals used in this study for solar wavelengths (band 1) was still substantially higher than the value of 0.75 generally matching observations, which might lead to an underestimation of the effects of crystal complexity on the radiative effect. In addition, Yi et al. (2017) implemented the severely roughened column aggregate model in the RRTMG RTM, which shows good agreement with CERES observation on CRE.

A similar investigation was performed by Järvinen et al. (2018a) who replaced the current parameterization for the bulk g in the ECHAM-HAM global climate model with a parameterization based on severely roughened column aggregate model (Yang et al., 2013). The ice crystal optical model was chosen since it was found to best reproduce in situ observations of ice cloud angular scattering function at 532 nm. The resulting global mean SW CRE change due to crystal complexity was found to be $-1.12\,\mathrm{W\,m^{-2}}$ but a regional change up to $-8\,\mathrm{W\ m^{-2}}$ was observed over the Indonesian maritime continent (Fig. 3.18).

Although the effect of crystal complexity on ice cloud (SW) CRE is smaller than the global mean cirrus CRE of about $5\,\mathrm{W\ m^{-2}}$ (Gasparini & Lohmann, 2016; Hong et al., 2016), it is large enough to cause noticeable uncertainty in estimations of the cirrus CRE. For climate model simulations, biases in radiative properties of ice clouds may lead to biases in modeled physical properties of clouds, such as cloud fraction, rain rates, thermodynamic phase, or of other components, such as water vapor profiles. This is because model parameters are generally calibrated ("tuned") within their uncertainty ranges so that the modeled present-day global radiative fluxes agree with observations and the models are generally energetically balanced in the pre-industrial era (Hourdin et al., 2017; Schmidt et al., 2017; Qian et al., 2018). Given the many parameters included in cloud microphysical schemes and their large uncertainties, such cloud variables are generally targeted as tuning parameters (Hourdin et al., 2017). Hence, relatively small but systematic biases in radiative properties of ice clouds caused by an unrealistic treatment of crystal complexity may have significant impacts on climate models' skill in representing current atmospheric variables, and in further predicting changes in climate by increasing greenhouse gasses.

3.6. CONCLUSIONS

The level of detail in natural ice crystals is only just starting to be unraveled. Improved measurement capabilities have made it possible to examine ice crystal structure in the micrometer and sub-micron level. Information on ice crystal complexity has been gained from laboratory and from atmospheric measurements using in situ and polarimetric remote sensing techniques. These observations have led to a consensus that most natural ice crystals deviate from pristine hexagonal geometry.

In this chapter, we have used the term "crystal complexity" to describe all possible deviations from that of a pristine hexagonal prism. Such crystal complexities are identified in a wide scale from nanometer-scale facet roughness to microscale phenomena, such as aggregation. CSEM studies and airborne cloud particle imaging probes have indicated that some degree of facet roughness, hollowness, air bubbles, and aerosol inclusions are more a rule than an exception. Frequently, several types of complexities are observed in the same crystal population or even combined in a single crystal.

Crystal complexity can contain a random component, such as random facet roughening, or it can have a non-random form that mimics the typical sixfold symmetry of the ice crystal, as is the case typically with hollowing. There is no uniform way to quantify morphological complexity but different (physical) complexity parameters are applied depending on the observational method used. To describe the optical effect of ice crystal complexity, a frequently used method in atmospheric sciences is the "tilted facet" method, where the normal of the crystal facet surface is tilted randomly from its nominal orientation during ray-tracing calculations. Other, more realistic, methods are also applied but these require more computational resources and, thus, are only applicable for smaller crystal sizes. Crossover between physical and optical complexity parameters is not yet established, although CSEM studies have shown that the statistical distribution of the square of the surface gradient of laboratory-grown ice can often be well described with gradient distributions assumed in the TF method, such as Gaussian and Weibull.

Cirrus ice crystals are influenced by a variety of environmental conditions during their life cycle, such as supersaturation, subsaturation, or they can interact with other particles in collisions. All these conditions affect the crystal morphology and degree of complexity. Laboratory studies with artificially generated ice crystals have indicated that physical complexity parameters tend to increase with increasing crystal growth rate. In cloud chamber studies, complexity parameters have been found to linearly correlate with the available condensable water vapor – defined by growth temperature and ice saturation ratio. Sublimation of ice crystals can either result in a rounding of the crystal's edges, as seen in cloud particle imaging probes, or yield random facet roughness and an increase in the degree of crystal complexity, as seen in laboratory studies. Laboratory studies have also shown that repeated growth and sublimation cycles can lead to a memory effect and gradual buildup of complexity.

Remote sensing and in situ measurements in natural cirrus have not been able to show a clear link between complexity parameters and simultaneously measured temperature or RH_{ice}. This may be due to the fact that the history of individual ice crystals cannot be quantified from crystal formation to point of measurement, and that such history likely contains multiple growth and sublimation cycles that cannot be easily mimicked in laboratory conditions. However, observations of multi-angle solar reflectances and polarimetry have shown that complexity parameters only weakly vary with cloud top temperatures below 230 K. Partly, this observation can be attributed to the fact that remote sensing techniques yield the maximum allowed complexity parameter for most observations, which also suggests that the optical complexity parameterizations probably do not adequately represent the full range of crystal complexity in natural ice clouds. In situ observations have also shown a weak variation in complexity parameters between 230 K and 210 K but a significant increase at temperatures colder than 210 K. Remote sensing and in situ measurements also indicate an increase of complexity in subsaturated conditions. Furthermore, both remote sensing and in situ measurements have shown that complexity parameters increase with increasing ice crystal effective radius, consistent with laboratory observations that have shown that a faster growth rate of large crystals leads to a higher degree of complexity. However, faster growth rates cannot explain the increase of complexity parameter at the coldest temperatures below 210 K, where ice crystals are expected to grow slower and reach smaller sizes. In turn, a slower growth rate also means it takes longer for a crystal to grow to a considerable size, potentially exposing it to a wider range of conditions, including subsaturated conditions, which promotes complexity.

The morphologies of ice crystals largely affect their single-scattering properties in the SW. Since there is no analytical solution to describe the interaction of electromagnetic radiation with aspherical ice crystals, numerical approximations are needed. In case of ice crystals, geometric ray-tracing is usually used to calculate light ray refractions and reflections. Additionally, diffraction needs to be accounted for. To account for crystal complexity, a random procedure is generally used for determining the outgoing direction of a traced ray in the ray-tracing process.

Calculated ice crystal optical properties are needed for remote sensing retrievals and for radiative transfer calculations. For radiative transfer, three ice crystal single-scattering properties are important: the extinction cross-section, SSA, and angular scattering function (or asymmetry parameter), of which the latter is highly sensitive to crystal complexity. Theoretical and laboratory studies as well as field observations have shown that complex crystals have a smooth and featureless angular scattering function with increased scattering intensity in the angular range from 100° to 150°, which leads to a decrease in the asymmetry parameter. Remote sensing and in situ observations indicate that at temperatures typical of cirrus cloud (below 235 K) a relatively stable asymmetry parameter below 0.79 in the visible represents most global ice clouds, although substantial variation is seen in individual cases that may depend on temperature, humidity, cloud type, and other factors. For ice clouds with cloud tops warmer than 240 K, remote sensing observations have shown an increase in g with increasing cloud top temperature. However, a limitation of space-borne remote sensing measurements is that g can be only retrieved for ice crystals within the first few optical depths from the cloud top.

Cirrus and ice cloud CRE is driven by cloud optical thickness. In general, optically thin cirrus is considered to have a net warming effect while optically thicker (in tropics optical thickness >3) cirrus is considered to have a net cooling effect. The global mean cirrus CRE is considered to be around 5 W m^{-2}. Crystal complexity creates a source of uncertainty in the estimations of ice cloud CRE for a given cloud optical thickness and height through its influence on the magnitude of SW g. Model studies implementing optical parameterizations of complex crystals have shown that taking into account crystal complexity makes ice clouds with a given optical thickness more reflective, with an estimated effect on global SW CRE between 1 and 2 W m^{-2}.

3.7. RECOMMENDATIONS FOR FUTURE WORK

In order to increase the fundamental understanding of crystal complexity, it is recommended that a complexity parameterization crossover should be aimed for. For example, CSEM work provides the most detailed information about the morphology of crystal complexity, but cannot be easily applied to atmospheric ice crystals, whereas in situ measured complexity parameters, such as k_e, cannot be used to discriminate between different complexity types. Thus, a crossover between k_e and CSEM observations is needed. Such crossover can be achieved, for example, by implementing measurement of spatial light scattering patterns in CSEM.

The same recommendations can be made for physical complexity parameters and optical complexity models. Whereas the TF method has established itself as a popular way to model the optical effects of ice crystal complexity, it hardly provides a good representation of the actual physical facet roughness, nor does it represent the vast scale of other types of complexities observed in cirrus ice crystals. More realistic facet roughness models incorporating the surface slope distributions observed by CSEM should be incorporated into optical particle models. Also, more work should go toward investigating the sensitivity of ice crystal single-scattering properties to different types and degrees of complexity.

Recent developments in ice microphysical schemes for cloud and climate models (e.g., Jensen et al., 2017; Zhang et al., 2014) are able to predict the evolution of ice particle shape and density as crystals are subjected to varying atmospheric conditions. Such schemes could be extended to include complexity metrics, for example, by using empirical relationships derived from in situ and laboratory measurements, while results could be constrained by in situ and remote sensing observations. Additional work is also needed to understand ice crystal growth and sublimation cycles in small-scale models with domains up to few cubic meters.

In general, our knowledge of crystal complexity and its magnitude in natural clouds is still very limited. Remote sensing observations presented here were restricted to optically thicker clouds (optical thickness >5) and to measurements of cloud tops, whereas in situ observations of complexity parameters (e.g., with the SID-3 instrument) have been rare. Open questions remain, such as if the conclusions presented for optically thicker clouds can be applied for optically thinner cirrus, or what is the vertical structure of crystal complexity. On the scale of individual ice crystals, not enough is known on the origin of crystal complexity and how the complexity correlates with other microphysical properties, such as habit, or with crystal formation mechanisms (heterogeneous or homogeneous nucleation, liquid-origin, or in situ origin). Therefore, it is recommended to implement in situ measurements of crystal complexity and ice crystal optical properties as part of the standard flight instrumentation payloads.

Direct measurements of ice crystal optical properties are important for improving radiative transfer calculations. We have shown that calculating g theoretically only based on crystal size or shape can lead to very different g values than are seen in natural clouds. Furthermore, airborne single particle measurement methods in atmospheric sciences have, for decades, focused on discrete detection of light scattering signals by measuring typically one and in some cases up to three signals.

Although detection of one signal is adequate to retrieve particle size, pseudo-continuous 2-D (e.g., SID-3) light scattering patterns are needed for more detailed information on the crystal morphology. Such experimental systems are already applied in other fields (Romanov & Yurkin, 2021) and they should also be more frequently incorporated into atmospheric measurement applications (both in laboratory and field studies).

The results presented in this chapter indicate that ice crystal complexity is relevant for ice cloud radiative properties. Ignoring the optical effects of crystal complexity can lead to systematic biases in radiative properties of ice clouds, and hence to biases in modeled physical cloud properties. Therefore, it is recommended to develop ice microphysical schemes to predict the variation of ice crystal complexity and combine these with flexible ice optical properties schemes to provide consistent radiative properties.

ACKNOWLEDGMENTS

WebPlotDigitizer was used to extract datapoints for Figure 3.17. Emma Järvinen acknowledges funding through the Helmholtz Association's Initiative and Networking Fund (grant agreement no. VH-NG-1531). Bastiaan van Diedenhoven acknowledges funding by NASA through grants no. 80NSSC20M0224 (PACE-SAT) and NNX14AJ28G (TASNPP). Steven Neshyba acknowledges support from NSF, grant CHE-1807898. Martin Schnaiter acknowledges support from the Helmholtz Association research program "Atmosphere and Climate" and funding from the German Research Foundation (grants SCHN 1140/3-1 and SCHN 1140/3-2). Fritz Waitz acknowledges funding by the German Research Foundation (DFG grant JA 2818/1-1).

REFERENCES

Abdelmonem, A., Järvinen, E., Duft, D., Hirst, E., Vogt, S., Leisner, T., & Schnaiter, M. (2016). PHIPS–HALO: The airborne particle habit imaging and polar scattering probe – Part 1: Design and operation'. *Atmospheric Measurement Techniques*, *9*(7), 3131–3144.

Avila, E. E., Castellano, N. E., Saunders, C. P. R., Bürgesser, R. E., & Aguirre Varela, G. G. (2009). Initial stages of the riming process on ice crystals. *Geophysical Research Letters*, *36*(9).

Bacon, N. J., & Swanson, B. D. (2000). Laboratory measurements of light scattering by single levitated ice crystals. *Journal of the Atmospheric Sciences*, *57*(13), 2094–2104.

Bailey, M., & Hallett, J. (2002). Nucleation effects on the habit of vapour grown ice crystals from- 18 to- 42° C. *Quarterly Journal of the Royal Meteorological Society: A Journal of the Atmospheric Sciences, Applied Meteorology and Physical Oceanography*, *128*(583), 1461–1483.

Bailey, M., & Hallett, J. (2004). Growth rates and habits of ice crystals between- 20 and- 70° C. *Journal of the Atmospheric Sciences*, *61*(5), 514–544.

Bailey, M. P., & Hallett, J. (2009). A comprehensive habit diagram for atmospheric ice crystals: Confirmation from the laboratory. AIRS II, and other field studies. *Journal of the Atmospheric Sciences*, *66*(9), 2888 – 2899.

Baran, A. J. (2009). A review of the light scattering properties of cirrus. *Journal of Quantitative Spectroscopy and Radiative Transfer*, *110*(14), 1239–1260. XI Conference on Electromagnetic and Light Scattering by Non-Spherical Particles: 2008.

Baran, A. J. (2012). From the single-scattering properties of ice crystals to climate prediction: A way forward. *Atmospheric Research*, *112*, 45–69.

Barkey, B., Paulson, S., & Liou, K.-N. (2012). Polar nephelometers for light scattering by ice crystals and aerosols: Design and measurements. In *Light scattering reviews* (Vol. 6, pp. 3–37). Springer.

Baum, B. A., Yang, P., Heymsfield, A. J., Schmitt, C. G., Xie, Y., Bansemer, A., et al. (2011). Improvements in shortwave bulk scattering and absorption models for the remote sensing of ice clouds. *Journal of Applied Meteorology and Climatology*, *50*(5), 1037–1056.

Bi, L., & Yang, P. (2017). Improved ice particle optical property simulations in the ultraviolet to far-infrared regime. *Journal of Quantitative Spectroscopy and Radiative Transfer*, *189*, 228–237.

Bi, L., Yang, P., Kattawar, G. W., & Mishchenko, M. I. (2013). Efficient implementation of the invariant imbedding T-matrix method and the separation of variables method applied to large nonspherical inhomogeneous particles. *Journal of Quantitative Spectroscopy and Radiative Transfer*, *116*, 169–183.

Bi, L., Yang, P., Liu, C., Yi, B., Baum, B., Van Diedenhoven, B., & Iwabuchi, H. (2014). Assessment of the accuracy of the conventional ray-tracing technique: Implications in remote sensing and radiative transfer involving ice clouds. *Journal of Quantitative Spectroscopy and Radiative Transfer*, *146*, 158–174.

Borovoi, A., Konoshonkin, A., & Kustova, N. (2014). Backscatter ratios for arbitrary oriented hexagonal ice crystals of cirrus clouds. *Optics Letters*, *39*(19), 5788.

Borovoi, A., Kustova, N., & Konoshonkin, A. (2015). Interference phenomena at backscattering by ice crystals of cirrus clouds. *Optics Express*, *23*(19), 24557.

Boucher, O., Randall, D., Artaxo, P., Bretherton, C., Feingold, G., Forster, P., et al. (2013). Clouds and aerosols. *In Climate change 2013: The physical science basis. Contribution of Working Group I to the Fifth Assessment Report of the Intergovernmental Panel on Climate Change* (pp. 571–657). Cambridge University Press.

Butterfield, N., Rowe, P. M., Stewart, E., Roesel, D., & Neshyba, S. P. (2017). Quantitative three-dimensional ice roughness from scanning electron microscopy. *Journal of Geophysical Research (Atmospheres)*, *122*(5), 3023–3041.

Campbell, J., Lolli, S., Lewis, J., Gu, Y., & Welton, E. (2016). Daytime cirrus cloud top-of-the-atmosphere radiative forcing properties at a midlatitude site and their global consequences. *Journal of Applied Meteorology and Climatology*, *55*(8), 1667–1679.

Campbell, J. R., Dolinar, E. K., Lolli, S., Fochesatto, G. J., Gu, Y., Lewis, J. R., et al. (2021). Cirrus cloud top-of-the-atmosphere net daytime forcing in the Alaskan Subarctic from ground-based MPLNET monitoring. *Journal of Applied Meteorology and Climatology*, *60*(1), 51–63.

Chauvigné, A., Jourdan, O., Schwarzenboeck, A., Gourbeyre, C., Gayet, J. F., Voigt, C., et al. (2018). Statistical analysis of contrail to cirrus evolution during the contrail and cirrus experiment (concert). *Atmospheric Chemistry and Physics*, *18*(13), 9803–9822.

Chepfer, H., Goloub, P., Riedi, J., De Haan, J., Hovenier, J., & Flamant, P. (2001). Ice crystal shapes in cirrus clouds derived from POLDER/ADEOS-1. *Journal of Geophysical Research: Atmospheres*, *106*, 7955–7966.

Cole, B. H., Yang, P., Baum, B. A., Riedi, J., & C.-Labonnote, L. (2014). Ice particle habit and surface roughness derived from PARASOL polarization measurements. *Atmospheric Chemistry and Physics*, *14*(7), 3739–3750.

Collier, C. T., Hesse, E., Taylor, L., Ulanowski, Z., Penttilä, A., & Nousiainen, T. (2016). Effects of surface roughness with two scales on light scattering by hexagonal ice crystals large compared to the wavelength: DDA results. *Journal of Quantitative Spectroscopy and Radiative Transfer*, *182*, 225–239.

Connolly, P., Saunders, C., Gallagher, M., Bower, K., Flynn, M., Choularton, T., et al. (2005). Aircraft observations of the influence of electric fields on the aggregation of ice crystals. *Quarterly Journal of the Royal Meteorological Society: A Journal of the Atmospheric Sciences, Applied Meteorology and Physical Oceanography*, *131*(608), 1695–1712.

Cox, C., & Munk, W. (1954). Measurement of the roughness of the sea surface from photographs of the sun's glitter. *Journal of the Optical Society of America 44*(11), 838–850.

del Genio, A. D., Yao, M.-S., Kovari, W., & Lo, K. K. W. (1996). A prognostic cloud water parameterization for global climate models. *Journal of Climate*, *9*(2), 270–304.

Draine, B. T., & Flatau, P. J. (1994). Discrete-dipole approximation for scattering calculations. *JoSAA*, *11*(4), 1491–1499.

Dugin, V., Volkovitskii, O., Mirumiants, S., & Nikiforova, N. (1977). Anisotropy of light scattering by artificial crystalline cloud formations. *Academy of Sciences, USSR, Izvestiya, Atmospheric and Oceanic Physics*, *13*, 22–25.

Forster, L., Seefeldner, M., Baumgartner, A., Kölling, T., & Mayer, B. (2020). Ice crystal characterization in cirrus clouds II: Radiometric characterization of HaloCam for the quantitative analysis of halo displays. *Atmospheric Measurement Techniques*, *13*(7), 3977–3991.

Freudenthaler, V., Homburg, F., & Jäger, H. (1996). Optical parameters of contrails from lidar measurements: Linear depolarization. *Geophysical Research Letters*, *23*(25), 3715–3718.

Fridlind, A., Atlas, R., Van Diedenhoven, B., Um, J., McFarquhar, G., Ackerman, A., et al. (2016). Derivation of physical and optical properties of mid-latitude cirrus ice crystals for a size-resolved cloud microphysics model. *Atmospheric Chemistry and Physics*, *16*(11), 7251–7283.

Garrett, T. J., Gerber, H., Baumgardner, D. G., Twohy, C. H., & Weinstock, E. M. (2003). Small, highly reflective ice crystals in low-latitude cirrus. *Geophysical Research Letters*, *30*(21).

Garrett, T. J., Hobbs, P. V., & Gerber, H. (2001). Shortwave, single-scattering properties of Arctic ice clouds. *Journal of Geophysical Research: Atmospheres*, *106*(D14), 15155–15172.

Gasparini, B., Blossey, P. N., Hartmann, D. L., Lin, G., & Fan, J. (2019). What drives the life cycle of tropical anvil clouds?. *Journal of Advances in Modeling Earth Systems*, *11*, 2586–2605.

Gasparini, B., & Lohmann, U. (2016). Why cirrus cloud seeding cannot substantially cool the planet. *Journal of Geophysical Research: Atmospheres*, *121*(9), 4877–4893.

Gayet, J.-F., Auriol, F., Oshchepkov, S., Schröder, F., Duroure, C., Febvre, G. et al. (1998). In situ measurements of the scattering phase function of stratocumulus, contrails and cirrus. *Geophysical Research Letters*, *25*(7), 971–974.

Gayet, J.-F., Ovarlez, J., Shcherbakov, V., Ström, J., Schumann, U., Minikin, A. et al. (2004). Cirrus cloud microphysical and optical properties at southern and northern midlatitudes during the inca experiment. *Journal of Geophysical Research: Atmospheres*, *109*(D20).

Gayet, J.-F., Shcherbakov, V., Voigt, C., Schumann, U., Schäuble, D., Jessberger, P., et al. (2012). The evolution of microphysical and optical properties of an A380 contrail in the vortex phase. *Atmospheric Chemistry and Physics*, *12*(14), 6629–6643.

Geogdzhayev, I., & van Diedenhoven, B. (2016). The effect of roughness model on scattering properties of ice crystals. *Journal of Quantitative Spectroscopy and Radiative Transfer*, *178*, 134–141.

Gerber, H., Takano, Y., Garrett, T. J., & Hobbs, P. V. (2000). Nephelometer measurements of the asymmetry parameter, volume extinction coefficient, and backscatter ratio in Arctic clouds. *Journal of the Atmospheric Sciences*, *57*(18), 3021–3034.

Gultepe, I., Heymsfield, A. J., Gallagher, M., Ickes, L., & Baumgardner, D. (2017). Ice fog: The current state of knowledge and future challenges. *Meteorological Monographs*, *58*, 4–1.

Hallett, J., & Mossop, S. C. (1974). Production of secondary ice particles during the riming process. *Nature*, *249*(5452), 26–28.

Hartmann, D. L., Gasparini, B., Berry, S. E., & Blossey, P. N. (2018). The life cycle and net radiative effect of tropical anvil clouds. *Journal of Advances in Modeling Earth Systems*, *10*, 3012–3029.

Hesse, E., Macke, A., Havemann, S., Baran, A., Ulanowski, Z., & Kaye, P. H. (2012). Modelling diffraction by facetted particles. *Journal of Quantitative Spectroscopy and Radiative Transfer*, *113*(5), 342–347.

Heymsfield, A. J., Krämer, M., Luebke, A., Brown, P., Cziczo, D. J., Franklin, C., et al. (2017). Cirrus clouds. *Meteorological Monographs*, *58*, 2–1.

Hioki, S., Yang, P., Baum, B. A., Platnick, S., Meyer, K. G., King, M. D., & Riedi, J. (2016). Degree of ice particle surface roughness inferred from polarimetric observations'. *Atmospheric Chemistry and Physics*, *16*(12), 7545–7558.

Holz, R. E., Platnick, S., Meyer, K., Vaughan, M., Heidinger, A., Yang, P., et al. (2016). Resolving ice cloud optical thickness biases between CALIOP and MODIS using infrared retrievals. *Atmospheric Chemistry and Physics*, *16*(8), 5075–5090.

Hong, G., & Minnis, P. (2015). Effects of spherical inclusions on scattering properties of small ice cloud particles. *Journal of Geophysical Research: Atmospheres*, *120*(7), 2951–2969.

Hong, G., Yang, P., Baum, B. A., Heymsfield, A. J., & Xu, K.-M. (2009). Parameterization of shortwave and longwave radiative properties of ice clouds for use in climate models. *Journal of Climate*, *22*(23), 6287–6312.

Hong, Y., Liu, G., & Li, J.-L. F. (2016). Assessing the radiative effects of global ice clouds based on CloudSat and CALIPSO measurements. *Journal of Climate*, *29*(21), 7651–7674.

Hourdin, F., Mauritsen, T., Gettelman, A., Golaz, J.-C., Balaji, V., Duan, Q., et al. (2017). The Art and Science of Climate Model Tuning. *Bulletin of the American Meteorological Society*, *98*(3), 589–602.

Huffman, P. (1970). Polarization of light scattered by ice crystals. *Journal of the Atmospheric Sciences*, *27*(8), 1207–1208.

Järvinen, E., Jourdan, O., Neubauer, D., Yao, B., Liu, C., Andreae, M. O. et al. (2018a). Additional global climate cooling by clouds due to ice crystal complexity. *Atmospheric Chemistry and Physics*, *18*(21), 15767–15781.

Jensen, A. A., Harrington, J. Y., Morrison, H., & Milbrandt, J. A. (2017). Predicting ice shape evolution in a bulk microphysics model. *Journal of the Atmospheric Sciences*, *74*(6), 2081–2104.

Johnson, B. R. (1988). Invariant imbedding T-matrix approach to electromagnetic scattering. *Applied Optics*, *27*(23), 4861–4873.

Jourdan, O., Oshchepkov, S., Shcherbakov, V., Gayet, J.-F., & Isaka, H. (2003). Assessment of cloud optical parameters in the solar region: Retrievals from airborne measurements of scattering phase functions. *Journal of Geophysical Research: Atmospheres*, *108*(D18).

Järvinen, E., Schnaiter, M., Mioche, G., Jourdan, O., Shcherbakov, V. N., Costa, A. et al. (2016). Quasi-spherical ice in convective clouds. *Journal of the Atmospheric Sciences*, *73*(10), 3885–3910.

Järvinen, E., Wernli, H., & Schnaiter, M. (2018b). Investigations of mesoscopic complexity of small ice crystals in midlatitude cirrus. *Geophysical Research Letters*, *45*(20), 11,465–11,472.

Kahnert, M., Nousiainen, T., Thomas, M. A., & Tyynelä, J. (2012). Light scattering by particles with small-scale surface roughness: Comparison of four classes of model geometries. *Journal of Quantitative Spectroscopy and Radiative Transfer*, *113*(18), 2356–2367.

Kato, S., Loeb, N. G., Rutan, D. A., & Rose, F. G. (2020). Effects of electromagnetic wave interference on observations of the Earth radiation budget. *Journal of Quantitative Spectroscopy and Radiative Transfer*, *253*, 107157.

Kato, S., Rose, F. G., Rutan, D. A., & Charlock, T. P. (2008). Cloud effects on the meridional atmospheric energy budget estimated from Clouds and the Earth's Radiant Energy System (CERES) data. *Journal of Climate*, *21*, 4223–4241.

Krämer, M., Rolf, C., Spelten, N., Afchine, A., Fahey, D., Jensen, E. et al. (2020). A microphysics guide to cirrus – Part 2: Climatologies of clouds and humidity from observations. *Atmospheric Chemistry and Physics*, *20*(21), 12569–12608.

Lauber, A., Kiselev, A., Pander, T., Handmann, P., & Leisner, T. (2018). Secondary ice formation during freezing of levitated droplets. *Journal of the Atmospheric Sciences*, *75*, 2815–2826.

Lawson, R. P., Woods, S., Jensen, E., Erfani, E., Gurganus, C., Gallagher, M. et al. (2019). A review of ice particle shapes in cirrus formed in situ and in anvils. *Journal of Geophysical Research: Atmospheres*, *124*(17–18), 10049–10090.

Letu, H., Ishimoto, H., Riedi, J., Nakajima, T. Y., -Labonnote, L. C., Baran, A. J., Nagao, T. M., et al. (2016). Investigation of ice particle habits to be used for ice cloud remote sensing for the GCOM-C satellite mission. *Atmospheric Chemistry & Physics*, *16*(18), 12287–12303.

Liou, K.-N., & Yang, P. (2016). *Light scattering by ice crystals: Fundamentals and applications*. Cambridge University Press.

Liu, C., Lee Panetta, R., & Yang, P. (2013). The effects of surface roughness on the scattering properties of hexagonal columns with sizes from the Rayleigh to the geometric optics regimes. *Journal of Quantitative Spectroscopy and Radiative Transfer*, *129*, 169–185.

Liu, C., Panetta, R. L., & Yang, P. (2012). Application of the pseudo-spectral time domain method to compute particle single-scattering properties for size parameters up to 200. *Journal of Quantitative Spectroscopy and Radiative Transfer*, *113*(13), 1728–1740.

Liu, C., Yang, P., Minnis, P., Loeb, N., Kato, S., Heymsfield, A., & Schmitt, C. (2014). A two-habit model for the microphysical and optical properties of ice clouds. *Atmospheric Chemistry and Physics*, *14*, 13719–13737.

Liu, Q. H. (1997). The PSTD algorithm: A time-domain method requiring only two cells per wavelength. *Microwave and Optical Technology Letters*, *15*(3), 158–165.

Loeb, N. G., Kato, S., Loukachine, K., & Smith, N. M. (2005). Angular distribution models for top-of-atmosphere radiative flux estimation from the Clouds and the Earth's Radiant Energy System instrument on the Terra satellite. Part I: Methodology. *Journal of Atmospheric and Oceanic Technology*, *22*, 338–351.

Loeb, N. G., Yang, P., Rose, F. G., Hong, G., Sun-Mack, S., Minnis, P., et al. (2018). Impact of ice cloud microphysics on satellite cloud retrievals and broadband flux radiative transfer model calculations. *Journal of Climate*, *31*, 1851–1864.

Lolli, S. (2021). Is the air too polluted for outdoor activities? check by using your photovoltaic system as an air-quality monitoring device. *Sensors (Basel, Switzerland)*, *21*(19).

Lolli, S., Campbell, J., Lewis, J., Gu, Y., Marquis, J., Chew, B., et al. (2017). Daytime top-of-the-atmosphere cirrus cloud radiative forcing properties at Singapore. *Journal of Applied Meteorology and Climatology*, *56*(5), 1249–1257.

Lolli, S., Di Girolamo, P., Demoz, B., Li, X., & Welton, E. (2017). Rain evaporation rate estimates from dual-wavelength lidar measurements and intercomparison against a model analytical solution. *Journal of Atmospheric and Oceanic Technology*, *34*(4), 829–839.

Lu, R.-S., Tian, G.-Y., Gledhill, D., & Ward, S. (2006). Grinding surface roughness measurement based on the co-occurrence matrix of speckle pattern texture. *Applied Optics*, *45*(35), 8839–8847.

L'Ecuyer, T. S., Wood, N. B., Haladay, T., Stephens, G. L., & Stackhouse, Jr., P. W. (2008). Impact of clouds on atmospheric heating based on the R04 CloudSat flux and heating rate data set. *Journal of Geophysical Research: Atmospheres*, *113*.

Mace, G. G., Zhang, Q., Vaughan, M., Marchand, R., Stephens, G., Trepte, C., & Winker, D. (2009). A description of hydrometeor layer occurrence statistics derived from the first year of merged CloudSat and CALIPSO data. *Journal of Geophysical Research: Atmospheres*, *114*(D8).

Macke, A., Francis, P. N., McFarquhar, G. M., & Kinne, S. (1998). The role of ice particle shapes and size distributions in the single scattering properties of cirrus clouds. *Journal of the Atmospheric Sciences*, *55*(17), 2874–2883.

Macke, A., Mishchenko, M. I., & Cairns, B. (1996). The influence of inclusions on light scattering by large ice particles. *Journal of Geophysical Research: Atmospheres*, *101*(D18), 23311–23316.

Macke, A., Mueller, J., & Raschke, E. (1996). Single scattering properties of atmospheric ice crystals. *Journal of the Atmospheric Sciences*, *53*(19), 2813–2825.

Maeno, N. (1967). Air bubble formation in ice crystals. *Physics of Snow and Ice: Proceedings*, *1*(1), 207–218.

Magee, N. B., Miller, A., Amaral, M., & Cumiskey, A. (2014). Mesoscopic surface roughness of ice crystals pervasive across a wide range of ice crystal conditions. *Atmospheric Chemistry and Physics*, *14*(22), 12357–12371.

Magee, N., Boaggio, K., Staskiewicz, S., Lynn, A., Zhao, X., Tusay, N. et al. (2020). Captured cirrus ice particles in high definition. *Atmospheric Chemistry and Physics*, *2020*, 1–27.

Magono, C., & Lee, C. W. (1966). Meteorological classification of natural snow crystals. *Journal of the Faculty of Science, Hokkaido University. Series 7, Geophysics*, *2*(4), 321–335.

Matus, A. V., & L'Ecuyer, T. S. (2017). The role of cloud phase in Earth's radiation budget. *Journal of Geophysical Research*, *122*(5), 2559–2578.

McFarquhar, G. M., Yang, P., Macke, A., & Baran, A. J. (2002). A new parameterization of single scattering solar radiative properties for tropical anvils using observed ice crystal size and shape distributions. *Journal of the Atmospheric Sciences*, *59*(16), 2458–2478.

McHardy, T. M., Campbell, J. R., Peterson, D. A., Lolli, S., Bankert, R. L., Garnier, A., et al. (2021). Advancing maritime transparent cirrus detection using the advanced baseline imager "cirrus" band. *Journal of Atmospheric and Oceanic Technology*, *38*(6), 1093–1110.

Min, M., Wang, P., Campbell, J. R., Zong, X., & Li, Y. (2010). Midlatitude cirrus cloud radiative forcing over China. *Journal of Geophysical Research: Atmospheres*, *115*(D20).

Mioche, G., Josset, D., Gayet, J.-F., Pelon, J., Garnier, A., Minikin, A., & Schwarzenboeck, A. (2010). Validation of the CALIPSO-CALIOP extinction coefficients from in situ observations in midlatitude cirrus clouds during the CIRCLE-2 experiment. *Journal of Geophysical Research: Atmospheres*, *115*(D4).

Mishchenko, M. I., & Macke, A. (1998). Incorporation of physical optics effects and computation of the Legendre expansion for ray-tracing phase functions involving δ-function transmission. *Journal of Geophysical Research: Atmospheres*, *103*(D2), 1799–1805.

Mishchenko, M. I., & Sassen, K. (1998). Depolarization of lidar returns by small ice crystals: An application to contrails. *Geophysical Research Letters*, *25*(3), 309–312.

Mishchenko, M. I., & Travis, L. D. (1998). Capabilities and limitations of a current FORTRAN implementation of the T-matrix method for randomly oriented, rotationally symmetric scatterers. *Journal of Quantitative Spectroscopy and Radiative Transfer*, *60*(3), 309–324.

Mishchenko, M., Rossow, W., Macke, A., & Lacis, A. (1995). Effect of particle nonsphericity on bidirectional reflectance of cirrus clouds. *In Proceedings of the 1995 ARM Science Meeting* (pp. 19–23). San Diego, CA, March.

Morrison, H., & Gettelman, A. (2008). A new two-moment bulk stratiform cloud microphysics scheme in the Community Atmosphere Model, Version 3 (CAM3). Part I: Description and numerical tests. *Journal of Climate*, *21*(15), 3642–3659.

Mosimann, L., Steiner, M., Collett, J., Henrich, W., Schmid, W., & Waldvogel, A. (1993). Ice crystal observations and the degree of riming in winter precipitation. *Water Air Soil Pollution*, *68*, 29–42.

Murray, B. J., Salzmann, C. G., Heymsfield, A. J., Dobbie, S., Neely, III. R. R., Cox, C. J., et al. (2015). Trigonal ice crystals in Earth's atmosphere. *Bulletin of the American Meteorological Society*, *96*(9): 1519–1531.

Nelson, J., & Swanson, B. D. (2019). Lateral facet growth of ice and snow–part 1: Observations and applications to secondary habits. *Atmospheric Chemistry and Physics*, *19*(24), 15285–15320.

Neshyba, S. P., Lowen, B., Benning, M., Lawson, A., & Rowe, P. M. (2013). Roughness metrics of prismatic facets of ice. *Journal of Geophysical Research (Atmospheres)*, *118*(8), 3309–3318.

Nousiainen, T., & McFarquhar, G. M. (2004). Light scattering by quasi-spherical ice crystals. *Journal of the Atmospheric Sciences*, *61*(18), 2229–2248.

Nousiainen, T., & Muinonen, K. (2007). Surface-roughness effects on single-scattering properties of wavelength-scale particles. *Journal of Quantitative Spectroscopy and Radiative Transfer*, *106*(1–3), 389–397.

Okamoto, H., Sato, K., Borovoi, A., Ishimoto, H., Masuda, K., Konoshonkin, A., & Kustova, N. (2020). Wavelength dependence of ice cloud backscatter properties for space-borne polarization lidar applications. *Optics Express*, *28*(20), 29178.

Ono, A. (1969). The shape and riming properties of ice crystals in natural clouds. *Journal of Atmospheric Sciences*, *26*(1), 138–147.

Pfalzgraff, W. C., Hulscher, R. M., & Neshyba, S. P. (2010). Scanning electron microscopy and molecular dynamics of surfaces of growing and ablating hexagonal ice crystals. *Atmospheric Chemistry and Physics*, *10*(6), 2927–2935.

Pfisterer, R. N. (2011). Approximated scatter models for stray light analysis. *Optics and Photonics News*, *22*(10), 16–17.

Purcell, E. M., & Pennypacker, C. R. (1973). Scattering and absorption of light by nonspherical dielectric grains. *The Astrophysical Journal*, *186*, 705–714.

Qian, Y., Wan, H., Yang, B., Golaz, J. C., Harrop, B., Hou, Z., et al. (2018). Parametric sensitivity and uncertainty quantication in the version 1 of E3SM Atmosphere Model based on short Perturbed Parameter Ensemble Simulations. *Journal of Geophysical Research: Atmospheres*, *123*(23), 13,046–13,073.

Ramanathan, V., Cess, R. D., Harrison, E. F., Minnis, P., Barkstrom, B. R., Ahmad, E., & Hartmann, D. (1989).

Cloud-radiative forcing and climate: Results from the Earth Radiation Budget Experiment. *Science*, *243*, 57–63.

Rimmer, J., & Saunders, C. (1997). Radiative scattering by artificially produced clouds of hexagonal plate ice crystals. *Atmospheric Research*, *45*(2), 153–164.

Riskilä, E., Lindqvist, H., & Muinonen, K. (2021). Light scattering by fractal roughness elements on ice crystal surfaces. *Journal of Quantitative Spectroscopy and Radiative Transfer*, *267*, 107561.

Romanov, A. V., & Yurkin, M. A. (2021). Single-particle characterization by elastic light scattering. *Laser and Photonics Reviews*, *2000368*, 1–18.

Saito, M., Yang, P., Heidinger, A. K., & Li, Y. (2020). An improved beta method for ice cloud property retrievals: Theory. *Journal of Geophysical Research (Atmospheres)*, *125*(14), e31863.

Sasaki, Y., Nishiyama, N., & Furukawa, Y. (1998). Experimental study on light scattering from an artificial ice cloud. *Polar Meteorology and Glaciology*, *12*, 130–139.

Sassen, K., Kayetha, V. K., & Zhu, J. (2012). Ice cloud depolarization for nadir and off-nadir CALIPSO measurements. *Geophysical Research Letters*, *39*(20).

Sassen, K., & Liou, K.-N. (1979). Scattering of polarized laser light by water droplet, mixed-phase and ice crystal clouds. Part I: Angular scattering patterns. *Journal of the Atmospheric Sciences*, *36*(5), 838–851.

Sassen, K., Wang, Z., & Liu, D. (2008). Global distribution of cirrus clouds from Cloudsat/Cloud-Aerosol Lidar and Infrared Pathfinder Satellite Observations (CALIPSO) measurements. *Journal of Geophysical Research: Atmospheres* *113*(D8).

Sassen, K., & Zhu, J. (2009). A global survey of CALIPSO linear depolarization ratios in ice clouds: Initial findings. *Journal of Geophysical Research: Atmospheres*, *114*(D4).

Schmidt, G. A., Bader, D., Donner, L. J., Elsaesser, G. S., Golaz, J.-C., Hannay, C. et al. (2017). Practice and philosophy of climate model tuning across six US modeling centers. *Geoscientific Model Development*, *10*(9), 3207–3223.

Schmitt, C. G., & Heymsfield, A. J. (2007). On the occurrence of hollow bullet rosette- and column-shaped ice crystals in midlatitude cirrus. *Journal of the Atmospheric Sciences*, *64*(12), 4514–4519.

Schmitt, C. G., & Heymsfield, A. J. (2014). Observational quantification of the separation of simple and complex atmospheric ice particles. *Geophysical Research Letters*, *41*(4), 1301–1307.

Schmitt, C. G., Heymsfield, A. J., Connolly, P., Järvinen, E., & Schnaiter, M. (2016). A global view of atmospheric ice particle complexity. *Geophysical Research Letters*, *43*(22), 11,913–11,920.

Schmitt, C. G., Stuefer, M., Heymsfield, A. J., & Kim, C. K. (2013). The microphysical properties of ice fog measured in urban environments of Interior Alaska. *Journal of Geophysical Research Atmospheres*, *118*(19), 11136–11147.

Schnaiter, M., Büttner, S., Möhler, O., Skrotzki, J., Vragel, M., & Wagner, R. (2012). Influence of particle size and shape on the backscattering linear depolarisation ratio of small ice crystals–cloud chamber measurements in the context of contrail and cirrus microphysics. *Atmospheric Chemistry and Physics*, *12*(21), 10465–10484.

Schnaiter, M., Järvinen, E., Abdelmonem, A., & Leisner, T. (2018). PHIPS_HALO: The airborne particle habit imaging and polar scattering probe – Part 2: Characterization and first results. *Atmospheric Measurement Techniques*, *11*(1), 341–357.

Schnaiter, M., Järvinen, E., Vochezer, P., Abdelmonem, A., Wagner, R., Jourdan, O. et al. (2016). Cloud chamber experiments on the origin of ice crystal complexity in cirrus clouds. *Atmospheric Chemistry and Physics*, *16*(8), 5091–5110.

Shcherbakov, V., Gayet, J.-F., Jourdan, O., Minikin, A., Ström, J., & Petzold, A. (2005). Assessment of cirrus cloud optical and microphysical data reliability by applying statistical procedures. *Journal of Atmospheric and Oceanic Technology*, *22*(4), 409–420.

Shcherbakov, V., Gayet, J.-F., Jourdan, O., Ström, J., & Minikin, A. (2006). Light scattering by single ice crystals of cirrus clouds. *Geophysical Research Letters*, *33*(15), L15809.

Soden, B. J., Broccoli, A. J., & Hemler, R. S. (2004). On the use of cloud forcing to estimate cloud feedback. *Journal of Climate*, *17*, 3661–3665.

Sohn, B. J., Nakajima, T., Satoh, M., & Jang, H.-S. (2010). Impact of different definitions of clear-sky flux on the determination of longwave cloud radiative forcing: NICAM simulation results. *Atmospheric Chemistry and Physics*, *10*, 11641–11646.

Stephens, G. L., Gabriel, P. M., & Partain, P. T. (2001). Parameterization of atmospheric radiative transfer. Part I: Validity of simple models. *Journal of the Atmospheric Sciences*, *58*(22), 3391–3409.

Stephens, G. L., Tsay, S.-C., Stackhouse, Paul W., J., & Flatau, P. J. (1990). The relevance of the microphysical and radiative properties of cirrus clouds to climate and climatic feedback. *Journal of the Atmospheric Sciences*, *47*(14), 1742–1754.

Stoelinga, M. T., Locatelli, J. D., Woods, C. P., Stoelinga, M. T., Locatelli, J. D., & Woods, C. P. (2007). The occurrence of "irregular" ice particles in stratiform clouds. *Journal of the Atmospheric Sciences*, *64*(7), 2740–2750.

Su, W., Corbett, J., Eitzen, Z., & Liang, L. (2015). Next-generation angular distribution models for top-of-atmosphere radiative flux calculation from CERES instruments: Validation. *Atmospheric Measurement Techniques*, *8*.

Sun, M., Doelling, D. R., Loeb, N. G., Scott, R. C., Wilkins, J., Nguyen, L. T., & Mlynczak, P. (2022). Clouds and the Earth's Radiant Energy System (CERES) FluxByCldTyp edition 4 data product. *Journal of Atmospheric and Oceanic Technology*, *39*(3), 303–318.

Sun, W., Fu, Q., & Chen, Z. (1999). Finite-difference time-domain solution of light scattering by dielectric particles with a perfectly matched layer absorbing boundary condition. *Applied Optics*, *38*(15), 3141–3151.

Taflove, A., & Umashankar, K. R. (1990). The finite-difference time-domain method for numerical modeling of electromagnetic wave interactions. *Electromagnetics*, *10*(1–2), 105–126.

Takahashi, T. (1993). High ice crystal production in winter cumuli over the Japan Sea. *Geophysical Research Letters*, *20*(6), 451–454.

Takano, Y., & Liou, K. (1995). Radiative transfer in cirrus clouds. Part III: Light scattering by irregular ice crystals. *Journal of Atmospheric Sciences*, *52*(7), 818–837.

Takano, Y., Liou, K., & Minnis, P. (1992). The effects of small ice crystals on cirrus infrared radiative properties. *Journal of Atmospheric Sciences*, *49*(16), 1487–1493.

Tang, G., Panetta, R. L., Yang, P., Kattawar, G. W., & Zhai, P.-W. (2017). Effects of ice crystal surface roughness and air bubble inclusions on cirrus cloud radiative properties from remote sensing perspective. *Journal of Quantitative Spectroscopy and Radiative Transfer*, *195*, 119–131. Laser-light and Interactions with Particles 2016.

Tape, W., & Moilanen, J. (2006). Atmospheric Halos and the search for Angle X. (Vol. 58). Washington, D. C: American Geophysical Union.

Thorsen, T. J., Fu, Q., & Comstock, J. M. (2013). Cloud effects on radiative heating rate profiles over darwin using arm and A-Train radar/lidar observations. *Journal of Geophysical Research: Atmospheres*, *118*(11), 5637–5654.

Twomey, S. (1977). *Atmospheric aerosols*. Elsevier Scientific publishing Co.

Ulanowski, Z., Hesse, E., Kaye, P. H., & Baran, A. J. (2006). Light scattering by complex ice-analogue crystals. *Journal of Quantitative Spectroscopy and Radiative Transfer*, *100*(1), 382–392. VIII Conference on Electromagnetic and Light Scattering by Nonspherical Particles.

Ulanowski, Z., Kaye, P. H., Hirst, E., Greenaway, R. S., Cotton, R. J., Hesse, E., & Collier, C. T. (2014). Incidence of rough and irregular atmospheric ice particles from Small Ice Detector 3 measurements. *Atmospheric Chemistry and Physics*, *14*(3), 1649–1662.

Um, J., & McFarquhar, G. M. (2007). Single-scattering properties of aggregates of bullet rosettes in cirrus. *Journal of Applied Meteorology and Climatology*, *46*(6), 757.

Um, J., & McFarquhar, G. M. (2011). Dependence of the single-scattering properties of small ice crystals on idealized shape models. *Atmospheric Chemistry and Physics*, *11*(7), 3159–3171.

Um, J., McFarquhar, G. M., Hong, Y. P., Lee, S.-S., Jung, C. H., Lawson, R. P., & Mo, Q. (2015). Dimensions and aspect ratios of natural ice crystals. *Atmospheric Chemistry and Physics*, *15*(7), 3933–3956.

van Diedenhoven, B. (2014). The prevalence of the 22° halo in cirrus clouds. *Journal of Quantitative Spectroscopy and Radiative Transfer*, *146*, 475–479.

van Diedenhoven, B. (2018). Remote sensing of crystal shapes in ice clouds, In: Kokhanovsky, A. (eds) *Springer Series in Light Scattering*. Springer Series in Light Scattering. Springer, Cham.

van Diedenhoven, B. (2021). Variation of ice microphysical properties with temperature and humidity at tops of convective clouds. *Geophysical Research Letters*, *48*(16), e2021GL093673.

van Diedenhoven, B., Ackerman, A. S., Fridlind, A. M., Cairns, B., & Riedi, J. (2020). Global statistics of ice microphysical and optical properties at tops of optically thick ice clouds. *Journal of Geophysical Research (Atmospheres)*, *125*(6), e31811.

van Diedenhoven, B., Cairns, B., Geogdzhayev, I. V., Fridlind, A. M., Ackerman, A. S., Yang, P., & Baum, B. A. (2012). Remote sensing of ice crystal asymmetry parameter using multi-directional polarization measurements - Part 1: Methodology and evaluation with simulated measurements. *Atmospheric Measurement Techniques*, *5*, 2361–2374.

van Diedenhoven, B., Fridlind, A. M., Ackerman, A. S., & Cairns, B. (2012). Evaluation of hydrometeor phase and ice properties in cloud-resolving model simulations of tropical deep convection using radiance and polarization measurements. *Journal of Atmospheric Sciences*, *69*(11), 3290–3314.

Vogelmann, A. M., & Ackerman, T. P. (1995). Relating cirrus cloud properties to observed fluxes: A critical assessment. *Journal of the Atmospheric Sciences*, *52*(23), 4285–4301.

Voigtländer, J., Chou, C., Bieligk, H., Clauss, T., Hartmann, S., Herenz, P. et al. (2018). Surface roughness during depositional growth and sublimation of ice crystals. *Atmospheric Chemistry and Physics*, *18*(18), 13687–13702.

Walden, V. P., Warren, S. G., & Tuttle, E. (2003). Atmospheric ice crystals over the Antarctic Plateau in winter. *Journal of Applied Meteorology*, *42*(10), 1391–1405.

Wang, Y., Yang, P., Hioki, S., King, M. D., Baum, B. A., Di Girolamo, L., & Fu, D. (2019). Ice cloud optical thickness, effective radius, and ice water path inferred from fused MISR and MODIS measurements based on a pixel-level optimal ice particle roughness model. *Journal of Geophysical Research (Atmospheres)*, *124*(22), 12,126–12,140.

Waterman, P. C. (1971). Symmetry, unitarity, and geometry in electromagnetic scattering, *Physical Review D*, *3*(4), 825.

Welton, E. J., Campbell, J. R., Spinhirne, J. D., & Stanley Scott III, V. (2001). Global monitoring of clouds and aerosols using a network of micropulse lidar systems, *SPIE Asia-Pacific Remote Sensing*.

Wielicki, B. A., Barkstrom, B. R., Harrison, E. F., Lee, R. B., Smith, G. L., & Cooper, J. E. (1996). Clouds and the Earth's Radiant Energy System (CERES): An Earth observing system experiment. *Bulletin of the American Meteorological Society*, *77*, 853–868.

Woods, S., Lawson, R. P., Jensen, E., Bui, T., Thornberry, T., Rollins, A., et al. (2018). Microphysical properties of tropical tropopause layer cirrus. *Journal of Geophysical Research: Atmospheres*, *123*(11), 6053–6069.

Xie, Y., Yang, P., Kattawar, G. W., Minnis, P., & Hu, Y. X. (2009). Effect of the inhomogeneity of ice crystals on retrieving ice cloud optical thickness and effective particle size. *Journal of Geophysical Research: Atmospheres*, *114*(D11).

Xu, G., Schnaiter, M., & Järvinen, E. (2022). Accurate retrieval of asymmetry parameter for large and complex ice crystals from in-situ polar nephelometer measurements. *Journal of Geophysical Research: Atmospheres*, *127*(3), e2021JD036071.

Yang, P., Bi, L., Baum, B. A., Liou, K.-N., Kattawar, G. W., Mishchenko, M. I., & Cole, B. (2013). Spectrally consistent scattering, absorption, and polarization properties of atmospheric ice crystals at wavelengths from 0.2 to 100 μm. *Journal of the Atmospheric Sciences*, *70*(1), 330–347.

Yang, P., Hioki, S., Saito, M., Kuo, C.-P., Baum, B., & Liou, K.-N. (2018). A review of ice cloud optical property models for passive satellite remote sensing. *Atmosphere*, *9*(12), 499.

Yang, P., & Liou, K. (1996a). Finite-difference time domain method for light scattering by small ice crystals in three-dimensional space. *JOSA A*, *13*(10), 2072–2085.

Yang, P., & Liou, K. (1996b). Geometric-optics–integral-equation method for light scattering by nonspherical ice crystals. *Applied Optics*, *35*(33), 6568–6584.

Yang, P., & Liou, K. (1998a). Single-scattering properties of complex ice crystals in terrestrial atmosphere. *Beitrage zur Physik der Atmosphare-Contributions to Atmospheric Physics*, *71*(2), 223–248.

Yang, P., & Liou, K.-N. (1998b). *Single-scattering properties of complex ice crystals in terrestrial atmosphere*. Germany.

Yang, P., Zhang, L., Hong, G., Nasiri, S. L., Baum, B. A., Huang, H., et al. (2007). Differences between collection 4 and 5 MODIS ice cloud optical/microphysical products and their impact on radiative forcing simulations. *IEEE Transactions on Geoscience and Remote Sensing*, *45*(9), 2886–2899.

Yi, B., Rapp, A. D., Yang, P., Baum, B. A., & King, M. D. (2017). A comparison of aqua MODIS ice and liquid water cloud physical and optical properties between collection 6 and collection 5.1: Cloud radiative effects. *Journal of Geophysical Research: Atmospheres*, *122*(8), 4550–4564.

Yi, B., Yang, P., Baum, B. A., L'Ecuyer, T., Oreopoulos, L., Mlawer, E. J., et al. (2013). Influence of ice particle surface roughening on the global cloud radiative effect. *Journal of the Atmospheric Sciences*, *70*(9), 2794–2807.

Yorks, J. E., Hlavka, D. L., Hart, W. D., & McGill, M. J. (2011). Statistics of cloud optical properties from airborne lidar measurements. *Journal of Atmospheric and Oceanic Technology*, *28*, 869–883.

Yurkin, M. A., & Hoekstra, A. G. (2011). The discrete-dipole-approximation code ADDA: Capabilities and known limitations. *Journal of Quantitative Spectroscopy and Radiative Transfer*, *112*(13), 2234–2247.

Yurkin, M. A., Hoekstra, A. G., Brock, R. S., & Lu, J. Q. (2007). Systematic comparison of the discrete dipole approximation and the finite difference time domain method for large dielectric scatterers. *Optics Express*, *15*(26), 17902–17911.

Zhang, C., Harrington, J. Y., Zhang, C., & Harrington, J. Y. (2014). Including surface kinetic effects in simple models of ice vapor diffusion. *Journal of the Atmospheric Sciences*, *71*(1), 372–390.

Zhang, M., Teng, S., Di, D., Hu, X., Letu, H., Min, M., & Liu, C. (2020). Information content of ice cloud properties from multi-spectral, -angle and -polarization observations. *Remote Sensing*, *12*(16), 2548.

Zhang, Y., Macke, A., & Albers, F. (1999). Effect of crystal size spectrum and crystal shape on stratiform cirrus radiative forcing. *Atmospheric Research*, *52*(1), 59–75.

Zhao, W., Peng, Y., Wang, B., Yi, B., Lin, Y., & Li, J. (2018). Comparison of three ice cloud optical schemes in climate simulations with community atmospheric model version 5. *Atmospheric Research*, *204*(Jan), 37–53.

Zhao, X., Lin, Y., Peng, Y., Wang, B., Morrison, H., & Gettelman, A. (2017). A single ice approach using varying ice particle properties in global climate model microphysics. *Journal of Advances in Modeling Earth Systems*, *9*(5), 2138–2157.

4

Cloud-Radiation Interactions and Cloud-Climate Feedbacks From an Active-Sensor Satellite Perspective

Grégory V. Cesana[1,2], Andrew S. Ackerman[2], Thibault Vaillant de Guélis[3,4], and David S. Henderson[5]

ABSTRACT

Clouds are ubiquitous in the troposphere. Their interactions with radiation may result in either a warming or a cooling of the Earth system and generate diverse climate feedbacks. The vertical structure of the radiative effects of clouds as well as the response of clouds to global warming (i.e., the cloud feedbacks) are inadequately constrained within the diversity of current climate models, which limits our ability to project the magnitude of future warming. In this chapter, we show how relatively recent active-sensor space-borne observations have narrowed constraints on cloud feedbacks. The value added beyond what can be retrieved from passive sensors is only just beginning to be exploited.

4.1. INTRODUCTION

The interactions between cloud particles and radiation modulate the amount of energy that is reflected, emitted, and absorbed by the Earth. Changes in cloud properties (e.g., areal and vertical extent, temperature, emissivity, opacity, and absorptivity profiles) modify their radiative effects, which, in turn, can profoundly affect the radiative balance of the Earth-atmosphere system (Kiehl & Trenberth, 1997; Wielicki et al., 1996) because clouds cover a large part of the troposphere (Cesana & Waliser, 2016; Stubenrauch et al., 2013). For example, liquid clouds are typically more opaque than their frozen counterparts, which corresponds to more reflection of shortwave (SW) radiation and also more absorption and emission of longwave (LW) radiation for a given profile of water content and temperature (e.g., Cesana & Storelvmo, 2017; Rogers & Yau, 1989).

For the past few decades, passive-sensor satellites have been monitoring cloud properties (e.g., Rossow & Schiffer, 1999) and the outgoing and incoming radiative fluxes at the top of the atmosphere (e.g., Wielicki et al., 1996). More recently, active sensors onboard the Cloud-Aerosol Lidar and Infrared Pathfinder Satellite Observations (CALIPSO; Winker et al., 2010) and CloudSat (Stephens et al., 2002) flying in the A-Train Constellation have provided detailed vertical distribution of cloud properties. Combining these new observations with passive-sensor information from other A-Train satellites has made it possible to characterize radiative fluxes at a high vertical resolution for different types of clouds (Hang et al., 2019; L'Ecuyer et al., 2019; Matus & L'Ecuyer, 2017) and has improved previous estimates of surface and top-of-atmosphere (TOA) radiative fluxes (Stephens et al., 2012). The use of such active and passive remote sensing platforms for precipitation is outlined in Chapter 15: Satellite Precipitation Measurements.

To better understand past, present, and future climates, we partly rely on Earth system models (ESMs), which

[1]*Center for Climate Systems Research, Columbia University, New York, NY, USA*

[2]*NASA Goddard Institute for Space Studies, New York, NY, USA*

[3]*Science Systems and Applications, Inc., Hampton, VA, USA*

[4]*NASA Langley Research Center, Hampton, VA, USA*

[5]*Space Science and Engineering Center, University of Wisconsin–Madison, Madison, WI, USA*

Clouds and Their Climatic Impacts: Radiation, Circulation, and Precipitation, Geophysical Monograph 281, First Edition.
Edited by Sylvia C. Sullivan and Corinna Hoose.

DOI: 10.1002/9781119700357.ch04

are a simplified version of nature and whose utility and plausibility can be evaluated to some extent through rigorous comparison with pertinent observations. Because of their coarse horizontal mesh of O(100 km), climate models by design do not resolve cloud processes on smaller scales, instead relying on parameterizations that represent the expressions of subgrid processes on the gridbox scale. Often times, such oversimplification results in large cloud biases (e.g., Klein et al., 2013; Zhang et al., 2005). With its more detailed vertical information, CloudSat-CALIPSO observations have exposed biases in climate models (e.g., Cesana & Waliser, 2016; Jiang et al., 2012; Nam et al., 2012) beyond the reach of previous satellite observations. However, it is assumed that simulating the current cloud climatology is a necessary but not sufficient condition to build confidence in the ability of the models to reliably project future climate change.

When subjected to "external" forcings, such as anthropogenic changes in atmospheric greenhouse gases, the atmosphere and surface warm at a rate determined not only by the forcing itself, but also by positive and negative feedbacks, defined as changes in the climate system in response to the forcing that respectively amplify and reduce the warming. Of all the feedbacks (i.e., Planck, water vapor, temperature lapse rate, surface albedo), those associated with changes in clouds in both the current and previous Coupled Models Intercomparison Project (CMIP) generations remain the largest source of uncertainty in Earth's equilibrium climate sensitivity (ECS) in ESMs, which is a measure of the warming resulting from a doubling of CO_2 levels (Andrews et al., 2012; Bony & Dufresne, 2005; Caldwell et al., 2016; Vial et al., 2013; Zelinka et al., 2016, 2020). Such diverse behavior limits our ability to project the magnitude of future climate change impacts. Being able to simulate cloud properties accurately is therefore critical to narrowing down the uncertainty in cloud feedbacks and their associated spread in models' ECS, and, ultimately, improving our confidence in projections of future climate change impacts.

In this chapter, we focus on recent advances made in monitoring cloud-radiation interactions from active-sensor satellite observations (section 4.2) and how these observations have been used to estimate and improve our understanding of cloud feedbacks (section 4.3).

4.2. CLOUD-RADIATION INTERACTIONS

4.2.1. Cloud Radiative Heating Rates

Although passive-sensor satellites have been monitoring the outgoing and incoming radiative fluxes at TOA for an extended period of time (e.g., Wielicki et al., 1996), vertically resolved profiles of such fluxes have been lacking. L'Ecuyer et al. (2009) first developed a new radiative flux retrieval based on CloudSat measurements and a radiative transfer model. A few years later, this product integrated other A-Train satellite measurements from CALIPSO (Winker et al., 2010) and Moderate Resolution Imaging Spectroradiometer (MODIS, King et al., 2013) to take into account the contribution of thin cirrus and near-surface shallow clouds and aerosols (the 2B-FLXHR-LIDAR product, Henderson et al., 2013). In parallel, another combined product was developed by an independent group also based on CloudSat-CALIPSO-MODIS but with the addition of Clouds and the Earth's Radiant Energy System (CERES) information: the CERES-CALIPSO-CloudSat-MODIS (CCCM, Kato et al., 2011).

Before the active-sensor satellite era, the evaluation of the radiative effect of clouds on a global scale was limited to the TOA (e.g., Wang & Su, 2013) and the surface, although with less confidence (e.g., Wild et al., 2015). In addition to providing a more complete view of the radiative fluxes at both the surface and TOA (Stephens et al., 2012), these aforementioned active-sensor observations provide new opportunities to document the radiative heating rate (RHR) contribution of the clouds at an unprecedented vertical resolution (Haynes et al., 2013) and thereby enable further evaluation of climate models on a global scale (Cesana et al., 2019a). Figure 4.1 shows zonal profiles of cloud RHR for the A-Train observations, the average of five ESMs and its bias (Cesana et al., 2019a). The simulated clouds produce too little warming in the lowest levels compared to the observations (blue shading in the bottom plot) as opposed to an excess of warming in the high levels, particularly in the tropics. Both of these biases are driven by the infrared component and explained by biases in cloud properties. In the low levels, the cloud amount and cloud top height are underestimated by the models, while in the high levels the models produce excessive cloud fraction and ice water content compared to the observations (Cesana et al., 2019a). The cloud RHR biases highlighted here are likely to cause an amplification of cloud biases in climate models. For example, the excess of warming in the high levels may modify the convection (Li et al., 2016), whereas the lack of warming near the surface may prevent the clouds from lifting up and reinforce the cloud biases (Brient & Bony, 2012).

4.2.2. Cloud Radiative Effects as a Function of the Cloud Phase

Combined A-Train observations make it possible to quantify cloud radiative effects (CREs) not only at the surface, within the atmosphere, and at the TOA but

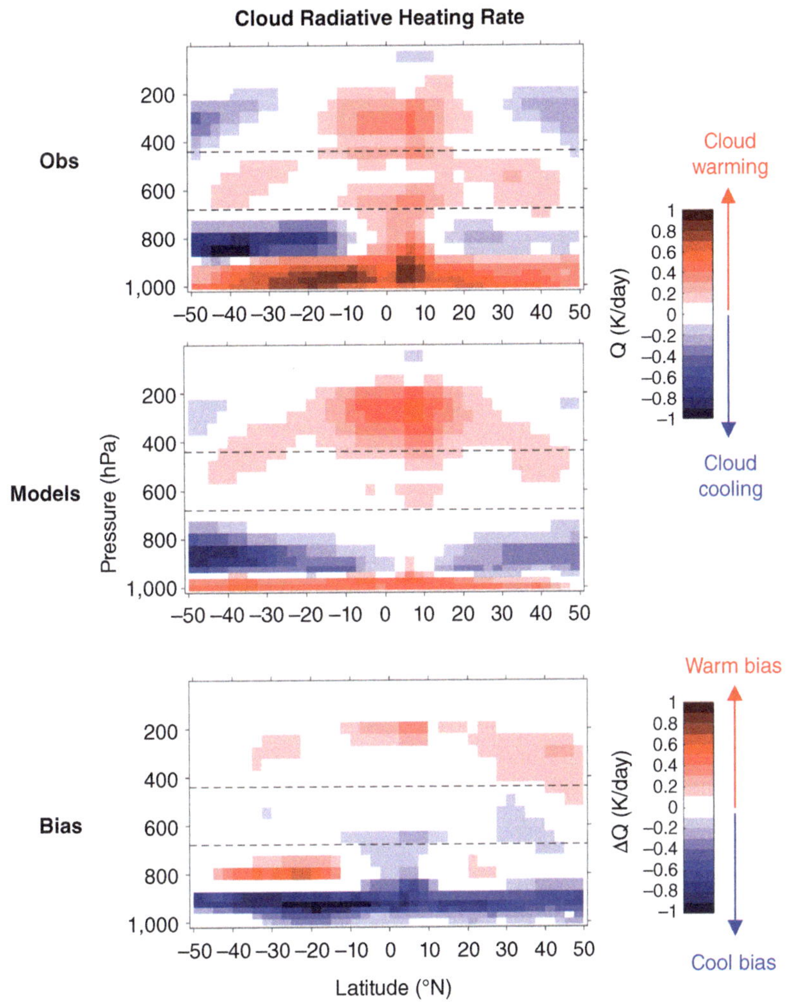

Figure 4.1 Cloud radiative heating rate biases in climate models. Zonal profiles (x-axis, latitude [°]; y-axis, pressure [hPa]) of annual mean total cloud radiative heating rate (K/day) for the A-Train observations (top, 2007–2010 daytime and nighttime, monthly files) for the multimodel mean (middle, 5 models, 1991–2008), and the multimodel mean bias (bottom). Horizontal black dashed lines separate the low- and mid-level clouds (680 hPa), and mid- and high-level clouds (440 hPa). The red and blue shadings designate cooling and warming, respectively. To highlight significant model error estimates, only biases larger than the observed maximum uncertainty estimates are shown in the bottom plot. Note that simulated data poleward 50° were not requested in this model experiment. Source: Modified from Cesana et al. (2019a).

also as a function of the cloud phase. Cloud phase is of particular interest since liquid and ice clouds generally have very distinct radiative properties. Clouds composed of liquid drops have greater numbers of cloud particles that are smaller in size than their frozen counterparts for a given water content. As a result, their optical depths are larger than that of the ice clouds (Rogers & Yau, 1989), which results in more reflection of solar radiation and also more emission of infrared radiation for a cloud at the same temperature (e.g., Cesana & Storelvmo, 2017).

At middle and high latitudes, when the temperature drops below 0 °C, clouds that contain both supercooled liquid water and ice are referred to as mixed-phase clouds (see also Chapter 5: A Review of the Factors Influencing Arctic Mixed-Phase Clouds). These clouds are usually topped by a thin supercooled layer with underlying ice crystals that continue to grow below the liquid cloud base where the air is supersaturated with respect to ice (e.g., Fridlind & Ackerman, 2018). Such clouds can be challenging to detect with passive-sensor satellites (e.g., Cesana et al., 2016; Stubenrauch et al., 2013) and climate models struggle to accurately represent them (Cesana et al., 2012, 2015; Klein et al., 2009; Komurcu et al., 2014; McIlhattan et al., 2020). While mixed-phase clouds are estimated to only account for 10% of total cloud cover (less than either type of single-phase cloud), their net radiative impact relative to all clouds is estimated to be around 20% (Matus & L'Ecuyer, 2017), based on 2B-FLXHR-LIDAR observations. This product uses lidar and radar signal intensity differences between liquid and ice particles, relying on the fact that the lidar is highly sensitive to small particles (i.e., droplets) while the radar is more sensitive to larger particles (i.e., ice crystals). In addition, mixed-phase clouds also influence atmospheric and even oceanic circulations (Matus & L'Ecuyer, 2017). Thus, assessing their presence and radiative effect in climate models is an important step toward improving climate simulations and how clouds could respond to climate warming, which now may be undertaken using these new active-sensor based observations.

For example, Figure 4.2 quantifies the TOA cloud radiative fluxes as observed by CERES observations and partitioned by phase as observed by CALIPSO-GOCCP in each grid box. The CALIPSO-GOCCP cloud-phase observations (Cesana & Chepfer, 2013) provide 333 m-along-track-resolution near-nadir lidar profiles for 480 m height intervals. Unlike 2B-FLXHR-LIDAR, which uses the different sensitivities of the lidar and radar to small and large particles (typically water droplets and ice crystals, respectively) to determine the cloud phase, CALIPSO-GOCCP utilizes the state of polarization of the lidar beam to separate between liquid and ice clouds, which results in a good agreement with airborne in situ measurements (Cesana et al., 2016). A non-spherical ice crystal changes the polarization state of the lidar return, unlike a spherical liquid droplet. Whether a gridbox is dominated by ice or liquid-containing clouds, which often correspond to mixed-phase clouds at subfreezing temperatures (Silber et al., 2020), is determined by the ratio of ice to ice plus liquid cloud frequencies (above and below one half for ice or liquid dominated clouds, respectively). The results clearly show distinct contribution of the two-phase regimes in terms of TOA radiative forcings, with differences up to 60 W/m^2 (Fig. 4.2, right).

4.2.3. Multilayered Clouds

Finally, by documenting nearly all vertical levels of the troposphere for every profile, the combined A-Train observations (e.g., 2B-FLXHR-LIDAR) also allow the distinction of radiative contributions from single-layer and multilayered clouds. Using this capability, recent studies (Hang et al., 2019; L'Ecuyer et al., 2019; Matus & L'Ecuyer, 2017) have shown that multilayered clouds, which correspond to multiple non-contiguous cloudy layers in the same profile, are ubiquitous and exert a strong influence on TOA energy balance and atmospheric heating. More specifically, they report that the largest contribution to global CRE at TOA and two third of atmospheric radiative heating is associated with multilayered clouds, which are often misclassified as geometrically thick mid-level clouds by passive-sensor instruments. This novel result gives raise to the question of the extent to which radiative biases in ESMs are related to poor representation of multilayered cloud systems, a problem that has been suggested as most likely attributable to the use of oversimplified vertical overlap assumptions (Barker, 2009; Cesana & Waliser, 2016; Hogan & Illingworth, 2000).

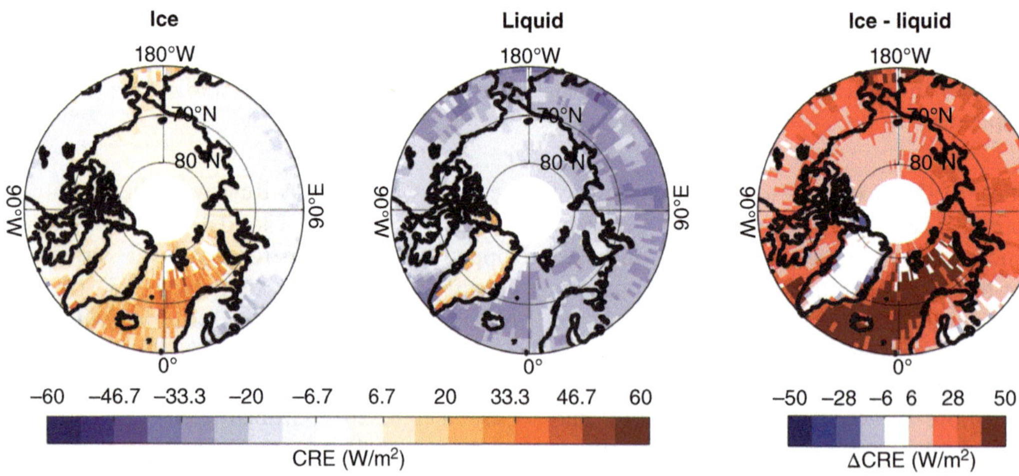

Figure 4.2 Cloud radiative effect as a function of the cloud phase regime. Monthly CERES-EBAF (Loeb et al., 2018) averaged net CRE at the TOA for ice-dominated (left), liquid-dominated gridboxes (middle), and the difference (right) based on CALIPSO-GOCCP frequency phase ratio (2007–2016 daytime and nighttime; Cesana et al., 2016).

4.3. CONSTRAINING CLOUD FEEDBACKS

4.3.1. Different Cloud Feedbacks

Within the context of global warming, the radiative impact of clouds in response to surface temperature variations, referred to as cloud feedback, remains the largest source of uncertainty in ESMs' climate sensitivity (e.g., Andrews et al., 2012; Caldwell et al., 2016; Zelinka et al., 2020). The cloud feedback can be decomposed into three main components (Fig. 4.3, middle column) coming mostly from high and low clouds (cloud top above ~7 km and below ~3 km, respectively). In the CMIP5 models, the first and largest component is the cloud cover feedback, mostly contributed by tropical and midlatitudes low clouds. On average, tropical low clouds are expected to dissipate somewhat, which would result in more incoming solar radiation, reinforcing the surface warming through a positive feedback (with an ensemble mean of +0.35 W m^{-2} K^{-1}). It is also the most uncertain one with values ranging roughly from –1 W m^{-2} K^{-1} to 2 W m^{-2} K^{-1} (e.g., Cesana & Del Genio, 2021; Klein et al., 2017). The second component is the altitude feedback (+0.2 W m^{-2} K^{-1}), which results from changes in altitude while keeping the cloud amount and optical depth fixed. This feedback is positive where high clouds rise higher in response to warming, hampering surface LW cooling. The third is the opacity feedback (–0.11 W m^{-2} K^{-1}), the only negative feedback out of the three, which is due to an increase in cloud optical depth from ice and mixed-phase clouds turning into liquid clouds with warmer temperatures. As a result, the amount of SW radiation reflected back to space is increased, thereby reducing the initial surface temperature warming (negative feedback).

4.3.2. Interannual Cloud Feedbacks

Capturing the mechanisms that govern the change of clouds in response to a surface warming is therefore necessary – but not sufficient – to reliably predict future climate. As a result, numerous studies have been devoted to evaluating the sensitivity of clouds to surface temperature in climate models against observations (e.g., Cesana et al., 2019b; Myers & Norris, 2016; McCoy et al., 2017; Qu et al., 2015). Most of these studies focus on the positive low-cloud feedbacks since multi-model spread of tropical low-cloud feedbacks dominates the uncertainty in model estimates of ECS (Andrews et al., 2012; Bony & Dufresne, 2005; Caldwell et al., 2016; Vial et al., 2013).

For example, the interannual variation of low clouds and their radiative effects (i.e., interannual low-cloud feedbacks) drive the low-cloud feedbacks through their physical relationships with sea surface temperature (SST) and estimated inversion strength (EIS; Qu et al., 2014), the two

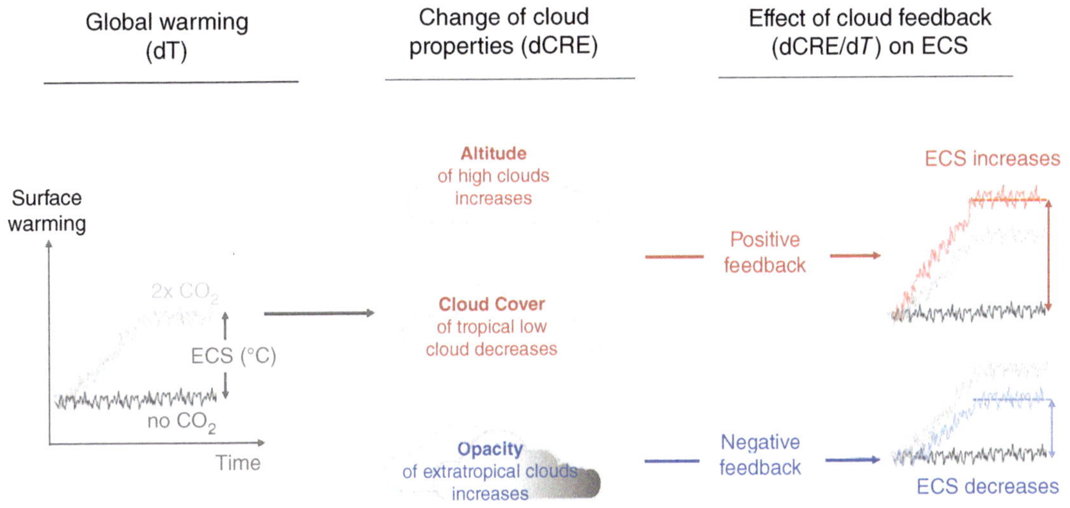

Figure 4.3 Illustration of the main cloud feedbacks and their impacts on climate sensitivity. The global mean surface temperature warms in response to an instantaneous doubling of CO_2 concentrations (light gray line, left column) relative to the pre-industrial atmosphere (dark gray line). At equilibrium, the temperature difference between the two curves defines the equilibrium climate sensitivity (ECS). That anthropogenic-induced warming alters cloud properties (middle column): high clouds rising higher, tropical low clouds dissipating, and extratropical clouds becoming optically thicker. As a consequence, their cloud radiative effects also change and feed back on the surface temperature by either enhancing the initial warming, a positive feedback (altitude and cloud cover feedbacks), or weakening it, a negative feedback (opacity feedback). Ultimately, the positive and negative feedbacks contribute to increasing and decreasing the ECS, respectively (right column).

main cloud-controlling factors (CCF), which appear to be timescale invariant (i.e., to hold in the future climate; Klein & Hall, 2015; Zhou et al., 2015a). These interannual cloud feedbacks can be derived from satellite observations in both the SW and the LW and used to evaluate models (e.g., Fig. 4.4e; Cesana et al., 2019b) and to constrain their low-cloud feedbacks (e.g., Cesana et al., 2019b; Myers & Norris, 2016).

A series of studies focused on the interannual variability of low clouds (described above) have shown that the low-cloud amount as observed from space substantially decreases in response to a surface warming. Such a decrease of low clouds generates a weakening of their negative SW CRE at TOA, which is defined as the difference between clear- and all-sky TOA SW fluxes. As a consequence, less SW radiation is reflected back to space

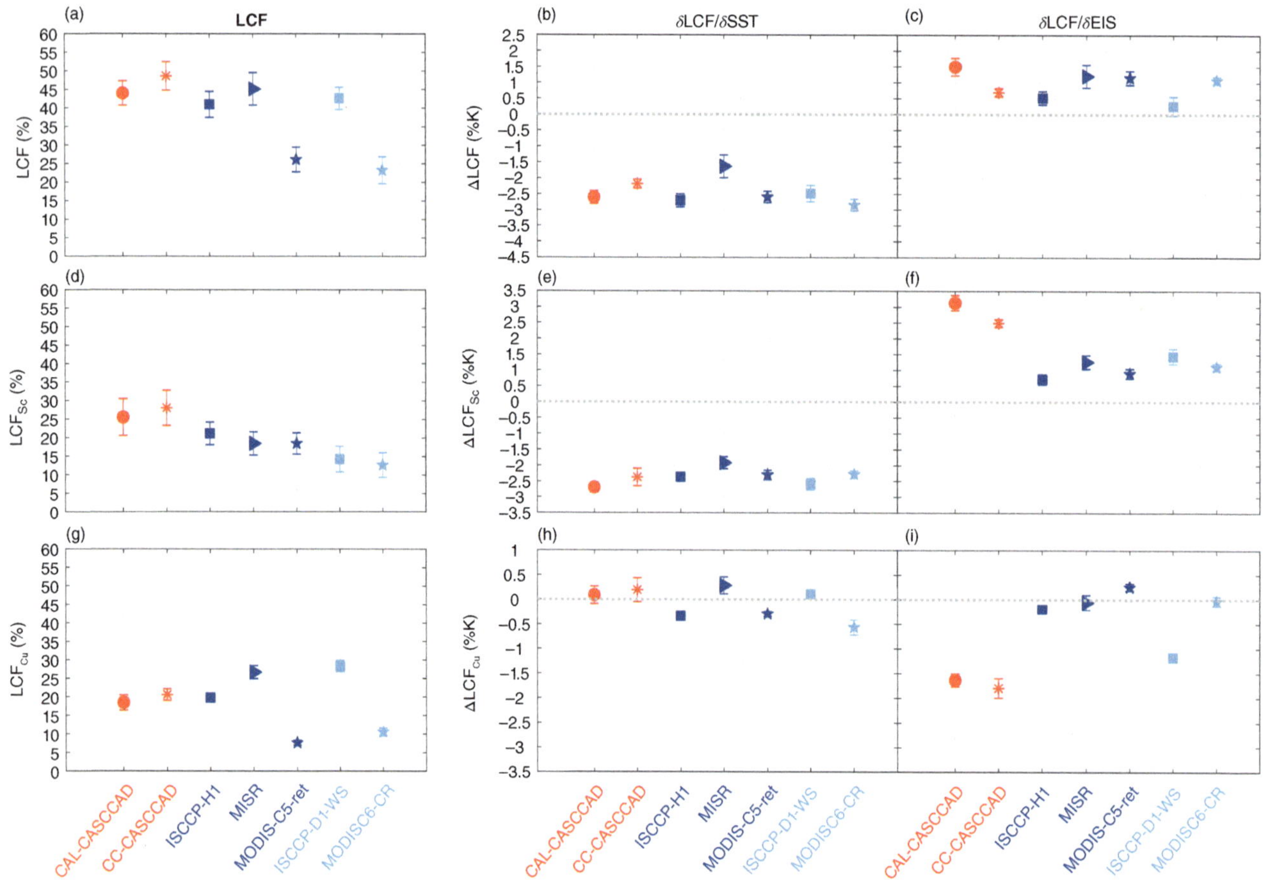

Figure 4.4 Low-cloud sensitivities of Sc- and shallow Cu-dominated to cloud-controlling factors. Observed low-cloud fraction (%) in subsidence regimes over the tropical oceans (ω_{500} > 10 hPa/day) for different satellite datasets (the active-sensor products in red and the passive-sensor products in blue shading) (a) for all low clouds, (d) for stratocumulus-dominated regions and (g) for shallow cumulus-dominated regions. Observed interannual low-cloud change (b, e, h) per K of SST warming with EIS held constant ($\partial LCC/\partial SST$, % K^{-1}), and (c, f, i) per K of EIS increase with SST held constant ($\partial LCC/\partial EIS$, % K^{-1}) for the same satellite products and using six observational and reanalysis products for the SST and EIS datasets. The uncertainty bars correspond to the interannual mean variability for the cloud fractions (left column), and the 10–90% confidence interval using the three SST and EIS datasets for the partial derivatives (middle and right columns). The method to compute these partial derivatives is described in Cesana et al. (2019a). See the text for the different methods used to discriminate Sc and Cu clouds in each product. In order of appearance in the plots, the satellite datasets used are (see also Table 4.1): CALIPSO-CASCCAD (2007–2016) and CloudSat-CALIPSO-CASCCAD (2007–2010; Cesana et al., 2019c), ISCCP-H1 (2007–2016; Rossow and Schiffer, 1999), MISR-V6 (2007–2016; Marchand et al., 2010), MODIS-C5 (2007–2015; Pincus et al., 2012), ISCCP-D1 weather state (1984–2007; Tselioudis et al., 2013), and MODIS-C6 cloud regimes (2003–2014; Oreopoulos et al., 2014). The dark and light blue shadings distinguish two different methods of Sc and Cu discrimination used in passive-sensor observations (see Table 4.1 and text for more details).

Table 4.1 Description of the datasets used in Figure 4.4.

Product	CAL-CASCCAD	CC-CASCCAD	ISCCP-H1	MODIS-C5-retrieval	MISR V6	ISCCP-D1 Weather State	MODIS-C6 Cloud Regime
Satellite	CALIPSO	CloudSat-CALIPSO	Multiple weather and polar-orbiting satellites	Aqua/Terra	Terra	Multiple weather and polar-orbiting satellites	Aqua/Terra
Active/passive sensor	Active	Active	Passive	Passive	Passive	Passive	Passive
Instrument	Lidar (532 nm)	Lidar (532 nm) and radar (94 GHz)	Multispectral imagers	Multispectral imagers	Multispectral + multiangle imagers	Multispectral imagers	Multispectral imagers
Sc-Cu discrimination	Cloud morphology (cloud height, horizontal cloud fraction and vertical cloud fraction variability)	Cloud morphology (cloud height, horizontal cloud fraction and vertical cloud fraction variability)	Cloud top pressure and optical depth	Cloud top pressure and optical depth	Cloud top height and optical depth	Clustering (weather state)	Clustering (cloud regime)
Time period	2007–2016	2007–2010	2007–2016	2007–2015	2007–2016	1984–2007	2003–2014
References	Cesana et al. (2019c)	Cesana et al. (2019c)	Rossow and Schiffer (1999)	Pincus et al. (2015)	Marchand et al. (2010)	Tselioudis et al. (2013)	Oreopoulos et al. (2014)

and low clouds contribute to further warming the system through a positive feedback. Figure 4.4 shows the tropical low-cloud fraction (LCF) in subsidence regimes (ω_{500} > 10 hPa/d) from various active- and passive-sensor satellite products. Although the LCF amplitude (Fig. 4.4a) can vary substantially from one product to another (mainly due to different detection sensitivities, cloud definitions, and retrieval method), its interannual variability with respect to SST and EIS (Fig. 4.4b,c) is robustly depicted by all products. In agreement with previous studies (Cesana et al., 2019b; Cesana & Del Genio, 2021; Myers & Norris, 2016; Zhou et al., 2015a), the LCF largely decreases with increasing SST (Fig. 4.4b; –2.4 %/K on average) whereas it slightly increases with increasing EIS (Fig. 4.4c; 1 %/K on average). On the one hand, an increasing SST corresponds to a stronger vertical gradient of specific humidity, which contributes to drying of the planetary boundary layer (PBL), therefore reducing the cloud amount. On the other hand, a stronger inversion leads to less mixing of the PBL with the free troposphere, which reduces the PBL height and favors cloud formation and persistence. In this example, the added value of active-sensor observations is not obvious, but Cesana et al. (2019b) showed that models with a similar low-cloud interannual variability (e.g., ∂LCF/∂SST) could be further constrained with the interannual variability of global cloud profiles using CALIPSO-GOCCP and the lidar simulator. Much uncertainty remains on how controlling factors affect different low-cloud types, but here again, active-sensor satellite can help to advance understanding of the mechanisms at play.

4.3.3. Sc and Shallow Cu Clouds

Low clouds may be separated into two main categories: stratocumulus (Sc, including stratus clouds) and shallow cumulus (Cu). Sc typically produce nearly overcast conditions off the west coasts of continents, where PBL depth is limited by large-scale subsidence and PBL mixing is primarily driven by cloud top LW cooling, and they reflect much more SW radiation than the underlying sea surface (see also Chapter 7. Tropical Marine Low Clouds). Downwind in the extensive open ocean trade-wind regions further west, the Sc decks transition to more scattered Cu cloud fields sustained by convective updrafts emerging from the loosely coupled surface layer, with a weaker radiative effect. Since the convection in these low clouds is driven by different processes, they are not expected to exhibit the same feedback. Satellite observations hint that low clouds in regions dominated by Sc show somewhat different interannual variations than in regions dominated by shallow Cu (Cesana, et al., 2019b). In addition, idealized large-eddy simulation (LES) studies show that feedbacks from Sc and Cu clouds can be substantially different (Bretherton, 2015). This has been confirmed in a recent satellite-based study over the whole tropics, the Sc cloud produce a much larger positive feedback than the Cu, regardless of the future SST and EIS trends (Cesana & Del Genio, 2021). Yet, understanding of the underlying reasons remains incomplete, particularly for trade-wind Cu, which may even produce negative feedbacks in response to surface warming in LES studies (Narenpitak & Bretherton, 2019).

Without having access to the vertical structure of the low clouds (i.e., with passive-sensor measurements), it is particularly challenging to tell Sc and Cu types apart. This caveat has been alleviated with the use of active-sensor satellites (Cesana et al., 2019c). Figure 4.4 also shows LCFs and interannual variations of Sc- and shallow Cu-dominated regimes from some active- and passive-sensor satellite observations (Table 4.1). The Sc and Cu regimes are determined using three different methods described into detail by Cesana et al. (2019c), and briefly introduced hereafter. Sc and Cu clouds have been commonly distinguished by joint distributions of cloud top pressure and optical thickness in passive-sensor observations (Rossow & Schiffer, 1999), although this method has been shown to misclassify Sc and Cu clouds that have moderate optical thickness (Pincus et al., 2012) and to mix different cloud types together (Mace & Wrenn, 2013). The newer passive-sensor clustering approaches, such as the International Satellite Cloud Climatology Project (ISCCP) weather states (Tselioudis et al., 2013) and MODIS cloud regimes (Oreopoulos et al., 2014), do not discriminate cloud types but rather regimes of prevalent cloud types. As such, they cannot be compared directly to Sc and Cu cloud types. Finally, the new Cumulus And Stratocumulus CloudSat-CALIPSO Dataset (CASCCAD)-based active-sensor observations (both CALIPSO-CASCCAD and CloudSat-CALIPSO-CASCCAD products) separate Sc from shallow Cu clouds at the orbital level (i.e., for every profile) based on their morphology using additional information on cloud height and vertical and horizontal cloud fraction variation, improving Sc-Cu classification compared to previous techniques using passive-sensor satellite observations (Cesana et al., 2019c).

Consistent with previous literature (Bretherton, 2015; Cesana et al., 2019b; Cesana & Del Genio, 2021), the Sc and shallow Cu responses to SST and EIS are substantially different regardless of the satellite observations considered. The Sc cloud fraction largely decreases and increases with increasing SST and EIS, respectively (Fig. 4.4e,f), while shallow Cu cloud fraction barely varies with SST and may either decrease (CASCCAD and ISCCP weather state observations) or remain almost invariant (other passive-sensor observations) with increasing EIS (Fig. 4.4h,i). Although all considered satellite observations more or less agree on the sign and amplitude of the low-cloud interannual variations (Fig. 4.4b,c), only active-sensor CASCCAD observations capture their variability with respect to EIS when decomposed into Sc and Cu cloud regimes. We note that the agreement between active- and passive-sensor observations is far better when Sc and Cu cloud regimes are determined using CALIPSO-CASCCAD for overlapping periods (not shown). Such results – made possible by active-sensor satellite observations – provide a robust observational target for model evaluation (e.g., Cesana et al., 2023) and pave the way for new studies exploring the cloud processes behind the distinct behaviors of Sc and shallow Cu clouds in response to SST and EIS forcings.

4.3.4. Inferring Low-Cloud Feedbacks from Observations

Assuming that the relationships (i.e., the partial derivatives such as Fig. 4.4, middle and right columns) between low clouds and their controlling factors remain effectively identical in a future climate, they may be used to estimate the future change of low clouds in combination with the change in each controlling factors derived from multi-model climate projections (i.e., the difference between future climate and control climate, e.g., Klein et al., 2017; Myers & Norris, 2016). As a result, one can infer an observationally based estimate of the low-cloud change by using the observed partial derivatives of each controlling factor multiplied by their change obtained from future climate projections of multiple models as in Myers and Norris (2016, their fig. 2). Then, the observationally inferred future cloud feedbacks can be found by simply changing the sign of the observationally inferred low-cloud change since the two quantities are almost perfectly negatively correlated (Cesana et al., 2019b; Klein et al., 2017). Using this method and multiple estimates of partial derivatives with respect to SST and EIS from past studies, Klein et al. (2017) found that the tropical global mean cloud feedback is $0.25 \pm 0.18\,W\,m^{-2}\,K^{-1}$ (90% confidence interval), consistent with previous evidence suggesting that the tropical low-cloud net feedback is positive. However, Cesana and Del Genio (2021) show that because Sc and shallow Cu clouds have different responses to controlling factors (see Fig. 4.4) and cover distinct geographic areas (Cesana et al., 2019c), their inferred feedback must be computed separately and added up to estimate the inferred low-cloud feedback. By doing so, the inferred low-cloud feedback and its uncertainty estimates are two times smaller than that computed without separating Sc and shallow Cu contributions as in past studies using passive-sensor observations. Using active-sensor satellite observations, we can now reliably determine the separate sensitivities of Sc and Cu clouds to their controlling factors (Fig. 4.4, middle and bottom rows) and subsequently infer more accurate, less uncertain low-cloud feedbacks (Cesana & Del Genio, 2021) than previous estimates (e.g., Klein et al., 2017).

4.3.5. Extratropical (Opacity) Cloud Feedbacks

Low clouds also produce substantial feedbacks in the extratropics, which can be either positive or negative, in response to changes in their amount and microphysical properties (Tsushima et al., 2006; Zelinka et al., 2016, 2020). Most of the increase in ECS between CMIP5 and CMIP6 models has been linked to extratropical low-cloud feedbacks (Zelinka et al., 2020), which is mainly determined by its SW component. The macrophysical processes leading to a change in cloud amount should be the same as those in the tropics, but the mechanisms involved in the change in microphysical properties remain unclear. Past studies showed that the extratropical cloud feedback greatly depends on the sensitivity of cloud phase and cloud opacity to climate warming (Ceppi et al., 2016; Terai et al., 2016) but these feedbacks remain largely unconstrained in the current generation of ESMs (Zelinka et al., 2020). In a warmer climate, more liquid water would form at the expense of ice in such clouds, thereby leading to a larger optical depth and a longer life cycle (e.g., Cesana & Storelvmo, 2017). As a result, the amount of solar radiation reflected back to space would be increased, weakening the surface warming (a negative feedback), described in the literature as "optical depth feedback." For the foregoing reasons, documenting separately liquid and ice clouds globally and over the polar regions appears essential to better constraining the optical depth feedback, but doing so from passive-sensor satellite observations is particularly challenging (Baum et al., 2012; Cesana et al., 2016). CloudSat-CALIPSO observations offer unique advantages over passive-sensor satellites by providing the vertical structure of clouds over highly reflective surfaces and by discriminating liquid- and ice-dominated clouds, all of which has led to substantial discoveries on extratropical clouds (Bodas-Salcedo et al., 2016; Cesana et al., 2022; Kay & Gettelman, 2009; Kay et al., 2016).

Previous attempts to better understand the cloud optical depth feedback using CALIPSO observation as a constraint on the supercooled liquid fraction in the present-day climate confirmed that the strength of the cloud optical depth feedback is tightly connected to the cloud-phase response to climate warming (Tan et al., 2016). However, although such constraint on present-day climate in ESMs is arguably necessary, it is not sufficient to guarantee a reliable optical depth feedback in ESM simulations of future climate. Therefore, more work is needed to complement traditional analyses of feedbacks by further breaking down the feedbacks into low-cloud regimes and directly evaluating their responses to climate warming through a process-level approach using global-scale satellite observations and reanalysis data. This could be done using the value-added set of CALIPSO observations that discriminate between Sc and shallow Cu clouds, liquid, ice, and mixed-phase clouds and also documents their opacities (Cesana & Chepfer, 2013; Cesana et al., 2019c; Guzman et al., 2017).

4.3.6. High-Level Cloud Feedbacks

With global warming, high-level cloud tops are expected to rise, according to climate models (e.g., Ceppi et al., 2016). Because this altitude increase is accompanied by warming of the troposphere, the outgoing LW emission from high-level clouds is expected to hold steady while the LW surface emission increases from warming of the surface (Hartmann & Larson, 2002), and the combined effect is a positive feedback that further warms the surface (Zelinka & Hartmann, 2010; Zelinka et al., 2016). Although it is possible to detect observational evidence of cloud top pressure change with passive-sensor observations (Norris et al., 2016), doing so requires a long record (25 years), complex data processing to minimize biases and artifacts, and, unfortunately, trends cannot be quantified. In contrast, for a similar record length, cloud top height information derived from active-sensor satellite observations may be used to precisely estimate such a trend (Chepfer et al., 2014; Takahashi et al., 2019). Moreover, active-sensor satellite observations allow for analysis of the interannual variability of cloud profiles, revealing the altitudes at which cloud fraction changes occur (Li et al., 2012; Saint-Lu et al., 2020) and possibly providing estimates of interannual cloud feedbacks. For example, Zhou et al. (2015b) used CALIPSO observations to derive a short-term global net feedback from cirrus clouds, which had been lacking for want of adequate cirrus observations from passive-sensor satellites, and show that this feedback is most likely positive and contributes substantially to total cloud feedback globally.

Another CALIPSO-based study analyzed how changes in cloud properties could affect cloud feedbacks in the LW (mostly driven by non-low clouds, altitude >3 km; e.g., Zelinka et al., 2016) using a novel technique (Vaillant de Guélis et al., 2018). They characterize LW CRE over oceans using five parameters: cloud altitude (Z_T in Fig. 4.5) and cover (C in Fig. 4.5) from opaque and thin clouds separately as well as emissivity (ε in Fig. 4.5) of thin clouds, where opaque clouds are diagnosed when no surface echo is detected (CALIPSO-GOCCP OPAQ, Guzman et al., 2017). This lidar-based method uniquely turns a lidar's limitation – its attenuation for optical thickness greater than 3 – into a strength, by separating these contributions to present-day (short-term) LW cloud feedbacks. Such a separation can also be derived from climate models via a lidar simulator, allowing comparison with observed interannual feedbacks and long-term feedbacks from climate simulations. By doing so over oceans only, Vaillant de Guélis et al. (2018) find that the CALIPSO-derived cloud altitude feedback is positive and dominates the present-day LW cloud feedback in a climate model, consistent with results using long-term LW cloud feedback in ESMs. However, in the observations, its contribution to the present-day LW feedback is smaller than that from changes in the opaque cloud fraction. Their results suggest that LW positive cloud feedback could be overestimated in climate model projections.

Following Vaillant de Guélis et al. (2018), we use the CALIPSO-GOCCP observational dataset to derive LW CRE over both oceans and continents and for a longer record. The top panels of Figure 4.5 show the short-term LW cloud feedbacks from CALIPSO-GOCCP, CloudSat-CALIPSO 2B-FLXHR-LIDAR, and CERES observations (top row), and the lower panels its respective decomposition as a function of the five aforementioned cloud properties for CALIPSO-GOCCP. To estimate the LW cloud feedback, we regress monthly anomalies of deseasonalized LW CRE against monthly anomalies of deseasonalized global mean surface temperatures using three different surface temperature datasets (Berkeley Earth Surface Temperature project, Rohde & Hausfather, 2020; GISTEMP v4, Lenssen et al., 2019; HadCRUT 4.6.0.0, Morice et al., 2012). Although very similar in terms of pattern, the active-sensor based cloud feedbacks are more positive at midlatitudes, particularly over the Southern Ocean and over continents where the detection of mixed-phase clouds and high-level clouds is challenging for passive-sensor instruments. Even though we use observations over oceans and continents and derive the short-term cloud feedback from interannual instead of seasonal variations as in Vaillant de Guélis et al. (2018), we also find that the main contributors to LW cloud feedback are changes in opaque cloud cover and altitude while the other components only modestly contribute. Such results provide additional information on short-term LW CRE and cloud feedback beyond previous passive-sensor based analysis (e.g., Dessler, 2010; Loeb et al., 2020) owing to the lidar's capabilities, determining cloud top heights more accurately than passive sensors and detecting optically thin clouds and broken clouds that are often undetected by passive-sensor instruments (e.g., Stubenrauch et al., 2013) despite being radiatively active in the LW.

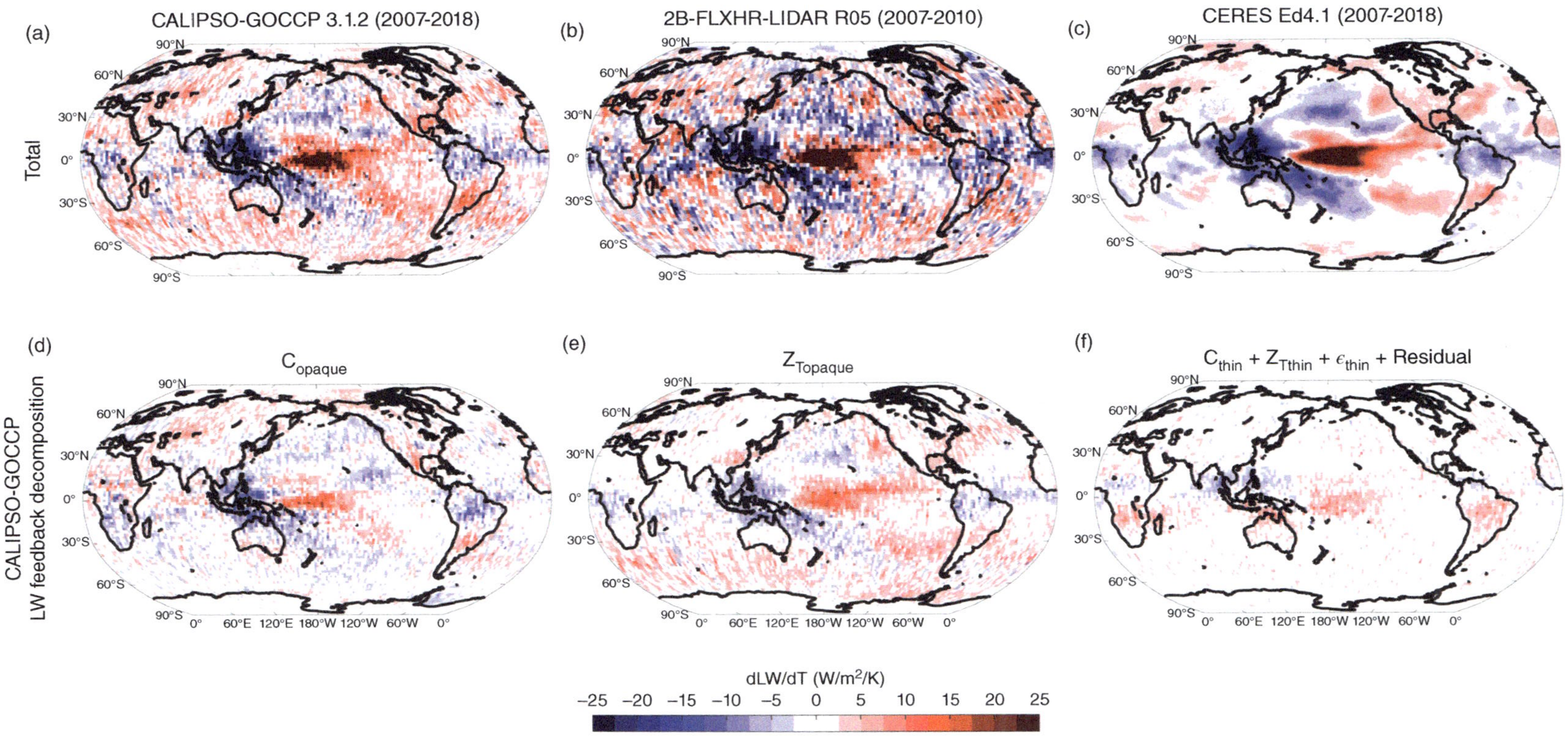

Figure 4.5 Short-term LW cloud feedback for (a) CALIPSO-GOCCP OPAQ (2007–2018; version 3.1.2), (b) 2B-FLXHR-LIDAR (2007–2010; R05), and (c) CERES (2007–2018; version Ed4.1). Contributions from (d) opaque cloud cover, (e) opaque cloud altitude, and (f) thin cloud cover, altitude, emissivity and the residual, all derived from CALIPSO-GOCCP OPAQ. The sum of (d–e–f) panels is equal to (a). Note that changes in opaque cloud cover and altitude are the main contributors to LW cloud feedback.

4.3.7. Radiative Kernel-Based Cloud Feedbacks

Radiative kernels quantify the radiative response of the system to small perturbations in a set of climate variables such as the temperature, water vapor, and albedo. The radiative kernel approach is often used to compute cloud feedbacks in climate models (Soden et al., 2008; Zelinka et al., 2016, 2020) rather than in observations because cloud information is required at a high vertical resolution. Kramer et al. (2019) took advantage of the CloudSat-CALIPSO ability to compile cloud profiles as well as radiative fluxes to observationally derive radiative kernel-based cloud feedbacks in both the LW and the SW, at the surface and TOA, which is a first to our knowledge. Although there are uncertainties related to sampling spatially and within the diurnal cycle, as well as from reliance on reanalysis, this study holds promise for future model evaluation and improved understanding of cloud feedback processes.

4.4. SUMMARY

Cloud-radiation interactions may result in either a warming or a cooling of the Earth system, generating changes in the large-scale circulation, vertical motions, and water cycle. So-called external forcings produce variations of the climate system temperature that modify cloud-radiation interactions by amplifying or reducing temperature changes induced by external forcings, which are called cloud feedbacks. These cloud-radiation interactions, in particular the RHR profiles of clouds as well as the cloud feedbacks, remain largely unconstrained in climate models, which cause a large spread in the multi-model climate sensitivity. Such diverse behavior limits our ability to project the magnitude of future climate change impacts.

Here, we show that the relatively recent active-sensor space-borne observations have improved our understanding of cloud-radiation interactions (L'Ecuyer et al., 2015; Stephens et al., 2012) by giving us access to the RHR profiles (Cesana et al., 2019a; Henderson et al., 2013; L'Ecuyer et al., 2009). It also paved the way to new possibilities such as documenting the atmospheric radiative heating and radiative effects of cloud as a function of the cloud properties. For example, the unique ability of CloudSat-CALIPSO observations to distinguish single-layer and multilayered clouds has revealed the dominant impact of multilayered clouds (Hang et al., 2019; L'Ecuyer et al., 2019; Matus & L'Ecuyer, 2017). Furthermore, these new observations have been used to assess cloud RHRs in climate models for the first time (Cesana et al., 2019a) and will be most likely further used to provide guidance for their future development.

Additionally, we show that active-sensor space-borne observations improve the discrimination of different cloud types compared to passive-sensor satellite observations (e.g., Sc vs. shallow Cu, liquid vs. ice, opaque vs. thin) and allow us to refine understanding of the response of clouds to environmental factors, which can be further used to infer cloud feedbacks with significantly less uncertainty (Cesana & Del Genio, 2021) or to further decompose contributions of cloud properties to short-term cloud feedbacks (e.g., Fig. 4.5; Vaillant de Guélis et al., 2018). These new active-sensor based estimates therefore provide novel constraints on ESM short-term and long-term cloud feedbacks (e.g., Cesana & Del Genio, 2021; Vaillant de Guélis et al., 2018), which have remained the main contributor to the spread in models' climate sensitivity.

In the future, new satellite missions (e.g., Aerosol, Cloud, Convection and Precipitation [ACCP]; Earth Clouds, Aerosols and Radiation Explorer [EarthCARE]) will address part of CloudSat-CALIPSO limitations by improving cloud and aerosol detection, particularly close to the surface, with a more sensitive cloud radar and a High Spectral Resolution Lidar (HSRL) and additional instruments (e.g., multi-angle polarimeter; Knobelspiesse et al., 2020). Moreover, these missions will provide additional useful information (e.g., microphysics, wind speed, diurnal cycle) and extend the current record, making it easier to quantify the impact of anthropogenic forcings on climate. We argue that existing CloudSat-CALIPSO observations, combined with those from future missions, when available, should be used at a process level to improve our understanding of cloud-radiation interactions and cloud feedbacks and therefore better quantify the impact of clouds on climate sensitivity.

ACKNOWLEDGMENTS

Grégory V. Cesana and Andrew S. Ackerman were supported by a NASA CloudSat-CALIPSO RTOP at the NASA Goddard Institute for Space Studies. Grégory V. Cesana was also supported by NOAA grant NA20OAR4310390. Grégory V. Cesana would like to thank NASA and CNES for giving access to CALIPSO and CloudSat observations, and Climserv for giving access to CALIPSO-GOCCP observations and for providing computing resources. Finally, the authors thank Pr. Corinna Hoose for editing the manuscript and the two anonymous reviewers for their helpful comments.

REFERENCES

Andrews, T., Gregory, J. M., Webb, M. J., & Taylor, K. E. (2012). Forcing, feedbacks and climate sensitivity in CMIP5 coupled atmosphere-ocean climate models. *Geophysical Research Letters*, *39*(9), 1–7. https://doi.org/10.1029/2012GL051607

Barker, H. W. (2009). Overlap of fractional cloud for radiation calculations in GCMs: A global analysis using CloudSat and CALIPSO data. *Journal of Geophysical Research Atmospheres*, *114*(8), 1–15. https://doi.org/10.1029/2007JD009677

Baum, B. A., Menzel, W. P., Frey, R. A., Tobin, D. C., Holz, R. E., Ackerman, S. A., et al. (2012). MODIS cloud-top property refinements for collection 6. *Journal of Applied Meteorology and Climatology*, *51*(6), 1145–1163. https://doi.org/10.1175/JAMC-D-11-0203.1

Bodas-Salcedo, A., Hill, P. G., Furtado, K., Williams, K. D., Field, P. R., Manners, J. C., et al. (2016). Large contribution of supercooled liquid clouds to the solar radiation budget of the Southern Ocean. *Journal of Climate*, *29*(11), 4213–4228. https://doi.org/10.1175/JCLI-D-15-0564.1

Bony, S., & Dufresne, J. L. (2005). Marine boundary layer clouds at the heart of tropical cloud feedback uncertainties in climate models. *Geophysical Research Letters*, *32*(20), L20806. https://doi.org/10.1029/2005GL023851

Bretherton, C. S. (2015). Insights into low-latitude cloud feedbacks from high-resolution models. *Philosophical Transactions of the Royal Society A: Mathematical, Physical and Engineering Sciences*, *373*(2054). https://doi.org/10.1098/rsta.2014.0415

Brient, F., & Bony, S. (2012). How may low-cloud radiative properties simulated in the current climate influence low-cloud feedbacks under global warming? *Geophysical Research Letters*, *39*(20), 1–6. https://doi.org/10.1029/2012GL053265

Caldwell, P. M., Zelinka, M. D., Taylor, K. E., & Marvel, K. (2016). Quantifying the sources of intermodel spread in equilibrium climate sensitivity. *Journal of Climate*, *29*(2), 513–524. https://doi.org/10.1175/JCLI-D-15-0352.1

Ceppi, P., Hartmann, D. L., & Webb, M. J. (2016). Mechanisms of the negative shortwave cloud feedback in middle to high latitudes. *Journal of Climate*, *29*(1), 139–157. https://doi.org/10.1175/JCLI-D-15-0327.1

Cesana, G., & Chepfer, H. (2013). Evaluation of the cloud thermodynamic phase in a climate model using CALIPSO-GOCCP. *Journal of Geophysical Research: Atmospheres*, *118*(14), 7922–7937. https://doi.org/10.1002/jgrd.50376

Cesana, G., & Waliser, D. E. (2016). Characterizing and understanding systematic biases in the vertical structure of clouds in CMIP5/CFMIP2 models. *Geophysical Research Letters*, *43*(19), 10,538–10,546. https://doi.org/10.1002/2016GL070515

Cesana, G., & Storelvmo, T. (2017). Improving climate projections by understanding how cloud phase affects radiation. *Journal of Geophysical Research*, *122*(8), 4594–4599. https://doi.org/10.1002/2017JD026927

Cesana, G., Kay, J. E., Chepfer, H., English, J. M., & de Boer, G. (2012). Ubiquitous low-level liquid-containing Arctic clouds: New observations and climate model constraints from CALIPSO-GOCCP. *Geophysical Research Letters*, *39*(20), 1–6. https://doi.org/10.1029/2012GL053385

Cesana, G., Waliser, D. E., Jiang, X., & Li, J. L. F. (2015). Multimodel evaluation of cloud phase transition using satellite and reanalysis data. *Journal of Geophysical Research*, *120*(15), 7871–7892. https://doi.org/10.1002/2014JD022932

Cesana, G., Chepfer, H., Winker, D., Getzewich, B., Cai, X., Jourdan, O., et al. (2016). Using in situ airborne measurements to evaluate three cloud phase products derived from CALIPSO. *Journal of Geophysical Research*, *121*(10), 5788–5808. https://doi.org/10.1002/2015JD024334

Cesana, G., Waliser, D. E., Henderson, D., L'Ecuyer, T. S., Jiang, X., & Li, J. -L. F. (2019a). The vertical structure of radiative heating rates: A multimodel evaluation using A-Train satellite observations. *Journal of Climate*, *32*(5), 1573–1590. https://doi.org/10.1175/JCLI-D-17-0136.1

Cesana, G., Del Genio, A. D., Ackerman, A. S., Kelley, M., Elsaesser, G., Fridlind, A. M., et al. (2019b). Evaluating models' response of tropical low clouds to SST forcings using CALIPSO observations. *Atmospheric Chemistry and Physics*, *19*(5), 2813–2832. https://doi.org/10.5194/acp-19-2813-2019

Cesana, G., Del Genio, A. D., & Chepfer, H. (2019c). The Cumulus and Stratocumulus CloudSat-CALIPSO Dataset (CASCCAD). *Earth System Science Data Discussions*, *2667637*(Nov), 1–33. https://doi.org/10.5194/essd-2019-73

Cesana, G., Khadir, T., Chepfer, H., & Chiriaco, M. (2022). Southern Ocean solar reflection biases in CMIP6 models linked to cloud phase and vertical structure representations. *Geophysical Research Letters*, *49*(22), e2022GL099777. doi:10.1029/2022GL099777

Cesana, G., Ackerman, A., Črnivec, N., Pincus, R., & Chepfer, H. (2023). An observation-based method to assess tropical stratocumulus and shallow cumulus clouds and feedbacks in CMIP6 and CMIP5 models. *Environmental Research Communications*, *5*, 045001. doi:10.1088/2515-7620/acc78a.

Cesana, G. V., & Del Genio, A. D. (2021). Observational constraint on cloud feedbacks suggests moderate climate sensitivity. *Nature Climate Change*, *11*(3), 213–218. https://doi.org/10.1038/s41558-020-00970-y

Chepfer, H., Noel, V., Winker, D., & Chiriaco, M. (2014). Where and when will we observe cloud changes due to climate warming? *Geophysical Research Letters*, *41*, 8387–8395. doi:10.1002/2014GL061792

Dessler, A. E. (2010). A determination of the cloud feedback from climate variations over the past decade. *Science*, *330*(6010), 1523–1527. https://doi.org/10.1126/science.1192546

Fridlind, A. M., & Ackerman, A. S. (2018). *Simulations of Arctic mixed-phase boundary layer clouds: Advances in understanding and outstanding questions. Mixed-phase clouds: observations and modeling*. Elsevier Inc. https://doi.org/10.1016/B978-0-12-810549-8.00007-6

Guzman, R., Chepfer, H., Noel, V., de Guélis, T. V., Kay, J. E., Raberanto, P., et al. (2017). Direct atmosphere opacity observations from CALIPSO provide new constraints on cloud-radiation interactions. *Journal of Geophysical Research*, *122*(2), 1066–1085. https://doi.org/10.1002/2016JD025946

Hang, Y., L'Ecuyer, T. S., Henderson, D. S., Matus, A. V., & Wang, Z. (2019). Reassessing the effect of cloud type on Earth's energy balance in the age of active spaceborne observations. Part II: Atmospheric heating. *Journal of Climate*, *32*(19), 6219–6236. https://doi.org/10.1175/JCLI-D-18-0754.1

Hartmann, D. L., & Larson, K. (2002). An important constraint on tropical cloud – climate feedback. *Geophysical Research Letters*, *29*(20), 1951. doi:10.1029/2002GL015835.

Haynes, J. M., Vonder Haar, T. H., L'Ecuyer, T., & Henderson, D. (2013). Radiative heating characteristics of Earth's cloudy atmosphere from vertically resolved active sensors. *Geophysical Research Letters*, *40*(3), 624–630. https://doi.org/10.1002/grl.50145

Henderson, D. S., L'ecuyer, T., Stephens, G., Partain, P., & Sekiguchi, M. (2013). A multisensor perspective on the radiative impacts of clouds and aerosols. *Journal of Applied Meteorology and Climatology*, *52*(4), 853–871. https://doi.org/10.1175/JAMC-D-12-025.1

Hogan, R. J., & Illingworth, A. J. (2000). Deriving cloud overlap statistics from radar. *Quarterly Journal of the Royal Meteorological Society*, *126*(569), 2903–2909. https://doi.org/10.1256/smsqj.56913

Jiang, J. H., Su, H., Zhai, C., Perun, V. S., Del Genio, A., Nazarenko, L. S., et al. (2012). Evaluation of cloud and water vapor simulations in CMIP5 climate models using NASA "A-Train" satellite observations. *Journal of Geophysical Research Atmospheres*, *117*(14), D14105. https://doi.org/10.1029/2011JD017237

Kato, S., Rose, F. G., Sun-Mack, S., Miller, W. F., Chen, Y., Rutan, D. A., et al. (2011). Improvements of top-of-atmosphere and surface irradiance computations with CALIPSO-, CloudSat-, and MODIS-derived cloud and aerosol properties. *Journal of Geophysical Research Atmospheres*, *116*(19), 1–21. https://doi.org/10.1029/2011JD016050

Kay, J. E., & Gettelman, A. (2009). Cloud influence on and response to seasonal Arctic sea ice loss. *Journal of Geophysical Research Atmospheres*, *114*(18), 1–18. https://doi.org/10.1029/2009JD011773

Kay, J. E., L'Ecuyer, T., Chepfer, H., Loeb, N., Morrison, A., & Cesana, G. (2016). Recent advances in Arctic cloud and climate research. *Current Climate Change Reports*, *2*(4), 159–169. https://doi.org/10.1007/s40641-016-0051-9

Kiehl, J. T., & Trenberth, K. E. (1997). Earth's annual global mean energy budget. *Bulletin of the American Meteorological Society*, *78*(2), 197–208. https://doi.org/10.1175/1520-0477(1997)078<0197:EAGMEB>2.0.CO;2

King, M. D., Platnick, S., Menzel, W. P., Ackerman, S. A., & Hubanks, P. A. (2013). Spatial and temporal distribution of clouds observed by MODIS onboard the Terra and Aqua satellites. *IEEE Transactions on Geoscience and Remote Sensing*, *51*(7), 3826–3852. https://doi.org/10.1109/TGRS.2012.2227333

Klein, S. A., & Hall, A. (2015). Emergent constraints for cloud feedbacks. *Current Climate Change Reports*, *1*(4), 276–287. https://doi.org/10.1007/s40641-015-0027-1

Klein, S. A., McCoy, R. B., Morrison, H., Ackerman, A. S., Avramov, A., Boer, G. de., et al. (2009). Intercomparison of model simulations of mixed-phase clouds observed during the ARM Mixed-Phase Arctic Cloud Experiment. I: Single-layer cloud. *Quarterly Journal of the Royal Meteorological Society*, *135*(641), 979–1002. https://doi.org/10.1002/qj.416

Klein, S. A., Zhang, Y., Zelinka, M. D., Pincus, R., Boyle, J., & Gleckler, P. J. (2013). Are climate model simulations of clouds improving? An evaluation using the ISCCP simulator. *Journal of Geophysical Research Atmospheres*, *118*(3), 1329–1342. https://doi.org/10.1002/jgrd.50141

Klein, S. A., Hall, A., Norris, J. R., & Pincus, R. (2017). Low-cloud feedbacks from cloud-controlling factors: A review. *Surveys in Geophysics*, *38*(6), 1307–1329. https://doi.org/10.1007/s10712-017-9433-3

Knobelspiesse, K., Barbosa, H. M. J., Bradley, C., Bruegge, C., Cairns, B., Chen, G., et al. (2020). The Aerosol Characterization from Polarimeter and Lidar (ACEPOL) airborne field campaign. *Earth System Science Data*, *12*(3), 2183–2208. https://doi.org/10.5194/essd-12-2183-2020

Komurcu, M., Storelvmo, T., Tan, I., Lohmann, U., Yun, Y., Penner, J. E., et al. (2014). Intercomparison of the cloud water phase among global climate models. *Journal of Geophysical Research: Atmospheres*, *119*(6), 3372–3400. https://doi.org/10.1002/2013JD021119

Kramer, R. J., Matus, A. V., Soden, B. J., & L'Ecuyer, T. S. (2019). Observation-based radiative kernels from CloudSat/CALIPSO. *Journal of Geophysical Research: Atmospheres*, *124*(10), 5431–5444. https://doi.org/10.1029/2018JD029021

L'Ecuyer, T. S., Wood, N. B., Haladay, T., Stephens, G. L., & Stackhouse, P. W. (2009). Impact of clouds on atmospheric heating based on the R04 CloudSat fluxes and heating rates data set. *Journal of Geophysical Research Atmospheres*, *114*(8), 1–15. https://doi.org/10.1029/2008JD009951

L'Ecuyer, T. S., Beaudoing, H. K., Rodell, M., Olson, W., Lin, B., Kato, S., et al. (2015). The observed state of the energy budget in the early twenty-first century. *Journal of Climate*, *28*(21), 8319–8346. https://doi.org/10.1175/JCLI-D-14-00556.1

L'Ecuyer, T. S., Hang, Y., Matus, A. V., & Wang, Z. (2019). Reassessing the effect of cloud type on Earth's energy balance in the age of active spaceborne observations. Part I: Top of atmosphere and surface. *Journal of Climate*, *32*(19), 6197–6217. https://doi.org/10.1175/JCLI-D-18-0753.1

Lenssen, N. J. L., Schmidt, G. A., Hansen, J. E., Menne, M. J., Persin, A., Ruedy, R., & Zyss, D. (2019). Improvements in the GISTEMP uncertainty model. *Journal of Geophysical Research: Atmospheres*, *124*, 6307– 6326. https://doi.org/10.1029/2018JD029522

Li, Y., Yang, P., North, G. R., & Dessler, A. (2012). Test of the fixed anvil temperature hypothesis. *Journal of the Atmospheric Sciences*, *69*, 2317–2328. https://doi.org/10.1175/JAS-D-11-0158.1

Li, J. L. F., Lee, W. L., Waliser, D., Wang, Y. H., Yu, J. Y., Jiang, X., et al. (2016). Considering the radiative effects of snow on tropical Pacific Ocean radiative heating profiles in contemporary GCMs using A-Train observations. *Journal of Geophysical Research*, *121*(4), 1621–1636. https://doi.org/10.1002/2015JD023587

Loeb, N. G., Wang, H., Allan, R. P., Andrews, T., Armour, K., Cole, J. N. S., et al. (2020). New generation of climate models track recent unprecedented changes in Earth's radiation budget observed by CERES. *Geophysical Research Letters*, *47*(5), 1–10. https://doi.org/10.1029/2019GL086705

Mace, G. G., & Wrenn, F. J. (2013). Evaluation of the hydrometeor layers in the East and West Pacific within ISCCP cloud-top pressure-optical depth bins using merged CloudSat and CALIPSO data. *Journal of Climate*, *26*(23), 9429–9444. https://doi.org/10.1175/JCLI-D-12-00207.1

Marchand, R., Ackerman, T., Smyth, M., & Rossow, W. B. (2010). A review of cloud top height and optical depth histograms from MISR, ISCCP, and MODIS. *Journal of Geophysical Research Atmospheres*, *115*(16), 1–25. https://doi.org/10.1029/2009JD013422

Matus, A. V., & L'Ecuyer, T. S. (2017). The role of cloud phase in Earth's radiation budget. *Journal of Geophysical Research*, *122*(5), 2559–2578. https://doi.org/10.1002/2016JD025951

McCoy, D. T., Eastman, R., Hartmann, D. L., & Wood, R. (2017). The change in low cloud cover in a warmed climate inferred from AIRS, MODIS, and ERA-Interim. *Journal of Climate*, *30*(10), 3609–3620. https://doi.org/10.1175/JCLI-D-15-0734.1

McIlhattan, E. A., Kay, J. E., & L'Ecuyer, T. S. (2020). Arctic clouds and precipitation in the Community Earth System Model Version 2. *Journal of Geophysical Research: Atmospheres*, *125*(22), e2020JD032521. https://doi.org/10.1029/2020JD032521

Morice, C. P., Kennedy, J. J., Rayner, N. A., & Jones, P. D. (2012). Quantifying uncertainties in global and regional temperature change using an ensemble of observational estimates: The HadCRUT4 data set. *Journal of Geophysical Research*, *117*, D08101. doi:10.1029/2011JD017187.

Myers, T. A., & Norris, J. R. (2016). Reducing the uncertainty in subtropical cloud feedback. *Geophysical Research Letters*, *43*(5), 2144–2148. https://doi.org/10.1002/2015GL067416

Nam, C., Bony, S., Dufresne, J., & Chepfer, H. (2012). The 'too few, too bright' tropical low-cloud problem in CMIP5 models. *Geophysical Research Letters*, *39*(Sep), 1–7. https://doi.org/10.1029/2012GL053421

Narenpitak, P., & Bretherton, C. S. (2019). Understanding negative subtropical shallow cumulus cloud feedbacks in a near-global aquaplanet model using limited area cloud-resolving simulations. *Journal of Advances in Modeling Earth Systems*, 1600–1626. https://doi.org/10.1029/2018MS001572

Oreopoulos, L., Cho, N., Lee, D., Kato, S., & Huffman, G. J. (2014). An examination of the nature of global MODIS cloud regimes. *Journal of Geophysical Research: Atmospheres*, *119*(13), 8362–8383. https://doi.org/10.1002/2013JD021409

Pincus, R., Platnick, S., Ackerman, S. A., Hemler, R. S., & Hofmann, R. J. P. (2012). Reconciling simulated and observed views of clouds: {MODIS}, {ISCCP}, and the limits of instrument simulators. *Journal of Climate*, *25*(13), 4699–4720. https://doi.org/10.1175/jcli-d-11-00267.1

Qu, X., Hall, A., Klein, S. A., & Caldwell, P. M. (2014). On the spread of changes in marine low cloud cover in climate model simulations of the 21st century. *Climate Dynamics*, *42*(9–10), 2603 2626. https://doi.org/10.1007/s00382-013-1945-z

Qu, X., Hall, A., Klein, S. A., & Deangelis, A. M. (2015). Positive tropical marine low-cloud cover feedback inferred from cloud-controlling factors. *Geophysical Research Letters*, *42*(18), 7767–7775. https://doi.org/10.1002/2015GL065627

Rogers, R. R., & Yau, M. K. (1989). *A short course in cloud physics*. (3rd Ed.). Pergamon Press.

Rohde, R. A., & Hausfather, Z. (2020). The Berkeley Earth land/ocean temperature record. *Earth System Science Data*, *12*, 3469–3479. https://doi.org/10.5194/essd-12-3469-2020, 2020.

Rossow, W. B., & Schiffer, R. A. (1999). Advances in understanding clouds from ISCCP. *Bulletin of the American Meteorological Society*, *80*(11), 2261–2287. https://doi.org/10.1175/1520-0477(1999)080<2261:AIUCFI>2.0.CO;2

Saint-Lu, M., Bony, S., & Dufresne, J.-L. (2020). Observational evidence for a stability Iris effect in the tropics. *Geophysical Research Letters*, *47*, e2020GL089059. https://doi.org/10.1029/2020GL089059

Silber, I., Fridlind, A., Verlinde, J., Ackerman, A., Cesana, G., & Knopf, D. (2020). The prevalence of precipitation from polar supercooled clouds. *Atmospheric Chemistry and Physics Discussions*, *21*, 1–32. https://doi.org/10.5194/acp-2020-993

Soden, B. J., Held, I. M., Colman, R. C., Shell, K. M., Kiehl, J. T., & Shields, C. A. (2008). Quantifying climate feedbacks using radiative kernels. *Journal of Climate*, *21*(14), 3504–3520. https://doi.org/10.1175/2007JCLI2110.1

Stephens, G. L., Vane, D. G., Boain, R. J., Mace, G. G., Sassen, K., Wang, Z., et al. (2002). The CloudSat mission and the A-Train. *Bulletin of the American Meteorological Society*, *83*(12), 1771–1790. https://doi.org/10.1175/BAMS-83-12-1771

Stephens, G. L., Li, J., Wild, M., Clayson, C. A., Loeb, N., Kato, S., et al. (2012). Latest global observations. *Nature Geoscience*, *5*(10), 691–696. https://doi.org/10.1038/ngeo1580

Stubenrauch, C. J., Rossow, W. B., Kinne, S., Ackerman, S., Cesana, G., Chepfer, H., et al. (2013). Assessment of global cloud datasets from satellites. *Bulletin of the American Meteorological Society*, *94*(7), 1031–1049. https://doi.org/10.1175/BAMS-D-12-00117.1

Takahashi, H., Lebsock, M. D., Richardson, M., Marchand, R., & Kay, J. E. (2019). When will spaceborne cloud radar detect upward shifts in cloud heights? *Journal of Geophysical Research: Atmospheres*, *124*, 7270– 7285. https://doi.org/10.1029/2018JD030242

Tan, I., Storelvmo, T., & Zelinka, M. D. (2016). Observational constraints on mixed-phase clouds imply higher climate sensitivity. *Science*, *352*(6282), 224–227. https://doi.org/10.1126/science.aad5300

Terai, C. R., Klein, S. A., & Zelinka, M. D. (2016). Constraining the low-cloud optical depth feedback at middle and high latitudes using satellite observations. *Journal of Geophysical Research*, *121*(16), 9696–9716. https://doi.org/10.1002/2016JD025233

Tselioudis, G., Rossow, W., Zhang, Y., & Konsta, D. (2013). Global weather states and their properties from passive and active satellite cloud retrievals. *Journal of Climate*, *26*(19), 7734–7746. https://doi.org/10.1175/JCLI-D-13-00024.1

Tsushima, Y., Emori, S., Ogura, T., Kimoto, M., Webb, M. J., Williams, K. D., et al. (2006). Importance of the mixed-phase cloud distribution in the control climate for assessing the response of clouds to carbon dioxide increase: A multi-model study. *Climate Dynamics*, *27*(2–3), 113–126. https://doi.org/10.1007/s00382-006-0127-7

Vaillant de Guélis, T., Chepfer, H., Guzman, R., et al. (2018). Space lidar observations constrain longwave cloud feedback. *Scientific Reports*, *8*, 16570. https://doi.org/10.1038/s41598-018-34943-1

Vial, J., Dufresne, J. L., & Bony, S. (2013). On the interpretation of inter-model spread in CMIP5 climate sensitivity estimates. *Climate Dynamics*, *41*(11–12), 3339–3362. https://doi.org/10.1007/s00382-013-1725-9

Wang, H., & Su, W. (2013). Evaluating and understanding top of the atmosphere cloud radiative effects in Intergovernmental Panel on Climate Change (IPCC) Fifth Assessment Report (AR5) Coupled Model Intercomparison Project Phase 5 (CMIP5) models using satellite observations. *JGR Atmospheres*, *118*, 683–699. https://doi.org/10.1029/2012JD018619

Wielicki, B. A., Barkstrom, B. R., Harrison, E. F., Lee, R. B., Louis Smith, G., & Cooper, J. E. (1996). Clouds and the Earth's Radiant Energy System (CERES): An Earth Observing System experiment. *Bulletin of the American Meteorological Society*, *77*(5), 853–868. https://doi.org/10.1175/1520-0477(1996)077<0853:CATERE>2.0.CO;2

Wild, M., Folini, D., Hakuba, M.Z. et al. (2015). The energy balance over land and oceans: An assessment based on direct observations and CMIP5 climate models. *Climate Dynamics*, *44*, 3393–3429. https://doi.org/10.1007/s00382-014-2430-z

Winker, D. M., Pelon, J., Coakley, J. A., Ackerman, S. A., Charlson, R. J., Colarco, P. R., et al. (2010). The CALIPSO mission. *Bulletin of the American Meteorological Society*, *91*(9), 1211–1230. https://doi.org/10.1175/2010BAMS3009.1

Zelinka, M. D., & Hartmann, D. L. (2010). Why is longwave cloud feedback positive? *Journal of Geophysical Research*, *115*, D16117. doi:10.1029/2010JD013817.

Zelinka, M. D., Zhou, C., & Klein, S. A. (2016). Insights from a refined decomposition of cloud feedbacks. *Geophysical Research Letters*, *43*(17), 9259–9269. https://doi.org/10.1002/2016GL069917

Zelinka, M. D., Myers, T. A., McCoy, D. T., Po-Chedley, S., Caldwell, P. M., Ceppi, P., et al. (2020). Causes of higher climate sensitivity in CMIP6 models. *Geophysical Research Letters*, *47*(1), 1–12. https://doi.org/10.1029/2019GL085782

Zhang, M. H., Lin, W. Y., Klein, S. A., Bacmeister, J. T., Bony, S., Cederwall, R. T., et al. (2005). Comparing clouds and their seasonal variations in 10 atmospheric general circulation models with satellite measurements. *Journal of Geophysical Research Atmosphere*, *110*, 1–18. https://doi.org/10.1029/2004JD005021

Zhou, C., Zelinka, M. D., Dessler, A. E., & Klein, S. A. (2015a). The relationship between interannual and long-term cloud feedbacks. *Geophysical Research Letters*, *42*(23), 10463–10469. https://doi.org/10.1002/2015GL066698

Zhou, C., Dessler, A. E., Zelinka, M. D., Yang, P., & Wang, T. (2015b). Cirrus feedback on interannual climate fluctuations. *Geophysical Research Letters*, *41*, 9166–9173. doi:10.1002/2014GL062095.

Part II
Cloud Types

5

A Review of the Factors Influencing Arctic Mixed-Phase Clouds: Progress and Outlook

Ivy Tan[1], Georgia Sotiropoulou[2,3], Patrick C. Taylor[4], Lauren Zamora[5,6], and Manfred Wendisch[7]

ABSTRACT

Mixed-phase clouds are ubiquitous in the Arctic and play a critical role in Earth's energy budget at the surface and top-of-the-atmosphere. These clouds typically occupy the lower and mid-level troposphere and are composed of purely supercooled liquid droplets or mixtures of supercooled liquid water droplets and ice crystals. Here, we review progress in our understanding of the factors that control the formation and dissipation of Arctic mixed-phase clouds, including the thermodynamic structure of the lower troposphere, warm and moist air intrusions into the Arctic, large-scale subsidence, and aerosol particles. We then provide a brief survey of numerous Arctic field campaigns that targeted local cloud-controlling factors and follow this with specific examples of how the Arctic Cloud Observations Using airborne measurements during polar Day (ACLOUD)/ Physical feedback of Arctic PBL, Sea ice, Cloud And AerosoL (PASCAL) and Airborne measurements of radiative and turbulent FLUXes of energy and momentum in the Arctic boundary layer (AFLUX) field campaigns that took place in the vicinity of Svalbard in 2019 were able to advance our understanding on this topic to demonstrate the value of field campaigns. Finally, we conclude with a discussion of the outlook of future research in the study of Arctic cloud-controlling factors and provide several recommendations for the observational and modeling community to advance our understanding of the role of Arctic mixed-phase clouds in a rapidly changing climate.

5.1. INTRODUCTION

The Earth's Arctic is warming at approximately twice the pace of the rest of the globe. This phenomenon, commonly known as Arctic amplification, is most pronounced during the late autumn and early winter (Serreze et al., 2009). Arctic amplification has been considered in various ways; previous studies have defined it as the ratio of warming in the Arctic to either global or tropical warming (Middlemas et al., 2020; Pithan & Mauritsen, 2014; Stuecker et al., 2018), where the Arctic has been

[1]*Department of Atmospheric and Oceanic Sciences, McGill University, Montreal, Canada; formerly at NASA Goddard Space Flight Center Greenbelt, MD, USA, and affiliated with University of Maryland, Baltimore County, Baltimore, MD, USA*

[2]*Department of Meteorology, and Bolin Center for Climate Research, Stockholm University, Stockholm, Sweden; currently at the Department of Physics, National and Kapodistrian University of Athens, Athens, Greece*

[3]*Laboratory of Atmospheric Processes and their Impacts, School of Architecture, Civil & Environmental Engineering, Ecole Polytechnique Fédérale de Lausanne, Lausanne, Switzerland*

[4]*Climate Science Branch, NASA Langley Research Center, Hampton, VA, USA*

[5]*Earth System Science Interdisciplinary Center, University of Maryland, College Park, MD, USA*

[6]*NASA Goddard Space Flight Center, Greenbelt, MD, USA*

[7]*Leipzig Institute for Meteorology, University of Leipzig, Leipzig, Germany*

Clouds and Their Climatic Impacts: Radiation, Circulation, and Precipitation, Geophysical Monograph 281, First Edition.
Edited by Sylvia C. Sullivan and Corinna Hoose.

DOI: 10.1002/9781119700357.ch05

defined to be poleward of latitudes ranging from 60 to 75 °N and different definitions of the ratio have been used (Hind et al., 2016). Furthermore, the timescale of Arctic amplification has also been studied on both transient and equilibrium timescales (Holland & Bitz, 2003; Tan & Storelvmo, 2019). Here, we use Arctic amplification as an umbrella term that encompasses amplified warming in the Arctic relative to either the tropics or the entire globe and on either transient or equilibrium timescales (Yoshimori et al., 2016).

A number of mechanisms have been proposed to explain Arctic amplification (Serreze & Barry, 2011; Taylor et al., 2013; Wendisch et al., 2017). An early mechanism proposed over half a century ago is the surface-albedo feedback, whereby warming induces the melting of snow and sea ice, which in turn induces further warming by reducing surface albedo (Budyko, 1969; Sellers, 1969). This positive feedback is among many others that have since been proposed. These include feedbacks related to the strength of surface temperature inversions (Boé et al., 2009; Bintanja et al., 2011), poleward energy transport (Graversen & Langen, 2019; Hwang et al., 2011; Merlis & Henry, 2018), and clouds (Cronin & Tziperman, 2015; Middlemas et al., 2020; Tan & Storelvmo, 2019; Vavrus, 2004; Wendisch et al., 2019) that may also interact nonlinearly with the surface-albedo feedback to further amplify or dampen Arctic warming. Several studies suggest that the sea ice albedo feedback is the leading contributor to Arctic amplification (Dai et al., 2019; Hall, 2004; Manabe & Wetherald, 1975). Furthermore, it has been shown that the sea ice minimum in summer and early fall both directly and indirectly contributes to Arctic amplification through surface heat flux exchange (Screen & Simmonds, 2010). However, simulations have indicated that Arctic amplification still occurs even when the surface-albedo feedback is locked (Graversen & Wang, 2009), and a combination of energy balance and coupled climate models suggest that although the surface-albedo feedback plays a contributing role to Arctic amplification, it does not play a dominating role in climate models (Pithan & Mauritsen, 2014; Winton, 2006).

This review focuses on the processes and factors that control the evolution and properties of Arctic mixed-phase clouds known to influence Arctic amplification. These clouds consist of a combination of liquid droplets and ice crystals within the temperature range extending from 0 °C to the homogeneous freezing temperature of approximately −38 °C (Korolev et al., 2017). Although mixed-phase clouds are the focus of this manuscript, we note that some of the processes and factors considered herein also influence pure supercooled liquid clouds as well, especially in the summer where they are common in the Arctic boundary layer (Nomokonova et al., 2019). Since clouds were identified as the largest source of uncertainty in the climate sensitivity in global climate models (Cess et al., 1989) decades ago, cloud feedbacks, that is, the response of clouds to changes in surface temperature warming, continue to remain the largest contributor to uncertainty in climate projections today (Zelinka et al., 2020). However, substantial progress in narrowing the uncertainty range and clarifying the dominant mechanisms involved has been made throughout the decades (Zelinka et al., 2017), in large part due to improvements to model parameterizations and innovative diagnostic techniques such as satellite simulators that enable consistent comparisons between satellite observations and large-scale models (Bodas-Salcedo et al., 2011). The Arctic exhibits the largest spread in near-surface air temperature projections in climate models across all latitudes (Boucher et al., 2013). Arctic clouds are also especially poorly represented in climate models (Klein et al., 2009; Karlsson & Svensson, 2011), with several models and even observations disagreeing on the annual cycle of cloud fraction (Boeke & Taylor, 2016; Taylor et al., 2019). Although the magnitude of the contribution of the surface temperature-mediated cloud feedback to Arctic amplification was shown to be relatively small in the previous generation of climate models (Pithan & Mauritsen, 2014), poor observational constraints on Arctic clouds combined with linear diagnostic techniques for highly nonlinear cloud feedbacks in the Arctic (Zhu et al., 2019) call into question the true sign and magnitude of the Arctic cloud feedback. Many of these low-level clouds are of the mixed-phase type that exhibit a unique vertical structure (Curry et al., 2000) and commonly exist as multilayer clouds, especially during the summer months. Due to the gradient in saturation vapor pressure over liquid and ice surfaces, these clouds are inherently unstable due to the Wegener-Bergeron-Findeisen (WBF) process (Korolev et al., 2003) and their maintenance and life cycle are difficult to represent in models of all scales (Korolev et al., 2017). In particular, climate models have a tendency to overestimate the proportion of ice crystals relative to the total cloud water in mixed-phase clouds (Cesana et al., 2015; Komurcu et al., 2014; McCoy et al., 2016), potentially due to the lack of a representation of subgrid-scale variability in cloud liquid and ice (Tan & Storelvmo, 2016; Zhang et al., 2019) that is commonly observed in nature as revealed by aircraft in situ and remote sensing measurements (Chylek & Borel, 2004; D'Alessandro et al., 2019; Korolev et al., 2003; Ruiz-Donoso et al., 2020). As such, climate models that parameterize the WBF process may be too active in climate models (Tan & Storelvmo, 2016; Zhang et al., 2019), which may also lead to an excessive production of snow compared to observations (McIlhattan et al., 2017).

The thermodynamic phase of Arctic clouds and how it is distributed spatially is of critical importance for climate and radiation because the radiative properties of ice crystals and liquid droplets differ substantially; for the same water path and solar zenith angle, the reflectivity of water clouds can be up to four times greater than that of ice clouds (Sun & Shine, 1994) (see also Chapter 4: Cloud-Radiation Interactions and Cloud-Climate Feedbacks). This is due to the greater abundance and smaller size of liquid droplets relative to their solid counterpart, which is facilitated through the amount of cloud condensation nuclei (CCN) and ice-nucleating particles (INPs) available to initiate droplet and ice crystal formation in the atmosphere, respectively (see also Chapter 2: An Overview of Aerosol-Cloud Interactions). Liquid clouds emit more downward longwave (LW) radiation to the surface compared to ice clouds. Here, LW radiation refers to radiation with wavelengths between 3 and 100 μm, where ~99% of Earth's outgoing LW radiation is emitted (Petty, 2006). Solar radiation wavelengths, on the other hand, range from ~0.25 to 4.0 μm, which includes the visible and shortwave infrared part of the electromagnetic spectrum. The LW radiation effect is particularly important during the polar night; however, it saturates when the cloud liquid water path reaches approximately 30 gm^{-2} (Shupe & Intrieri, 2004), at which point clouds act as efficient blackbody emitters, although the exact value for a given cloud depends on the average droplet effective radius (Wang et al., 2005). Downward LW radiation from clouds is also typically masked by increases in water vapor in a warmer climate; however, at constant relative humidity the LW cloud radiative effect (CRE), defined as the difference between all-sky and clear-sky radiative flux, was found to remain constant in the Arctic and, therefore, the sensitivity of the Arctic surface to LW CRE may increase in the future (Cox et al., 2015). Statistically significant multidecadal trends of cloud cover during spring and autumn based on data from ground stations in the Arctic Ocean also show increasing cloud cover that is conducive to sea ice melt via surface warming from enhanced downward LW radiation (Eastman & Warren, 2010). Ice crystals also tend to precipitate more efficiently as a result of their larger sizes, which affects cloud fraction and therefore the radiative impact of mixed-phase clouds. Therefore, the efficiency of the WBF process, which depends on how supercooled liquid and ice are spatially distributed within mixed-phase clouds, and in turn, the partitioning of the thermodynamic phase of mixed-phase clouds, can therefore greatly impact the Earth's radiation budget and ultimately climate sensitivity (Frey & Kay, 2018; Mitchell et al., 1989; Tsushima et al., 2006; Tan et al., 2016) and Arctic amplification as a result (Tan & Storelvmo, 2019).

A better process-level understanding of Arctic mixed-phase clouds is required to reduce the spread in the representation of these clouds in large-scale and high-resolution models. Here, we review a number of studies that have approached the problem using various observations, controlled experiments in large-eddy simulations and cloud-resolving models. We first provide an overview of Arctic mixed-phase cloud formation mechanisms (section 5.2). This is followed by an up-to-date overview of the influence of the dominant factors that control Arctic clouds from the microscale to the synoptic scale, and from the interannual to the decadal timescale and beyond using multiple tools and observations in section 5.3. These factors include temperature and moisture inversions, moisture intrusions, large-scale subsidence, and aerosol particles. We emphasize the importance of taking into account the different surface types in the Arctic when considering the impact of the factors on clouds, classifying the interactions as belonging to either sea ice or open ocean categories. We argue that a comprehensive understanding of these factors is necessary to better constrain the impact of clouds on Arctic climate change and point out weaknesses that require more attention in future studies. Section 5.4 provides a broad overview of a number of Arctic field observations that studied the targeted role of various cloud- controlling factors and demonstrates the utility of airborne and ship-borne in situ observations. We conclude with a discussion of the outlook for future research on Arctic clouds and put forward several recommendations for improving our understanding of Arctic mixed-phase clouds in section 5.5.

5.2. THE FORMATION AND CHARACTERISTICS OF ARCTIC MIXED-PHASE CLOUDS

Arctic mixed-phase clouds occur year-round with a minimum frequency of occurrence of 30% in the winter and 50% during the rest of the year based on space-borne active remote sensing instruments limited to approximately 82°N (Mioche et al., 2015). Year-round ground-based remote sensing instruments in the Beaufort Sea observed mixed-phase clouds 41% of the time, with seasonal maxima in spring and autumn and cloud bases ranging from 0–2 km (Shupe et al., 2006). Their rather unique vertical structure often consists of a geometrically thin layer of supercooled liquid water droplets with a depth of approximately 0.5 km and a layer of ice virga beneath it that commonly precipitates down to the surface (Curry et al., 1997; de Boer et al., 2009; Mioche et al., 2017; Shupe et al., 2006; Silber et al., 2020a; Verlinde et al., 2007). Liquid water-topped clouds are particularly common in the Arctic. Active satellite observations have shown that this distinct vertical structure of mixed-phase clouds is observed in 55–70% of all mixed- phase clouds in the entire Arctic domain, however, with the caveat that these satellite observations inaccurately quantify the bottommost kilometer closest to the surface (Mioche et al., 2015). These clouds also occur at other latitudes; active space-borne lidar and radar instruments observed their global frequency of occurrence to be ~8% based on Zhang et al., (2010).

Advection of warm and moist air over the Arctic cold sea ice surface initiates mixed-phase cloud formation and can lead to extensive stratus clouds (Herman & Goody, 1976). LW radiation is then emitted to space from cloud top (Pinto, 1998), which decreases static stability and leads to the formation of eddies and a negatively buoyant overturning circulation (Nicholls, 1984), analogous to the process occurring in subtropical stratocumulus clouds (Wood, 2012). When supersaturation exceeds the equilibrium water vapor pressure of liquid water and ice surfaces, which is established through sufficient updraft velocities in the eddies, these turbulent eddies promote the growth of both thermodynamic phases – liquid water and ice, rather than ice solely growing at the expense of the supercooled liquid droplets through the WBF process (Korolev, 2007). The interactions of the cloud top layer driven by radiative cooling with the surface below and/or the advected air aloft play a critical role in sustaining the liquid condensate and preventing the cloud from immediate glaciation (Solomon et al., 2018). Overall, a complex web of interactions between turbulent, radiative, and dynamical processes collectively contribute to sustaining them, causing Arctic mixed-phase clouds to ultimately persist for days despite their inherent thermodynamic instability due to the WBF process (Morrison et al., 2012).

Multilayer clouds are also common in the Arctic, especially during summer (Curry et al., 1996; Shupe et al., 2011). These structures often consist of a low-level liquid or mixed-phase cloud and higher level mixed-phase or cirrus clouds (Vassel et al., 2019). These clouds usually form when large-scale horizontal advection associated with low- and high-pressure synoptic systems brings warm and moist air into the Arctic, which can condense in the presence of sufficient CCN and INPs at various levels. However, multiple cloud layers can also form within the lower troposphere (Verlinde et al., 2013), when sublimation of ice precipitation generated by the upper clouds results in cooling and moistening of the subcloud layer; this can lead to the formation of secondary inversion and a lower cloud (Harrington et al., 1999). The upper clouds of a multilayer system not only interact with radiation directly but also impact the shape, microphysical and radiative properties of the lower clouds. Overlying clouds shield lower layers from cloud- radiative cooling (Verlinde et al., 2013), which limits cloud-generated turbulent motions in the latter and thus condensation (Shupe et al., 2013). Moreover, ice precipitation from the higher layers into lower mixed-phase clouds can act as a sink for liquid water droplets through both vapor growth of the ice and rime collection (Verlinde et al., 2013). Enhanced riming can further reinforce secondary ice production and eventually trigger cloud glaciation (Lawson et al., 2015; Lloyd et al., 2015). Despite the critical impact of multilayer systems on the structure of the atmospheric radiative heating profile, the interactions of the individual layers are poorly quantified and understood. As a result, the microphysical structure of these systems is inadequately represented in models, which often overestimate (underestimate) the overall cloud liquid (ice) content (Morrison et al., 2009).

5.3. FACTORS THAT CONTROL ARCTIC CLOUDS

We contend that the main factors that influence Arctic mixed-phase clouds include the thermodynamic structure of the Arctic atmosphere, which is determined by the frequent presence of temperature and moisture inversions, the oscillation of moisture intrusions that bring large bouts of moisture to the Arctic and the presence of CCN and INPs. Although large-scale subsidence is an important factor controlling subtropical clouds (Myers & Norris, 2013), evidence suggests that large-scale subsidence plays a secondary role in controlling the evolution of Arctic mixed-phase clouds. A discussion of each of these cloud-controlling factors and the interconnection between them follows.

5.3.1. Thermodynamic Structure of the Arctic Atmosphere

The vertical profiles of temperature and humidity define the thermodynamic structure of the atmosphere. The Arctic atmosphere's thermodynamic structure is strongly influenced by the frequent presence of low-level temperature inversions (Curry et al., 1996). In addition, the Arctic frequently experiences coincident specific humidity inversions (Devasthale et al., 2011). The degree to which temperature and moisture inversions impact Arctic cloud properties strongly depends on surface type, that is, the extent to which the underlying surface is covered by sea ice, open ocean, or land, whether the cloud layer interacts with the surface layer, and the degree to which the clouds are coupled to the surface (sections 5.3.2 and 5.3.3). Arctic clouds are deemed "coupled" to the Arctic surface when cloud-driven turbulence interacts with surface turbulence (Fig. 5.1d) and "decoupled" when the subcloud mixed layer remains disconnected from the surface layer (Fig. 5.1b,c). Field measurements indicate that decoupled clouds dominate both in winter

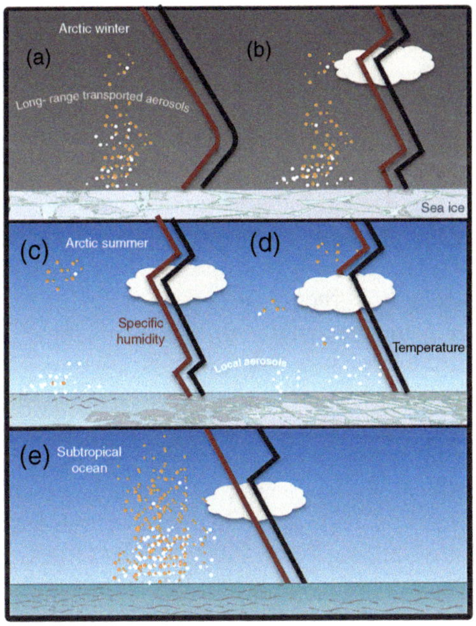

Figure 5.1 Schematic of typical thermodynamic structures of the Arctic atmosphere during (a) winter in the presence of clear-sky or thin ice clouds, (b) winter in the presence of mixed-phase clouds, (c) summer when clouds are decoupled from the surface, (d) summer when clouds are coupled to the surface, and (e) subtropical stratocumulus clouds. The dashed (solid) lines indicate specific humidity (temperature) profiles and the triangles indicate local aerosol particles that are overall less abundant than long-range transported aerosol particles (dots) in the winter. Coupling of the clouds to the surface facilitates interactions with more local aerosol particles. Overall, aerosols are generally less abundant in the Arctic compared to the lower latitudes, such as the subtropics.

(Gierens et al., 2020) and summer (Shupe et al., 2013; Sotiropoulou et al., 2014). Cloud-surface interactions can have important implications for a cloud's life cycle (Shupe et al., 2013), especially from late spring to early October, when both ice-free ocean and melting sea ice can supply clouds with enhanced moisture (Pinto & Curry, 1995). Along with thermodynamic effects, clouds coupled to the surface may also be more affected by local sources of CCN (Mahmood et al., 2019) and INPs (Creamean et al., 2019), which may result in additional interactions and impacts on the Arctic's surface radiative energy budget and hydrological cycle (section 5.3.4).

5.3.2. Temperature Inversions

Temperature soundings from the year-long Surface Heat Budget of the Arctic Ocean (SHEBA) field campaign that took place in the Beaufort Sea in 1997 (Uttal et al., 2002) first revealed two preferred thermodynamic states over sea ice and snow-covered surfaces during the Arctic winter (Stramler et al., 2011). The most prevalent preferred state is characterized by strong surface inversions and cloud-free conditions or optically thin ice clouds (Fig. 5.1a) and is referred to as the "clear state." The second state is less preferred in the winter and dominates the summer and autumn seasons when clouds occur for 70–90% of the time and can persist for days to weeks (Nomokonova et al., 2019; Shupe et al., 2011; Zygmuntowska et al., 2012). The latter state is characterized by weaker, usually elevated temperature inversions (Tjernström & Graversen, 2009) and the formation of optically thick mixed-phase clouds (Fig. 5.1b) and is referred to as the "cloudy state." Both clear and cloudy states have also been identified over open ocean and sea ice, such that altogether four states of typical atmospheric conditions in the ice-covered and open-ocean Arctic prevail (Wendisch et al., 2019).

The clear state is characterized by large net upward surface LW radiative energy flux densities of at least 40 Wm^{-2}. Downward LW emission from clouds in the cloudy state tends to offset surface cooling (Zuidema et al., 2005), often resulting in near-zero surface net LW radiation (Graham et al., 2017; Stramler et al., 2011). While the accurate representation of these two preferred states is critical for the correct representation of the Arctic surface radiative energy budget, climate models partly fail to reproduce the cloudy state and this causes systematically cold surface temperature biases and a stronger winter surface inversion (Pithan et al., 2014; Pithan et al., 2016).

Satellite remote sensing and radiosonde observations show that temperature inversions tend to be stronger and occur closer to the surface over sea ice than open ocean (Nygård et al., 2014; Pavelsky et al., 2011). Although the temperature inversions are typically stronger during the winter months (Ganeshan & Wu, 2015), averaging over 5 K, they tend to be weaker during summer, averaging approximately 2 K (Devasthale et al., 2011). The strength of the temperature inversions, as commonly quantified by the lower tropospheric stability (LTS), exerts a strong influence on Arctic low-level cloud properties, which in turn depends on whether the clouds are coupled or decoupled to the surface. Whether clouds are decoupled (Fig. 5.1c) or coupled (Fig. 5.1d) from the surface in turn is strongly influenced by surface type (Kay & Gettelman, 2009). Over sea ice, atmospheric columns with stronger LTS tend to have fewer clouds, less cloud liquid water content, more cloud ice water content, and are also closer to the surface with a tendency for more frequent multi layer clouds (Taylor et al., 2015; Taylor et al., 2019). Here, clouds tend to be decoupled from the surface since it is generally colder than the overlying atmospheric boundary layer, which is influenced by the episodic advection of warm air from lower latitudes. Decoupling of the clouds from the surface leaves radiation exchanges as the only coupling mechanism. Previous studies found that climate models tend to overestimate the decoupling and the related Arctic LTS relative to observations (Barton et al., 2014; Medeiros et al., 2011). This was due to cold surface temperature biases over sea ice that arise from a lack of cloud liquid water path (Barton et al., 2014). On the other hand, a study using active satellite remote sensing observations showed that the opposite correlation holds over the ocean – stronger LTS favors more cloud cover with more liquid water and less ice (Yu et al., 2019). Over areas of open ocean, that are common during autumn, warmer surface temperatures relative to the atmosphere enhance surface turbulent fluxes to the atmosphere and couple the clouds more often to the surface. This warming and moistening of the lower atmosphere can promote larger cloud fraction and liquid water content, and strong LTS can restrict boundary layer deepening and enhance low-level liquid clouds. The opposite effect occurs in the presence of weak LTS that reduces low-level liquid clouds over a coupled boundary layer through enhanced entrainment effects that are well studied in subtropical marine stratocumulus clouds (Klein & Hartmann, 1993; Wood, 2012). For example, sensible heat fluxes that deepen the boundary layer were found to reduce cloud cover above Arctic wintertime leads (Li et al., 2020). A notable exception to the coupling of Arctic clouds to the surface over partially open ocean occurs during the summer, when clouds are decoupled from the surface due to a lag in sea ice melting that keeps the surface cooler than the overlying atmosphere. In general, over both surface types, stronger LTS promotes lower cloud bases. Moreover, in addition to the direct impact of LTS on cloud macrophysical properties, it can also indirectly influence aerosol-cloud interactions through the entrainment, transport and recycling of aerosol particles (section 5.3.6).

5.3.3. Moisture Inversions

Specific humidity or "moisture" inversions are often observed to accompany temperature inversions in humidity soundings and satellite profiles of water vapor at multiple layers in the Arctic atmosphere (Devasthale et al., 2011; Nygård et al., 2014). Over Arctic land, moisture inversions coincide with temperature inversions roughly 50% of the time on average. These moisture inversions form via two mechanisms: (1) surface or cloud top radiative cooling and (2) large-scale moisture transport (Naakka et al., 2018). In the former mechanism, LW radiative cooling at either cloud top or at the surface can decrease the local moisture supply by promoting the condensation of water vapor and simultaneously cool the cloud top or surface, thereby forming coincident moisture and temperature inversions. In the latter mechanism, moisture inversions are formed via the advection of moist air over a dry air mass. Over mountainous regions with large slopes such those that exist in Greenland, katabatic winds are commonly responsible for moisture inversions (Vihma et al., 2011). Note that katabatic winds over Greenland and wind-induced mixing have the opposite effect on the strength of moisture inversions; although the latter typically tends to erode moisture inversions, the former strengthens them.

The frequency, strength, and mechanism of formation of Arctic moisture inversions vary with season and with surface type. Arctic moisture inversions tend to be stronger during summer. Annually, their strength varies from 0.2 to 2 gkg^{-1}. However, moisture inversions are more frequent in the winter (Devasthale et al., 2011; Naakka et al., 2018), occurring up to 80% of the time, with depths ranging from 200 to 900 m (Nygård et al., 2014; Sotiropoulou et al., 2016). While the surface radiative cooling mechanism controls their frequency in the winter, the moisture advection mechanism is considered the dominant formation mechanism in summer (Naakka et al., 2018) when the Arctic Ocean is frequently affected by the transport of southerly warm and moist air (Naakka et al., 2018; Tjernstro et al., 2019), although episodes of intense poleward moisture transport, known as moisture intrusions (section 5.3.2), have been reported year-round (Pithan et al., 2018). Over sea ice, the frequency of coexistent moisture and temperature inversions is higher (Sedlar et al., 2012; Sotiropoulou et al., 2016) due to the fact that surface sensible and latent heat fluxes are generally small and of similar magnitude, which limits the instantaneous differences in boundary-layer heat and moisture transport (Nygård et al., 2014).

Although moisture inversions are not unique to the polar regions, they do occur much more frequently in the polar regions relative to warmer latitudes and they are also structurally different in the Arctic compared to other regions. The increasing moisture supplies above the temperature inversion base often promotes condensation within the stable layer, allowing Arctic low-level clouds to extend into the inversion (Fig. 5.1b,c). This unique feature of Arctic clouds is commonly found over sea ice (Sedlar et al., 2012) but not over open water (Sotiropoulou et al., 2016). Moreover, the extension of the liquid layer into the inversion alters the effective cloud emission temperature; although it has a weak positive impact on LW irradiances at the surface ($\sim$1.5 Wm^{-2}) the increase in outgoing LW radiation at the top-of-the-atmosphere (TOA) can be up to 10 Wm^{-2} (Sedlar et al., 2012).

If the cloud layer is decoupled from the surface, elevated moisture inversions in the Arctic can play an important role in sustaining the lifetime of liquid clouds by entraining moist air from the inversion into the cloud and therefore increasing their optical thickness and radiative properties (Egerer et al., 2021). Idealized large-eddy simulations have shown that elevated moisture inversions can serve as a sufficient moisture source to maintain a decoupled cloud for days to a week, although this resulted in small differences in the properties of liquid and ice within a single mixed-phase cloud layer compared to the case with a coupled cloud without an elevated moisture inversion (Solomon et al., 2014). Thus, the existence of elevated moisture inversions implies that surface coupling is not the only source of moisture for Arctic clouds and that the cloud layer can evolve independently of whether it is coupled to the surface and is independent of the surface type. There is currently no consensus on the impact of surface coupling as a moisture source to maintain Arctic mixed-phase clouds based on the analysis of field observations; while some observational studies suggest that additional surface moisture sources enhance liquid condensate (Gierens et al., 2020; Shupe et al., 2013), others do not find a significant impact (Sotiropoulou et al., 2014). However, footprint-level satellite observations have shown that the influence of surface coupling on the evolution of cloud properties depends on the local atmospheric meteorological regime partitioned by LTS and mid-tropospheric vertical velocity (Taylor et al., 2015). The situation contrasts with cloud-surface decoupling in the midlatitudes where decoupling promotes cloud break-up (Wood, 2012). Entrained air at cloud top is usually dry air over the midlatitude counterpart (Fig. 5.1e) and thus cloud top entrainment, evaporation, and precipitation lead to cloud dissipation when the cloud system is disconnected from the surface vapor supply.

5.3.4. Warm and Moist Air Intrusions

Several recent studies indicate that anomalously large moisture and heat transport from the south into the Arctic plays a critical role in Arctic amplification (Kapsch et al., 2013; Woods & Caballero, 2016; Johansson et al., 2017; Messori et al., 2018). Such episodes bring warm and moist air into the Arctic and are often linked to extreme surface and sea ice melting (Woods et al., 2013; Tjernström et al., 2015; Park et al., 2015a, 2015b, 2015c, Boisvert et al., 2015; Hegyi & Taylor, 2018). This phenomenon is sometimes referred to as warm and moist air intrusions or simply “moisture intrusions.”

Wintertime moisture intrusions are supported by a synoptic blocking pattern to the east of the region affected (Woods et al., 2013). Moreover, a blocking system over the Ural Mountains can induce significant sea ice decline in the Barents and Kara Seas when combined with the positive phase of the North Atlantic Oscillation, as this circulation pattern favors southerly moisture transport into the basins (Gong & Luo, 2017). Several studies have linked the transport of enhanced moisture to poleward-propagating planetary-scale Rossby waves triggered by tropical convection (Baggett et al., 2016; Lee, 2014; Park et al., 2015c; Yoo et al., 2012), which is referred to as the “tropically excited Arctic warming mechanism.”

Moisture intrusions exert a substantial influence on Arctic cloud conditions. When transported air masses originate from open water, they cool and condense when advected over sea ice, resulting in cloud formation (Ali & Pithan, 2020). Moisture intrusions are also one of the large- scale moisture transport mechanisms that lead to moisture inversions described in section 5.3.3 that sustain clouds that are decoupled from the surface. Many studies have linked the occurrence of moisture intrusions with increased cloudiness (Johansson et al., 2017; Liu et al., 2018; Messori et al., 2018; Persson et al., 2017) and enhanced downward LW radiation (Messori et al., 2018; Park et al., 2015a; Park et al., 2015b; Park et al., 2015c; Woods et al., 2013) and thus surface warming. Large-eddy simulations reveal that advected heat has a relatively weak impact on cloud properties whereas moisture is crucial for cloud maintenance in the Arctic boundary layer (Sotiropoulou et al., 2018). In contrast, the advection of warm and moist air within oceanic boundary layers is associated with a negative impact on cloud fraction (Eirund et al., 2019; Knudsen et al., 2018). Heat and moisture increase within the boundary layer promote destabilization and convection in the lower atmosphere (Eirund et al., 2019). As a result, the stratocumulus cloud breaks up, the cloud top is lifted and cloud fraction is substantially reduced. Decreases in cloud fraction and cloud liquid water content result in enhanced outgoing LW radiation and surface cooling. However, when advection occurs above the boundary layer, it can promote the formation of multilayer structures and thus an overall increase in liquid water path (Eirund et al., 2019).

While infrequent, moisture intrusions are responsible for the bulk of the poleward moisture transport in the Arctic during both winter and summer (Liu & Barnes, 2015). Moreover, the associated anomalies in moisture content and cloudiness have been linked to accelerated onset of the sea ice melting period (Kapsch et al., 2013; Mortin et al., 2016). Yet, despite their climatic significance (Pithan et al., 2018), moisture intrusions are poorly represented in climate models, which fail to reproduce their regional characteristics (Woods et al., 2017). These deficiencies can have a substantial impact on Arctic cloud representation in climate models, as models with enhanced poleward heat and moisture transport produce improved cloud fractions and cloud liquid properties (Baek et al., 2020). Resolving these model issues requires dedicated measurement campaigns in a Lagrangian air parcel framework (Pithan et al., 2018; Wendisch et al. 2021).

5.3.5. Large-Scale Subsidence

While several recent studies have focused on the impact of large-scale advection on Arctic clouds, less is known about the impact of subsidence, which often accompanies poleward atmospheric transport (Neggers et al., 2019; Tjernström et al., 2019). Large-scale subsidence is weaker in the Arctic compared to the subtropics and thus potentially plays a lesser role in the Arctic. Dedicated measurements of synoptic-scale divergence and derived vertical pressure velocity, such as those performed in the subtropics (Stevens et al., 2021), are scarce in the Arctic; however, there are plans to conduct appropriate samplings (Wendisch et al., 2021). In the Arctic, large-scale subsidence is weak and can be thought of as being correlated with the generation of a surface temperature inversion – as the air is advected over the central Arctic from lower latitudes it slowly sinks as the air radiatively cools (Tjernström et al., 2019). Thus, unlike the case in the subtropics, subsidence is not an active driver of the inversion strength and is merely correlated with stronger LTS and inversion strength. In this section, we discuss our current limited knowledge of the role of subsidence in Arctic clouds based on a limited set of field observations, large-eddy simulations, as well as large-scale climate models and satellite observations.

Reanalysis data and field observations from the Arctic Clouds in Summer Experiment (ACSE) have revealed that air parcels were higher in altitude and further south a few days before the presence of a surface inversion (Tjernström et al., 2019). Although subsidence is generally linked with the presence of a surface temperature inversion over melting sea ice, moisture and cloud characteristics are more variable (Tjernström et al., 2019). In particular, the formation of moisture inversions and low clouds or fog was found to be associated with weaker subsidence, compared to cases where the stratified boundary layer is drier and often cloud-free. While reanalysis data suggest that subsidence weakly enhances the fraction and liquid water path of Arctic clouds (Zhao & Wang, 2010), field observations suggest that subsidence is correlated with the existence of optically thinner Arctic clouds, potentially by impacting the entrainment of aerosol particles into the boundary layer (Zuidema et al., 2005).

The limited number of large-eddy simulations has shown contradictory results in terms of the role of subsidence in Arctic low-level cloud properties. In a study employing large-eddy simulations, increases in large-scale subsidence result in more turbulent clouds with enhanced liquid condensate over open water (Young et al., 2018). The enhanced subsidence reinforces the boundary-layer inversion strength and reduces entrainment of warmer air aloft. This allows for greater cloud liquid, thus more efficient precipitation, cloud top radiative cooling, and downdraft turbulent production. The combination of strong cloud top radiative cooling, subcloud rain evaporative cooling, and latent heat release from snow growth at cloud base destabilize the boundary layer, resulting in more convective structures (Young et al., 2018). However, other studies showed that enhanced subsidence resulted in an overall decrease in cloud liquid water content over sea ice due to entrainment of drier air (Dimitrelos et al., 2020). In line with this study, strong and sudden subsidence led to cloud dissipation when the boundary-layer top was pushed below the lifting condensation level (Neggers et al., 2019). The cloud response to variations in large-scale vertical forcing for different surface, thermodynamic, and microphysical conditions has not been comprehensively explored. Additional studies with cloud-resolving simulations, preferably in a Lagrangian framework (Dimitrelos et al., 2020; Neggers et al., 2019), are necessary to improve the understanding of the role of subsidence in Arctic cloud evolution and radiation (Wendisch et al., 2021).

Finally, the role of subsidence as an Arctic cloud-controlling factor is also limited to a few studies employing large-scale climate models and satellite observations. Synergistic CloudSat/CALIPSO observations and reanalysis-based regimes have shown that the Arctic atmosphere produces a wide range of 500 hPa vertical pressure velocity values (ω500) ranging from weak ascent to strong sinking motion for each of the LTS-based regimes (Barton et al., 2012; Taylor et al., 2015). Under conditions of large-scale subsidence, the altitude of low-level clouds was very sensitive to LTS. Satellite observations and a climate model also show that stronger subsidence may also increase relative humidity in the lower troposphere (Curry et al., 1988), and in turn trigger Arctic sea ice melt in the summer via enhanced downward LW cloud radiation at the surface (Huang et al., 2021). However, the dependence of cloud properties on ω500 was shown to be much weaker than that on LTS in the CMIP5 models (Taylor et al., 2019), suggesting a relatively minor role compared to other cloud-controlling factors in the context of large spatial and temporal scales.

5.3.6. Aerosol Particles

The influence of aerosol particles on the cloud microphysical properties that drive cloud radiative effects is poorly quantified yet is of fundamental importance to Earth's climate and its future change (Fan et al., 2016). Although Arctic low- and mid-level cloud properties and radiative effects can be highly susceptible to aerosol effects on a local scale (Garrett & Zhao, 2006; Lubin & Vogelmann, 2006), regional-scale impacts have not been thoroughly explored. A series of recent reviews on topics related to Arctic aerosol distributions, mixed-phase cloud modeling, microphysics, and aerosol interactions show that regional uncertainty in aerosol-cloud interactions (ACIs) is in large part due to (1) CCN and INP levels being difficult to predict, and (2) aerosol impacts on clouds being microphysically complex and linked to meteorology (Fridlind et al., 2007; Fan et al., 2016; Kanji et al., 2017; Lohmann, 2017; Morrison et al., 2012; Schmale et al., 2021; Willis et al., 2018). This section complements these reviews by linking the larger-scale meteorological influences discussed in the sections above with what is known about Arctic-specific aerosol sources and microphysical cloud impacts. Specifically, we discuss the factors affecting the concentrations and activity of Arctic CCN and INPs, the robust and uncertain mechanisms by which these CCN and INPs impact radiation-relevant cloud properties, and how aerosol-cloud interactions may be changing with a warming Arctic.

Sources, Concentrations, and Activity of Arctic CCN and INPs

CCN and INPs are derived from marine, terrestrial, and anthropogenic sources. Combustion- derived aerosols are sporadically transported to the Arctic from lower latitudes (Soja et al., 2008) and there are local near-surface aerosol particle sources from exposed glacial till dust (Tobo et al., 2019; Zwaaftink et al., 2016), shipping and oil extraction (Schmale et al., 2018), and thawing permafrost (Creamean et al., 2020). However, more commonly, summertime Arctic aerosol particles are produced from local marine primary and secondary sources. These sources can provide at least half of the CCN supply to the Arctic atmosphere via primary sea spray emissions (Quinn et al., 2017) and new particle formation (Dunne et al., 2016; Heintzenberg et al., 2017; Merikanto et al., 2009; Yu & Luo, 2009). Marine aerosols may also supply more than half of all INPs at high latitudes (Burrows et al., 2013; Wilson et al., 2015). In contrast, during winter and spring, marine biogenic aerosol particle concentrations are lower when the open ocean is less exposed to the atmosphere due to both greater sea ice cover and lower biological productivity, and aerosol particles derived from long-range transport, dust and combustion sources, are more prevalent (Arrigo, 2008; Barrie, 1986; Croft et al., 2016; Engvall et al., 2008; Stohl, 2006; Quinn et al., 2007), although marine sources of INPs are still present to some extent (Hartmann et al., 2019b). Sea salt aerosol particle concentrations can be larger in winter, potentially due to upward migration of brine from underlying sea ice and subsequently lifting and sublimation in the atmosphere through a strong influence from blowing snow (Huang & Jaeglé, 2017). Long- range transport of aerosol particles to the Arctic from sources such as dust and smoke (Bullard et al., 2016) as well as biomass burning aerosols derived from boreal forest fires especially during the spring season (Marelle et al., 2015) are also important sources of Arctic CCN and INPs. Long-range transported wintertime aerosols can accumulate and form Arctic haze due to the combination of a cold, stable boundary layer and reduced particle and gas removal rates (Shaw, 1995). In particular, black carbon is an important contributor of Arctic haze and its wintertime peak is controlled by its hydrophilic fraction and weak wet deposition rate (Shen et al., 2017).

Concentrations of CCN typically range between 1 and 100 cm^{-3} at supersaturations between 0.3 and 0.8%, but "tenuous" regimes with CCN concentrations <10 cm^{-3} have been observed during multiple field campaigns (Bigg & Leck, 2001; Bigg et al., 1996; Lannefors et al., 1983; Leaitch et al. 2016; Leck & Svensson, 2015; Leck et al., 2002; Mauritsen et al., 2011; Tjernström et al., 2014). CCN levels are lowest during the summer, when midlatitude aerosol transport is inefficient and midlatitude wet deposition is likely to scavenge long-range transported aerosols before they reach the Arctic (Bourgeois & Bey, 2011; Di Pierro et al., 2013; Law & Stohl, 2007; Quinn et al., 2007). Unlike CCN and aerosols in general, Arctic INPs appear to be more prevalent during the summer (Wex et al., 2019), although overall their concentrations tend to be quite low, ranging from 10^{-4} to 10^{-2} L^{-1} at $-15\,°C$ (Creamean et al., 2019; Hartmann et al., 2019a; Irish et al., 2019; Mason et al., 2016; Si et al., 2018; Wex et al., 2019), although concentrations have been observed as high as 0.25 L^{-1} (Bigg, 1996). Ice core data suggest that the summertime INP peak is caused by biological aerosols of marine and possibly terrestrial origin (Hartmann et al., 2019b). Ship-based CCN observations from (Wendisch et al., 2019) were lowest when their research vessel was surrounded by sea ice and highest during open ocean conditions, which supports the role of local marine emissions. However, in those locations where dust is present, dust may be an equal or more significant source of INPs than biogenic aerosols (Abbatt et al., 2019; Huang et al., 2018; Irish et al., 2019; Si et al., 2018; Tobo et al., 2019; Vergara- Temprado et al., 2017). Local dust is more exposed during the summer, but long-range transported dust may be more common in the winter.

The distributions and ability of CCN and INPs to impact Arctic clouds are quite heterogeneous in space and time (Moore et al., 2011; Willis et al., 2018) and are difficult to characterize because both CCN and INPs can be modified during atmospheric transport. For example, polluted aerosols are not thought to be major sources of INPs at the temperature ranges important to mixed-phase clouds (Borys, 1989; Hartmann et al., 2019b). Moreover, if they mix with INPs from other sources, the INP activity of these other particles may be reduced after being coated with sulfuric acid, and they may not freeze until colder temperatures (Borys, 1989; Cziczo et al., 2009; Coopman et al., 2018b; Eastwood et al., 2009; Grenier & Blanchet, 2010; Girard & Asl, 2014; Tan et al., 2014). However, it is important to note that while some studies found that pollution coatings can decrease the ice-nucleating ability of certain INPs, others found no impact (Archuleta et al., 2005; Knopf & Koop, 2006). Given the generally low concentrations of aerosols in the region, the impact of pollution coatings may be particularly important for Arctic aerosol- cloud interactions (Coopman et al., 2018a).

Neither CCN nor INPs can be accurately measured in the Arctic with current satellite remote sensors, so much of our information on their distributions relies on field data. Arctic INP field data are rare and are associated with non-negligible uncertainties (Demott et al., 2015; Garimella et al., 2018). The link between CCN and INP distributions and ambient meteorological conditions adds another level of challenge to CCN and INP prediction. At temperatures below ~−15 °C in the Arctic, INPs can be activated from sources that otherwise might not have been important at lower latitudes (Kanji et al., 2017; Wilson et al., 2015) where INPs cannot get lofted to high enough altitudes with sufficiently cold temperatures outside regions of deep convection. At the same time, Arctic INPs are able to nucleate ice at temperatures as high as −5 °C (Wex et al., 2019), although activation at these warm temperatures are typical for INPs composed of bacteria that can also be found at other latitudes (Murray et al., 2012). Moreover, INPs become less effective at warmer temperatures, and more INPs and CCN become active at higher water vapor supersaturations. CCN are easier to sample than INPs, but their distribution is also not well known. For example, in very clean Arctic conditions, even particles as small as 20–30 nm can nucleate cloud droplets at high water vapor supersaturations (Burkart et al., 2017; Croft et al., 2016; Koike et al., 2019; Leaitch et al., 2016). These potentially cloud-active aerosol particles are too small to be detectable from current spaceborne instruments (Hallen & Philbrick, 2018), and must be sampled by in situ measurements. Moreover, the aforementioned tenuous aerosol layers consisting of low concentrations of CCN and INPs that are very optically thin are also not detectable from space-borne instruments (Cho et al., 2013; Winker et al., 2013). For these reasons, CCN and INP distributions are least well-constrained outside of clouds, at high altitudes and over remote regions where few CCN and INP field data exist.

Aerosol-Cloud-Radiation Interactions

Models consistently show that aerosol impacts in Arctic mixed-phase clouds are large enough to potentially affect sea ice melt (Alterskjær et al., 2010; Dalsøren et al., 2013; Gagné et al., 2017; Jiang et al., 2000; Koch et al., 2009; Mahmood et al., 2019; Regayre et al., 2015; Shindell & Faluvegi, 2009). For example, INP levels have a large impact on modeled cloud phase, cloud fraction, and precipitation (Fridlind et al., 2012; Morrison et al., 2011; Ovchinnikov et al., 2014; Prenni et al., 2007), with potentially important impacts on surface CREs (Shupe & Intrieri, 2004), TOA CREs (English et al., 2014; Xie et al., 2013), and, ultimately, Arctic amplification (Tan & Storelvmo, 2019). Enhanced INP levels will affect heterogeneous ice crystal formation and growth processes, for example, via immersion freezing (de Boer et al., 2010). INP-driven glaciation could affect cloud lifetime and precipitation through either liquid removal in the cloud or through the WBF process referred to as the "glaciation effect" (Curry, 1995; Lohmann, 2002) and/or associated secondary ice multiplication (Field et al., 2017; Korolev & Leisner, 2020). Conversely, deactivation of preexisting INPs when pollution aerosols are present can reduce ice nuclei levels and glaciation (Girard et al., 2005; Lohmann, 2017). However, clear in situ evidence is still lacking for how often these processes occur or how important they might be in part because many past aircraft campaigns (Table 5.1) have missed key parameters, such as INP or CCN levels, aerosol composition, or ice cloud particle habit. Also, parameterizations for ice-phase processes in mixed-phase clouds lead to large uncertainties in modeled cloud properties that have not yet been resolved (Boucher et al., 2013; Klein et al., 2009; Liu et al., 2011; Morrison et al., 2011; Tan &

Table 5.1 List of Arctic field campaigns and the cloud-controlling factors that were observed.

Meteorological variable	Surface type	Relevant field campaign
Thermodynamic structure		BASE, FIRE-ACE, M- PACE, ASCOS, ACSE, AFLUX, SHEBA
Moisture intrusions		SHEBA, ACCACIA, ACLOUD, PASCAL, MOSAiC
Aerosol particles		M-pace, AMISA, ASTAR, ISDAC, ACCACIA, ARCTAS, ARCPAC, POLARCAT, VERDI, ICEALOT, RACEPAC, ACME-V, NETCARE
	Marginal Ice Zone (MIZ) Melt ponds Polynya Leads	ACSE, ACLOUD, PASCAL, MACCSIMIZE, NETCARE MELTEX, NETCARE PAMARCMiP ASCOS, MOCCHA, PAMARCMiP

Storelvmo, 2016; Taylor et al., 2019; Xie et al., 2013). Moreover, while INP-related processes appear to drive cloud microphysical and radiative responses to aerosols in some conditions and locations (Costa et al., 2017; Jouan et al., 2012; Solomon et al., 2018), CCN-related processes may be more important in other situations (Lance et al., 2011; Norgren et al., 2018). For example, the role of CCN-driven processes seems to be particularly important in tenuous mixed-phase cloud regimes with very low cloud droplet number concentrations (Loewe et al., 2017; Stevens et al., 2018).

There are also various other ways that CCN can impact mixed-phase clouds. Enhanced levels of CCN lead to smaller and more numerous liquid cloud droplets in Arctic mixed-phase clouds on average over large spatial and timescales (Coopman et al., 2018a). Smaller droplets can affect cloud lifetime and precipitation by (1) impeding liquid droplet precipitation (Albrecht, 1989), (2) reducing liquid droplet collection from falling ice particles (Borys et al., 2000; Lohmann et al., 2003) known as the "riming indirect effect" (Lohmann, 2017), and (3) reducing secondary ice production from collision and splintering processes (Rosenfeld, 2000). Cloud lifetime and droplet size in thin clouds affect cloud LW radiative emissivity (Shupe & Intrieri, 2004), which in turn impacts moisture and surface turbulent fluxes, cloud top cooling, and mixed layer depth (Garrett & Zhao, 2006; Garrett et al., 2009; Lubin & Vogelmann, 2006; Solomon et al., 2018). In summer, smaller and more numerous droplets at constant liquid water content will also cause more radiation to be scattered back to space (Twomey, 1977). Multilayer clouds are commonly observed in this region (Liu et al., 2012), and changes to mixed-phase cloud CCN-driven precipitation could also affect seeding of lower level clouds. Subsequent seeding-related changes to cloud ice and precipitation formation (Luo et al., 2008; Silber et al., 2020a; Vassel et al., 2019) may then affect cloud dissipation and surface albedo. In some conditions, both CCN and INPs might drive co-occurring processes that can behave nonlinearly.

Currently, the concentrations of CCN and INPs are a major source of uncertainties in models of Arctic mixed-phase clouds. These concentrations are particularly poorly constrained within clouds, where they can be entrained (Avramov et al., 2011; Igel et al., 2017), redistributed (Solomon et al., 2018), scavenged, and precipitated (Morrison et al., 2005; Willis et al., 2018). INPs may also be sublimated and then re-entrained (Fan et al., 2009; Possner et al., 2017; Solomon et al., 2015; Verlinde et al., 2007) and they may become more efficient when CCN-related processes like LW cloud top radiative cooling lower cloud temperatures and increase immersion freezing rates (Fu & Xue, 2017; Possner et al., 2017). The uncertainty in aerosol impacts on cloud phase is another major issue in models. Cloud phase, along with cloud fraction, exerts a large influence on CREs. Aerosol impacts on cloud phase are poorly constrained in global climate models not solely in the Arctic but also globally (Cesana et al., 2015; Karlsson & Svensson, 2011; McCoy et al., 2016; Tan et al., 2016; Taylor et al., 2019).

Field and remote sensing data offer complementary insights to models of aerosol interactions in Arctic mixed-phase clouds, but must be viewed in light of their own uncertainties. A main challenge is that aerosols often co-vary with meteorological factors that control clouds as discussed in the previous sections. For example, across Arctic sea ice regions during polar night, satellite observations showed that combustion aerosols are associated with an average 10 Wm^{-2} difference in surface LW CREs, but that between 57 and 91% of this signal is caused by changes in meteorological conditions associated with aerosol transport (Zamora et al., 2018). Other challenges with interpreting field data include that ice concentrations within clouds are difficult to measure accurately (Fridlind et al., 2007), and microphysical processes can be impacted by co-occurring cold-weather phenomena, including secondary ice production (Field et al., 2017; Korolev & Leisner, 2020) and seeding from either Arctic multilayer clouds above the mixed- phase cloud layer (Luo et al., 2008) or frost flowers (Xu et al., 2016). Moreover, the sign, magnitude, and mechanisms by which aerosols impact Arctic mixed-phase cloud precipitation and radiative effects vary, depending not only on CCN and INP levels, but also on incoming radiation and surface albedo, multilayer cloud radiative shielding, and cloud properties such as height, temperature, and liquid water content (Morrison et al., 2012; Quinn et al., 2008; Shupe & Intrieri, 2004; Sedlar et al., 2011; Sedlar & Shupe, 2014; Stofferahn & Boybeyi, 2017; Willis et al., 2018). For example, at Utqiaġvik, Alaska, aerosol microphysical effects on clouds may lead to surface heating as large as 12 Wm^{-2} in winter, and surface cooling as large as 12 Wm^{-2} in the summer (Zhao & Garrett, 2015). Therefore, to quantify aerosol-cloud effects across the Arctic, observations are needed over large spatial and temporal scales, with attention paid to verifiably accounting for meteorological co-variability with aerosols.

When data are compared over large temporal and spatial scales and across cloud types, most remote sensing-based observations seem to agree that combustion and dust aerosols are associated with some combination of higher glaciation temperatures, more cloud ice, more precipitation, and reduced cloud fraction in the Arctic (Coopman et al., 2018b; Coopman et al., 2020; Filioglou et al., 2019; Villanueva et al., 2020; Zhang et al., 2018; Zamora et al., 2018). Aerosols have been associated with less cloud ice and less precipitation at specific locations or in specific cloud types (Norgren et al., 2018;

Zamora et al., 2017), but these trends seem to be associated with combustion aerosols (Filioglou et al., 2019) and might be caused by either a CCN-related process or by a deactivation effect. Even presuming that the glaciation effect is dominant now (and more work is still needed to verify this hypothesis), it is unclear whether this process will remain dominant in a future warmer, wetter, and more aerosol-laden Arctic environment. Either way, for the reasons discussed above, aerosol-related uncertainties contribute to major uncertainties in climate projections (Bellouin et al., 2020), especially for the Arctic where rapid changes to historic aerosol, moisture, and heat fluxes are expected.

Aerosol-Meteorology Interactions and Their Impact on Clouds and Radiation

Aerosol impacts on radiation-relevant cloud properties have a strong relationship with meteorological conditions, such as temperature. Warmer temperatures reduce INP effectiveness and glaciation and riming processes (Eirund et al., 2019). As Arctic INPs are thought to have a large influence on mixed-phase cloud processes (section 5.3.4.2), this warming could become increasingly important to cloud dynamics. Warmer temperatures can also affect the microphysical environment in which aerosols are suspended, influencing the degree to which aerosols contribute to the Twomey effect, the WBF process, precipitation, splintering, and riming, and more generally, the potential importance of CCN compared to INP-dominated aerosol microphysical processes. Thus, although many uncertainties remain, INP-driven glaciation processes might become less influential in the future warmer Arctic.

Besides temperature, other related meteorological parameters can influence Arctic ACIs as well, such as decoupling with the surface (Creamean et al., 2021), as well as stability and moisture levels. Besides affecting temperature, decoupling limits the influence of surface aerosol sources on clouds at higher altitudes, but promotes recycling of INPs that are released during sublimation of precipitating ice particles at the base of the subcloud layer (Fan et al., 2015; Solomon et al., 2018). This process, which plays a critical role in maintaining cloud-phase partitioning (Solomon et al., 2015; Solomon et al., 2018), is likely not favored in coupled clouds (Kalesse et al., 2016). A more stable atmosphere (such as over sea ice) promotes weaker cloud top entrainment of free- tropospheric air that can serve as moisture and a source of CCN and INP concentrations in the cloud layer (Coopman et al., 2018b; Fridlind et al., 2007; Morrison et al., 2012; Solomon et al., 2011; Solomon et al., 2014). It may also concentrate aerosols emitted from local sources (Willis et al., 2018). Atmospheric moisture content, which is impacted by temperatures, atmospheric stability, and moisture intrusions, influences aerosol deposition and loss processes (Browse et al., 2012), and in turn impacts a CCN/INP's lifetime potential for impacting clouds. More moist and less stable conditions over open ocean can also activate smaller CCN particles, which might then affect cloud droplet feedbacks with mixed-phase cloud vertical mixing and radiative cooling (Silber et al., 2020b).

Moisture intrusions may also influence ACIs, particularly in areas decoupled from the surface. These intrusions are often aerosol-laden, and they produce not only in more aerosol transport (Thomas et al., 2019), but also in more frequent precipitation and aerosol loss. The extension of cloud top into the inversion layer modulates aerosol fluxes into the cloud as well as moisture entrainment fluxes and can thus impact cloud life cycles (Egerer et al., 2020; Igel et al., 2017; Solomon et al., 2011). Other changes, for example, in large-scale subsidence, might also affect the cloud microphysical environment upon which aerosols operate, for example, impacting cloud top radiative cooling rates and ice and liquid water paths, as well as aerosol entrainment rates and precipitation loss rates (Brooks et al., 2017; Dimitrelos et al., 2020; Young et al., 2018).

Multiple studies have found Arctic cloud responses to non-marine aerosols to be clearly reduced over open ocean compared to sea ice (Filioglou et al., 2019; Eirund et al., 2019; Zamora et al., 2017; Zamora et al., 2018). That there would be a difference between open ocean and sea ice ACIs is not surprising, given that the two surface types produce very different levels of stability and aerosol, heat, and moisture fluxes and aerosol emissions (Schmale et al., 2021; Willis et al., 2018; Wendisch et al., 2019). For example, not only are marine aerosol levels much larger over the open ocean than over sea ice, but clouds over open ocean generally experience warmer and wetter conditions compared to those over sea ice. Aerosols may at times also impact meteorology, as when aerosol-driven increases in thin cloud LW radiative emissivity warm the surface and thereby increase moisture and heat fluxes (Morrison et al., 2012). Although the dominant causes for the observed differences between open ocean and sea ice are unknown, the trend of reduced aerosol influence over open ocean regions suggests that in the absence of significant new aerosol sources or pathways, the impacts of non-marine aerosols may become less influential in the future as the Arctic warms.

5.4. A BRIEF SURVEY OF ARCTIC FIELD CAMPAIGNS TARGETING CLOUD-CONTROLLING FACTORS

Despite limitations in their temporal and spatial coverage, in situ field observations are an indispensable

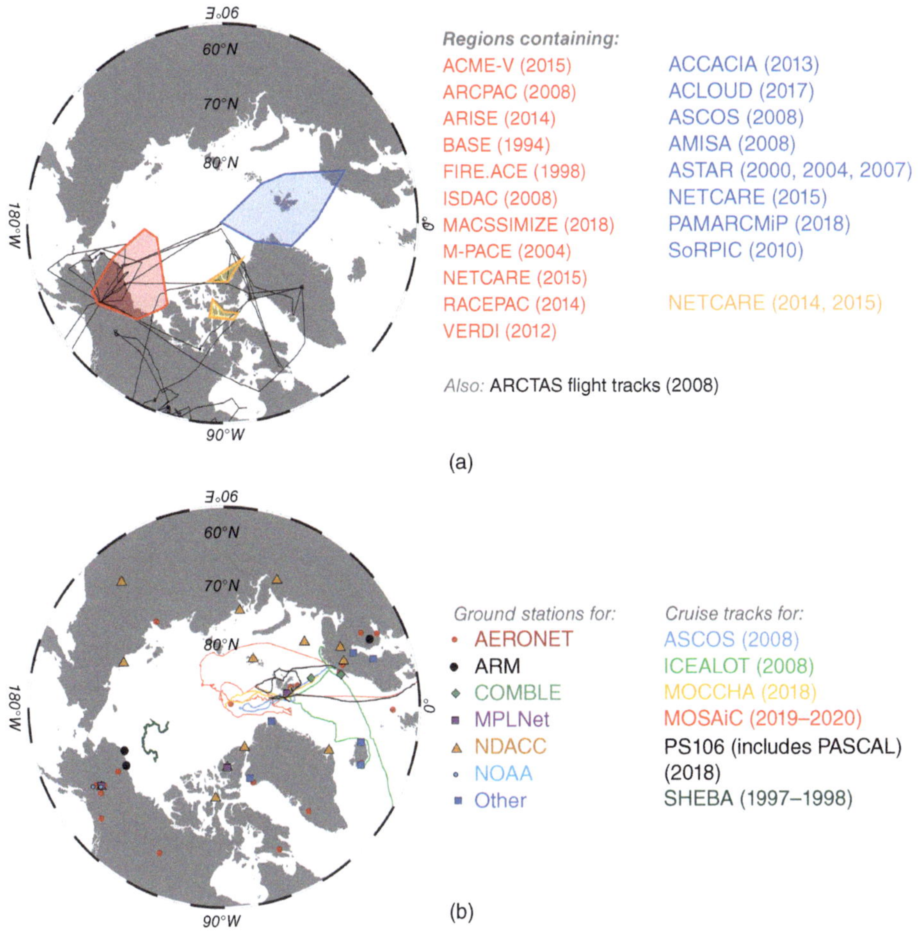

Figure 5.2 Locations of selected Arctic campaigns and datasets, including (a) aircraft datasets, and (b) ground and ship-based datasets.

tool for climate science by virtue of their relatively high accuracy and frequency of measurements relative to global satellite observations that can be used to validate regional models and develop model parameterizations. A number of field campaigns over the Arctic that took place mostly during the non-winter months have been performed over the past few decades (Fig. 5.2a) and have been combined with ground-based stations in the Arctic (Fig. 5.2b) to compensate for the limitations of space-borne remote sensing. This section begins with a brief overview of a number of these field campaigns and some of the studies that have applied observations from them to gain insight on the influence of several factors that influence Arctic clouds. This is followed by a short discussion of some lessons learned from three examples of the numerous field campaigns conducted in the past, namely, the combined airborne and ship-based Arctic Cloud Observations Using airborne measurements during polar Day (ACLOUD), the Airborne measurements of radiative and turbulent FLUXes of energy and momentum in the Arctic boundary layer (AFLUX), and the Physical feedback of Arctic PBL, Sea ice, Cloud And AerosoL (PASCAL) field campaigns.

5.4.1. Overview

Several aspects of Arctic clouds, aerosols, radiation, and their interactions were targeted by field campaigns. Selected examples that aimed to study the interaction of the thermodynamic and turbulent boundary-layer structure with clouds include the Beaufort and Arctic Storms Experiment (BASE) (Gultepe et al., 2000), the First International Satellite Cloud Climatology Project (ISCCP) Regional Experiment-Arctic Cloud Experiment (FIRE-ACE) (Curry et al., 2000), the Mixed-Phase Arctic Cloud Experiment (M-PACE) (Verlinde et al., 2007), the Arctic Summer Cloud Ocean Study (ASCOS) (Tjernström et al., 2014), the Arctic Clouds in Summer Expedition (ACSE) (Tjernström et al., 2015), AFLUX, and Surface Heat Budget of the Arctic Ocean (SHEBA)

(Uttal et al., 2002). Taken together with data collected from ground-based remote sensing observations at Ny-Ålesund, cloud liquid and ice water contents appear to be strongly influenced by synoptic conditions such as wind direction and the degree of thermodynamic coupling to the surface (Gierens et al., 2020).

The influence of aerosols on Arctic clouds was also documented based on observations from a large number of campaigns such as M-PACE (Prenni et al., 2009), the Arctic Study of Aerosol, Clouds and Radiation (ASTAR) (Yamagata et al., 2009), the Indirect and Semi-Direct Aerosol Campaign (ISDAC) (McFarquhar et al., 2011), the Aerosol-Cloud Coupling and Climate Interactions in the Arctic (ACCACIA) (Lloyd et al., 2015), the Aerosol, Radiation and Cloud Processes affecting Arctic Climate (ARCPAC) (Brock et al., 2011), PASCAL (Griesche et al., 2020), and the Polar Study using Aircraft, Remote Sensing, Surface Measurements and Models, of Climate, Chemistry, Aerosols, and Transport (POLARCAT) (Law et al., 2014), Radiation- Aerosol-Cloud Experiment in the Arctic (RACEPAC) (Herenz et al., 2018), the Vertical Distribution of Ice in Arctic Clouds (VERDI) (Klingebiel et al., 2015), and The Arctic Research of the Composition of the Troposphere from Aircraft and Satellites (ARCTAS) (Jacob et al., 2010), which was one of the largest airborne field campaigns to study the impact of air pollution on Arctic climate. Network on Climate and Aerosols: Addressing Key Uncertainties in Remote Canadian Environments (NETCARE) (Abbatt et al., 2019) was a highly interdisciplinary field campaign that was able to observe melt ponds as a source of dimethyl sulfide and long-range mineral dust as a prominent springtime source of INPs and local mineral dust as a local source in the summer. Arctic Mechanisms for the Interaction of the Surface and Atmosphere (AMISA) (Persson et al., 2017) was also a field campaign that complemented ASCOS with its information on the impact of synoptic and mesoscale flow and vertical mixing of aerosol particles. Additionally, International Chemistry Experiment in the Arctic LOwer Troposphere (ICEALOT) was a field campaign dedicated to determining the influence of local and transported aerosol particles on clouds among other effects such as haze and ozone over open ocean in the Arctic (Huang & Jaeglé, 2017; Quinn et al., 2017; Russell et al., 2010). Rare high-resolution airborne measurements of INPs and air temperature in the high Arctic were measured by the Polar Airborne Measurements and Arctic Regional Climate Model Simulation Project (PAMARCMiP) (Hartmann et al., 2019a), while evidence from airborne measurements during the Fifth Airborne Carbon Measurements (ACME-V) field campaign in the Alaska revealed that the Arctic is not always as pristine in the summer as once thought due to wildfires and local oil extraction activities (Creamean et al., 2018).

A number of field campaigns have also studied cloud-radiation impacts. Among the earliest to study Arctic surface energy fluxes were the Radiation and Eddy Flux Experiment (REFLEX) (Hartmann et al., 1991) and the Arctic Radiation and Turbulence Interaction STudy (ARTIST) (Hartmann et al., 1999). Other studies with goals to better understand the impact of clouds on atmospheric radiation followed, including SHEBA (Stramler et al., 2011), ASCOS (Sedlar et al., 2011), the Solar Radiation and Phase Discrimination of Arctic Clouds (SoRPIC) (Bierwirth et al., 2013), the Arctic Radiation-IceBridge Sea & Ice Experiment (ARISE) (Smith et al., 2017), and ACLOUD/PASCAL (Stapf et al., 2021).

The influence of various surface types such as the Marginal Ice Zone (MIZ), melt ponds, leads, and polynyas on cloud properties was also the interest of several Arctic field campaigns such as Measurements of Arctic Clouds, Snow, and Sea Ice nearby the Marginal Ice ZonE (MACSSIMIZE), ACLOUD/PASCAL (Stapf et al., 2021), the recently completed Multidisciplinary Drifting Observatory for the Study of Arctic Climate (MOSAiC) campaign, melt ponds on energy and momentum fluxes between atmosphere and ocean (MELTEX) (Rösel & Kaleschke, 2012), Microbiology-Ocean-Cloud-Coupling in the High Arctic (MOCCHA), which aimed to quantify influences of new aerosol particle formation (Baccarini et al., 2020), particularly over leads, and PAMARCMiP. A summary of the impact of these various factors on clouds and the relevant field campaigns are summarized in Table 5.1.

5.5. INSIGHTS ON ARCTIC CLOUD-CONTROLLING FACTORS GAINED FROM ACLOUD, PASCAL, AND AFLUX

The influence of a number of cloud-controlling factors on Arctic low clouds was investigated during ACLOUD, PASCAL (Wendisch et al., 2019), and ALFUX, which were components of Phase I of the German ArctiC Amplification: Climate Relevant Atmospheric and SurfaCe Processes and Feedback Mechanisms $(AC)^3$ field campaign (Wendisch et al., 2017). ACLOUD and PASCAL concurrently took place in May and June 2017 in and around Svalbard, while ALFUX took place between mid-March and mid-April 2019 in the same area. Here, we present a brief description of results from the campaigns to demonstrate the effectiveness of Arctic field campaigns in advancing our understanding of the factors driving cloud formation and cloud properties.

The influence of INPs on clouds and their dependence on the coupling state of the clouds to the surface was observed using ship-based remote sensing instruments taken during PASCAL. A high occurrence of surface-coupled ice-containing clouds with cloud top temperatures warmer than −10 °C suggests the influence of near-surface INPs on Arctic boundary-layer clouds at relatively warm supercooled temperatures when the cloud layer is coupled to the surface.

The combination of ACLOUD/PASCAL and AFLUX, both of which took place in a MIZ, revealed that the familiar cloud-free and cloudy atmospheric states observed during SHEBA over sea ice and snow-covered regions also occur over open ocean (Wendisch et al., 2019). The differences in the surface temperature and lapse rates between the ACLOUD/PASCAL and AFLUX field campaigns, which took place during different months, influence the clear and cloudy states. While the horizontal surface temperature gradient between sea ice and open ocean was 25 K in AFLUX, it was only 6 K in ACLOUD. The horizontal surface temperature gradients in turn affects the vertical lapse rate, that is, thermodynamic stability of the atmosphere, which consequently affects downward LW emission profile in cloud-free conditions. Less stable atmospheric conditions decrease the net irradiances because less downward LW radiation is emitted from the atmosphere. Thus, the cloud-free modes over sea ice and open ocean may both shift in response to thermodynamic stability. Cloud-base temperature remains almost unchanged whether over sea ice or open ocean and, thus, downward LW radiation emitted by the cloud-base stays nearly constant. These shifts were revealed during the early-spring AFLUX (very stable) and the summer ACLOUD/PASCAL (less stable) campaigns.

ACLOUD and PASCAL also raised several open questions related to Arctic clouds and the factors that control them. Although clouds are clearly impacted by surface type, the degree to which sea ice and open water impact cloud properties still remains poorly quantified. Moreover, observations revealed stronger turbulence between clouds at high altitudes than expected, raising the question of the dominant contributing physical processes leading to the enhanced turbulence. On the other hand, atmospheric thermodynamic stability was also observed to be weaker than previous studies suggest (Stapf et al., 2021).

5.6. OUTLOOK

We have outlined and discussed the influence of various meteorological factors and aerosols, and the mechanisms by which they influence Arctic mixed-phase cloud properties. In so doing, we have identified several outstanding questions that remain to be addressed in the future to improve our understanding of the factors controlling the behavior of Arctic mixed-phase clouds.

Progress in resolving these major questions in Arctic cloud evolution and their radiative effects is hindered by the limited number of high-quality observations of cloud and aerosol processes. Satellite measurements have substantially advanced our current state of knowledge of Arctic cloud properties and their interaction with sea ice; however, both passive and active satellite observations suffer from a number of limitations. Current passive satellite data of cloud properties are limited by inaccurate retrievals at steep solar zenith angles that are exacerbated in the polar regions (Grosvenor & Wood, 2014). While the development of new algorithms to correct for these biases related to the lack of three-dimensional radiative transfer effects has led to promising improvements (Lebsock & Su, 2014; Khanal et al., 2020), there is still non-trivial disagreement in the various cloud properties among satellite instruments. The common supercooled liquid-topped structure of Arctic mixed-phase clouds also presents a challenge for active space-borne lidar that cannot penetrate entire cloud layers with optical thicknesses greater than approximately 5 (Winker et al., 2009). Although the synergistic use of collocated measurements with space-borne radar can remediate this shortcoming, the combination of instruments still fails to observe the bottommost kilometer of the atmosphere due to the combination of radar ground clutter and lidar beam attenuation (Liu et al., 2017). The horizontal and vertical spatial resolution of satellite observations is also insufficient to accurately determine the spatial distribution of clusters of liquid and ice structures that comprise mixed-phase clouds, which in turn impact the efficiency of the WBF process. Furthermore, space-borne remote sensing instruments cannot reliably retrieve CCN and INP concentrations at a spatial resolution that is sufficient for cloud process modeling; this is particularly true for very small aerosol particles <0.1 μm. To evaluate these issues dedicated validation exercises using ground and airborne measurements are required. While in situ observations are a more suitable tool for this purpose, they suffer from a lack of spatial coverage for the widespread low-level stratiform mixed-phase clouds that are ubiquitous in the Arctic (Eastman & Warren, 2010). The lack of in situ observations is especially problematic during Arctic winter when harsh weather conditions prevail that can lead to aircraft icing during in situ measurements. In addition to clouds, tenuous aerosol layers are common in the Arctic and preclude measurements from space-borne observations. Moreover, while the TOA radiative fluxes are better observed by satellite observations compared to surface fluxes (Kato et al., 2018), it is crucial to characterize surface radiative energy fluxes for the important surface radiative energy budget and the related near- surface warming in the Arctic, which are difficult to retrieve from satellite data. Finally, although space-borne infrared sounders have improved our knowledge of temperature and moisture inversions in the Arctic, their coarse vertical resolution precludes observations of shallow inversions. As a result of the limited high-quality observations, the precise mechanisms relating lower tropospheric stratification, cloud dynamics, and vertical velocity are still poorly understood. We emphasize the need for reliable and comprehensive data of the response of Arctic mixed-phase clouds under a broad range of relevant meteorological and surface conditions. In this regard, the validation of models using data from dedicated measurement campaigns has a powerful potential to unravel model deficiencies in parameterizations, subscale process representation, and other issues. The validated models then reveal critical processes determining the evolution and effects of clouds.

Regarding the impact of aerosols on Arctic mixed-phase clouds, it is clear that Arctic meteorology and aerosol levels are continually undergoing dramatic changes. Local CCN and INP emissions will likely increase due to shipping and oil and gas development, mining, exposed soil from irreversible loss of snow, permafrost, and sea ice (Meredith et al., 2019), and altered sea spray and biogenic emissions from changes in sea ice cover, wind intensity, and warmer temperatures (Arrigo, 2008; Ardyna et al., 2014; Deslippe et al., 2012). A better understanding of continually changing natural aerosol emissions and the fundamental physical processes involved was emphasized to better constrain Arctic ACIs (Schmale et al., 2021). Aerosol transport from lower latitudes will also change with shifting wind and precipitation patterns, and there will likely be increasing sub-Arctic wildfire emissions and changing anthropogenic aerosol particle emissions as well. Drawing from the previous sections, we put forth 10 recommendations to improve our understanding of cloud-controlling factors in the Arctic that would ideally involve the development of an overall community-wide strategic plan to improve on this front as follows:

- Targeted field campaigns, dedicated model validations, and model intercomparisons of synoptic influences such as cyclones, moisture intrusions, and large-scale subsidence on clouds. In particular, cloud evolution and airmass transformations over Arctic sea ice and open ocean during moist intrusions, particularly from a Lagrangian perspective based on in situ observations are lacking yet important for model evaluation (Dimitrelos et al., 2020; Neggers et al., 2019; Pithan et al., 2018; Wendisch et al., 2021). This last point is the target of the upcoming HALO-(AC)^3 field campaign planned to take place in 2022.
- Detailed investigations using high-resolution models to quantify the impact of surface aerosol and moisture sources versus cloud top entrainment fluxes under various meteorological and surface conditions. High-resolution models are also needed to clarify the dominant planetary boundary-layer processes affecting the in-cloud redistribution of CCN and INPs in mixed-phase clouds.
- Improved methods for observing INPs, CCN, and ice particles in situ and to the extent possible, also from remote sensing measurements. The former task requires higher sensitivity to tenuous aerosol layers typical of the Arctic and accurate distinctions between INP and CCN types. Although limited accurate high-latitude (poleward of 82 °N) aerosol particle measurements are available from passive satellite observations, they are currently unavailable from active space-borne remote sensing instruments, despite their increasing importance in a warming Arctic with decreasing sea ice extent. There has recently been active progress on in situ INP measurements. Year-long surface-based INP measurements at Oliktok that use techniques described in McCluskey et al., (2018) and Suski et al., (2018) will soon be available and MOSAiC will provide the first year-round observations of Arctic INPs in the Arctic Ocean. However, these observations are limited to the surface and may not represent the cloud layer. The latter task requires improved retrieval algorithms for ice number concentration and ice crystal effective radius with reduced uncertainties in stratiform mixed-phase clouds. While such algorithms have been explored for ice number concentration using ground-based observations (Zhang et al., 2014), they are completely lacking using current satellite observations. This also includes further reduced shattering effects of ice crystals on aircraft measurements (Korolev et al., 2013).
- Improved understanding of ice formation and growth in Arctic mixed-phase clouds and representations of these processes in climate models. For example, representing subgrid- scale variability in the liquid and ice partitioning in mixed-phase clouds in climate models (Tan & Storelvmo, 2016; Zhang et al., 2019) can result in more accurate rates of the WBF process in climate models and requires continuous and high spatial resolution observations of mixed-phase clouds. Detailed observations of snowflakes using three- dimensional ground-based cameras, for example, the Multi-Angle Snowflake Camera (Garrett et al., 2012) in the Arctic, are expected to aid in the development of more sophisticated parameterizations of ice cloud microphysical processes.
- Long-term observations that can be linked to multiscale models, including in multilayer cloud conditions. Furthermore, existing fair and consistent comparisons between models and satellite remote observations via the satellite simulator approach (Bodas-Salcedo et al., 2011) should not only be continued given their previous success in identifying model biases (Cesana et al., 2015; Nam et al., 2012), but also expanded to include

other types of remote sensing instruments and a larger variety of observables. Ground-based satellite simulators (Kuma et al., 2021) are an example of a recent advance that has taken us one step closer to closing the gap between model and observation comparisons from the surface perspective, and the development of scale-aware and definition-aware diagnostics for near-surface precipitation frequency is also another example (Kay et al., 2018). The full potential of model and satellite comparisons is critical to reducing model biases but has yet to be fully exploited.

- Upgraded sophisticated methods to isolate the aerosol in observational studies from confounding factors such as co-varying meteorology and secondary ice formation; these methods are needed in the current and changing Arctic climate, including in response to sea ice decline.
- Boosted development and testing of Arctic aerosol transport models of dust and other aerosol particles, particularly over remote regions and in the presence of precipitation, along with better techniques for integrating satellite and suborbital data with models. These efforts could benefit from focused field campaigns that aim to validate Arctic aerosol transport models.

In summary, we have highlighted and reviewed a number of important Arctic cloud-controlling factors. The influences of these cloud-controlling factors share similarities yet are also markedly different from the impacts of tropical cloud-controlling factors (Klein et al., 2017). We contend that a better understanding of the various controls over Arctic clouds is contingent on improved observations of clouds and aerosols in terms of both quality and quantity. Some of the shortcomings in the satellite instruments of the past and present are currently being considered in NASA's ATMosphere Observing System (ATMOS) mission resulting from the National Academies of Sciences, Engineering and Medicine's 2017 Decadal Survey (National Academies of Sciences and Medicine, 2018). The launch of a space-borne high spectral resolution lidar will enable higher sensitivity to tenuous aerosol layers. Additionally, while currently still under development, coincident observations of aerosols and clouds in the Arctic by exploiting far-infrared measurements as well as improved observations of cloud ice microphysics and snowfall are being considered. Due to the previous success of active satellite instruments in improving our understanding of cloud processes and better constraining cloud feedbacks (Winker et al., 2017), particularly in the Arctic (Kay & Gettelman, 2009; Morrison et al., 2018; Taylor et al., 2015), active satellite observations are being considered in the ACCP mission. Combining these observations with targeted field campaigns presents a path forward to closing the gap in our knowledge of the controls over Arctic clouds and therefore enables a better understanding of the role of clouds in the changing Arctic climate system.

ACKNOWLEDGMENTS

Ivy Tan acknowledges support from NASA Grant 80NSSC18K1599. Georgia Sotiropoulou acknowledges support from the Swedish Research Council for Sustainable Development (FORMAS) – project-ID 2018-01760. Patrick C. Taylor acknowledges support from the NASA Interdisciplinary Studies Program (NNH19ZDA001N-IDS) and Radiation Budget Science Project. Lauren Zamora acknowledges support from the NASA Aerosol-Cloud Modeling and Analysis Program (Grant 80NSSC19K0978) under Richard Eckman. Manfred Wendisch appreciates the funding by the Deutsche Forschungsgemeinschaft (DFG, German Research Foundation) – Project-ID 268020496 – TRR 172. The authors thank Ralph Kahn at NASA GSFC and three anonymous reviewers for their helpful comments and suggestions.

REFERENCES

Abbatt, J. P. D., Leaitch, W. R., Aliabadi, A. A., Bertram, A. K., Blanchet, J.-P., Boivin-Rioux, A., et al. (2019). New insights into aerosol and climate in the Arctic. *Atmospheric Chemistry and Physics.*, *19*, 2527–2650.

Albrecht, B. A. (1989). Aerosols, cloud microphysics, and fractional cloudiness. *Science*, *45*(4923), 1227–1230.

Ali, S. M., & Pithan, F. (2020). Following moist intrusions into the Arctic using SHEBA observations in a Lagrangian perspective. *Quarterly Journal of the Royal Meteorological Society*, *146*(732), 3522–3533.

Alterskjær, K., Kristánsson, J. E., & Hoose, C. (2010). Do anthropogenic aerosols enhance or suppress the surface cloud forcing in the Arctic? *Journal of Geophysical Research*, *115*(D22).

Archuleta, C. M., DeMott, P. J., & Kreidenweis, S. M. (2005). Ice nucleation by surrogates for atmospheric mineral dust and mineral dust/sulfate particles at cirrus temperatures. *Atmospheric Chemistry and Physics*, *5*, 2617–2634.

Ardyna, M., Babin, M., Gosselin, M., Devred, E., Rainville, L., & Tremblay, J. E. (2014). Recent Arctic Ocean sea ice loss triggers novel fall phytoplankton blooms. *Geophysical Research Letters*, *41*(17), 6207–6212.

Arrigo, K. R., van Dijken, G., & Pabi, S. (2008). Impact of a shrinking Arctic ice cover on marine primary production. *Geophysical Research Letters*, *35*(19).

Avramov, A., Ackerman, A. S., Fridlind, A., Diedenhoven, B. V., Botta, G., Aydin, K., et al. (2011). Toward ice formation closure in Arctic mixed-phase boundary layer clouds during ISDAC. *Journal of Geophysical Research*, *116*(D1).

Baccarini, A., Karlsson, L., Dommen, J., Duplessis, P., Vullers, J., Brooks, I. M., et al. (2020). Frequent new particle formation over the high Arctic pack ice by enhanced iodine emissions. *Nature Communications*, *11*(1), 1–11.

Baek, E.-H., Kim, J. H., Park, S., Kim, B. M., & Jeong, J. H. (2020). Impact of poleward heat and moisture transports on Arctic clouds and climate simulation. *Atmospheric Chemistry and Physics*, *20*(5), 2953–2966.

Baggett, C., Lee, S., & Feldstein, S. (2016). An investigation of the presence of atmospheric rivers over the North Pacific during planetary-scale wave life cycles and their role in Arctic warming. *Journal of the Atmospheric Sciences*, *73*, 4329–4347.

Barrie, L. A. (1986). Arctic air pollution: An overview of current knowledge. *Atmospheric Environment*, *20*(4), 643–663.

Barton, N. P., Klein, S. A., & Boyle, J. S. (2014). On the contribution of longwave radiation to global climate model biases in Arctic lower tropospheric stability. *Journal of Climate*, *27*(19), 7250–7269.

Barton, N. P., Klein, S. A., Boyle, J. S., & Zhang, Y. Y. (2012). Arctic synoptic regimes: Comparing domain-wide Arctic cloud observations with CAM4 and CAM5 during similar dynamics. *Journal of Geophysical Research*, *117*(D15).

Bellouin, N., Quaas, J., Gryspeerdt, E., Kinne, S., Stier, P., Watson-Parris, D., et al. (2020). Bounding global aerosol radiative forcing of climate change. *Reviews of Geophysics*, *58*(1), e2019RG000660.

Bierwirth, E., Ehrlich, A., Wendisch, M., Gayet, J.-F., Gourbeyre, C., Dupuy, R., et al. (2013). Optical thickness and effective radius of Arctic boundary-layer clouds retrieved from airborne nadir and imaging spectrometry. *Atmospheric Measurement Techniques*, *6*(5), 1189–1200.

Bigg, E. K. (1996). Ice forming nuclei in the high Arctic. *Tellus B: Chemical and Physical Meteorology*, *48*(2), 223–233.

Bigg, E. K., & Leck, C. (2001). Cloud-active particles over the central Arctic Ocean. *Journal of Geophysical Research*, *106*(D23), 32155–32166.

Bigg, K. E., Leck, C., & Nilsson, D. E. (1996). Sudden changes in Arctic atmospheric aerosol concentrations during summer and autumn. *Tellus B: Chemical and Physical Meteorology*, *48*(2), 252–271.

Bintanja, R., Graversen, R. G., & Hazeleger, W. (2011). Arctic winter warming amplified by the thermal inversion and consequent low infrared cooling to space. *Nature Geoscience*, *4*(11), 758–761.

Bodas-Salcedo, A., Web, M. J., Bony, S., Chepfer, H., Dufresne, J. L., Klein, S. A., et al. (2011). COSP: Satellite simulation software for model assessment. *Bulletin of the American Meteorological Society*, *92*(8), 1023–1043.

Boé, J., Hall, A., & Qu, X. (2009). Current GCMs' unrealistic negative feedback in the Arctic. *Journal of Climate*, *22*, 4682–4695.

Boeke, R. C., & Taylor, P. C. (2016). Evaluation of the Arctic surface radiation budget in CMIP5 models. *Journal of Geophysical Research*, *121*, 8525–8548.

Boisvert, L. N., Petty, A. A., & Stroeve, J. C. (2015). The impact of the extreme winter 2015/16 Arctic cyclone on the Barents-Kara Seas. *Monthly Weather Review*, *144*(11), 4279–4287.

Borys, R. D. (1989). Studies of ice nucleation by Arctic aerosol on AGASP-II. *Journal of Atmospheric Chemistry*, *9*(1–3), 169–185.

Borys, R. D., Lowenthal, D. H., & Mitchell, D. L. (2000). The relationships among cloud microphysics, chemistry, and precipitation rate in cold mountain clouds. *Atmospheric Environment*, *34*(16), 2593–2602.

Boucher, O., Randall, D., P. Artaxo, et al. (2013). Clouds and aerosols. In T. F. Stocker, D. Qin, & G. K. Plattner, et al. (Eds.), *Climate Change 2013: The Physical Science Basis. Contribution of Working Group I to the Fifth Assessment Report of the Intergovernmental Panel on Climate Change*. Cambridge University Press.

Bourgeois, Q., & Bey, I. (2011). Pollution transport efficiency toward the Arctic: Sensitivity to aerosol scavenging and source regions. *Journal of Geophysical Research*, *116*(D8).

Brock, C. A., Cozic, J., Bahreini, R., Froyd, K. D., Middlebrook, A. M., McComiskey, A., et al. (2011). Characteristics, sources, and transport of aerosols measured in spring 2008 during the Aerosol, Radiation, and Cloud Processes affecting Arctic Climate (ARCPAC) project. *Atmospheric Chemistry and Physics*, *11*(6), 2423–2453.

Brooks, I. M., Tjernström, M., Persson, P. O. G., Shupe, M. D., Atkinson, R. A., Canut, G., et al. (2017). The turbulent structure of the Arctic summer boundary layer during the Arctic summer cloud-ocean study. *Journal of Geophysical Research*, *122*(18), 9685–9704.

Browse, J., Carslaw, K. S., Arnold, S. R., Pringle, K., & Boucher, O. (2012). The scavenging processes controlling the seasonal cycle in Arctic sulphate and black carbon aerosol. *Atmospheric Chemistry and Physics*, *12*(15), 6775–6798.

Budyko, M. I. (1969). The effect of solar radiation variations on the climate of the Earth. *Tellus*, *21*(5), 611–619.

Bullard, J. E., Baddock, M., Bradwell, T., Crusius, J., Darlington, E., Gaiero, D., et al. (2016). High-latitude dust in the Earth system. *Reviews of Geophysics*, *54*, 447–485.

Burkart, J., Willis, M. D., Bozem, H., Thomas, J. L., Law, K., Hoor, P., et al. (2017). Summertime observations of elevated levels of ultrafine particles in the high Arctic marine boundary layer. *Atmospheric Chemistry and Physics*, *17*(8), 5515–5535.

Burrows, S. M., Hoose, C., Pöschl, U., & Lawrence, M. G. (2013). Ice nuclei in marine air: Biogenic particles or dust. *Atmospheric Chemistry and Physics*, *13*(1), 245–267.

Cesana, G., Waliser, D. E., Jiang, X., & Li, J.-L. F. (2015). Multimodel evaluation of cloud phase transition using satellite and reanalysis data. *Journal of Geophysical Research*, *120*, 7871–7892.

Cess, R. D., Potter, G. L., Blanchet, J. P., Boer, G. J., Ghan, S. J., Kiehl, J. T., et al (1989). Interpretation of cloud-climate feedback as produced by 14 atmospheric general circulation models. *Science*, *245*(4917), 513–516.

Cho, H.-M., Nasiri, S. L., Yang, P., Laszlo, I., & Zhao, X. (2013). Detection of optically thin mineral dust aerosol layers over the ocean using MODIS. *Journal of Atmospheric and Oceanic Technology Atmospheric and Oceanic Technology*, *30*(5), 896–916.

Chylek, P., & Borel, C. (2004). Mixed phase cloud water/ice structure from high spatial reso- lution satellite data. *Geophysical Research Letters*, *31*(14), L14104.

Coopman, Q., Garrett, T. J., Finch, D. P., & Riedi, J. (2018a). High sensitivity of Arctic liquid clouds to long-range anthropogenic aerosol transport. *Geophysical Research Letters*, *45*(1), 372–381.

Coopman, Q., Riedi, J., Finch, D. P., & Garrett, T. J. (2018b). Evidence for changes in Arctic cloud phase due to long-range pollution transport. *Geophysical Research Letters*, *45*(19), 10709–10718.

Coopman, Q., Riedi, J., Zeng, S., & Garrett, T. J. (2020). Space-based analysis of the cloud thermodynamic phase transition for varying microphysical and meteorological regimes. *Geophysical Research Letters*, *47*(6), e2020GL087122.

Costa, A., Meyers, M., Afchine, A., Luebke, A., Gu nther, G., Dorsey, J. R., et al (2017). Classification of Arctic, midlatitude and tropical clouds in the mixed-phase temperature regime. *Atmospheric Chemistry and Physics*, *17*(19), 12219–12238.

Cox, C. J., Walden, V. P., Rowc, P. M., & Shupe, M. D. (2015). Humidity trends imply increased sensitivity to clouds in a warming Arctic. *Nature Communications*, *6*(1), 1–8.

Creamean, J., de Boer, G., Telg, H., Mei, F., Dexheimer, D., Shupe, M. D., et al. (2021). Assessing the vertical structure of Arctic aerosols using balloon-borne measurements. *Atmospheric Chemistry and Physics*, *21*(3), 1737–1757.

Creamean, J., Hill, T., DeMott, P. J., Uetake, J., Kreidenweis, S., & Douglas, T. A. (2018). The influence of local oil exploration and regional wildfires on summer 2015 aerosol over the North Slope of Alaska. *Atmospheric Chemistry and Physics*, *18*, 555–570.

Creamean, J., Hill, T., DeMott, P. J., Uetake, J., Kreidenweis, S., & Douglas, T. A. (2020). Thawing permafrost: An overlooked source of seeds for Arctic cloud formation. *Environmental Research Letters*, *15*(8).

Creamean, J. M., Cross, J. N., Pickart, R., McRaven, L., Lin, P., Pacini, A., et al. (2019). Ice nucleating particles carried from below a phytoplankton bloom to the Arctic atmosphere. *Geophysical Research Letters*, *46*, 8572–8581.

Croft, B., Martin, R. V., Leaitch, W. R., Tunved, P., Breider, T. J., D'Andrea, S. D., & Pierce, J. R. (2016). Processes controlling the annual cycle of Arctic aerosol number and size distributions. *Atmospheric Chemistry and Physics*, *16*(6), 3665–3682.

Cronin, T. W., & Tziperman, E. (2015). Low clouds suppress Arctic air formation and amplify high-latitude continental winter warming. *Proceedings of the National Academy of Sciences*, *112*(37), 11490–11495.

Curry, J. A. (1995). Interactions among aerosols, clouds, and climate of the Arctic ocean. *Science of the Total Environment*, *160*, 777–791.

Curry, J. A., Ebert, E. E., & Herman, G. F. (1988). Mean and turbulence structure of the summertime Arctic cloudy boundary layer. *Quarterly Journal of the Royal Meteorological Society*, *144*, 715–746.

Curry, J. A., Hobbs, P. V., King, M. D., Randall, D. A., Minnis, P., Isaac, G. A., et al. (2000). FIRE Arctic clouds experiment. *Bulletin of the American Meteorological Society*, *81*(1), 5–30.

Curry, J. A., Pinto, J. O., Benner, T., & Tschudi, M. (1997). Evolution of the cloudy boundary layer during the autumnal freezing of the Beaufort Sea. *Journal of Geophysical Research*, *102*(D12), 13851–13860.

Curry, J. A., Rossow, W. B., Randall, D., & Schramm, J. L. (1996). Overview of Arctic cloud and radiation characteristics. *Journal of Climate*, *9*(8), 1731–1764.

Cziczo, D. J., Froyd, K. D., Gallavardin, S. J., Moehler, O., Benz, S., Saathoff, H., & Murphy, D. M. (2009). Deactivation of ice nuclei due to atmospherically relevant surface coatings. *Environmental Research Letters*, *4*(4), 044013.

Dai, A., Luo, M., Song, M., & Liu, J. (2019). Arctic amplification is caused by sea-ice loss under increasing CO2. *Nature Communications*, *10*(1), 1–13.

D'Alessandro, J. J., Diao, M., Wu, C., Liu, X., Jensen, J. B., & Stephens, B. B. (2019). Cloud phase and relative humidity distributions over the Southern Ocean in Austral summer based on in situ observations and CAM5 simulations. *Journal of Climate*, *32*(10), 2781–2805.

Dalsøren, S., Samset, B. H., Myhre, G., Corbett, J. J., Minjares, R., Lack, D., & Fuglestvedt, J. S. (2013). Environmental impacts of shipping in 2030 with a particular focus on the Arctic region. *Atmospheric Chemistry and Physics*, *13*(4), 1941–1955.

de Boer, G., Eloranta, E. W., & Shupe, M. D. (2009). Arctic mixed-phase stratiform cloud properties from multiple years of surface-based measurements at two high-latitude locations. *Journal of the Atmospheric Sciences*, *66*(9), 2874–2887.

de Boer, G., Hashino, T., & Tripoli, G. J. (2010). Ice nucleation through immersion freezing in mixed-phase stratiform clouds: Theory and numerical simulations. *Atmospheric Research*, *96*(2–3), 315–324.

Demott, P. J., Prenni, A. J., McMeeking, G. R., Sullivan, R. C., Petters, M. D., Tobo, Y., et al. (2015). Integrating laboratory and field data to quantify the immersion freezing ice nucleation activity of mineral dust particles. *Atmospheric Chemistry and Physics*, *15*(1), 393–409.

Deslippe, J. R., Hartmann, M., Simard, S. W., & Mohn, W. W. (2012). Long-term warming alters the composition of Arctic soil microbial communities. *FEMS Microbiology Ecology*, *82*(2), 303–315.

Devasthale, A., Sedlar, J., & Tjernström, M. (2011). Characteristics of water-vapour inversions observed over the Arctic by Atmospheric Infrared Sounder (AIRS) and radiosondes. *Atmospheric Chemistry and Physics*, *11*, 9813–9823.

Di Pierro, M., Jaeglé, L., Eloranta, E. W., & Sharma, S. (2013). Spatial and seasonal distribution of Arctic aerosols observed by the CALIOP satellite instrument (2006–2012). *Atmospheric Chemistry and Physics*, *13*(14), 7075–7095.

Dimitrelos, A., Ekman, A. M. L., Caballero, R., & Savre, J. (2020). A sensitivity study of Arctic air-mass transformation using large eddy simulation. *Journal of Geophysical Research*, *125*(6).

Dunne, E.M., Gordon, H., Kürten, A., Almeida, J., Duplissy, J., Williamson, C., et al. (2016). Global atmospheric particle formation from CERN CLOUD measurements. *Science*, *354*(6316), 1119–1124.

Eastman, R., & Warren, S. G. (2010). Interannual variations of Arctic cloud types in relation to sea ice. *Journal of Climate*, *23*(15), 4216–4232.

Eastwood, M., Cremel, S., Wheeler, M., Murray, B. J., Girard, E., & Bertram, A. K. (2009). Effects of sulfuric acid and ammonium sulfate coatings on the ice nucleation properties of kaolinite particles. *Geophysical Research Letters.*, *36*(2).

Egerer, U., Ehrlich, A., Gottschalk, M., Neggers, R. A., Siebert, H., & Wendisch, M. (2021). Case study of a humidity layer above Arctic stratocumulus using balloon-borne turbulence and radiation measurements and large eddy simulations. *Atmospheric Chemistry and Physics*, 1–27.

Eirund, G. K., Possner, A., & Lohmann, U. (2019). Response of Arctic mixed-phase clouds to aerosol perturbations under different surface forcings. *Atmospheric Chemistry and Physics*, *19*(15), 9847–9864.

English, J., Kay, J. E., Gettelman, A., Liu, X., Wang, Y., Zhang, Y., & Chepfer, H. (2014). Contributions of clouds, surface albedos, and mixed-phase ice nucleation schemes to Arctic radiation biases in CAM5. *Journal of Climate*, *27*(13), 5174–5197.

Engvall, A.-C., Krejci, R., Ström, J., Treffeisen, R., Scheele, R., Hermansen, O., & Paatero, J. (2008). Changes in aerosol properties during spring-summer period in the Arctic troposphere. *Atmospheric Chemistry and Physics*, *8*(3), 445–462.

Fan, J., Wang, Y., Rosenfeld, D., & Liu, X. (2016). Review of aerosol–cloud interactions: Mechanisms, significance, and challenges. *Journal of the Atmospheric Sciences*, *73*(11), 4221–4252.

Fan, Y., Ovtchinnikov, M., Comstock, J. M., McFarlane, S. A., & Khain, A. (2009). Ice formation in Arctic mixed-phase clouds: Insights from a 3-D cloud-resolving model with size-resolved aerosol and cloud microphysics. *Journal of Geophysical Research*, *114*(D4).

Fan, Y., Ovtchinnikov, M., & Shaw, R. A. (2015). Long-lifetime ice particles in mixed-phase stratiform clouds: Quasi-steady and recycled growth. *Journal of Geophysical Research*, *120*(22), 11617–11635.

Field, P. R., Lawson, R. P., Brown, P. R. A., Lloyd, G., Westbrook, C., Moisseev, D., et al. (2017). Secondary ice production: Current state of the science and recommendations for the future. *Meteorological Monographs*, *58*.

Filioglou, M., Mielonen, T., Balis, D., Giannakaki, E., Arola, A., Kokkola, H., et al. (2019). Aerosol effect on the cloud phase of low-level clouds over the Arctic. *Journal of Geophysical Research*, *124*(14), 7886–7899.

Frey, W. R., & Kay, J. E. (2018). The influence of extratropical cloud phase and amount feedbacks on climate sensitivity. *Climate Dynamics*, *50*(7), 3097–3116.

Fridlind, A. M., & Ackerman, A. S. (2018). *Simulations of Arctic mixed-phase boundary layer clouds: Advances in understanding and outstanding questions*. In: Mixed-Phase Clouds (pp. 153–183). Elsevier.

Fridlind, A. M., Ackerman, A. S., McFarquhar, G., Zhang, G., Poellot, M. R., DeMott, P. J., et al. (2007). Ice properties of single-layer stratocumulus during the Mixed-Phase Arctic Cloud Experiment: 2. model results. *Journal of Geophysical Research*, *112*(D24).

Fridlind, A. M., Diedenhoven, B. V., Ackerman, A. S., Avramov, A., Mrowiec, A., Morrison, H., et al. (2012). A FIRE-ACE/SHEBA case study of mixed-phase Arctic boundary layer clouds: Entrainment rate limitations on rapid primary ice nucleation processes. *Journal of the Atmospheric Sciences*, *69*(1), 365–389.

Fu, S., & Xue, H. (2017). The effect of ice nuclei efficiency on Arctic mixed-phase clouds from large-eddy simulations. *Journal of the Atmospheric Sciences.*, *74*(12), 3901–3913.

Gagné, M. E., Fyfe, J. C., Gillett, N. P., Polyakov, I. V., & Flato, G. M. (2017). Aerosol-driven increase in Arctic sea ice over the middle of the twentieth century. *Geophysical Research Letters*, *44*(14), 7338–7346.

Ganeshan, M., & Wu, D. L. (2015). An investigation of the Arctic inversion using COSMIC RO observations. *Journal of Geophysical Research*, *120*(18), 9338–9351.

Garimella, S., Rothenberg, D. A., Wolf, M. J., Wang, C., & Cziczo, D. J. (2018). How uncertainty in field measurements of ice nucleating particles influences modeled cloud forcing. *Journal of the Atmospheric Sciences*, *75*(1), 179–187.

Garrett, T., & Zhao, C. (2006). Increased Arctic cloud longwave emissivity associated with pollution from mid-latitudes. *Nature*, *440*(7085), 787–789.

Garrett, T. J., Fallgatter, C., Shkurko, K., & Howlett, D. (2012). Fall speed measurement and high-resolution multi-angle photography of hydrometeors in free fall. *Atmospheric Measurement Techniques*, *5*(11), 2625–2633.

Garrett, T. J., Maestas, M. M., Krueger, S. K., & Schmidt, C. T. (2009). Acceleration by aerosol of a radiative-thermodynamic cloud feedback influencing Arctic surface warming. *Geophysical Research Letters*, *36*(19).

Gierens, R., Kneifel, S., Shupe, M. D., Ebell, K., Maturilli, M., & Löhnert, U. (2020). Low- level mixed-phase clouds in a complex Arctic environment. *Atmospheric Chemistry and Physics*, *20*, 3459–3481.

Girard, E., & Asl, N. S. (2014). Relative importance of acid coating on ice nuclei in the deposition and contact modes for wintertime Arctic clouds and radiation. *Meteorology and Atmospheric Physics*, *123*(1), 81–92.

Girard, E., Blanchet, J.-P., & Dubois, Y. (2005). Effects of arctic sulphuric acid aerosols on wintertime low-level atmospheric ice crystals, humidity and temperature at Alert, *Nunavut. Atmospheric Research*, *73*(1–2), 131–148.

Gong, T., & Luo, D. (2017). Ural blocking as an amplifier of the Arctic sea ice decline in winter. *Journal of Climate*, *30*, 2639–2654.

Graham, R. M., Rinke, A., Cohen, L., Hudson, S. R., Walden, V. P., Granskog, M. A., et al. (2017). A comparison of the two Arctic atmospheric winter states observed during N-ICE2015 and SHEBA. *Journal of Geophysical Research*, *122*(11), 5716–5737.

Graversen, R. G., & Langen, P. L. (2019). On the role of the atmospheric energy transport in 2xCO2-induced polar amplification in CESM1. *Journal of Climate*, *32*(13), 3941–3956.

Graversen, R. G., & Wang, M. (2009). Polar amplification in a coupled climate model with locked albedo. *Climate Dynamics*, *33*(5), 629–643.

Grenier, P., & Blanchet, J.-P. (2010). Investigation of the sulphate-induced freezing inhibition effect from CloudSat and CALIPSO measurements. *Journal of Geophysical Research*, *115*(D22).

Griesche, H. J., Ohneiser, K., Seifert, S., Ansmann, A., & Engelmann, R. (2020). Contrasting ice formation in Arctic clouds: Surface coupled vs decoupled clouds. *Atmospheric Chemistry and Physics Discussions*, 1–22.

Grosvenor, D. P., & Wood, R. (2014). The effect of solar zenith angle on MODIS cloud optical and microphysical retrievals within marine liquid water clouds. *Atmospheric Chemistry and Physics*, *14*(14), 7291–7321.

Gultepe, I., Isaac, G., Hudak, D., Nissen, R., & Strapp, J. W. (2000). Dynamical and microphysical characteristics of Arctic clouds during BASE. *Journal of Climate.*, *13*(7), 1225–1254.

Hall, A. (2004). The role of surface albedo feedback in climate. *Journal of Climate.*, *17*(7), 1550–1568.

Hallen, H. D., & Philbrick, C. R. (2018). Lidar detection of small aerosol size distribution. *Laser Radar Technology and Applications,* XXIII, 10636.

Harrington, J., Reisin, T., Cotton, W. R., & Kreidenweis, S. M. (1999). Cloud resolving simulations of Arctic stratus: Part II: Transition-season clouds. *Atmospheric Research*, *51*(1), 45–75.

Hartmann, J., Albers, F., Argentini, S., Bochert, A., Bonafe, U., Cohrs, W., et al. (1999). Arctic Radiation and Turbulence Interaction Study (ARTIST). *Berichtezur Polarforschung (Reports on Polar Research) Alfred Wegener Institute for Polar and Marine Research, Bremerhaven*, 305.

Hartmann, J. N. C., Kottmeier, C., & Wamser, C. (1991). Radiation and Eddy Flux Experiment 1991 (REFLEX I). *Berichtezur Polarforschung*, *105*.

Hartmann, M., Adachi, K., Eppers, O., Haas, C., Herber, A., Holzinger, R., et al. (2019a). Wintertime airborne measurements of ice nucleating particles in the high Arctic: A hint to a marine, biogenic source for ice nucleating particles. *Geophysical Research Letters*, *47*(13), e2020GL087770.

Hartmann, M., Blunier, T., Brügger, S. O., Schmale, J., Schwikowski, M., Vogel, A., et al. (2019b). Variation of ice nucleating particles in the European Arctic over the last centuries. *Geophysical Research Letters*, *46*(7), 4007–4016.

Hegyi, B. M., & Taylor, P. C. (2018). The unprecedented 2016–2017 Arctic sea ice growth season: The crucial role of atmospheric rivers and longwave fluxes. *Geophysical Research Letters*, *45*(10), 5204–5212.

Heintzenberg, J., Tunved, P., Gali, M., & Leck, C. (2017). New particle formation in the Svalbard region 2006–2015. *Atmospheric Chemistry and Physics*, *17*(10), 6153–6175.

Herenz, P., Wex, H., Henning, S., Bjerring, T., Kristensen, F. R., Roth, A., et al. (2018). Measurements of aerosol and CCN properties in the Mackenzie River Delta (Canadian Arctic) during spring–summer transition in may 2014. *Atmospheric Chemistry and Physics*, *18*(7), 4477–4496.

Herman, G., & Goody, R. (1976). Formation and persistence of summertime Arctic stratus clouds. *Journal of the Atmospheric Sciences*, *33*(8), 1537–1553.

Hind, A., Zhang, Q., & Brattstrom, G. (2016). Problems encountered when defining Arctic amplification as a ratio. *Scientific Reports*, *6*(30469).

Holland, M. M., & Bitz, C. M. (2003). Polar amplification of climate change in coupled models. *Climate Dynamics*, *21*(2–3), 221–232.

Huang, J., & Jaeglé, L. (2017). Wintertime enhancements of sea salt aerosol in polar regions consistent with a sea ice source from blowing snow. *Atmospheric Chemistry and Physics*, *17*, 3699–3712.

Huang, W. T. K., Ickes, L., Tegen, I., Rinaldi, M., Ceburnis, D., & Lohmann, U. (2018). Global relevance of marine organic aerosol as ice nucleating particles. *Atmospheric Chemistry and Physics*, *18*(15), 11423–11445.

Huang, Y., Ding, Q., Dong, X., Xi, B., & Baxter, I. (2021). Summertime low clouds mediate the impact of the large-scale circulation on Arctic sea ice. *Communications Earth & Environment*, *2*(1), 1–10.

Hwang, Y.-T., Frierson, D. M. W., & Kay, J. E. (2011). Coupling between Arctic feedbacks and changes in poleward energy transport. *Geophysical Research Letters*, *38*(17).

Igel, A., Ekman, A. M. L., Leck, C., Tjernström, M., Savre, J., & Sedlar, J. (2017). The free troposphere as a potential source of Arctic boundary layer aerosol particles. *Geophysical Research Letters*, *44*(13), 7053–7060.

Irish, V. E., Hanna, S. J., Willis, M. D., China, W., Thomas, J. L., Wentzell, J. B., et al. (2019). Ice nucleating particles in the marine boundary layer in the Canadian Arctic during summer 2014. *Atmospheric Chemistry and Physics*, *19*(2), 1027–1039.

Jacob, D. J., Crawford, J. H., Maring, H., Clarke, A. D., Dibb, J. E., Emmons, L. K., et al. (2010). The Arctic Research of the Composition of the Troposphere from Aircraft and Satellites (ARCTAS) mission: Design, execution, and first results. *Atmospheric Chemistry and Physics*, *10*, 5191–5212.

Jiang, H., Cotton, W. R., Pinto, J. O., Curry, J. A., & Weissbluth, M. J. (2000). Cloud resolving simulations of mixed-phase Arctic stratus observed during BASE: Sensitivity to concentration of ice crystals and large-scale heat and moisture advection. *Journal of the Atmospheric Sciences*, *57*(13), 2105–2117.

Johansson, E., Devasthale, A., Tjernström, M., Ekman, A. M. L., & L'Ecuyer, T. (2017). Response of the lower troposphere to moisture intrusions into the Arctic. *Geophysical Research Letters*, *44*, 2527–2536.

Jouan, C., Girard, E., Pelon, J., Gultepe, I., Delanoë, J., & Blanchet, J. P. (2012). Characterization of Arctic ice cloud properties observed during ISDAC. *Journal of Geophysical Research*, *117*(D23).

Kalesse, H., de Boer, G., Solomon, A., Oue, M., Ahlgrimm, M., Zhang, D., et al. (2016). Understanding rapid changes in phase partitioning between cloud liquid and ice in stratiform mixed-phase clouds: An Arctic case study. *Monthly Weather Review*, *144*(12), 4805–4826.

Kanji, Z., Ladino, L. A., Wex, H., Boose, Y., Burkert-Kohn, M., Cziczo, D. J., & Krämer, M. (2017). Overview of ice nucleating particles. *Meteorological Monographs*, *58*(1).

Kapsch, M.-L., Graversen, R.G., & Tjernström, M. (2013). Springtime atmospheric energy transport and the control of Arctic summer sea-ice extent. *Nature Climate Change*, *3*, 744–748.

Karlsson, J., & Svensson, G. (2011). The simulation of Arctic clouds and their influence on the winter surface temperature in present-day climate in the CMIP3 multi-model dataset. *Climate Dynamics*, *36*, 623–635.

Kato, S., Rose, F. G., Rutan, D. A., Thorsen, T. J., Loeb, N. G., Doelling, D. R., et al. (2018). Surface irradiances of edition 4.0 Clouds and the Earth's Radiant Energy System (CERES) Energy Balanced and Filled (EBAF) data product. *Journal of Climate*, *31*(11), 4501–4527.

Kay, J. E., & Gettelman, A. (2009). Cloud influence on and response to seasonal Arctic sea ice loss. *Journal of Geophysical Research, 114*(D18204).

Kay, J. E., L'Ecuyer, T., Pendergrass, A., Chepfer, H., Guzman, R., & Yettella, V. (2018). Scale- aware and definition-aware evaluation of modeled near-surface precipitation frequency using CloudSat observations. *Journal of Geophysical Research, 123*(8), 4294–4309.

Khanal, S., Wang, Z., & French, J. R. (2020). Improving middle and high latitude cloud liquid water path measurements from MODIS. *Atmospheric Research, 243*(105033).

Klein, S. A., Hall, A., Norris, J. R., & Pincus, R. (2017). Low-cloud feedbacks from cloud- controlling factors: A review. *Surveys in Geophysics, 38*, 1307–1329.

Klein, S. A., & Hartmann, D. L. (1993). The seasonal cycle of low stratiform clouds. *Journal of Climate, 6*(8), 1587–1606.

Klein, S. A., McCoy, R. B., Morrison, H., Ackerman, A. S., Avramov, A., de Boer, G., et al. (2009). Intercomparison of model simulations of mixed-phase clouds observed during the ARM mixed-phase Arctic cloud experiment. I: Single-layer cloud. *Quarterly Journal of the Royal Meteorological Society, 135*(641), 979–1002.

Klingebiel, M., de Lozar, A., Molleker, S., Weigel, R., Roth, A., Schmidt, L., et al. (2015). Arctic low-level boundary layer clouds: In situ measurements and simulations of mono-and bimodal supercooled droplet size distributions at the top layer of liquid phase clouds. *Atmospheric Chemistry and Physics, 15*(2), 617–631.

Knopf, D. A., & Koop, T. (2006). Heterogeneous nucleation of ice on surrogates of mineral dust. *Journal of Geophysical Research, 111*(D12).

Knudsen, E. M., Heinold, B., Dahlke, S., Bozem, H., Crewell, S., Gorodetskaya, I. V., et al. (2018). Meteorological conditions during the ACLOUD/PASCAL field campaign near Svalbard in early summer 2017. *Atmospheric Chemistry and Physics, 18*, 17995–18022.

Koch, D., Menon, S., Genio, A. D. D., Ruedy, R., Alienov, I., & Schmidt, G. A. (2009). Distinguishing aerosol impacts on climate over the past century. *Journal of Climate, 22*(10), 2659–2677.

Koike, M., Ukita, J., Stro m, J., Tunved, P., Shiobara, M., Vitale, V., et al. (2019). Year-round in situ measurements of Arctic low-level clouds: Microphysical properties and their relationships with aerosols. *Journal of Geophysical Research, 124*(3), 1798–1822.

Komurcu, M., Storelvmo, T., Tan, I., Lohmann, J., Yun, Y., Penner, J. E., et al. (2014). Intercomparison of the cloud water phase among global climate models. *Journal of Geophysical Research, 119*(6), 3372–3400.

Korolev, A. V. (2007). Limitations of the Wegener–Bergeron–Findeisen mechanism in the evolution of mixed-phase clouds. *Journal of the Atmospheric Sciences, 64*(9), 3372–3375.

Korolev, A. V., Emery, E. F., Strapp, J. W., Cober, S. G., & Isaac, G. A. (2013). Quantification of the effects of shattering on airborne ice particle measurements. *Journal of Atmospheric and Oceanic Technology, 30*(11), 2527–2553.

Korolev, A. V., Isaac, G. A., Cober, S. G., Strapp, J. W., & Hallett, J. (2003). Microphysical characterization of mixed-phase clouds. *Quarterly Journal of the Royal Meteorological Society, 129*(587), 39–65.

Korolev, A. V., & Leisner, T. (2020). Review of experimental studies of secondary ice production. *Atmospheric Chemistry and Physics, 20*(20), 11767–11797.

Korolev, A. V., McFarquhar, G., Field, P. R., Franklin, C., Lawson, P., Wang, Z., et al. (2017). Mixed-phase clouds: Progress and challenges. *Meteorological Monographs, 58*(5–1).

Kuma, P., McDonald, A. J., Morgenstern, O., Querel, R., Silber, I., & Flynn, C. J. (2021). Ground- based lidar processing and simulator framework for comparing models and observations (ALCF 1.0). *Geoscientific Model Development, 14*(1), 43–72.

Lance, S., Shupe, M. D., Feingold, G., Brock, C. A., Cozic, J., Holloway, J. S., et al. (2011). Cloud condensation nuclei as a modulator of ice processes in Arctic mixed-phase clouds. *Atmospheric Chemistry and Physics, 11*(15), 8003–8015.

Lannefors, H., Heintzenberg, J., & Hansson, H.-C. (1983). A comprehensive study of physical and chemical parameters of the Arctic summer aerosol: Results from the Swedish expedition Ymer-80. *Tellus B: Chemical and Physical Meteorology, 35*(1), 40–54.

Law, K. S., & Stohl, A. (2007). Arctic air pollution: Origins and impacts. *Science, 315*(5818), 1537–1540.

Law, K. S., Stohl, A., Quinn, P. K., Brock, C. A., Burkhart, J. F., Paris, J. D., et al. (2014). Arctic air pollution: New insights from POLARCAT-IPY. *Bulletin of the American Meteorological Society, 95*(12), 1873–1895.

Lawson, R. P., Woods, S., & Morrison, H. (2015). The microphysics of ice and precipitation development in tropical cumulus clouds, *Journal of the Atmospheric Sciences, 72*(6), 2429–2445.

Leaitch, W. R., Korolev, A., Aliabadi, A. A., Burkart, J., Willis, M. D., Abbatt, J. P., et al. (2016). Effects of 20–100nm particles on liquid clouds in the clean summertime Arctic. *Atmospheric Chemistry and Physics, 16*(17), 11107–11124.

Lebsock, M., & Su, H. (2014). Application of active spaceborne remote sensing for understanding biases between passive cloud water path retrievals. *Journal of Geophysical Research, 119*(14), 8962–8979.

Leck, C., Norman, M., Bigg, E. K., & Hillamo, R. (2002). Chemical composition and sources of the high Arctic aerosol relevant for cloud formation. *Journal of Geophysical Research, 107*(D12).

Leck, C., & Svensson, E. (2015). Importance of aerosol composition and mixing state for cloud droplet activation over the Arctic pack ice in summer. *Atmospheric Chemistry and Physics, 15*(5), 2545–2568.

Lee, S. (2014). A theory for polar amplification from a general circulation perspective. *Asia-Pacific Journal of Atmospheric Sciences, 50*, 31–43.

Li, X., Krueger, S. K., Strong, C., Mace, G. G., & Benson, S. (2020). Midwinter Arctic leads form and dissipate low clouds. *Nature Communications, 11*(1), 1–8.

Liu, C., & Barnes, E. A. (2015). Extreme moisture transport into the Arctic linked to Rossby wave breaking. *Journal of Geophysical Research, 120*, 3774–3788.

Liu, X., Boyle, S., Klein, S. A., Shi, X., Wang, Z., Lin, W., et al. (2011). Testing cloud microphysics parameterizations in NCAR CAM5 with ISDAC and M-PACE observations. *Journal of Geophysical Research*, *116*(D1).

Liu, Y., Key, J. R., Ackerman, S. A., Mace, G. G., & Zhang, Q. (2012). Arctic cloud macrophysical characteristics from CloudSat and CALIPSO. *Remote Sensing of Environment*, *124*, 159–173.

Liu, Y., Key, J. R., Vavrus, S., & Woods, C. (2018). Time evolution of the cloud response to moisture intrusions into the Arctic during winter. *Journal of Climate*, *31*(22), 9389–9405.

Liu, Y., Shupe, M. D., Wang, Z., & Mace, G. (2017). Cloud vertical distribution from combined surface and space radar-lidar observations at two Arctic atmospheric observatories. *Atmospheric Chemistry and Physics*, *17*(9), 5973–5989.

Lloyd, G., Choularton, T. W., Bower, K. N., Crosier, J., Jones, H., Dorsey, J. R., et al. (2015). Observations and comparisons of cloud microphysical properties in spring and summertime Arctic stratocumulus clouds during the ACCACIA campaign. *Atmospheric Chemistry and Physics*, *15*, 3719–3737.

Loewe, K., Ekman, A. M., Paukert, M., Sedlar, J., Tjernström, M., & Hoose, C. (2017). Modelling micro- and macrophysical contributors to the dissipation of an Arctic mixed-phase cloud during the Arctic Summer Cloud Ocean Study (ASCOS). *Atmospheric Chemistry and Physics*, *17*(11), 6693–6704.

Lohmann, U. (2002). A glaciation indirect aerosol effect caused by soot aerosols. *Geophysical Research Letters*, *29*(4).

Lohmann, U. (2017). Anthropogenic aerosol influences on mixed-phase clouds. *Current Climate Change Reports*, *3*(1), 32–44.

Lohmann, U., Zhang, J., & Pi, J. (2003). Sensitivity studies of the effect of increased aerosol concentrations and snow crystal shape on the snowfall rate in the Arctic. *Journal of Geophysical Research*, *108*(D11).

Lubin, D., & Vogelmann, A. M. (2006). A climatologically significant aerosol longwave indirect effect in the Arctic. *Nature*, *439*(7075), 453–456.

Luo, Y., Xu, K.-M., Morrison, H., & McFarquhar, G. (2008). Arctic mixed-phase clouds simulated by a cloud-resolving model: Comparison with ARM observations and sensitivity to microphysics parameterizations. *Journal of the Atmospheric Sciences*, *65*(4), 1285–1303.

Mahmood, R., von Salzen, K., Norman, A.-L., Gali, M., & Levasseur, M. (2019). Sensitivity of Arctic sulfate aerosol and clouds to changes in future surface seawater dimethylsulfide concentrations. *Atmospheric Chemistry and Physics*, *19*, 6419–6435.

Manabe, S., & Wetherald, R. T. (1975). The effects of doubling the CO2 concentration on the climate of a general circulation model. *Journal of the Atmospheric Sciences*, *32*(1), 3–15.

Marelle, L., Raut, J.-C., Thomas, J. L., Law, K. S., Quennehen, B., Ancellet, G., et al. (2015). Transport of anthropogenic and biomass burning aerosols from Europe to the Arctic during spring 2008. *Atmospheric Chemistry and Physics*, *15*(7), 3831–3850.

Mason, R. H., Si, M., Chou, C., Irish, V. E., Dickie, R., Elizondo, P., et al. (2016). Size-resolved measurements of ice-nucleating particles at six locations in North America and one in Europe. *Atmospheric Chemistry and Physics*, *16*(3), 1637–1651.

Mauritsen, T., Sedlar, J., Tjernstro m, M., Leck, C., Martin, M., Shupe, M., et al. (2011). An Arctic CCN-limited cloud-aerosol regime. *Atmospheric Chemistry and Physics*, *11*, 165–173.

McCluskey, C. S., Hill, T. C. J., Sultana, C. M., Laskina, O., Trueblood, J., Santander, M. V., et al. (2018). A mesocosm double feature: Insights into the chemical makeup of marine ice nucleating particles. *Journal of the Atmospheric Sciences*, *75*(7), 2405–2423.

McCoy, D. T., Tan, I., Hartmann, D. L., Zelinka, M. D., & Storelvmo, T. (2016). On the relationships among cloud cover, mixed-phase partitioning, and planetary albedo in GCMs. *Journal of Advances in Modeling Earth Systems*, *8*(2), 650–668.

McFarquhar, G. M., Ghan, S., Verlinde, J., Korolev, A., Strapp, J. W., Schmid, B., et al. (2011). Indirect and semi-direct aerosol campaign: The impact of Arctic aerosols on clouds. *Bulletin of the American Meteorological Society*, *92*(2), 183–201.

McIlhattan, E. A., L'Ecuyer, T. S., & Miller, N. B. (2017). Observational evidence linking Arctic supercooled liquid cloud biases in CESM to snowfall processes. *Journal of Climate*, *30*(12), 4477–4495.

Medeiros, B., Deser, C., Tomas, R. A., & Kay, J. E. (2011). Arctic inversion strength in climate models. *Journal of Climate*, *24*, 4733–4740.

Meredith, M., Sommerkorn, M., Cassotta, S., Derksen, C., Ekaykin, A., Hollwed, A., et al. (2019). Polar regions. In *IPCC Special Report on the Ocean and Cryosphere in a Changing Climate*.

Merikanto, J., Spracklen, D. V., Mann, G. W., Pickering, S. J., & Carslaw, K. S. (2009). Impact of nucleation on global CCN. *Atmospheric Chemistry and Physics*, *9*(21), 8601–8616.

Merlis, T. M., & Henry, M. (2018). Simple estimates of polar amplification in moist diffusive energy balance models. *Journal of Climate*, *31*(15), 5811–5824.

Messori, G., Woods, C., & Caballero, R. (2018). On the drivers of wintertime temperature extremes in the high Arctic. *Journal of Climate*, *31*, 1597–1618.

Middlemas, E. A., Kay, J. E., Medeiros, B. M., & Maroon, E. A. (2020). Quantifying the influence of cloud radiative feedbacks on Arctic surface warming using cloud locking in an Earth system model. *Geophysical Research Letters*, *47*(e2020GL089207).

Mioche, G., Jourdan, O., Ceccaldi, M., & Delanoe, J. (2015). Variability of mixed-phase clouds in the Arctic with a focus on the Svalbard region: A study based on spaceborne active remote sensing. *Atmospheric Chemistry and Physics*, *15*, 2445–2461.

Mioche, G., Jourdan, O., Delanoë, J., Gourbeyre, C., Febvre, G., Dupuy, R., et al. (2017). Vertical distribution of microphysical properties of Arctic springtime low-level mixed-phase clouds over the Greenland and Norwegian seas. *Atmospheric Chemistry and Physics*, *17*(20), 12845–12869.

Mitchell, J. F. B., Senior, C. A., & Ingram, W. J. (1989). CO_2 and climate: A missing feedback? *Nature*, *341*(6238), 132–134.

Moore, R. H., Bahreini, R., Brock, C. A., Froyd, K. D., Cozic, J., Holloway, J. S., et al. (2011). Hygroscopicity and composition of Alaskan Arctic CCN during April 2008. *Atmospheric Chemistry and Physics*, *11*(22), 11807–11835.

Morrison, A. L., Kay, J. E., Chepfer, H., Guzman, R., & Yettella, V. (2018). Isolating the liquid cloud response to recent Arctic sea ice variability using spaceborne lidar observations. *Journal of Geophysical Research*, *123*(1), 473–490.

Morrison, H., Boer, G. D., Feingold, G., Harrington, J., Shupe, M. D., & Sulia, K. (2012). Resilience of persistent Arctic mixed-phase clouds. *Nature Geoscience*, *5*(1), 11–17.

Morrison, H., Shupe, M. D., Pinto, J. O., & Curry, J. A. (2005). Possible roles of ice nucleation mode and ice nuclei depletion in the extended lifetime of Arctic mixed-phase clouds. *Geophysical Research Letters*, *32*(18).

Morrison, H., Zuidema, P., Ackerman, A. S., Avramov, A., de Boer, G., Fan, J., et al. (2011). Intercomparison of cloud model simulations of Arctic mixed-phase boundary layer clouds observed during SHEBA/FIRE- ACE. *Journal of Advances in Modeling Earth Systems*, *3*(2).

Morrison, M., McCoy, R. B., Klein, S. A., Xie, S., Luo, Y., Avramov, A., et al. (2009). Intercomparison of model simulations of mixed- phase clouds observed during the ARM mixed-phase Arctic cloud experiment. II: Multilayer cloud. *Quarterly Journal of the Royal Meteorological Society*, *135*, 1003–1019.

Mortin, J., Svensson, G., Graversen, R. G., Kapsch, M. L., Stroeve, J. C., & Boisvert, L. N. (2016). Melt onset over Arctic sea ice controlled by atmospheric moisture transport. *Geophysical Research Letters*, *43*(12), 6636–6642.

Murray, B. J., O'Sullivan, D., Atkinson, J. D., & Webb, M. E. (2012). Ice nucleation by particles immersed in supercooled cloud droplets. *Chemical Society Reviews*, *41*(19), 6519–6554.

Myers, T. A., & Norris, J. R. (2013). Observational evidence that enhanced subsidence reduces subtropical marine boundary layer cloudiness. *Journal of Climate*, *26*(19), 7507–7524.

Naakka, T., Nygård, T., & Vihma, T. (2018). Arctic humidity inversions: Climatology and processes. *Journal of Climate*, *31*(10), 3765–3787.

Nam, C., Bony, S., Dufresne, J. L., & Chepfer, H. (2012). The 'too few, too bright' tropical low-cloud problem in CMIP5 models. *Geophysical Research Letters*, *39*(21).

National Academies of Sciences, E., & Medicine (2018). *Thriving on our changing planet: A decadal strategy for Earth observation from space*. National Academies Press.

Neggers, R. A. J., Chylik, J., Egerer, U., Griesche, H., Schemann, V., Seifert, P., et al. (2019). Local and remote controls on Arctic mixed-layer evolution. *Journal of Advances in Modeling Earth Systems*, *11*, 2214–2237.

Nicholls, S. (1984). The dynamics of stratocumulus: Aircraft observations and comparisons with a mixed layer model. *Quarterly Journal of the Royal Meteorological Society*, *110*(466), 783–820.

Nomokonova, T., Ebell, K., Löhnert, U., Maturilli, M., Ritter, C., & O'Connor, E. (2019). Statistics on clouds and their relation to thermodynamic conditions at Ny-Alesund using ground-based sensor synergy. *Atmospheric Chemistry and Physics*, *19*, 4105–4126.

Norgren, M. S., de Boer, G., & Shupe, M. D. (2018). Observed aerosol suppression of cloud ice in low-level Arctic mixed-phase clouds. *Atmospheric Chemistry and Physics*, *18*(18), 13345–13361.

Nygård, T., Valkonen, T., & Vihma, T. (2014). Characteristics of Arctic low-tropospheric humidity inversions based on radio soundings. *Atmospheric Chemistry and Physics*, *14*, 1959–1971.

Ovchinnikov, M., Ackerman, A. S., Avramov, A., Cheng, A., Fan, J., Fridlind, A. M., et al. (2014). Intercomparison of large-eddy simulations of Arctic mixed-phase clouds: Importance of ice size distribution assumptions. *Journal of Advances in Modeling Earth Systems*, *6*(1), 223–248.

Park, D.-S. R., Lee, S., & Feldstein, S. B. (2015a). Attribution of the recent winter sea ice decline over the Atlantic sector of the Arctic ocean. *Journal of Climate*, *28*, 4027–4033.

Park, H.-S., Lee, S., Son, S.-W., Feldstein, S. B., & Kosaka, Y. (2015b). The impact of poleward moisture and sensible heat flux on Arctic winter sea ice variability. *Journal of Climate*, *28*, 5030–5040.

Park, H.-S., Lee, Y., Kosaka, S.-S., & Kim, S.-W. (2015c). The impact of Arctic winter infrared radiation on early summer sea ice. *Journal of Climate*, *28*, 6281–6296.

Pavelsky, T. M., Boé, J., Hall, A., & Fetzer, E. J. (2011). Atmospheric inversion strength over polar oceans in winter regulated by sea ice. *Climate Dynamics*, *36*(5–6), 945–955.

Persson, P. O. G., Shupe, M. D., & Perovich, D. (2017). Linking atmospheric synoptic transport, cloud phase, surface energy fluxes, and sea-ice growth: Observations of midwinter SHEBA conditions. *Climate Dynamics*, *49*, 1341–1364.

Petty, G. W. (2006). *A first course in atmospheric radiation*. Sundog Publication.

Pinto, J. O. (1998). Autumnal mixed-phase cloudy boundary layers in the Arctic. *Journal of the Atmospheric Sciences*, *55*(11), 2016–2038.

Pinto, J. O., & Curry, J. A. (1995). Atmospheric convective plumes emanating from leads: Microphysical and radiative processes. *Journal of Geophysical Research*, *100*(C3), 4633–4642.

Pithan, F., Ackerman, A., Angevine, W. M., Hartung, K., Ickes, L., Kelley, M., et al. (2016). Select strengths and biases of models in representing the Arctic winter boundary layer over sea ice: The Larcform 1 single column model intercomparison. *Journal of Advances in Modeling Earth Systems*, *8*(3), 1345–1357.

Pithan, F., & Mauritsen, T. (2014). Arctic amplification dominated by temperature feedbacks in contemporary climate models. *Nature Geoscience*, *7*(3), 181–184.

Pithan, F., Medeiros, B. M., & Mauritsen, T. (2014). Mixed-phase clouds cause climate model biases in Arctic wintertime temperature inversions. *Climate Dynamics*, *43*, 289–303.

Pithan, F., Svensson, G., Caballero, R., Chechin, D., Cronin, T. W., Ekman, A. M. L., et al. (2018). Role of air-mass transformations in exchange between the Arctic and mid-latitudes. *Nature Geoscience*, *11*(11), 805–812.

Possner, A., Ekman, A. M. L., & Lohmann, U. (2017). Cloud response and feedback processes in stratiform mixed-phase clouds perturbed by ship exhaust. *Geophysical Research Letters*, *44*(4), 1964–1972.

Prenni, A. J., DeMott, P. J., Rogers, D. C., Kreidenweis, S. M., McFarquhar, G., Zhang, G., & Poellot, M. R. (2009). Ice nuclei characteristics from M-PACE and their relation to ice formation in clouds. *Tellus B: Chemical and Physical Meteorology*, *61*(2), 436–448.

Prenni, A. J., Harrington, J. Y., Tjernström, M., DeMott, P. J., Avramov, A., Long, C. N., et al. (2007). Can ice-nucleating aerosols affect Arctic seasonal climate? *Bulletin of the American Meteorological Society*, *88*(4), 541–550.

Quinn, P. K., Bates, T. S., Baum, E., Doubleday, N., Fiore, A. M., Flanner, M., et al. (2008). Short-lived pollutants in the Arctic: Their climate impact and possible mitigation strategies. *Atmospheric Chemistry and Physics*, *8*(6), 1723–1735.

Quinn, P. K., Coffamn, D. J., Johnson, J. E., Upchurch, L. M., & Bates, T. S. (2017). Small fraction of marine cloud condensation nuclei made up of sea spray aerosol. *Nature*, *10*(9), 674–679.

Quinn, P. K., Shaw, G., Andrews, E., Dutton, E. G., Ruoho-Airoloa, T., & Gong, S. L. (2007). Arctic Haze: Current trends and knowledge gap. *Chemical and Physical Meteorology*, *59*(1), 99–114.

Regayre, L. A., Pringle, K. J., Lee, L. A., Rap, A., Browse, J., Mann, G. W., et al. (2015). The climatic importance of uncertainties in regional aerosol–cloud radiative forcings over recent decades. *Journal of Climate*, *28*(17), 6589–6607.

Rösel, A., & Kaleschke, L. (2012). Melt ponds on Arctic sea ice determined from MODIS satellite data using an artificial neural network. *The Cryosphere*, *6*(2), 431–446.

Rosenfeld, D. (2000). Suppression of rain and snow by urban and industrial air pollution. *Science*, *287*(5459), 1793–1796.

Ruiz-Donoso, E., Ehrlich, A., Schäfer, M., Jäkel, E., Schemann, V., Crewell, S., et al. (2020). Small-scale structure of thermodynamic phase in Arctic mixed-phase clouds observed by airborne remote sensing during a cold air outbreak and a warm air advection event. *Atmospheric Chemistry and Physics*, *20*, 5487–5511.

Russell, L. M., Hawkins, L. N., Frossard, A. A., Quinn, P. K., & Bates, T. S. (2010). Carbohydrate- like composition of submicron atmospheric particles and their production from ocean bubble bursting. *Proceedings of the National Academy of Sciences*, *107*(15), 6652–6657.

Schmale, J., Zieger, P., & Ekman, A. M. L. (2021). Aerosols in current and future Arctic climate. *Nature Climate Change*, *11*, 95–105.

Schmale, J. S. R., Arnold, K. S., Law, T., Thorp, T., Anenberg, S., Simpson, W. R., et al. (2018). Local Arctic air pollution: A neglected but serious problem. *Earth's Future*, *6*, 1385–1412.

Screen, J. A., & Simmonds, I. (2010). The central role of diminishing sea ice in recent Arctic temperature amplification. *Nature*, *464*(7293), 1334–1337.

Sedlar, J., & Shupe, M. D. (2014). Characteristic nature of vertical motions observed in Arctic mixed-phase stratocumulus. *Atmospheric Chemistry and Physics*, *14*(6), 3461–3478.

Sedlar, J., Shupe, M. D., & Tjernström, M. (2012). On the relationship between thermodynamic structure, cloud top, and climate significance in the Arctic. *Journal of Climate*, *25*(7).

Sedlar, J., Tjernström, M., Mauritsen, T., Shupe, M. D., Brooks, I. M., Persson, P. O. G., et al. (2011). A transitioning Arctic surface energy budget: The impacts of solar zenith angle, surface albedo and cloud radiative forcing. *Climate Dynamics*, *37*(7–8), 1643–1660.

Sellers, W. D. (1969). A global climatic model based on the energy balance of the Earth- atmosphere system. *Journal of Applied Meteorology and Climatology*, *8*(3), 392–400.

Serreze, M., Barrett, A. P., Stroeve, J. C., Kindig, D. N., & Holland, M. M. (2009). The emergence of surface-based Arctic amplification. *The Cryosphere*, *3*(11–19).

Serreze, M. C., & Barry, R. G. (2011). Processes and impacts of Arctic amplification: A research synthesis. *Global and Planetary Change*, *77*, 85–96.

Shaw, G. E. (1995). The Arctic haze phenomenon. *Bulletin of the American Meteorological Society*, *76*(12), 2403–2414.

Shen, Z., Ming, Y., Horowitz, L. W., Ramaswamy, V., & Lin, M. (2017). On the seasonality of Arctic black carbon. *Journal of Climate*, *30*(12), 4429–4441.

Shindell, D., & Faluvegi, G. (2009). Climate response to regional radiative forcing during the twentieth century. *Nature Geoscience*, *2*(4), 294–300.

Shupe, M. D., & Intrieri, J. M. (2004). Cloud radiative forcing of the Arctic surface: The influence of cloud properties, surface albedo, and solar zenith angle. *Journal of Climate*, *17*(3), 616–628.

Shupe, M. D., Matrosov, S. Y., & Uttal, T. (2006). Arctic mixed-phase cloud properties derived from surface-based sensors at SHEBA. *Journal of the Atmospheric Sciences*, *63*(2), 697–711.

Shupe, M. D., Persson, P. O. G., Brooks, I. M., Tjernström, M., Sedlar, J., Mauritsen, T., et al. (2013). Cloud and boundary layer interactions over the Arctic sea-ice in late summer. *Atmospheric Chemistry and Physics*, *13*, 9379–9399.

Shupe, M. D., Walden, V. P., Eloranta, E., Uttal, T., Campbell, J. R., Starkweather, S. M., & Shiobara, M. (2011). Clouds at Arctic atmospheric observatories. part I: Occurrence and macrophysical properties. *Journal of Applied Meteorology and Climatology*, *50*, 626–644.

Si, M., Irish, V. E., Mason, R. H., Vergara-Temprado, J., Hanna, S. J., Ladino, L. A., et al. (2018). Ice- nucleating ability of aerosol particles and possible sources at three coastal marine sites. *Atmospheric Chemistry and Physics*, *18*(21), 15669–15685.

Silber, I., Fridlind, A. M., Verlinde, J., Ackerman, A. S., Cesana, G. V., & Knopf, D. A. (2020a). The prevalence of precipitation from polar supercooled clouds. *Atmospheric Chemistry and Physics*, *21*, 3949–3971.

Silber, I., Fridlind, A. M., Verlinde, J., Russell, L. M., & Ackerman, A. S. (2020b). Nonturbulent liquid-bearing polar clouds: Observed frequency of occurrence and simulated sensitivity to gravity waves. *Geophysical Research Letters*, *47*(10), e2020GL087099.

Smith, W. L., Hansen, C., Bucholtz, A., Anderson, B. E., Beckley, M., Corbett, J. G., et al. (2017). Arctic radiation-icebridge sea and ice experiment: The Arctic radiant energy system during the critical seasonal ice transition. *Bulletin of the American Meteorological Society*, *98*(7), 1399–1426.

Soja, A. J., Stocks, B., Maczek, P., Fromm, M., Servrancks, R., & Turetsky, M. (2008). ARCTAS: The perfect smoke. *The Canadian Smoke Newsletter*, 2–7.

Solomon, A., de Boer, G., Creamean, J. M., McComiskey, A., Shupe, M. D., Maahn, M., & Cox, C. (2018). The relative impact of cloud condensation nuclei and ice nucleating particle concentrations on phase partitioning in Arctic mixed-phase stratocumulus clouds. *Atmospheric Chemistry and Physics*, *18*, 17047–17059.

Solomon, A., Feingold, G., & Shupe, M. D. (2015). The role of ice nuclei recycling in the maintenance of cloud ice in Arctic mixed-phase stratocumulus. *Atmospheric Chemistry and Physics*, *15*(18), 10631–10643.

Solomon, A., Shupe, M. D., Persson, O., Morrison, H., Yamaguchi, T., Caldwell, P. M., & de Boer, G. (2014). The sensitivity of springtime Arctic mixed-phase stratocumulus clouds to surface-layer and cloud-top inversion-layer moisture sources. *Journal of the Atmospheric Sciences*, *71*(2), 574–595.

Solomon, A., Shupe, M. D., Persson, P. O. G., & Morrison, H. (2011). Moisture and dynamical interactions maintaining decoupled Arctic mixed-phase stratocumulus in the presence of a humidity inversion. *Atmospheric Chemistry and Physics*, *11*, 10127–10148.

Sotiropoulou, G., Sedlar, J., Tjernström, M., Shupe, M. D., Brooks, I. M., & Persson, P. O. G. (2014). The thermodynamic structure of summer Arctic stratocumulus and the dynamic coupling to the surface. *Atmospheric Chemistry and Physics*, *14*(22), 12573–12592.

Sotiropoulou, G., Tjernström, M., Savre, J., Ekman, A. M., Hartung, K., & Sedlar, J, (2018). Large-eddy simulation of a warm-air advection episode in the summer Arctic. *Journal of the Royal Meteorological Society*, *144*(717), 2449–2462.

Sotiropoulou, G., Tjernström, M., Sedlar, J., Achtert, P., Brooks, B. J., Brooks, I. M., et al. (2016). Atmospheric conditions during the Arctic Clouds in Summer Experiment (ACSE): Contrasting open water and sea ice surfaces during melt and freeze-up seasons. *Journal of Climate*, *29*(24), 8721–8744.

Stapf, J., Ehrlich, A., & Wendisch, M. (2021). Influence of thermodynamic state changes on surface cloud radiative forcing in the Arctic: A comparison of two approaches using data from AFLUX and SHEBA. *Journal of Geophysical Research*, *126*(5), e2020JD033589.

Stevens, R. G., Loewe, K., Dearden, C., Dimitrelos, A., Possner, A., Eirund, G. K., et al. (2018). A model intercomparison of CCN-limited tenuous clouds in the high Arctic. *Atmospheric Chemistry and Physics*, *18*(15), 11041–11071.

Stevens, B. e. a. (2021). EUREC^{4}A. *Journal of Earth System Science Data Discuss*.

Stofferahn, E., & Boybeyi, Z. (2017). Investigation of aerosol effects on the Arctic surface temperature during the diurnal cycle: Part 2–separating aerosol effects. *International Journal of Climatology*, *37*, 775–787.

Stohl, A. (2006). Characteristics of atmospheric transport into the Arctic troposphere. *Journal of Geophysical Research*, *111*(D11).

Stramler, S., Genio, A. D. D., & Rossow, W. B. (2011). Synoptically driven Arctic winter states. *Journal of Climate*, *24*(6), 1747–1762.

Stuecker, M. F., Bitz, C., Armour, K. M., Proistosescu, C., Kang, S. M., Xie, S.-P., et al. (2018). Polar amplification dominated by local forcing and feedbacks. *Nature Climate Change*, *8*(12), 1076–1081.

Sun, Z., & Shine, K. P. (1994). Studies of the radiative properties of ice and mixed-phase clouds. *Quarterly Journal of the Royal Meteorological Society*, *120*(515), 111–137.

Suski, K. J., Hill, T. C. J., Levin, E. J. T., Miller, A., DeMott, P. J., & Kreidenweis, S. M. (2018). Agricultural harvesting emissions of ice-nucleating particles. *Atmospheric Chemistry and Physics*, *18*, 13755–13771.

Tan, I., & Storelvmo, T. (2016). Sensitivity study on the influence of cloud microphysical parameters on mixed-phase cloud thermodynamic phase partitioning in CAM5. *Journal of the Atmospheric Sciences*, *73*, 709–728.

Tan, I., & Storelvmo, T. (2019). Evidence of strong contributions from mixed-phase clouds to Arctic climate change. *Geophysical Research Letters*, *46*(6), 2894–2902.

Tan, I., Storelvmo, T., & Choi, Y.-S. (2014). Spaceborne lidar observations of the ice- nucleating potential of dust, polluted dust, and smoke aerosols in mixed-phase clouds. *Journal of Geophysical Research*, *119*(11), 6653–6665.

Tan, I., Storelvmo, T., & Zelinka, M. D. (2016). Observational constraints on mixed-phase clouds imply higher climate sensitivity. *Science*, *352*, 224–227.

Taylor, P. C., Boeke, R. C., Li, Y., & Thompson, D. W. J. (2019). Arctic cloud annual cycle biases in climate models. *Atmospheric Chemistry and Physics*, *19*, 8759–8782.

Taylor, P. C., Cai, M., Hu, A., Meehl, J., Washington, W., & Zhang, G. J. (2013). A decomposition of feedback contributions to polar warming amplification. *Journal of Climate*, *26*, 7023–7043.

Taylor, P. C., Kato, S., Xu, K.-M., & Cai, M. (2015). Covariance between Arctic sea ice and clouds within atmospheric state regimes at the satellite footprint level. *Journal of Geophysical Research*, *120*(24), 12656–12678.

Thomas, M. A., Devasthale, A., Tjermström, M., & Ekman, A. M. (2019). The relation between aerosol vertical distribution and temperature inversions in the Arctic in winter and spring. *Geophysical Research Letters*, *46*(5), 2836–2845.

Tjernström M., & Graversen, R. G. (2009). The vertical structure of the lower Arctic troposphere analysed from observations and the ERA-40 reanalysis. *Quarterly Journal of the Royal Meteorological Society*, *135*(639), 431–443.

Tjernström, M., Leck, C., Birch, C. E., Bottenheim, J. W., Brooks, B. J., Brooks, I. M., et al. (2014). The Arctic Summer Cloud Ocean Study (ASCOS): Overview and experimental design. *Atmospheric Chemistry and Physics*, *14*(6), 2823–2869.

Tjernström, M., Shupe, M. D., Brooks, I. M., Achtert, P., Prytherch, J., & Sedlar, J. (2019). Arctic summer airmass transformation, surface inversions, and the surface energy budget. *Journal of Climate*, *32*(3), 769–789.

Tjernström, M., Shupe, M. D., Brooks, I. M., Persson, P. O. G., Prytherch, J., Salisbury, D., et al. (2015). Warm-air advection, air mass transformation and fog causes rapid ice melt. *Geophysical Research Letters*, *42*, 5594–5602.

Tobo, Y., Adachi, K., Hill, P. J. D., Hamilton, D. S., Mahowald, N. M., Nagatsuka, N., et al. (2019). Glacially sourced dust as a potentially significant source of ice nucleating particles. *Nature Geoscience*, *12*(4), 253–258.

Tsushima, Y., Emori, S., Ogura, T., Kimoto, K., Webb, M. J., Williams, K. D., et al. (2006). Importance of the mixed-phase cloud distribution in the control climate for assessing the response of clouds to carbon dioxide increase: A multi- model study. *Climate Dynamics*, *27*, 113–126.

Twomey, S. (1977). The influence of pollution on the shortwave albedo of clouds. *Journal of the Atmospheric Sciences*, *34*(7), 1149–1152.

Uttal, T., Curry, J. A., McPhee, M. G., Perovich, D. K., Moritz, R. E., Maslanik, J. A., et al. (2002). Surface heat budget of the Arctic Ocean. *Bulletin of the American Meteorological Society*, *83*(2), 255–276.

Vassel, M., Ickes, L., Maturilli, M., & Hoose, C. (2019). Classification of Arctic multilayer clouds using radiosonde and radar data in Svalbard. *Atmospheric Chemistry and Physics*, *19*(7), 5111–5126.

Vavrus, S. (2004). The impact of cloud feedbacks on Arctic climate under greenhouse forcing. *Journal of Climate*, *17*(3), 603–615.

Vergara-Temprado, J., Murray, B. J., Wilson, T. W., O'Sullivan, D., Pringle, K. J., Ardon- Dryer, K., & Bertram, A. K. (2017). Contribution of feldspar and marine organic aerosols to global ice nucleating particle concentrations. *Atmospheric Chemistry and Physics*, *17*(5), 3637–3658.

Verlinde, J., Harrington, J. Y., McFarquhar, G. M., Yunnuzzi, V. T., Avramov, A., Greenbert, S., et al. (2007). The mixed-phase Arctic cloud experiment. *Bulletin of the American Meteorological Society*, *88*(2), 205–222.

Verlinde, J., Rambukkange, M. P., Clothiaux, E. E., McFarquhar, G. M., & Eloranta, E. W. (2013). Arctic multi-layered, mixed-phase cloud processes revealed in millimeter-wave cloud radar doppler spectra. *Journal of Geophysical Research*, *118*(23), 13199–13213.

Vihma, T., Kilpeläinen, T., Manninen, M., Sjöblom, A., Jakobson, E., Palo, T., et al. (2011). Characteristics of temperature and humidity inversions and low-level jets over Svalbard fjords in spring. *Advances in Meteorology.*

Villanueva, D., Heinold, G., Seifert, P., Deneke, H., Radenz, M., & Tegen, I. (2020). The day- to-day co-variability between mineral dust and cloud glaciation: A proxy for heterogeneous freezing. *Atmospheric Chemistry and Physics*, *20*, 2177–2199.

Wang, Z., Sassen, K., Whiteman, D. N., & Demoz, B. B. (2005). Studying altocumulus with ice virga using ground-based active and passive remote sensors. *Journal of Applied Meteorology and Climatology*, *43*(3), 449–460.

Wendisch, M., M. Brückner, J. P. Burrows, S. Crewell, K. Dethloff, K. Ebell, et al. (2017). Understanding causes and effects of rapid warming in the Arctic. *Eos*, *98*(8), 22–26, doi:10.1029/2017EO064803. https://eos.org/project-updates/understanding-causes-and-effects-of-rapid-warming-in-the-arctic

Wendisch, M., D. Handorf, I. Tegen, R. A. J. Neggers, and G. Spreen (2021), Glimpsing the ins and outs of the Arctic atmospheric cauldron, *Eos, 102,* https://doi.org/10.1029/2021EO155959. Published on 16 March 2021.

Wendisch, M., A. Macke, A. Ehrlich, C. Lüpkes, M. Mech, D. Chechin, et al. (2019): The Arctic Cloud Puzzle: Using ACLOUD/PASCAL Multi-Platform Observations to Unravel the Role of Clouds and Aerosol Particles in Arctic Amplification. *Bull. Amer. Meteor. Soc.*, *100*(5), 841–871, doi: 10.1175/BAMS-D-18-0072.1

Wex, H., Huang, L., Zhang, W., Hung, H., Traversi, R., Becagli, S., et al. (2019). Annual variability of ice-nucleating particle concentrations at different Arctic locations. *Atmospheric Chemistry and Physics*, *19*(7), 5293–5311.

Willis, M. D., Leaitch, W. R., & Abbatt, J. P. (2018). Processes controlling the composition and abundance of Arctic aerosol. *Reviews of Geophysics*, *56*(4), 621–671.

Wilson, T. W., Ladino, L. A., Alpert, P. A., Breckels, M. N., Brooks, I. M., Browse, J., et al. (2015). A marine biogenic source of atmospheric ice-nucleating particles. *Nature*, *525*(7568), 234–238.

Winker, D., Chepfer, H., Noel, V., & Cai, X. (2017). Observational constraints on cloud feedbacks: The role of active satellite sensors. *Surveys in Geophysics*, *38*(6), 1483–1508.

Winker, D., Vaughan, M. A., Omar, A., Hu, Y., Powell, K. A., Liu, Z., et al. (2009). Overview of the CALIPSO mission and CALIOP data processing algorithms. *Journal of Atmospheric and Oceanic Technology*, *26*(11), 2310–2323.

Winker, D. M., Tackett, J. L., Getzewich, B. J., Liu, Z., Vaughan, M. A., & Rogers, R. R. (2013). The global 3-D distribution of tropospheric aerosols as characterized by CALIOP. *Atmospheric Chemistry and Physics*, *13*(6), 3345–3361.

Winton, M. (2006). Amplified Arctic climate change: What does surface albedo feedback have to do with it? *Geophysical Research Letters*, *33*(3).

Wood, R. (2012). Stratocumulus clouds. *Monthly Weather Review*, *140*(8), 2373–2423.

Woods, C., & Caballero, R. (2016). The role of moist intrusions in winter Arctic warming and sea ice decline. *Journal of Climate*, *29*, 4473–4485.

Woods, C., Caballero, R., & Svensson, G. (2013). Large-scale circulation associated with moisture intrusions into the Arctic during winter. *Geophysical Research Letters*, *40*(4717–4721).

Woods, C., Caballero, R., & Svensson, G. (2017). Representation of Arctic moist intrusions in CMIP5 models and implications for winter climate biases. *Journal of Climate*, *30*(11), 4083–4102.

Xie, S., Liu, X., Zhao, C., & Zhang, Y. (2013). Sensitivity of CAM5-simulated Arctic clouds and radiation to ice nucleation parameterization. *Journal of Climate*, *26*(16), 5981–5999.

Xu, L., Russell, L. M., & Burrows, S. M. (2016). Potential sea salt aerosol sources from frost flowers in the pan-Arctic region. *Journal of Geophysical Research*, *121*(18).

Yamagata, S., Kobayashi, D., Ohta, S., Murao, N., Shiobara, M., Wada, M., et al. (2009). Properties of aerosols and their wet deposition in the Arctic spring during ASTAR2004 at Ny-Alesund, Svalbard. *Atmospheric Chemistry and Physics*, *9*, 261–270.

Yoo, C., Lee, S., & Feldstein, S. B. (2012). Mechanisms of Arctic surface air temperature change in response to the Madden–Julian Oscillation. *Journal of Climate*, *25*(17), 5777–5790.

Yoshimori, M., Watanabe, M., Shiogama, H., Oka, A., Abe-Ouchi, A., Ohgaito, R., & Kamae, Y. (2016). A review of progress towards understanding the transient global mean surface temperature response to radiative perturbation. *Progress in Earth and Planetary Science*, *3*(1), 1–14.

Young, G., Connolly, P. J., Dearden, C., & Choularton, T. W. (2018). Relating large-scale subsidence to convection development in Arctic mixed-phase marine stratocumulus. *Atmospheric Chemistry and Physics*, *18*(3), 1475–1494.

Yu, F., & Luo, G. (2009). Simulation of particle size distribution with a global aerosol model: Contribution of nucleation to aerosol and CCN number concentrations. *Atmospheric Chemistry and Physics*, *9*(20), 7691–7710.

Yu, Y., Taylor, P. C., & Cai, M. (2019). Seasonal variations of Arctic low-level clouds and its linkage to sea ice seasonal variations. *Journal of Geophysical Research*, *124*(22), 12206–12226.

Zamora, L. M., Kahn, R. A., Eckhardt, S., McComiskey, A., Sawamura, P., Moore, R., & Stohl, A. (2017). Aerosol indirect effects on the nighttime Arctic Ocean surface from thin, predominantly liquid clouds. *Atmospheric Chemistry and Physics*, *17*, 7311–7332.

Zamora, L. M., Kahn, R. A., Huebert, K. B., Stohl, A., & Eckhardt, S. (2018). A satellite-based estimate of combustion aerosol cloud microphysical effects over the Arctic Ocean. *Atmospheric Chemistry and Physics*, *18*(20), 14949–14964.

Zelinka, M. D., Myers, T. A., McCoy, D. T., Po-Chedley, S., Caldwell, P. M., Ceppi, P., et al. (2020). Causes of higher climate sensitivity in CMIP6 models. *Geophysical Research Letters*, *47*(1), e2019GL085782.

Zelinka, M. D., Randall, D. A., Webb, M. J., & Klein, S. A. (2017). Clearing clouds of uncertainty. *Nature Climate Change*, *7*(10), 674–678.

Zhang, D., Wang, Z., Heymsfield, A., Fan, J., & Luo, T. (2014). Ice concentration retrieval in stratiform mixed-phase clouds using cloud radar reflectivity measurements and 1D ice growth model simulations. *Journal of the Atmospheric Sciences*, *71*(10), 3613–3635.

Zhang, D., Wang, Z., Kollias, P., Vogelmann, A. M., Yang, K., & Luo, T. (2018). Ice particle production in mid-level stratiform mixed-phase clouds observed with collocated A-Train measurements. *Atmospheric Chemistry and Physics*, *18*, 4317–4327.

Zhang, D., Wang, Z., & Liu, D. (2010). A global view of midlevel liquid-layer topped stratiform cloud distribution and phase partition from CALIPSO and CloudSat measurements. *Journal of Geophysical Research*, *115*(D00H13).

Zhang, M., Liu, X., Diao, M., D'Alessandro, J. J., Wong, Y., Wu, C., et al. (2019). Impacts of representing heterogeneous distribution of cloud liquid and ice on phase partitioning of Arctic mixed-phase clouds with NCAR CAM5. *Journal of Geophysical Research*, *124*(23), 13071–13090.

Zhao, C., & Garrett, T. J. (2015). Effects of Arctic haze on surface cloud radiative forcing. *Geophysical Research Letters*, *42*(2), 557–564.

Zhao, M., & Wang, Z. (2010). Comparison of Arctic clouds between European Center for Medium-range Weather Forecasts simulations and atmospheric radiation measurement climate research facility long-term observations at the north slope of Alaska barrow site. *Journal of Geophysical Research*, *115*(D23202).

Zhu, T., Huang, Y., & Wei, H. (2019). Estimating climate feedbacks using a neural network. *Journal of Geophysical Research*, *124*(6), 3246–3258.

Zuidema, P., Baker, B., Han, Y., Intrieri, J., Key, J., Lawson, P., et al. (2005). An Arctic springtime mixed-phase cloudy boundary layer observed during SHEBA. *Journal of the Atmospheric Sciences*, *62*(1), 160–176.

Zwaaftink, G., Grythe, C. D. H., Skov, H., & Stohl, A. (2016). Substantial contribution of northern high-latitude sources to mineral dust in the Arctic. *Journal of Geophysical Research*, *121*(22), 13678–13697.

Zygmuntowska, M., Mauritsen, T., Quaas, J., & Kaleschke, L. (2012). Arctic clouds and surface radiation–A critical comparison of satellite retrievals and the ERA-interim reanalysis. *Atmospheric Chemistry and Physics*, *12*(14), 6667–6677.

6

Extratropical Cloud Feedbacks

Daniel T. McCoy[1], Michelle E. Frazer[2,3], Johannes Mülmenstädt[4], Ivy Tan[5], Christopher R. Terai[6], and Mark D. Zelinka[6]

ABSTRACT

The extratropics are the cloudiest region on Earth. Changes in clouds in this region in response to warming have the potential to substantially affect global mean cloud feedback and by extension climate sensitivity. Global climate models (GCMs) predict a relatively small, but consistent positive longwave (LW) cloud feedback throughout much of the extratropics. The bulk of GCMs transition from positive subtropical shortwave (SW) cloud feedback to negative extratropical SW cloud feedback is driven by increasing cloud optical depth. However, the strength of the negative feedback in the extratropics is not agreed upon by GCMs. Recent shifts in extratropical SW cloud feedback toward more positive values in the most recent generation of GCMs have led to the emergence of several high ECS (> 4.5K) GCMs. Thus, understanding and constraining the processes that drive extratropical cloud feedback has global implications and constraint of the SW extratropical cloud feedback has garnered significant attention in the literature. In this chapter, we summarize recent literature and present the processes important for extratropical cloud feedback in the context of meteorological regimes and global climate.

6.1. INTRODUCTION

The extratropical atmosphere is characterized by the persistent surface westerly winds of the Ferrel cell that begin at roughly 30° and extend poleward. Moisture and energy exported from the tropics by transient atmospheric eddies converge in this region (Hartmann, 2015), and clouds cover over 75% of the ocean surface (Marchand et al., 2009). Changes in cloud cover or opacity and trapping of escaping heat by clouds have the potential to substantially affect the Earth's radiation budget (Hartmann & Short, 1980; Hartmann and Larson, 2002). Extratropical clouds present a unique set of challenges to simulation and observation. Cloud structures span scales from open-cellular convection following fronts with a lengthscale of kilometers to synoptic systems spanning thousands of kilometers (Fig. 6.1). Meteorological regimes shift from low pressure centers surrounding cyclones to high pressure in anticyclones on a timescale of hours to days, leading to larger swings in temperature, humidity, and circulations than typically seen in the tropics or subtropics. In addition to the diversity of processes and environmental conditions in this region, there are numerous observational challenges. The multilayered and complex cloud topography in the extratropics leads to

[1]Department of Atmospheric Science, University of Wyoming, Laramie, WY, USA
[2]Program in Atmospheric and Oceanic Sciences, Princeton University, Princeton, NJ, USA
[3]Department of Geosciences, Pennsylvania State University, University Park, PA, USA
[4]Department of Atmospheric Sciences & Global Change, Pacific Northwest National Laboratory, Richland, WA, USA
[5]Department of Atmospheric and Oceanic Sciences, McGill University, Montreal, Canada
[6]Atmospheric, Earth, and Energy Division, Lawrence Livermore National Laboratory, Livermore, CA, USA

Clouds and Their Climatic Impacts: Radiation, Circulation, and Precipitation, Geophysical Monograph 281, First Edition.
Edited by Sylvia C. Sullivan and Corinna Hoose.

DOI: 10.1002/9781119700357.ch06

Figure 6.1 Himawari-8 image of the North Pacific 31 December 2020. Cloud structures over a variety of scales are visible over the North Pacific. Source: CIRA/CSU & NOAA.

challenges for both passive and active remote sensing instruments (Haynes et al., 2011; Marchand et al., 2009; Sourdeval et al., 2016). Furthermore, the large seasonal changes in insolation closer to the poles make cloud retrievals in the visible wavelengths impossible during high-latitude winter, and challenging in many cases when the Sun is low in the sky (Grosvenor & Wood, 2014; Grosvenor et al., 2018).

Given the challenges to our understanding of extratropical cloud processes, it is reasonable to be concerned as to the effects of this region on our predictions of future climate. Global climate models (GCMs) tend to predict similar, modest positive longwave (LW) cloud feedback in this region. It is likely that this feedback is driven by processes that are similar to those that drive the positive LW feedback in the tropics set by anvil clouds maintaining a fixed temperature while the underlying surface warms (Ceppi et al., 2017; Hartmann & Larson, 2002; Thompson et al., 2019, 2017). Given the strong consensus between GCMs regarding the extratropical LW feedback and its modest strength, it has received little attention relative to the highly uncertain shortwave (SW) feedback in this region. As this monograph surveys existing literature this chapter focuses on the SW cloud feedback, which has dominated publications.

Early studies noted the possibility of negative SW cloud feedbacks in this region linked to an ice to liquid transition (Mitchell et al., 1989; Senior & Mitchell, 1993). Analysis of Southern Ocean energy fluxes in Trenberth and Fasullo (2010) found correlation between mean-state radiative fluxes in this region and climate sensitivity. In particular, analysis of cyclone cold sectors shows linkages between cloud feedback and mean-state cloud radiative effect (CRE) (Bodas-Salcedo, Andrews et al., 2016; Kelleher & Grise, 2019). Assessment of overall climate sensitivity in Sherwood et al. (2020) assigns a best estimate and standard deviation for optical depth feedback poleward of 60° of 0 ± 0.1 $W/m^2/K$ and a range of 0.12 ± 0.12 $W/m^2/K$ for midlatitude (30°–60°) low-cloud amount feedback. All other feedbacks in the extratropics were assigned 0 ± 0 $W/m^2/K$ due to lack of process level understanding or constraint. Based on this assessed range of extratropical cloud feedback, combined with other lines of evidence, the most probable range of effective climate sensitivity (ECS) is calculated at 2.3–4.7 K (in the baseline case shown in Sherwood et al. (2020)).

Uncertainty in changes in cloud properties across the 40 °S-40 °N band has a larger impact on SW cloud feedback and ECS relative to the extratropics due to the larger area and stronger insolation in this region. As discussed in Chapter 7: Tropical Marine Low Clouds, substantial progress has been made in recent years in constraining the cloud feedback in this region (Klein et al., 2017). The narrowing of the probable range of subtropical cloud feedback has left the extratropics as a substantial contributor to uncertainty in ECS, and as shown in Zelinka et al. (2020), shifts in this quantity between climate model generations can lead to substantial differences in ensemble ECS. Thus, extratropical cloud feedback bears further investigation.

In this chapter, we will present a synthesis of observational and modeling studies aimed at better understanding the extratropical SW cloud feedback (GCM SW cloud feedback is shown in Fig. 6.2). As discussed above, the positive extratropical LW cloud feedback due to increased cloud top height is non-negligible, differences

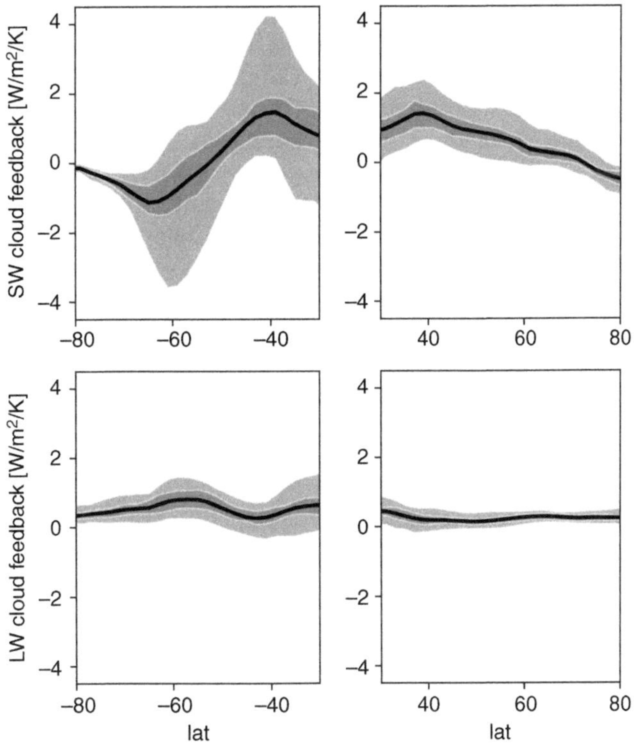

Figure 6.2 SW (top) and LW (bottom) cloud feedback calculated following Zelinka et al. (2020) for the GCMs listed in appendix. The mean of GCM values is shown as a black line. The 50th and 95th percentiles are shown using shading. Source: Adapted from Zelinka et al. (2020).

between GCMs are relatively small (Fig. 6.2) and less research has been devoted to reducing this spread (Ceppi et al., 2017; Thompson et al., 2017, 2019; Zelinka et al., 2016, 2020). We, therefore, focus our attention on the SW cloud feedback, while presenting existing LW feedback in GCMs. As noted above, here, we define the lower latitude bound of the extratropics as near 30° while the upper bound is around 70°. This latitude range is somewhat consistent with the latitude range where substantial SW cloud feedback exists. In this monograph, we will focus on this latitude range, although the precise bounds of different studies vary. As an extension of the geographic region, we must also consider the contrast between feedback over maritime and continental regions. The 30 °S-70 °S region is nearly all maritime and many studies focus on this region. Strong contrasts exist in the SW cloud feedback between land and ocean in the 30 °N-70 °N region and where applicable we will discuss these differences. The overall structure of this chapter is as follows: we will first discuss what inferences can be drawn from the observational record (section 6.2) and robust elements of the predicted response of clouds and radiation to warming (section 6.3). In the context of the observational and simulation results, we will present a summary of processes that are hypothesized to underpin the extratropical cloud feedback in section 6.4. In section 6.5, we discuss potential avenues toward better constraining extratropical SW cloud feedback.

6.2. CONSISTENT FEATURES OF THE OBSERVATIONAL RECORD

Diverse observational studies attempting to infer the extratropical SW cloud feedback from present-day variability have been performed. It is important to note here that these inferences of cloud feedback from observations are based on the assumption that we can learn about the long-term cloud feedback in response to warming from cloud responses to temperature that occur at much shorter timescales (daily – interannual). While some of these studies examine long-term trends in extratropical clouds to relate observed warming trends over the observational time period to future warming (Manaster et al., 2017; McCoy et al., 2019; Norris et al., 2016), many are based on observed covariability between clouds and their environment (so-called cloud controlling factor analysis; Stevens and Brenguier, 2009). These studies evaluate the timescale invariance (Klein et al., 2017) of various cloud controlling factors through comparison to GCMs and their representation of warming scenarios. The exact techniques used to evaluate timescale invariance (whether observations in the present-day can be used to predict the response of cloud to increased global mean temperature) vary between studies. Generally, the evaluation of timescale invariance relies on using GCMs to infer this quantity, or examining whether long-term (decades or longer) trends in liquid water path (LWP) can be predicted by observations of variability at a short timescale (days-to-months).

In this section, we synthesize existing research in the context of cloud top pressure regime (whether or not the study focused on boundary-layer [BL] cloud) and whether cloud optical depth, coverage, or an aggregate (e.g., area-mean LWP or CRE) were examined. The height distinction is important because of the different mechanisms that drive cloud feedbacks in those clouds (discussed below). A summary table of select observational constraint studies is given in Table 6.1. Table 6.1 includes many differences between instrumentation, regions of examination, and technique. The strength of extratropical SW cloud feedback assessed by the studies in Table 6.1 varies. However, when examined in the context of whether the clouds being examined are limited to low-topped clouds in the BL or include some higher topped clouds (such as those occurring along cyclone fronts), the general consensus is that low-topped cloud changes result in a positive SW feedback and upper level cloud changes result in a negative SW feedback with an overall inferred negative feedback. It is also important to examine whether cloud thickness, cloud extent, or an aggregate of both (as contained in microwave LWP [Elsaesser et al., 2017; Wentz & Meissner, 2000] or CRE) are being analyzed, with strong consensus among studies inferring a decrease in low-topped cloud τ with warming (Gordon & Klein, 2014; Sherwood et al., 2020; Terai et al., 2019, 2016), but possible increases in low-topped cloud fraction (Ceppi, McCoy, et al., 2016; Norris et al., 2016). There is also broad consensus in projecting an increase in upper level cloud fraction and optical depth in response to warming across studies (Ceppi, McCoy, et al., 2016; McCoy, Field,

Table 6.1 Summary of findings of observational constraints on changes in shortwave-relevant extratropical cloud properties from the literature sorted by cloud top height. The cells are colored by the approximate inferred feedback on warming (Red=Positive, Blue=Negative). Sign is noted in the text as (+/−). Latitude ranges of interest vary between studies, but are approximately poleward of 30° and equatorward of 70°. Studies using of boundary layer (generally $CTP < 680$) are denoted BL and studies examining clouds with tops in the free troposphere (generally $CTP > 680$) are denoted FT. Where appropriate the Cloud Controlling Factors (CCFs) used are provided. Further details are provided in Table 6.2.

	Regime → Study ↓	BL	FT	Total	CCFs	Notes
(T92)	Tselioudis et al. (1992)	[+]			Low cloud T	Constraint on low cloud over land and ocean τ
(G14)	Gordon and Klein (2014)	[+]			T_{2m}	Constraint on low cloud τ
(T16)	Terai et al. (2016)	[+]		[−]	T_{2m}, EIS	Constraint on low cloud τ feedback
(C16)	Ceppi, McCoy et al., (2016)	[−]	[−]	[−]	500–850 hPa T	Maritime, LWP, cloud area, and τ
(N16)	Norris et al. (2016)	[+]	[−]	[+/−]	N/A	Trends in cloud fraction and albedo
(M17)	Manaster et al. (2017)			[−]	N/A	Extratropical maritime LWP trends
(TA19)	Tan et al. (2019)	[+]	[−]			Constraint on thermodynamic phase shifts over land and ocean
(TE19)	Terai et al. (2019)	[+]				Constraint on mechanisms of BL clouds using surface retrievals
(M19)	McCoy et al. (2019)	[+]	[−]		q_{conv}, SST	Maritime LWP composited around cyclones
(M20)	McCoy, Field, Bodas-Salcedo, et al. (2020)	[+]	[−]		q_{conv}, EIS, ω_{500}, SST	Maritime LWP, composited around cyclones and anticyclones
(M22)	McCoy et al. (2022)			[−]	q_{conv}, SST	Maritime LWP and upwelling SW radiation

Note: SST = Sea Surface Temperature, EIS = Estimated Inversion Strength, LWP = Liquid Water Path, SW = Shortwave.

Table 6.2 As in Table 6.1, but with additional explanation of inferred feedback.

Regime → Study ↓	BL	FT	Total
(G92)	[+] τ decreases with warming over oceans and in warm continental clouds. τ only increases in cold, continental clouds.		
(G14)	[+] Decreased τ in response to surface warming.		
(T16)	[+] Stability increases τ and τ decreases with increased SST. Total effect is decreased τ with warming in majority of satellite data records.		[−] Total SW cloud feedback is inferred to be negative with observational constraint on τ_{BL}.
(C16)	[−] Increased cloud fraction with low-to-mid tropospheric warming (driven by $\tau > 3.6$ regime).	[−] Increased cloud fraction with low-to-mid tropospheric warming (driven by $\tau > 1.3$ regime).	[−] Increased cloud fraction (driven by $\tau > 3.6$ regime) and increased area-mean (microwave) LWP with low-to-mid tropospheric warming.
(N16)	[+] Decreasing trends in CF detected by PATMOS-X and ISCCP ($\tau > 3.6$).	[−] Increasing trend in CF detected by ISCCP ($\tau > 3.6$, $CTP > 300\ hPa$) and ambiguous trends in PATMOS-X.	[+/−] Increased cloud albedo detected by CERES-ERBS and decreased cloud fraction detected by ISCCP.
(M17)			[−] Examination of warming trends in microwave data record shows decadal increase in Southern Ocean LWP.
(TA19)	[+] Maritime τ decreases with warming due to liquid cloud processes.	[−] Maritime τ increases due to phase shifts in response to warming. (CTT of −20 °C approximate a CTP of 600 hPa).	
(TE19)	[+] τ decreases or remains constant (increases) in response to warming in warm (cold) stratiform clouds due to multiple mechanisms affecting liquid water path.		
(M19 & M20)	[+] Decreased area-mean (microwave) LWP in response to surface warming in subsiding regions of cyclones dominated by BL cloud.	[−] Increased LWP in response to increased moisture convergence (primarily along cold front).	[−] Increased LWP. Inferred feedback calculated by scaling ΔLWP by $dSW_{\uparrow}/dLWP$ is negative.
(M22)			[−] Increased LWP. Inferred feedback calculated by scaling ΔLWP by $dSW_{\uparrow}/dLWP$ is negative.

Note: FT = Free Troposphere, BL = boundary layer, SST = Sea Surface Temperature, EIS = Estimated Inversion Strength, LWP = Liquid Water Path, SW = Shortwave, CTT = Cloud top temperature, CTP = Cloud top pressure, CF = cloud fraction, τ = optical depth.

Bodas-Salcedo, et al., 2020; Norris et al., 2016; Tan et al., 2019).

Analysis in the dynamical context of jet shift as opposed to warming is consistent with the analyses described above, showing enhanced upper level cloud in the region where the jet shifts to and enhanced low-level cloud in equatorward regions where boundary layer stability increases (Grise & Medeiros, 2016). Overall, the extratropical CRE in CMIP5 GCMs has been shown to be sensitive to mid-tropospheric vertical velocity and lower atmospheric stability; excessive sensitivity of the GCMs to vertical velocity has led to overestimated CRE in the frontal region of cyclones and underestimated CRE in the center of anticyclones (Kelleher & Grise, 2021). At the same time, the sensitivity of the CMIP5 SW CRE to stability was shown to be underestimated (Grise & Medeiros, 2016).

How competing effects from upper and lower level clouds on cloud feedback combine is not certain. Broadly, studies support a negative extratropical feedback when changes in extent and thickness are aggregated over all cloud top height regimes. This has been evaluated by either constraining components of GCM extratropical SW feedback (Terai et al., 2016) or by constraining aggregate changes in clouds and radiation (Ceppi, McCoy, et al., 2016; McCoy, Field, Bodas-Salcedo, et al., 2020; McCoy et al., 2022).

As shown in Table 6.1, observations tend to point toward a negative SW cloud feedback due to increased liquid cloud in the extratropics when aggregated across all cloud regimes. In the following section, we will contrast the observationally inferred cloud property response to warming and overall SW cloud feedback to simulations of idealized CO_2 quadrupling experiments conducted in GCMs.

6.3. GCM RESPONSES TO WARMING

As discussed in the previous section, observations indicate an aggregate negative SW cloud feedback in the extratropics. This is consistent with predictions of SW cloud feedback by GCMs (Zelinka et al., 2020) (Fig. 6.2). We now examine the response of CMIP5 and CMIP6 GCMs to warming in the Southern Ocean. This analysis examines pre-industrial control (piControl) and abrupt CO_2 quadrupling (abrupt4XCO2) simulations (Gregory et al., 2004; Zelinka et al., 2020). The analysis focuses on the output from 57 GCMs spanning CMIP5 and CMIP6 with available simulation output at a monthly resolution (see appendix). Our analysis focuses on changes in flows of moisture, condensed water mass, and radiative flux. This analysis expands on the analysis shown in McCoy et al. (2022).

Ultimately, the SW cloud feedback is driven by changes in cloud condensate and apportioning of cloud condensate between phases. We will now discuss the climate mean-state cloud condensate and response to warming of cloud condensate. We will also discuss the representation of moisture in GCMs as this is a fairly robust feature (Held & Soden, 2006) that underpins clouds. That is to say, atmospheric moisture is one of the key factors influencing cloud formation. We will go into detail regarding the processes relating moisture convergence in cloud in section 6.4 and how it relates to other changes in atmospheric state like stability and subsidence. We would prefer to use moisture convergence, but as that is not available across Climate Model Intercomparison Project (CMIP) models we approximate moisture convergence as the difference between precipitation and evaporation: $P - E$ (Held & Soden, 2006). As noted in earlier intermodel comparisons (Jiang et al., 2012), GCMs have strongly varying estimates of mean-state LWP and IWP (ice water path) (Fig. 6.3 top, middle) although their mean-state moisture convergence is generally consistent (Fig. 6.3 bottom). Extratropical atmospheric temperature rises in step with global temperature (Fig. 6.4 top) throughout the GCM experiments discussed here, and moisture convergence increases (Figs. 6.5 bottom and 6.6 bottom). Moisture convergence into the extratropics is driven by local flux and eddy transport from lower latitudes (Dacre et al., 2015; Lorenz & DeWeaver, 2007; Lavers et al., 2015). It can be seen that moisture convergence has a weaker dependence on local atmospheric temperature than global temperature (Figs. 6.5, bottom and 6.4), which is consistent with previous studies showing an increase in moisture flux into the extratropics at slightly less than the rate dictated by the Clausius-Clapeyron (Lorenz & DeWeaver, 2007).

The most salient feature of the extratropical response to global warming is increased liquid cloud at higher latitudes (Tsushima et al., 2006) (Figs. 6.5a (top) and 6.6a (top)), generally accompanied by a small increase in IWP (Figs. 6.5b (middle) and 6.6b (middle)). Broadly, these increases occur in the same region as the extratropical moistening predicted by GCMs (Fig. 6.6 bottom) (Held & Soden, 2006). This is a robust feature of GCM cloud response to warming, although the sensitivity of extratropical LWP to global mean temperature (GMT) varies dramatically across GCMs (Fig. 6.5 top). It is particularly intriguing to note that of GCMs from CMIP5 and CMIP6 that are examined in Figure 6.5a (top), only CESM2 (and variants) and E3SM-1-0 display a non-monotonic response of LWP to GMT and begin to flatten and even curve back down toward the end of the first 150 years of the abrupt4XCO2 simulation (cyan lines in Fig. 6.5 top). These models share a common set of cloud physics packages, and this behavior has been examined in the literature (Bjordal et al., 2020; Gettelman

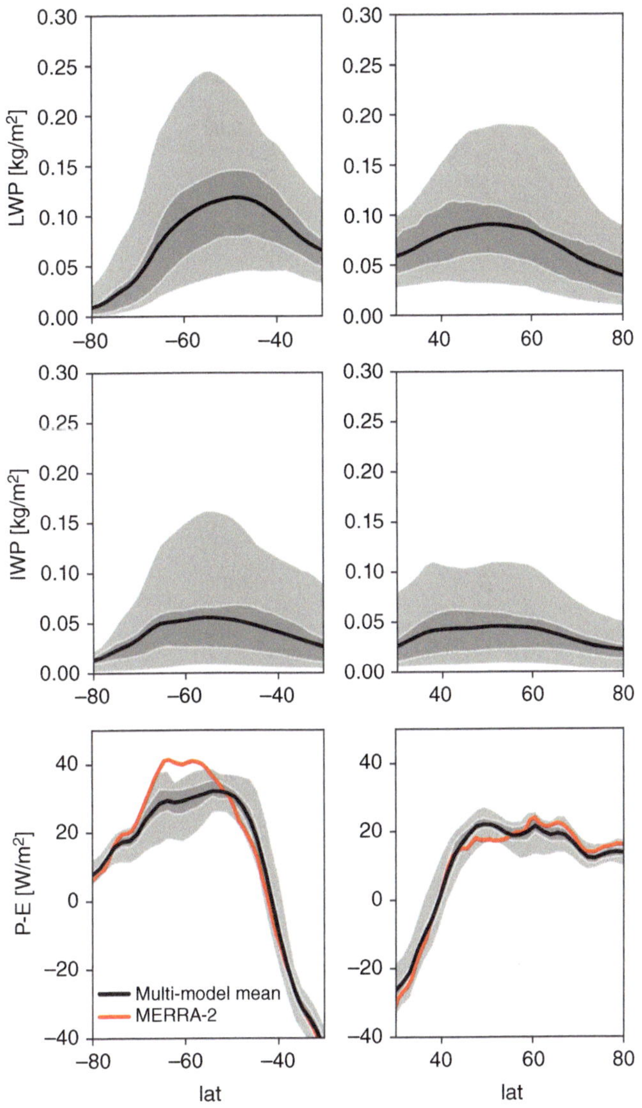

Figure 6.3 LWP (top), IWP (middle), and P-E (bottom) in piControl simulations from GCMs (see appendix). The solid line is the multi-model mean. Shaded regions correspond to the 50th and 95th percentiles of the model means. MERRA-2 reanalysis P-E is shown.

et al. 2019; Y. Zhang et al., 2019). This may be partially related to the enhancement in supercooled liquid between CAM5 and CAM6, meaning that there is less ice that may be shifted to liquid as the atmosphere warms in these GCMs (Bjordal et al., 2020). With the exception of these models, GCMs all display a monotonic trend in LWP. This robust response of extratropical liquid cloud to warming is a strong driver of the emergence of a negative extratropical SW cloud feedback that is shared by many GCMs (Bodas-Salcedo, 2018; Zelinka et al., 2012, 2016, 2020). In contrast, GCMs do not consistently increase or decrease extratropical IWP with warming (Fig. 6.5 middle). Having presented some of the salient features of the GCM response to warming in the extratropics, we will discuss some of the processes discussed in the literature that may give rise to these features.

6.4. PROCESSES CONTRIBUTING TO EXTRATROPICAL SW CLOUD FEEDBACK

6.4.1. Summary of Processes

As discussed in section 6.1, a complex array of cloud feedback mechanisms may operate in the extratropics. As shown in section 6.3, GCMs are highly uncertain with regard to the magnitude of the SW cloud feedback

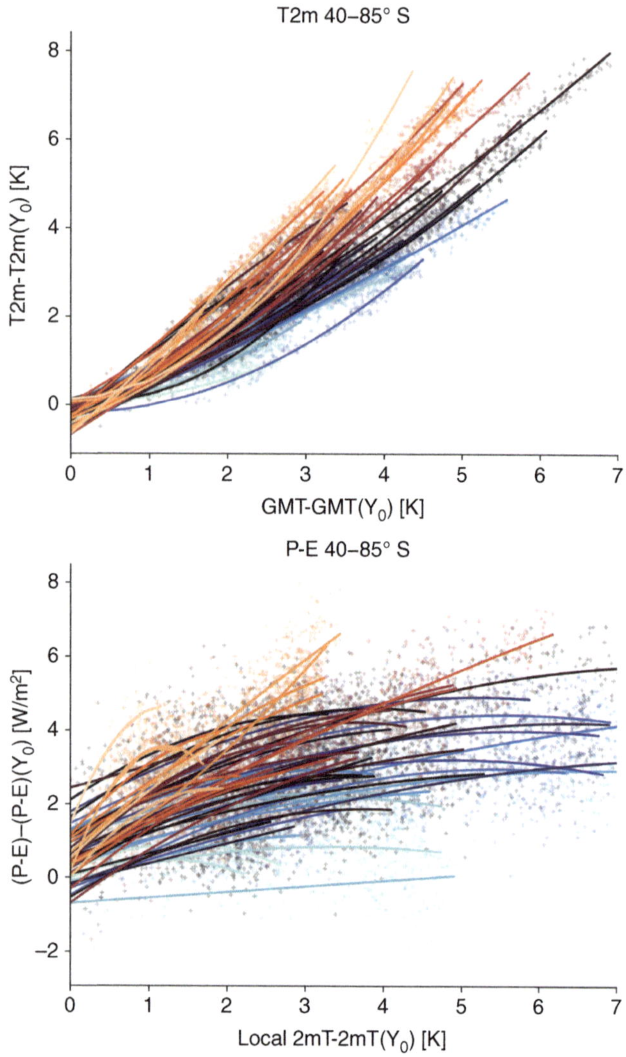

Figure 6.4 Two-meter temperature for the 40–85 °S region versus global mean 2m temperature (GMT) (top) and P-E versus 40–85 °S 2m temperature. Data are shown during the first 150 years of the abrupt4XCO2 simulations for CMIP5 and CMIP6 GCMs. The anomalies are shown relative to the start of the abrupt4xCO2 simulation (Y_0). Second-order polynomial fits to annual mean values are shown for visual clarity. Each line corresponds to a single GCM.

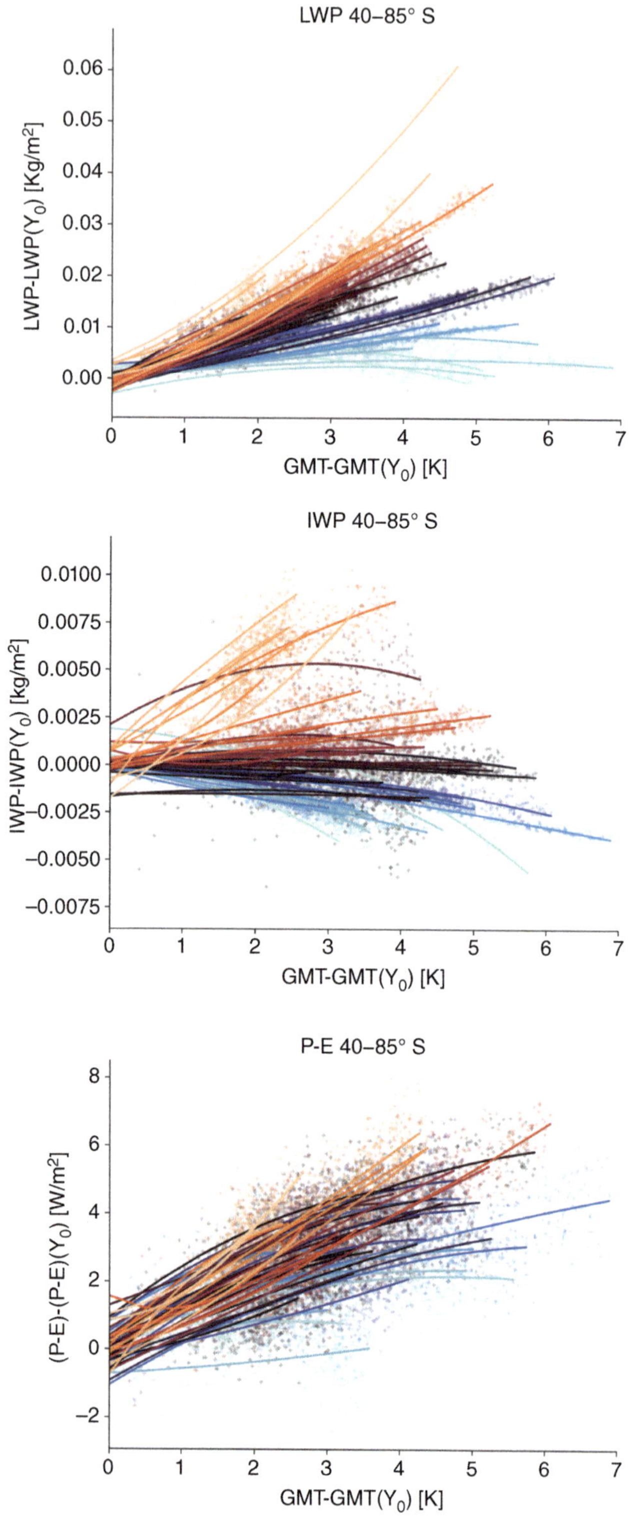

Figure 6.5 As in Figure 6.4, but for LWP (top), IWP (middle), and P-E (bottom) averaged 40 S– 85 °S versus global mean 2m temperature (GMT).

and do not agree on the sign of the feedback. However, as discussed in section 6.2, observations tend to predict a positive SW cloud feedback related to BL clouds and a negative feedback related to upper level clouds whose combined effect is a negative SW cloud feedback. Similarly, GCMs tend to predict an overall negative SW cloud feedback. We will now discuss the cloud processes that are potentially behind the extratropical SW cloud feedback. A summary of feedback mechanisms from

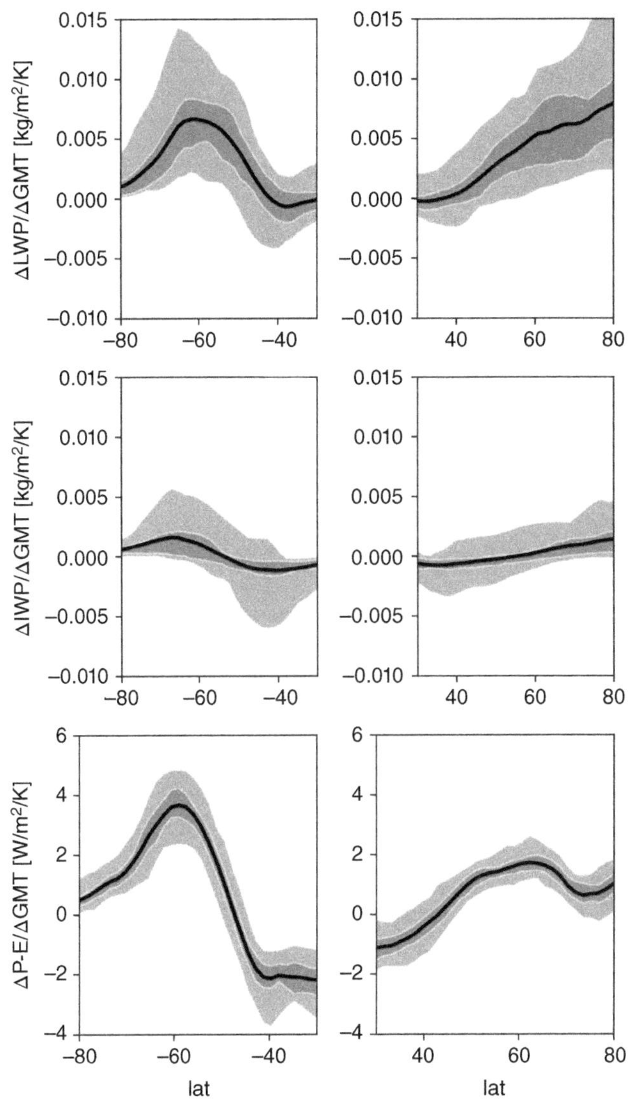

Figure 6.6 As in Figure 6.3, but showing the difference between piControl and abrupt4XCO2 simulations, normalized by change in GMT.

the literature is shown in Figure 6.7 (see discussion in Terai et al., (2019), from which this figure is adapted). Broadly, these processes apply to BL ($CTP > 680\,\text{hPa}$) and mid-topped cloud ($680\,\text{hPa} < CTP < 440\,\text{hPa}$), with the exception of (c) efficiency of cloud top drying, (d) strengthening inversions, and (e) decoupling, which are relevant to BL cloud cover. We introduce each process as follows:

(a) *Increasing liquid water content (LWC) from warmer moist adiabat.* Shifts in the moist adiabat as the atmosphere warms should result in an increase in the LWC of clouds at constant geometric thickness (Betts & Harshvardhan, 1987; Paltridge, 1980; Tselioudis et al., 1998) with stronger effects at colder temperatures (Betts & Harshvardhan, 1987; Frazer & Ming, 2022).

(b) *Increasing LWC from shift in cloud phase.* As atmospheric temperature rises, the frequency of occurrence of ice-phase condensate will decrease and liquid will increase (Frey et al., 2017; Senior & Mitchell 1993; Storelvmo et al., 2015; Tsushima et al., 2006; Tan et al., 2019), particularly near where the freezing isotherm shifts (Frazer and Ming, 2022). This can result in an increase in optical depth due to the small relative particle size of liquid compared to ice (McCoy et al., 2014) and result in an overall increase in condensate due to the relatively high precipitation efficiency of ice (Ceppi, Hartmann, & Webb, 2016; Hoose et al., 2008; McCoy et al., 2015; Mülmenstädt et al., 2021; Tsushima et al., 2006).

(c) *Decreased cloud condensate from efficient cloud top drying.* Increased vertical humidity gradient between the BL and free troposphere in step with atmospheric warming and approximately constant relative humidity promotes enhanced cloud top entrainment, reducing cloud condensate (Bretherton et al., 2013; Frey et al., 2017; Rieck et al., 2012).

(d) *Increased cloud cover from strengthening BL inversion.* The capping inversion at the top of the cloudy BL suppresses entrainment, leading to a moister BL and more cloud (Bretherton et al., 2013; Myers and Norris, 2013; Qu et al., 2015).

(e) *Decreased cloud from increased BL decoupling.* Decoupling in the BL inhibits the movement of moisture from the surface to the cloud layer, decreasing cloud. Decoupling may increase in a warming climate as increased latent heat flux overwhelms the cloud top radiative cooling (Bretherton & Wyant, 1997; Bretherton et al., 2013; Zheng et al., 2020).

(f) *Increased cloud condensate from enhanced moisture convergence.* As atmospheric temperature in the tropics and extratropics rise, both local surface fluxes and tropical export of moisture into the extratropics will increase moisture convergence (Held & Soden, 2006; Lorenz & DeWeaver, 2007; Lavers et al., 2015). As the convergence of moisture strengthens more clouds will form. More clouds will increase the precipitation sink, keeping the system in pseudo steady state (McCoy, Field, Bodas-Salcedo, et al., 2020; McCoy et al., 2019).

(g) *Increased cloud condensation nuclei from enhanced air sea exchange and increased cloud optical depth. Potentially increased cloud condensate in response to suppressed precipitation.* The extratropical storm track strengthens and shifts poleward in response

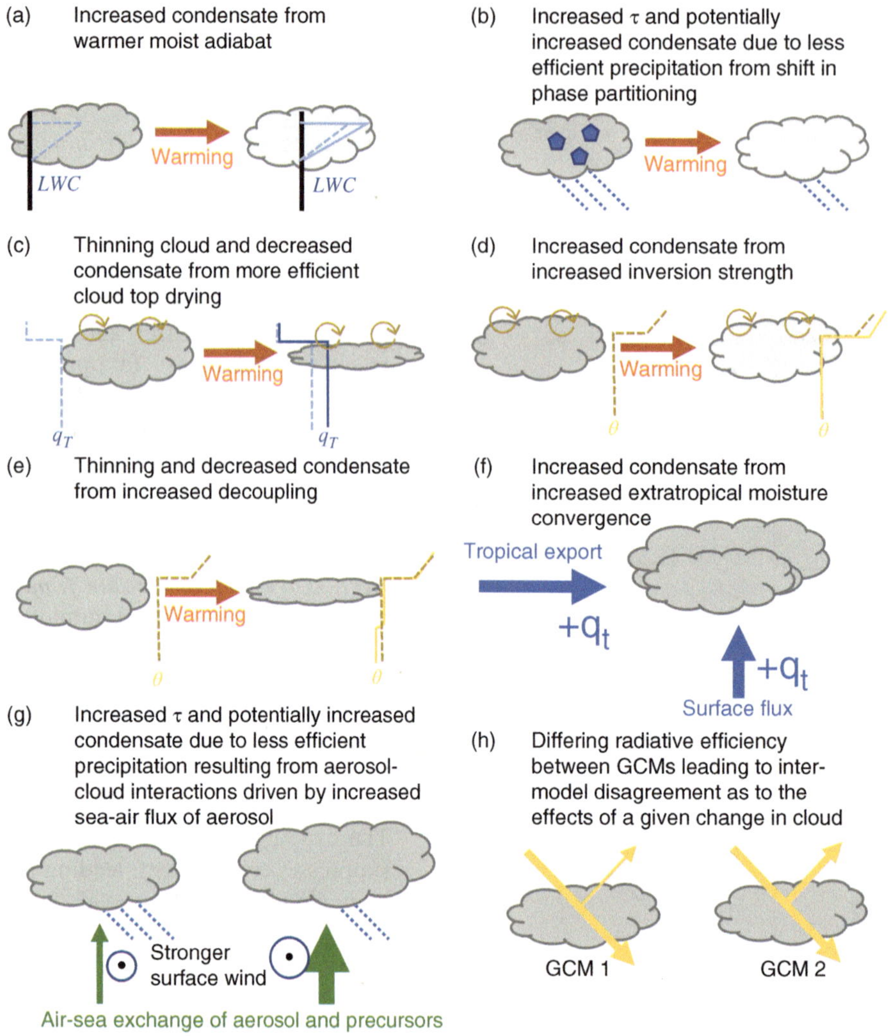

Figure 6.7 Mechanisms from the literature relevant to extratropical cloud feedback. Source: Adapted from Terai et al. (2019) with the addition of mechanisms proposed in McCoy, Field, Bodas-Salcedo, et al. (2020), Korhonen et al. (2010), and Bodas-Salcedo et al. (2019). The fixed tropopause temperature hypothesis relating to LW cloud feedback is not shown. Readers should consult Ceppi et al. (2017) for a discussion of this process.

to warming (Bender et al., 2011; Barnes & Polvani, 2013). Enhanced surface wind speed increases air sea exchange of sea-spray and aerosol precursor gases such as dimethylsulfide (Charlson et al., 1987), which in turn increase cloud condensation nuclei and cloud droplet number concentration. Increased cloud droplet number concentration increases scattering (Twomey, 1977) and potentially decreases precipitation (Albrecht, 1989) leading to increased LWC (Bodas-Salcedo et al., 2019; Korhonen et al., 2010).

(h) *Differing radiative efficiency among GCMs.* While not technically a process driving cloud changes between mean-state and warming climates, the efficiency with which clouds can affect the SW flux is critical to setting the SW cloud feedback. As demonstrated in Bender et al. (2017), GCMs differ substantially in how increasing cloud fraction increases upwelling SW. Thus, intermodel differences in how clouds and radiation interact may contribute strongly to intermodel differences in SW cloud feedback.

(i) High-topped ($CTP < 440$ hPa) clouds are likely to trap enhanced outgoing surface LW as the climate warms due to constant cloud detrainment temperature, consistent with the mechanisms governing the so-called fixed anvil temperature (FAT) mechanisms, which is critical to setting tropical LW cloud feedback (Ceppi et al., 2017; Hartmann & Larson 2002; Thompson et al., 2017, 2019; Wetherald & Manabe, 1988).

These hypothesized mechanisms are not an exhaustive set of all possible cloud feedback mechanisms in the extratropics. It should also be noted that they exhibit a dependence on sea ice in the Arctic as discussed in Chapter 5: A Review of the Factors Influencing Arctic Liquid-Containing Clouds. As can be seen, they are interwoven and difficult to disentangle. In the following section, we expand on several processes discussed in this summary.

6.4.2. Boundary-Layer Cloud Changes

The mechanisms shown in Figure 6.7 are all broadly applicable to BL cloud. In particular, changes in in-cloud LWP ($LWP_{in-cloud}$) and optical depth have been investigated closely (Eitzen et al., 2011; Gordon & Klein, 2014; Terai et al., 2016, 2019). Terai et al. (2019) analyzed contributions to change in $LWP_{in-cloud}$ in stratiform clouds from mechanisms (a–e) in Figure 6.7. The relative dominance of different mechanisms splits roughly between cold clouds (clouds below 0 °C) and warm clouds (above 0 °C). For cold clouds, changes in the moist adiabatic lapse rate and increased adiabatic ratio (the ratio between $LWP_{in-cloud}$ and maximum $LWP_{in-cloud}$ of an undiluted, adiabatic ascending cloud parcel) contribute roughly equally to increased $LWP_{in-cloud}$ along with ice to liquid phase shifts (Terai et al., 2019). However, changes in cloud geometric thickness counteract this behavior in some regimes (Terai et al., 2019). Complementary analysis of satellite observations of maritime clouds agrees in inferring a strong contribution of phase shifts in cold clouds to increased optical depth as well as revealing differing responses to warming for maritime and continental cloud τ (Tan et al., 2019).

In contrast to a general increase in $LWP_{in-cloud}$ in cold clouds, warm BL clouds tend to thin with warming in observations and models (Terai et al., 2019, 2016; Tan et al., 2019); however, there are marked differences between land and ocean potentially due to the impact of aerosols on cloud properties and precipitation (Tselioudis et al., 1992, 1998; Tan et al., 2019). Over land, passive satellite instruments have shown that cloud optical depth appears to consistently increase with temperature at cold temperatures while cloud optical depth consistently decreases with temperature over relatively warm ocean temperatures (Tselioudis et al., 1992; Tan et al., 2019). The decrease in low-cloud optical depth with temperature over oceans may be associated with fewer cloud condensation nuclei (CCN) that allow clouds to grow to larger sizes and form precipitation (Tselioudis et al., 1992). In mixed-phase clouds, thermodynamic phase transitions from ice to liquid may occur at warmer temperatures over land due to the increase presence of ice nucleating particles (INPs) in the Northern Hemisphere (Tan et al., 2019) that lower the temperature required for the formation of ice crystals thereby producing more susceptible ice. Satellite observations have noted a strong contrasts in INP concentrations in the Northern and Southern Hemispheres and their subsequent impact on thermodynamic phase partitioning in mixed-phase clouds (Coopman et al., 2020; Filioglou et al., 2019; Tan et al., 2014). In single-layer stratus clouds over land, this appears to be related to decreased adiabaticity overwhelming increases owing to changes in the moist adiabat (Terai et al., 2019). The reasons for this decrease in adiabaticity are likely to be related to decoupling and inversion strength that may potentially result in an increase in cloud base and therefore geometric thinning of the cloud (Chang & Coakley, 2007). Further investigation is required. Overall, the contrast in response to warm and cold BL clouds is consistent with modeling studies (Tselioudis et al., 1998; Tsushima et al., 2006), and observations in the Southern Ocean where supercooled liquid is frequently present (Huang et al., 2016). It is also consistent with differing sensitivities of LWP in poleward and equatorward low-cloud dominated portions of cyclones (McCoy et al., 2019) and regions outside of cyclone systems (McCoy, Field, Bodas-Salcedo, et al., 2020). Decreased τ_{BL} in warm clouds is also consistent with analyses of maritime low clouds by (Tan et al., 2019).

Overall, analysis of observations and global models is broadly consistent in finding increased optical depth in cold BL clouds with warming and a decrease in warm clouds, although uncertainty exists in their exact magnitudes (Gordon & Klein 2014; Terai et al., 2016). Further stratification of Table 6.1 into cloud top pressure (CTP) as well as cloud temperature could enhance consistency among empirical studies. However, the studies summarized in Table 6.1 agree in assigning an overall decrease in BL cloud optical depth and associated decrease in LWP in response to warming when summed over cloud temperature and by extension an overall positive SW cloud feedback issuing from BL cloud.

6.4.3. Ice, Liquid, and Mixed-Phase

Many studies have focused on how extratropical SW cloud feedbacks are affected by decisions regarding the treatment of mixed-phase and supercooled liquid (at or below 0° C) clouds in GCMs (Bjordal et al., 2020; Frey & Kay 2018; Komurcu et al., 2014; McCoy et al., 2015, 2016; Tsushima et al., 2006; Tan et al., 2016) (mechanism b). In several of these studies the dominant driver of extratropical SW cloud feedback was inferred to be related to ice-to-liquid transitions (Bjordal et al., 2020; Tan et al., 2016). In particular, the strength of the Wegener–Bergeron–Findeisen (WBF) process has been singled out as strongly affecting SW cloud feedback (Ceppi, Hartmann, & Webb, 2016; Storelvmo & Tan, 2015; Tan et al., 2016). One way in which these studies can be reconciled with an extratropical SW cloud feedback driven by moisture convergence (discussed in the previous section) is through the strong control of precipitation by snow (Field & Heymsfield, 2015; Mülmenstädt et al., 2015). This can occur in two ways: first, through ice setting the mean-state precipitation efficiency of a given GCM, or, second, reduced ice with warming rendering precipitation less efficient (Mülmenstädt et al., 2021). A negative extratropical SW cloud feedback due to decreased precipitation efficiency through diminished ice amount is consistent with process-based studies, which show a strong effect from cold-rain precipitation on how liquid clouds respond to warming (Ceppi, Hartmann & Webb, 2016; Mülmenstädt et al., 2021). Furthermore, the reduced precipitation of ice was also shown to be linked to enhanced Arctic amplification (Tan & Storelvmo, 2019). However, it is important to note that not all GCMs agree on this effect, as shown in model intercomparison (Ceppi, Hartmann et al., 2016). Dissimilarity between GCMs in terms of ice and liquid response to warming is shown schematically in Figure 6.8. The extratropical response of ice and LWP to warming in the abrupt4XCO2 simulations is shown. Models generally fall into two camps: models where ice decreases a small amount and liquid increases (potentially following reduced precipitation efficiency as ice is diminished) and where ice and liquid both increase (potentially following an overall driving of enhanced condensate as moisture converges). Very few models approach the behavior where ice is replaced with liquid, although we note that IWP includes contributions from cirrus and precipitating ice clouds and is therefore not a perfect representation of thermodynamic shifts associated with mixed-phase cloud feedbacks. This is generally consistent with previous surveys showing a 1-1 replacement of ice with liquid was insufficient to explain GCM response (Ceppi, Hartmann & Webb, 2016; McCoy et al., 2015).

Cloud top phase has been measured utilizing lidar observations and passive spectroradiometers (Hogan et al., 2004; Hu et al., 2010; Nasiri & Kahn, 2008; Naud & Kahn 2015; Tan et al., 2014; Thompson et al., 2018). Passive measures of cloud top phase may be more compatible with the scale of ice and liquid regions in clouds (Tan & Storelvmo, 2016) and high-resolution modeling.

Evaluation of cloud phase in GCMs has the potential to strongly constrain model behavior (Komurcu et al., 2014; Tan et al., 2016), and contrasting these satellite retrievals with GCMs is attractive. However, it is extremely difficult to directly compare observations of cloud phase at cloud top to GCMs. Analysis utilizing GCM satellite simulator output may be compared to lidar (Cesana & Chepfer, 2013; Tan et al., 2016). Even with careful utilization of a satellite simulator this comparison is fraught with difficulty. Supercooled liquid often exists in a thin layer at cloud top above an ice cloud layer (Chubb et al., 2013) and this vertical cloud structure is particularly commonly

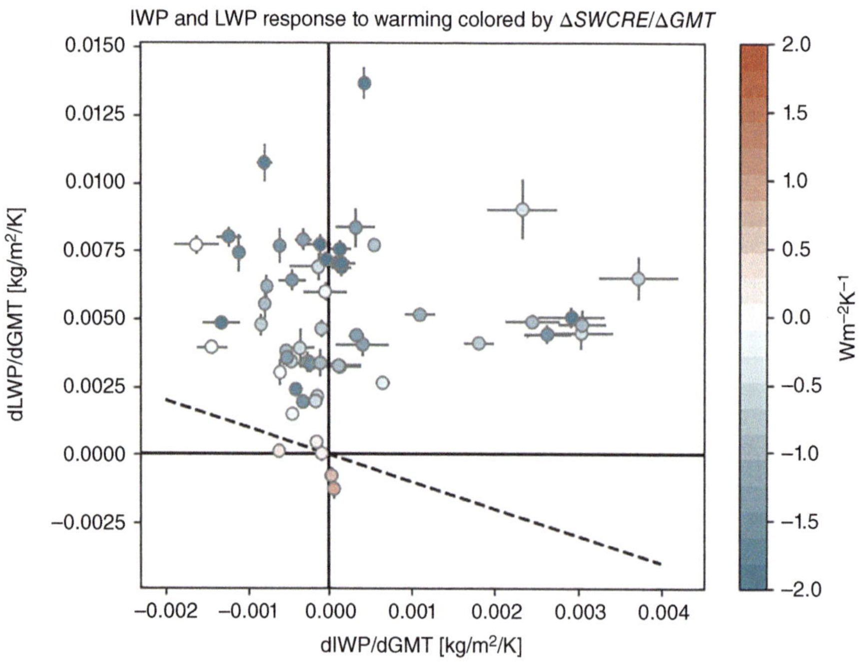

Figure 6.8 Extratropical ice and liquid water path response averaged over 40 S–85 °S in GCMs normalized by change in GMT. The 1:−1 line is shown to signify a perfect replacement of ice with liquid following warming. The figure shows points colored by $\Delta CRE_{SW}/\Delta GMT$ (approximately equivalent to SW cloud feedback).

observed in the Arctic (Mioche et al., 2017), although also observed in all regions of the globe (Zhang et al., 2010). If this layer is thinner than the height of the relevant model layer, then comparison between observed and modeled supercooled liquid is not robust as the simulator output from the GCM is likely to underestimate the cloud top liquid fraction. Some studies have suggested a simple examination of ice and liquid probability stratified by atmospheric temperature allows rapid evaluation of many models (McCoy et al., 2015, 2016), but doing this on anything less than instantaneous model output risks reducing the traceability to actual cloud phase and failure to use a satellite simulator further muddies comparison.

6.4.4. Hydrological Cycle Changes

As discussed in section 6.3, GCM cloud response to warming varies substantially between models (Figs. 6.5 top and middle and 6.6 top and middle), but the change in moisture convergence is fairly consistent (Figs. 6.5 bottom and 6.6 bottom).

Observations and models agree on a robust present-day trend in liquid water over the Southern Hemisphere (Manaster et al., 2017). The LWP response to warming as a function of latitude in GCMs reveals a robust increase in LWP toward the pole, and also shows a decrease in LWP in regions of drying toward the equator (Fig. 6.6a). This may be linked to the expansion of the Hadley cell and poleward jet shift (Bender et al., 2011; Kelleher & Grise, 2021; Sousa et al., 2018; Tselioudis et al., 2016). The region where LWP decreases is broadly consistent with the region where the atmosphere dries ($\Delta(P - E) < 0$, Fig. 6.6c (bottom)). The latitudinal transition between drying and moistening across the region where GCMs generally transition from a positive subtropical feedback to a negative extratropical feedback is intriguing. Understanding shifts in cloud properties in this region brought on by the expanding Hadley cell (Bender et al., 2011; Kelleher & Grise, 2021) and tightening area of moistening may offer insight into the cloud and circulation processes at work across the extratropics (Miyamoto et al., 2018). Overall, analysis of shifts in the position of the midlatitude jet shows little effect on cloud feedback (Ceppi & Hartmann, 2015) and it may be that this feature is of more use as a steppingstone to insight into the processes underlying extratropical SW cloud feedback than a strong determinant of extratropical SW feedback in and of itself (Kelleher & Grise, 2021).

Across models there appears to be consistency between regions of large-scale atmospheric moistening in the extratropics ($\Delta(P - E) > 0$ in Fig. 6.6c) (Held & Soden, 2006; Lorenz & DeWeaver, 2007) and increasing liquid cloud (Fig. 6.6a (top)). It is predicted that intensification of the hydrological cycle will lead to strengthened extratropical precipitation from storm systems (Lorenz & DeWeaver, 2007; Trenberth, 1998; Yettella & Kay, 2017). Most precipitation in the extratropics originates in fronts (Catto et al., 2012) and intensification of extratropical storms may occur by either enhancing atmospheric convergence, or enhancing the atmospheric moisture that can be converged (Field and Wood, 2007; Tselioudis & Rossow, 2006). Analysis of trends in cyclone behavior over the last two decades in the Southern Ocean suggests that the dominant contributor is moisture increases, as opposed to convergence (McCoy et al., 2019). This appears to be consistent with GCM predictions of cyclone behavior (Bengtsson et al., 2006; Yettella & Kay, 2017) as well as predictions of atmospheric river moisture flux (Algarra et al., 2020; Gonzales et al., 2020; Payne et al., 2020; Ramos et al., 2016). However, some caution is needed in the interpretation of the satellite record of cyclone precipitation as analysis of centennial-scale proxies in the North Atlantic region by Feser et al. (2015) agrees in showing a strong increase in regional storms, but over longer periods this increase is within natural variability.

Enhanced convergence of moisture must be converted into precipitation which happens in cloud. Analyses of regime composited behavior have found that enhanced extratropical LWP over the last decades (Manaster et al., 2017) can be explained by trends in convergence and moisture in cyclones and that the response of LWP to warming predicted by GCMs can be explained by these same factors (Field & Wood 2007; Field et al., 2008; McCoy et al., 2019; McCoy, Field, Bodas-Salcedo, et al., 2020). Generally, it is found that moisture convergence into the extratropical atmosphere strongly affects the increase in LWP in GCMs, although this effect is modulated by changes in BL cloud possibly following mechanisms (c)–(e) (Fig. 6.7). The extratropical LWP response to warming in many GCMs can be explained to within 30% by enhanced moisture convergence due to increased import from the subtropics and local evaporation (McCoy et al., 2019; McCoy, Field, Bodas-Salcedo, et al., 2020). Dynamical effects such as strengthening storms may further enhance moisture convergence driven by increased global mean temperature and appear to overwhelm predicted decreases in extratropical cyclone frequency in their aggregate effect on reflected SW (Tselioudis & Rossow, 2006).

6.4.5. Aerosol-Cloud Interactions

Our discussion of cloud-precipitation-radiation processes and their effect on setting extratropical cloud feedback has occurred in a vacuum with respect to how liquid droplets nucleate on aerosol particles and how ice forms. Given the effect of particle habit on setting radiative efficacy and precipitation processes, it is important that we discuss pathways linking aerosol to cloud feedback. This is presented in the context of mechanism (g) (Fig. 6.7).

In the case of ice nucleation, the relationship to the preceding discussion is clear. Ice plays a key role in setting precipitation flux (Field & Heymsfield, 2015; Mülmenstädt et al., 2015). However, in the absence of an aerosol that can act as an INP on which to crystallize, ice will not necessarily form above −37 °C (Hoose & Möhler (2012; Murray et al., 2021; Storelvmo, 2017). This leaves a sizable population of clouds that may be affected by INP and in turn affect both precipitation efficiency (Mülmenstädt et al., 2021) and radiative flux (condensate in the form of ice tends to be less reflective than liquid) (McCoy et al., 2014). One of the more effective INP is dust (primarily feldspar) (Atkinson et al., 2013), and supercooled liquid is relatively rare in dust outflow regions (Lohmann & Diehl, 2006; Tan et al., 2014), but other aerosol such as marine organics may play a role in regions far from dust sources, such as the remote Southern Ocean (Burrows et al., 2013; Vergara-Temprado, Miltenberger et al., 2018). Although the efficiency of black carbon as an INP is currently highly uncertain, its sheer abundance in Earth's atmosphere (Schill, Froyd et al., 2020) implies that it may potentially be an important INP (Tan et al., 2014); however, recent studies suggest that it is not an efficient immersion INP as previously thought (Schill, DeMott et al., 2020; Vergara-Temprado, Holden et al., 2018).

Given the uncertainty associated with how INPs affect clouds, modeling efforts linking cloud feedback to INPs are highly uncertain. However, high-resolution modeling of Southern Ocean cyclones coupled with implementation of state-of-the-art INP parameterization suggests a potentially sizable effect from INP on Southern Ocean cloud radiative effect and liquid cloud prevalence (Vergara-Temprado, Miltenberger et al., 2018). In addition to modeling the effects of ice nucleation in extratropical clouds accurately, it is likely that understanding sources of INP to the extratropics and how these sources will change in a warming world will be important to setting ice cloud frequency and extratropical precipitation efficiency and by extension cloud feedback (Murray et al., 2021).

The relationship between CCN acting on liquid clouds and extratropical SW cloud feedback is also relevant. Increased aerosol that can act as cloud condensation nuclei in warm clouds enhances the number concentration of cloud droplets (N_d) at fixed cloud liquid content (see Chapter 2: An Overview of Aerosol-Cloud Interactions). Scattering strengthens with enhanced N_d and cloud albedo increases (Twomey, 1977), warm rain becomes less efficient (Albrecht, 1989), and entrainment evaporation becomes more efficient (Ackerman et al., 2004; Bretherton et al., 2007). Thus, the manifestation of warm-cloud aerosol-cloud interactions in a given model may act to set the base-state reflectivity as well as warm-rain precipitation efficiency with knock-on effects to the SW cloud feedback (Gettelman & Sherwood, 2016). There is substantial uncertainty in the representation of liquid aerosol-cloud interaction across GCMs (Bellouin et al., 2019; Storelvmo, 2017) and this may result in aerosol-cloud interactions strongly affecting extratropical cloud feedback in some GCMs (Gettelman et al., 2016, 2019). Analysis of the HadGEM3 GCM and development models in Bodas-Salcedo et al. (2019) found that dimethylsulfide emissions increased with warming as the jet strengthened and moved poleward. An increase in N_d and cloud reflectivity was driven by this increase in dimethylsulfide leading to a more negative SW cloud feedback. This effect was weakened in the version of HadGEM3 contributed to CMIP6 due to a reduction in the strength of aerosol-cloud interactions relative to development versions (Mulcahy et al., 2018). Aerosol-cloud interactions driven by air-sea exchange in this way may contribute substantially to the extratropical cloud feedback in other GCMs where aerosol-cloud interactions are strong. However, this effect may be artificially too strong in the Southern Ocean extratropics of many GCMs given that most GCMs underestimate N_d in those regions (McCoy et al., 2020) and thus even small increases in dimethylsulfide flux and N_d may lead to large increases in cloud optical depth due to the nonlinear relationship between these quantities (Carslaw et al., 2013).

Disentangling aerosol-cloud interactions and cloud feedbacks may also play an important role in interrogating the observational record for trends that can constrain extratropical cloud feedback. In the Northern Hemisphere, trends in aerosol emissions over the last decade have led to decreases in N_d in continental outflow regions (Bennartz et al., 2011; Li et al., 2018; McCoy, Bender et al., 2018). If decreased N_d in the North Atlantic and Pacific decreases liquid cloud amount in the extratropical storm tracks as indicated by modeling and analysis of observations (Malavelle et al., 2017; McCoy, Field, et al., 2018; McCoy, Field, Gordon, et al., 2020; Wang et al., 2014), the resultant trend in cloudiness would be of the opposite sign to the trend in LWP consistent with a negative SW cloud feedback. Thus, it is likely that trends induced in cloudiness through aerosol-cloud adjustments make a clean evaluation of meteorology-induced trends, and by extension cloud feedback, difficult to assess in anthropogenically dominated regions.

6.4.6. Radiative Response to Cloud Changes

We have discussed shifts in cloud macrophysical properties in response to warming. We will briefly discuss coupling between cloud changes and radiative fluxes. While not a process-level mechanism shown in Figure 6.7, cloud-radiation interactions are key to setting the variable of interest: cloud feedback, in particular the SW component.

The expectation is that cloud changes will contribute to changes in albedo (α), which in turn will reflect incoming solar radiation. There is a strong seasonal cycle of insolation in the extratropics, making the season in which cloud changes occur important to the overall change in radiation. However, there is also the expectation that changes in variables such as cloud optical depth and cloud fraction will be related to α in a roughly consistent way across GCMs. This is not necessarily the case, as has been shown for the cloud fraction-α relationship. Evaluation of the sensitivity of extratropical α to cloud fraction ($d\alpha/dCF$) found substantial intermodel spread ranging from 0.25 to 0.55 (units of α over cloud fraction) in CMIP5 GCMs, relative to an observed value of 0.4 (Bender et al., 2017). Similar evaluation of the relationship between α and area-mean LWP, which contains information about cloud fraction as well as liquid content of clouds, found substantial differences across CMIP5 and CMIP6 GCMs with sensitivity of α to LWP in December ranging from 0.25 to 1.75 m^2/kg (where α is a unitless quantity). Observations of the α sensitivity to LWP were estimated from CERES and MAC-LWP at 0.5 m^2/kg (McCoy et al., 2022). One possible reason underlying the diversity in sensitivities of α and by extension radiative flux to area-mean LWP is that models with larger mean-state cloud cover and condensate mass in the extratropics will be nearer to radiative saturation (McCoy et al., 2022). Examination of mean-state LWP in CMIP5 and CMIP6 GCMs shows nearly an order of magnitude difference between the highest and lowest LWP models. Another source of spread in relationships between cloud macrophysical state (e.g., extent and LWP) may be cloud microphysical state (e.g., number concentration of cloud droplets). This is discussed in the following section in the context of aerosol.

6.5. PROSPECTS

In this chapter, we have outlined some of the mechanisms of the extratropical feedback and discussed what information observed present-day behavior can yield. Over many generations of GCMs the extratropical SW cloud feedback has stayed stubbornly uncertain and the processes underpinning its manifestation have remained ambiguous, making it difficult to focus on a specific process and reduce model uncertainty. In particular, the scientific understanding of several processes remains elusive: ice nucleation (Murray et al., 2021), biogenic aerosol processes in remote regions (Gordon et al., 2017; Korhonen et al., 2008; McCoy et al., 2020, 2021), the representation of WBF process in climate models (Storelvmo & Tan, 2015; Tan et al., 2016), and warm and cold rain (Mülmenstädt et al., 2021).

The dependence of climate sensitivity on extratropical SW cloud feedback (Zelinka et al., 2020) continues to motivate study and constraint. The interconnected nature of process uncertainty in extratropical clouds means that no single process can be fairly singled out for analysis. However, the following areas may return significant reductions in the probable range of extratropical SW cloud feedback and by extension climate sensitivity.

6.5.1. Precipitation Efficiency

Precipitation efficiency in the extratropics has been shown to affect cloud feedback (Mülmenstädt et al., 2021) and potentially Arctic amplification (Tan & Storelvmo, 2019). Precipitation efficiency rests on complex pillars: how moisture converges into the extratropical atmosphere and responds to warming (Held & Soden, 2006; Lavers et al., 2015) and how ice-phase processes set the aggregate precipitation efficiency (Field & Heymsfield, 2015; Frazer & Ming, 2022). The former links this problem to the general circulation of the atmosphere (Bony et al., 2015). The latter links it to the representation of a suite of complex ice-phase microphysical and aerosol processes (Murray et al., 2021; Vergara-Temprado, Miltenberger, et al., 2018).

Precipitation efficiency also depends on the ability of models to resolve fronts and represent the subgrid variability of mixed-phase properties (Tan & Storelvmo, 2016). A turbulence-based approach to representing the persistence of supercooled liquid in GCMs, which reduces process representation dependence on resolution, has shown promising results in improving supercooled liquid persistence (Furtado et al., 2016). Fronts are the dominant precipitation sink in the extratropics (Catto et al., 2012). Examination of precipitation efficiency in cyclones in convection-permitting simulations has shown convergence between observations and models at resolutions that resolve fronts (McCoy et al., 2019). This result suggests that models that are able to resolve frontal passages may yield more realistic representations of extratropical cloud feedback. In particular, GCM frontal structure may be confronted with existing observations (Naud et al., 2010, 2012) to better understand modeled extratropical precipitation processes.

Precipitation efficiency may also be an important unknown in Southern Ocean postfrontal BL clouds, in addition to cyclone frontal clouds. Using a CloudSat and CALIPSO cloud classification (Mace & Zhang, 2014; Mace et al., 2021) report that 40% of single-layer BL clouds precipitate (mostly clouds with >2 km geometrical thickness). This indicates that postfrontal clouds may need to be further subdivided into different cloud regimes according to the processes – precipitation versus entrainment drying – that act as the predominant sinks for cloud condensate, as these processes may lead to opposite-sign sensitivity to warming (Mülmenstädt et al., 2021). However, there is still much that needs to be better understood. This includes what role (supercooled) liquid and mixed-phase precipitation processes play in these clouds. Supercooled liquid can be the predominant phase in Southern Ocean BL clouds to very low temperatures, and such clouds commonly contain supercooled drizzle droplets (Chubb et al., 2013; Silber et al., 2021). However, surface-reaching precipitation usually originates as ice (Mace and Protat, 2018), again indicating the efficiency differential between "warm" (liquid) and "cold" (mixed-phase) rain processes. How sensitive cold rain formation is to the usually low INP concentrations over the Southern Ocean (McCluskey et al., 2018) has significant implications. If INP become less active with temperature but precipitation depends on INP, then the warmer Southern Ocean clouds of the future will precipitate less. But if cold rain can initiate relatively independently of INP concentration – say, from efficient ice multiplication mechanisms starting with a small number of freezing drizzle droplets – then the precipitation sink may be fairly insensitive to warming. Relatively little is known about the relevant cloud physics because Northern Hemisphere clouds, embedded in higher INP concentrations, are not necessarily reliable analogs, and Southern Ocean field campaigns are only now beginning to be exploited.

6.5.2. Aerosol-Cloud Interactions

A more holistic examination of aerosol-cloud interactions in relation to cloud feedback, as suggested by Gettelman and Sherwood (2016) and Gettelman et al. (2016), is likely to yield constraints on both forcing and feedback within GCMs.

The processes driving aerosol-cloud interactions in warm clouds are broadly understood, but the source term for CCN in the remote Southern Ocean is still poorly understood (Gordon et al., 2017) and GCMs tend toward low-cloud droplet concentrations in this region relative to observations (McCoy et al., 2020). Thus, part of accurately representing the warm-rain contribution to precipitation efficiency (Mülmenstädt et al., 2021) is accurate representation of pristine sources of aerosol and nucleation processes. As pointed out by Gettelman and Sherwood (2016), aerosol-cloud interactions also act to set mean-state properties such as brightness, which in turn impacts cloud feedbacks by setting radiative efficiency.

Relative to warm clouds, the representation of ice-nucleation by aerosols is much more complex (Hoose & Möhler, 2012), especially given the distance of the Southern Hemisphere extratropics from sources of dust, which act as an efficient ice nuclei (Atkinson et al., 2013; Murray et al., 2021; Tan et al., 2014). As discussed above, ice creation directly impacts precipitation efficiency (Field & Heymsfield, 2015) as well as the radiative efficacy of condensate (McCoy et al., 2014; Vergara-Temprado, Miltenberger, et al., 2018). In particular, improved constraint of remote, marine sources of INPs (Burrows et al., 2013) and of the aggregate impact of those INPs on IWP and precipitation are likely to improve model behavior.

6.5.3. Boundary-Layer Clouds in Cold Air Outbreaks

Several studies have drawn attention to the representation of BL cloud cover in relation to solar radiation biases and cloud feedback (Bodas-Salcedo et al., 2014; Bodas-Salcedo, Andrews, et al., 2016; Bodas-Salcedo, Hill, et al., 2016; Bodas-Salcedo, 2018; Kelleher & Grise, 2019). Accurately reproducing this regime in simulations, even at very high resolution is difficult (Field et al., 2017) and sources of sensitivity include both ice and BL processes (Field et al., 2014). Open and closed cellular convection are strongly dependent on these processes and have differing effects on SW radiation (McCoy et al., 2017).

The complexity of BL clouds in cold air outbreaks is daunting. Given the frequent occurrence of cold air outbreaks behind cyclone cold fronts, there already exists a substantial in situ and remotely sensed observational record to draw on. Planned field campaigns (Arctic Cold Air Outbreak [ACAO] and Cold-Air outbreak Experiment in the Sub-Arctic Region [CAESAR]) in the coming years in the Northern Hemisphere combined with analysis of recent field campaigns in both hemispheres (SOCRATES and COMBLE) (Geerts, 2019; McFarquhar et al., 2021) may lead to rapid improvements in our understanding of clouds and BL processes in cold air outbreaks, their representation in GCMs, and resultant contribution to GCM response and mean-state behavior.

6.5.4. Improving Comparisons Between Models and Observations

In the context of cloud feedbacks and the hydrological cycle, satellite observations are a useful tool for both directly constraining cloud feedbacks and precipitation and for identifying shortcomings at the process level in GCMs (DeAngelis et al., 2015; Tan et al., 2016; Myers & Norris, 2016; Mülmenstädt et al., 2021; Su et al., 2017; Scott et al., 2020; Tan et al., 2015). While useful comparisons between GCMs and ground-based and satellite instruments can be conducted with knowledge of the algorithm and sampling limitations of the instruments (see for instance Qu et al. (2015), McCoy et al. (2019), or Myers et al. (2021)), a consistent framework to compare GCMs and observations can be a powerful and convenient tool. These comparisons between models and observations require a consistent framework to allow for fair comparisons between the two tools. Satellite observations examine clouds biased toward the cloud top in contrast to GCM output. They also suffer spatial and temporal sampling biases. Furthermore, retrieval algorithms differ greatly between various satellite remote sensing instruments. Satellite simulators, such as the Cloud Feedback Model Intercomparison Project (CFMIP) Observational Simulator Package (COSP) (Bodas-Salcedo et al., 2011) that has been implemented in a number of CMIP5 and CMIP6 models, address this issue by consistently comparing model output with satellite retrievals. CFMIP cfSites output, included in a subset of CMIP5 and CMIP6 models and experiments, also provides high frequency and height-resolved temperature and humidity variables and tendencies at 120 locations that make it possible to evaluate models with in situ measurements and ground-based remote sensing (Webb et al., 2017). A wide adoption of these tools will help improve our ability to consistently and systematically compare models with the wide array of available observations. More recently, ground-based simulators in COSP have also been implemented to facilitate comparisons between ground-based lidar and radar observations (Kuma et al., 2021).

While COSP is extensively used to diagnose biases in modeling cloud properties consistently with satellite observations and has been very useful in revealing model biases (Zhang et al. 2005), instrument simulators have yet to be thoroughly evaluated and suffer from known limitations related to discrepancies in the retrieval of cloud optical depth (Mace, 2010). Instrument simulators also suffer from a lack of representation of precipitation processes. The use of simulators to diagnose biases in precipitation in GCMs is much less common and has been limited to the evaluation of precipitation frequency against NASA's CloudSat instrument as opposed to precipitation intensity (Kay et al., 2018). Although high-resolution models have implemented radar simulators for comparisons with additional satellite simulators including those for Global Precipitation Measurement (GPM) (Matsui et al., 2013) and Tropical Rainfall Measuring Mission (TRMM) (Chern et al., 2020; Roh et al., 2017), advances in the representations of cloud microphysics would enable a comparison with existing satellite observations of precipitation in large-scale GCMs. Simulators are particularly relevant for comparing precipitation in models and observations, because many current-generation satellite instruments are unable to observe the drizzle that commonly forms over the extratropical latitudes (Mace & Protat, 2018; Silber et al., 2019; Wang et al., 2015). Furthermore, previous studies suggest that low-level clouds in GCMs drizzle too frequently (Bodas-Salcedo et al., 2008; Stephens et al., 2012; Suzuki et al., 2015). A recent study noted that accounting for snow in the radiation scheme of a GCM also improved the comparison of mixed-phase clouds with observations that may have a large impact on cloud feedbacks (Cesana et al., 2021). Because the strength and importance of a number of the mechanisms outlined in section 6.4 hinge on an accurate characterization of precipitation, the future expansion of precipitation diagnostics and simulators (Kay et al., 2018; Michibata et al., 2019) in large-scale GCMs can potentially provide large advances in understanding.

ACKNOWLEDGMENTS

Daniel T. McCoy acknowledges the support of the University of Wyoming and NASA-PMMST Grant #80NSSC22K0599. Michelle E. Frazer's efforts were supported by the National Science Foundation under award AWD1005319. The efforts of Christopher R. Terai and Mark D. Zelinka were performed under the auspices of the U.S. Department of Energy (DOE) by Lawrence Livermore National Laboratory under contract DE-AC52-07NA27344 and were supported by the Regional and Global Model Analysis program area of DOE's Earth and Environmental Systems Science Division (EESSD). Johannes Mülmenstädt was supported by the Atmospheric System Research program of DOE EESSD. The Pacific Northwest National Laboratory is operated for the U.S. Department of Energy by Battelle Memorial Institute under contract DE-AC05-76RL01830. We thank two anonymous reviewers for their insightful comments and suggestions.

APPENDIX

Table S1 List of GCMs

Number	Model	CMIP
1	ACCESS1-0	cmip5
2	ACCESS1-3	cmip5
3	CCSM4	cmip5
4	CNRM-CM5	cmip5
5	CSIRO-Mk3-6-0	cmip5
6	CanESM2	cmip5
7	FGOALS-g2	cmip5
8	FGOALS-s2	cmip5
9	GFDL-CM3	cmip5
10	GFDL-ESM2G	cmip5
11	GFDL-ESM2M	cmip5

(continued)

Table S1 *(continued)*

Number	Model	CMIP
12	GISS-E2-H	cmip5
13	GISS-E2-R	cmip5
14	PSL-CM5A-LR	cmip5
15	IPSL-CM5A-MR	cmip5
16	IPSL-CM5B-LR	cmip5
17	MIROC-ESM	cmip5
18	MIROC5	cmip5
19	MPI-ESM-LR	cmip5
20	MPI-ESM-MR	cmip5
21	MPI-ESM-P	cmip5
22	MRI-CGCM3	cmip5
23	NorESM1-M	cmip5
24	bcc-csm1-1-m	cmip5
25	bcc-csm1-1	cmip5
26	inmcm4	cmip5
27	BCC-CSM2-MR	cmip6
28	BCC-ESM1	cmip6
29	CAMS-CSM1-0	cmip6
30	CESM2-FV2	cmip6
31	CESM2-WACCM-FV2	cmip6
32	CESM2-WACCM	cmip6
33	CESM2	cmip6
34	CMCC-CM2-SR5	cmip6
35	CNRM-CM6-1-HR	cmip6
36	CNRM-CM6-1	cmip6
37	CNRM-ESM2-1	cmip6
38	CanESM5	cmip6
39	E3SM-1-0	cmip6
40	EC-Earth3-AerChem	cmip6
41	EC-Earth3-Veg	cmip6
42	FGOALS-f3-L	cmip6
43	FGOALS-g3	cmip6
44	GFDL-CM4	cmip6
45	GFDL-ESM4	cmip6
46	GISS-E2-1-H	cmip6
47	GISS-E2-2-G	cmip6
48	INM-CM4-8	cmip6
49	IPSL-CM6A-LR	cmip6
50	MIROC-ES2L	cmip6
51	MIROC6	cmip6
52	MPI-ESM-1-2-HAM	cmip6
53	MPI-ESM1-2-HR	cmip6
54	MPI-ESM1-2-LR	cmip6
55	MRI-ESM2-0	cmip6
56	NorESM2-MM	cmip6
57	UKESM1-0-LL	cmip6

The number reflects the number listed in main-text figures.

REFERENCES

Ackerman, A. S., Kirkpatrick, M. P., Stevens, D. E., & Toon, O. B. (2004, Dec). The impact of humidity above stratiform clouds on indirect aerosol climate forcing. *Nature*, *432*(7020), 1014–7. doi: 10.1038/nature03174

Albrecht, B. A. (1989, Sept). Aerosols, cloud microphysics, and fractional cloudiness. *Science*, *245*(4923), 1227–1230. doi: 10.1126/science.245.4923.1227

Algarra, I., Nieto, R., Ramos, A. M., Eiras-Barca, J., Trigo, R. M., & Gimeno, L. (2020, Oct). Significant increase of global anomalous moisture uptake feeding landfalling Atmospheric Rivers. *Nature Communications*, *11*(1), 5082. doi: 10.1038/s41467-020-18876-w

Atkinson, J. D., Murray, B. J., Woodhouse, M. T., Whale, T. F., Baustian, K. J., Carslaw, K. S., et al. (2013, June). The importance of feldspar for ice nucleation by mineral dust in mixed-phase clouds. *Nature*, *498*(7454), 355–358. doi: 10.1038/nature12278

Barnes, E. A., & Polvani, L. (2013, Sept). Response of the midlatitude jets, and of their variability, to increased greenhouse gases in the CMIP5 models. *Journal of Climate*, *26*(18), 7117–7135. doi: 10.1175/jcli-d-12-00536.1

Bellouin, N., Quaas, J., Gryspeerdt, E., Kinne, S., Stier, P., Watson-Parris, D., et al. (2019, Nov). Bounding global aerosol radiative forcing of climate change. *Reviews of Geophysics*, *58*(1), e2019RG000660. doi: 10.1029/2019rg000660

Bender, F. A. M., Engström, A., Wood, R., & Charlson, R. J. (2017, June). Evaluation of hemispheric asymmetries in marine cloud radiative properties. *Journal of Climate*, *30*(11), 4131–4147. doi: 10.1175/JCLI-D-16-0263.1

Bender, F. A. M., Ramanathan, V., & Tselioudis, G. (2011, May). Changes in extratropical storm track cloudiness 1983–2008: Observational support for a poleward shift. *Climate Dynamics*, *38*(9–10), 2037–2053. doi: 10.1007/s00382-011-1065-6

Bengtsson, L., Hodges, K. I., & Roeckner, E. (2006). Storm tracks and climate change. *Journal of Climate*, *19*(15), 3518–3543.

Bennartz, R., Fan, J., Rausch, J., Leung, L. R., & Heidinger, A. K. (2011). Pollution from China increases cloud droplet number, suppresses rain over the East China Sea. *Geophysical Research Letters*, *38*(9). doi: 10.1029/2011gl047235

Betts, A. K., & Harshvardhan. (1987, July). Thermodynamic constraint on the cloud liquid water feedback in climate models. *Journal of Geophysical Research-Atmospheres*, *92*(D7), 8483–8485. doi: 10.1029/JD092iD07p08483

Bjordal, J., Storelvmo, T., Alterskjær, K., & Carlsen, T. (2020, Nov). Equilibrium climate sensitivity above 5 °C plausible due to state-dependent cloud feedback. *Nature Geoscience*, *13*(11), 718–721. doi: 10.1038/s41561-020-00649-1

Bodas-Salcedo, A. (2018). Cloud condensate and radiative feedbacks at midlatitudes in an aquaplanet. *Geophysical Research Letters*, *45*(8), 3635–3643. doi: 10.1002/2018gl077217

Bodas-Salcedo, A., Andrews, T., Karmalkar, A. V., & Ringer, M. A. (2016). Cloud liquid water path and radiative feedbacks over the Southern Ocean. *Geophysical Research Letters*, *43*, 10,938–10,946. doi: 10.1002/2016GL070770

Bodas-Salcedo, A., Hill, P. G., Furtado, K., Williams, K. D., Field, P. R., Manners, J. C., et al. (2016). Large contribution of supercooled liquid clouds to the solar radiation budget of the Southern Ocean. *Journal of Climate*, *29*(11), 4213–4228. doi: 10.1175/jcli-d-15-0564.1

Bodas-Salcedo, A., Mulcahy, J. P., Andrews, T., Williams, K. D., Ringer, M. A., Field, P. R., & Elsaesser, G. S. (2019). Strong dependence of atmospheric feedbacks on mixed-phase microphysics and aerosol-cloud interactions in HadGEM3. *Journal of Advances in Modeling Earth Systems*, *11*(6), 1735–1758. doi: 10.1029/2019ms001688

Bodas-Salcedo, A., Webb, M. J., Bony, S., Chepfer, H., Dufresne, J. L., Klein, S. A., et al. (2011, Aug). COSP: Satellite simulation software for model assessment. *Bulletin of the American Meteorological Society*, *92*(8), 1023–1043. doi: 10.1175/2011BAMS2856.1

Bodas-Salcedo, A., Williams, K. D., Ringer, M. A., Beau, I., Cole, J. N. S., Dufresne, J. L., et al. (2014, Jan). Origins of the solar radiation biases over the Southern Ocean in CFMIP2 models. *Journal of Climate*, *27*(1), 41–56. doi: 10.1175/jcli-d-13-00169.1

Bodas-Salcedo, A., Webb, M., Brooks, M., Ringer, M., Williams, K., Milton, S., & Wilson, D. (2008). Evaluating cloud systems in the Met Office global forecast model using simulated CloudSat radar reflectivities. *Journal of Geophysical Research: Atmospheres*, *113*(D8).

Bony, S., Stevens, B., Frierson, D. M. W., Jakob, C., Kageyama, M., Pincus, R., et al. (2015). Clouds, circulation and climate sensitivity. *Nature Geoscience*, *8*(4), 261–268. doi: 10.1038/ngeo2398

Bretherton, C. S., Blossey, P. N., & Jones, C. R. (2013, June). Mechanisms of marine low cloud sensitivity to idealized climate perturbations: A single-LES exploration extending the CGILS cases. *Journal of Advances in Modeling Earth Systems*, *5*(2), 316–337. doi: 10.1002/jame.20019

Bretherton, C. S., Blossey, P. N., & Uchida, J. (2007). Cloud droplet sedimentation, entrainment efficiency, and subtropical stratocumulus albedo. *Geophysical Research Letters*, *34*(3). doi: 10.1029/2006gl027648

Bretherton, C. S., & Wyant, M. C. (1997, Jan). Moisture transport, lower-tropospheric stability, and decoupling of cloud-topped boundary layers. *Journal of the Atmospheric Sciences*, *54*(1), 148–167. doi: 10.1175/1520-0469(1997)054<0148:mtltsa>2.0.co;2

Burrows, S. M., Hoose, C., Poschl, U., & Lawrence, M. G. (2013). Ice nuclei in marine air: Biogenic particles or dust? *Atmospheric Chemistry and Physics*, *13*(1), 245–267. doi: 10.5194/acp-13-245-2013

Carslaw, K. S., Lee, L. A., Reddington, C. L., Pringle, K. J., Rap, A., Forster, P. M., et al. (2013). Large contribution of natural aerosols to uncertainty in indirect forcing. *Nature*, *503*(7474), 67–71. doi: 10.1038/ nature12674

Catto, J. L., Jakob, C., Berry, G., & Nicholls, N. (2012). Relating global precipitation to atmospheric fronts. *Geophysical Research Letters*, *39*(10). doi: 10.1029/2012GL051736

Ceppi, P., Brient, F., Zelinka, M. D., & Hartmann, D. L. (2017). Cloud feedback mechanisms and their representation in global climate models. *Wiley Interdisciplinary Reviews: Climate Change*, *8*(4), e465.

Ceppi, P., & Hartmann, D. (2015, June). Connections between clouds, radiation, and midlatitude dynamics: A review. *Current Climate Change Reports*, *1*(2) 94–102. doi: 10.1007/s40641-015-0010-x

Ceppi, P., Hartmann, D. L., & Webb, M. J. (2016). Mechanisms of the negative shortwave cloud feedback in middle to high latitudes. *Journal of Climate*, *29*(1), 139–157. doi: doi:10.1175/JCLI-D-15-0327.1

Ceppi, P., McCoy, D. T., & Hartmann, D. L. (2016). Observational evidence for a negative shortwave cloud feedback in middle to high latitudes. *Geophysical Research Letters*, *43*(3), 1331–1339. doi: 10.1002/2015gl067499

Cesana, G., & Chepfer, H. (2013). Evaluation of the cloud thermodynamic phase in a climate model using CALIPSO-GOCCP. *Journal of Geophysical Research: Atmospheres*, *118*(14), 7922–7937. doi: 10.1002/jgrd.50376

Cesana, G. V., Ackerman, A. S., Fridlind, A. M., Silber, I., & Kelley, M. (2021, Oct). Snow reconciles observed and simulated phase partitioning and increases cloud feedback. *Geophysical Research Letters*, *48*(20), e2021GL094876. doi: 10.1029/2021GL094876

Chang, F.-L., & Coakley, J. A. (2007, May). Relationships between marine stratus cloud optical depth and temperature: inferences from AVHRR observations. *Journal of Climate*, *20*(10), 2022–2036. doi: 10.1175/JCLI4115.1

Charlson, R. J., Lovelock, J. E., Andreae, M. O., & Warren, S. G. (1987, Apr). Oceanic phytoplankton, atmospheric sulfur, cloud albedo and climate. *Nature*, *326*(6114), 655–661. doi: 10.1038/326655a0

Chern, J., Tao, W., Lang, S. E., Li, X., & Matsui, T. (2020). Evaluating precipitation features and rainfall characteristics in a multi-scale modeling framework. *Journal of Advances in Modeling Earth Systems*, *12*(8), e2019MS002007. (Publisher: Wiley Online Library)

Chubb, T. H., Jensen, J. B., Siems, S. T., & Manton, M. J. (2013). In situ observations of supercooled liquid clouds over the Southern Ocean during the HIAPER Pole-to-Pole Observation campaigns. *Geophysical Research Letters* *40*(19), 5280–5285. doi: 10.1002/grl.50986

Coopman, Q., Riedi, J., Zeng, S., & Garrett, T. J. (2020, March). Space-based analysis of the cloud thermodynamic phase transition for varying microphysical and meteorological regimes. *Geophysical Research Letters*, *47*(6), e2020GL087122.

Dacre, H. F., Clark, P. A., Martinez-Alvarado, O., Stringer, M. A., & Lavers, D. A. (2015, Aug). How do atmospheric rivers form? *Bulletin of the American Meteorological Society*, *96*(8), 1243–1255. doi: 10.1175/BAMS-D-14-00031.1

DeAngelis, A. M., Qu, X., Zelinka, M. D., & Hall, A. (2015). An observational radiative constraint on hydrologic cycle intensification. *Nature*, *528*(7581), 249–253.

Eitzen, Z. A., Xu, K.-M., & Wong, T. (2011). An estimate of low-cloud feedbacks from variations of cloud radiative and physical properties with sea surface temperature on interannual time scales. *Journal of Climate*, *24*(4), 1106–1121.

Elsaesser, G. S., O'Dell, C. W., Lebsock, M. D., Bennartz, R., Greenwald, T. J., & Wentz, F. J. (2017). The Multi-Sensor Advanced Climatology of Liquid Water Path (MAC-LWP). *Journal of Climate*, *30*(24), 10193–10210. doi: 10.1175/jcli-d-16-0902.1

Feser, F., Barcikowska, M., Krueger, O., Schenk, F., Weisse, R., & Xia, L. (2015, Jan). Storminess over the North Atlantic and northwestern Europe—A review. *Quarterly Journal of the Royal Meteorological Society*, *141*(687), 350–382. doi: 10.1002/qj.2364

Field, P. R., Brozková, R., Chen, M., Dudhia, J., Lac, C., Hara, T., et al. (2017). Exploring the convective grey zone with regional simulations of a cold air outbreak. *Quarterly Journal of the Royal Meteorological Society*, *143*(707), 2537–2555. doi: 10.1002/qj.3105

Field, P. R., Cotton, R. J., McBeath, K., Lock, A. P., Webster, S., & Allan, R. P. (2014). Improving a convection-permitting model simulation of a cold air outbreak. *Quarterly Journal of the Royal Meteorological Society*, *140*(678), 124–138. doi: 10.1002/qj.2116

Field, P. R., Gettelman, A., Neale, R. B., Wood, R., Rasch, P. J., & Morrison, H. (2008). Midlatitude cyclone compositing to constrain climate model behavior using satellite observations. *Journal of Climate*, *21*(22), 5887–5903. doi: doi:10.1175/2008JCLI2235.1

Field, P. R., & Heymsfield, A. J. (2015). Importance of snow to global precipitation. *Geophysical Research Letters*, *42*(21), 9512–9520. doi: 10.1002/ 2015GL065497

Field, P. R., & Wood, R. (2007, Jan). Precipitation and cloud structure in midlatitude cyclones. *Journal of Climate*, *20*(2), 233–254. doi: 10.1175/jcli3998.1

Filioglou, M., Mielonen, T., Balis, D., Giannakaki, E., Arola, A., Kokkola, H., et al. (2019, July). Aerosol effect on the cloud phase of low-level clouds over the Arctic. *Journal of Geophysical Research: Atmospheres*, *124*(14), 7886–7899. doi: 10.1029/2018JD030088

Frazer, M. E., & Ming, Y. (2022, Jan). Understanding the extratropical liquid water path feedback in mixed-phase clouds with an idealized global climate model. *Journal of Climate*, 1–48. doi: 10.1175/JCLI-D-21-0334.1

Frey, W. R., & Kay, J. E. (2018, Apr). The influence of extratropical cloud phase and amount feedbacks on climate sensitivity. *Climate Dynamics*, *50*(7), 3097–3116. doi: 10.1007/s00382-017-3796-5

Frey, W. R., Maroon, E. A., Pendergrass, A. G., & Kay, J. E. (2017). Do southern ocean cloud feedbacks matter for 21st century warming? *Geophysical Research Letters*, *44*(24), 12,447–12,456. doi: 10.1002/2017gl076339

Furtado, K., Field, P. R., Boutle, I. A., Morcrette, C. J., & Wilkinson, J. M. (2016). A physically based subgrid parameterization for the production and maintenance of mixed-phase clouds in a general circulation model. *Journal of the Atmospheric Sciences*, *73*(1), 279–291. doi: doi:10.1175/JAS-D-15-0021.1

Geerts, B. (2019). *Cold-Air Outbreaks in the Marine Boundary Layer Experiment (COMBLE) Science Implementation Plan* (Technical Report). DOE Office of Science Atmospheric Radiation Measurement (ARM) Program, United States.

Gettelman, A., Hannay, C., Bacmeister, J. T., Neale, R. B., Pendergrass, A. G., Danabasoglu, G., et al. (2019). High climate sensitivity in the Community Earth System Model version 2 (CESM2). *Geophysical Research Letters*, *46*(14), 8329–8337. doi: 10.1029/2019gl083978

Gettelman, A., Lin, L., Medeiros, B., & Olson, J. (2016, Sept). Climate feedback variance and the interaction of aerosol forcing and feedbacks. *Journal of Climate*, *29*(18), 6659–6675. doi: 10.1175/JCLI-D-16-0151.1

Gettelman, A., & Sherwood, S. C. (2016). Processes responsible for cloud feedback. *Current Climate Change Reports*, 1–11. doi: 10.1007/s40641-016-0052-8

Gonzales, K. R., Swain, D. L., Barnes, E. A., & Diffenbaugh, N. S. (2020, Dec). Moisture- versus wind-dominated flavors of atmospheric rivers. *Geophysical Research Letters*, *47*(23), e2020GL090042. doi: 10.1029/2020GL090042

Gordon, H., Kirkby, J., Baltensperger, U., Bianchi, F., Breitenlechner, M., Curtius, J., et al. (2017). Causes and importance of new particle formation in the present-day and preindustrial atmospheres. *Journal of Geophysical Research: Atmospheres*, *122*(16), 8739–8760. doi: 10.1002/2017JD026844

Gordon, N. D., & Klein, S. A. (2014). Low-cloud optical depth feedback in climate models. *Journal of Geophysical Research: Atmospheres*, *119*(10), 6052–6065. doi: 10.1002/2013JD021052

Gregory, J. M., Ingram, W. J., Palmer, M. A., Jones, G. S., Stott, P. A., Thorpe, R. B., et al. (2004, Feb). A new method for diagnosing radiative forcing and climate sensitivity. *Geophysical Research Letters*, *31*(3). doi: https://doi.org/10.1029/2003GL018747

Grise, K. M., & Medeiros, B. (2016, Dec). Understanding the varied influence of midlatitude jet position on clouds and cloud radiative effects in observations and global climate models. *Journal of Climate*, *29*(24), 9005–9025. doi: 10.1175/JCLI-D-16-0295.1

Grosvenor, D. P., Sourdeval, O., Zuidema, P., Ackerman, A., Alexandrov, M. D., Bennartz, R., et al. (2018, June). Remote sensing of droplet number concentration in warm clouds: A review of the current state of knowledge and perspectives. *Reviews of Geophysics*, *56*(2), 409–453. doi: 10.1029/2017rg000593

Grosvenor, D. P., & Wood, R. (2014). The effect of solar zenith angle on MODIS cloud optical and microphysical retrievals within marine liquid water clouds. *Atmospheric Chemistry and Physics*, *14*(14), 7291–7321. doi: 10.5194/acp-14-7291-2014

Hartmann, D. L. (2015). *Global physical climatology* (Vol. 103). Newnes.

Hartmann, D. L., & Larson, K. (2002, Oct). An important constraint on tropical cloud - climate feedback. *Geophysical Research Letters*, *29*(20), 12–1. doi: 10.1029/2002GL015835

Hartmann, D. L., & Short, D. A. (1980, June). On the use of Earth radiation budget statistics for studies of clouds and climate. *Journal of the Atmospheric Sciences*, *37*(6), 1233–1250. doi: 10.1175/1520-0469(1980)037<1233:OTUOER>2.0.CO;2

Haynes, J. M., Jakob, C., Rossow, W. B., Tselioudis, G., & Brown, J. (2011, Oct). Major characteristics of Southern Ocean cloud regimes and their effects on the energy budget. *Journal of Climate*, *24*(19), 5061–5080. doi: 10.1175/2011jcli4052.1

Held, I. M., & Soden, B. J. (2006, Nov). Robust responses of the hydrological cycle to global warming. *Journal of Climate*, *19*(21), 5686–5699. doi: 10.1175/jcli3990.1

Hogan, R. J., Behera, M. D., O'Connor, E. J., & Illingworth, A. J. (2004, Mar). Estimate of the global distribution of stratiform supercooled liquid water clouds using the LITE lidar. *Geophysical Research Letters*, *31*(5). doi: 10.1029/2003GL018977

Hoose, C., Lohmann, U., Bennartz, R., Croft, B., & Lesins, G. (2008, Dec). Global simulations of aerosol processing in clouds. *Atmospheric Chemistry and Physics*, *8*(23), 6939–6963. doi: 10.5194/acp-8-6939-2008

Hoose, C., & Möhler, O. (2012). Heterogeneous ice nucleation on atmospheric aerosols: A review of results from laboratory experiments. *Atmospheric Chemistry and Physics*, *12*(20), 9817–9854. doi: 10.5194/acp-12-9817-2012

Hu, Y. X., Rodier, S., Xu, K. M., Sun, W. B., Huang, J. P., Lin, B., et al. (2010, Oct). Occurrence, liquid water content, and fraction of supercooled water clouds from combined CALIOP/IIR/MODIS measurements. *Journal of Geophysical Research-Atmospheres*, *115*, 13. doi: 10.1029/2009jd012384

Huang, Y., Siems, S. T., Manton, M. J., Rosenfeld, D., Marchand, R., McFarquhar, G. M., & Protat, A. (2016, Oct). What is the role of sea surface temperature in modulating cloud and precipitation properties over the Southern Ocean? *Journal of Climate*, *29*(20), 7453–7476. doi: 10.1175/JCLI-D-15-0768.1

Jiang, J. H., Su, H., Zhai, C., Perun, V. S., Del Genio, A., Nazarenko, L. S., et al. (2012). Evaluation of cloud and water vapor simulations in CMIP5 climate models using NASA "A-Train" satellite observations. *Journal of Geophysical Research: Atmospheres*, *117*(D14), D14105. doi: 10.1029/2011JD017237

Kay, J. E., L'Ecuyer, T., Pendergrass, A., Chepfer, H., Guzman, R., & Yettella, V. (2018). Scale-aware and definition-aware evaluation of modeled near-surface precipitation frequency using CloudSat observations. *Journal of Geophysical Research: Atmospheres*, *123*(8), 4294–4309.

Kelleher, M. K., & Grise, K. M. (2019). Examining Southern Ocean cloud controlling factors on daily time scales and their connections to midlatitude weather systems. *Journal of Climate*, *32*(16), 5145–5160. doi: 10.1175/jcli-d-18-0840.1

Kelleher, M. K., & Grise, K. M. (2021, Oct). Varied midlatitude shortwave cloud radiative responses to Southern Hemisphere circulation shifts. *Atmospheric Science Letters*, *23*(1), e1068. doi: 10.1002/asl.1068

Klein, S. A., Hall, A., Norris, J. R., & Pincus, R. (2017, Nov). Low-cloud feedbacks from cloud-controlling factors: A review. *Surveys in Geophysics*, *38*(6), 1307–1329. doi: 10.1007/s10712-017-9433-3

Komurcu, M., Storelvmo, T., Tan, I., Lohmann, U., Yun, Y. X., Penner, J. E., et al. (2014, Mar). Intercomparison of the cloud water phase among global climate models. *Journal of Geophysical Research-Atmospheres*, *119*(6), 3372–3400. doi: 10.1002/2013jd021119

Korhonen, H., Carslaw, K. S., Forster, P. M., Mikkonen, S., Gordon, N. D., & Kokkola, H. (2010, Jan). Aerosol climate feedback due to decadal increases in Southern Hemisphere wind speeds. *Geophysical Research Letters*, *37*(2). doi: 10.1029/2009GL041320

Korhonen, H., Carslaw, K. S., Spracklen, D. V., Mann, G. W., & Woodhouse, M. T. (2008, Aug). Influence of oceanic dimethyl sulfide emissions on cloud condensation nuclei concentrations and seasonality over the remote Southern Hemisphere oceans: A global model study. *Journal of Geophysical Research-Atmospheres*, *113*(D15). doi: D1520410.1029/2007jd009718

Kuma, P., McDonald, A. J., Morgenstern, O., Querel, R., Silber, I., & Flynn, C. J. (2021). Ground-based lidar processing and simulator framework for comparing models and observations (ALCF 1.0). *Geoscientific Model Development*, *14*(1), 43–72.

Lavers, D. A., Ralph, F. M., Waliser, D. E., Gershunov, A., & Dettinger, M. D. (2015, July). Climate change intensification of horizontal water vapor transport in CMIP5. *Geophysical Research Letters*, *42*(13), 5617–5625. doi: 10.1002/2015GL064672

Li, J., Jian, B., Huang, J., Hu, Y., Zhao, C., Kawamoto, K., et al. (2018, Aug). Long-term variation of cloud droplet number concentrations from space-based Lidar. *Remote Sensing of Environment*, *213*, 144–161. doi: https://doi.org/10.1016/j.rse.2018.05.011

Lohmann, U., & Diehl, K. (2006, Mar). Sensitivity studies of the importance of dust ice nuclei for the indirect aerosol effect on stratiform mixed-phase clouds. *Journal of the Atmospheric Sciences*, *63*(3), 968–982. doi: 10.1175/jas3662.1

Lorenz, D. J., & DeWeaver, E. T. (2007). The response of the extratropical hydrological cycle to global warming. *Journal of Climate*, *20*(14), 3470–3484. doi: 10.1175/JCLI4192.1

Mace, G. G. (2010, May). Cloud properties and radiative forcing over the maritime storm tracks of the Southern Ocean and North Atlantic derived from A-Train. *Journal of Geophysical Research-Atmospheres*, *115*, D10201. doi: D1020110.1029/2009jd012517

Mace, G. G., & Protat, A. (2018). Clouds over the Southern Ocean as observed from the R/V Investigator during CAPRICORN. Part I: Cloud occurrence and phase partitioning. *Journal of Applied Meteorology and Climatology*, *57*(8), 1783–1803.

Mace, G. G., Protat, A., Humphries, R. S., Alexander, S. P., McRobert, I. M., Ward, J., et al. (2021, Feb). Southern Ocean cloud properties derived From CAPRICORN and MARCUS Data. *Journal of Geophysical Research: Atmospheres*, *126*(4), e2020JD033368. doi: 10.1029/2020JD033368

Mace, G. G., & Zhang, Q. (2014, Aug). The CloudSat radar-lidar geometrical profile product (RL-GeoProf): Updates, improvements, and selected results. *Journal of Geophysical Research: Atmospheres*, *119*(15), 9441–9462. doi: 10.1002/2013JD021374

Malavelle, F. F., Haywood, J. M., Jones, A., Gettelman, A., Clarisse, L., Bauduin, S., et al. (2017). Strong constraints on aerosol–cloud interactions from volcanic eruptions. *Nature*, *546*(7659), 485–491. doi: 10.1038/nature22974

Manaster, A., O'Dell, C. W., & Elsaesser, G. (2017). Evaluation of cloud liquid water path trends using a multidecadal record of passive microwave observations. *Journal of Climate*, *30*(15), 5871–5884. doi: 10.1175/jcli-d-16-0399.1

Marchand, R., Haynes, J., Mace, G. G., Ackerman, T., & Stephens, G. (2009, Jan). A comparison of simulated cloud radar output from the multiscale modeling framework global climate model with CloudSat cloud radar observations. *Journal of Geophysical Research-Atmospheres*, *114*, 18. doi: D00a2010.1029/2008jd009790

Matsui, T., Iguchi, T., Li, X., Han, M., Tao, W.-K., Petersen, W., et al. (2013). GPM satellite simulator over ground validation sites. *Bulletin of the American Meteorological Society*, *94*(11), 1653–1660.

McCluskey, C. S., Hill, T. C. J., Humphries, R. S., Rauker, A. M., Moreau, S., Strutton, P. G., et al. (2018, Nov). Observations of ice nucleating particles over Southern Ocean waters. *Geophysical Research Letters*, *45*(21), 11,989–11,997. doi: 10.1029/2018GL079981 &

McCoy, D. T., Bender, F. A. M., Grosvenor, D. P., Mohrmann, J. K., Hartmann, D. L., Wood, R., & Field, P. R. (2018). Predicting decadal trends in cloud droplet number concentration using reanalysis and satellite data. *Atmospheric Chemistry and Physics*, *18*(3), 2035–2047. doi: 10.5194/acp-18-2035-2018

McCoy, D. T., Field, P., Bodas-Salcedo, A., Elsaesser, G. S., & Zelinka, M. D. (2020). A regime-oriented approach to observationally constraining extratropical shortwave cloud feedbacks. *Journal of Climate*, 1–55. doi: 10.1175/JCLI-D-19-0987.1

McCoy, D. T., Field, P., Frazer, M. E., Zelinka, M. D., Elsaesser, G. S., Mülmenstädt, J., et al. (2022, Mar). Extratropical shortwave cloud feedbacks in the context of the global circulation and hydrological cycle. *Geophysical Research Letters*, *49*(8), e2021GL097154. doi: 10.1029/2021GL097154

McCoy, D. T., Field, P., Gordon, H., Elsaesser, G. S., & Grosvenor, D. P. (2020). Untangling causality in midlatitude aerosol–cloud adjustments. *Atmospheric Chemistry and Physics*, *20*(7), 4085–4103. doi: 10.5194/acp-20-4085-2020

McCoy, D. T., Field, P. R., Elsaesser, G. S., Bodas-Salcedo, A., Kahn, B. H., Zelinka, M. D., et al. (2019). Cloud feedbacks in extratropical cyclones: Insight from long-term satellite data and high-resolution global simulations. *Atmospheric Chemistry and Physics*, *19*(2), 1147–1172. doi: 10.5194/acp-19-1147-2019

McCoy, D. T., Field, P. R., Schmidt, A., Grosvenor, D. P., Bender, F. A. M., Ship-way, B. J., et al (2018). Aerosol midlatitude cyclone indirect effects in observations and high-resolution simulations. *Atmospheric Chemistry and Physics*, *18*(8), 5821–5846. doi: 10.5194/acp-18-5821-2018

McCoy, D. T., Hartmann, D. L., & Grosvenor, D. P. (2014, Dec). Observed Southern Ocean cloud properties and shortwave reflection. Part II: Phase changes and low cloud feedback*. *Journal of Climate*, *27*(23), 8858–8868. doi: 10.1175/jcli-d-14-00288.1

McCoy, D. T., Hartmann, D. L., Zelinka, M. D., Ceppi, P., & Grosvenor, D. P. (2015). Mixed-phase cloud physics and Southern Ocean cloud feedback in climate models. *Journal of Geophysical Research: Atmospheres*, *120*(18), 9539–9554. doi: 10.1002/2015jd023603

McCoy, D. T., Tan, I., Hartmann, D. L., Zelinka, M. D., & Storelvmo, T. (2016). On the relationships among cloud cover, mixed-phase partitioning, and planetary albedo in GCMs. *Journal of Advances in Modeling Earth Systems*, *8*(2), 650–668. doi: 10.1002/2015ms000589

McCoy, I. L., Bretherton, C. S., Wood, R., Twohy, C. H., Gettelman, A., Bardeen, C. G., & Toohey, D. W. (2021, Apr). Influences of recent particle formation on Southern Ocean aerosol variability and low cloud properties. *Journal of Geophysical Research: Atmospheres*, *126*(8), e2020JD033529. doi: 10.1029/2020JD033529

McCoy, I. L., McCoy, D. T., Wood, R., Regayre, L., Watson-Parris, D., Grosvenor, D. P., et al. (2020). The hemispheric contrast in cloud microphysical properties constrains aerosol forcing. *Proceedings of the National Academy of Sciences*, 201922502. doi: 10.1073/pnas.1922502117

McCoy, I. L., Wood, R., & Fletcher, J. K. (2017). Identifying meteorological controls on open and closed mesoscale cellular convection associated with marine cold air outbreaks. *Journal of Geophysical Research: Atmospheres*, *122*(21), 11,678–11,702. doi: 10.1002/2017JD027031

McFarquhar, G. M., Bretherton, C. S., Marchand, R., Protat, A., DeMott, P. J., Alexander, S. P., et al. (2021, Apr). Observations of clouds, aerosols, precipitation, and surface radiation over the Southern Ocean: An overview of CAPRICORN, MARCUS, MICRE, and SOCRATES. *Bulletin of the American Meteorological Society*, *102*(4), E894–E928. doi: 10.1175/BAMS-D-20-0132.1

Michibata, T., Suzuki, K., Ogura, T., & Jing, X. (2019, Oct). Incorporation of inline warm rain diagnostics into the COSP2 satellite simulator for process-oriented model evaluation. *Geoscientific Model Development*, *12*(10), 4297–4307. doi: 10.5194/gmd-12-4297-2019

Mioche, G., Jourdan, O., Delanoë, J., Gourbeyre, C., Febvre, G., Dupuy, R., et al. (2017, Oct). Vertical distribution of microphysical properties of Arctic springtime low-level mixed-phase clouds over the Greenland and Norwegian seas. *Atmospheric Chemistry and Physics*, *17*(20), 12845–12869. doi: 10.5194/acp-17-12845-2017

Mitchell, J. F. B., Senior, C. A., & Ingram, W. J. (1989, Sept). CO2 and climate - A missing feedback. *Nature*, *341*(6238), 132–134. doi: 10.1038/341132a0

Miyamoto, A., Nakamura, H., & Miyasaka, T. (2018). Influence of the subtropical high and storm track on low-cloud fraction and its seasonality over the South Indian Ocean. *Journal of Climate*, *31*(10), 4017–4039. doi: 10.1175/jcli-d-17-0229.1

Mulcahy, J. P., Jones, C., Sellar, A., Johnson, B., Boutle, I. A., Jones, A., et al. (2018). Improved aerosol processes and effective radiative forcing in HadGEM3 and UKESM1. *Journal of Advances in Modeling Earth Systems*, *10*(11), 2786–2805. doi: 10.1029/2018MS001464

Murray, B. J., Carslaw, K. S., & Field, P. R. (2021, Jan). Opinion: Cloudphase climate feedback and the importance of ice-nucleating particles. *Atmospheric Chemistry and Physics*, *21*(2), 665–679. doi: 10.5194/ acp-21-665-2021

Myers, T. A., & Norris, J. R. (2013, Oct). Observational evidence that enhanced subsidence reduces subtropical marine boundary layer cloudiness. *Journal of Climate*, *26*(19), 7507–7524. doi: 10.1175/JCLI-D-12-00736.1

Myers, T. A., & Norris, J. R. (2016). Reducing the uncertainty in subtropical cloud feedback. *Geophysical Research Letters*, *43*(5). doi: 10.1002/ 2015GL067416

Myers, T. A., Scott, R. C., Zelinka, M. D., Klein, S. A., Norris, J. R., & Cald-well, P. M. (2021, May). Observational constraints on low cloud feedback reduce uncertainty of climate sensitivity. *Nature Climate Change 11*, 501–507. doi: 10.1038/s41558-021-01039-0

Mülmenstädt, J., Salzmann, M., Kay, J. E., Zelinka, M. D., Ma, P.-L., Nam, C., et al. (2021). An underestimated negative cloud feedback from cloud lifetime changes. *Nature Climate Change*, *11*(6), 508–513.

Mülmenstädt, J., Sourdeval, O., Delanoë, J., & Quaas, J. (2015, Aug). Frequency of occurrence of rain from liquid-, mixed-, and ice-phase clouds derived from A-Train satellite retrievals. *Geophysical Research Letters*, *42*(15), 6502–6509. doi: 10.1002/2015GL064604

Nasiri, S. L., & Kahn, B. H. (2008). Limitations of bispectral infrared cloud phase determination and potential for improvement. *Journal of Applied Meteorology and Climatology*, *47*(11), 2895–2910.

Naud, C. M., Del Genio, A. D., Bauer, M., & Kovari, W. (2010, June). Cloud vertical distribution across warm and cold fronts in CloudSat-CALIPSO data and a general circulation model. *Journal of Climate*, *23*(12), 3397–3415. doi: 10.1175/2010jcli3282.1

Naud, C. M., & Kahn, B. H. (2015). Thermodynamic phase and ice cloud properties in Northern Hemisphere winter extratropical cyclones observed by aqua AIRS. *Journal of Applied Meteorology and Climatology*, *54*(11), 2283–2303. doi: 10.1175/jamc-d-15-0045.1

Naud, C. M., Posselt, D. J., & van den Heever, S. C. (2012, July). Observational analysis of cloud and precipitation in midlatitude cyclones: Northern versus Southern Hemisphere warm fronts. *Journal of Climate*, *25*(14), 5135–5151. doi: 10.1175/jcli-d-11-00569.1

Norris, J. R., Allen, R. J., Evan, A. T., Zelinka, M. D., O'Dell, C. W., & Klein, S. A. (2016). Evidence for climate change in the satellite cloud record. *Nature*, *536*(7614), 72–75. doi: 10.1038/nature18273

Paltridge, G. W. (1980). Cloud-radiation feedback to climate. *Quarterly Journal of the Royal Meteorological Society*, *106*(450), 895–899. doi: 10.1256/smsqj.45017

Payne, A. E., Demory, M.-E., Leung, L. R., Ramos, A. M., Shields, C. A., Rutz, J. J., et al. (2020, Mar). Responses and impacts of atmospheric rivers to climate change. *Nature Reviews Earth and Environment*, *1*(3), 143–157. doi: 10.1038/s43017-020-0030-5

Qu, X., Hall, A., Klein, S. A., DeAngelis, & Anthony, M. (2015). Positive tropical marine low-cloud cover feedback inferred from cloud-controlling factors. *Geo-physical Research Letters*. doi: 10.1002/2015GL065627

Ramos, A. M., Tomé, R., Trigo, R. M., Liberato, M. L. R., & Pinto, J. G. (2016, Sept). Projected changes in atmospheric rivers affecting Europe in CMIP5 models. *Geophysical Research Letters*, *43*(17), 9315–9323. doi: 10.1002/ 2016GL070634

Rieck, M., Nuijens, L., & Stevens, B. (2012, Aug). Marine boundary layer cloud feedbacks in a constant relative humidity atmosphere. *Journal of the Atmospheric Sciences*, *69*(8), 2538–2550. doi: 10.1175/JAS-D-11-0203.1

Roh, W., Satoh, M., & Nasuno, T. (2017). Improvement of a cloud microphysics scheme for a global nonhydrostatic model using TRMM and a satellite simulator. *Journal of Atmospheric Sciences*, *74*(1), 167–184.

Schill, G. P., DeMott, P. J., Emerson, E. W., Rauker, A. M. C., Kodros, J. K., Suski, K. J., et al. (2020, Sept). The contribution of black carbon to global ice nucleating particle concentrations relevant to mixed-phase clouds. *Proceedings of the National Academy of Sciences*, *117*(37), 22705. doi: 10.1073/pnas.2001674117

Schill, G. P., Froyd, K. D., Bian, H., Kupc, A., Williamson, C., Brock, C. A., et al. (2020, June). Widespread biomass burning smoke throughout the remote troposphere. *Nature Geoscience*, *13*(6), 422–427. doi: 10.1038/s41561-020-0586-1

Scott, R. C., Myers, T. A., Norris, J. R., Zelinka, M. D., Klein, S. A., Sun, M., & Doelling, D. R. (2020). Observed sensitivity of low-cloud radiative effects to meteorological perturbations over the global oceans. *Journal of Climate*, *33*(18), 7717–7734.

Senior, C. A., & Mitchell, J. F. B. (1993, Mar). Carbon-dioxide and climate - the impact of cloud parameterization. *Journal of Climate*, *6*(3), 393–418. doi: 10.1175/1520-0442(1993) 006<0393:cdacti>2.0.co;2

Sherwood, S. C., Webb, M. J., Annan, J. D., Armour, K. C., Forster, P. M., Hargreaves, J. C., et al. (2020). An assessment of Earth's climate sensitivity using multiple lines of evidence. *Reviews of Geophysics*, *58*(4), e2019RG000678.

Silber, I., Fridlind, A. M., Verlinde, J., Ackerman, A. S., Cesana, G. V., & Knopf, D. A. (2021, Mar). The prevalence of precipitation from polar supercooled clouds. *Atmospheric Chemistry and Physics*, *21*(5), 3949–3971. doi: 10.5194/acp-21-3949-2021

Silber, I., Fridlind, A. M., Verlinde, J., Ackerman, A. S., Chen, Y.-S., Bromwich, D. H., et al. (2019, Oct). Persistent supercooled drizzle at temperatures below 25 °C observed at McMurdo Station, Antarctica. *Journal of Geophysical Research: Atmospheres*, *124*(20), 10878–10895. doi: 10.1029/2019JD030882

Sourdeval, O., C.-Labonnote, L., Baran, A. J., Mülmenstädt, J., & Brogniez, G. (2016, Oct). A methodology for simultaneous retrieval of ice and liquid water cloud properties. Part 2: Near-global retrievals and evaluation against

A-Train products. *Quarterly Journal of the Royal Meteorological Society*, *142*(701), 3063–3081. doi: 10.1002/qj.2889

Sousa, P. M., Blamey, R. C., Reason, C. J. C., Ramos, A. M., & Trigo, R. M. (2018, Dec). The 'Day Zero' Cape Town drought and the poleward migration of moisture corridors. *Environmental Research Letters*, *13*(12), 124025. doi: 10.1088/1748-9326/aaebc7

Stephens, G. L., Wild, M., Stackhouse, P. W., L'Ecuyer, T., Kato, S., & Henderson, D. S. (2012, Apr). The global character of the flux of downward longwave radiation. *Journal of Climate*, *25*(7), 2329–2340. doi: 10.1175/JCLI-D-11-00262.1

Stevens, B., & Brenguier, J. L. (2009). *Cloud controlling factors: Low clouds*. Massachusetts: MIT Press Cambridge.

Storelvmo, T. (2017). Aerosol effects on climate via mixed-phase and ice clouds. *Annual Review of Earth and Planetary Sciences*, *45*(1), 199–222. doi: 10.1146/annurev-earth-060115-012240

Storelvmo, T., & Tan, I. (2015). The Wegener-Bergeron-Findeisen process? Its discovery and vital importance for weather and climate. *Meteorologische Zeitschrift*, *24*(4), 455–461. doi: 10.1127/metz/2015/0626

Storelvmo, T., Tan, I., & Korolev, A. (2015, Oct). Cloud phase changes induced by CO2 warming—a powerful yet poorly constrained cloud-climate feedback. *Current Climate Change Reports*, *1*(4), 1–9. doi: 10.1007/s40641-015-0026-2

Su, H., Jiang, J. H., Neelin, J. D., Shen, T. J., Zhai, C., Yue, Q., et al. (2017, June). Tightening of tropical ascent and high clouds key to precipitation change in a warmer climate. *Nature Communications*, *8*(1), 15771. doi: 10.1038/ncomms15771

Suzuki, K., Stephens, G., Bodas-Salcedo, A., Wang, M., Golaz, J.-C., Yokohata, T., & Koshiro, T. (2015, Oct). Evaluation of the warm rain formation process in global models with satellite observations. *Journal of the Atmospheric Sciences*, *72*(10), 3996–4014. doi: 10.1175/JAS-D-14-0265.1

Tan, I., Oreopoulos, L., & Cho, N. (2019). The role of thermodynamic phase shifts in cloud optical depth variations with temperature. *Geophysical Research Letters*, *46*(8), 4502–4511. doi: 10.1029/2018gl081590

Tan, I., & Storelvmo, T. (2016). Sensitivity study on the influence of cloud microphysical parameters on mixed-phase cloud thermodynamic phase partitioning in CAM5. *Journal of the Atmospheric Sciences*, *73*(2), 709–728. doi: 10.1175/JAS-D-15-0152.1

Tan, I., & Storelvmo, T. (2019, Mar). Evidence of strong contributions from mixed-phase clouds to Arctic climate change. *Geophysical Research Letters*, *46*(5), 2894–2902. doi: 10.1029/2018GL081871

Tan, I., Storelvmo, T., & Choi, Y.-S. (2014). Spaceborne lidar observations of the ice-nucleating potential of dust, polluted dust, and smoke aerosols in mixedphase clouds. *Journal of Geophysical Research: Atmospheres*. doi:10.1002/2013JD021333

Tan, I., Storelvmo, T., & Zelinka, M. D. (2016). Observational constraints on mixed-phase clouds imply higher climate sensitivity. *Science*, *352*(6282), 224–227. doi: 10.1126/science.aad5300

Tan, J., Jakob, C., Rossow, W. B., & Tselioudis, G. (2015, Mar). Increases in tropical rainfall driven by changes in frequency of organized deep convection. *Nature*, *519*(7544), 451–454. doi: 10.1038/nature14339

Terai, C. R., Klein, S. A., & Zelinka, M. D. (2016). Constraining the low-cloud optical depth feedback at middle and high latitudes using satellite observations. *Journal of Geophysical Research: Atmospheres*. doi:10.1002/2016JD025233

Terai, C. R., Zhang, Y., Klein, S. A., Zelinka, M. D., Chiu, J. C., & Min, Q. (2019). Mechanisms behind the extratropical stratiform low-cloud optical depth response to temperature in ARM site observations. *Journal of Geophysical Research: Atmospheres*, *124*(4), 2127–2147. doi: 10.1029/2018jd029359

Thompson, D. R., Kahn, B. H., Green, R. O., Chien, S. A., Middleton, E. M., & Tran, D. Q. (2018). Global spectroscopic survey of cloud thermodynamic phase at high spatial resolution, 2005–2015. *Atmospheric Measurement Techniques*, *11*(2), 1019.

Thompson, D. W. J., Bony, S., & Li, Y. (2017, Aug). Thermodynamic constraint on the depth of the global tropospheric circulation. *Proceedings of the National Academy of Sciences*, *114*(31), 8181. doi: 10.1073/pnas.1620493114

Thompson, D. W. J., Ceppi, P., & Li, Y. (2019, Jan). A robust constraint on the temperature and height of the extratropical tropopause. *Journal of Climate*, *32*(2), 273–287. doi: 10.1175/JCLI-D-18-0339.1

Trenberth, K. E. (1998). Atmospheric moisture residence times and cycling: Implications for rainfall rates and climate change. *Climatic Change*, *39*(4), 667–694.

Trenberth, K. E., & Fasullo, J. T. (2010, Jan). Simulation of present-day and twenty-first-century energy budgets of the Southern Oceans. *Journal of Climate*, *23*(2), 440–454. doi: 10.1175/2009jcli3152.1

Tselioudis, G., DelGenio, A. D., Kovari, W., & Yao, M.-S. (1998, Dec). Temperature dependence of low cloud optical thickness in the GISS GCM: Contributing mechanisms and climate implications. *Journal of Climate*, *11*(12), 3268–3281. doi: 10.1175/1520-0442(1998)011<3268:TDOLCO>2.0.CO;2

Tselioudis, G., Lipat, B. R., Konsta, D., Grise, K. M., & Polvani, L. M. (2016, May). Midlatitude cloud shifts, their primary link to the Hadley cell, and their diverse radiative effects. *Geophysical Research Letters*, *43*(9), 4594–4601. doi: https://doi.org/10.1002/2016GL068242

Tselioudis, G., & Rossow, W. B. (2006). Climate feedback implied by observed radiation and precipitation changes with midlatitude storm strength and frequency. *Geophysical Research Letters*, *33*(2). doi: 10.1029/2005GL024513

Tselioudis, G., Rossow, W. B., & Rind, D. (1992). Global patterns of cloud optical thickness variation with temperature. *Journal of Climate*, *5*(12), 1484–1495. doi: 10.1175/1520-0442(1992)005<1484:GPOCOT>2.0.CO;2

Tsushima, Y., Emori, S., Ogura, T., Kimoto, M., Webb, M. J., Williams, K. D., et al. (2006, Aug). Importance of the mixed-phase cloud distribution in the control climate for assessing the response of clouds to carbon dioxide increase: A multi-model study. *Climate Dynamics*, *27*(2–3), 113–126. doi: 10.1007/s00382-006-0127-7

Twomey, S. (1977). Influence of pollution on shortwave albedo of clouds. *Journal of the Atmospheric Sciences*, *34*(7), 1149–1152. doi: 10.1175/1520-0469(1977)034⟨1149:tiopot⟩2.0.co;2

Vergara-Temprado, J., Holden, M. A., Orton, T. R., O'Sullivan, D., Umo, N. S., Browse, J., et al. (2018, Apr). Is black carbon an unimportant ice-nucleating particle in mixed-phase clouds? *Journal of Geophysical Research: Atmospheres*, *123*(8), 4273–4283. doi: 10.1002/2017JD027831

Vergara-Temprado, J., Miltenberger, A. K., Furtado, K., Grosvenor, D. P., Shipway, B. J., Hill, A. A., et al. (2018). Strong control of Southern Ocean cloud reflectivity by ice-nucleating particles. *Proceedings of the National Academy of Sciences*, *115*(11), 2687–2692.

Wang, Y., Wang, M., Zhang, R., Ghan, S. J., Lin, Y., Hu, J., et al. (2014, May). Assessing the effects of anthropogenic aerosols on Pacific storm track using a multiscale global climate model. *Proceedings of the National Academy of Sciences*, *111*(19), 6894–6899. doi: 10.1073/pnas.1403364111

Wang, Z., Siems, S. T., Belusic, D., Manton, M. J., & Huang, Y. (2015). A climatology of the precipitation over the Southern Ocean as observed at Macquarie Island. *Journal of Applied Meteorology and Climatology*, *54*(12), 2321–2337.

Webb, M. J., Andrews, T., Bodas-Salcedo, A., Bony, S., Bretherton, C. S., Chadwick, R., et al. (2017). The cloud feedback model intercomparison project (CFMIP) contribution to CMIP6. *Geoscientific Model Development*, *10*(1), 359–384.

Wentz, F. J., & Meissner, T. (2000). AMSR Ocean Algorithm, Version 2. *Remote Sensing Systems, Santa Rosa, CA.*, *Report Number 121599A-1*, p. 66.

Wetherald, R. T., & Manabe, S. (1988, Apr). Cloud feedback processes in a general circulation model. *Journal of Atmospheric Sciences*, *45*(8), 1397–1416. doi: 10.1175/1520-0469(1988)045⟨1397:CFPIAG⟩2.0.CO;2

Yettella, V., & Kay, J. E. (2017). How will precipitation change in extratropical cyclones as the planet warms? Insights from a large initial condition climate model ensemble. *Climate Dynamics*, *49*(5–6), 1765–1781.

Zelinka, M. D., Klein, S. A., & Hartmann, D. L. (2012, June). Computing and partitioning cloud feedbacks using cloud property histograms. Part I: Cloud radiative kernels. *Journal of Climate*, *25*(11), 3715–3735. doi: 10.1175/jcli-d-11-00248.1

Zelinka, M. D., Myers, T. A., McCoy, D. T., Po-Chedley, S., Caldwell, P. M., Ceppi, P., et al. (2020, Jan). Causes of higher climate sensitivity in CMIP6 models. *Geophysical Research Letters*, 47(1). doi:10.1029/2019GL085782

Zelinka, M. D., Zhou, C., & Klein, S. A. (2016). Insights from a refined decomposition of cloud feedbacks. *Geophysical Research Letters*, *43*(17), 9259–9269. doi: 10.1002/2016GL069917

Zhang, D., Wang, Z., & Liu, D. (2010, Feb). A global view of midlevel liquid-layer topped stratiform cloud distribution and phase partition from CALIPSO and CloudSat measurements. *Journal of Geophysical Research: Atmospheres*, *115*(D4). doi:10.1029/2009JD012143

Zhang, M. H., Lin, W. Y., Klein, S. A., Bacmeister, J. T., Bony, S., Cederwall, R. T., Del Genio, A. D., et al. (2005). Comparing Clouds and Their Seasonal Variations in 10 Atmospheric General Circulation Models with Satellite Measurements. *Journal of Geophysical Research-Atmospheres*, *110*, no. D15.

Zhang, Y., Xie, S., Lin, W., Klein, S. A., Zelinka, M., Ma, P.-L., et al. (2019, May). Evaluation of clouds in version 1 of the E3SM atmosphere model with satellite simulators. *Journal of Advances in Modeling Earth Systems*, *11*(5), 1253–1268. doi: 10.1029/2018MS001562

Zheng, Y., Rosenfeld, D., & Li, Z. (2020, July). A more general paradigm for understanding the decoupling of stratocumulus-topped boundary layers: The importance of horizontal temperature advection. *Geophysical Research Letters*, *47*(14), e2020GL087697. doi: 10.1029/2020GL087697

7

Tropical Marine Low Clouds: Feedbacks to Warming and on Climate Variability

Timothy A. Myers[1], Raphaela Vogel[2,3], Florent Brient[2], Hossein Parishani[4], and Ryan C. Scott[5]

ABSTRACT

Stratocumulus and shallow cumulus clouds over the tropical oceans are critical components of the mean-state and present-day variability of Earth's climate and the climate's response to increasing greenhouse gases. Observational analyses and high-resolution models agree that, on aggregate, these tropical marine low clouds will act as a positive feedback to climate warming, driven largely in direct response to increasing sea surface temperature (SST). However, it is likely that the trade-wind cumulus response to warming will be weaker than that of stratocumulus. From an observational perspective, this feature is suggested by the weaker sensitivity of trade cumulus cloud amount and optical depth to interannual perturbations in SST than that of stratocumulus. Shallow cumulus in the trade winds are organized into several different mesoscale patterns that may evolve as the climate warms. These possible mesoscale organization feedbacks are very uncertain and require further investigation. A conceptual view of the morphology of low-cloud feedback would help reduce this uncertainty. Tropical marine low clouds also exert important feedbacks on climate variability, owing to their coupling to SST anomalies and their patterns. These feedbacks occur on regional and global scales, with influences on surface and top-of-atmosphere energy budgets. The role of ocean heat transport in modifying tropical low-cloud feedbacks to climate warming and variability remains largely unexplored. High-resolution atmospheric models have been instrumental in understanding tropical low-cloud physics and the relevant processes. As these evolve to global scale and into high-resolution Earth system models, new challenges and opportunities will emerge in the study and simulation of low-cloud processes and feedbacks.

7.1. INTRODUCTION

7.1.1. Climatological Characteristics of Tropical Low Clouds and Their Impacts on Earth's Radiation Budget

Low clouds occurring in the marine boundary layer (MBL) are a critical aspect of the coupled atmosphere-ocean system. On a near-global, annual mean basis, they cool the climate by strongly reflecting incoming solar radiation while only weakly restricting the emission of thermal radiation. This cooling cloud radiative effect (CRE) provides a majority contribution

[1]*Lawrence Livermore National Laboratory, Livermore, CA, USA; currently at Cooperative Institute for Research in Environmental Sciences, University of Colorado, Boulder, CO, USA; and Physical Science Laboratory, National Oceanic and Atmospheric Administration, Boulder, CO, USA*

[2]*Laboratoire de Météorologie Dynamique/Institut Pierre Simon Laplace, Sorbonne University, CNRS, Paris, France*

[3]*Currently at: Meteorological Institute, Department of Earth System Sciences, University of Hamburg, Hamburg, Germany*

[4]*Department of Earth System Science, University of California Irvine, Irvine, CA, USA; currently at Ansys, Inc., Irvine, CA, USA*

[5]*NASA Langley Research Center, Hampton, VA, USA*

Clouds and Their Climatic Impacts: Radiation, Circulation, and Precipitation, Geophysical Monograph 281, First Edition.
Edited by Sylvia C. Sullivan and Corinna Hoose.

DOI: 10.1002/9781119700357.ch07

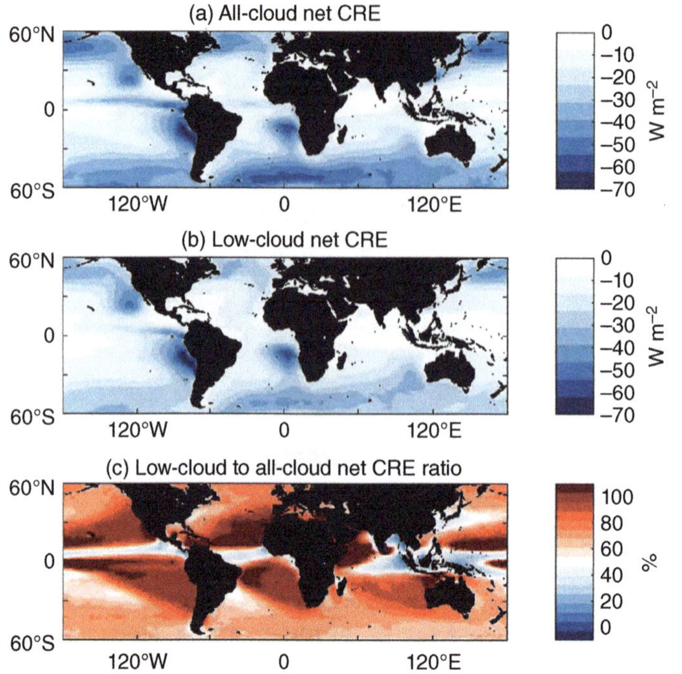

Figure 7.1 Long-term annual mean (a) net CRE at TOA from all cloud types, (b) net CRE at TOA from just low clouds, and (c) the annual mean low cloud to annual mean all-cloud net CRE ratio, based on July 2002 to December 2019 data from the CERES Flux-by-Cloud-Type dataset (Eitzen et al., 2017), available at https://ceres .larc.nasa.gov/data/. CERES-FBCT fluxes are constrained by satellite measurements of radiative fluxes at the top of the atmosphere and cloud retrievals from the Moderate Resolution Imaging Spectroradiometer (MODIS) (Pincus et al., 2012). CERES-FBCT provides monthly cloud fraction and radiative flux in seven cloud top pressure levels and six cloud optical depth categories. For a given month, the all-cloud net CRE is the product of cloud fraction and the difference between clear-sky and overcast fluxes within respective cloud top pressure and optical depth bins, summed over all bins. The low-cloud net CRE is computed similarly, but just over the two pressure levels greater than 680 hPa.

(~60%) to the net CRE of all cloud types on Earth's top-of-atmosphere (TOA) energy budget (Hartmann et al., 1992). Perturbations in low-cloud spatial extent or opacity due to anthropogenic climate change or internal climate variability can exert marked radiative effects and hence feedbacks on local and global climate, both at the TOA and at the ocean surface.

The climatological distribution of marine low-level cloud fraction, to first order, mirrors that of net CRE (from all cloud types) at the TOA (Fig. 7.1a). The cooling CRE is strongest where low-level cloud fraction is largest, such as over the subtropical oceans off the west coasts of continents, where there are extensive and persistent stratus and stratocumulus. The cooling CRE is weakest where cloud fraction is smallest, such as in trade-wind regions of the tropical oceans characterized by scattered shallow cumulus. Recent advances in satellite-based estimates of the radiative contribution of various cloud types to Earth's radiation budget enable a robust estimate of the low-level (lower than 680 hPa in elevation) cloud contribution to net CRE, as illustrated in Figure 7.1b. The pattern of low-cloud net CRE is nearly identical to that of all-cloud net CRE, and low clouds provide around 100% of the all-cloud net CRE within ocean regions experiencing climatological subsidence (Fig. 7.1c). In general, with the exception of regions of deep convection in the tropics, low-level clouds provide a majority contribution to all-cloud net CRE throughout the global oceans, including in the middle latitudes.

Subtropical stratus and stratocumulus locally reflect up to 100 W m^{-2} of incident solar radiation back to space. This exerts a strong imprint on the energy budget at TOA as well as on that at the surface, where the presence of these clouds leads to a strong cooling net CRE. This strong surface CRE is especially important in light of recent findings that sea surface temperature (SST) anomaly patterns produce decadal changes in tropical low clouds (Zhou et al., 2016), and, more generally, in light of the highly coupled nature of the atmosphere-ocean system in regions of the world's major stratocumulus decks. Examination of the contributions to the surface energy budget under the stratocumulus deck off the Peruvian coast provides an illustrative example of this coupled system. Here, the surface net radiative flux heats the ocean mixed layer (de Szoeke et al., 2010). To maintain approximately constant and sufficiently cool SSTs in steady state, this radiative heating is balanced by cooling by surface sensible and latent heat fluxes and by residual oceanic processes. These oceanic processes include vertical mixing or upwelling and cold horizontal advection. Oceanic heat transport is therefore a fundamental aspect of climate in stratocumulus regions, yet with rare exceptions its role in stratocumulus feedbacks remains largely unexplored (e.g., Middlemas et al., 2019; Tan et al., 2017).

Subtropical stratocumulus preferentially occur within large-scale meteorological conditions that determine the structure of the MBL, including its depth, temperature, and moisture content. These clouds are located on the eastern flank of the semi-permanent subtropical high-pressure systems, under the subsiding branches of the Hadley Circulation. Here, equatorward winds induce oceanic Ekman transport that helps to maintain cool SSTs, relative to the zonal mean, via coastal upwelling. Cold advection from higher latitudes driven by the eastern boundary currents also helps to generate these cool surface waters. Subsidence warming and drying above the MBL, in combination with the cool SSTs, sustains a sharp temperature inversion and sudden decrease in moisture that separate the MBL and the cloud top from the free troposphere. Strong subsidence limits the depth of the MBL, allowing turbulent fluxes generated via cloud top radiative cooling and at the surface to sustain a well-mixed MBL.

As subtropical stratocumulus are advected equatorward by trade winds, there is a gradual transition of stratocumulus to scattered shallow cumulus as the large-scale meteorological environment is altered. SSTs increase and subsidence weakens, which erodes the inversion and allows the MBL and the clouds to deepen. This deeper MBL is often decoupled, such that a stable transition layer "decouples" well-mixed subcloud and cloud layers. This weakly stable layer inhibits the mixing of heat and moisture between the two adjacent layers.

These brief descriptions of the patterns of cloudiness and large-scale climatological conditions over the tropical oceans highlight the interconnectedness between marine low clouds and the environmental conditions in which they occur. A large body of research based on observations of low-cloud variability, theory, and high-resolution models has identified several large-scale factors that control variations in subtropical stratocumulus and trade cumulus (see reviews by Bretherton, 2015, Klein et al., 2017, Nuijens and Siebesma, 2019, and Vial et al., 2017). These cloud-controlling factors act as boundary conditions that the MBL responds to. Subtropical stratocumulus and trade cumulus respond in similar ways to perturbations in some cloud-controlling factors and in dissimilar ways to perturbations in other factors. These similarities and differences have major implications for how these cloud types are likely to respond to climate change, which is a main topic of section 7.2. We note also that low clouds are coupled strongly to circulations over different scales; this coupling is discussed in more detail in Chapter 10: Interactions Between the Tropical Atmospheric Overturning Circulation and Clouds in Present and Future Climates and Chapter 11: Clouds and Radiatively Induced Circulations.

7.1.2. The Simplest Model of the Marine Boundary Layer

The simplest model of the MBL assumes that it is well mixed with respect to heat and moisture and non-precipitating (Lilly, 1968; Stevens, 2006), conditions that are often observed for stratocumulus. Equations for the conservation of mass, liquid water potential temperature θ, and total water mixing ratio q within the boundary layer are then formulated as

$$\frac{dz_i}{dt} = w_{ent} - w_{sub} \tag{7.1}$$

$$\frac{d\theta}{dt} = (V\Delta\theta_0 + w_{ent}\Delta\theta_+ - \Delta F)/z_i \tag{7.2}$$

$$\frac{dq}{dt} = (V\Delta q_0 + w_{ent}\Delta q_+)/z_i \tag{7.3}$$

where the operator d/dt describes the time rate of change following a Lagrangian trajectory. Moreover, $\Delta\theta_0 = \theta_0 - \theta$ and $\Delta\theta_+ = \theta_+ - \theta$, where θ_0 and θ_+ are the liquid water potential temperatures at the sea surface and just above the MBL top, with the same notation used for the mixing ratio jumps Δq_0 and Δq_+. z_i is the depth of the MBL; w_{ent} is the rate of entrainment of free-tropospheric air into the boundary layer; w_{sub} is the subsidence rate at the MBL top; V is the surface horizontal windspeed scaled by a non-dimensional constant; and ΔF is the radiative flux difference between the surface and top of the MBL.

Assuming steady state, such that the right-hand sides of equations (7.1)–(7.3) are equal to zero, some fundamental dynamics of the cloudy, well-mixed MBL become apparent. In this case, equation (7.1) states that the entrainment rate must balance the subsidence rate at the top of the MBL. Equation (7.2) states that warming by surface turbulent heat flux $V\Delta\theta_0$ and warming induced by entrainment at the MBL top $w_{ent}\Delta\theta_+$ are balanced by cloud top radiative cooling ΔF. And equation (7.3) states that surface turbulent moisture flux $V\Delta q_0$ is balanced by drying induced by entrainment at the MBL top $w_{ent}\Delta q_+$. A different set of equations is required to model the decoupled MBL in regions of trade cumulus (Stevens, 2006).

7.1.3. How do Cloud-Controlling Factors Impact Tropical Low Clouds?

Figure 7.2 provides a schematic, adopted from a recent observational study and supported by additional observational, model, and theoretical evidence, of how tropical marine low clouds respond to interannual perturbations in individual cloud-controlling factors in the current climate when all other factors are held fixed. This latter constraint is crucial in order to distinguish between the effects of individual cloud-controlling factors, which can co-vary on daily-to-interannual timescales and thus confound their impacts on cloudiness. Qualitatively and on average, greater marine low-cloud coverage and optical thickness is favored by anomalously cool SST, strong inversions, cold horizontal surface temperature advection, high free-tropospheric humidity, weak subsidence, and faster surface winds (including, but not limited to: Blossey et al., 2013; Bretherton et al., 2013; Bretherton & Blossey 2014; Brueck et al., 2015; Myers & Norris, 2013; Myers & Norris, 2016; Mieslinger et al., 2019; Norris & Iacobellis, 2005; Qu et al., 2015a; Scott et al., 2020; van der Dussen et al., 2015; Wood & Bretherton, 2006). The effects of these cloud-controlling factors on surface turbulent moisture flux and cloud top entrainment drying are particularly important in explaining their relationships to cloud coverage and optical thickness. Since surface moisture flux and entrainment are turbulent processes, they cannot be directly simulated by global climate models (GCMs) and thus must be parameterized, making the simulation of low clouds in GCMs a challenging problem. High-resolution models attempt to directly simulate these turbulent processes, though their simulation at a near-global scale and their coupling to the ocean remain a challenge. Recent advances in and prospects for high-resolution modeling are discussed in section 7.5.

The effects of these cloud-controlling factors are described extensively in the aforementioned recent

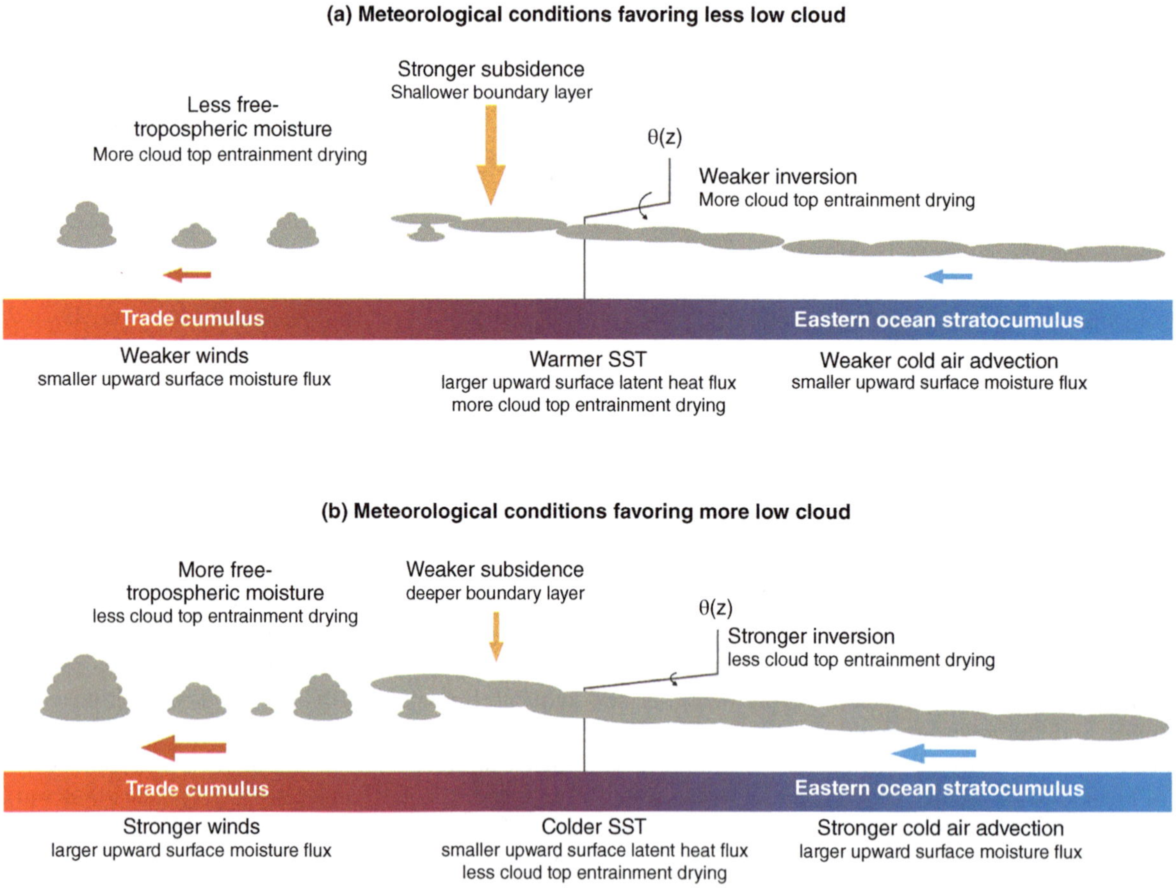

Figure 7.2 Schematic summary of how subtropical stratocumulus and trade cumulus respond to changes in individual large-scale meteorological cloud-controlling factors. Source: Scott et al. (2020)/with permission of American Meteorological Society.

reviews. Here, we focus on a few recent advances in our understanding of them. Changes in free-tropospheric moisture can have two opposing effects on MBL cloudiness. Free-tropospheric moistening reduces entrainment drying by increasing the humidity of the entrained air, which moistens the MBL and promotes cloudiness. Simultaneously, free-tropospheric moistening reduces cloud top radiative cooling as downward longwave radiation increases. This reduces turbulent mixing of the MBL by sinking air parcels at the cloud top, thus suppressing cloudiness by promoting a more decoupled MBL. As illustrated in Figure 7.2, evidence from recent years increasingly suggests that, in the current climate, the dominant effect of increasing free-tropospheric moisture on cloudiness is the moistening of the MBL through reduced entrainment drying (Eastman & Wood, 2018; Myers & Norris, 2015; Myers & Norris, 2016; Scott et al., 2020; van der Dussen et al., 2015).

Observational evidence also has emerged that cloud fraction, cloud optical depth, and CRE in regions of trade cumulus are less sensitive to independent changes in SST than in regions of subtropical stratocumulus (Cesana et al., 2019; Scott et al., 2020). This is corroborated by a recent satellite-based analysis of mesoscale patterns of trade cumulus that found that interannual anomalies in SST have a weak relationship with different organization patterns (Bony et al., 2020). The topic of mesoscale organization and its possible implications for cloud feedbacks is discussed in section 7.4. The differing responses of trade cumulus and stratocumulus to perturbations in SST provide evidence that their feedbacks to climate warming will be dissimilar, which will be further discussed in section 7.2.

7.2. RESPONSE OF TRADE CUMULUS AND STRATOCUMULUS TO WARMING

Global cloud feedbacks are the largest contributor to intermodel spread in equilibrium climate sensitivity (ECS) among GCMs (Flato et al., 2014; Soden & Held, 2006; Zelinka et al., 2020). The decomposition of cloud feedbacks into circulation regimes and altitude ranges reveals that the differing warming response of tropical marine low clouds is a major source of uncertainty in ECS (Bony et al., 2004; Bony & Dufresne, 2005; Caldwell et al., 2016; Cesana & Del Genio, 2021; Lutsko et al., 2021; Vial et al., 2013; Zelinka et al., 2012; Zelinka et al., 2020). The model spread in the low-cloud feedback largely depends on model differences in the parameterization of the unresolved convective, turbulent, radiative, and microphysical processes governing the dynamics of low clouds (Qu et al., 2014; Vial et al., 2016; Webb et al., 2015).

The limited confidence in GCMs' ability to represent low clouds requires other frameworks to constrain tropical low-cloud feedbacks. Two important frameworks to increase process-level understanding and to help assess the sign and magnitude of the feedback are (1) high-resolution simulations to test the low-cloud response to specified perturbations in the large-scale forcings, and (2) constraints from observed natural variability.

7.2.1. Forcing-Feedback Studies

For the first framework (also called forcing-feedback studies in Nuijens and Siebesma (2019)), models that resolve boundary layer and cumulus cloud processes, such as large-eddy simulations (LES), are run on small doubly periodic domains at grid spacings of 10–250 m in the horizontal and 5–50 m in the vertical. The LES are performed with different prescribed large-scale forcings, representing either the current climate or scenarios for the future climate. The prescribed large-scale forcings include the most important cloud-controlling factors such as large-scale subsidence, SST, and free-tropospheric temperature and humidity, with the climate perturbations representing the predicted changes by GCMs (Blossey et al., 2013, 2016; Bretherthon et al., 2013; Shen et al., 2022) or more idealized versions thereof (Rieck et al., 2012; Radtke et al., 2021; Vogel et al., 2016).

For the stratocumulus regime, Bretherton et al., (2013) classify the process understanding gained from these forcing-feedback studies into four feedback mechanisms (see Bretherton (2015) for a detailed review of these mechanisms). The *thermodynamic* mechanism refers to cloudiness reductions due to a larger vertical gradient of specific humidity in a warmer climate, which increases the entrainment drying of the boundary layer. The *radiative* mechanism represents cloudiness reductions due to reduced longwave radiative cooling at the top of the boundary layer, caused by increased CO_2 and water vapor in the free troposphere. This decreases turbulence production and cloud top entrainment, lowering cloud tops and thinning the clouds. The *stability* mechanism refers to the strengthening of the inversion due to the stabilization of the lower troposphere, which increases cloudiness due to reduced dry air entrainment. Finally, the *dynamic* mechanism also leads to a slight increase in cloudiness due to reduced subsidence and a thickening of clouds resulting from an increase in cloud top height. The first two cloud reduction mechanisms are more important than the latter two cloud increase mechanisms (Blossey et al., 2013, 2016; Bretherthon et al., 2013), supporting the positive low-cloud feedback simulated by previous and current-generation GCMs (Zelinka et al., 2020).

While variants of these mechanisms are also at play in the shallow trade cumulus regime, the trade cumulus feedbacks in forcing-feedback studies tend to be weakly positive or near-zero (Blossey et al., 2013; Bretherton et al., 2013; Rieck et al., 2012; Radtke et al., 2021; Shen et al., 2022; Vogel et al., 2016). The cloud reductions are mostly caused by stronger penetrative entrainment in response to increased surface evaporation, which dries the cloud layer and decreases cloudiness (similar to the *thermodynamic* mechanism above). Also for trade cumuli, stronger inversions tend to enhance cloud cover through increased cloud fraction near the inversion due to the formation of detrained stratiform layers (Lock, 2009; Stevens et al., 2001; Vogel et al., 2020a). However, the increased stability in a warmer climate will likely not lead to larger fractions of stratiform layers, because the local inversion strength in the trade cumulus regime is strongly influenced by shallow convection itself, through its propensity to deepen and organize (Vogel et al., 2020a). Cloud depth and inversion strength in the trade cumulus regime are also regulated by precipitation, which functions as an important buffer that limits changes in cloud top height with warming (Bretherton et al., 2013; Vogel et al., 2016).

The trade cumulus feedback mechanisms described above mostly induce cloud changes in the upper cloud layer above 1 km height, which accounts for about one third of total cloud cover in this regime (Nuijens et al., 2014). Cloudiness near the base of the cumulus layer instead is relatively insensitive to changes in the large-scale forcings both in forcing-feedback studies (Blossey et al., 2013; Vogel et al., 2016) and in observations (Nuijens et al., 2014, 2015). As clouds near the lifting condensation level (LCL) make up around two thirds of total cloud cover (Nuijens et al., 2014), their insensitivity to warming buffers feedbacks and leads to the much weaker low-cloud feedback in the trade cumulus regime compared to the stratocumulus regime.

7.2.2. Constraints from Observed Natural Variability

The second framework – constraints from observed natural variability – uses observed relationships between low clouds and their controlling factors and combines them with the GCM simulated changes in these cloud-controlling factors under climate change to estimate the sign and magnitude of the cloud feedback (Brient & Schneider, 2016; Cesana & Del Genio, 2021; Ceppi & Nowack, 2021; McCoy et al., 2017a; Myers & Norris, 2016; Myers et al., 2021; Qu et al., 2015a; Zhai et al., 2015). See Klein et al., (2017) for a review of this framework and meta-analysis of the original studies that applied it. The initial studies applying this framework provide observational evidence that the *thermodynamic* cloud reduction mechanism explained above will largely dominate the *stability* and *dynamic* cloud increase mechanisms. These observational analyses therefore predict a positive tropical marine low-cloud feedback. The most modern studies to apply this framework have advanced the method to predict the spatial pattern of tropical marine cloud feedback (Ceppi & Nowack, 2021; Cesana & Del Genio, 2021; Myers et al., 2021). They provide observational evidence from independent satellite datasets for a positive stratocumulus feedback and a much weaker or near-zero trade cumulus feedback.

The observed response of low-level cloud fraction, cloud optical depth, and CREs to SST perturbations in the current climate shows major differences in the response of trade cumulus compared to that of stratocumulus (Ceppi & Nowack, 2021; Cesana et al., 2019; Cesana & Del Genio, 2021; Myers et al., 2021; Scott et al., 2020). This explains why their feedbacks will be similarly different. In particular, for current climate variability, trade cumulus cloud amount and optical depth are less sensitive to perturbations in SST than stratocumulus cloud amount and optical depth. This is clearly seen in Figure 7.3, which shows the spatially resolved observed sensitivities of low-cloud amount and optical depth-induced radiative flux to interannual changes in SST when five other cloud-controlling factors are held fixed. Since SST is likely to have the largest climate change response relative to the magnitude of its natural variability compared to that of any other cloud-controlling factor, this provides observational evidence that the response of trade cumulus to surface warming in the future climate will be weaker than that of stratocumulus (Ceppi & Nowack, 2021; Cesana & Del Genio, 2021; Myers et al., 2021), corroborating model evidence. Indeed, the pattern exhibited in Figure 7.3 closely tracks that of the cloud feedback estimated using constraints from observed natural variability (Myers et al., 2021).

7.2.3. Why the Trade Cumulus Feedback likely Differs from that of Stratocumulus

The insights from both forcing-feedback studies and observational constraints emphasize the question why the response of trade cumulus to climate change is likely to be much weaker than the response of stratocumulus. As mentioned above, cloudiness near the LCL makes up for about two thirds of total cloud cover in the trade cumulus regime and is nearly invariant on timescales longer than a day (Nuijens et al., 2014). The variability in cloudiness near the LCL is likely constrained by the venting function of the cumulus mass flux, which is thought to act like a valve that vents air out of the subcloud mixed layer as soon as the subcloud layer top grows appreciably beyond the LCL (Neggers et al., 2006; Nuijens et al., 2015; Vial et al., 2017). Such a cumulus valve would keep the subcloud layer top close to the LCL, potentially constraining

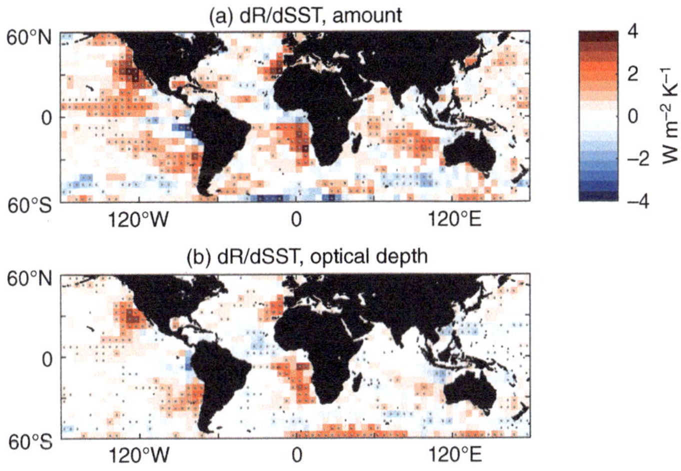

Figure 7.3 Observed sensitivity of low-cloud (a) amount-induced and (b) optical depth-induced TOA radiative flux to a local 1 K increase in SST, with five other local meteorological parameters held constant. Positive values indicate a net gain of energy by the local climate system, and vice versa. Stippling indicates where the coefficients are statistically significant at the 90% confidence level. The sensitivity is estimated as one of the resulting multi-linear regression coefficients after regressing monthly resolved radiative anomalies, derived from July 2002 to December 2018 cloud fraction histogram data from the Moderate Resolution Imaging Spectroradiometer (MODIS) collection 6.1, onto anomalies of six cloud-controlling factors, derived from the ECMWF atmospheric reanalysis of the global climate (ERA5) and the National Oceanic and Atmospheric Administration (NOAA) Optimum Interpolation (OI) SST dataset V2. Source: Adapted from Scott et al., (2020).

variability in humidity and likely also cloudiness near the base of the cumulus layer (Nuijens et al., 2015).

In contrast to LES and observations, near-base cloudiness in many GCMs is very sensitive to climate change. This could explain why most GCMs tend to produce an inflated trade cumulus feedback relative to what the current observational and LES evidence suggests (Myers et al., 2021). Model differences in the strength of lower tropospheric mixing by convective and large-scale circulations explained a large portion of the uncertainty in climate sensitivity in CMIP5 models (Brient et al., 2016; Sherwood et al., 2014; Zhao, 2014). Presumably, models with a stronger increase in mixing with warming tend to have a higher climate sensitivity because stronger mixing leads to enhanced drying and reduced cloudiness near cloud base (Sherwood et al., 2014). However, lower tropospheric mixing is not strongly correlated to climate sensitivity in CMIP6 models and appears uncorrelated to shallow cumulus cloud fraction in observations, calling into question whether such a mechanism explains model behavior (Hu et al., 2022; Schlund et al., 2020). Nonetheless, how much a GCM reduces near-base cloudiness with warming depends on both the coupling between convective mixing, surface turbulent fluxes, and low-cloud radiative effects (Vial et al., 2016), and on the present-day cloud fraction profile (Brient et al., 2016). Observations collected in the EUREC^{4}A campaign, which took place in January and February 2020 upstream of Barbados, will allow for a test of whether a cumulus-valve mechanism is at play in nature and if the GCM behavior is indeed unrealistic (Bony et al., 2017; Stevens et al., 2021; Vogel et al., 2020b).

7.2.4. Outlook

The substantial progress in analysis and simulation techniques in recent years has significantly improved our understanding of the mechanisms governing low-cloud changes and reduced the intermodel spread in tropical low-cloud amount feedbacks in the newest generation of GCMs (Zelinka et al., 2020). Nevertheless, substantial uncertainties remain regarding the magnitude of the tropical low-cloud feedback. Important factors modifying the low-cloud response that have not received enough attention in previous forcing-feedback studies include the mesoscale organization of low clouds and interactions with the underlying ocean surface. For example, when LES are run on larger domains, trade cumulus fields spontaneously organize into larger and deeper clusters with more precipitation and developing cold pool structures (Seifert & Heus, 2013; Vogel et al., 2016). LES with organized shallow convection show some reductions in near-base cloudiness in response to warming (Vogel et al., 2016), motivating a more in-depth study of this deeper cumulus regime (see also section 7.4). Allowing for an interactive SST (with prescribed ocean heat uptake) also influences the low-cloud feedbacks (Tan et al., 2017). Another factor that has often been neglected especially for the trade cumulus regime is the diurnal cycle. The trade cumulus diurnal cycle is distinct for near-base cloudiness and cloudiness further aloft, and mediated by the occurrence frequency of distinct patterns of mesoscale cloud organization (Vial et al., 2019, Vial et al., 2021). The diurnal cycle likely impacts cloud feedbacks, as storm-resolving simulations showed that including a diurnal cycle of insolation influences the magnitude of the low-cloud feedback in the trade cumulus regime (Blossey et al., 2009). Observations from the EUREC^{4}A field campaign as well as recent advances in high-resolution modeling (see section 7.5) will be important to further reduce uncertainty in tropical marine low-cloud feedbacks.

7.3. ROLE OF LOW CLOUDS IN VARIATIONS IN CLIMATE

Low clouds can also exert important radiative feedbacks on climate variability. As low-cloud coverage or its optical thickness deviates from its climatological mean, the TOA and surface radiation budget is perturbed from its quasi steady state. These perturbations can occur on regional and global scales, with influences on patterns and extremes in SST variability, temporal variations in the Earth's energy budget, and the evolution of the historical rate of global warming.

7.3.1. Cloud-Ocean Coupling

When SST is perturbed by some forcing process, such as a change in upper ocean vertical mixing induced by anomalous surface winds, it tends to relax back to its initial state by means of longwave radiative damping and turbulent heat flux damping. In regions of the world with persistent and widespread low clouds, such as the eastern subtropical oceans, the rate of this damping process, however, is reduced due to a positive feedback between low-cloud radiative effect and SST. This reinforcing atmosphere-ocean interaction increases the SST anomaly decorrelation timescale, amplifying the SST anomaly relative to a situation in which this positive cloud radiative feedback was absent. Mechanistically, the positive feedback arises because anomalously warm SST tends to weaken the inversion at cloud top. As discussed in previous sections, the individual effects of warm SST and a weak inversion suppress low-cloud fraction and optical depth. Consequently, downward solar radiation at the surface increases, heating the upper ocean.

The effects of such a feedback on interannual-to-interdecadal patterns of SST variability have been quantified in several idealized modeling experiments using distinct methods (Bellomo et al., 2014; Bellomo et al., 2015; Brown et al., 2016; Burgman et al., 2017; Middlemas et al., 2019). In one method, the feedback strength (defined as a linear relationship between SST and CRE) was artificially increased in targeted subtropical stratocumulus regions in an atmospheric GCM coupled to a slab ocean model (Bellomo et al., 2014, 2015). It was found that the enhanced feedback strength increases the amplitude and persistence of leading modes of interdecadal SST variability in various ocean basins. These modes include those resembling the Pacific Decadal Oscillation (PDO), the El Nino Southern Oscillation, and the Atlantic Nino. Interestingly, SST amplitude and persistence changes were not confined to the regions where the feedback was altered. This suggests that the positive cloud-SST feedback has remote effects, potentially communicated equatorward along the trade winds from subtropical stratocumulus regions through the wind-evaporation-SST (WES) feedback that occurs when tropical SST gradients are altered. The link between subtropical clouds and remote SST was also explored in a pioneering study by Ma et al., (1996), who prescribed a realistic climatological stratocumulus deck off the Peruvian coast in a fully coupled climate model. The SST over the eastern equatorial Pacific was cooler compared to that in the model with a smaller stratocumulus deck, which they attributed to a WES feedback and enhanced cold oceanic advection.

A more recent study using a hierarchy of climate model configurations finds an important role for ocean dynamics in mediating the impact of low-cloud feedbacks on interannual SST variability in the subtropical northeastern oceans (Middlemas et al., 2019). Specifically, in a slab ocean-atmospheric GCM configuration in which clouds freely vary and thus cloud radiative feedbacks are active, SST variance in the subtropical northeastern oceans is amplified relative to a simulation in which cloud feedbacks are disabled via "cloud-locking" (a method in which the annual cycle of clouds in the radiation code is specified to repeat identically each year), consistent with expectation. This is illustrated in Figure 7.4 (upper panel). Interestingly, the introduction of ocean dynamics modifies this response such that, in a fully coupled GCM, the positive low cloud-SST feedback has a much smaller impact on SST variability in the northeast Atlantic than that in the northeast Pacific, as shown in Figure 7.4 (lower panel). This is corroborated by an observational analysis of variations in the oceanic mixed layer energy budget that finds a very small contribution of surface CREs on interannual SST variability in the subtropical northeast Atlantic (Myers & Mechoso, 2020). In contrast, in that study, surface CREs were found to contribute as much as surface turbulent heat fluxes or upper oceanic processes to periods of amplifying SST anomalies in the subtropical northeast Pacific.

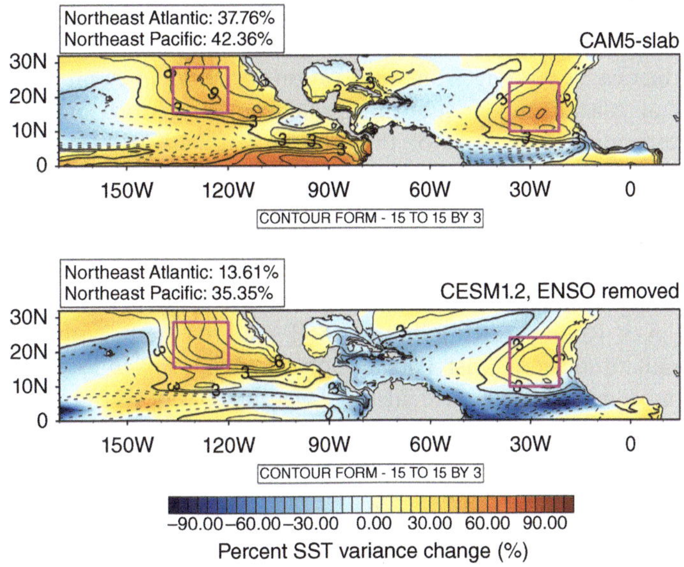

Figure 7.4 Effect of cloud feedback on SST variance in models, taken from Middlemas et al. (2019)/ U.S Department of Energy/CC BY 4.0. Percent change SST variance (% K K^{-1}) due to cloud-locking in CAM5-slab (top) and fully coupled CESM1.2 (bottom) in colored contours, and shortwave cloud radiative feedback overlaid in black contours (W m^{-2} K^{-1}). Shortwave cloud radiative feedback is calculated as the pointwise regression of surface shortwave cloud forcing onto local SST anomalies from the control simulations. Regions that fall within 75% of maximum shortwave cloud radiative feedback are outlined in magenta. The average percent change in variance in either of these boxes are indicated in text above each panel. Source: Figure reproduced under the terms of the Creative Commons license, https://creativecommons.org/licenses/by/4.0/.

However, another "cloud-locking" analysis using a fully coupled GCM concludes that low-cloud radiative feedbacks strongly amplify North Atlantic SST anomalies associated with the Atlantic Multidecadal Oscillation (AMO) and are needed to produce its basin-scale SST anomaly pattern (Brown et al., 2016). This is supported by ship- and satellite-based cloud observations, which reveal reduced subtropical northeast Atlantic cloud fraction during decades of anomalously warm North Atlantic SST and enhanced cloud fraction during decades of anomalously cool North Atlantic SST (Bellomo et al., 2016; Yuan et al., 2016).

How regional low-cloud feedbacks differ across timescales and between ocean basins, and how they relate to ocean dynamics are new and emerging topics that require more investigation[1]. Differing spatial scales of ocean basins and their associated ocean circulation patterns and differing climatological mean coverages of stratocumulus might explain some of the geographical and temporal differences in the role of low clouds in patterns of SST variability documented in the literature.

Finally, intermodel analysis of CMIP5 models reveals that fully coupled GCMs differ with respect to their typical amplitudes of interannual-to-interdecadal subtropical SST variability associated with the PDO and AMO (Myers et al., 2018a; Yuan et al., 2016). Across models, these amplitudes tend to increase in tandem with the diagnosed strength of their local subtropical low cloud-SST feedbacks, thus providing additional evidence that stratocumulus interact with the upper ocean so as to amplify patterns of SST variability. Such intermodel relationships preferentially occur during Northern Hemisphere summer, when stratocumulus are most pervasive and when shallow subtropical ocean mixed layer depths are more sensitive to anomalous surface CREs than in winter, when deeper mixed layers occur (Myers et al., 2018a; Norris et al., 1998).

Marine low-cloud variations associated with extremely warm SST anomalies recently observed in the northeast Pacific Ocean also provide insight into their radiative feedbacks. Between 2013 and 2016, this region experienced the warmest SST anomalies to date (Amaya et al., 2016; Bond et al., 2015; Myers et al., 2018b; Schmeisser

1 We note that climate models tend to simulate a warm SST bias in eastern ocean stratocumulus regions (e.g., Zuidema et al., 2016), which arises in part due to the models' insufficient or lack of coverage of low-level clouds that would otherwise reflect more solar radiation back to space (e.g., Mechoso et al., 1995). This indicates the importance of a positive cloud radiative feedback for generating model biases. Yet, a recent analysis demonstrates that the processes generating a model's SST errors in the subtropical southeast Pacific change dramatically when coastal upwelling is resolved (da Silveira et al., 2019). This further highlights that additional work is needed to better understand the relationship between low clouds and ocean dynamics.

et al., 2019; Zaba & Rudnick 2016). Warm anomalies of this "marine heatwave" occurred along and offshore the west coast of North America, extending from the midlatitudes, near the Gulf of Alaska, to the subtropics, off Baja California. Observational analyses of the oceanic mixed layer energy budget show that an increase in downward solar radiation attributable to a reduction in MBL clouds increased the persistence of the warm anomalies in the midlatitudes and strongly amplified the warm anomalies in the subtropics (Myers et al., 2018b; Schmeisser et al., 2019; Zaba & Rudnick 2016). More recently, extremely warm midlatitude northeast Pacific SST anomalies occurring in summer 2019 were also associated with a large local reduction in low-level clouds and commensurate increase in absorbed solar radiation at the ocean surface (Amaya et al., 2020). Collectively, these studies provide convincing observational evidence that low clouds can exert surface radiative feedbacks that increase the intensity and duration of extreme SST anomalies in the northeast Pacific Ocean.

7.3.2. Low-Cloud Feedbacks and Patterns of Sea Surface Temperature

It has become increasingly apparent that low-cloud radiative feedbacks also exert a strong imprint on interannual-to-interdecadal changes in Earth's TOA energy budget and on the historical rate of greenhouse gas-induced global warming (Andrews et al., 2015; Ceppi & Fueglistaler, 2021; Fueglistaler, 2019; Gregory & Andrews, 2016; Loeb et al., 2018; Loeb et al., 2020; Zhou et al., 2016). Climate model simulations forced by historical SSTs show that the global mean TOA cloud feedback varies on interdecadal timescales (Andrews et al., 2015; Gregory & Andrews, 2016; Zhou et al., 2016). These changes are linked to decadal variations in the pattern of SST anomalies over the Pacific Ocean that cause corresponding perturbations in stratocumulus cloud coverage over the subtropical northeast and southeast Pacific Oceans. For example, even as global mean surface temperature increased from the 1980s to early 2000s, a transition to the cool phase of the PDO occurred. SST decreased in eastern tropical Pacific subsidence regions and increased in western tropical Pacific ascent regions, where warm free-tropospheric temperature anomalies were propagated to the rest of the tropics. The result was a strengthening of the temperature inversion capping stratocumulus clouds in the subsiding regions, increasing their coverage and cooling radiative effect. These conditions produced a globally averaged negative low-cloud feedback and may have contributed to the slowdown or "hiatus" in the rate of global mean warming in the early 2000s (Zhou et al., 2016).

State-of-the-art satellite observations indicate that a decrease in globally averaged outgoing shortwave radiation contributed dominantly to an acceleration of the rate of planetary warming in the years following the end of the hiatus in 2014 (Loeb et al., 2018). This decrease in outgoing shortwave radiation was attributable to a reduction in marine low-cloud coverage over the subtropical eastern Pacific Oceans and was associated with a pattern of SST anomalies resembling the warm phase of the PDO. Current-generation models forced with these recent SST patterns are broadly able to simulate the associated observed changes in local and global low cloud-induced TOA fluxes (Loeb et al., 2020). In summary, temporally varying Pacific SST patterns produce changes in subtropical marine low clouds that trigger strong global radiative feedbacks on the climate system.

7.4. MESOSCALE ORGANIZATION AND POSSIBLE IMPLICATIONS FOR CLOUD FEEDBACKS

Much of the discussion in previous sections referred to macrophysical properties of clouds such as their coverage, optical depth, and radiative effects. Yet, it is manifestly obvious that clouds, including stratocumulus and trade cumulus, are characterized by much more complicated features than these statistical descriptions. From a rudimentary perspective, a cloud can either be considered as a single entity or as part of a larger spatial organization. When analyzing feedback, cloudiness is often studied as one single layer or as disorganized individual cloud entities. But this does not represent the observed spatial organization of low clouds, which is varied and complex. In climate models, the horizontal scale of model grids does not explicitly resolve the mesoscale spatial organization of low clouds. Parameterizations represent the grid-averaged characteristics, while implicitly defining the subgrid organization of the convective plumes (Cheinet, 2003; Suselj et al., 2012). Similarly, observed cloudiness and radiation are generally averaged over a given geographical region or a weather regime; the mesoscale heterogeneity is implicitly included in each "grid mean." Therefore, to date, observational analyses of the low-cloud feedback have not explicitly or systematically taken into account the mesoscale organization of clouds.

7.4.1. Observed Mesoscale Organization of Low Clouds

In the tropics, low clouds organize on the mesoscales, that is, with spatial ranges from ~20 to 2,000 km. Historical aircraft observations (LeMone, 1973), then satellite observations (Agee, 1987), recognized organizations of stratiform convective low clouds, such as stratocumulus closed cells (Agee et al., 1973), stratocumulus open cells (Agee, 1984), and pockets of open cells surrounded by closed cells (Stevens et al., 2005). A first-order view of the spatial organization of cumulus clouds is also recognized, which consists of horizontal convective rolls. These are often named cloud streets, as they organize along lines. However, this only reflects a small part of the diversity of the cumulus organization. Deeper convective clouds also organize at the mesoscale and the mechanisms and climatic impacts are discussed in Chapter 8: Mechanisms for the Self-Organization of Tropical Deep Convection and Chapter 9: An Overview of Mesoscale Convective Systems.

Over the last two decades, the considerable improvement in the quality of the instruments on board satellites has allowed for a more rigorous examination of low-cloud patterns. In the context of a field campaign of the tropical Atlantic trade winds, Stevens et al. (2020a) define four new additional mesoscale configurations for the organization of shallow cumuliform clouds, ranging from an area of small individual clouds, with no obvious sign of self-organization, to large elongated skeletal-like features that encompass a scale of greater than 1,000 km. Observations of these four trade cumulus types are shown in Figure 7.5. Contrary to the stratocumulus structure, the spatial characterization of trade cumulus is less obvious to define rigorously and leads to some empirical definition of the observed regimes (Stevens et al., 2020a). For the most part, the definitions of cloud structures are based on visual inspection, even when machine-learning algorithms are used (Rasp et al., 2020). Unsupervised approaches to disentangling cloud patterns on satellites images could, however, reduce this human bias (Denby, 2020). More rigorous classifications of cloud structures, as performed by machine-learning techniques, could reduce the significant occurrence of disorganized or random cloud regimes suggested by other studies (e.g., Wood & Hartmann, 2006).

7.4.2. Unknown Mesoscale Organization Feedbacks

In investigations of global low-cloud feedback, the focus should be on the change in global-mean albedo. In that context, the albedo difference between stratocumulus and shallow cumulus is tremendous, and a change in the relative occurrence of these two regimes would have a major influence on the climate system (Schneider et al., 2019). However, the physical mechanisms that could explain such a change have yet to be thoroughly assessed.

Cloud regimes with a stratocumulus top exist as the closed-cell stratocumulus structure (Wood, 2012) or as some cumuliform regimes, such as the "flower" regime defined by Stevens et al. (2020a). Therefore, a reduction of global mean albedo with global warming (as suggested by a majority of models and observations) implies moving from closed cells to either open cells or cumulus, or changing the relative occurrence of cumulus regimes. Recently, Bony et al. (2020) use the new definition of trade cumulus regimes to quantify the observational response of cumulus to environmental variations. They show that the frequency of occurrence of regimes is sensitive to changes in vertical stability. Structures with larger spatial scales and higher albedo, corresponding to the "fish" and "flower" cloud types shown in two panels of Figure 7.5, occur more often with increasing stability. In a warmer climate in which stability robustly increases (Qu et al., 2015b), this suggests a possible negative morphology feedback of tropical trade wind cumulus. Bony et al. (2020) also show that increasing SST has a small effect on the frequency of different cumulus regimes, suggesting a weak cloud morphology feedback in response to surface warming alone. Sensitivity of cloud organization over higher temporal variability and shorter periods appears to corroborate these conclusions (Schulz et al., 2021; Vial et al., 2021). These results remain to be replicated with different methodologies (e.g., Denby, 2020; Janssens et al., 2021), and to be understood with a better theoretical framework (Vogel et al. 2020a). Similarly, the observed morphological sensitivity of stratocumulus regimes needs to be quantified, for example, through analyzing the low-frequency temporal evolution of the relative occurrence of closed and open cells (Muhlbauer et al., 2014; McCoy et al., 2017b; Wood & Hartmann, 2006). In that regard, deriving a robust definition of observed cloud regimes is crucial. This would be the essential step for separating possible changes in regime occurrence from cloud changes within a regime in response to climate warming. Fortunately, both the recent improvement of satellite sensors and the ability to run more easily highly resolved models provide the ability to analyze and test the relative importance of these separate cloud changes for low-cloud feedbacks.

Taking mesoscale organization into account would help build novel and more physically based mechanisms of cloud feedback. Since tropical low-level clouds represent the condensed liquid water part of the boundary-layer dynamics, process-oriented analyses of cloud feedback may be understood as pertaining to the structural changes of the boundary layer. Given the long history of studying the formation and the evolution of the atmospheric boundary layer (LeMone et al., 2019), it is hence somewhat surprising that uncertainty in the magnitude of the tropical low-cloud feedback persists. Investigating this challenge from the perspective of cloudy structural changes could be a major step forward. Indeed, the sensitivity of the organization of deep convection has recently been extensively studied, especially through

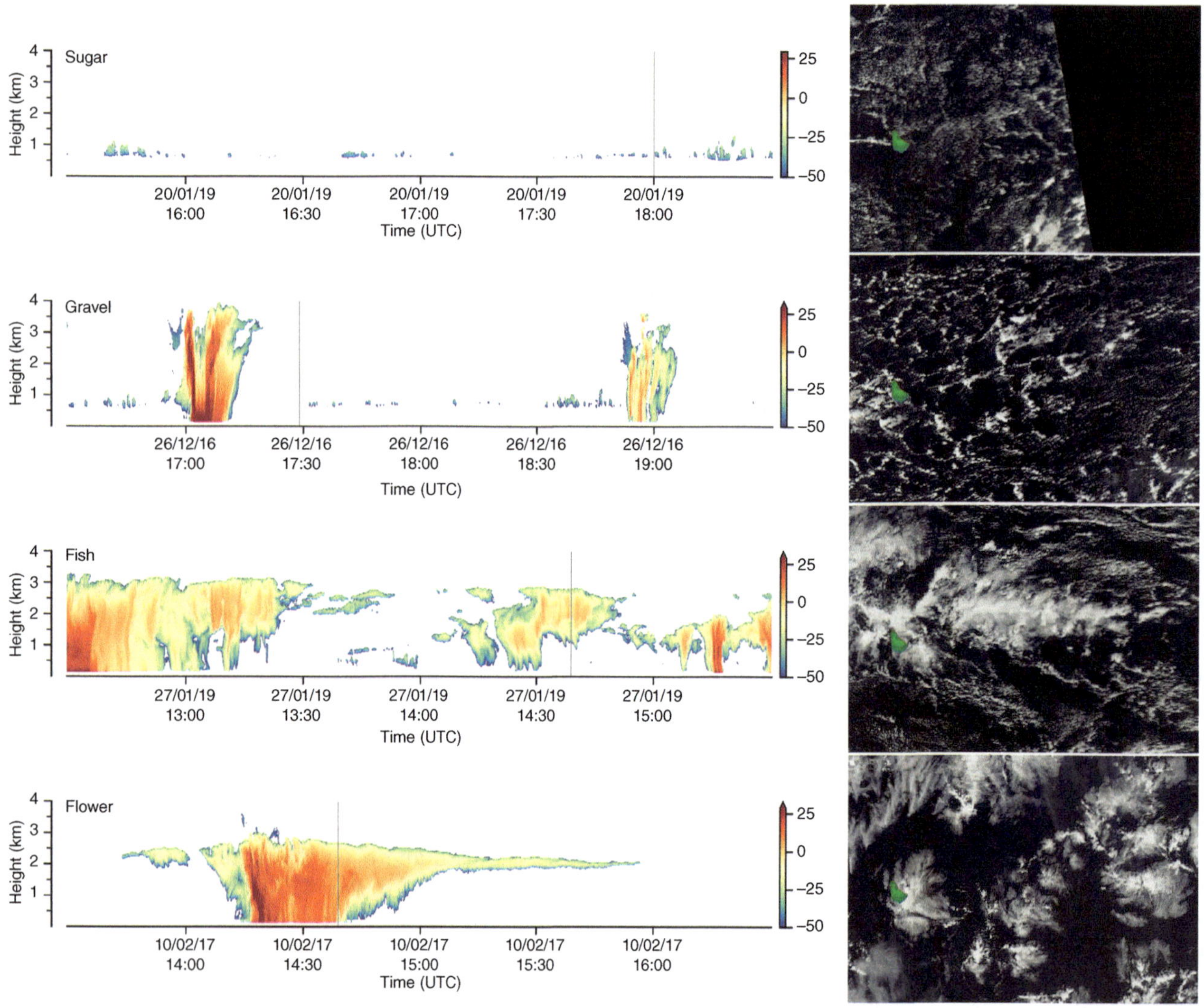

Figure 7.5 Radar retrievals, from the cloud radar at the Barbados Cloud Observatory, and passive satellite retrievals, from MODIS, of four different trade cumulus types, taken from Stevens et al. (2020a)/John Wiley & Sons/CC BY 4.0. The types are named sugar, gravel, fish, and flowers (from top to bottom). Radar cross-sections each span 3 h which for typical wind speeds of 7 m·s^{-1} correspond to a spatial scale of about 75 km, and the color shading is the radar reflectivity (dBZ). Precipitation is indicated by a reflectivity signature extending to the surface. Each MODIS image is over the same geographic region, with Barbados in the left quartile below the centerline, in green. The gray vertical line in the time series indicates the time of the satellite overpass. Source: Figure reproduced under the terms of the Creative Commons license, https://creativecommons.org/licenses/by/4.0/.

three-dimensional radiative convective equilibrium, and provides a better picture of the mechanisms controlling the tropical climate (Wing, 2019). For low clouds, LES that resolve convective structures are already able to reproduce ideal and realistic conditions of shallow convective organization (e.g., Narenpitak et al., 2021), promising major advances in our understanding of the physical mechanisms driving low-cloud organizations.

Finally, the recent geostationary satellites may now be able to describe new and detailed characteristics of the high-frequency evolution of cloud structures (e.g., GOES-16), which would certainly lead to a better understanding of the life cycle of clouds. This assessment through modeling and observations must be combined with a theoretical view of the physics of boundary-layer structures. For example, stratocumulus models are generally understood via the canonical view of Rayleigh-Bénard's cell organization, which partly explains the plume structures by vertical boundary-layer thermal instabilities (Agee, 1987). This helps explain some features of low-cloud organization (Atkinson & Wu Zhang, 1996; Schmidt & Schumann 1989). Yet, a conceptual view of the low-cloud morphology feedback is missing and requires further investigation. Combining observations, modeling, and theory to better understand mesoscale cloud organization might lead us to a potential breakthrough on the as yet incompletely resolved issue of tropical low-cloud feedback.

7.5. RECENT ADVANCES IN AND PROSPECTS FOR HIGH-RESOLUTION MODELING OF LOW CLOUDS

7.5.1. Limitations of GCMs

"Physics parameterization" is at the heart of uncertainties in GCMs, including that of the low-cloud feedback. Some of the small-scale mechanisms relevant to low clouds, which represent a significant scientific and computational challenge to GCMs include: entrainment, subgrid closure of turbulence, aerosols, and microphysics. In all of these processes, the length and timescales of the process are much smaller than those that can be represented in GCMs, since GCMs have resolutions on the order of 1–10 km at best and, more typically, of 100 km. For instance, a sharp boundary-layer inversion and the associated gradients cannot be captured in GCMs simply because of their coarse vertical and horizontal resolutions. Furthermore, in GCMs, boundary-layer turbulence is traditionally parameterized using lower order closure schemes. Newer and higher order boundary layer parameterizations have been investigated, yet have only been tested in models with grid spacing on the order of 10 km at finest (Golaz et al., 2002; Guo et al., 2014). On top of that, grid dependency of the results in the "gray zone" of shallow convection (Tomassini et al., 2017) has made it clear that the grid spacing in the current GCMs must be significantly refined for more accuracy. Therefore, high-resolution models (which are capable of resolving much smaller scales of turbulence) have emerged as an alternative, with the goal of reducing the extent of physics parameterization in the boundary layer.

7.5.2. Current State of High-Resolution Models

Nuijens and Siebesma (2019) review a host of low-cloud modeling techniques and conclude that, among all, turbulence-resolving models – LES – are paving the way forward both for better understanding of physical mechanisms involved and also a more accurate prediction of low-cloud feedbacks. There have been a significant number of limited area high-resolution analyses performed with LES (see Bretherton, 2015). More recently, Stevens et al. (2020b) studied the impact of grid spacing in large-domain simulations. They show that resolving the sub-kilometer scale not only improves the qualitative representation of cumulus versus stratiform clouds, but that it also helps in reducing precipitation biases, particularly over the tropical oceans. These studies typically focus on scales and processes that cannot be resolved in global high-resolution models. In contrast, simulations applying global high-resolution models are narrow in scope and variety. Some of the recent global high-resolution models that have shown promise in advancing the low-cloud modeling capability are as follows:

1. NICAM (Noda et al., 2014; Satoh et al., 2018; Stevens et al., 2019), a global cloud resolving model that resolves eddies down to about a 3 km scale.
2. Global System for Atmospheric Modeling (https://www.esiwace.eu/services/dyamond), a near-global LES model with the caveat of excluded poles (Stevens et al., 2019).
3. Ultra-parameterization (Parishani et al., 2017, 2018), a mixed approach that builds upon strengths of super-parameterization by refining the horizontal and vertical resolutions of the embedded limited area LES cloud resolving models down to a scale of 100 m.

High-resolution models have been instrumental in understanding the low-cloud physics and the relevant processes. Of particular interest here, a positive tropical marine low-cloud feedback is predicted by LES models using the forcing-feedback framework described in section 7.2 (Nuijens & Siebesma, 2019). However, uncertainties still exist, especially with respect to the representation of the cloudy boundary layer (Heim et al., 2021) and differing feedbacks simulated by cloud resolving models. For example, Narenpitak and Bretherton (2019) found a negative shallow cumulus feedback in response to surface warming using a near-global aqua-planet cloud resolving model, which they attributed to an increase in clear-sky radiative cooling being balanced by a stronger updraft cloud mass flux.

7.5.3. New Challenges

There are some major infrastructural scientific and computational challenges facing high-resolution models. For high-resolution models to be more accurate, the involved physical phenomena (e.g., cloud top entrainment) need to be looked at and studied from new perspectives. Here, as an example, we would like to go over two physical processes that need to be rethought for high-resolution models. First, in embedded boundary layers, the cross-grid flow is traditionally ignored. However, Wyant et al., (2018) used an LES and showed that including the boundary layer cross-grid flow can modify the cloud top entrainment, and, as a result, may significantly increase or decrease the low-cloud cover and liquid water path depending on the local vertical grid resolution. Second, subgrid processes in the GCMs are traditionally parameterized and tuned against a limited number of observational data. Use of the same subgrid parameterization in higher resolution models may lead to significant model biases. For instance, Terai et al., (2020) showed that, in the ultra-parameterization framework, resolving sub-kilometer scales dampens the liquid water path increase with higher aerosol concentration compared to the coarser model setup. This shows that a simple embedding of an individually verified small-scale physics solver in a GCM by itself does not lead to a more accurate prediction of low clouds or a better understanding of the underlying physics. In essence, solvers of the small-scale physics need to be re-evaluated, retuned, or completely rethought for the purpose of high-resolution models to compensate for the scale disparity.

Another critical challenge that persists in testing and validation of high-resolution models (and does not seem to go away at least in the near future) is a general lack of whole-earth vertically high-resolution observational profiles of quantities like temperature, liquid water content, and radiative properties that are essential for tuning and validation of high-resolution models. In the absence of these observational data, one typically resorts to the following:

1. Use of vertically integrated properties, which most of the times defeats the purpose of running a high-resolution model. Parishani et al. (2017) and others have shown that, for instance, liquid water path in the low-cloud regions is highly underpredicted in an ultra-parameterization framework (obtaining liquid-rich low clouds has not been easier in LES models either). An accurate prediction of liquid water path (as the most representative column-integrated property) still remains a big challenge in high-resolution models;
2. Vertical profiles from field campaigns through which only a limited model validation can be performed for a small geographical area, or
3. Vertically lower resolution reanalysis data through which only a partial insight can be gained since some of the most important features (e.g., high vertical gradients around boundary-layer inversion) are typically muted or at best diffused in reanalysis data.

High vertical resolution reanalysis that is augmented with machine learning can be a possible solution. Only the initial steps have been taken in this direction, and the authors are not aware of such infrastructure for global reanalysis. For instance, Gentine et al. (2018) and Rasp et al. (2018) demonstrated that a well-tuned machine-learning algorithm can replace the subgrid scheme processes in a GCM at a fraction of the cost, and yet reproduce the key features of the global climate and also the smaller scales cloud features. Machine learning/artificial intelligence can be a potential solution for high-resolution reanalysis, but the question of training the baseline algorithm still remains open.

Other than physics, there are also computational and numerical issues that have to be tackled for a consistent high-resolution simulation. For instance, numerical artificial diffusion can plague the high-resolution simulations, thereby rendering them inaccurate (Wyant et al., 2018). On the other hand, Pressel et al. (2017) hypothesize that numerical diffusion inherent in an advective scheme can replace the subgrid scale closure model, and the resulting numerical diffusion produces an effect equivalent or quantitatively better than a separate subgrid scheme for stratocumulus clouds. These apparently opposing effects of numerical diffusion call for a coherent study of more advanced subgrid (Matheou & Chung, 2014; Shi et al., 2018) and also advective schemes in global high-resolution models. Moreover, Lamaakel and Matheou (2021) show that the boundary-layer flow configuration and the choice of frame of reference (whether it is stationary or translating with the flow) may introduce substantial errors in flow statistics that are considered inherently invariant in an inertial frame of reference. This error is shown to be reduced using a higher order advection scheme. Finally, some less diffusive algorithms have been developed and used for lower resolution (Bretherton et al., 1999) or limited area (e.g., Pressel et al., 2017) simulations.

In another front, data handling at the PetaByte scale becomes a real challenge. Large amounts of data will be presented to the community from global high-resolution models that need to be stored, filtered, and processed efficiently with a high throughput. The current computational infrastructure (hardware, parallel data processing capability, network, storage capacity) has to significantly leap forward for these challenges to be properly addressed.

The challenges of high-resolution simulation will only grow in the foreseeable future, as high-resolution oceanic, land, and ice models will eventually be coupled to the high-resolution atmospheric models, forming the first all-integrated global high-resolution Earth system models. However, for the near term, GCMs still remain in use to study the global-scale phenomena, with the high-resolution models complementing the GCMs.

ACKNOWLEDGMENTS

We thank Hideaki Kawai and two anonymous reviewers for their comments that improved the manuscript. This work was performed under the auspices of the U.S. Department of Energy (DOE) by Lawrence Livermore

National Laboratory under Contract DE-AC52-07NA27344. IM Release: LLNL-BOOK-815033. Timothy A. Myers was supported by the NOAA Cooperative Agreements with CIRES, NA17OAR4320101 and NA22OAR4320151, and by the NOAA/ESRL Atmospheric Science for Renewable Energy program. Raphaela Vogel has received funding from the European Research Council (ERC) under the European Union's Horizon 2020 research and innovation programme (grant agreement No 694768).

REFERENCES

Agee, E. M. (1984). Observations from space and thermal convection: A historical perspective. *Bulletin of the American Meteorological Society*, *65*(9), 938–949.

Agee, E. M. (1987). Mesoscale cellular convection over the oceans. *Dynamics of Atmospheres and Oceans*, *10*(4), 317–341.

Agee, E. M., Chen, T. S., & Dowell, K. E. (1973). A review of mesoscale cellular convection. *Bulletin of the American Meteorological Society*, *54*(10), 1004–1012.

Amaya, D. J., Bond, N. E., Miller, A. J., & DeFlorio, M. J. (2016). The evolution and known atmospheric forcing mechanisms behind the 2013–2015 North Pacific warm anomalies. *US Clivar Variations*, *14*(2), 1–6.

Amaya, D. J., Miller, A. J., Xie, S. P., & Kosaka, Y. (2020). Physical drivers of the summer 2019 North Pacific marine heatwave. *Nature Communications*, *11*(1), 1–9.

Andrews, T., Gregory, J. M., & Webb, M. J. (2015). The dependence of radiative forcing and feedback on evolving patterns of surface temperature change in climate models. *Journal of Climate*, *28*(4), 1630–1648.

Atkinson, B. W., & Wu Zhang, J. (1996). Mesoscale shallow convection in the atmosphere. *Reviews of Geophysics*, *34*(4), 403–431.

Bellomo, K., Clement, A., Mauritsen, T., Rädel, G., & Stevens, B. (2014). Simulating the role of subtropical stratocumulus clouds in driving Pacific climate variability. *Journal of climate*, *27*(13), 5119–5131.

Bellomo, K., Clement, A. C., Mauritsen, T., Rädel, G., & Stevens, B. (2015). The influence of cloud feedbacks on equatorial Atlantic variability. *Journal of Climate*, *28*(7), 2725–2744.

Bellomo, K., Clement, A. C., Murphy, L. N., Polvani, L. M., & Cane, M. A. (2016). New observational evidence for a positive cloud feedback that amplifies the Atlantic Multidecadal Oscillation. *Geophysical Research Letters*, *43*(18), 9852–9859.

Blossey, P. N., Bretherton, C. S., & Wyant, M. C. (2009). Subtropical low cloud response to a warmer climate in a superparameterized climate model. Part II: Column modeling with a cloud resolving model. *Journal of Advances in Modeling Earth Systems*, *1*, 8. https://doi.org/10.3894/JAMES.2009.1.8

Blossey, P. N., Bretherton, C.S., Zhang, M., Cheng, A., Endo, S., Heus, T., et al. (2013). Marine low cloud sensitivity to an idealized climate change: The CGILS LES intercomparison. *Journal of Advances in Modeling Earth Systems*, *5*(2), 234–258.

Blossey, P., Bretherton, C., Cheng, A., Endo, S., Heus, T., Lock, A. P., et al. (2016). CGILS Phase 2 LES intercomparison of response of subtropical marine low cloud regimes to CO2 quadrupling and a CMIP3 composite forcing change. *JAMES*, *8*, 1714–1726. https://doi.org/10.1002/2016MS000765

Bond, N. A., Cronin, M. F., Freeland, H., & Mantua, N. (2015). Causes and impacts of the 2014 warm anomaly in the NE Pacific. *Geophysical Research Letters*, *42*(9), 3414–3420.

Bony, S., Dufresne, J.-L., Le Treut, H., Morcrette, J.-J., & Senior, C. (2004). On dynamic and thermodynamic components of cloud changes. *Climate Dynamics*, *22*, 71–86. doi:10.1007/s00382-003-0369-6.

Bony, S., & Dufresne, J. L. (2005). Marine boundary layer clouds at the heart of tropical cloud feedback uncertainties in climate models. *Geophysical Research Letters*, *32*(20), L20806.

Bony, S., Stevens, B., Ament, F., Bigorre, S., Chazette, P., Crewell, S., et al. (2017). EUREC4A: A field campaign to elucidate the couplings between clouds, convection and circulation. *Surveys in Geophysics*, *38*(6), 1529–1568.

Bony, S., Schulz, H., Vial, J., & Stevens, B. (2020). Sugar, gravel, fish, and flowers: Dependence of mesoscale patterns of trade-wind clouds on environmental conditions. *Geophysical Research Letters*, *47*(7), e2019GL085988.

Bretherton, C.S., Macvean, M.K., Bechtold, P., Chlond, A., Cotton, W., Cuxart, J., et al. (1999). An intercomparison of radiatively driven entrainment and turbulence in a smoke cloud, as simulated by different numerical models. *Quarterly Journal of the Royal Meteorological Society*, *125*, 391–423. https://doi.org/10.1002/qj.49712555402

Bretherton, C. S. (2015). Insights into low-latitude cloud feedbacks from high-resolution models. *Philosophical Transactions of the Royal Society A: Mathematical, Physical and Engineering Sciences*, *373*(2054), 20140415.

Bretherton, C. S., & Blossey, P. N. (2014). Low cloud reduction in a greenhouse-warmed climate: Results from Lagrangian LES of a subtropical marine cloudiness transition. *Journal of Advances in Modeling Earth Systems*, *6*(1), 91–114.

Bretherton, C. S., Blossey, P. N., & Jones, C. R. (2013). Mechanisms of marine low cloud sensitivity to idealized climate perturbations: A single-LES exploration extending the CGILS cases. *Journal of Advances in Modeling Earth Systems*, *5*(2), 316–337.

Brient, F., & Schneider, T. (2016). Constraints on climate sensitivity from space-based measurements of low-cloud reflection. *Journal of Climate*, *29*(16), 5821–5835.

Brient, F., Schneider, T., Tan, Z., Bony, S., Qu, X., & Hall, A. (2016). Shallowness of tropical low clouds as a predictor of climate models' response to warming. *Climate Dynamics*, *47*(1–2), 433–449.

Brown, P. T., Lozier, M. S., Zhang, R., & Li, W. (2016). The necessity of cloud feedback for a basin-scale Atlantic Multidecadal Oscillation. *Geophysical Research Letters*, *43*(8), 3955–3963.

Brueck, M., Nuijens, L., & Stevens, B. (2015). On the seasonal and synoptic time-scale variability of the North Atlantic trade wind region and its low-level clouds. *Journal of the Atmospheric Sciences*, *72*(4), 1428–1446.

Burgman, R. J., Kirtman, B. P., Clement, A. C., & Vazquez, H. (2017). Model evidence for low-level cloud feedback driving persistent changes in atmospheric circulation and regional hydroclimate. *Geophysical Research Letters*, *44*(1), 428–437.

Caldwell, P. M., Zelinka, M. D., Taylor, K. E., & Marvel, K. (2016). Quantifying the sources of intermodel spread in equilibrium climate sensitivity. *Journal of Climate*, *29*(2), 513–524. doi:10.1175/jcli-d-15-0352.1.

Ceppi, P., & Fueglistaler, S. (2021). The El Niño–Southern Oscillation pattern effect. *Geophysical Research Letters*, *48*(21), e2021GL095261.

Ceppi, P., & Nowack, P. (2021). Observational evidence that cloud feedback amplifies global warming. *Proceedings of the National Academy of Sciences*, *118*(30), e2026290118.

Cesana, G., Del Genio, A.D., Ackerman, A.S., Kelley, M., Elsaesser, G., Fridlind, A.M., et al. (2019). Evaluating models' response of tropical low clouds to SST forcings using CALIPSO observations. *Atmospheric Chemistry and Physics*, *19*(5), 2813–2832.

Cesana, G. V., & Del Genio, A. D. (2021). Observational constraint on cloud feedbacks suggests moderate climate sensitivity. *Nature Climate Change*, *11*(3), 213–218.

Cheinet, S. (2003). A multiple mass-flux parameterization for the surface-generated convection. Part I: Dry plumes. *Journal of the Atmospheric Sciences*, *60*(18), 2313–2327.

da Silveira, I. P., Zuidema, P., & Kirtman, B. P. (2019). Fast SST error growth in the southeast Pacific Ocean: Comparison between high and low-resolution CCSM4 retrospective forecasts. *Climate Dynamics*, *53*(9), 5237–5251.

De Szoeke, S. P., Fairall, C. W., Wolfe, D. E., Bariteau, L., & Zuidema, P. (2010). Surface flux observations on the southeastern tropical Pacific Ocean and attribution of SST errors in coupled ocean–atmosphere models. *Journal of Climate*, *23*(15), 4152–4174.

Denby, L. (2020). Discovering the importance of mesoscale cloud organization through unsupervised classification. *Geophysical Research Letters*, *47*(1), e2019GL085190.

Eastman, R., & Wood, R. (2018). The competing effects of stability and humidity on subtropical stratocumulus entrainment and cloud evolution from a Lagrangian perspective. *Journal of the Atmospheric Sciences*, *75*(8), 2563–2578.

Eitzen, Z. A., Su, W., Xu, K. M., Loeb, N., Sun, M., Doelling, D., et al. (2017). Evaluation of a general circulation model by the CERES Flux-by-cloud type simulator. *Journal of Geophysical Research: Atmospheres*, *122*(20), 10–655.

Flato, G., Marotzke, J., Abiodun, B., Braconnot, P., Chou, S. C., Collins, W., et al. (2014). Evaluation of climate models. In *Climate change 2013: The physical science basis. Contribution of Working Group I to the Fifth Assessment Report of the Intergovernmental Panel on Climate Change* (pp. 741–866). Cambridge University Press.

Fueglistaler, S. (2019). Observational evidence for two modes of coupling between sea surface temperatures, tropospheric temperature profile, and shortwave cloud radiative effect in the tropics. *Geophysical Research Letters*, *46*(16), 9890–9898.

Gentine, P., Pritchard, M., Rasp, S., Reinaudi, G., & Yacalis, G. (2018). Could machine learning break the convection parameterization deadlock? *Geophysical Research Letters*, *45*, 5742–5751. https://doi.org/10.1029/2018GL078202

Golaz, J.-C., Larson, V. E., & Cotton, W. R. (2002). A PDF-based model for boundary layer clouds. Part II: Model results. *Journal of the Atmospheric Sciences*, *59(24)*, 3552–3571. https://doi.org/10.1175/1520-0469(2002)059h3552:APBMFBi2.0.CO;2

Gregory, J. M., & Andrews, T. (2016). Variation in climate sensitivity and feedback parameters during the historical period. *Geophysical Research Letters*, *43*(8), 3911–3920.

Guo, Z., Wang, M., Qian, Y., Larson, V. E., Ghan, S., Ovchinnikov, M., et al. (2014). A sensitivity analysis of cloud properties to CLUBB parameters in the Single-column Community Atmosphere Model (SCAM5). *Journal of Advances in Modeling Earth Systems*, *6*, 829–858. https://doi.org/10.1002/2014MS000315

Hartmann, D. L., Ockert-Bell, M. E., & Michelsen, M. L. (1992). The effect of cloud type on Earth's energy balance: Global analysis. *Journal of Climate*, *5*(11), 1281–1304.

Heim, C., Hentgen, L., Ban, N., & Schär, C. (2021). Inter-model variability in convection-resolving simulations of subtropical marine low clouds. *Journal of the Meteorological Society of Japan. Ser. II*, *99*(5), 1271–1295.

Hu, J., Bailey, A., Nusbaumer, J., Dee, S., Sasser, C., & Worden, J. (2022). Tracking shallow convective mixing and its influence on low-level clouds with stable water isotopes in vapor. *Journal of Geophysical Research: Atmospheres*, *127*(5), e2021JD035355.

Janssens, M., de Arellano, J. V.-G., Scheffer, M., Antonissen, C., Siebesma, A. P., & Glassmeier, F. (2021). Cloud patterns in the trades have four interpretable dimensions. *Geophysical Research Letters*, *48*(5), e2020GL091001. https://doi.org/10.1029/2020GL091001

Klein, S. A., Hall, A., Norris, J. R., & Pincus, R. (2017). Low-cloud feedbacks from cloud-controlling factors: A review. In *Shallow Clouds, Water Vapor, Circulation, and Climate Sensitivity* (pp. 135–157). Springer, Cham.

Lamaakel, O., & Matheou, G. (2021). Galilean invariance of shallow cumulus convection large-eddy simulations. *Journal of Computational Physics*, *427*, 110012.

LeMone, M. A. (1973). The structure and dynamics of horizontal roll vortices in the planetary boundary layer. *Journal of the Atmospheric Sciences*, *30*(6), 1077–1091.

LeMone, M. A., Angevine, W., Bretherton, C., Chen, F., Dudhia, J., Fedorovich, E., et al. (2018). 100 years of progress in boundary layer meteorology. *Meteorological Monographs*, *59*, 9–1.

Lilly, D. K. (1968). Models of cloud-topped mixed layers under a strong inversion. *Quarterly Journal of the Royal Meteorological Society*, *94*(401), 292–309.

Lock, A. P. (2009). Factors influencing cloud area at the capping inversion for shallow cumulus clouds. *Quarterly Journal of the Royal Meteorological Society*, *135*, 941–952. https://doi.org/10.1002/qj.424

Loeb, N. G., Thorsen, T. J., Norris, J. R., Wang, H., & Su, W. (2018). Changes in Earth's energy budget during and after the "pause" in global warming: An observational perspective. *Climate*, *6*(3), 62.

Loeb, N. G., Wang, H., Allan, R. P., Andrews, T., Armour, K., Cole, J. N. S., et al. (2020). New generation of climate models track recent unprecedented changes in Earth's radiation budget observed by CERES. *Geophysical Research Letters*, *47*(5), e2019GL086705.

Lutsko, N. J., Popp, M., Nazarian, R. H., & Albright, A. L. (2021). Emergent constraints on regional cloud feedbacks. *Geophysical Research Letters*, *48*(10), e2021GL092934.

Ma, C. C., Mechoso, C. R., Robertson, A. W., & Arakawa, A. (1996). Peruvian stratus clouds and the tropical Pacific circulation: A coupled ocean-atmosphere GCM study. *Journal of Climate*, *9*(7), 1635–1645.

Matheou, G., & Chung, D. (2014). Large-eddy simulation of stratified turbulence. Part II: Application of the stretched-vortex model to the atmospheric boundary layer. *Journal of the Atmospheric Sciences*, *71*(12), 4439–4460.

McCoy, D. T., Eastman, R., Hartmann, D. L., & Wood, R. (2017a). The change in low cloud cover in a warmed climate inferred from AIRS, MODIS, and ERA-Interim. *Journal of Climate*, *30*(10), 3609–3620.

McCoy, I. L., Wood, R., & Fletcher, J. K. (2017b). Identifying meteorological controls on open and closed mesoscale cellular convection associated with marine cold air outbreaks. *Journal of Geophysical Research: Atmospheres*, *122*(21), 11–678.

Mechoso, C. R., Robertson, A. W., Barth, N., Davey, M. K., Delecluse, P., Gent, P. R., et al. (1995). The seasonal cycle over the tropical Pacific in coupled ocean–atmosphere general circulation models. *Monthly Weather Review*, *123*(9), 2825–2838.

Middlemas, E., Clement, A., & Medeiros, B. (2019). Contributions of atmospheric and oceanic feedbacks to subtropical northeastern sea surface temperature variability. *Climate Dynamics*, *53*(11), 6877–6890.

Mieslinger, T., Horváth, Á., Buehler, S. A., & Sakradzija, M. (2019). The dependence of shallow cumulus macrophysical properties on large-scale meteorology as observed in ASTER imagery. *Journal of Geophysical Research: Atmospheres*, *124*(21), 11477–11505.

Muhlbauer, A., McCoy, I. L., & Wood, R. (2014). Climatology of stratocumulus cloud morphologies: Microphysical properties and radiative effects. *Atmospheric Chemistry & Physics*, *14*(5), 6695–6716.

Myers, T. A., & Norris, J. R. (2013). Observational evidence that enhanced subsidence reduces subtropical marine boundary layer cloudiness. *Journal of Climate*, *26*(19), 7507–7524.

Myers, T. A., & Norris, J. R. (2015). On the relationships between subtropical clouds and meteorology in observations and CMIP3 and CMIP5 models. *Journal of Climate*, *28*(8), 2945–2967.

Myers, T. A., & Norris, J. R. (2016). Reducing the uncertainty in subtropical cloud feedback. *Geophysical Research Letters*, *43*(5), 2144–2148.

Myers, T. A., Mechoso, C. R., & DeFlorio, M. J. (2018a). Coupling between marine boundary layer clouds and summer-to-summer sea surface temperature variability over the North Atlantic and Pacific. *Climate Dynamics*, *50*(3–4), 955–969.

Myers, T. A., Mechoso, C. R., Cesana, G. V., DeFlorio, M. J., & Waliser, D. E. (2018b). Cloud feedback key to marine heatwave off Baja California. *Geophysical Research Letters*, *45*(9), 4345–4352.

Myers, T. A., & Mechoso, C. R. (2020). Relative contributions of atmospheric, oceanic, and coupled processes to North Pacific and North Atlantic variability. *Geophysical Research Letters*, *47*(5), e2019GL086321.

Myers, T. A., Scott, R. C., Zelinka, M. D., Klein, S. A., Norris, J. R., & Caldwell, P. M. (2021). Observational constraints on low cloud feedback reduce uncertainty of climate sensitivity. *Nature Climate Change*, *11*(6), 501–507.

Narenpitak, P., & Bretherton, C. S. (2019). Understanding negative subtropical shallow cumulus cloud feedbacks in a near-global aquaplanet model using limited area cloud-resolving simulations. *JAMES*, *11*, 1600–1626. https://doi.org/10.1029/2018MS001572

Narenpitak, P., Kazil, J., Yamaguchi, T., Quinn, P., & Feingold, G. (2021). From sugar to flowers: A transition of shallow cumulus organization during ATOMIC. *Journal of Advances in Modeling Earth Systems*, *13*(10), e2021MS002619.

Neggers, R., Stevens, B., & Neelin, J. D. (2006). A simple equilibrium model for shallow-cumulus-topped mixed layers. *Theoretical and Computational Fluid Dynamics*, *20*(5–6), 305–322.

Noda, A., Satoh, M., Yamada, Y., Kodama, C., & Seiki, T. (2014). Responses of tropical and subtropical high-cloud statistics to global warming. *Journal of Climate*, *27*, 7753–7768. https://doi.org/10.1175/JCLI-D-14-00179.1

Norris, J. R., Zhang, Y., & Wallace, J. M. (1998). Role of low clouds in summertime atmosphere–ocean interactions over the North Pacific. *Journal of Climate*, *11*(10), 2482–2490.

Norris, J. R., & Iacobellis, S. F. (2005). North Pacific cloud feedbacks inferred from synoptic-scale dynamic and thermodynamic relationships. *Journal of Climate*, *18*(22), 4862–4878.

Nuijens, L., Serikov, I., Hirsch, L., Lonitz, K., & Stevens, B. (2014). The distribution and variability of low-level cloud in the North Atlantic trades. *Quarterly Journal of the Royal Meteorological Society*, *140*(684), 2364–2374.

Nuijens, L., Medeiros, B., Sandu, I., & Ahlgrimm, M. (2015). Observed and modeled patterns of covariability between low-level cloudiness and the structure of the trade-wind layer. *Journal of Advances in Modeling Earth Systems*, *7*(4), 1741–1764.

Nuijens, L., & Siebesma, A. P. (2019). Boundary layer clouds and convection over subtropical oceans in our current and in a warmer climate. *Current Climate Change Reports*, *5*(2), 80–94.

Parishani, H., Pritchard, M., Bretherton, C. S., Wyant, M. C., & Khairoutdinov, M. (2017). Toward low-cloud-permitting cloud superparameterization with explicit boundary layer turbulence. *JAMES*, *9*, 1542–1571. https://doi.org/10.1002/2017MS000968

Parishani, H., Pritchard, M. S., Bretherton, C. S., Terai, C. R., Wyant, M. C., Khairoutdinov, M., et al. (2018). Insensitivity of the cloud response to surface warming under radical changes to boundary layer turbulence and cloud microphysics: Results from the ultraparameterized CAM. *JAMES*, *10*, 3139–3158. https://doi.org/10.1029/2018MS001409

Pincus, R., Platnick, S., Ackerman, S. A., Hemler, R. S., & Patrick Hofmann, R. J. (2012). Reconciling simulated and observed views of clouds: MODIS, ISCCP, and the limits of instrument simulators. *Journal of Climate*, *25*(13), 4699–4720.

Pressel, K. G., Mishra, S., Schneider, T., Kaul, C. M., & Tan, Z. (2017). Numerics and subgrid-scale modeling in large eddy simulations of stratocumulus clouds. *Journal of Advances in Modeling Earth Systems*, *9*, 1342–1365. doi:10.1002/2016MS000778.

Qu, X., Hall, A., Klein, S. A., & Caldwell, P. M. (2014). On the spread of changes in marine low cloud cover in climate model simulations of the 21st century. *Climate Dynamics*, *42*(9–10), 2603–2626.

Qu, X., Hall, A., Klein, S. A., & DeAngelis, A. M. (2015a). Positive tropical marine low-cloud cover feedback inferred from cloud-controlling factors. *Geophysical Research Letters*, *42*(18), 7767–7775.

Qu, X., Hall, A., Klein, S. A., & Caldwell, P. M. (2015b). The strength of the tropical inversion and its response to climate change in 18 CMIP5 models. *Climate Dynamics*, *45*(1–2), 375–396.

Radtke, J., Mauritsen, T., & Hohenegger, C. (2021). Shallow cumulus cloud feedback in large eddy simulations–bridging the gap to storm-resolving models. *Atmospheric Chemistry and Physics*, *21*(5), 3275–3288.

Rasp, S., Pritchard, M., & Gentine, P. (2018). Deep learning to represent subgrid processes in climate models. *PNAS*, *115*, 9684-9689. https://doi.org/10.1073/pnas.1810286115

Rasp, S., Schulz, H., Bony, S., & Stevens, B. (2020). Combining crowdsourcing and deep learning to explore the mesoscale organization of shallow convection. *Bulletin of the American Meteorological Society*, *101(11)*, E1980–E1995.

Rieck, M., Nuijens, L., & Stevens, B. (2012). Marine boundary layer cloud feedbacks in a constant relative humidity atmosphere. *Journal of the Atmospheric Sciences*, *69*(8), 2538–2550.

Satoh, M., Noda, A., Seiki, T., Chen, Y.-W., Kodama, C., Yamada, Y., et al. (2018). Toward reduction of the uncertainties in climate sensitivity due to cloud processes using a global non-hydrostatic atmospheric model. *Progress in Earth and Planetary Science*, *5*, 67. https://doi.org/10.1186/s40645-018-0226-1

Schmeisser, L., Bond, N. A., Siedlecki, S. A., & Ackerman, T. P. (2019). The role of clouds and surface heat fluxes in the maintenance of the 2013–2016 Northeast Pacific Marine heatwave. *Journal of Geophysical Research: Atmospheres*, *124*(20), 10772–10783.

Schmidt, H., & Schumann, U. (1989). Coherent structure of the convective boundary layer derived from large-eddy simulations. *Journal of Fluid Mechanics*, *200*, 511–562.

Schneider, T., Kaul, C. M., & Pressel, K. G. (2019). Possible climate transitions from breakup of stratocumulus decks under greenhouse warming. *Nature Geoscience*, *12*(3), 163–167.

Schlund, M., Lauer, A., Gentine, P., Sherwood, S. C., & Eyring, V. (2020). Emergent constraints on equilibrium climate sensitivity in CMIP5: Do they hold for CMIP6? *Earth System Dynamics*, *11*(4), 1233–1258.

Schulz, H., Eastman, R., & Stevens, B. (2021). Characterization and evolution of organized shallow convection in the downstream North Atlantic trades. *Journal of Geophysical Research: Atmospheres*, *126*(17), e2021JD034575.

Scott, R. C., Myers, T. A., Norris, J. R., Zelinka, M. D., Klein, S. A., Sun, M., et al. (2020). Observed sensitivity of low-cloud radiative effects to meteorological perturbations over the global oceans. *Journal of Climate*, *33*(18), 7717–7734.

Seifert, A., & Heus, T. (2013). Large-eddy simulation of organized precipitating trade wind cumulus clouds. *Atmospheric Chemistry and Physics*, *13*(11), 5631–5645. doi:10.5194/acp-13-5631-2013.

Shen, Z., Sridhar, A., Tan, Z., Jaruga, A., & Schneider, T. (2022). A library of large-eddy simulations forced by global climate models. *Journal of Advances in Modelling Earth Systems*, *14*(3), e2021MS002631.

Sherwood, S. C., Bony, S., & Dufresne, J. L. (2014). Spread in model climate sensitivity traced to atmospheric convective mixing. *Nature*, *505*, 37–42. doi:10.1038/nature12829.

Shi, X., Hagen, H. L., Chow, F. K., Bryan, G. H., & Street, R. L. (2018). Large-eddy simulation of the stratocumulus-capped boundary layer with explicit filtering and reconstruction turbulence modeling. *Journal of the Atmospheric Sciences*, *75*(2), 611–637.

Soden, B. J., & Held, I. M. (2006). An assessment of climate feedbacks in coupled ocean–atmosphere models. *Journal of climate*, *19*(14), 3354–3360.

Stevens, B. (2006). Bulk boundary-layer concepts for simplified models of tropical dynamics. *Theoretical and Computational Fluid Dynamics*, *20*(5–6), 279–304.

Stevens, B., Ackerman, A.S., Albrecht, B.A., Brown, A. R., Chlond, A., Cuxart, J., et al. (2001). Simulations of trade wind cumuli under a strong inversion. *Journal of the Atmospheric Sciences*, *58*(14), 1870–1891. https://doi.org/10.1175/1520-0469(2001)058<1870:SOTWCU>2.0.CO;2

Stevens, B., Vali, G., Comstock, K., Wood, R., van Zanten, M. C., Austin, P. H., et al. (2005). Pockets of open cells and drizzle in marine stratocumulus. *Bulletin of the American Meteorological Society*, *86*(1), 51–58.

Stevens, B., Satoh, M., Auger, L., Biercamp, J., Bretherton, C. S., Chen, X., et al. (2019). DYAMOND: The dynamics of the atmospheric general circulation modeled on non-hydrostatic domains. *Progress in Earth and Planetary Science*, *6*(1), 61.

Stevens, B., Bony, S., Brogniez, H., Hentgen, L., Hohenegger, C., Kiemle, C., et al. (2020a). Sugar, gravel, fish and flowers: Mesoscale cloud patterns in the trade winds. *Quarterly Journal of the Royal Meteorological Society*, *146*, 141–152.

Stevens, B., Acquistapace, C., Hansen, A., Heinze, R., Klinger, C., Klocke, D., et al. (2020b). The added value of large-Eddy and storm-resolving models for simulating clouds and precipitation. *Journal of the Meteorological Society of Japan. Ser. II*, *98*(2), 395–435. https://doi.org/10.2151/jmsj.2020-021

Stevens, B., Bony, S., Farrell, D., Ament, F., Blyth, A., Fairall, C., et al. (2021). EUREC[4]A. *Earth System Science Data*, *13*(8), 4067–4119.

Sušelj, K., Teixeira, J., & Matheou, G. (2012). Eddy diffusivity/mass flux and shallow cumulus boundary layer: An updraft PDF multiple mass flux scheme. *Journal of the Atmospheric Sciences*, *69*(5), 1513–1533.

Tan, Z., Schneider, T., Teixeira, J., & Pressel, K. G. (2017). Large-eddy simulation of subtropical cloud-topped boundary layers: 2. Cloud response to climate change. *Journal of Advances in Modeling Earth Systems*, *9*(1), 19–38.

Terai, C. R., Pritchard, M. S., Blossey, P., & Bretherton, C. S. (2020). The impact of resolving sub-kilometer processes on aerosol-cloud interactions in global model simulations. *JAMES*, *12*, e2020MS002274. https://doi.org/10.1029/2020MS002274

Tomassini, L., Field, P. R., Honnert, R., Malardel, S., McTaggart-Cowan, R., Saitou, K., et al. (2017). The "Grey Zone" cold air outbreak global model intercomparison: A cross evaluation using large-eddy simulations. *Journal of Advances in Modeling Earth Systems*, *9*, 39–64. doi:10.1002/2016MS000822.

Van der Dussen, J. J., De Roode, S. R., Dal Gesso, S., & Siebesma, A. P. (2015). An LES model study of the influence of the free tropospheric thermodynamic conditions on the stratocumulus response to a climate perturbation. *Journal of Advances in Modeling Earth Systems*, *7*(2), 670–691.

Vial, J., Bony, S., Stevens, B., & Vogel, R. (2017). Mechanisms and model diversity of trade-wind shallow cumulus cloud feedbacks: A review. *Surveys in Geophysics*, *38*(6), 1331–53. https://doi.org/10.1007/s10712-017-9418-2

Vial, J., Dufresne, J. L., & Bony, S. (2013). On the interpretation of inter-model spread in CMIP5 climate sensitivity estimates. *Climate Dynamics*, *41*(11–12), 3339–3362.

Vial, J., Bony, S., Dufresne, J. L., & Roehrig, R. (2016). Coupling between lower-tropospheric convective mixing and low-level clouds: Physical mechanisms and dependence on convection scheme. *Journal of Advances in Modeling Earth Systems*, *8*(4), 1892–1911.

Vial, J., Vogel, R., Bony, S., Stevens, B., Winker, D. M., Cai, X., et al. (2019). A new look at the daily cycle of trade wind cumuli. *Journal of advances in modeling earth systems*, *11*(10), 3148–3166.

Vial, J., Vogel, R., & Schulz, H. (2021). On the daily cycle of mesoscale cloud organization in the winter trades. *Quarterly Journal of the Royal Meteorological Society*, *147*(738), 2850–2873.

Vogel, R., Nuijens, L., & Stevens, B. (2016). The role of precipitation and spatial organization in the response of trade-wind clouds to warming. *Journal of Advances in Modeling Earth Systems*, *8*(2), 843–862.

Vogel, R., Nuijens, L., & Stevens, B. (2020a). Influence of deepening and mesoscale organization of shallow convection on stratiform cloudiness in the downstream trades. *Quarterly Journal of the Royal Meteorological Society*, *146*(726), 174–185.

Vogel, R., Bony, S., & Stevens, B. (2020b). Estimating the shallow convective mass flux from the subcloud-layer mass budget. *Journal of the Atmospheric Sciences*, *77*(5), 1559–1574.

Webb, M. J., Lock, A. P., Bretherton, C. S., Bony, S., Cole, J. N. S., Idelkadi, A., et al. (2015). The impact of parametrized convection on cloud feedback. *Philosophical Transactions of the Royal Society A: Mathematical, Physical and Engineering Sciences*, *373*(2054), 20140414.

Wing, A. A. (2019). Self-aggregation of deep convection and its implications for climate. *Current Climate Change Reports*, *5*(1), 1–11.

Wood, R., & Bretherton, C. S. (2006). On the relationship between stratiform low cloud cover and lower-tropospheric stability. *Journal of Climate*, *19*(24), 6425–6432.

Wood, R., & Hartmann, D. L. (2006). Spatial variability of liquid water path in marine low cloud: The importance of mesoscale cellular convection. *Journal of Climate*, *19*(9), 1748–1764.

Wyant, M. C., Bretherton, C. S., & Blossey, P. N. (2018). The sensitivity of numerical simulations of cloud-topped boundary layers to cross-grid flow. *Journal of Advances in Modeling Earth Systems*, *10*(2), 466–480.

Yuan, T., Oreopoulos, L., Zelinka, M., Yu, H., Norris, J. R., Chin, M., et al. (2016). Positive low cloud and dust feedbacks amplify tropical North Atlantic Multidecadal Oscillation. *Geophysical Research Letters*, *43*(3), 1349–1356.

Źaba, K. D., & Rudnick, D. L. (2016). The 2014–2015 warming anomaly in the Southern California Current System observed by underwater gliders. *Geophysical Research Letters*, *43*(3), 1241–1248.

Zelinka, M. D., Klein, S. A., & Hartmann, D. L. (2012). Computing and partitioning cloud feedbacks using cloud property histograms. Part I: Cloud radiative kernels. *Journal of Climate*, *25*(11), 3715–3735.

Zelinka, M. D., Myers, T. A., McCoy, D. T., Po-Chedley, S., Caldwell, P. M., Ceppi, P., et al. (2020). Causes of higher climate sensitivity in CMIP6 models. *Geophysical Research Letters*, *47*, e2019GL085782. https://doi.org/10.1029/2019GL085782

Zhai, C., Jiang, J. H., & Su, H. (2015). Long-term cloud change imprinted in seasonal cloud variation: More evidence of high climate sensitivity. *Geophysical Research Letters*, *42*(20), 8729–8737.

Zhao, M. (2014). An investigation of the connections among convection, clouds, and climate sensitivity in a global climate model. *Journal of Climate*, *27*, 1845–1862. https://doi.org/10.1175/JCLI-D-13-00145.1.

Zhou, C., Zelinka, M. D., & Klein, S. A. (2016). Impact of decadal cloud variations on the Earth's energy budget. *Nature Geoscience*, *9*(12), 871–874.

Zuidema, P., Chang, P., Medeiros, B., Kirtman, B. P., Mechoso, R., Schneider, E. K., et al. (2016). Challenges and prospects for reducing coupled climate model SST biases in the eastern tropical Atlantic and Pacific Oceans: The US CLIVAR eastern tropical oceans synthesis working group. *Bulletin of the American Meteorological Society*, *97*(12), 2305–2328.

8

Mechanisms for the Self-Organization of Tropical Deep Convection

Jan O. Haerter[1,2,3] and Caroline Muller[4,5]

ABSTRACT

Organization – or departure from a random pattern – in tropical deep convection is heavily studied due to its immediate relevance to climate sensitivity and extremes. Low-latitude convection has motivated numerical model idealizations, where the Coriolis force is removed and boundary conditions are simplified spatially and temporally. One of the most stunning aspects of such idealized simulated cloud organization is the spontaneous clumping of convection that can occur without any predetermining external perturbation, such as inhomogeneous surface boundary conditions or large-scale waves. Whereas individual convective rain cells measure only few kilometers in horizontal diameter, the clusters they form can often span hundreds or even thousands of kilometers. Hence, organization may emerge from the very small scales but can show effects at the synoptic scale. We refer to such emergent organization as convective self-organization. Convective self-organization thus features characteristics of emergence, such as non-trivial system-scale pattern formation or hysteresis. We summarize observational evidence for large-scale organization and briefly recap classical idealized modeling studies that yield convective self-aggregation – emergent organization under strongly idealized boundary conditions. We then focus on developing research, where temporal variation, such as the diurnal cycle, or two-way interactive surface properties yield distinct organizational modes. Convectively generated cold pools and mesoscale convective systems, both ubiquitous in nature, are thereby found to potentially play key roles in promoting – rather than suppressing – sustained system-scale organization.

8.1. INTRODUCTION

When a homogeneous fluid is captured between two vertically stacked horizontally infinite plates, and the temperature of the lower plate is sufficiently large compared to that of the upper, the fluid, initially at rest, will develop a state of spontaneously broken symmetry under gravity, known as *Rayleigh-Bénard convection* (Turner, 1979). In this state, the sign of the fluid's vertical velocity will alternate horizontally and heat will be transported more efficiently between the lower and upper plates than in the state of rest – where heat must flow through conduction.

In planetary atmospheres, the situation is far more complicated, as a number of additional processes cannot be excluded. Pressure inevitably decreases with height, typically causing the fluid's temperature to decrease, for example, due to adiabatic expansion in convective updrafts in Earth's troposphere. There, under *moist convection*, this vertical temperature gradient requires vertically rising water vapor to locally release latent heat through condensation, causing clouds. The presence of

[1]*Complexity and Climate, Leibniz Centre for Tropical Marine Research, Bremen, Germany*
[2]*Constructor University, Bremen, Germany*
[3]*Niels Bohr Institute, Copenhagen University, Copenhagen, Denmark*
[4]*Information and System Sciences, Institute of Science and Technology Austria, Klosterneuburg, Austria*
[5]*Laboratoire de Météorolgie Dynamique, Paris Sciences Lettres/Institut Pierre Simon Laplace, Paris, France*

Clouds and Their Climatic Impacts: Radiation, Circulation, and Precipitation, Geophysical Monograph 281, First Edition.
Edited by Sylvia C. Sullivan and Corinna Hoose.

DOI: 10.1002/9781119700357.ch08

clouds strongly affects the *radiative heating and cooling* of the atmosphere as well as that of the liquid or solid surface at the lower boundary. Radiative effects include reflection of shortwave, but absorption and re-emission of both short wave and longwave radiation – with both positive and negative feedbacks possible.

When cloud condensate particles further grow into hydrometeors (such as rain, snow, or intermediate phases), these can locally sediment (fall) toward the surface – yet, usually in the subsaturated boundary layer beneath the cloud, a fraction of these condensate particles may re-evaporate. They thereby take up latent heat and cool and moisten the subcloud air volume, forming a so-called *convective cold pool (CP)*.

Away from the equator and at large horizontal scales – beyond the Rossby deformation radius – Earth's rotation cannot be ignored and the Coriolis force must be considered. Whereas elements of Rayleigh-Bénard convection are indeed observed, such as in subtropical non-precipitating shallow cumulus convection (Glassmeier & Feingold, 2017; Krishnamurti, 1975; Koren & Feingold, 2013), circumstances with intense precipitation and strong cloud-radiation interactions often present an even larger variety of competing effects.

Whereas Earth's atmosphere is never in complete steady state, the assumption of a balanced heat flux budget is of large theoretical modeling convenience and, in reality, indeed approximately justified over sufficiently large spatial ($\mathcal{O}(10^3\,\text{km})$, Jakob et al. (2019)) and temporal scales. *Radiative convective equilibrium* (RCE) is a simplifying conceptual framework which describes an idealized atmosphere, where radiative and convective heat fluxes are balanced, that is, the radiative cooling is equal to the convective heating (Bretherton et al., 2005; Held et al., 1993; Manabe & Strickler, 1964; Rennó et al., 1994). RCE hence represents a statistical steady state, where large-scale, system-averaged, observables are approximately constant over time. In RCE simulations, boundary conditions, such as surface temperature and moisture as well as insolation, are typically defined to be uniform (spatially constant) and are often even held temporally constant. The RCE framework is a modeling paradigm for the atmosphere over the tropical ocean – there, sea surface temperature (SST) variations may be relatively small, both in space and throughout the year, and there are fewer external ("large-scale") perturbations to the atmosphere than, say, in the midlatitudes, where convective activity is often perturbed by frontal dynamics, resulting from baroclinic instabilities.

Tropical deep convection can theoretically self-organize into a state referred to as *convective self-aggregation (CSA)*, where the horizontal translational symmetry provided by RCE boundary conditions is broken in favor of a horizontal segregation into strongly convecting and dry, subsiding, subregions. CSA is often quantified by variance measures, such as of column water vapor or moist static energy. When running a numerical simulation in RCE over an extended time period ($\mathcal{O}(10-100\,\text{days})$) within a sufficiently large domain ($\mathcal{O}(10^3\,\text{km})$), the initial small-scale ($\mathcal{O}(10\,\text{km})$) regular organization of clouds, precipitation, and moisture spontaneously segregates into one or more strongly convecting subregions, whereas the remainder of the model domain is nearly cloud free with subsiding air motion (Fig. 8.1). CSA hence can be viewed as an additional system-scale symmetry breaking of the initial disaggregated (near random) convective pattern. Beyond the systematic patterning of convective rain cells, CSA also modifies domain-average properties (Bretherton et al., 2005; Bony et al., 2015, 2016; Muller & Held, 2012; Muller & Bony, 2015; Wing & Cronin, 2016; Wing, 2019; Wing et al., 2020) such as reducing moisture and increasing outgoing longwave radiation – thus making CSA a potentially relevant climate feedback. In line with overall increased free tropospheric cooling compared to a disaggregated state, simulated CSA yielded increased system-average precipitation and vertical heat fluxes. Dry regions are characterized by overall subsidence, induced by stronger-than-domain-average radiative cooling.

Assuming a steady state, where the spatial extent of the moist patch does not change and the circulation is stationary, it is possible to estimate the subsidence velocity w_s within the dry area. In steady state, w_s can then be obtained from equating the warming induced by subsiding air masses along the vertical potential temperature gradient, $d\theta/dz$, with the differential cooling from the stronger thermal radiative emission Q_{rad} in the dry region. Assuming a large domain and a small convective domain fraction, Tompkins and Craig (1998) estimated the steady state subsidence velocity as

$$w_s = -\frac{\langle Q_{rad} \rangle}{\rho c_p H \frac{d\theta}{dz}} \approx 0.5\,\text{m}\,\text{s}^{-1}, \tag{8.1}$$

assuming reasonable values for $\langle Q_{rad} \rangle = 125\,\text{W}\,\text{m}^{-2}$, $\rho H = 9{\cdot}10^3\,\text{kg}\,\text{m}^{-2}$, and $d\theta/dz = 3\,\text{K}\,\text{km}^{-1}$.

Within the transient phase before reaching a steady state, initial subsidence can feed back on drying and thus suppression of convection in the already dry region. In the course of the simulation, dry regions are initially small, but tend to expand and merge, in many cases leading to only one persistent dry region that covers most of the domain (typically ~80% of it, see Fig. 8.1). The remainder of the domain is covered by an intensely convecting moist patch, featuring strong convective updrafts, reduced outgoing longwave radiation and heavy precipitation. In summary, the approach to the fully aggregated state can be described as "dry gets drier" and "moist gets moister."

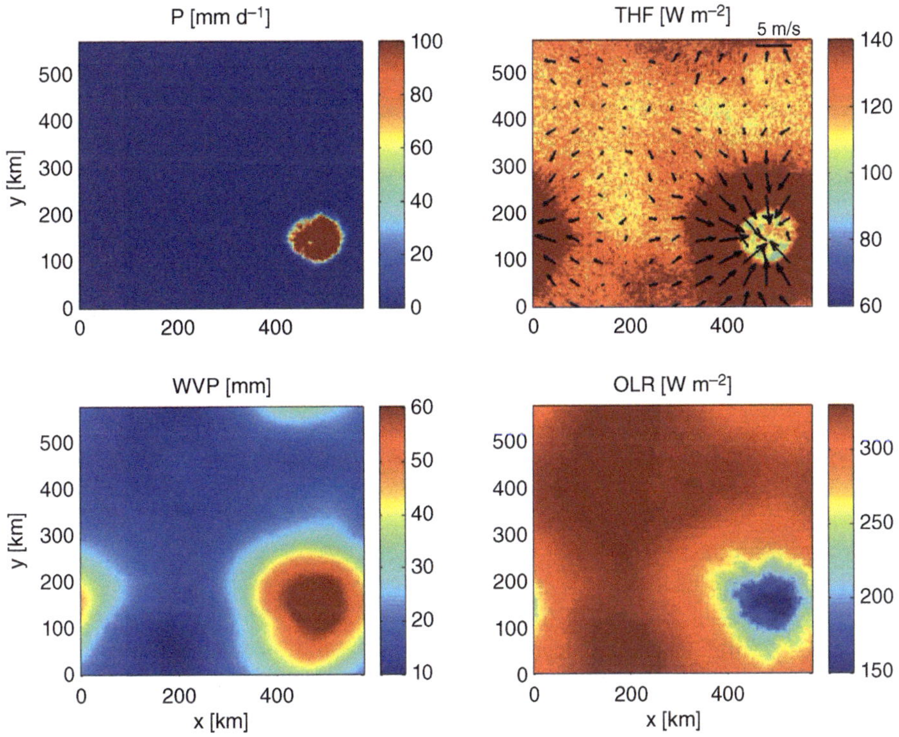

Figure 8.1 A typical final state of convective self-aggregation. Precipitation rate (*P*), thermal surface heat fluxes (*THF*), water vapor path (*WVP*), and outgoing longwave radiation (*OLR*) during day 50 in an RCE simulation run at 3 km horizontal grid resolution and sea surface temperature fixed to 301K. Source: Adapted from Bretherton et al. (2005).

Several thorough reviews exist on CSA (Holloway, 2017; Wing et al., 2017; Wing, 2019) cold pools (Zuidema et al., 2017), and the implications for extreme events (Muller & Takayabu, 2020). In the following, we therefore only briefly summarize the main concepts from this vast body of literature, at the benefit of allowing for a reflection on departure from the strictly idealized assumptions typically made in studies on CSA, and to highlight the ways in which cold pools may disrupt or even facilitate self-organization.

8.2. OBSERVING LARGE-SCALE TROPICAL CLOUD CLUSTERING

While elegant, the RCE modeling framework comes with various idealizations and one may ask, to which extent convective cloud clustering can actually be observed. The large spatial scales involved necessitate near-global remote sensing measurements. Fortunately, improvements in satellite imagery make more and more detailed observations from space possible. Pioneering observational analysis already employed infrared and visible images from geostationary satellites to observe extended cloud clusters (Miller & Fritsch, 1991; Mapes & Houze Jr, 1993; Velasco & Fritsch, 1987; Zuidema, 2003), whereas satellite radar was used to characterize precipitation rates (Nesbitt et al., 2000; Schumacher & Houze Jr, 2003) and compare continental to oceanic convection (Futyan & Del Genio, 2007). Microwave measurements can further probe column water vapor (Mapes et al., 2009), which is sometimes used to quantify CSA.

Mapes and Houze Jr (1993) analyzed infrared satellite imagery to study tropical cloud clusters over the oceanic warm-pool region. The cloud cluster size distribution obtained was broad, approximately log-normal. Such long-tailed distributions can hint at non-random organizing mechanisms (Christensen & Moloney, 2005). For the distribution obtained, approximately half of the total cloud area is contained in large clusters for different cloud top height – that is, both for cold (<208 K) and moderately cold (<235 K) cloud tops.

Building on the notion of non-random emergent organization, Peters and Neelin (2006) in seminal work plotted the precipitation rate and its variance against column water vapor and found characteristics similar to those measured in critical phenomena, such as divergent susceptibilities at the critical point. Identifying precipitation as a relaxation process toward a critical water vapor value w_c, they compared tropical cloud organization to self-organized critical systems (SOC, Bak et al. (1987)), which are characterized by a timescale separation between the slow driver, such as surface moisture flux, and the fast relaxation mechanism, rainfall. Exploring the spatial implications of SOC-like phenomena, Peters et al. (2009) analyzed Tropical Rainfall Measurement Mission (TRMM) data, finding scale-free cluster size distributions and a non-trivial scaling of the cluster circumference. They interpreted these findings using percolation theory and concluded that the statistical properties found were commensurate with a continuous phase transition.

From this idealized equilibrium framework it remains to be explained, how, in the past 30 years, the observed frequency of organized convection might have increased transiently across the tropics (Tselioudis et al., 2010; Tan et al., 2015) and why organized convection might correlate with increased precipitation intensity (Dai & Soden, 2020; Semie & Bony, 2020; Tan et al., 2015) (see also Chapter 14: Observed Scaling of Precipitation Extremes).

To measure organization, Tobin et al. (2012) defined a Simple Convective Aggregation Index (SCAI) for $10^\circ \times 10^\circ$ satellite data and found that the degree of clustering is inversely related to the number of clusters detected (Fig. 8.2). Cloud clusters were thereby defined as spatially contiguous cold cloud shields. Tobin's work indeed confirmed one of the key predictions made from CSA, namely, that outgoing longwave radiation increased significantly (by $30\,\mathrm{W\,m^{-2}}$) for the more clustered cases. But in contrast to simulations, such increased thermal emission was almost entirely compensated for by reduced reflection of incident solar shortwave radiation – a finding that might suggest a relatively weak overall climate response caused by the clustering. However, in other satellite observations, Bony et al. (2020) find a correlation between monthly anomalies of aggregation and the tropically averaged radiation budget. Satellite observations may thus be consistent with the suggestion from modeling studies that the spatial organization of deep convection substantially influences the Earth's radiative balance at large scales.

To probe idealized numerical predictions directly, Holloway et al. (2017) compared simulations on horizontally square domains with observations from Nauru Island, finding that the simulated CSA produces humidity distributions that are broader than those observed, and that simulated dry regions are drier than the observed ones. On rectangular domains, however, moisture variabilities tended to be more consistent with the observations at Nauru.

8.3. ORGANIZATIONAL MECHANISMS

Before discussing some of the different feedbacks involved using a range of conceptual models, we first give a brief overview. The simplified RCE boundary conditions and the resultant CSA make for a more transparent starting point in discussing convective self-organization. Yet, even CSA is the result of a complex interplay between cloud processes, moisture redistribution, radiation, and circulation (see also Chapter 10: Interactions Between the Tropical Atmospheric Overturning Circulation and Clouds in Present and Future Climates). The relative roles of these different processes are currently not fully settled and vary from one simulation setup to another. Many works seem to have singled out interactive radiation, and the atmosphere's dynamical response to it, as essential in maintaining CSA (Emanuel et al., 2014; Muller & Held, 2012; Muller & Bony, 2015). Whereas Muller & Held (2012) found the feedback from low cloud within dry regions to be critical in the initial formation of CSA, later work (Emanuel et al., 2014; Muller & Bony, 2015) relaxed this requirement for higher surface temperatures, where water vapor feedbacks alone were seen as a necessity. Dry regions, with their strong surface latent heat fluxes and divergent low-level flow (Fig. 8.5d), serve to export moisture to already moist regions, often referred to as "upgradient moisture transport," thus resulting in a positive feedback. Shortwave radiation feedbacks tend to act against CSA, at least in the mature stage, as they more strongly cool deep convective regions, which have highly reflecting clouds (Muller & Held, 2012). Other conceptual models do not invoke spatially non-uniform radiation processes in order to reach CSA, with local moisture reinforcement (Craig & Mack, 2013) or local cold pool effects (Haerter, 2019; Windmiller, 2017) reported as sufficient in bringing about the spatial segregation into convectively active and quiescent regions.

There is no final consensus as to the effect of (sea) surface temperature on convective self-aggregation (Wing et al., 2020). A number of studies suggest that convective self-aggregation might be favored by high SST (Coppin & Bony, 2015; Emanuel et al., 2014; Held et al., 1993; Khairoutdinov & Emanuel, 2010; Wing & Emanuel, 2014), although some studies do not find this temperature dependence (Abbot, 2014; Coppin & Bony, 2015; Wing & Cronin, 2016). Given that such high SSTs are usually only reached near the equator, these findings would, in addition to the typically non-rotating RCE boundary conditions, support the notion of CSA as a model idealization corresponding mainly to the tropical ocean.

8.3.1. Feedbacks Between Radiation, Moisture, and Circulation

Founded in radiation feedbacks, Emanuel et al. (2014) proposes a two-layer model, which assumes a fixed temperature in each of the two layers, in accordance with the weak-temperature gradient approximation

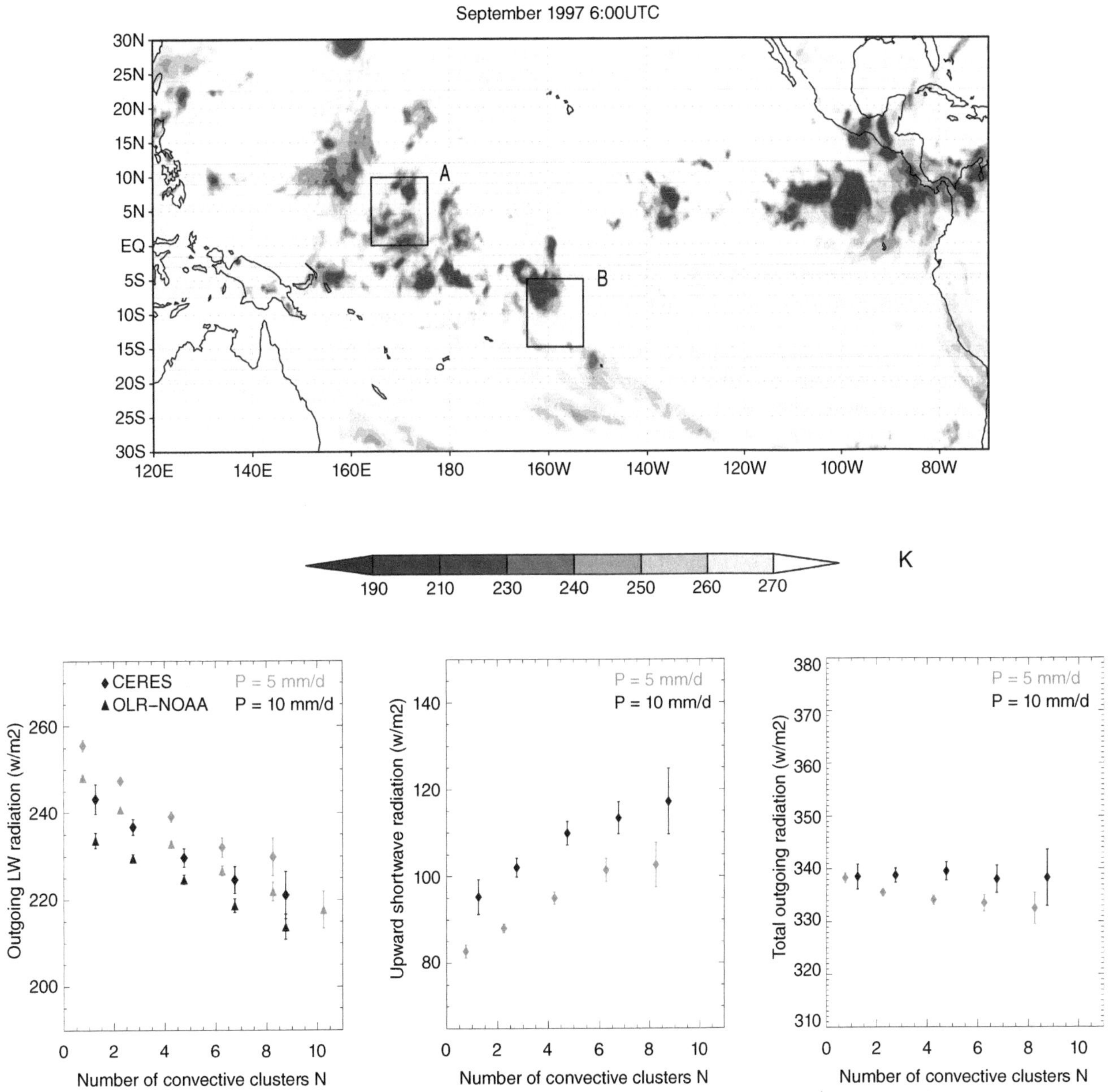

Figure 8.2 Defining clusters in satellite imagery. (top) Snapshot of brightness temperature (Cloud Archive User Service, CLAUS), employed to detect deep convective clouds (0600 UTC 29 September 1997. Boxes (labeled A and B) exemplify subdomains considered by Tobin et al. (bottom) Outgoing longwave, shortwave, and total radiation versus the number of convective clusters, conditioned on precipitation intensity (*see* legend). Source: Adapted from Tobin et al. (2012).

(Sobel et al., 2001). By relating the convective mass flux to radiative emission and absorption within the two layers, they obtain two coupled ordinary differential equations for the moisture in each of the two layers, that is,

$$L_v \frac{\partial q_i'}{\partial t} = c_{i1} q_1' + c_{i2} q_2', \text{ where } i \in \{1,2\}. \qquad (8.2)$$

The dynamical variables, q_i', are the moisture fluctuations in the two layers and the coefficients, c_{ij}, encode the dependence of radiative heating rates on moisture content (Fig. 8.3). Examining the stability of equation (8.2) in each layer by considering the absorption of radiation emitted from the other layer and the surface results in

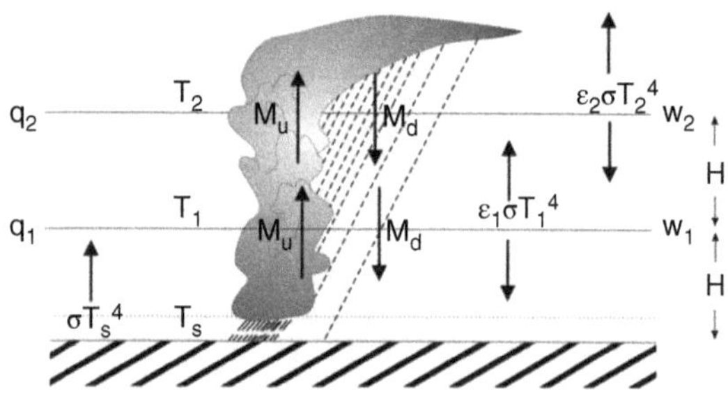

Figure 8.3 A simplified two-layer model for radiative-convective instability. The model assumes that surface temperature (T_s) and the temperatures (T_1 and T_2) of the two layers are specified and constant. Emissivities, ϵ_1 and ϵ_2, the mass fluxes corresponding to updrafts and downdrafts, M_u and M_d, respectively, as well as vertical velocities, w_1 and w_2, and specific humidities, q_1 and q_2, vary. Source: Figure adapted from Emanuel et al. (2014).

a stability criterion, which depends on the layer temperatures. Instability is suggested to be facilitated, when the upper tropospheric emissivity depends strongly on water vapor concentration, whereas that of the lower troposphere depends little on water vapor. An overall high emissivity of the lower layer and high precipitation efficiency further foster the instability. The combination of lower layer high emissivity but weak low-layer dependence on water vapor is achieved, when overall specific humidity is increased at higher SST. The authors therefore argue that there might be a transition between scattered and aggregated convection as a function of SST. Notably, this simplified model neglects shortwave radiative and surface flux feedbacks, assumes a thin boundary layer, and utilizes the quasi-equilibrium assumption. It further assumes that turbulence as well as condensation and evaporation occur exclusively within clouds and that cloud fraction is very small.

The perspective of higher SST leading to a transition to CSA is intriguing: a state of CSA – in simulations found to feature increased outgoing radiative fluxes – could then be a natural negative feedback to increased surface temperature (Khairoutdinov & Emanuel, 2010). In this way, CSA would counteract a likely future increase in SST, thus acting to naturally mitigate some of the consequences of such warming. Mathematically, such self-regulated restoration toward a phase transition could be seen as a natural realization of self-organized criticality (Bak et al., 1987).

Challenging this notion, at least several studies reported that CSA could also occur at temperatures well below those of tropical sea surfaces (Abbot, 2014; Coppin & Bony, 2015; Holloway & Woolnough, 2016; Holloway, 2017; Wing & Cronin, 2016). It has further been argued that strong vertical wind shear may hamper CSA (Abbot, 2014; Bretherton et al., 2005; Held et al., 1993; Wing & Emanuel, 2014), whereas mesoscale SST anomalies (cold or warm) accelerate the aggregation process (Müller & Hohenegger, 2020; Shamekh et al., 2020b). Indeed, it is well known that SST gradients can spatially organize convection, but self-aggregation feedbacks further modulate and strengthen this SST-induced organization (as quantified in detail in Shamekh 2020). For the CRM and configuration used, the presence of a warm SST anomaly also extends the range of SSTs for which aggregation occurs. A large convective instability over the warm sea surface leads to stronger convection and generates a large-scale circulation which forces subsidence drying outside the warm anomaly, thus reinforcing the drying in dry regions and the convergence of moisture into the moist convecting region.

8.3.2. Coarsening dynamics through local moisture feedbacks

A different view is taken by Craig and Mack (2013), where radiative cooling takes place uniformly and only sets the domain total precipitation rate; thus, radiation feedbacks are not required to yield segregation into a moist and a dry subregion. Craig and Mack (2013) implement a type of reaction-diffusion equation by assuming a local moisture feedback, where precipitation increases exponentially with humidity, an assumption commensurate with observations (Bretherton et al., 2004). Their model is formulated in terms of vertically integrated moisture content, I_v,

$$\frac{\partial I_v}{\partial t} = -\alpha I_v + f(I_v) + K\nabla^2 I_v, \tag{8.3}$$

where the constant α represents subsidence drying, $f(I_v)$ encodes monotonically increasing convective moistening, that is, "moist-gets-moister" dynamics, and K is an eddy diffusivity, representing horizontal transport. $f(I_v)$ is taken as proportional to the precipitation rate, but weakly decreasing with moisture due to reductions in rain re-evaporation. The combination of the local reinforcing term $f(I_v)$, diffusive smoothening, and the large-scale precipitation constraint makes for a bistable dynamics, yielding coarsening (Bray, 2002). The system eventually organizes in a way that is qualitatively similar to CSA, where a single moist patch is formed and surrounded by otherwise dry conditions, that is, dry regions become drier and moist regions moister.

8.3.3. Cold pools

The above conceptual models leave out any explicit encoding of boundary-layer processes. Yet, thunderstorm rain cells produce a direct, negative buoyancy feedback on themselves, once rainfall evaporates within the subsaturated boundary layer, thus causing a so-called convective CP (Droegemeier & Wilhelmson, 1985; Simpson, 1980). CPs are volumes of denser air, cooled by hydrometeor evaporation, that then spread along the surface as gravity currents. CPs can be delimited from their surroundings by the *gust front*, a narrow area of strong near-surface horizontal convergence, hence pronounced vertical velocity (Droegemeier & Wilhelmson, 1987; Rotunno et al., 1988). Thermally, CPs are often dry in their interior but sometimes moistened along their edges (Tompkins, 2001). In turn, the speed of spreading depends on the initial temperature perturbation, which increases with rain evaporated. Rain evaporation increases for deeper and drier subsaturated boundary layers and depends on the drop size spectrum. CPs act to "shut down" convective activity directly, by depleting the buoyancy that drives it, with recovery times of hours to days. Additionally, surface heat fluxes can be enhanced underneath spreading CPs due to the increased horizontal wind speed and temperature or moisture depressions. Recent numerical simulations suggest that the dynamics of spreading CPs is well captured by the initial potential energy of the density anomaly (Meyer & Haerter 2020). The simulated radius, $R(t)$, shows algebraic dependence on time, t, with results roughly in line with those from similarity theory ($R(t) \sim t^{1/2}$; (Rooney, 2015)) or vorticity conservation ($R(t) \sim t^{2/3}$; (Rooney, 2018)).

8.3.4. Surface heat fluxes

In numerical models, surface fluxes are often approximated by the so-called bulk formula (Fairall et al., 2003), which is based on Monin-Obukhov similarity theory (MOST) and represents the fluxes through mean quantities:

$$\overline{w'x'} = C_x S \Delta X \tag{8.4}$$

where C_x is the total transfer coefficient, $\Delta X = X_{surf} - X(z)$ is the difference between the mean of a quantity X in the air and surface, respectively, and S is the mean horizontal wind speed. In the case of latent heat fluxes over sea, $X(z)$ would represent atmospheric specific humidity, whereas X_{surf} would be the saturation water vapor mixing ratio at the sea surface. The transfer coefficient C_x itself is a complicated function of wind speed, height, and atmospheric stability and is sometimes simply taken as a constant. Hence, surface heat fluxes can either be increased by large horizontal wind speed or large vertical gradients in the respective quantity between the air and the surface.

Several studies highlight the potential positive feedback of interactive surface fluxes on CSA (Bretherton et al., 2005), notably from enhanced winds and WISHE effects (Wing & Emanuel, 2014). The relative role of the surface flux feedback, compared to the aforementioned radiative feedback, is still debated and may depend on the SST (Coppin & Bony, 2015). In particular, the extent to which spatial variation in surface heat fluxes is required in order to initiate CSA is debated (Muller & Bony, 2015; Wing et al., 2017), but surface heat fluxes may well help stabilize convective self-aggregation, once it is formed: in the later stage of self-aggregation the extremely dry conditions outside of the convecting region lead to enhanced surface heat fluxes (see equation (8.4), limiting further drying and helping CSA to reach an equilibrium.

8.3.5. Effects of cold pools on organization

In simulations, moisture rings can be formed by the combination of increased latent heat fluxes (Langhans et al., 2015; Schlemmer & Hohenegger, 2016) as the CP spreads, moisture anomalies already present, as well as advection of moisture. Observationally, moisture rings have also been detected in some cases (Chandra et al., 2018; de Szoeke et al., 2017; Zuidema et al., 2017), but were found to lie outside of the gust front.

Convective CPs have a complex internal vortical circulation (Fig. 8.4), which can cause mechanical lifting, in particular at locations where CPs collide (Moncrieff & Liu, 1999; Meyer & Haerter, 2020; Torri et al., 2015). CP collisions have recently been studied in a range of cloud-resolving simulation frameworks (Böing et al., 2012; Haerter & Schlemmer, 2018; Schlemmer & Hohenegger, 2016), as well as through conceptual modeling (Böing, 2016; Grandpeix & Lafore, 2010; Haerter et al., 2019). Feedbacks between precipitation intensity, CP strength, and subsequent updraft generation and thus the resulting self-organization of convective cells in space are however far from settled. Recent conceptual modeling brings forward a mechanism for phase segregation akin to CSA, which is entirely based on CP collisions and a global energy constraint. The latter acts to restrict the total number of new rain cells – and thus CPs – formed (Haerter, 2019). In this model, new rain cells are set off at the points where several gust fronts collide. When starting from random initial conditions, where the number density of CP centers is spread homogeneously across the domain, the model dynamics gradually causes organization into larger and larger clusters of convective activity – compensated by regions without activity. The dynamics of the rain cell number density ρ is described by

$$\begin{aligned}\dot{\rho}(\mathbf{r}, t) = {} & p_0(1 - \overline{\rho})\rho(\mathbf{r})^2(1 - \rho(\mathbf{r})) \\ & + f_{sg}(1 - \overline{\rho})(1 - \rho(\mathbf{r})) - f_d\rho(\mathbf{r}),\end{aligned} \tag{8.5}$$

where $\rho(\mathbf{r}, t)$ is the local number density of rain cells, $\overline{\rho}$ is its domain mean, the parameter p_0 describes the interaction strength between the CP gust fronts, and the

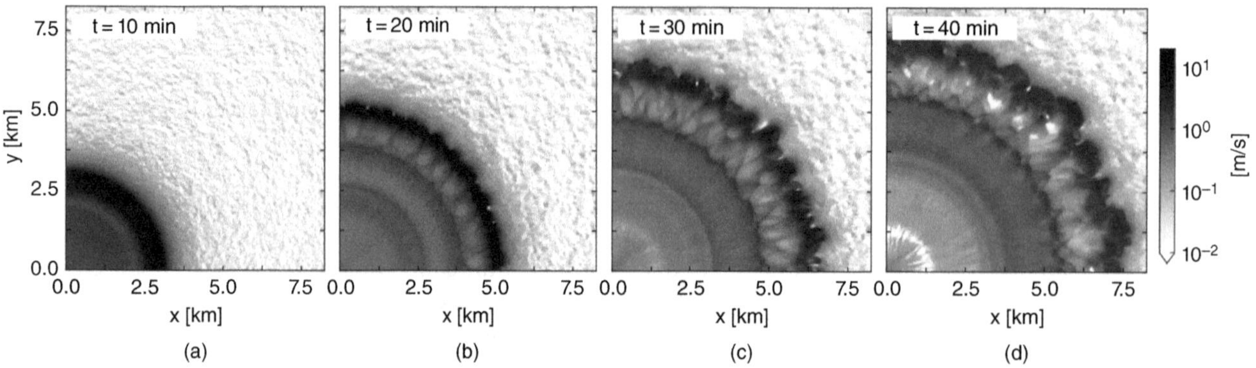

Figure 8.4 Internal structure of a simulated cold pool. Panels show the radial velocity at the lowest model level at several times (as marked in panels) during the cold pool expansion. The cold pool was initialized as a 3 K temperature reduction relative to the surrounding atmosphere, the numerical resolution was 25 m, and the linear domain size was 20 km horizontally. Note the lobe-and-cleft instabilities developing near the cold pool gust front. Source: Reproduced from Meyer and Haerter (2020) / John Wiley & Sons.

parameter f_{sg} allows for spontaneous generation of new rain cells, even without CP interaction. The factor $1 - \overline{\rho}$ accounts for the global energy constraint, whereby overall high activity feeds back negatively on the production of new rain cells, and $1 - \rho(\mathbf{r})$ models a local suppression effect through CPs. This term ensures that the negative buoyancy produced by existing CPs in the vicinity of $\mathbf{r}$ acts to suppress further activity locally. The factor $\rho(\mathbf{r})^2$ results from two rain cells being required within a given subregion, in order for a new rain cell to be triggered.

The spatial dynamics of $\rho(\mathbf{r})$ undergoes a form of gradual coarsening. In the steady state, an analytical approximation to equation (8.5) was presented (Haerter, 2019) and backed by simulations. A key parameter is the interaction strength p_0, and, as a function of this parameter, the model exhibits two continuous transitions between spatially homogeneous and segregated phases. The second parameter, f_{sg}, acts as a form of noise, where large f_{sg} diminishes the parameter range of p_0, where CSA occurs. For very large f_{sg} CSA is no longer possible. Recent work adds to this, by making the notion of circular CP spreading more explicit (Nissen & Haerter, 2020). It reaches the conclusion that CPs, which only grow to small maximal radii, facilitate the emergence of CSA.

Whereas the conceptual models by Craig and Mack (2013) and Emanuel et al. (2014) build entirely on free-tropospheric moisture and radiation feedbacks, and thus neglect CP effects, the approach in Haerter (2019) disregards larger scale circulation and radiation feedbacks, but emphasizes boundary-layer processes that act to redistribute moisture through CPs. One might speculate that processes at the CP scale are relevant in enabling or hampering the initial onset of CSA, but that larger scale radiation and circulation feedbacks dominate, once disturbances, such as dry patches, exceed the typical scale of individual CPs.

8.3.6. Interaction of cold pools and system scale circulations

Large-scale subsidence regions, which develop with the onset of evolution of CSA (Figs. 8.1 and 8.5a and equation (8.1), have recently been referred to as "radiatively driven dry pools" (Coppin & Bony, 2015; Zuidema et al., 2017), in contrast to the convective CPs under precipitation events. The role of circulations in convective aggregation is also discussed in Chapter 11: Clouds and Radiatively Induced Circulations. Common to convective CPs and radiatively driven dry pools, in a steady state approach the radial speed, C_{cp}, at which a pool spreads along the surface can be approximated by (Grandpeix & Lafore, 2010; Rotunno et al., 1988) $C_{cp} = k_*(2\mathcal{B})^{1/2}$ where $k_* = \mathcal{O}(1)$ is a parameter and $\mathcal{B} = -g\int_0^{h_{bl}} \frac{\theta_{vp}-\theta_{ve}}{\theta_{ve}} dz$ is the vertically integrated buoyancy anomaly resulting from the mean virtual potential temperatures within the pool, θ_{vp}, and outside of it, θ_{ve}, as well as the boundary layer height, h_{bl}. Note that both temperature and moisture anomalies (the so-called virtual effect, related to the smaller molecular mass of water vapor molecules compared to that of dry air molecules), contribute to the buoyancy anomaly. To analyze radiatively-driven dry pools, Coppin and Bony (2015) distinguish several contributing processes, such as longwave and shortwave radiational, convection, large-scale condensation, and vertical diffusion, and attribute the buildup of $\mathcal{B}$ mainly to the large longwave radiative cooling of the dry patch relative to the environment, in particular when low-level clouds are present.

Again, CP effects should be considered explicitly: Recent works specifically describe the stationary interface, which – in the final steady state of CSA – builds up between (1) a subsidence-associated dry pool and

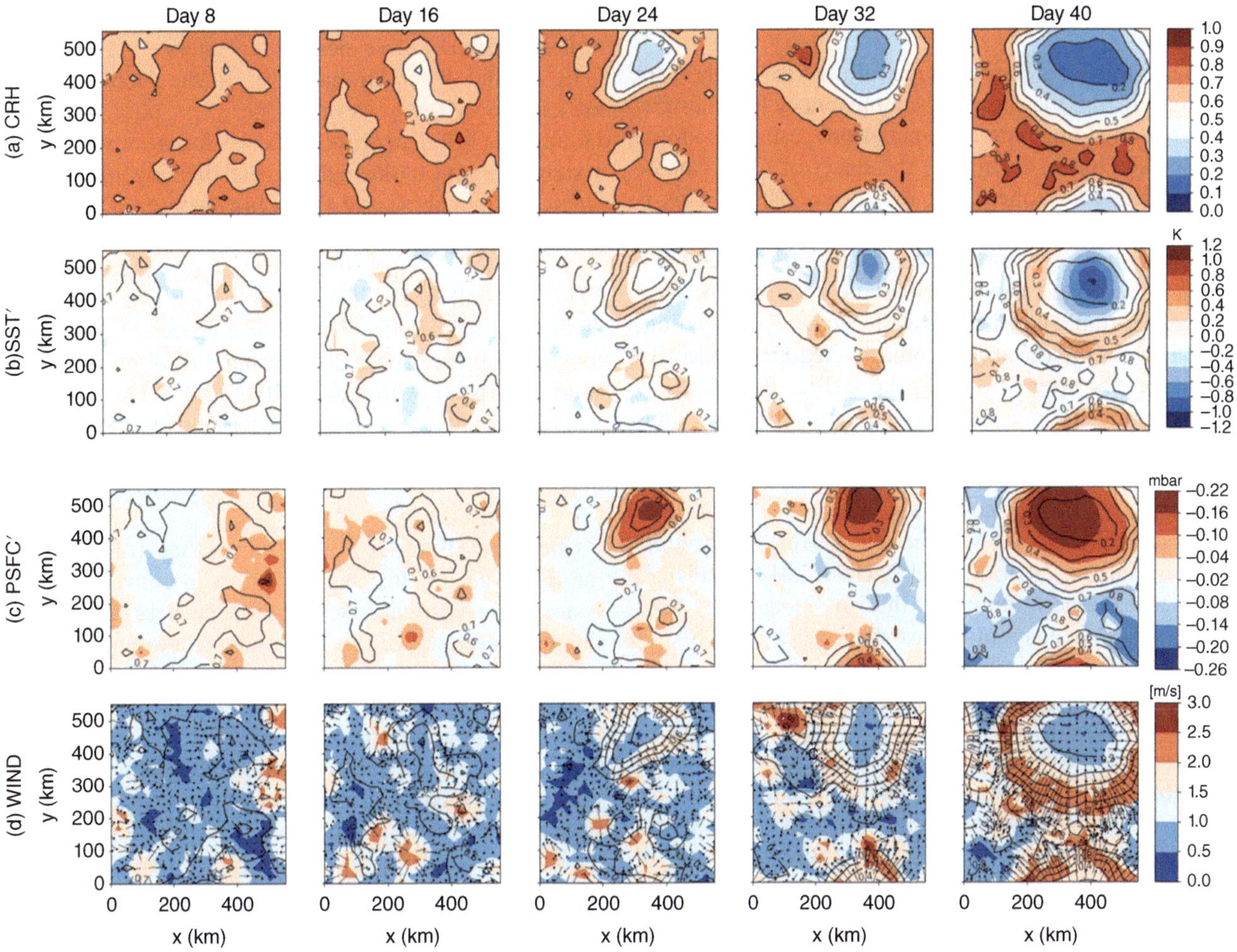

Figure 8.5 Aggregation with interactive SST. Panels show the column relative humidity (*CRH*), the SST anomaly (*SST′*), the surface pressure anomaly (*PSFC′*), and the wind speed (colors) and direction (arrows) at various stages of aggregation with interactive SST. The SST anomaly is at first positive in dry regions (a,b), thus reducing the high surface pressure (c) and concomitant low-level divergence (d), slowing CSA. But it eventually becomes positive (from day 24) and accelerates CSA. Source: Adapted from Shamekh et al. (2020a).

(2) a moist combined CP composed of many convective CPs (Windmiller & Hohenegger, 2019; Yanase et al., 2020). Windmiller and Hohenegger (2019) find that when the moist combined CP is large, convective rainfall is enhanced at the interface, an effect they refer to as *edge intensification* and attribute to the horizontal convergence between the two system-scale pools. They further estimate the velocity at which the combined CP expands and find that it approximately equals the opposing velocity. This result is supported by Yanase et al. (2020), who estimate the horizontal wind speed u_s of the combined CP as

$$u_s(r) = \frac{w_{sub}}{2 h_{bl}} r, \tag{8.6}$$

where r is the radius of the dry patch, and w_{sub} is subsidence wind speed within the dry patch. By inserting meaningful values of w_{sub} and h_{bl}, and equating u_s to typical wind speeds of convective CPs within the moist subdomain, they obtain a rough length scale $r \approx 500$ km, beyond which the dry patch might be assumed to persist.

Yanase et et al. also make contact to Jeevanjee and Romps (2013), who found that weakening convectively generated CPs facilitates the emergence of CSA. Yanase et al. (2020) support a reasoning, where convectively generated CPs would redistribute moisture but argue that a radiatively driven dry pool would counteract this effect. Yanase et al. (2020) follow this line of thought by coarsening resolution, confirming that this fosters CSA. They explain this effect by increased low-level cloudiness, which might increase outgoing longwave radiation. This link between increased low-level cloudiness, enhanced longwave cooling, and CSA is also consistent with results from Muller & Bony (2015).

8.3.7. Interactive surface temperatures

When surface temperatures are allowed to locally react to radiation and the state of the atmosphere, for example, by cloud-shielding effects, simulations using a simple slab surface have shown that the onset of CSA can be further delayed (Hohenegger & Stevens, 2016). This is because the surface temperature anomaly in dry regions is positive at first, thus opposing the high surface pressure and diverging shallow circulation known to favor self-aggregation, consistent with slower aggregation. But surprisingly, the driest columns acquire a negative SST anomaly after about 20 days (Fig. 8.5a,b), thus strengthening the pressure anomaly and diverging shallow circulation (Fig. 8.5c,d), and favoring aggregation (Shamekh et al., 2020a).

Hysteresis effects have been reported by several authors (Holloway & Woolnough, 2016; Khairoutdinov & Emanuel, 2010; Muller & Held, 2012; Muller & Bony, 2015). One finding was that in simulation setups that do not give spontaneous emergence of CSA, a self-aggregated state is nonetheless preserved when imposed as an initial condition. It might hence be argued that there could often be a sufficient degree of spatial inhomogeneity to induce moisture segregation, such as through local temperature anomalies (Shamekh et al., 2020b), and that the resulting variation of moisture would lead to CSA, sustainable even under then uniform boundary conditions.

8.4. TEMPORALLY VARYING SURFACE CONDITIONS AND THE DIURNAL CYCLE

Whereas the debate is not settled on the effect of changes in mean surface temperature on convective self-organization (Wing, 2019), profound spatiotemporal clustering appears to be consistently brought about by diurnal temporal variations in surface temperatures. Measured globally, satellite data analysis has shown that convection produced over continents has different characteristics from that over the sea, with continental regions typically featuring more frequent occurrence of mesoscale convective systems (MCS) (Nesbitt et al., 2006). Mapes et al. (2006) highlights the role of mesoscale clusters, suggesting that "the ultimate building block of deep convective weather is a multicellular cumulonimbus event." The observed amplitudes of oceanic surface temperatures can be as large as (Johnson et al., 1999; Weller & Anderson, 1996) 2–5 K (Kawai & Wada, 2007), and oscillations in so-called diurnal warm layers can persist for up to 5 days (Bellenger & Duvel, 2009), with a measurable impact on atmospheric properties (Bellenger & Duvel, 2009; Bellenger et al., 2010; Kawai & Wada, 2007). Chapter 9: An Overview of Mesoscale Convective Systems discusses climatology, observation, and model representation of these mesoscale clusters in greater detail.

In seminal observational work on the deep convective life cycle over the western Pacific warm pool, Chen and Houze Jr (1997) provide evidence for afternoon cloud system formation through diurnal heating of both the tropical atmosphere and the ocean surfaces. The authors acknowledge that the afternoon timing might be unexpected in light of previous work (Gray & Jacobson Jr, 1977; Randall & Dazlich, 1991), which argues solely in favor of radiative effects. Radiative processes would tend to stabilize the upper troposphere through afternoon solar absorption within clouds (Randall & Dazlich, 1991) and additional radiative cooling within cloud-free regions at night – resulting in additional subsidence there, much like the radiatively driven dry pools, discussed above for CSA at constant SST.

Chen and Houze Jr (1997) reason that, to understand the observed daytime buildup of deep convective cloud systems, the entire life cycle of these clusters must be considered. In pursuit of this hypothesis, they analyze hourly infrared satellite images at 10 km resolution, microwave soundings, and buoy surface measurements to capture cloud top temperature, precipitation rate, and surface data, respectively. The derived high cloudiness percentage shows a stark contrast between land and sea areas, with afternoon peaks over land but early morning peaks for the sea.

To better understand differences resulting from cloud clustering, the authors further track spatially extended cloud systems over time and condition on the spatial extent of these clusters. They find that, whereas small systems fall in line with the overall oceanic diurnal cycle of rainfall, larger systems follow aggregate diurnal timeseries with buildup of convection during the afternoon, maxima at night and decay during the morning of the subsequent day. Chen and Houze Jr (1997) find these large convective systems to have relatively long-lasting impacts on the thermal state of the boundary layer, with recovery times ranging between 9 and 18 hours (Parsons et al., 1994; Young et al., 1995), especially under weak winds. The resulting CPs, likely in combination with cloud shielding, act to measurably lower the near-surface surface temperature, documented from buoy measurements (Chen & Houze Jr, 1997; Zuidema et al., 2017). Chen and Houze Jr (1997) then further identify a spatiotemporal structure over the western Pacific as a "checkerboard pattern in time-longitude domain," and refer to the dynamics of individual mesoscale cloud clusters as "diurnal dancing." In diurnal dancing, one mesoscale system leaves behind a region of suppressed, unfavorable, conditions the subsequent day, but induces new mesoscale systems in the vicinity. Again one day later, the initial area has recovered and can again allow for a fresh mesoscale system to evolve. They find this dynamics to be measurable in a bi-diurnal cycle of cloud top and near-surface air temperature.

Recent idealized high-resolution simulations under RCE-like lower boundary conditions as well as with diurnally varying surface temperature conditions build on that notion by finding qualitatively distinct behavior for oceanic versus continental conditions (Haerter et al., 2020). In the former case, mimicked by a small-magnitude surface temperature forcing, dis-aggregated convection appears, with no apparent spatial structure beyond the scale of single convective CPs, typically $\mathcal{O}(10\,\text{km})$. In the latter, with larger magnitude of surface temperature forcing, MCSs, extending more than 100 km in the horizontal, self-organize at several locations within the domain, leaving other regions all but rainfree (Fig. 8.6). This process, termed "diurnal self-aggregation," is explained by the higher spatial number density of convective updrafts during the convectively active times within the diurnal cycle. Under such higher number density, the convective CPs under convective cells are found to merge, forming a combined CP of much larger spatial (and temporal) extent. This combined CP, then acting autonomously in instigating additional convection along its deep gust front, is capable of setting off a cascade of subsequent convective rain cells, again adding to the combined CP. In line with the observational results by Chen and Houze Jr (1997), the thermodynamic anomalies thus produced within the boundary layer through these combined CPs recover relatively slowly – suppressing further convection in the same spatial area during the subsequent day. However, activity then resumes in other locations, establishing a checkerboard-like dynamics, both in space and time (Fig. 8.6).

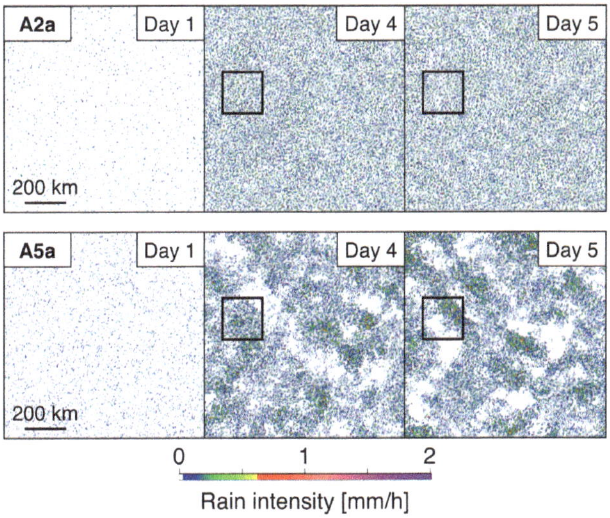

Figure 8.6 Diurnal self-aggregation. Comparison of two cloud-resolving simulations, with a 2 K temperature forcing amplitude (A2a, top row) compared to 5 K amplitude (A5a, bottom row). Note the pronounced spatial and temporal cloud organization for the latter case. To exemplify this, the bold squares highlight subregions of a 200 km horizontal extent, where in A5a a notable alternation occurs between overall rainfall and partially dry conditions. For A2a, there is no notable difference in organization from day to day, and the pattern remains featureless throughout the simulation domain. Source: Reproduced from (Haerter et al., 2020) / Springer Nature / CC BY 4.0.

More recently, Jensen et al. (2022) extended such diurnally forced simulations to incorporate a longer timeseries. They find that apart from the temporally alternating day-to-day pattern a persistent super-structure of dry patches emerges. These dry patches were attributed to the action of vigorous MCSs and their associated combined CPs. These strong combined CPs were thereby shown capable of clearing large-diameter $\mathcal{O}(100\,\text{km})$ areas from moisture and thereby suppressing subsequent convective activity. A resulting subsiding circulation over these already dry regions was shown to be sufficient to permanently drive moisture out toward already moist regions – thus establishing a CSA-like circulation and feedbacks commensurate with traditional CSA. Notably, the CSA-like effect disappeared, when horizontal model resolution was low (2 km) but strengthened, when resolution was increased, such as to 0.5 km. In this way, the establishment of CSA takes a different route than in the traditional case, where higher resolution and strong CPs were detrimental to the buildup of the segregated state (Jeevanjee & Romps, 2013). Further, the effect also disappeared, when surface temperature was held constant from the outset. Jensen et al. (2022) then explored, whether removing the temporal oscillation would cause the dry patches to disappear, once they were established through the diurnal oscillation in surface temperature. This is not the case: Once established, the dry patches persist, even when surface temperature is then set constant.

Whereas Emanuel et al. (2014) bring an interpretation forward, where the onset of CSA in RCE is caused by a linear moisture instability, Jensen et al. (2022) argue that for varying forcing, an abrupt, strong perturbation can instigate the initial dry anomaly, which can then become permanent. Whereas Emanuel et al. (2014) assume homogeneous boundary-layer conditions, those building up in Jensen et al. (2022) are strongly inhomogeneous and drive the subsequent route to persistent segregation.

8.4.1. The quest for the origin of observed convective clustering

The complexity of convective self-organization arises from the breadth of spatial and temporal scales involved, ranging from cloud microphysical processes and turbulent energy dissipation at the molecular scale to the near-planetary scale of the Madden-Julian Oscillation. It is intriguing to describe the apparent organization on the largest scales as *emergent* from the very small scales, that is, taking the viewpoint of a fully self-organized system. For this reason, many numerical studies on CSA will start with homogeneous initial conditions, which are only infinitesimally perturbed to break complete translational symmetry. In the real world, spatial inhomogeneity is always present and extended relatively dry or warm areas will always be found. It may therefore be more important to better understand the timescales at which self-organization takes place (Holloway et al., 2017; Wing et al., 2017) and initial perturbations amplify to grow to the size of poorly explained planetary-scale organizational patterns – and how organization, once present, can resist further perturbations, such as temporal fluctuations in boundary conditions.

Yet, so far such an approach has required substantial idealization and brought about a gap between simulations, which allow for such idealization, and observations – which come with a range of real-world complications. Even under the idealizations made, interpreting simulation results is far from straightforward, as in particular model grid resolution and the interaction with the surface pose unsolved questions: reality requires arbitrarily high grid resolution, yet, simulated constant-SST organization effects have often been found to be hampered by increased resolution – raising questions as to obtaining proper grid convergence. Especially boundary-layer moisture, temperature and momentum fluxes appear to be greatly influenced by grid resolution, and strongly feed back on convective organization.

Further, simulations often simplify the surface condition to be an infinite heat capacity ocean surface. Yet, given the lifetime of typical mesoscale convective clusters, both their direct effect on the surface temperature through cloud shielding as well as the resultant feedback on the cloud cluster are likely relevant. Additionally, with large-scale advection allowing cloud clusters to migrate from land to sea (or vice versa), constant surface conditions can hardly be considered sufficiently realistic. Already single tropical islands (Cronin et al., 2015) or spatial temperature anomalies (Shamekh et al., 2020b) were shown to profoundly or even sustainably affect the local climate and cloud clustering. Overall, actual tropical convective organization is likely a combination of deep convective updrafts within the Intertropical Convergence Zone (ITCZ) and emergent organization, which is driven through the feedbacks discussed above. More work is thus desirable to make quantitative comparisons between simulated and observed convective clustering, and to clarify the role of cold pools, as well as to grasp the implications for precipitation extremes and climate sensitivity.

ACKNOWLEDGMENTS

The authors thank the anonymous reviewers and Chris Holloway for useful comments on the manuscript. Jan O. Haerter gratefully acknowledges funding from the Villum Foundation (grant no. 13168) and from the European Research Council under the European Union's Horizon 2020 Research and Innovation programme (grant no. 771859) as well as the Novo Nordisk Foundation Interdisciplinary Synergy Program (grant no. NNF19OC0057374). He gratefully acknowledges high-performance computing resources from the German Climate Computing Center (DKRZ) as well as the Danish Climate Computing Center (DC3). Caroline Muller gratefully acknowledges funding from the European Research Council (ERC) under the European Union's Horizon 2020 research and innovation program (Project CLUSTER, grant agreement 805041). She also thanks Grand Equipement National de Calcul Intensif (GENCI), France, for providing access to their computing platforms at Tres Grand Centre de Calcul (TGCC).

REFERENCES

Abbot, D. S. (2014). Resolved snowball Earth clouds. *Journal of Climate*, *27*(12), 4391–4402.

Bak, P., Tang, C., & Wiesenfeld, K. (1987). Self-organized criticality: An explanation of the 1/f noise. *Physical Review Letters*, *59*(4), 381.

Bellenger, H., & Duvel, J.-P. (2009). An analysis of tropical ocean diurnal warm layers. *Journal of Climate*, *22*(13), 3629–3646.

Bellenger, H., Takayabu, Y., Ushiyama, T., & Yoneyama, K. (2010). Role of diurnal warm layers in the diurnal cycle of convection over the tropical Indian Ocean during MISMO. *Monthly Weather Review*, *138*(6), 2426–2433.

Böing, S. J. (2016). An object-based model for convective cold pool dynamics. *Mathematics of Climate and Weather Forecasting*, *2*(1), 43–60.

Böing, S. J., Jonker, H. J., Siebesma, A. P., & Grabowski, W. W. (2012). Influence of the subcloud layer on the development of a deep convective ensemble. *Journal of the Atmospheric Sciences*, *69*(9), 2682–2698. doi: https://doi.org/10.1175/JAS-D-11-0317.1.

Bony, S., Semie, A., Kramer, R., Soden, B., Tompkins, A., & Emanuel, K. (2020). Observed modulation of the tropical radiation budget by deep convective organization and lower-tropospheric stability. *AGU Advances*, *1*(3), e2019AV000155.

Bony, S., Stevens, B., Frierson, D. M. W., Jakob, C., Kageyama, M., Pincus, R., et al. (2015). Clouds, circulation and climate sensitivity. *Nature Geoscience*, *8*(4), 261–268.

Bony, S., Stevens, B., Coppin, D., Becker, T., Reed, K. A., Voigt, A., et al. (2016). Thermodynamic control of anvil cloud amount. *Proceedings of the National Academy of Sciences (USA)*.

Bray, A. J. (2002). Theory of phase-ordering kinetics. *Advances in Physics*, *51*(2), 481–587.

Bretherton, C. S., Blossey, P. N., & Khairoutdinov, M. (2005). An energy-balance analysis of deep convective self-aggregation above uniform SST. *Journal of the Atmospheric Sciences*, *62*(12), 4273–4292.

Bretherton, C. S., Peters, M. E., & Back, L. E. (2004). Relationships between water vapor path and precipitation over the tropical oceans. *Journal of Climate*, *17*(7), 1517–1528.

Chandra, A. S., Zuidema, P., Krueger, S., Kochanski, A., de Szoeke, S. P., & Zhang, J. (2018).Moisture distributions in tropical cold pools from equatorial Indian Ocean observations and cloud-resolving simulations. *Journal of Geophysical Research: Atmospheres*, *123*(20), 11–445.

Chen, S. S., & Houze Jr, R. A. (1997). Diurnal variation and life-cycle of deep convective systems over the tropical Pacific warm pool. *Quarterly Journal of the Royal Meteorological Society*, *123*(538), 357–388.

Christensen, K., & Moloney, N. R. (2005). *Complexity and criticality*, (Vol. 1). World Scientific Publishing Company.

Coppin, D., & Bony, S. (2015). Physical mechanisms controlling the initiation of convective self-aggregation in a general circulation model. *Journal of Advances in Modeling Earth Systems*, *7*. doi:10.1002/2015MS000571.

Craig, G. C., & Mack, J. M. (2013). A coarsening model for self-organization of tropical convection. *Journal of Geophysical Research*, *118*(16), 8761–8769.

Cronin, T. W., Emanuel, K. A., & Molnar, P. (2015). Island precipitation enhancement and the diurnal cycle in radiative-convective equilibrium. *Quarterly Journal of the Royal Meteorological Society*, *141*(689), 1017–1034.

Dai, N., & Soden, B. J. (2020). Convective aggregation and the amplification of tropical precipitation extremes. *AGU Advances*, *1*(4), e2020AV000201.

de Szoeke, S. P., Skyllingstad, E. D., Zuidema, P., & Chandra, A. S. (2017). Cold pools and their influence on the tropical marine boundary layer. *Journal of the Atmospheric Sciences*, *74*(4), 1149–1168.

Droegemeier, K., & Wilhelmson, R. (1985). Three-dimensional numerical modeling of convection produced by interacting thunderstorm outflows. Part I: Control simulation and low-level moisture variations. *Journal of the Atmospheric Sciences*, *42*(22), 2381–2403. doi: 10.1175/1520-0469(1985)042⟨2381: TDNMOC⟩2.0.CO;2

Droegemeier, K., & Wilhelmson, R. (1987). Numerical simulation of thunderstorm outflow dynamics. Part I: Outflow sensitivity experiments and turbulence dynamics. *Journal of the Atmospheric Sciences*, *44*(8), 1180–1210. doi: 10.1175/1520-0469(1987)044⟨1180:NSOTOD⟩2.0CO;2

Emanuel, K. A., Wing, A. A., & Vincent, E. M. (2014). Radiative-convective instability. *Journal of Advances in Modeling Earth Systems*, *6*(1), 75–90.

Fairall, C. W., Bradley, E. F., Hare, J., Grachev, A. A., & Edson, J. B. (2003). Bulk parameterization of air–sea fluxes: Updates and verification for the COARE algorithm. *Journal of Climate*, *16*(4), 571–591.

Futyan, J. M., & Del Genio, A. D. (2007). Deep convective system evolution over Africa and the tropical Atlantic. *Journal of Climate*, *20*(20), 5041–5060.

Glassmeier, F., & Feingold, G. (2017). Network approach to patterns in stratocumulus clouds. *Proceedings of the National Academy of Sciences*, *114*(40), 10578–10583.

Grandpeix, J.-Y., & Lafore, J.-P. (2010). A density current parameterization coupled with Emanuel's convection scheme. Part I: The models. *Journal of the Atmospheric Sciences.*, *67*(4), 881–897.

Gray, W. M., & Jacobson Jr, R. W. (1977). Diurnal variation of deep cumulus convection. *Monthly Weather Review*, *105*(9), 1171–1188.

Haerter, J. O. (2019). Convective self-aggregation as a cold pool-driven critical phenomenon. *Geophysical Research Letters*, *46*(7), 4017–4028. doi: https://doi.org/10.1029/2018GL081817.

Haerter, J. O., Böing, S. J., Henneberg, O., & Nissen, S. B. (2019). Circling in on convective organization. *Geophysical Research Letters*, *46*(12), 7024–7034. doi: https://doi.org/10.1029/2019GL082092.

Haerter, J. O., Meyer, B., & Nissen, S. B. (2020). Diurnal self-aggregation. *NPJ Climate and Atmospheric Science*, *3*, 30.

Haerter, J. O., & Schlemmer, L. (2018). Intensified cold pool dynamics under stronger surface heating. *Geophysical Research Letters*, *45*(12), 6299–6310. doi: https://doi.org/10.1029/2017GL076874.

Held, I. M., Hemler, R. S., & Ramaswamy, V. (1993). Radiative-convective equilibrium with explicit two-dimensional moist convection. *Journal of the Atmospheric Sciences*, *50*(23), 3909–3909.

Hohenegger, C., & Stevens, B. (2016). Coupled radiative convective equilibrium simulations with explicit and parameterized convection. *Journal of Advances in Modeling Earth Systems*, *8*. DOI:10.1002/2016MS000666.

Holloway, C. E. (2017). Convective aggregation in realistic convective-scale simulations. *Journal of Advances in Modeling Earth Systems*, *9*(2), 1450–1472.

Holloway, C. E., Wing, A. A., Bony, S., Muller, C. J., Masunaga, H., L'Ecuyer, T. S., & et al. (2017). Observing convective aggregation. *Surveys in Geophysics*, *38*(6), 1199–1236.

Holloway, C. E., & Woolnough, S. J. (2016). The sensitivity of convective aggregation to diabatic processes in idealized radiative-convective equilibrium simulations. *Journal of Advances in Modeling Earth Systems*, *8*(1), 166–195.

Jakob, C., Singh, M., & Jungandreas, L. (2019). Radiative convective equilibrium and organized convection: An observational perspective. *Journal of Geophysical Research: Atmospheres*, *124*(10), 5418–5430.

Jeevanjee, N., & Romps, D. M. (2013). Convective self-aggregation, cold pools, and domain size. *Geophysical Research Letters*, *40*(5), 994–998.

Jensen, G. G., Fiévet, R., & Haerter, J. O. (2022). The diurnal path to persistent convective self-aggregation. *Journal of Advances in Modeling Earth Systems (JAMES)*, *14*(5):e2021MS002923.

Johnson, R. H., Rickenbach, T. M., Rutledge, S. A., Ciesielski, P. E., & Schubert, W. H. (1999). Trimodal characteristics of tropical convection. *Journal of Climate*, *12*(8), 2397–2418.

Kawai, Y., & Wada, A. (2007). Diurnal sea surface temperature variation and its impact on the atmosphere and ocean:

A review. *Journal of Oceanography*, *63*(5), 721–744. doi: https://doi.org/10.1007/s10872-007-0063-0.

Khairoutdinov, M. F., & Emanuel, K. A. (2010). Aggregated convection and the regulation of tropical climate. In *29th Conference on Hurricanes and Tropical Meteorology*. Arizona

Koren, I., & Feingold, G. (2013). Adaptive behavior of marine cellular clouds. *Scientific Reports*, *3*, 2507.

Krishnamurti, R. (1975). On cellular cloud patterns. Part 1: Mathematical model. *Journal of the Atmospheric Sciences*, *32*(7), 1353–1363.

Langhans, W., Yeo, K., & Romps, D. M. (2015). Lagrangian investigation of the precipitation efficiency of convective clouds. *Journal of the Atmospheric Sciences*, *72*(3), 1045–1062.

Manabe, S., & Strickler, R. F. (1964). Thermal equilibrium of the atmosphere with a convective adjustment. *Journal of the Atmospheric Sciences*, *21*(4), 361–385.

Mapes, B. E., & Houze Jr, R. A. (1993). Cloud clusters and superclusters over the oceanic warm pool. *Monthly Weather Review*, *121*(5), 1398–1416.

Mapes, B. E., Milliff, R., & Morzel, J. (2009). Composite life cycle of maritime tropical mesoscale convective systems in scatterometer and microwave satellite observations. *Journal of the Atmospheric Sciences*, *66*(1), 199–208.

Mapes, B. E., Tulichb, S., Linb, J., & Zuidemaa, P. (2006). The mesoscale convection life cycle: Building block or prototype for large-scale tropical waves? *Dynamics of Atmospheres and Oceans*, *42*, 3–29.

Meyer, B., & Haerter, J. O. (2020). Mechanical forcing of convection by cold pools: Collisions and energy scaling. *Journal of Advances in Modeling Earth Systems*, e2020MS002281.

Miller, D., & Fritsch, J. (1991). Mesoscale convective complexes in the western Pacific region. *Monthly Weather Review*, *119*(12), 2978–2992.

Moncrieff, M. W., & Liu, C. (1999, Oct). Convection initiation by density currents: Role of convergence, shear, and dynamical organization. *Monthly Weather Review*, *127*, 2455–2464.

Muller, C. J., & Bony, S. (2015). What favors convective aggregation and why? *Geophysical Research Letters*, *42*, 5626–5634.

Muller, C. J., & Held, I. M. (2012). Detailed investigation of the self-aggregation of convection in cloud-resolving simulations. *Journal of the Atmospheric Sciences*, *69*, 2551–2565.

Muller, C. J., & Takayabu, Y. (2020). Response of precipitation extremes to warming: What have we learned from theory and idealized cloud-resolving simulations, and what remains to be learned? *Environmental Research Letters*, *15*(3), 035001.

Müller, S. K., & Hohenegger, C. (2020). Self-aggregation of convection in spatially varying sea surface temperatures. *Journal of Advances in Modeling Earth Systems*, *12*(1), e2019MS001698.

Nesbitt, S. W., Cifelli, R., & Rutledge, S. A. (2006). Storm morphology and rainfall characteristics of trmm precipitation features. *Monthly Weather Review*, *134*(10), 2702–2721.

Nesbitt, S. W., Zipser, E. J., & Cecil, D. J. (2000). A census of precipitation features in the tropics using TRMM: Radar, ice scattering, and lightning observations. *Journal of Climate*, *13*(23), 4087–4106.

Nissen, S. B., & Haerter, J. O. (2020). How weakened cold pools open for convective self-aggregation, *EGU General Assembly 2021*, online, 19–30 Apr 2021, EGU21-1520.

Parsons, D., Dabberdt, W., Cole, H., Hock, T., Martin, C., Barrett, A.-L., et al. (1994). The integrated sounding system: Description and preliminary observations from TOGA COARE. *Bulletin of the American Meteorological Societ*, *75*(4), 553–568.

Peters, O., & Neelin, J. D. (2006). Critical phenomena in atmospheric precipitation. *Nature Physics*, *2*(6), 393.

Peters, O., Neelin, J. D., & Nesbitt, S. W. (2009). Mesoscale convective systems and critical clusters. *Journal of the Atmospheric Sciences*, *66*(9), 2913–2924.

Randall, D. A., & Dazlich, D. A. (1991). Diurnal variability of the hydrologic cycle in a general circulation model. *Journal of the Atmospheric Sciences*, *48*(1), 40–62.

Rennó, N. O., Emanuel, K. A., & Stone, P. H. (1994). Radiative-convective model with an explicit hydrologic cycle: 1. Formulation and sensitivity to model parameters. *Journal of Geophysical Research: Atmospheres*, *99*(D7), 14429–14441.

Rooney, G. G. (2015). Descent and spread of negatively buoyant thermals. *Journal of Fluid Mechanics*, *780*, 457–479. doi: 10.1017/jfm.2015.484

Rooney, G. G. (2018, Oct). Similarity-based approximations for the evolution of a gravity current. *Quarterly Journal of the Royal Meteorological Society*, *144*(716), 2302–2310. doi: 10.1002/qj.3371

Rotunno, R., Klemp, J. B., & Weisman, M. L. (1988, Feb). A theory for strong, long-lived squall lines. *Journal of the Atmospheric Sciences*, *45*(3), 463–485. doi: 10.1175/1520-0469(1988)045⟨0463:ATFSLL⟩2.0.CO;2

Schlemmer, L., & Hohenegger, C. (2016). Modifications of the atmospheric moisture field as a result of cold-pool dynamics. *Quarterly Journal of the Royal Meteorological Society*, *142*(694), 30–42.

Schumacher, C., & Houze Jr, R. A. (2003). Stratiform rain in the tropics as seen by the TRMM precipitation radar. *Journal of Climate*, *16*(11), 1739–1756.

Semie, A. G., & Bony, S. (2020). Relationship between precipitation extremes and convective organization inferred from satellite observations. *Geophysical Research Letters*, *47*(9), e2019GL086927.

Shamekh, S., Muller, C. J., Duvel, J.-P., & D'Andrea, F. (2020a). Self-aggregation of convective clouds with interactive sea surface temperature. *Journal of Advances in Modeling Earth Systems*, e2020MS002164.

Shamekh, S., Muller, C. J., Duvel, J.-P., & d'Andrea, F. (2020b). How do ocean warm anomalies favor the aggregation of deep convective clouds? *Journal of the Atmospheric Sciences*, *77*(11), 3733–3745.

Simpson, J. (1980, Apr). Downdrafts as linkages in dynamic cumulus seeding effects. *Journal of Applied Meteorology*, *19*(4), 477–487. doi: 10.1175/1520-0450(1980)019⟨0477:DALIDC⟩2.0.CO;2

Sobel, A. H., Nilsson, J., & Polvani, L. M. (2001). The weak temperature gradient approximation and balanced tropical moisture waves. *Journal of the Atmospheric Sciences*, *58*(23), 3650–3665.

Tan, J., Jakob, C., Rossow, W. B., & Tselioudis, G. (2015). Increases in tropical rainfall driven by changes in frequency of organized deep convection. *Nature*, *519*(7544), 451–454. doi: https://doi.org/10.1038/nature14339.

Tobin, I., Bony, S., & Roca, R. (2012). Observational evidence for relationships between the degree of aggregation of deep convection, water vapor, surface fluxes, and radiation. *Journal of Climate*, *25*(20), 6885–6904.

Tompkins, A. M. (2001). Organization of tropical convection in low vertical wind shears: The role of cold pools. *Journal of the Atmospheric Sciences*, *58*, 1650–1672.

Tompkins, A. M., & Craig, G. C. (1998). Radiative–convective equilibrium in a three-dimensional cloud-ensemble model. *Quarterly Journal of the Royal Meteorological Society*, *124*(550), 2073–2097. doi: https://doi.org/10.1002/qj.49712455013.

Torri, G., Kuang, Z., & Tian, Y. (2015, Feb). Mechanisms for convection triggering by cold pools. *Geophysical Research Letters*, *42*(6), 1943–1950. doi: 10.1002/2015GL063227

Tselioudis, G., Tromeur, E., Rossow, W. B., & Zerefos, C. (2010). Decadal changes in tropical convection suggest effects on stratospheric water vapor. *Geophysical Research Letters*, *37*(14).

Turner, J. S. (1979). *Buoyancy effects in fluids*. Cambridge university press.

Velasco, I., & Fritsch, J. M. (1987). Mesoscale convective complexes in the Americas. *Journal of Geophysical Research: Atmospheres*, *92*(D8), 9591–9613.

Weller, R., & Anderson, S. (1996). Surface meteorology and air-sea fluxes in the western equatorial Pacific warm pool during the TOGA coupled oceanatmosphere response experiment. *Journal of Climate*, *9*(8), 1959–1990.

Windmiller, J. M. (2017). Organization of tropical convection (Unpublished doctoral dissertation). lmu.

Windmiller, J. M., & Hohenegger, C. (2019). Convection on the edge. *Journal of Advances in Modeling Earth Systems*, *11*(12), 3959–3972.

Wing, A. A. (2019). Self-aggregation of deep convection and its implications for climate. *Current Climate Change Reports*, *5*(1), 1–11.

Wing, A. A., & Cronin, T. W. (2016). Self-aggregation of convection in long channel geometry. *Quarterly Journal of the Royal Meteorological Society*. doi:10.1002/qj.2628.

Wing, A. A., Emanuel, K., Holloway, C., & Muller, C. J. (2017). Convective self-aggregation in numerical simulations: A review. *Surveys in Geophysics*, *142*(694), 1–15. doi: 10.1007/978-3-319-77273-8_1.

Wing, A. A., & Emanuel, K. A. (2014). Physical mechanisms controlling self-aggregation of convection in idealized numerical modeling simulations. *Journal of Advances in Modeling Earth Systems*. *5*, doi:10.1002/2013MS000269.

Wing, A. A., Stauffer, C. L., Becker, T., Reed, K. A., Ahn, M.-S., Arnold, N. P., et al. (2020). Clouds and convective self-aggregation in a multimodel ensemble of radiative-convective equilibrium simulations. *Journal of Advances in Modeling Earth Systems*, *12*(9), e2020MS002138.

Yanase, T., Nishizawa, S., Miura, H., Takemi, T., & Tomita, H. (2020). New critical length for the onset of self-aggregation of moist convection. *Geophysical Research Letters*, *47*(16), e2020GL088763.

Young, G. S., Perugini, S. M., & Fairall, C. (1995). Convective wakes in the equatorial western Pacific during TOGA. *Monthly Weather Review*, *123*(1), 110–123.

Zuidema, P. (2003). Convective clouds over the Bay of Bengal. *Monthly Weather Review*, *131*(5), 780–798.

Zuidema, P., Torri, G., Muller, C. J., & Chandra, A. (2017). A survey of precipitation-induced atmospheric cold pools over oceans and their interactions with the larger-scale environment. *Surveys in Geophysics*, *38*(6), 1283–1305.

9

An Overview of Mesoscale Convective Systems: Global Climatology, Satellite Observations, and Modeling Strategies

Sudip Chakraborty[1,2], Sylvia C. Sullivan[3,4], and Zhe Feng[5]

ABSTRACT

Deep convection is responsible for redistributing water as well as energy and providing vital water resources for many regions of the world. Cumulonimbus clouds aggregation into a single storm system develop mesoscale convective systems (MCS). MCSs have precipitation covering a horizontal region on the scale of 100 km or more. After more than the last seven decades, MCS has been gradually received wide attention in the scientific community as an important component of the Earth's hydrologic cycle and energy balance. MCSs differ from ordinary deep convection because of the mesoscale circulations consisting of a layer of air overturning on a scale much larger than those created by individual convective updrafts or downdrafts. Moreover, a key feature that separates MCS from ordinary deep convection is the significant portion of stratiform precipitation that results in the production and detrainment of a large number of ice particles from their organized convective cores and continues the mesoscale ascent that supports ice growth and aggregation as the crystals fall. This significant stratiform precipitation gives rise to top-heavy latent heating profiles that have substantial impacts on the upper-level global circulations, especially for those MCS that have a larger portion of stratiform rainfall. Top-heavy latent heating profiles are more frequent over the ocean than over the land. These unique features of MCS suggest that they are important for the hydrologic cycle, general circulation, and radiative balance of the climate system. In this chapter, we focus on the recent advances in the climatological perspective of MCS as well as the research tools used to better understand MCS after the advent of the satellite era. We will discuss both tropical and midlatitude MCS, the influence of modes of tropical climate variability on MCS, and the new insight from satellite observations on MCS interaction with the land surface.

9.1. AN OVERVIEW OF MESOSCALE CONVECTIVE SYSTEMS

Deep convection is a critical component of our Earth system, redistributing atmospheric moisture and energy and providing vital water resources for many regions of the world. The largest form of deep convective storms, known as mesoscale convective systems (MCSs), develops when cumulonimbus clouds aggregate, producing precipitation over regions of 100 km or more (Houze, 2004, 2018). Through research spanning 70 years, MCSs have emerged as a crucial component of the Earth's hydrologic

[1]Institute for Harnessing Data and Model Revolution in the Polar Regions, Department of Information Systems, University of Maryland, Baltimore County, Baltimore, MD, USA

[2]NASA Jet Propulsion Laboratory, California Institute of Technology, Pasadena, CA, USA

[3]Department of Chemical and Environmental Engineering, and Department of Hydrology & Atmospheric Science, University of Arizona, Tucson, AZ, USA

[4]Institute for Meteorology and Climate Research, Karlsruhe Institute of Technology, Karlsruhe, Germany

[5]Atmospheric Sciences and Global Change Division, Pacific Northwest National Laboratory, Richland, WA, USA

Clouds and Their Climatic Impacts: Radiation, Circulation, and Precipitation, Geophysical Monograph 281, First Edition.
Edited by Sylvia C. Sullivan and Corinna Hoose.

DOI: 10.1002/9781119700357.ch09

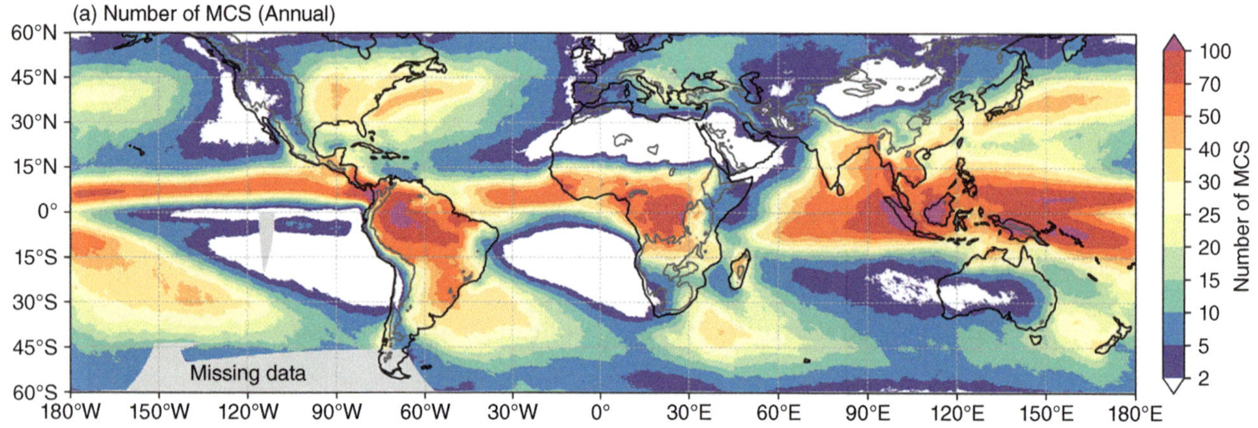

Figure 9.1 Climatology of annual MCS occurrence as in Feng et al. (2021a) (their Fig. 10a). Data are taken from 2001 to 2019. Source: Adapted from Feng et al. (2021).

cycle and energy balance. MCSs occur globally, especially along the Intertropical Convergence Zone (ITCZ), over the Tropical Warm Pool and Bay of Bengal, and Amazon but also off eastern seaboards throughout the midlatitudes (Fig. 9.1). An interesting historical perspective of MCS research can be found in Houze (2018).

Several elements distinguish MCSs from ordinary deep convection. Mesoscale circulations exist within MCSs, consisting of a layer of air overturning on a scale much larger than those created by individual convective updrafts or downdrafts (Houze, 2018; Moncrieff, 1992). Such circulations may arise as a gravity wave response to the latent heating associated with the aggregated convection (Pandya & Durran, 1996). MCSs produce more weak, stratiform precipitation than ordinary convection, as a result of ice detrainment from convective cores and the mesoscale ascent supporting ice growth and aggregation (Schumacher & Houze, 2003). This stratiform precipitation in MCSs results in top-heavy latent heating profiles relative to the bottom-heavy heating profiles characterizing ordinary convection (Liu et al., 2015). Such top-heavy latent heating profiles have substantial impacts on the upper level global circulation (Schumacher et al., 2004). When tropical deep convection is more aggregated, that is, developing into MCSs or clusters of multiple MCSs, the surrounding free troposphere also becomes drier, possibly related to a larger compensating subsidence and strengthening of the overturning circulations (Bony et al., 2020). Lastly, wind shear influences MCS evolution via convective updraft and downdraft structures, vorticity distribution, rainfall evaporation, and cold pool formation (Fritsch et al., 1994; Moncrieff, 1992; Rotunno et al., 1988; Weisman & Rotunno, 2004).

A distinct life cycle also characterizes MCSs relative to ordinary deep convection. This life cycle is generally understood in three phases (e.g., Houze, 1982; Roca et al., 2017): (1) initiation of isolated convective cores; (2) aggregation and upscale growth leading to detrainment of condensate into stratiform anvils; and (3) dissipation as condensate evaporates, sublimates, or sediments. These phases correspond to particular changes to the vertical depth, horizontal extent, and precipitation amount and intensity of the MCSs. For example, the initial formation of cores is associated with the minimum cloud top brightness temperature and the highest rain rates (e.g., Kondo et al., 2006). Various linear or polynomial growth-decay models have been proposed to describe these structural changes over time, using the time of maximum MCS extent (e.g., Feng et al., 2012; Roca et al., 2017). Due to this more complicated evolution, the duration and cold cloud shield extent of MCSs both tend to be much larger than that of ordinary deep convection. We note that mechanisms and metric of convective organization in the tropics are outlined in Chapter 8: Mechanisms for the Self-Organization of Tropical Deep Convection.

A number of reviews exist already, emphasizing the historical context or the meteorological aspects of MCSs (e.g., Houze, 2018; Markowski & Richardson, 2010; Schumacher & Rasmussen, 2020). In this chapter, we focus on the climatological perspective of MCSs and the research tools to understand their interactions with the environments, broadly construed. We discuss interactions of MCSs with modes of tropical climate variability, midlatitude synoptic conditions, the land surface, and atmospheric aerosol, as well as the effect of these factors on MCS characteristics (e.g., intensity and duration). The chapter presents both tropical and midlatitude MCS climatologies (sections 9.2 and 9.3), recent advances in satellite observations of MCSs that provide insight into their interactions with the land surface and atmospheric aerosols (section 9.4), and a discussion of numerical simulations of MCSs (section 9.5).

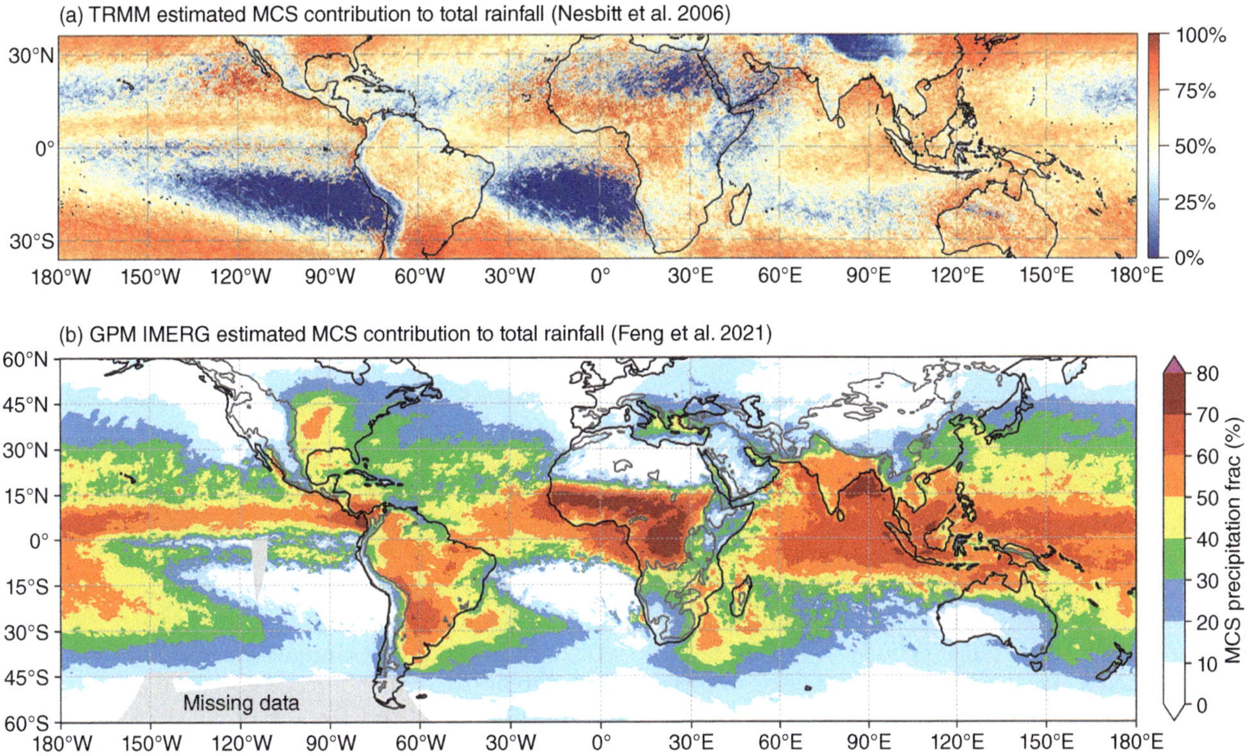

Figure 9.2 Climatology of the MCS contribution to total rainfall. Top panel shows rainfall statistics from 1997 to 2014 as in Schumacher and Rasmussen (2020) *Nature Reviews* (their Fig. 2a). Bottom panel shows rainfall statistics from 2001 to 2019 as in Feng et al. (2021a) (their Fig. 10c). We note the relatively good agreement in this MCS-associated precipitation from two different observational products in the tropics. Source: (a) Schumacher et al. (2020) / Springer Nature. (b) Adapted from Feng et al. (2021).

9.2. MESOSCALE CONVECTIVE SYSTEMS IN THE TROPICS

MCSs are commonly observed in the tropics, especially over the Indo-Pacific warm pool, the ITCZ, tropical Africa, and the Amazon (Huang et al., 2018; Laing & Fritsch, 1997; Yuan & Houze, 2010). Satellite precipitation measurements, particularly from the Tropical Rainfall Measurement Mission (TRMM), estimate that MCSs produce half of tropical precipitation on average and more than 80% in certain regions and seasons over land (Fig. 9.2, Nesbitt et al., 2006; Yuan & Houze, 2010). The clustering of deep convection associated with MCS formation also generates extreme precipitation events in the tropics with long-lived MCSs playing a particularly important role (Latos et al., 2021; Semie & Bony, 2020) (see also Chapter 14: Observed Scaling of Precipitation Extremes). Convective tracking data in combination with an ensemble of precipitation products indicate that MCSs with duration greater than 24 hours occur on almost half of the days with 250 mm of precipitation or more over tropical land (Roca & Fiolleau, 2020).

9.2.1. Interaction of MCSs with the El Niño Southern Oscillation

The El Niño Southern Oscillation (ENSO) refers to the periodic warming of sea surface temperatures (SSTs) off the coast of Peru and Ecuador in a region called the warm tongue (e.g., Trenberth, 1997). Climatologies from geostationary satellite data show a concentration of MCS occurrence over this warm tongue during El Niño periods, as convective initiation migrates toward these warmest SSTs (Sullivan et al., 2019; Zolman et al., 2000). Eastern Pacific warming reduces the zonal SST gradient and overlying sea-level pressure (SLP) gradient and easterlies. This ocean-atmosphere coupling through SST and SLP gradients is called the Bjerknes feedback and results in MCSs with larger maximum horizontal extent and vertical depth (quantified by cloud top brightness temperatures) over the central Pacific during El Niño periods (Fig. 9.3, Henderson et al., 2018; Sullivan et al., 2019). In contrast, in the tropics-wide trend, MCS maximum extent decreases with increasing SSTs, in agreement with burgeoning evidence that anvil coverage decreases with warming (e.g., Bony et al., 2020; Saint-Lu et al., 2020; Su et al., 2017; Wing

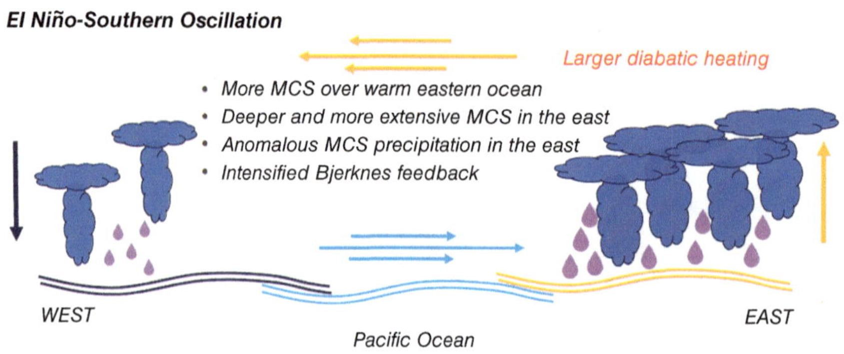

Figure 9.3 A schematic of the ENSO effects on MCS, as well as their feedbacks onto ENSO anomalies in sea surface warming and wind speeds.

et al., 2020). Over the full life cycle, MCS vertical depth also tends to be insensitive to anomalies in SST in line with the Fixed-Anvil Temperature hypothesis (Hartmann & Larson, 2002).

Different metrics exist to classify El Niño periods and the countervailing La Niña periods in which eastern Pacific SSTs cool. The Oceanic Niño Index uses only the SST anomaly between 5°S and 5°N and between 120°W and 170°W, whereas the Multivariate ENSO Index incorporates SST, SLP, zonal and meridional surface winds, and outgoing longwave radiation (OLR) in a larger tropical domain between 30°S and 30°N and 100°E and 70°W. The horizontal spreading and vertical deepening of MCSs over the central Pacific during El Niño are quantitatively larger when using the Multivariate ENSO Index, as the use of more environmental variables omits transient El Niños (Sullivan et al., 2019).

While the surface warming and shifting circulation of El Niño affect the occurrence and structure of MCSs, the diabatic heating within MCSs can also feed back on El Niño events. Increased occurrence of MCSs over the warm tongue during El Niño increases the upper tropospheric latent and radiative heating in the eastern Pacific. Several changes to circulation occur with upper level heating: (1) vertical velocities intensify and shift upward; (2) zonal circulation tilts vertically; and (3) the upper level anticyclonic flow over southeastern Asia weakens and expands out over the central Pacific (Schumacher et al., 2004). Mid-level and upper level cloud heating is associated with an increase in the Bjerknes feedback and El Niño variance (Rädel et al., 2016). Shortwave reflection by clouds also appears to be decisive in setting ENSO frequency, as it permits stronger upper oceanic heat discharge (Middlemas et al., 2019). Although MCSs have not been explicitly isolated in these studies, satellite retrievals indicate that tropical diabatic heating is dominated by that of MCSs (Liu et al., 2021; Takayabu & Tao, 2020).

Another important and directly felt component of ENSO-MCS coupling is alterations to surface precipitation. Indeed, most previous work on MCSs and El Niño has focused on these rainfall shifts. The deeper ocean mixed layer and increased convection over the eastern Pacific during El Niño promote disequilibrium between the sea surface and the overlying boundary layer, which in turn fuels precipitation (Chiang & Sobel, 2002). A dipole of precipitation increases over the central and eastern Pacific and decreases over the western Pacific results during El Niño (Trenberth, 1997). More precipitation also tends to occur over the western Indian Ocean and northern Atlantic, and less over the Amazon. Again, climatologies indicate that the majority of this tropical precipitation comes from MCSs (Fig. 9.2, Nesbitt et al., 2006; Yuan & Houze, 2010).

A few studies have also focused explicitly on MCS rainfall during El Niño. With deviations of the passive and active microwave measurements in the TRMM as a starting point, Henderson et al. (2017) found more stratiform precipitation associated with organized convection during El Niño relative to La Niña. Also with TRMM data, Liu et al. (2019) found increased MCS precipitation over the Gulf of Mexico and Argentina during El Niño with stronger response in boreal winter. With 25 years of convective tracking data from the International Satellite Cloud Climatology Project (ISCCP) and collocated rain rates from the Multi-Source Weighted-Ensemble Precipitation, Sullivan et al. (2020) showed how the frequency of heavy rainfall ($>10\,\text{mm h}^{-1}$) from the deepest MCSs increases twofold in El Niño versus La Niña (Fig. 9.4). This intensification of extreme precipitation can be explained via increases in ascent, and hence condensation rate over the central Pacific. Otherwise, regional observations indicate more frequent quasi-linear MCSs and positive rainfall anomalies over the continental United States (Anderson & Arritt, 2001), increased eastern but suppressed western Sahelian precipitation (Parhi et al., 2016), and Amazonian drought but flooding in Argentina and Uruguay (Cai et al., 2020; Liu et al., 2019) during El Niño.

Finally, El Niño comes in different flavors. Along with the canonical east-Pacific type discussed above, a Modoki

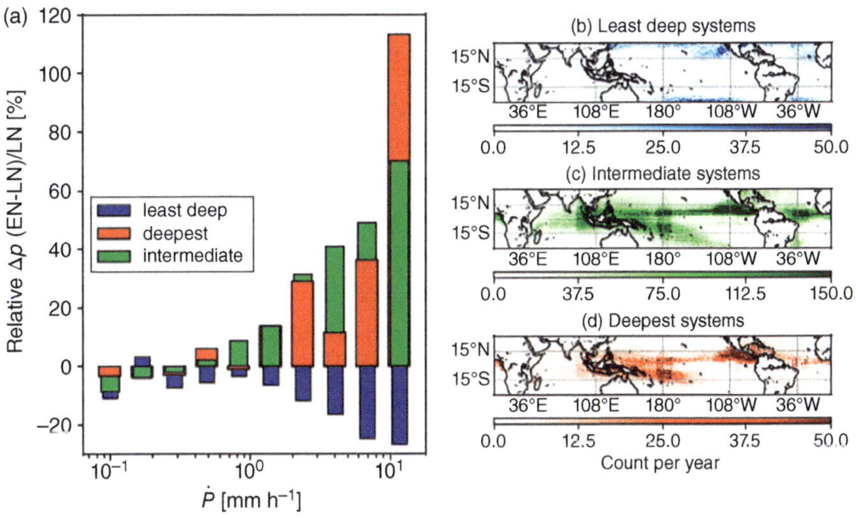

Figure 9.4 Relative changes in the probability of precipitation intensities between El Niño and La Niña (panel a). Climatology of MCS of different depth from the ISCCP convective tracking database (panels b–d). Source: Sullivan et al. (2020) / John Wiley & Sons.

central-Pacific type El Niño can form in which the warm SST anomalies are limited to the central Pacific with colder SST to both the east and west (Ashok et al., 2007). Studies have looked at how precipitation in different regions changes with Modoki-type El Niño relative to a canonical one. For example, Feng et al. (2011a) found that El Niño in its decaying phase brings increased rainfall to southern China, while El Niño Modoki has no effect during its decaying phase. Reduced precipitation over southern India, as well as northwestern and northern Australia, has also been documented during El Niño Modoki (Dandi et al., 2020; Taschetto & England, 2009). But the more general effect of different El Niño flavors on MCS occurrence, structure, and life cycle has not been explored to our knowledge.

9.2.2. Interactions of MCSs with Monsoons

Seasonal temperature changes in the tropics are far less pronounced than in the extratropics. Instead, the wet and dry season contrast dominates in monsoon systems that form in regions with significant topography and land-sea contrasts. The strongest of these is the Asian monsoon, a system that affects India, southeastern Asia, and much of eastern China, where more than half of the global population lives. Monsoon systems also exist over the Sahel in West Africa and to a lesser extent over Malaysia and northern Australia, eastern South America, and northwestern Mexico, and the Southwest United States.

The switch between monsoon wet and dry seasons corresponds to zonal migrations of the ITCZ, the region where meridional circulations in the Northern and Southern Hemisphere converge. But the representation of monsoons, as more regional phenomena, also requires that we take zonal asymmetries into account (Biasutti et al., 2018). An important source of such energetic zonal asymmetry is the diabatic heating associated with MCSs embedded in the migrating ITCZ. The lofting and condensation of water vapor within the convective cores of MCSs generates a net latent heating in the atmosphere when this condensed water falls out as precipitation, and the vertical distribution of this latent heating sets the dynamical response (Chiang et al., 2001; Hartmann et al., 1984; Schumacher et al., 2004). Work with both climate and cloud-resolving models (CRMs) indicates the importance of MCS latent heating to generate monsoon depressions over the Bay of Bengal (Choudhury & Krishnan, 2011; Fujinami et al., 2020), whereas precipitation radar observations from TRMM indicate the large interannual and vertical variability in this latent heating around the Himalayas (Magagi & Barros, 2004). A recent study using convection-permitting simulations showed that MCSs greatly contribute to the abrupt increase of rainfall and diabatic moist static energy sources during the south Asian monsoon onset and dominate the convection-associated positive feedbacks on diabatic energy enhancements (Chen et al., 2021c). While condensation heating dominates the diabatic heating from MCS, followed by riming heating and deposition heating in the upper troposphere (Fan et al., 2017), convective systems also influence the radiative budget, particularly through their extensive ice cloud outflow. Aqua-planet simulations show that ultraviolet reflection by deep cloud systems can accelerate the monsoon onset by 2 weeks, whereas infrared absorption by the same cloud systems can warm and moisten the monsoon core and intensify its flow (Byrne & Zanna, 2020). MCSs are both embedded within and affect the evolution of the monsoon flow.

Along with their diabatic heating, MCSs can also affect large-scale monsoon flow through momentum transport, primarily in the vertical. Indeed, compared to regular deep convection, the vertical transport of total mass and energy by MCSs over the south Asian summer monsoon region is at least 1.5 times stronger throughout the troposphere, and MCSs occur ~1.6 times more frequently during the active monsoon phase than the break phase, suggesting the crucial role of MCSs in the Asian monsoon overturning circulation and energetics (Chen et al., 2021b). Deriving the vorticity equation from the horizontal momentum budget, MCS diabatic heating appears as a first "vortex stretching" term that amplifies vorticity in regions of convergence, and a second momentum transport term that generates vorticity couplets at upper altitudes (Mapes & Houze, 1992). The "vortex stretching" from localized low-level convergence near MCSs can accelerate development of the monsoon circulation. Within a second set of momentum transport terms are the mean momentum fluxes, and also the eddy momentum fluxes that modulate the circulation shifts defining monsoons (Bordoni & Schneider, 2008). As more complex flow structures than simple upright convective plumes, MCSs have non-negligible eddy tendencies that can act as a "deep convective friction" to slow the shear flow speed (Mapes & Wu, 2001). But complicating these effects further, different MCS structures generate very different damping rates (Mapes & Wu, 2001).

Along with their role in the monsoon energy and momentum budgets, the most intense MCSs in the world, as defined by cold cloud top temperatures, high reflectivity aloft, and lightning flash rate, occur primarily in continental regions, including those affected by monsoons like the Himalayan foothills, the Sahel, and northern Australia (Zipser et al., 2006). The seasonality of this storm intensity also follows that of the monsoon. For example, deep and icy convective tops migrate northward from the Guinea-Congolian region over the Sahelian region between boreal spring and boreal summer. During the Asian monsoon, the deepest and broadest storms form near the southern ridge of the Himalayas, where moisture from the Bay of Bengal encounters strong topographical uplift but weak environmental shear (Houze et al., 2007; Zipser et al., 2006). MCSs generally have larger horizontal extent relative to isolated convection prior to the monsoon, but during this pre-monsoon period, deep storms with high reflectivity aloft can still form along the eastern coast of India and Bangladesh, where depressions from the Bay of Bengal often migrate (Virts & Houze, 2016). The seasonality of storm intensity is less pronounced over South America relative to Africa or Asia, but MCSs still tend to be deepest and have the highest lightning flash rates after monsoon onset in September, October, and November (Chen et al., 2021c; Kodama et al., 2005).

Many of the monsoon regions are also hotspots of MCS-associated precipitation. MCSs bring 70% of annual Sahelian rainfall and roughly a third of extreme precipitation events in East China (Nesbitt et al., 2006; Zhang et al., 2019). Synoptic conditions that facilitate moisture advection over land surfaces can generally be used to forecast extreme monsoonal rainfall from MCSs a week or more ahead of time. For example, high geopotential heights over the Tibetan Plateau and warm Eastern European temperatures were identified as driving factors in multiple flooding events over Pakistan between 2010 and 2012 (Rasmussen et al., 2015). Recent work with convection-permitting models (CPMs) also indicates that extreme rainfall from MCSs over Africa and Asia should intensify in coming years (Fitzpatrick et al., 2020; Schumacher & Rasmussen, 2020). Indeed, even coarse-resolution simulations of the Coupled Model Intercomparison Project (CMIP) show that monsoon rainfall over land will increase by 2.8% per degree Celsius (Wang et al., 2020a).

9.2.3. Interactions of MCSs with the Madden-Julian Oscillation

The Madden-Julian Oscillation (MJO) is an eastward-propagating dipole of enhanced and suppressed precipitation that generally last between 30 and 60 days and dominates intra-seasonal tropical variability. The MJO typically initiates over the western Indian Ocean, and its eastward movement around 5 m s^{-1} can be tracked by an area of reduced OLR associated with enhanced convective activity (Fig. 9.5). Eight phases characterize MJO progression from the western Indian Ocean over the west Pacific. Along with its hydrological and radiative signatures, the MJO is associated with strong surface westerlies at its western edge, as convergence fuels the region of increased precipitation. The production of regions of strong ascent by the MJO means it may facilitate tropical cyclogenesis in some instances, even remotely. Maloney and Hartmann (2000), for example, find a fourfold increase in hurricane occurrence over the Caribbean and the Gulf of Mexico when the MJO produces east Pacific low-level westerlies. The MJO is also characterized by strong seasonality, peaking in boreal winter and spring over the Indian and western Pacific Oceans and in boreal summer over the eastern Pacific (Zhang & Dong, 2004).

MCSs and the MJO are closely intertwined. The internal flow of an MCS consists of a deep, ascending front-to-rear inflow and an underlying descending rear inflow, and an analogous kind of layered overturning exists at a scale 10 times larger in the MJO (Houze, 2018;

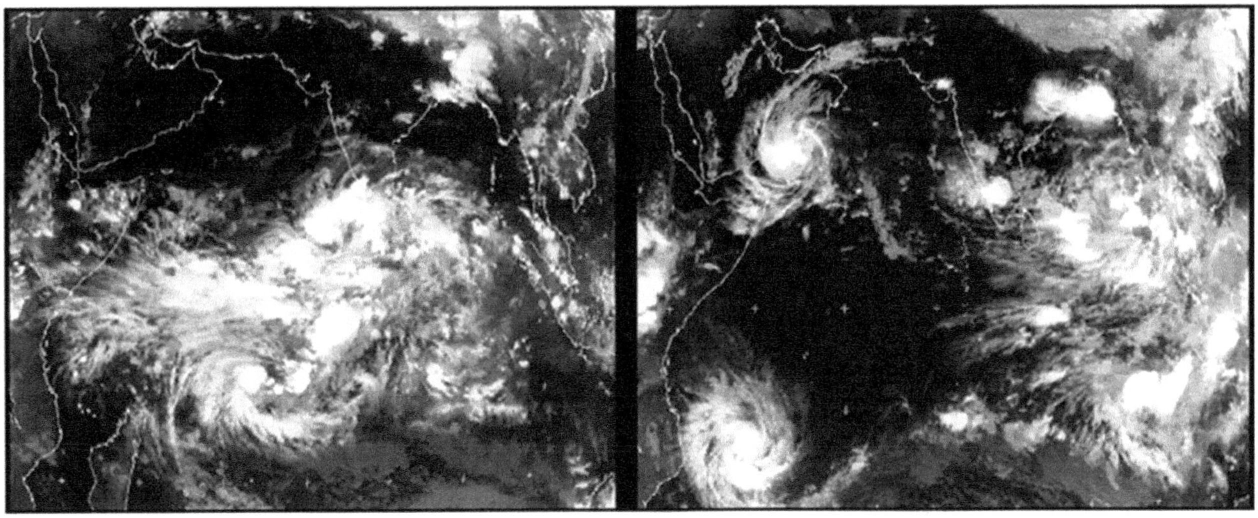

Figure 9.5 Infrared images of organized convection taken by the METEOSAT-5 satellite operated by EUMETSAT on 2 May 2002 on the left and 9 May 2002 on the right. The eastward propagation of MCSs over the Indian Ocean within the MJO can be seen distinctly. In the right panel, tropical cyclones can be seen over Indonesia. Source: Shapiro et.al. (2010) / American Meteorological Society.

Moncrieff, 2004). The MCS population in a given region changes as the MJO evolves through its eight phases; for example, more organized storms with larger horizontal extent and fewer isolated storms form over the Indian Ocean in the early MJO phases (Fig. 9.4, Barnes & Houze, 2013). Similar evolution occurs in the MCS-associated latent heating with MJO phase. Once the MJO has propagated eastward away from the Indian Ocean, upper level heating there is suppressed due to reduction of stratiform heating from MCSs (Barnes et al., 2015).

The wave nature of the MJO also suggests that we could understand it as a superposition of higher frequency convective activity, particularly MCS. Mapes et al. (2006) hypothesized that a "stretched building-block" conceptual model with changing proportion of shallow, deep isolated, and deep organized convective elements during different phases of large-scale waves could represent phenomena like the MJO. More recently, Toms et al. (2020) illustrated how mesoscale element morphology evolves over the MJO in cloud-resolving simulations, from anomalously long-lived convective elements initially to anomalously broad elements in later phases. These kinds of MJO decompositions imply that poor climate model representation of the MJO could be attributed to unresolved MCS (Yang et al., 2019). Indeed, recent work incorporating a parameterization of MCS, called the Multiscale Coherent Structure Parameterization (MCSP), into an Earth system model minimizes precipitation biases, particularly over the tropical west Pacific and improves the propagation speed of precipitation and zonal wind anomalies associated with the MJO (Chen et al., 2021a; Moncrieff et al., 2017).

A particularly important contribution of observational data on MCSs and the MJO came from the Dynamics of the Madden-Julian Oscillation (DYNAMO), an international field campaign over the Indian Ocean in 2011. Studies using the DYNAMO data or simulations during the field campaign period have given insight into phenomena as diverse as the diurnal cycle of precipitation during different phases of the MJO (Ruppert & Johnson, 2015), the importance of intersecting cold pools to organize and enhance triggering of secondary convection during the MJO initiation (Feng et al., 2015), and the influence of surface fluxes on convective organization (Sakaeda et al., 2018). Use of a dual-polarization S-band Doppler radar during the campaign also allowed the study of microphysical structure within MCSs during MJO initiation (Barnes & Houze, 2013; Rowe & Houze, 2014). With the ability to probe these smallest scales, we are a step closer to representing the complex multiscale interactions of organized tropical convection, as well as their coupling to modes of climate variability like the MJO (Moncrieff, 2004).

The modes of tropical climate variability discussed above also strongly interact. For example, a strong MJO may accelerate the onset of a strong ENSO by establishing a negative SST anomaly over the west Pacific during its early phases (Bergman et al., 2001), or the weakened descending branch of the Walker Cell over the west Pacific and Southeast Asia can weaken Asian monsoon rainfall (Slingo & Annamalai, 2000). Studies examining how MCSs affect and are affected by such mode interactions have been limited. Future research should examine such coupling for a more complete understanding of multiscale tropical climate variability.

9.3. MESOSCALE CONVECTIVE SYSTEMS OVER THE MIDLATITUDES

Many previous studies examining MCS climatology are focused on the tropics and use the mesoscale dimension of cloud systems or precipitation features to identify MCSs in satellite measurements (e.g., Houze et al., 2015; Huang et al., 2018; Nesbitt et al., 2006; Roca et al., 2014; Yuan & Houze, 2010). While this method works well in the tropics where MCSs are associated with convective clustering, applying such methodology in the midlatitudes, particularly beyond the summer season when baroclinic forcing becomes stronger, is problematic. Midlatitude synoptic systems such as fronts and extratropical cyclones often produce large cloud systems and rain bands that exceed the mesoscale dimension, especially during transitional and cool seasons. MCSs can be embedded in these synoptic systems when sufficient moisture, convective instability, and persistent lifting mechanisms exist. Along with these distinct morphological differences, midlatitude MCSs are also distinguished from tropical ones by their strong seasonality. Most previous regional studies of midlatitude MCSs focus on the warm season (Chen et al., 2019; Machado et al., 1998; Morel & Senesi, 2002; Yang et al., 2015). Although midlatitude MCSs are more frequent during the warm season, they can occur year-round under certain favorable large-scale environments, as demonstrated by ground-based radar observations in the United States (Feng et al., 2019; Geerts, 1998; Haberlie & Ashley, 2019; Rickenbach et al., 2015). Future work should be devoted to elucidating the full seasonal variability of midlatitude MCSs and linking this variability to midlatitude circulations and hydroclimate.

As more sophisticated, multi-datastream methods have emerged to identify midlatitude MCSs, their preferential occurrence in certain regions has become clear. Over the midlatitude oceans, MCSs occur frequently offshore of the east coasts of North America, South America, East Asia, and South Africa (Feng et al., 2021a). Over land, their occurrence is favored in prominent baroclinic zones downstream of major mountain ranges, such as the Rocky Mountains in the United States, the Andes in South America, Ethiopian Highlands in Africa, and Tibetan Plateau in Asia (Laing & Fritsch, 1997; Laing & Fritsch, 2000). MCSs also occur primarily over land in Europe between May and August, and their triggering is strongly correlated with orography as maximum MCS initiation is found near all mountain ranges (Morel & Senesi, 2002).

The central United States is another important and well-studied hotspot of midlatitude MCS occurrence. The advent of the Next Generation radar (NEXRAD) operated by the National Weather Service has contributed greatly to MCS studies over the United States in the past two decades. With the nearly continuous coverage of the NEXRAD network and long-term records collected since 1997, extensive climatology can be constructed to characterize MCS behavior over the central United States. Recent studies that leverage automated feature tracking techniques applied to the long-term NEXRAD data reveal that MCS frequency and precipitation is largest in the Great Plains and Midwest during the warm season, while cool-season occurrences peak in the southeast (Feng et al., 2019; Haberlie & Ashley, 2019). These NEXRAD-based MCS records also show that large-scale environments across the seasons strongly impact the characteristics of MCSs. Feng et al. (2019) found that transitional season (spring/fall) MCSs commonly initiate under strong baroclinic forcing, with the Great Plains Low-level Jet (GPLLJ) transporting large anomalous moisture into the region, and low-level convergence ahead of the mid-level trough helping to initiate MCSs. Spring MCSs in the United States have large and deep convective features, large stratiform rainfall area, and high volumetric rainfall. In contrast, summer MCSs often occur under or ahead of a high-pressure ridge, with high low-level humidity and weak low-level convergence (Song et al., 2019). Under synoptically unfavorable environments in summer, propagating sub-synoptic scale disturbances, as shown in multiple meteorological fields including potential vorticity, surface pressure, and equivalent potential temperature upstream, seem to provide favorable mesoscale environments to support MCS initiation (Song et al., 2021). Different from spring, summer MCSs have the largest and deepest convective features but smallest stratiform rain area and lowest total rainfall volume (TRV) compared to other seasons (Feng et al., 2019).

Upscale growth of deep convection also frequently results in MCS formation in subtropical South America, particularly east of the Andes and the central plains of southeastern South America. The large-scale environments supporting MCS genesis bear significant resemblance to those in the United States. Several essential ingredients have been identified in previous studies, including (1) upper level synoptic disturbances traversing the Andes, providing dynamical support through upper level divergence associated with the jet stream (Romatschke & Houze, 2010; Salio et al., 2007); (2) lee cyclogenesis inducing a pressure gradient that enhances the South American Low-level Jet (SALLJ), transporting anomalous moisture and convective instability from the Amazon closer to the foothills of the Andes (Rasmussen

& Houze, 2016); and (3) interaction between the SALLJ and steep mountains east of the Andes such as the Sierras de Córdoba, providing triggering mechanisms for deep convection and upscale growth (Mulholland et al., 2020; Romatschke & Houze, 2010; Rasmussen & Houze, 2016; Zhang et al., 2021). Similar to the United States, MCS initiation in subtropical South America also exhibits a strong diurnal cycle during the warm season (October to March), peaking in late afternoon to early evening, while the mature stage spans the nocturnal to early morning hours (Cancelada et al., 2020; Salio et al., 2007).

Warm season MCSs, particularly over the United States and subtropical South America, can produce a variety of hazardous weather, including extreme precipitation and floods (Doswell et al., 1996; Kunkel et al., 2012; Rasmussen et al., 2016), wind damage (Ashley & Mote, 2005; Ashley et al., 2019), hail (Kumjian et al., 2020; Smith et al., 2012), lightning (Cecil & Blankenship, 2012), and even tornadoes (Haberlie et al., 2021; Smith et al., 2012). Compared to non-MCS rainfall, MCS rainfall in the central United States during the warm season (April to August) is about seven times more intense, but it occurs less frequently in time and space (Hu et al., 2020a). Extreme precipitation in this region is more likely to occur from MCSs than ordinary deep convection or non-convective rain, both because of their higher rainfall rates and their more extensive precipitating area (Prein et al., 2017a). In the eastern two thirds of the United States, over half of the daily extreme precipitation events (defined by 50- or 100-year return periods) are associated with MCSs (Schumacher & Johnson, 2006; Stevenson & Schumacher, 2014), although over the Mid-Atlantic region east of the Appalachian Mountains, isolated deep convection also contributes significantly to summer extreme rainfall (Li et al., 2021). While MCSs are responsible for the majority of the warm season extreme precipitation events, the relationship between MCSs and flooding is more complex due to factors such as terrestrial storage, watershed characteristics, and human interventions (Kundzewicz et al., 2014; Stephens et al., 2015). By linking a long-term multi-sensor MCS tracking database (Feng, 2019) and the Storm Events database of flooding events, Hu et al. (2021b) found that warm season MCSs account for the majority of slow-rising and hybrid (mixture of flash and slow-rising) floods, while non-MCSs explain about half of flash floods in July and August.

High-resolution observational records of precipitation show that long-lived and intense MCSs have become more frequent in the central United States during spring over the past several decades (Feng et al., 2016). MCSs also explain the observed increases in frequency and amount of extreme spring rainfall in this region, due to statistically significant increases in their lifetime and associated rainfall intensity. These trends are driven by surface warming over the Rocky Mountains that enhances the pressure gradient across the central United States, strengthening the GPLLJ and moisture transport that moisten the inland region. These findings have significant implications for future extreme rainfall occurrence in this region and season and in other regions frequented by MCSs, because climate model projections suggest that the GPLLJ will intensify and extend northward in spring and autumn (Zhou et al., 2021), enhancing the low-level moisture flux in a warming climate (Cook et al., 2008; Harding & Snyder, 2014). While convection-resolving simulations can capture these increases in MCS precipitation rate and duration, they underestimate the increase in frequency of occurrence (Prein et al., 2017a).

9.4. SATELLITE OBSERVATIONS OF MESOSCALE CONVECTIVE SYSTEMS

Over the last few decades, advanced sensors onboard various satellites have made it possible to study different aspects of MCSs nearly globally. Satellite observations have been particularly helpful in elucidating the multiscale interactions of MCSs with their environments. For example, cloud microphysics at the smallest scales both affects and is impacted by MCS structure, and remote sensing measurements reveal the diabatic heating and microphysical properties that underlie this two-way coupling. The role of soil moisture on MCSs constitutes another example of interactions; higher soil moisture can amplify or suppress further MCS rainfall but this soil moisture is also dictated by the preceding rainfall. We discuss how satellite observations are helping us unravel these interactions at micro-, meso-, and macroscales in the next three subsections. We note also that Chapter 15: Satellite Precipitation Measurements focuses specifically on remote sensing of the full array of precipitating systems.

9.4.1. Remote Sensing of Mesoscale Convective System-Microphysical Interactions

In their convective region, MCSs are constituted of cloud droplets and raindrops with sizes from a few microns up to 0.5 mm in diameter (Hodson, 1986; Marshall & Palmer, 1948). Ice crystals also make up the upper regions of MCSs, with effective sizes spanning a few microns up to a few millimeters. These ice crystals tend to be much larger in the convective region due to aggregation and rimming than in the more quiescent stratiform layer (Heymsfield et al., 2013). In the most intense MCSs over land, large hail and graupel also form in the convective updrafts. A variety of remote sensing measurements is necessary to observe this wide range of hydrometeor types and sizes in MCSs. For example, CloudSat, a 94 GHz W band radar can measure reflectivity, vertical cloud structure, as well as infer cloud microphysical properties; TRMM and its successor, the Global Precipitation Measurement (GPM), contain 13.8 GHz radars to measure 3-D structures of precipitating clouds and rainfall; and the Micro-Limb Sounder onboard the Aura satellite, a microwave radiometer, measures the thermal emission from broad spectral bands near 118–640 and 2,500 GHz and provides information about upper level ice properties in storm systems.

Aerosol particles are generally required to form the cloud droplets and ice crystals within MCSs, and the interactions of these aerosols with MCSs via cloud microphysics remain highly uncertain (Fan et al., 2016; Rosenfeld, Andreae, et al., 2014) (see also Chapter 2: An Overview of Aerosol-Cloud Interactions). Satellite measurements from the Cloud-Aerosol Lidar and Infrared Pathfinder Satellite Observations (CALIPSO) and CloudSat provide closely collocated aerosol and cloud datasets to try to untangle such aerosol-cloud interactions (ACI). These data confirm certain indirect effects of aerosols on convection, for example, a reduction in the effective size and an enhancement in the number concentration of cloud particles that can delay precipitation (e.g., Chakraborty et al., 2016; Fan et al., 2016; Rosenfeld et al., 2008, 2013, 2014, 2016; Stevens & Feingold, 2009; Seinfeld et al., 2016). Correlations between the aerosol optical depth (AOD) or other aerosol proxies and MCS convective precipitation, latent heat release, cloud ice water content, horizontal extent, and convective lifetime may also be inferred from the satellite record, especially during MCS decay.

But inference of MCS-aerosol interactions from remote sensing data faces a number of challenges. Causality behind correlations of aerosol proxies and MCS properties cannot be established from remote sensing alone. Passive satellite sensors often are unable to see through clouds, and hence cannot provide information on overlapping layers of aerosol and cloud. Satellites cannot provide chemical information about cloud condensation nuclei or ice-nucleating particles, the subsets of aerosol on which cloud droplets and ice crystals form (Rosenfeld, Andreae, et al., 2014; Rosenfeld, Sherwood, et al., 2014). Information about aerosol properties near the boundary of clouds is also unreliable (Tackett & Di Girolamo, 2009; Várnai & Marshak, 2009). Given these limitations, a multi-platform approach integrating satellite measurements with ground-based and in situ observations will be necessary to make further progress on MCS-microphysics and -aerosol interactions.

9.4.2. Remote Sensing of the Mesoscale Convective System Life Cycle

The increasing availability of remote sensing data has provided insights into MCS initiation, growth, propagation, morphology, as well as their latent heating and precipitation characteristics (e.g., Chakraborty et al., 2018a; Chakraborty et al., 2020; Dolores et al., 2019, Fan et al., 2013; Feng et al., 2021a; Jackson et al., 2009; Schumacher & Houze, 2003; Taylor et al., 2010; Wang et al., 2000; Yuan & Houze, 2010). The generation of a global, homogenized long-term infrared dataset from geostationary satellites of various meteorological agencies has been crucial to this progress (Fiolleau et al., 2020). As a single example, from ISCCP radiance measurements, MCS tracks have been inferred over 35 years between 50°N and 50°S (Machado et al., 1998). Many other studies have been conducted using this ISCCP convective tracking dataset in combination with other data like that of TRMM, GPM, and sensors on board the A-Train constellation (Chakraborty et al., 2016; Sullivan et al., 2020). Indeed, the fusion of the infrared data of low-Earth orbiting geostationary satellites or ground-based observations of radiative, microphysical, and environmental properties has also contributed greatly to understanding MCS evolution (e.g., Futyan & Del Genio 2007; Feng et al., 2011b, Feng et al., 2012; Fiolleau & Roca 2013; Vant-Hull et al., 2016).

A recent study developed a novel approach to use high-resolution geostationary satellite infrared images collocated with satellite-retrieved surface precipitation features to identify and track MCSs (Feng, et al., 2021a). The new method improves upon previous detection algorithms for midlatitude MCSs for all seasons and delivers MCS statistics comparable to ground-based radar observations. This long-term high-resolution MCS tracking database provides the first global picture of MCS occurrence over all seasons, and numerous climatologically relevant conclusions can be drawn. For example, the database confirms that MCSs account for over 50% of annual total rainfall across most of the tropical belt and reveal their frequencies and contributions to rainfall in the midlatitudes for the first time (Fig. 9.2). A strong seasonality in MCS characteristics and rainfall is found over many regions of the globe. MCSs with the longest lifetime are primarily concentrated over subtropical oceans (Feng et al., 2021a). Continental MCSs are convectively more intense with higher cloud tops, but oceanic MCSs have much higher rainfall production. By combining this MCS tracking database with GPM satellite retrievals, Liu et al. (2021) examined the global distribution and seasonal variation of MCS latent heating profiles. They found that midlatitude MCS latent heating profiles are top-heavier than tropical ones, indicating greater stratiform heating in the midlatitudes. In line with these latent heating results, the majority of midlatitude MCSs have relatively high (>70%) stratiform fraction, in contrast to the higher convective rain fraction in tropical MCSs. These results come from a single tracking dataset, and many such datasets are now emerging with different identification algorithms, durations, and coverages (e.g., Chen et al., 2019; Cancelada et al., 2020; Fiolleau & Roca, 2013; Hayden et al., 2021; Núñez Ocasio et al., 2020). A more detailed review of these datasets and tracking tools would be an asset to the MCS research community.

Satellite measurements also provide insight into environmental conditions and macroscopic properties associated with MCS. Moisture convergence and entrainment mixing significantly modulate MCS updraft speed and precipitation efficiency, and recent observational advances allow us to be more quantitative about how environmental moisture affects MCS lifetime and rainfall. The Atmospheric Infrared Sounder and Tropospheric Emission Spectrometer (TES), in combination with various reanalysis products, have provided valuable information about the ambient meteorological conditions surrounding MCS. Some studies evaluated the changes in the convective lifetime and rainfall amount for MCSs over the Congo and Amazon rainforests and Indian subcontinent. For example, recent research using TRMM and ISCCP data along with MERRA-2 reanalysis data evaluates change in convective lifetime and rainfall for MCSs over the Congo and Amazon rainforests and Indian subcontinent. Increases in the lower (free) tropospheric relative humidity (RH) by one standard deviation can increase MCS lifetime by 3–21 (3–9) hours (Chakraborty et al., 2016). RH also strongly influences TRV over this MCS lifetime: when RH exceeds 80% or total precipitable water surrounding MCSs exceeds 60 mm, TRV exceeds 12 km^3, whereas TRV in a drier environment (RH < 40% and TPW < 30 mm) is less than 1 km^3 (Chakraborty et al., 2018b). They have also observed that more than 90% of those MCSs form and spend a significant amount of their mean convective lifetime (~30 and ~15 hours for MCSs over the Congo and Amazon rainforest, respectively) over land; however, a significant number of MCSs over the Indian subcontinent originate over the ocean (~60%) and spend a major part of their mean lifetime over the ocean (~80 hours).

9.4.3. Remote Sensing of Mesoscale Convective System-Land Surface Interactions

While the interaction with ambient meteorological conditions has been intensively studied, interactions between the land surface and MCSs have received less attention, especially from an observational perspective. The coupling between land and atmosphere involves parameters like land surface type, soil moisture, surface temperature, and vegetation types. Previous research has demonstrated that these above-mentioned land surface parameters influence energy fluxes in the boundary layer, destabilize it, and make it suitable for convective initiation (Betts & Ball, 1998; Gentine et al., 2013; Konings et al., 2010; Siqueira et al., 2009). In the Amazon rainforest, MCSs initiate 47% of the time over the evergreen forest, 31% over the savanna and contact vegetation, 19% over the deforested area, and 3% over the Pantanal wetlands and seasonal forest, highlighting the influence of vegetation type on MCS formation (Laurent et al., 2002).

It is well known that for surface-based convection, the surface temperature plays an important role in the intensity and organization of MCSs by modulating convective available potential energy (CAPE), convective inhibition (CIN), and environmental lapse rate. In addition, the temperature gradient at the surface, especially over the African continent, plays an important role in the location and strength of the African easterly jets (AEJ) that produce numerous MCSs over the Sahel region and the Congo rainforest (Adebiyi & Zuidema, 2016; Cook, 1999; Chakraborty et al., 2021a; Jackson et al., 2009; Nicholson & Grist, 2003). Very intense MCSs are associated with the presence of the AEJ-South and bring rainfall during boreal autumn (Jackson et al., 2009; Vondou et al., 2010). AEJ-South is a thermal wind formed due to the meridional temperature gradient between the Congo rainforest and the Kalahari Desert. It creates a strong converging environment in the equatorward direction and promotes the formation of intense MCSs over the Congo rainforest during the boreal autumn wet season.

Over the Amazon rainforest, afternoon shallow to evening deep convective transition, an important step in MCS formation, also depends on higher environmental free tropospheric as well as boundary-layer moisture (Chakraborty, et al., 2018a) and CAPE (Chakraborty et al., 2020). A medium-high column water vapor content (>30 mm) promotes the transition of shallow clouds (with a height from CloudSat data of less than 4 km) to deep clouds (with a height greater than 12 km). Shallow convection over the Amazon rainforest moistens the dry mid-troposphere during the transition period (August to November) and makes the environment suitable for mesoscale development (Wright et al., 2017) and the wet season onset over Amazonia. In contrast, over the Congo rainforest, such a moistening effect is not noticed (Chakraborty et al., 2020). Moreover, the shallow-deep convective evolutionary mechanisms over these two rainforests are different. Afternoon shallow clouds over the Congo region are associated with suppressed and weakened evening-time deep convection, whereas shallow clouds over the Amazon region are associated with the growth of the evening-time deep convection (Chakraborty et al., 2020).

The land surface is an important moisture source throughout the entire MCS life cycle. Soil moisture and vegetation play a crucial role in MCS formation, strength, and intensity over various regions of the world (Adler et al., 2011; Baker et al., 2001; Birch et al., 2013; Dolores et al., 2019; Siqueira et al., 2009; Taylor et al., 2010), and where local soils are dry, external moisture advection is needed to initiate convection (Siqueira et al., 2009). Rainforest evapotranspiration (ET) can also be an important local moisture source to MCS, for example, during the wet season in the Amazon and Congo rainforests (Wright et al., 2017; Worden et al., 2021). Different products like water vapor isotope measurements by the TES and vegetative properties like the solar-induced fluorescence (SIF) from satellites like Scanning Imaging Absorption Spectrometer for Atmospheric Chartography (SCHYMACHI), the Global Ozone Monitoring Experiment-**2** (GOME-2), and Orbiting Carbon Observatory **2** (OCO2) have widely been used to detect SIF and ET (Shi et al., 2019; Wright et al., 2017; Worden et al., 2021). These remote sensing measurements show that a significant and positive correlation exists between the concentration of the heavy water isotope (HDO) in the atmosphere (measured from TES and Atmospheric Infrared Sounder or AIRS) and the atmospheric moisture content when MCSs are fueled by ET moisture (Aron et al., 2019; Shi et al., 2019; Wright et al., 2017). Over the Congo rainforests, ET generates moisture for MCSs during the boreal spring wet season, whereas the long-range transport of moisture from the Atlantic Ocean is the primary source of moisture during the boreal autumn wet season (Worden et al., 2021). On the other hand, over the Sahel region, multiple studies show that dry soils within synoptic convergence zones can initiate and intensify MCSs by increasing moisture convergence and the convective instability, especially during the afternoon-evening transition (Birch et al., 2013; Klein & Taylor, 2020; Taylor et al., 2010). Hu et al. (2021a) found that early warm season (April to June) MCS rainfall in the central United States, characterized by high intensity and large coverage, produces coherent mesoscale soil moisture heterogeneity that promotes summer (July) afternoon rainfall dominated by non-MCS events. On the other hand, soil moisture sourced from early MCS rainfall contributes to July MCS rainfall with a longer lead time because higher intensity MCS rainfall percolates into deeper soil that has a longer memory. Their findings demonstrate the critical role of early warm season MCS rainfall in soil moisture-precipitation feedback and motivate future studies to examine the predictability of summer MCS rainfall, using early warm season MCS rainfall and soil moisture anomalies.

Generally speaking, water vapor supply to MCSs has two distinct sources: advection and/or convergence from remote regions by the large-scale circulation and surface evaporation locally near the MCS. But the relative roles of these remote versus local moisture sources cannot be quantified without ambiguity, mainly because of a lack of reliable observations and the large uncertainties in convective and turbulence parameterizations utilized in coarsely gridded global climate models. In this regard, accurately measuring and simulating the spatiotemporal variation of land surface characteristics (e.g., soil moisture, land use, surface winds, and evaporation) are critical. The ability to tag and trace moisture sources transported from the soil to the atmosphere through evapotranspiration (Hu et al., 2018, 2020b) and from atmospheric transport to precipitation (Dominguez et al., 2016, 2022) in high-resolution simulations is useful in quantifying contributions of various water vapor sources to MCS.

Several land surface-associated factors may also influence MCS structure and intensity. Changes in land use patterns such as urbanization and deforestation can alter the meteorology surrounding MCSs. Urbanization affects hydrometeorology through altered thermodynamics, atmospheric dynamics, and aerosol concentration (Yang et al., 2011; Zhong et al., 2017). Surface heat release associated with the urban heat island (UHI) effect could affect CAPE (Seeley & Romps, 2015; Zhong et al., 2017; Zhang et al., 2018). Besides the thermal perturbations, the UHI effect has been documented to modulate wind patterns (Hjelmfelt, 1982), evaporation (Wienert & Kuttler, 2005), and atmospheric circulations (Lei et al., 2008; Shepherd & Burian, 2003), all factors that will influence MCS. Increased urban roughness increases the frictional velocity and the drag over the urban areas, thus enhancing the low-level convergence as well as upper level divergence, and the updraft speed. As a result, increased roughness length in urban areas favors atmospheric instability (Shepherd et al., 2010; Zhang et al., 2018), which may increase cloud water content and precipitation from the lower to the middle troposphere (Zhong et al., 2017; Zhang et al., 2018). Such impacts have increased precipitation and exacerbated flooding by hurricane Harvey and tropical storm Allison (Zhang et al., 2018). The impact of deforestation on MCSs is also an area of active research (Laurent et al., 2002; Silva Dias et al., 2002). The frequency of occurrences of shallow clouds increases and deep convective clouds become less frequent over the deforested regions in the Amazonia due to a reduction in CAPE and moisture owing to deforestation – potentially suggesting to influence MCSs in the future (Wang et al., 2009). With Amazonian deforestation, CAPE and low-level moisture are reduced. Hence, shallow clouds are more frequent and deep clouds are less frequent. Given the importance of the shallow-to-deep transition for MCS initiation in this region, these shifts suggest that MCS occurrence may reduce in the future. However, observational evidence using satellite measurements is still sparse relative to modeling studies investigating deforestation-MCS relationships (Garcia-Carreras & Parker, 2011; Wang et al., 2000). For example, a recent modeling study shows that expansion in cropland and reduction in the forest lead to a more favorable thermodynamic environment for convection (e.g., increase CAPE and latent heat flux, reduces lifted condensation level), resulting in an increased MCS number and precipitation in the warm season (Devanand et al., 2020).

9.5. MODELING OF MESOSCALE CONVECTIVE SYSTEMS

Most traditional general circulation models (GCM) with coarse horizontal resolution (grid spacing of 100 km or coarser) and parameterized convection cannot simulate MCS. This is partly because representing the multiscale interactions between convective-scale dynamics and microphysics and the upscale effects and feedback of MCS through latent heating (Feng et al., 2018; Yang et al., 2017) is particularly challenging. Most convection parameterizations in GCMs use scale-separation assumptions, that is, cumulus convection is in a statistical equilibrium with the large-scale environments. In other words, the environment slowly adjusts to the fast convection processes that remove instability at small subgrid scales almost instantaneously, with little memory or internal variability (Arakawa & Schubert, 1974). However, such assumptions are violated in MCSs that occupy at or larger than a GCM grid box and last much longer than isolated convection (Moncrieff, 2010). Critical processes in MCS, including multiple convective updrafts in close proximity that interact with each other, mesoscale circulations, and organized downdrafts induced by top-heavy latent heating (warming above the melting layer and cooling below) associated with broad areas of stratiform rain, and vertical momentum transports, are neither resolved nor parameterized.

There are various model development approaches being pursued in the climate modeling community to improve the representation of convective clouds, such as MCS, including improving cumulus parameterizations with "super-parameterization" in which moist physics are simulated by embedding explicit CRMs within GCM grids. Refining resolution from the mesoscale down to the kilometer scale in CPMs is another strategy. Indeed, while CRMs have historically been used for case studies by the weather research community, increasing computational power means that climate can now begin to be studied at these meteorologically and MCS-relevant scales. Each approach has certain advantages and drawbacks, and recent progress and future challenges are discussed briefly below.

9.5.1. Parameterization and Resolution Improvements

Improving convection parameterization in GCMs to better represent organized convection appears to be the most logical choice. However, progress has been slow in the past several decades (Randall et al., 2003; Randall, 2013). A recent study implemented an MCSP (see also section 9.2.3) to represent the top-heavy mesoscale heating associated with MCSs in a GCM (Chen et al., 2021a). A simple sinusoidal heating profile formulation was used, in which an upper tropospheric heating and a lower tropospheric cooling that represent MCS stratiform latent heating profile were added to the convective heating from parameterized deep convection. The ratio of stratiform to convective heating (ranging from 0.3 to 0.5) was determined from a CPM simulation during an MJO event over the Indian Ocean and west Pacific. An additional vertical wind shear criteria was added such that MCSP is only applied when zonal wind shear between surface and 600 hPa was above a certain threshold, with the assumption that MCSs are more likely to develop in a shear environment. They showed that with MCSP the simulated eastward propagation of the MJO is significantly improved, while the mean state excessive tropical western Pacific precipitation bias is reduced, thus mitigating the common MJO-mean state trade-off (Ahn et al., 2019; Jiang et al., 2020; Moncrieff et al., 2017). However, it is unclear if MCSs are better represented in the model with the new parameterization, as the study did not examine individual cloud system characteristics. There are also various conceptual models to parameterize stochastic convective cloud population dynamics (Hagos et al., 2018; Khouider et al., 2010; Plant, 2012) and associated stratiform cloud area (Hagos et al., 2020). These stochastic parameterizations provide another potential path forward to incorporate nonequilibrium processes associated with observed convective cloud system transitions such as formation of MCSs and the diurnal cycle to improve representation of convection and precipitation in future GCMs.

Besides parameterization improvements, increasing model resolution has been shown to produce improved representation of the hydrologic cycle, particularly related to hydrologic hazards and impacts that are critical for climate risk assessments (Roberts et al., 2018). There are several complementary efforts to increase model resolutions: (1) uniform grid refinements across the globe such as the High Resolution Model Intercomparison Project (HighResMIP) for the Coupled Model Intercomparison

Project 6 (CMIP6) (Haarsma et al., 2016); (2) unstructured meshes that provide variable-resolution (VR) capacity in which increased resolution can be applied to a region of interest; and (3) improved dynamical mesh designs that support convective-permitting simulations both regionally and globally. Compared to increasing resolution globally, the VR approach has been explored to study regional processes at a much-reduced computational cost (Gettelman et al., 2018; Sakaguchi et al., 2015). While models with higher resolution have been shown to produce more intense and less frequent precipitation that agrees better with observations (Chen & Dai, 2019; Wehner et al., 2014), it is unclear if the improvements reflect a correct representation of the underlying physical processes. In order to critically address the uncertainty in climate model representation of precipitation, phenomena-based metrics that focus on evaluating model precipitation associated with weather phenomena such as MCSs can help better assess model skill in simulating the multifaceted and multiscale nature of precipitation, understand the relationships of MCSs with their environment, and elucidate precipitation generation mechanisms (Leung et al., 2022).

A number of very recent studies evaluate the ability of high-resolution GCMs to simulate MCSs by applying feature tracking techniques to model simulations and observations. Dong et al. (2021) track cloud systems in the tropics with low infrared brightness temperature following previous methods to identify MCSs and show that a global model with 50 km horizontal grid spacing can simulate organized cloud clusters that resemble observed MCS. However, the simulated MCSs are much longer lived, larger, with warmer cloud tops, and the mean MCS precipitation amount is much larger than observed. A recent study developed a new approach that uses both infrared brightness temperature and surface precipitation features to identify and track MCSs in observations and models with mesoscale resolutions ranging from 10 to 50 km (Feng et al., 2021b). They found that at mesoscale resolutions, GCMs with untuned convection parameterizations designed for coarse resolution consistently underestimate MCS frequency, precipitation amount and intensity in the United States, particularly in summer when baroclinic forcing is much weaker. Although the models can simulate large-scale environments favorable for MCS development in spring, the simulated precipitation is too weak, and the diurnal cycle amplitude and phase are incorrect. Further, Leung et al. (2022) show that MCS tracking using infrared brightness temperature alone and including precipitation features (e.g., contiguous area with rain rate exceeding a threshold) results in drastically different assessment of the simulated MCS numbers in a mesoscale resolution GCM. By including MCS precipitation feature criteria that are derived from observations, the model tends to underestimate MCSs in most regions, as the simulated large cloud clusters often fail to meet the precipitation feature criteria. These studies highlight the importance of using multiple metrics for comprehensive evaluation of MCSs and their associated precipitation in models.

9.5.2. Model Physics Improvements

Cloud microphysics parameterizations can also affect model simulation of MCS, yet representing microphysical processes remains a major challenge in GCMs and numerical weather forecasts (Morrison et al., 2020a). Microphysics becomes increasingly important with higher resolution when synoptic systems and mesoscale dynamics are better resolved. Most existing cloud microphysics in GCMs is rather simple, such as using single-moment hydrometeor species (only predicting the mass but not the number concentration), no consideration of rimed ice particles (graupel or hail), and artificial partitioning of frozen particles into non-precipitating cloud ice and precipitating snow (Morrison & Milbrandt, 2015). When implementing a more physics-based microphysics parameterization in a GCM with 25 km horizontal grid spacing in VR configuration, Wang et al. (2021) showed that heavy precipitation associated with MCSs under large frontal systems are better simulated in the United States during spring. They attributed the improvements to the added representation of rimed ice processes and modified depositional growth, which leads to a positive feedback whereby more latent heat release enhances large-scale ascent. Such processes are expected to become even more important as model resolution continues to increase, because cloud microphysics is better coupled with the increasingly resolved dynamics.

The multiscale modeling framework (MMF) approach, also known as super-parameterization (SP), uses a different approach to improve model representations of moist physics. MMF embeds a CRM within each column of a GCM to replace cloud and convection parameterizations (Grabowski & Smolarkiewicz, 1999; Grabowski, 2001). Moist physics is explicitly resolved by the CRM within a GCM grid; hence, they are "super-parameterized." Compared to GCMs with traditional convection parameterization, MMF produces significant improvements in the diurnal cycle of precipitation over land (Khairoutdinov et al., 2005; Pritchard & Somerville, 2009), and more realistic eastward propagation of precipitation in the Great Plains of the United States in the summer that is analogous to observed MCS behavior (Kooperman et al., 2013; Pritchard et al., 2011). Chen and Kirtman (2018) showed that SP better simulates stratiform cooling and moistening at low levels associated with tropical MCSs compared to traditional GCM stratiform microphysics parameterizations. However, the coarse host GCM resolution (1.25° grid spacing) may create artifacts in identification and tracking of individual MCS, resulting in very long-lived cloud clusters identified in the model. When the host model grid spacing is too coarse (e.g., 100 km), MMF failed to capture the mesoscale dynamical environments that are critical for initiating midlatitude MCSs (Lin et al., 2019). Higher resolution (e.g., 25 km) alleviates the problem, but the computational expense is significantly higher at such resolutions. More recent efforts in the development of the SP version of the Energy Exascale Earth System Model (SP-E3SM) supported by the U.S. Department of Energy leverage the architecture of Graphics Processing Units (GPU) to significantly increase the model efficiency (Hannah et al., 2020). Compared to standard E3SM at mesoscale grid-spacing (25 km), SP-E3SM substantially improves simulation of various MCS features in both spring and summer, such as MCS number, precipitation amount and intensity, diurnal cycle, and propagation. The improvement is partly related to better simulation of the large-scale environments including low-level moisture and moisture transport to the central United States (Lin et al., 2022). However, SP-E3SM still underestimates MCS frequencies and precipitation amount in summer, suggesting further improvements of the SP approach are needed.

9.5.3. Convection-Permitting Models for Climate Simulations

CPMs have been playing an increasingly important role to address our need to understand and simulate climate processes and phenomena unresolved in GCMs (Gutowski et al., 2021). At grid-spacings of a few kilometers, CPMs no longer need convection parameterizations and instead resolve important convective processes nearly explicitly. As a result, CPM produces significant improvements in simulating MCSs in regional climate simulations that are comparable to observations in many aspects (Feng et al., 2018; Prein et al., 2017a; Yang et al., 2017). However, certain long-standing biases remain in regional CPM climate simulations, particularly the warm and dry bias in the central United States during summer (Prein et al., 2017a; Qian et al., 2020). The exact cause of such biases remains unclear, but it is possibly related to land-atmosphere feedbacks from early season MCS precipitation (Hu et al., 2021a), missing land surface processes such as groundwater processes (Barlage et al., 2021) and irrigation from farming activities (Qian et al., 2020), and cloud-radiation feedbacks (Van Weverberg et al., 2018). The impact of model resolution, particularly when coupled with cloud microphysics feedback in moist convective updrafts, remains a large uncertainty in CPMs. Entrainment mixing with the environment in convective updrafts at kilometer scale is under-resolved (Lebo & Morrison, 2015; Morrison et al., 2020b; Peters et al., 2020; Varble et al., 2020), resulting in overestimated convective updraft intensity that produces large rimmed ice particles that fall out too quickly (Fan et al., 2017; Wang et al., 2020b), and detrainment altitudes that are too high, limiting ice growth in the stratiform cloud region (Han et al., 2019). These biases are manifested in CPM simulations that consistently overestimate convective rainfall and underestimate stratiform rainfall regardless of convective regimes (Hagos et al., 2014; Varble et al., 2014a, 2014b; Zhang et al., 2021). Despite these biases, which remain an active area of research, CPM is an indispensable tool that complements global models for understanding physical processes controlling the evolution of MCSs and their potential changes in future climate (Prein et al., 2017a; Prein et al., 2021).

The advent of the global convection-permitting model (GCPM) brings unprecedented opportunities to improve understanding of global weather and climate. In addition to the same advantages offered by the aforementioned regional CPMs, GCPM provides additional benefits that are particularly relevant for the representation of MCSs and their roles in the climate system: (1) representation of global mesoscale circulations that drive weather systems without constraints from lateral boundary conditions; and (2) multiscale interactions of convection, such as MCS and MJO (Satoh et al., 2019). The global modeling community recently organized efforts to design an experimental protocol for the first intercomparison project of GCPMs, known as the DYnamics of the Atmospheric general circulation Modeled On Non-hydrostatic Domains (DYAMOND; Stevens et al., 2019). While the scientific use of these simulations is still in early stages, they have been shown to improve representation of the diurnal cycle and spatial distribution of tropical precipitation, the vertical structure of tropical convection and stratocumulus, and important weather events like atmospheric rivers, cold air outbreaks, and tropical cyclones (Fig. 9.6, Caldwell et al. 2021; Judt et al., 2021). The key challenges for GCPM include significant computational expense and storage requirements, both of which limit the integration length of the simulations. In addition, these models are mostly atmospheric-only, with the atmosphere-ocean coupling capability being developed toward global storm-resolving Earth system models. The ability of GCPM in representing observed mean characteristics and variabilities of MCSs globally is currently being explored.

Lastly, advances in understanding processes critical to MCSs and their roles in a changing climate system require concurrent improvements in modeling and observational capabilities. While high-resolution models have

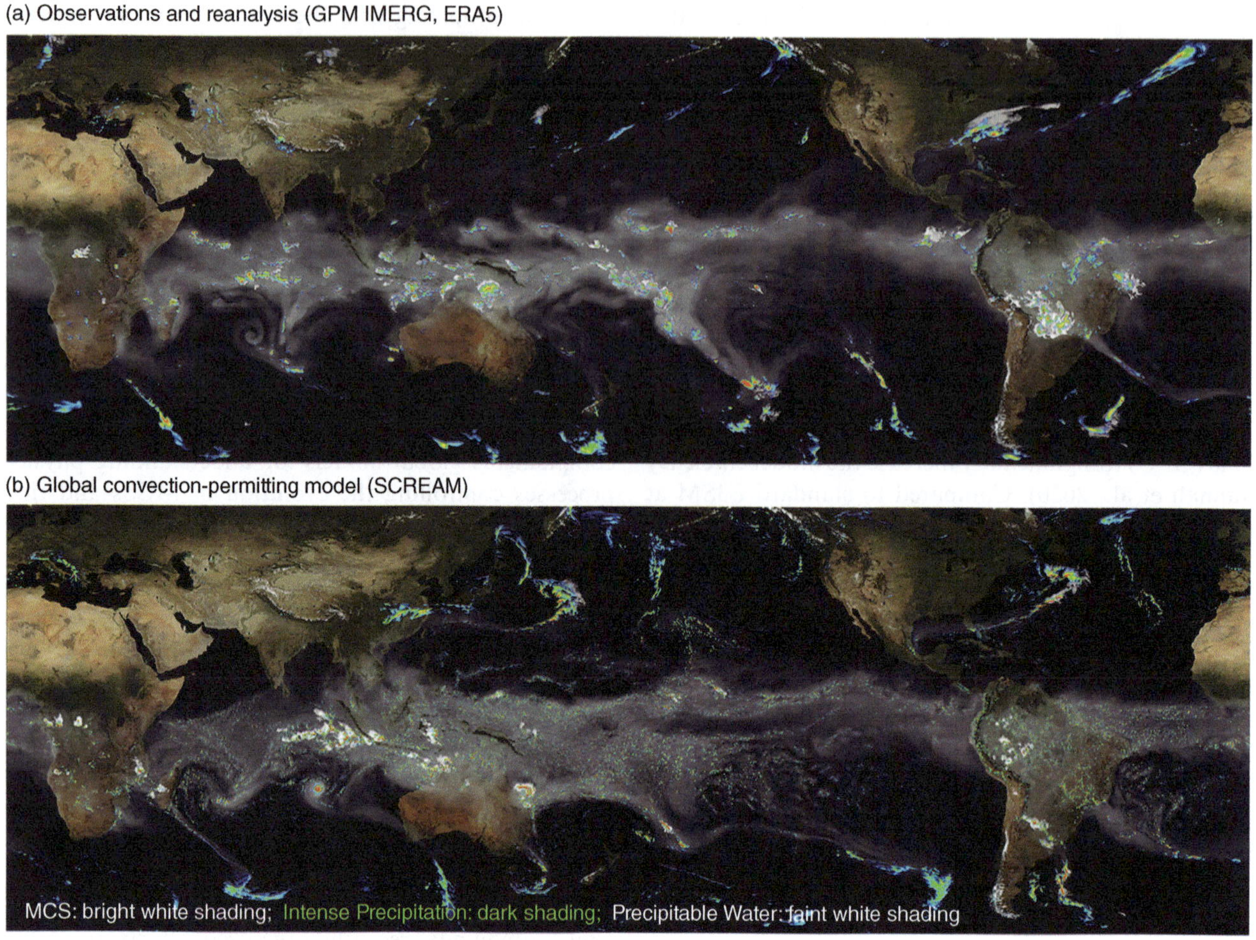

Figure 9.6 Examples of global MCSs (bright white shading) and precipitable water (faint white shading) from (a) observations and reanalysis, and (b) E3SM SCREAM (Caldwell et al., 2021). Source: Zhe Feng (Chapter Author). Precipitable water in (a) is from ERA5.

provided an unprecedented view of many processes of MCSs that are not directly observable and improve our fundamental understanding, they have many known biases and processes that still require parameterizations such as microphysics and turbulence. As refinement in resolution beyond kilometer scale in GCPMs and increase in domain size beyond 1,000 km in Large-eddy simulations (LES) may become the norm in the next several decades (Schalkwijk et al., 2015), improvements to our modeling capabilities require continue investments in our observational capabilities. At LES resolutions (grid-spacing ~100 m), our ability to provide sufficiently accurate observational constraints on critical convective processes, such as how the size, shape, and strength of convective updraft/downdraft in MCS influence entrainment, detrainment, and cloud microphysics, is extremely limited. Microphysics and thermodynamic measurements within MCS convective updrafts, which require storm penetrating aircrafts with appropriate instrumentations, are nearly non-existent. Convective updraft measurements can be made from vertical profilers (Giangrande et al., 2016; Wang et al., 2020b) or multi-Doppler retrievals (North et al., 2017), but both of them have very limited sampling, and neither has the capability to observe updraft evolution at relevant scales. While our observational sampling will continue to have major limitations in retrieving critical variables and representativeness, a better observation-model integration strategy is needed in which comprehensive and complementary high-resolution observations (e.g., from satellite, radar, surface network, and field campaign) and model simulations are better fused via instrument simulators/emulators and data assimilation techniques to lwink poorly observed variables (e.g., vertical motion, microphysical properties) and unobservable processes (e.g., entrainment mixing, hydrometeor production rates), and ultimately guiding the design of new observational strategies needed to constrain specific convective processes.

9.6. SUMMARY

We have reviewed progress on MCS research primarily from a climatological perspective in this chapter. Other review efforts presented MCSs from a meteorological perspective, as they are associated with extreme weather events, or from a historical perspective, as their research has been influenced by progress in radar technology, large-scale field campaigns, and satellite launches. But satellite data from missions like TRMM and GPM have now made it possible to consider MCSs in a climate context. Indeed, MCSs, as a form of convective aggregation, are associated with climatically relevant changes to the atmosphere, such as large-scale drying of the free troposphere and enhanced radiative cooling to space. MCSs have larger regions of stratiform precipitation and higher intensity precipitation than disorganized convection, supported in large part by mesoscale circulations that favor depositional and aggregational growth of ice from detrainment of their convective cores.

In the tropics, seasonality is muted and MCS frequency, structure, and attributable precipitation more closely follow intraseasonal to seasonal fluctuations like the ENSO or the MJO. Tropics-wide trends with these modes of variability often differ from regional ones. For example, the horizontal extent of MCSs shrinks in the tropical mean but expands over the central Pacific with warm SST anomalies of El Niño. While attributable precipitation does not change much in total throughout the tropics, the west Pacific receives roughly 10% more rainfall during La Niña and the central Pacific receives roughly 10% more during El Niño. Not only do MCS frequency and rainfall shift with modes of tropical climate variability, but their occurrences also feed back on these modes. For example, monsoon onset can be accelerated by solar reflection in the thick anvil outflow of MCS. Mid-level and upper level diabatic heating, dominated by MCSs also intensifies the Bjerknes feedback whereby the weaker Pacific SST gradient during El Niño generates a weaker overlying SLP gradient which further weakens the SST gradient.

The complexity of midlatitude dynamics and variety of weather phenomena observed make it somewhat harder to identify MCSs there, particularly over midlatitude oceans without ground-based observations. In contrast to the tropics, MCS properties and precipitation are strongly determined by seasonality. Owing to extensive weather radar coverage, for example, from the U.S. NEXRAD network, MCS seasonality has been thoroughly studied and connected to factors like baroclinic forcing and synoptic disturbances downstream of the Rocky Mountains. Similar factors seem to be responsible for supporting MCS development downstream of other major mountain ranges. Increases in observed MCS frequency, lifetime, and rainfall intensity have been tied to increases in extreme rainfall in the central United States during springtime. However, it is currently unclear if other regions of the world have experienced similar changes in MCSs with global warming, and more research is needed.

Along with these kinds of climatological pictures, increased sensor frequency, resolution and capability of satellite observations, and novel analysis techniques have allowed us to examine MCS life cycle and characteristics nearly globally. Newly developed MCS tracking databases have recently revealed climatically relevant patterns like more upper level latent heating from midlatitude MCSs than tropical ones and longest-lived systems over the subtropical oceans. From the burgeoning satellite record, we are also learning how to extract information on MCS cloud microphysics and MCS-land interactions. Multi-frequency radars and microwave sensors are able to retrieve ice crystal properties in the upper troposphere of MCSs or liquid precipitation in the lower part of the system. Absorption and emission spectrometers, on the other hand, are able to measure SIF and water vapor isotopes, giving us insights into evapotranspiration, soil moisture, and other land-based moisture sources to MCS development.

Finally, a wide variety of exciting modeling approaches for MCSs are underway. Typical climate model resolutions are insufficient to capture MCS, and methodologies like unstructured meshes or super-parameterization (in which a cloud-resolving model is embedded in a GCM grid box) allow us to effectively increase resolution. Improving model parameterizations is another route. The MCSP is one example of improving convective parameterization, in this case by employing a characteristic latent heating profile to represent the MCS effects on large-scale circulation. Improvements in microphysics parameterizations become increasingly relevant as model resolution continues to increase. More physical representation of ice hydrometeor properties (e.g., P3 scheme), Lagrangian particle tracking, and machine learning of microphysical relationships are all promising avenues. Lastly, as computational power continues to increase, global convection-permitting and storm-resolving models with horizontal grid-spacings of a few kilometers are becoming feasible. In the years ahead, these computational advances will go hand in hand with those in remote sensing to provide further insight into the crucial role of MCSs in the Earth system and how they may respond in a changing climate.

REFERENCES

Adebiyi, A. A., & Zuidema, P. (2016). The role of the southern African easterly jet in modifying the southeast Atlantic aerosol and cloud environments. *Quarterly Journal of the*

Royal Meteorological Society, *142*, 1574–1589. https://doi.org/10.1002/qj.2765

Adler, B., Kalthoff, N., & Gantner, L. (2011, July). The impact of soil moisture inhomogeneities on the modification of a mesoscale convective system: An idealised model study. *Atmospheric Research*, *101*(1–2), 354–372. https://doi.org/10.1016/j.atmosres.2011.03.013

Ahn, M.-S., Kim, D., Park, S., & Ham, Y.-G. (2019). Do we need to parameterize mesoscale convective organization to mitigate the MJO-mean state trade-off? *Geophysical Research Letters*, *46*, 2293–2301. https://doi.org/10.1029/2018GL080314

Anderson, C. J., & Arritt, R. W. (2001). Mesoscale convective systems over the United State during the 1997–98 El Niño. *Monthly Weather Review*, *129*, 2443–2457. https://doi.org/10.1175/1520-0493(2001)129<2443:MCSOTU>2.0.CO;2

Arakawa, A., & Schubert, W. H. (1974). Interaction of a cumulus cloud ensemble with the large-scale environment: Part I. *Journal of the Atmospheric Sciences*, *31*, 674–701. https://doi.org/10.1175/1520-0469(1974)031<0674:ioacce>2.0.co;2

Aron, P. G., Poulsen, C. J., Fiorella, R. P., & Matheny, A. M. (2019). Stable water isotopes reveal effects of intermediate disturbance and canopy structure on forest water cycling. *Journal of Geophysical Research: Biogeosciences*, *124*(10), 2958–2975. https://doi.org/10.1029/2019JG005118

Ashley, W. S., & Mote, T. L. (2005). Derecho hazards in the United States. *Bulletin of the American Meteorological Society*, *86*, 1577–1592. https://doi.org/10.1175/BAMS-86-11-1577

Ashley, W. S., Haberlie, A. M., & Strohm, J. (2019). A climatology of quasi-linear convective systems and their hazards in the United States. *Weather and Forecasting*, *34*, 1605–1631. https://doi.org/10.1175/WAF-D-19-0014.1

Ashok, K., Behera, S. K., Rao, S. A., Weng, H., & Yamagata, T. (2007). El Niño Modoki and its possible teleconnection. *Journal of Geophysical Research*, *112*, C11007. doi:10.1029/2006JC003798.

Baker, R. D., Lynn, B. H., Boone, A., Tao, W. K., & Simpson, J. (2001). The influence of soil moisture, coastline curvature, and land-breeze circulations on sea-breeze-initiated precipitation. *Journal of Hydrometeorology*, *2*, 193–211. https://doi.org/10.1175/1525-7541(2001)002<0193:TIOSMC>2.0.CO;2

Barlage, M., Chen, F., Rasmussen, R., Zhang, Z., & Miguez-Macho, G. (2021). The importance of scale-dependent groundwater processes in land-atmosphere interactions over the central United States. *Geophysical Research Letters*, *48*, e2020GL092171. https://doi.org/10.1029/2020GL092171

Barnes, H. C., & Houze, R. A. (2013). The precipitating cloud population of the Madden-Julian Oscillation over the Indian and west Pacific Oceans. *Journal of Geophysical Research: Atmospheres, 118*, 6996–7023. doi:10.1002/jgrd.50375.

Barnes, H. C., Zuluaga, M. D., & Houze, R. A. (2015). Latent heating characteristics of the MJO computed from TRMM Observations. *Journal of Geophysical Research: Atmospheres*, *120*, 1322–1334. doi: 10.1002/2014JD022530.

Bergman, J. W., Hendon, H. H., & Weikmann, K. M., (2001). Intraseasonal air–sea interactions at the onset of El Niño. *J. Climate*, *14*, 1702–1919.

Betts, A. K., & Ball, J. H. (1998). FIFE surface climate and site-average dataset 1987–89. *Journal of the Atmospheric Sciences*, *55*, 1091–1108. https://doi.org/10.1175/1520-0469(1998)055<1091:FSCASA>2.0.CO;2

Biasutti, M., Voigt, A., Boos, W. R., Braconnot, P., Hargreaves, J. C., Harrison, S. P., et al. (2018). Global energetics and local physics as drivers of past, present and future monsoons. *Nature Geoscience*, *11*, 392–400. https://doi.org/10.1038/s41561-018-0137-1

Birch, C.E., Parker, D.J., O'Leary, A., Marsham, J.H., Taylor, C.M., Harris, P.P., & Lister, G.M.S. (2013). Impact of soil moisture and convectively generated waves on the initiation of a West African mesoscale convective system. *Quarterly Journal of the Royal Meteorological Society*, *139*, 1712–1730. https://doi.org/10.1002/qj.2062

Bony, S., Semie, A., Kramer, R. J., Soden, B., Tompkins, A. M., & Emanuel, K. A. (2020). Observed modulation of the tropical radiation budget by deep convective organization and lower-tropospheric stability. *AGU Advances*, *1*, e2019AV000155. https://doi.org/10.1029/2019AV000155

Bordoni, S., & Schneider, T. (2008). Monsoons as eddy-mediated regime transitions of the tropical overturning circulation. *Nature Geoscience*, *1*, 515–519. https://doi.org/10.1038/ngeo248

Byrne, M. P., & Zanna, L. (2020). Radiative effects of clouds and water vapor on an axisymmetric monsoon. *Journal of Climate*, *33*, 8789–8811. https://doi.org/10.1175/JCLI-D-19-0974.1

Cai, W., McPhaden, M. J., Grimm, A. M., Rodrigues, R. R., Taschetto, A. S., Garreaud, R. D., et al. (2020). Climate impacts of the El Niño–Southern Oscillation on South America. *Nature Reviews Earth and Environment*, *1*, 215–231. https://doi.org/10.1038/s43017-020-0040-3

Caldwell, P. M., Terai, C. R., Hillman, B., Keen, N. D., Bogenschutz, P., Lin, W., et al. (2021). Convection-permitting simulations with the E3SM global atmosphere model. *Journal of Advances in Modeling Earth Systems*, *13*, e2021MS002544. https://doi.org/10.1029/2021MS002544

Cancelada, M., Salio, P., Vila, D., Nesbitt, S. W., & Vidal, L. (2020). Backward adaptive brightness temperature threshold technique (BAB3T): A methodology to determine extreme convective initiation regions using satellite infrared imagery. *Remote Sensing*, *12*, 337. https://doi.org/10.3390/rs12020337

Cecil, D. J., & Blankenship, C. B. (2012). Toward a global climatology of severe hailstorms as estimated by satellite passive microwave imagers. *Journal of Climate*, *25*, 687–703. https://doi.org/10.1175/JCLI-D-11-00130.1

Chakraborty, S., Jiang, J. H., Su, H., & Fu, R. (2021a). On the role of aerosol radiative effect in the wet season onset timing over the Congo rainforest during boreal autumn. *Atmospheric Chemistry and Physics Discussions*, *21*, 12855–12866. https://doi.org/10.5194/acp-2020-1138

Chakraborty, S., Fu, R., Massie, S. T., & Stephens, G. (2016). Relative influence of meteorological conditions and aerosols on the lifetime of mesoscale convective systems. *Proceedings of the National Academy of Sciences of the United States of America*, *113*(27), 7426–7431. https://doi.org/10.1073/pnas.1601935113

Chakraborty, S., Schiro, K. A., Fu, R., & David Neelin, J. (2018a). On the role of aerosols, humidity, and vertical wind shear in the transition of shallow-to-deep convection at the Green Ocean Amazon 2014/5 site. *Atmospheric Chemistry and Physics*, *18*, 11135–11148. https://doi.org/10.5194/acp-18-11135-2018

Chakraborty, S., Fu, R., Rosenfeld, D., & Massie, S. T. (2018b). The Influence of aerosols and meteorological conditions on the total rain volume of the mesoscale convective systems over tropical continents. *Geophysical Research Letters*, *45*(23), 13,099–13,106. https://doi.org/10.1029/2018GL080371

Chakraborty, S., Jiang, J. H., Su, H., & Fu, R. (2020). Deep convective evolution from shallow clouds over the Amazon and Congo rainforests. *Journal of Geophysical Research: Atmospheres*, *125*, e2019JD030962. https://doi.org/10.1029/2019JD030962

Chen, C.-C., Richter, J. H., Liu, C., Moncrieff, M. W., Tang, Q., Lin, W., et al. (2021a). Effects of organized convection parameterization on the MJO and precipitation in E3SMv1. Part I: Mesoscale heating. *Journal of Advances in Modeling Earth Systems*, *13*, e2020MS002401. https://doi.org/10.1029/2020MS002401

Chen, D., Guo, J., Yao, D., Lin, Y., Zhao, C., Min, M., et al. (2019). Mesoscale convective systems in the Asian monsoon region from advanced Himawari Imager: Algorithms and preliminary results. *Journal of Geophysical Research: Atmospheres*, *124*, 2210–2234. https://doi.org/10.1029/2018JD029707

Chen, D., & Dai, A. (2019). Precipitation characteristics in the Community Atmosphere Model and their dependence on model physics and resolution. *Journal of Advances in Modeling Earth Systems*, *11*, 2352–2374. https://doi.org/10.1029/2018MS001536

Chen, G., & Kirtman, B. P. (2018). Long-lived mesoscale convective systems of superparameterized CAM and the response of CAM. *Journal of Advances in Modeling Earth Systems*, *10*, 2269–2286. https://doi.org/10.1029/2018MS001339

Chen, X., Leung, L. R., Feng, Z., & Song, F. (2021b). Critical role of mesoscale convective systems in the vertical mass, water, and energy transports of the South Asian summer monsoon. *Journal of Climate*, *35*, 91–108. https://doi.org/10.1175/JCLI-D-21-0124.1

Chen, X., Leung, L. R., Feng, Z., Song, F., & Yang, Q. (2021c). Mesoscale convective systems dominate the energetics of the South Asian summer monsoon onset. *Geophysical Research Letters*, *48*, e2021GL094873. https://doi.org/10.1029/2021GL094873

Chiang, J. C. H., Zebiak, S. E., & Cane, M. A. (2001). Relative roles of elevated heating and surface temperature gradients in driving anomalous surface winds over tropical oceans. *Journal of the Atmospheric Sciences*, *58*, 1371–1394. https://doi.org/10.1175/1520-0469(2001)058<1371:RROEHA>2.0.CO;2

Chiang, J. C. H., & Sobel, A. H. (2002). Tropical tropospheric temperature variations caused by ENSO and their influence on the remote tropical climate. *Journal of Climate*, *15*, 2616–2631. https://doi.org/10.1175/1520-0442(2002)015<2616:TTTVCB>2.0.CO;2

Choudhury, A. D., & Krishnan, R. (2011). Dynamical response of the South Asian monsoon trough to latent heating from stratiform and convective precipitation. *Journal of the Atmospheric Sciences*, *68*, 1347–1363. https://doi.org/10.1175/2011JAS3705.1

Cook, K. H. (1999). Generation of the African easterly jet and its role in determining West African precipitation. *Journal of Climate*, *12*, 1165–1184. https://doi.org/10.1175/1520-0442(1999)012<1165:GOTAEJ>2.0.CO;2

Cook, K. H., Vizy, E. K., Launer, Z. S., & Patricola, C. M. (2008). Springtime intensification of the Great Plains Low-Level jet and Midwest precipitation in GCM Simulations of the twenty-first century. *Journal of Climate*, *21*, 6321–6340. https://doi.org/10.1175/2008JCLI2355.1

Dandi, R. A., Chowdary, J. S., Pillai, P. A., Sidhan N. S., Koteswararao, K., & Ramakrishna, S.S.V.S. (2020). Impact of El Niño Modoki on Indian summer monsoon rainfall: Role of western north Pacific circulation in observations and CMIP5 models. *International Journal of Climatology*, *40*, 2117–2133. https://doi.org/10.1002/joc.6322

Devanand, A., Huang, M., Lawrence, D. M., Zarzycki, C. M., Feng, Z., & Lawrence, P. J., et al. (2020). Land use and land cover change strongly modulates land-atmosphere coupling and warm-season precipitation over the central United States in CESM2-VR. *Journal of Advances in Modeling Earth Systems*, *12*, e2019MS001925. https://doi.org/10.1029/2019MS001925

Dolores, E., Caetano, E., López-Bravo, L. C., & Calheiros, A. (2019). Influence of soil moisture on mesoscale convective initiation in central Mexico. *European Journal of Remote Sensing*, *640–652*. https://doi.org/10.1080/22797254.2019.1700397

Dominguez, F., Miguez-Macho, G., & Hu, H. (2016). WRF with water vapor tracers: A study of moisture sources for the North American Monsoon. *Journal of Hydrometeorology*, *17*, 1915–1927. https://doi.org/10.1175/JHM-D-15-0221.1

Dominguez, F., Eiras-Barca, J., Yang, Z., Bock, D., Nieto, R., & Gimeno, L. (2022). Amazonian moisture recycling revisited using WRF with water vapor tracers. *Journal of Geophysical Research: Atmospheres*, *127*, e2021JD035259. https://doi.org/10.1029/2021JD035259

Dong, W., Zhao, M., Ming, Y., & Ramaswamy, V. (2021). Representation of tropical mesoscale convective systems in a general circulation model: Climatology and response to global warming. *Journal of Climate*, *34*, 5657–5671. https://doi.org/10.1175/JCLI-D-20-0535.1

Doswell, C. A., Brooks, H. E., & Maddox, R. A. (1996). Flash flood forecasting: An ingredients-based methodology. *Weather and Forecasting*, *11*, 560–581. https://doi.org/10.1175/1520-0434(1996)011<0560:FFFAIB>2.0.CO;2

Fan, J., Leung, L. R., Rosenfeld, D., Chen, Q., Li, Z., Zhang, J., & Yan, H. (2013). Microphysical effects determine macrophysical response for aerosol impacts on deep convective clouds. *Proceedings of the National Academy of Sciences of the United States of America*, *110*(48), E4581–E4590. https://doi.org/10.1073/pnas.1316830110

Fan, J., Wang, Y., Rosenfeld, D., & Liu, X. (2016). Review of aerosol-cloud interactions: Mechanisms, significance, and challenges. *Journal of the Atmospheric Sciences*, *73*, 4221–4252. https://doi.org/10.1175/JAS-D-16-0037.1

Fan, J., Han, B., Varble, A., Morrison, H., North, K., Kollias, P., et al. (2017). Cloud-resolving model intercomparison of an MC3E squall line case: Part I—Convective updrafts. *Journal of Geophysical Research: Atmospheres*, *122*, 9351–9378. doi:10.1002/2017JD026622.

Feng, J., Chen, W., Tam, C.-Y. & Zhou, W. (2011a) Different impacts of El Niño and El Niño Modoki on China rainfall in the decaying phases. *International Journal of Climatology*, *31*, 2091–2101. https://doi.org/10.1002/joc.2217

Feng, Z., Dong, X., Xi, B., Schumacher, C., Minnis, P., & Khaiyer, M. (2011b). Top-of-atmosphere radiation budget of convective core/stratiform rain and anvil clouds from deep convective systems. *Journal of Geophysical Research*, *116*, D23202. doi:10.1029/2011JD016451.

Feng, Z., Dong, X., Xi, B., McFarlane, S. A., Kennedy, A., Lin, B., & Minnis, P. (2012). Life cycle of midlatitude deep convective systems in a Lagrangian framework. *Journal of Geophysical Research*, *117*, D23201. doi:10.1029/2012JD018362.

Feng, Z., Hagos, S., Rowe, A. K., Burleyson, C. D., Martini, M. N., & de Szoeke, S. P. (2015). Mechanisms of convective cloud organization by cold pools over tropical Warm Ocean during the AMIE/DYNAMO field campaign. *Journal of Advances in Modeling Earth Systems*, *7*, 357–381. doi:10.1002/2014MS000384.

Feng, Z., Leung, L. R., Hagos, S., Houze, R. A., Burleyson, C. D., & Balaguru, K. (2016). More frequent intense and long-lived storms dominate the springtime trend in central US rainfall. *Nature Communications*, *7*, 13429. https://doi.org/10.1038/ncomms13429

Feng, Z., Leung, L. R., Houze, R. A., Hagos, S., Hardin, J., Yang, Q., et al. (2018). Structure and evolution of mesoscale convective systems: Sensitivity to cloud microphysics in convection-permitting simulations over the United States. *Journal of Advances in Modeling Earth Systems*, *10*, 1470–1494. https://doi.org/10.1029/2018MS001305

Feng, Z., Houze, R. A., Leung, L. R., Song, F., Hardin, J. C., Wang, J., et al. (2019). Spatiotemporal characteristics and large-scale environments of mesoscale convective systems east of the Rocky Mountains. *Journal of Climate*, *32*, 7303–7328. https://doi.org/10.1175/JCLI-D-19-0137.1

Feng, Z. (2019). FLEXTRKR algorithm for Mesoscale Convective System MCS, Convection Cell Tracking (FLEXTRKR). Atmospheric Radiation Measurement (ARM) User Facility. https://doi.org/10.5439/1571643.

Feng, Z., Leung, L. R., Liu, N., Wang, J., Houze, R. A., Li, J., et al. (2021a). A global high-resolution mesoscale convective system database using satellite-derived cloud tops, surface precipitation, and tracking. *Journal of Geophysical Research: Atmospheres*, *126*, e2020JD034202. https://doi.org/10.1029/2020JD034202

Feng, Z., Song, F., Sakaguchi, K., & Leung, L. R. (2021b). Evaluation of mesoscale convective systems in climate simulations: Methodological development and results from MPAS-CAM over the United States. *Journal of Climate*, *34*, 2611–2633. https://doi.org/10.1175/JCLI-D-20-0136.1

Fiolleau, T., & Roca, R. (2013). Composite life cycle of tropical mesoscale convective systems from geostationary and low Earth orbit satellite observations: Method and sampling considerations. *Quarterly Journal of the Royal Meteorological Society*, *139*, 941–953. https://doi.org/10.1002/qj.2174

Fiolleau, T., Roca, R., Cloche, S., Bouniol, D., & Raberanto, P. (2020). Homogenization of geostationary infrared imager channels for cold cloud studies using Megha-Tropiques/ScaRaB. *IEEE Transactions on Geoscience and Remote Sensing*, *58*(9), 6609–6622. https://doi.org/10.1109/TGRS.2020.2978171

Fitzpatrick, R. G. J., Parker, D. J., Marsham, J. H., Rowell, D. P., Guichard, F. M., Taylor, C. M., et al. (2020). What drives the intensification of mesoscale convective systems over the West African Sahel under climate change? *Journal of Climate*, *33*, 3151–3172. https://doi.org/10.1175/JCLI-D-19-0380.1

Fritsch, J. M., Murphy, J. D., & Kain, J. S. (1994). Warm core vortex amplification over land. *Journal of the Atmospheric Sciences*, *51*, 1780–1807. https://doi.org/10.1175/1520-0469(1994)051<1780:wcvaol>2.0.co;2

Fujinami, H., Hirata, H., Kato, M., & Tsuboki, K. (2020). Mesoscale precipitation systems and their role in the rapid development of a monsoon depression over the Bay of Bengal. *Quarterly Journal of the Royal Meteorological Society*, *146*, 267–283. https://doi.org/10.1002/qj.3672

Futyan, J. M., & Del Genio, A. D. (2007). Deep convective system evolution over Africa and the tropical Atlantic. *Journal of Climate*, *20*(20), 5041–5060. http://dx.doi.org/10.1175/JCLI4297.1

Garcia-Carreras, L., & Parker, D. J. (2011). How does local tropical deforestation affect rainfall? *Geophysical Research Letters*, *38*, L19802. doi:10.1029/2011GL049099.

Geerts, B. (1998). Mesoscale convective systems in the southeast United States during 1994–95: A survey. *Weather and Forecasting*, *13*, 860–869. https://doi.org/10.1175/1520-0434(1998)013<0860:mcsits>2.0.co;2

Gentine, P., Holtslag, A. A. M., D'Andrea, F., & Ek, M. (2013). Surface and atmospheric controls on the onset of moist convection over land. *Journal of Hydrometeorology*, *14*, 1443–1462. https://doi.org/10.1175/JHM-D-12-0137.1

Gettelman, A., Callaghan, P., Larson, V. E., Zarzycki, C. M., Bacmeister, J. T., Lauritzen, P. H., et al. (2018). Regional climate simulations with the Community Earth System Model. *Journal of Advances in Modeling Earth Systems*, *10*, 1245–1265. https://doi.org/10.1002/2017MS001227

Giangrande, S. E., Toto, T., Jensen, M. P., Bartholomew, M. J., Feng, Z., Protat, A., et al. (2016). Convective cloud vertical velocity and mass-flux characteristics from radar wind profiler observations during GoAmazon2014/5. *Journal of Geophysical Research: Atmospheres*, *121*, 12,891–12,913. doi:10.1002/2016JD025303.

Grabowski, W. W. (2001). Coupling cloud processes with the large-scale dynamics using the clouds-resolving convection parameterization (CRCP). *Journal of the Atmospheric Sciences*, *58*, 978–997. https://doi.org/10.1175/1520-0469(2001)058<0978:CCPWTL>2.0.CO;2

Grabowski, W. W., & Smolarkiewicz, P. K. (1999). CRCP: A Cloud Resolving Convection Parameterization for modeling the tropical convecting atmosphere. *Physica D: Nonlinear Phenomen, 133,* 171–178. https://doi.org/10.1016/S0167-2789(99)00104-9

Gutowski, W. J., Ullrich, P. A., Hall, A., Leung, L. R., O'Brien, T. A., Patricola, C. M., et al. (2021). The ongoing need for high-resolution regional climate models: Process understanding and stakeholder information. *Bulletin of the American Meteorological Society*, *101*, E664–E683. https://doi.org/10.1175/BAMS-D-19-0113.1

Haarsma, R. J., Roberts, M. J., Vidale, P. L., Catherine, A., Bellucci, A., Bao, Q., et al. (2016). High Resolution Model Intercomparison Project (HighResMIP v1.0) for CMIP6. *Geoscientific Model Development*, *9*, 4185–4208. https://doi.org/10.5194/gmd-9-4185-2016

Haberlie, A. M., & Ashley, W. S. (2019). A radar-based climatology of mesoscale convective systems in the United States. *Journal of Climate*, *32*, 1591–1606. https://doi.org/10.1175/JCLI-D-18-0559.1

Haberlie, A. M., Ashley, W. S., & Karpinski, M. R. (2021). Mean storms: Composites of radar reflectivity images during two decades of severe thunderstorm events. *International Journal of Climatology*, *41*(Suppl. 1), E1738–E1756. https://doi.org/10.1002/joc.6804

Hagos, S., Feng, Z., Burleyson, C. D., Lim, K.-S. S., Long, C. N., Wu, D., & Thompson, G. (2014). Evaluation of convection-permitting model simulations of cloud populations associated with the Madden-Julian Oscillation using data collected during the AMIE/DYNAMO field campaign. *Journal of Geophysical Research: Atmospheres*, *119*, 12,052–12,068. doi:10.1002/2014JD022143.

Hagos, S., Feng, Z., Plant, R. S., Houze, R. A., & Xiao, H. (2018). A stochastic framework for modeling the population dynamics of convective clouds. *Journal of Advances in Modeling Earth Systems*, *10*, 448–465. https://doi.org/10.1002/2017MS001214

Hagos, S., Feng, Z., Plant, R. S., & Protat, A. (2020). A machine learning assisted development of a model for the populations of convective and stratiform clouds. *Journal of Advances in Modeling Earth Systems*, *12*, e2019MS001798. https://doi.org/10.1029/2019MS001798

Han, B., Fan, J., Varble, A., Morrison, H., Williams, C. R., Chen, B., et al. (2019). Cloud-resolving model intercomparison of an MC3E squall line case: Part II. Stratiform precipitation properties. *Journal of Geophysical Research: Atmospheres*, *124*, 1090–1117. https://doi.org/10.1029/2018JD029596

Hannah, W. M., Jones, C. R., Hillman, B. R., Norman, M. R., Bader, D. C., Taylor, M. A., et al. (2020). Initial results from the super-parameterized E3SM. *Journal of Advances in Modeling Earth Systems*, *12*, e2019MS001863. https://doi.org/10.1029/2019MS001863

Harding, K. J., & Snyder, P. K. (2014). Examining future changes in the character of Central U.S. warm-season precipitation using dynamical downscaling. *Journal of Geophysical Research: Atmospheres*, *119*, 13,116–13,136. doi:10.1002/2014JD022575.

Hartmann, D. L., Hendon, H. H., & Houze, R. A. (1984). Some implications of the mesoscale circulations in tropical cloud clusters for large-scale dynamics and climate (Walker circulation). *Journal of the Atmospheric Sciences*, *41*, 113–121. https://doi.org/10.1175/1520-0469(1984)041<0113:SIOTMC>2.0.CO;2

Hartmann, D. L., & Larson, K. (2002). An important constraint on tropical cloud - climate feedback. *Geophysical Research Letters*, *29*(20), 1951. doi:10.1029/2002GL015835

Hayden, L., Liu, C., & Liu, N. (2021). Properties of mesoscale convective systems throughout their lifetimes using IMERG, GPM, WWLLN, and a simplified tracking algorithm. *Journal of Geophysical Research: Atmospheres*, *126*, e2021JD035264. https://doi.org/10.1029/2021JD035264

Henderson, D. S., Kummerow, C. D., Marks, D. A., & Berg, W. (2017). A regime-based evaluation of TRMM oceanic precipitation biases. *Journal of Atmospheric and Oceanic Technology*, *34*, 2613–2635. https://doi.org/10.1175/JTECH-D-16-0244.1.

Henderson, D. S., Kummerow, C. D., & Berg, W. (2018). ENSO influence on TRMM tropical oceanic precipitation characteristics and rainrates. *Journal of Climate*, *31*, 3979–3998. doi: 10.1175/JCLI-D-11-00050.1

Heymsfield, A. J., Schmitt, C., & Bansemer, A. (2013). Ice cloud particle size distributions and pressure-dependent terminal velocities from in situ observations at temperatures from 0° to 286°c. *Journal of the Atmospheric Sciences*, *70*(12), 4123–4154. https://doi.org/10.1175/JAS-D-12-0124.1

Hjelmfelt, M. R. (1982). Numerical simulation of the effects of St. Louis on mesoscale boundary-layer airflow and vertical air motion: Simulations of urban vs non-urban effects. *Journal of Applied Meteorology*, *21*, 1239–1257. https://doi.org/10.1175/1520-0450(1982)021<1239:NSOTEO>2.0.CO;2

Hodson, M. C. (1986). Raindrop size distribution. *Journal of Climate & Applied Meteorology*, *25*, 1070–1074 . https://doi.org/10.1175/1520-0450(1986)025<1070:RSD>2.0.CO;2

Houze, R. A. (1982). Cloud clusters and large-scale vertical motions in the tropics. *Journal of the Meteorological Society of Japan*, *60*, 396–410. https://doi.org/10.2151/jmsj1965.60.1_396

Houze, R. A. (2004). Mesoscale convective systems. *Reviews of Geophysics*, *42*. https://doi.org/10.1029/2004RG000150

Houze, R. A. (2018). 100 years of research on mesoscale convective systems. *Meteorological Monographs*, *59*, 17.1–17.54. https://doi.org/10.1175/amsmonographs-d-18-0001.1

Houze, R. A., Jr., Wilton, D. C., & Smull, B. F. (2007). Monsoon convection in the Himalayan region as seen by the TRMM Precipitation Radar. *Quarterly Journal of the Royal Meteorological Society*, *133*, 1389–1411. https://doi.org/10.1002/qj.106

Houze, R. A., Rasmussen, K. L., Zuluaga, M. D., & Brodzik, S. R. (2015). The variable nature of convection in the tropics and subtropics: A legacy of 16 years of the Tropical Rainfall Measuring Mission satellite. *Reviews of Geophysics*, *53*, 994–1021. https://doi.org/10.1002/2015RG000488

Hu, H., Dominguez, F., Kumar, P., Mcdonnell, J., & Gochis, D. (2018). A numerical water tracer model for understanding event-scale hydrometeorological phenomena. *Journal of Hydrometeorology*, *19*, 947–967. https://doi.org/10.1175/JHM-D-17-0202.1

Hu, H., Leung, L. R., & Feng, Z. (2020a). Observed warm-season characteristics of MCS and non-MCS rainfall and their recent changes in the Central United States. *Geophysical Research Letters*, *47*, e2019GL086783. https://doi.org/10.1029/2019GL086783

Hu, H., Leung, L. R., & Feng, Z. (2020b). Understanding the distinct impacts of MCS and non-MCS rainfall on the surface water balance in the Central United States using a numerical water-tagging technique. *Journal of Hydrometeorology*, *21*, 2343–2357. https://doi.org/10.1175/JHM-D-20-0081.1

Hu, H., Leung, L. R., & Feng, Z. (2021a). Early warm-season mesoscale convective systems dominate soil moisture–precipitation feedback for summer rainfall in central United States. *Proceedings of the National Academy of Sciences of the United States of America*, *118*(43), e21052601. https://doi.org/10.1073/pnas.2105260118

Hu, H., Feng, Z., & Leung, L.-Y. R. (2021b). Linking flood frequency with mesoscale convective systems in the US. *Geophysical Research Letters*, *48*, e2021GL092546. https://doi.org/10.1029/2021GL092546

Huang, X., Hu, C., Huang, X., Chu, Y., Tseng, Y. H., Zhang, G. J., & Lin, Y. (2018). A long-term tropical mesoscale convective systems dataset based on a novel objective automatic tracking algorithm. *Climate Dynamics*, *51*, 3145–3159. https://doi.org/10.1007/s00382-018-4071-0

Jackson, B., Nicholson, S. E., & Klotter, D. (2009). Mesoscale convective systems over western equatorial Africa and their relationship to large-scale circulation. *Monthly Weather Review*, *137*, 1272–1294. https://doi.org/10.1175/2008MWR2525.1

Jiang, X., Adames, Á. F., Kim, D., Maloney, E. D., Lin, H., Kim, H., et al. (2020). Fifty years of research on the Madden-Julian Oscillation: Recent progress, challenges, and perspectives. *Journal of Geophysical Research: Atmospheres*, *125*, e2019JD030911. https://doi.org/10.1029/2019JD030911

Judt, F., Klocke, D., Rios-Berrios, R., Vanniere, B., Ziemen, F., Auger, L., et al. (2021). Tropical cyclones in global storm-resolving models. *Journal of the Meteorological Society of Japan*, *99*, 579–602. https://doi.org/10.2151/jmsj.2021-029

Khairoutdinov, M., Randall, D., & DeMott, C. (2005). Simulations of the atmospheric general circulation using a cloud-resolving model as a superparameterization of physical processes. *Journal of the Atmospheric Sciences*, *62*, 2136–2154. https://doi.org/10.1175/JAS3453.1

Khouider, B., Biello, J., & Majda, A. J. (2010). A stochastic multicloud model for tropical convection. *Communications in Mathematical Sciences*, *8*(1): 187–216. https://doi.org/10.4310/cms.2010.v8.n1.a10

Klein, C., & Taylor, C. M. (2020). Dry soils can intensify mesoscale convective systems. *Proceedings of the National Academy of Sciences of the United States of America*, *117*(35), 21132–21137. https://doi.org/10.1073/pnas.2007998117

Kodama, Y.-M., Ohta, A., Katsumata, M., Mori, S., Satoh, S., & Ueda, H. (2005). Seasonal transition of predominant precipitation type and lightning activity over tropical monsoon areas derived from TRMM observations. *Geophysical Research Letters*, *32*, L14710. doi:10.1029/2005GL022986.

Konings, A. G., Katul, G. G., & Porporato, A. (2010). The rainfall-no rainfall transition in a coupled land-convective atmosphere system. *Geophysical Research Letters*, *37*, L14401. doi:10.1029/2010GL043967.

Kondo, Y., Higuchi, A., & Nakamura, K. (2006). Small-scale cloud activity over the maritime continent and the Western Pacific as revealed by satellite data. *Monthly Weather Review*, *134*, 1581–1599. https://doi.org/10.1175/MWR3132.1.

Kooperman, G. J., Pritchard, M. S., & Somerville, R. C. J. (2013). Robustness and sensitivities of central U.S. summer convection in the super-parameterized CAM: Multi-model intercomparison with a new regional EOF index. *Geophysical Research Letters*, *40*, 3287– 3291. doi:10.1002/grl.50597.

Kumjian, M. R., Gutierrez, R., Soderholm, J. S., Nesbitt, S. W., Maldonado, P., Luna, L. M., et al. (2020). Gargantuan Hail in Argentina. *Bulletin of the American Meteorological Society*, *101*(8), E1241–E1258. https://doi.org/10.1175/BAMS-D-19-0012.1

Kundzewicz, Z. W., Kanae, S., Seneviratne, S. I., Handmer, J., Nicholls, N., Peduzzi, P., et al. (2014). Flood risk and climate change: Global and regional perspectives. *Hydrological Sciences Journal*, *59*(1), 1–28. https://doi.org/10.1080/02626667.2013.857411

Kunkel, K. E., Easterling, D. R., Kristovich, D. A. R., Gleason, B., Stoecker, L., & Smith, R. (2012). Meteorological causes of the secular variations in observed extreme precipitation events for the conterminous United States. *Journal of Hydrometeorology*, *13*(3), 1131–1141. https://doi.org/10.1175/JHM-D-11-0108.1

Laing, A. G., & Fritsch, J. M. (2000). The large-scale environments of the global populations of mesoscale convective complexes. *Monthly Weather Review*, *128*(8), 2756–2776. https://doi.org/10.1175/1520-0493(2000)128<2756:tlseot>2.0.co;2

Laing, A.G., & Michael Fritsch, J. (1997). The global population of mesoscale convective complexes. *Quarterly Journal of the Royal Meteorological Society*, *123*, 389–405. https://doi.org/10.1002/qj.49712353807

Latos, B., Lefort, T., Flatau, M. K., Flatau, P. J., Permana, D. S., Baranowski, D. B., et al. (2021). Equatorial waves triggering extreme rainfall and floods in Southwest Sulawesi, Indonesia. *Monthly Weather Review*, *149*, 1381–1401. https://doi.org/10.1175/MWR-D-20-0262.1

Laurent, H., Machado, L. A. T., Morales, C. A., & Durieux, L. (2002). Characteristics of the Amazonian mesoscale convective systems observed from satellite and radar during the WETAMC/LBA experiment. *Journal of Geophysical Research*, *107*(D20), 8054. doi:10.1029/2001JD000337

Lebo, Z. J., & Morrison, H. (2015). Effects of horizontal and vertical grid spacing on mixing in simulated squall lines and implications for convective strength and structure. *Monthly Weather Review*, *143*, 4355–4375. https://doi.org/10.1175/MWR-D-15-0154.1

Lei, M., Niyogi, D., Kishtawal, C., Pielke, R. A., Beltrán-Przekurat, A., Nobis, T. E., & Vaidya, S. S. (2008). Effect of explicit urban land surface representation on the simulation of the 26 July 2005 heavy rain event over Mumbai, India. *Atmospheric Chemistry and Physics*, *8*, 5975–5995. https://doi.org/10.5194/acp-8-5975-2008

Leung, L. R., Boos, W. R., Catto, J. L., DeMott, C., Martin, G. M., Neelin, J. D., et al. (2022). Exploratory precipitation metrics: Spatiotemporal characteristics, process-oriented, and phenomena-based. *Journal of Climate*, *1–55*. https://doi.org/10.1175/JCLI-D-21-0590.1

Li, J., Qian, Y., Leung, L. R., & Feng, Z. (2021). Summer mean and extreme precipitation over the Mid-Atlantic region:

Climatological characteristics and contributions from different precipitation types. *Journal of Geophysical Research: Atmospheres, 126*, e2021JD035045. https://doi.org/10.1029/2021JD035045

Lin, G., Fan, J., Feng, Z., Gustafson, W. I., Ma, P.-L., & Zhang, K. (2019). Can the multiscale modeling framework (MMF) simulate the MCS-associated precipitation over the Central United States? *Journal of Advances in Modeling Earth Systems, 11*, 4669–4686. https://doi.org/10.1029/2019MS001849

Lin, G., Jones, C. R., Leung, L. R., Feng, Z., & Ovchinnikov, M. (2022). Mesoscale convective systems in a superparameterized E3SM simulation at high resolution. *Journal of Advances in Modeling Earth Systems, 14*, e2021MS002660. https://doi.org/10.1029/2021MS002660

Liu, C., Shige, S., Takayabu, Y. N., & Zipser, E. (2015). Latent heating contribution from precipitation systems with different sizes, depths, and intensities in the tropics. *Journal of Climate, 28*, 186–203. https://doi.org/10.1175/JCLI-D-14-00370.1

Liu, N., Liu, C., & Lavigne, T. (2019). The variation of the intensity, height, and size of precipitation systems with El Niño-Southern oscillation in the tropics and subtropics. *Journal of Climate, 32*, 4281–4297. https://doi.org/10.1175/JCLI-D-18-0766.1

Liu, N., Leung, L. R., & Feng, Z. (2021). Global Mesoscale Convective system latent heating characteristics from GPM retrievals and an MCS tracking dataset. *Journal of Climate, 34*, 8599–8613. https://doi.org/10.1175/JCLI-D-20-0997.1

Machado, L. A. T., Rossow, W. B., Guedes, R. L., & Walker, A. W. (1998). Life cycle variations of mesoscale convective systems over the Americas. *Monthly Weather Review, 126*, 1630–1654. https://doi.org/10.1175/1520-0493(1998)126<1630:LCVOMC>2.0.CO;2

Magagi, R., & Barros, A. P. (2004). Estimation of latent heating of rainfall during the onset of the Indian Monsoon using TRMM PR and radiosonde data. *Journal of Applied Meteorology, 43*, 328–349. https://doi.org/10.1175/1520-0450(2004)043<0328:EOLHOR>2.0.CO;2

Maloney, E. D., & Hartmann, D. L. (2000). Modulation of hurricane activity in the Gulf of Mexico by the Madden-Julian oscillation. *Science, 287*, 2002–2004. https://doi.org/10.1126/science.287.5460.2002

Mapes, B., & Houze, R.A., Jr. (1992). An integrated view of the 1987 Australian monsoon and its mesoscale convective systems. I: Horizontal structure. *Quarterly Journal of the Royal Meteorological Society, 119*, 733–754. https://doi.org/10.1002/qj.49711951207

Mapes, B., Tulich, S., Lin, J., & Zuidema, P. (2006). The mesoscale convection life cycle: Building block or prototype for large-scale tropical waves? *Dynamics of Atmospheres and Oceans, 42*, 3–29. https://doi.org/10.1016/j.dynatmoce.2006.03.003

Mapes, B. E., & Wu, X. (2001). Notes and correspondence convective eddy momentum tendencies in long cloud-resolving model simulations. *Journal of the Atmospheric Sciences, 58*, 517–526. https://doi.org/10.1175/1520-0469(2001)058<0517:NACCEM>2.0.CO;2

Markowski, P., & Richardson, Y. (2010). Chapter 9: Mesoscale convective systems. In *Mesoscale meteorology in midlatitudes deep moist convection* (pp. 245–272). Wiley & Sons, Ltd. doi: 10.1002/978047068104.ch9

Marshall, J. S., & Palmer, W. M. K. (1948). The distribution of raindrops with size. *Journal of Meteorology, 5*, 165–166. https://doi.org/10.1175/1520-0469(1948)005<0165:tdorws>2.0.co;2

Middlemas, E. A., Clement, A. C., Medeiros, B., & Kirtman, B. (2019). Cloud radiative feedbacks and El niño-southern oscillation. *Journal of Climate, 32*(15), 4661–4680. https://doi.org/10.1175/JCLI-D-18-0842.1

Moncrieff, M.W. (1992). Organized convective systems: Archetypal dynamical models, mass and momentum flux theory, and parametrization. *Quarterly Journal of the Royal Meteorological Society, 118*, 819–850. https://doi.org/10.1002/qj.49711850703

Moncrieff, M. W. (2004). Analytic representation of the large-scale organization of tropical convection. *Journal of Atmospheric Sciences, 61*(13).

Moncrieff, M. W., Liu, C., & Bogenschutz, P. (2017). Simulation, modeling, and dynamically based parameterization of organized tropical convection for global climate models. *Journal of the Atmospheric Sciences, 74*, 1363–1380. https://doi.org/10.1175/JAS-D-16-0166.1

Moncrieff, M. W. (2010). The multiscale organization of moist convection and the intersection of weather and climate. In climate dynamics: Why does climate vary? *Geophysical Monograph: American Geophysical Union, 189*, 3–26. https://doi.org/10.1029/2008GM000838

Morel, C., & Senesi, S. (2002). A climatology of mesoscale convective systems over Europe using satellite infrared imagery. II: Characteristics of European mesoscale convective systems. *Quarterly Journal of the Royal Meteorological Society, 128*, 1973–1995. https://doi.org/10.1256/003590002320603494

Morrison, H., & Milbrandt, J. A. (2015). Parameterization of cloud microphysics based on the prediction of bulk ice particle properties. Part I: Scheme description and idealized tests. *Journal of the Atmospheric Sciences, 72*, 287–311. https://doi.org/10.1175/JAS-D-14-0065.1

Morrison, H., van Lier-Walqui, M., Fridlind, A. M., Grabowski, W. W., Harrington, J. Y., Hoose, C., et al. (2020a). Confronting the challenge of modeling cloud and precipitation microphysics. *Journal of Advances in Modeling Earth Systems, 12*, e2019MS001689. https://doi.org/10.1029/2019MS001689

Morrison, H., Peters, J. M., Varble, A. C., Hannah, W. M., & Giangrande, S. E. (2020b). Thermal chains and entrainment in cumulus updrafts. Part I: Theoretical description. *Journal of the Atmospheric Sciences, 77*, 3637–3660. https://doi.org/10.1175/JAS-D-19-0243.1

Mulholland, J. P., Nesbitt, S. W., Trapp, R. J., & Peters, J. M. (2020). The influence of terrain on the convective environment and associated convective morphology from an idealized modeling perspective. *Journal of the Atmospheric Sciences, 72*, 3929–3949. https://doi.org/10.1175/JAS-D-19-0190.1

Nesbitt, S. W., Cifelli, R., & Rutledge, S. A. (2006). Storm morphology and rainfall characteristics of TRMM precipitation features. *Monthly Weather Review, 134*, 2702–2721. https://doi.org/10.1175/MWR3200.1

Nicholson, S. E., & Grist, J. P. (2003). The seasonal evolution of the atmospheric circulation over West Africa and equatorial Africa. *Journal of Climate*, *16*, 1013–1030. https://doi.org/10.1175/1520-0442(2003)016<1013:TSEOTA>2.0.CO;2

North, K. W., Oue, M., Kollias, P., Giangrande, S. E., Collis, S. M., & Potvin, C. K. (2017). Vertical air motion retrievals in deep convective clouds using the ARM scanning radar network in Oklahoma during MC3E. *Atmospheric Measurement Techniques*, *10*, 2785–2806. https://doi.org/10.5194/amt-10-2785-2017

Núñez Ocasio, K. M., Evans, J. L., & Young, G. S. (2020). Tracking mesoscale convective systems that are potential candidates for tropical cyclogenesis. *Monthly Weather Review*, *148*(2), 655–669. https://doi.org/10.1175/MWR-D-19-0070.1

Pandya, R. E., & Durran, D. R. (1996). The Influence of convectively generated thermal forcing on the mesoscale circulation around squall lines. *Journal of the Atmospheric Sciences*, *53*, 2924–2951. https://doi.org/10.1175/1520-0469(1996)053<2924:TIOCGT>2.0.CO;2

Parhi, P., Giannini, A., Gentine, P., & Lall, U. (2016). Resolving contrasting regional rainfall responses to EL Niño over tropical Africa. *Journal of Climate*, *29*, 1461–1476. https://doi.org/10.1175/JCLI-D-15-0071.1

Peters, J. M., Morrison, H., Varble, A. C., Hannah, W. M., & Giangrande, S. E. (2020). Thermal chains and entrainment in cumulus updrafts. Part II: Analysis of idealized simulations. *Journal of the Atmospheric Sciences*, *77*, 3661–3681. https://doi.org/10.1175/JAS-D-19-0244.1

Plant, R. S. (2012). A new modelling framework for statistical cumulus dynamics. *Philosophical Transactions of the Royal Society A: Mathematical, Physical and Engineering Sciences*, *3701041–3701060*. https://doi.org/10.1098/rsta.2011.0377

Prein, A. F., Rasmussen, R. M., Wang, D., & Giangrande, S. E. (2021). Sensitivity of organized convective storms to model grid spacing in current and future climates. *Philosophical Transactions of the Royal Society A: Mathematical, Physical and Engineering Sciences*, *379*. https://doi.org/10.1098/rsta.2019.0546

Prein, A. F., Liu, C., Ikeda, K., Bullock, R., Rasmussen, R. M., Holland, G. J., & Clark, M. (2017a). Simulating North American mesoscale convective systems with a convection-permitting climate model. *Climate Dynamics*, *55*, 95–110. https://doi.org/10.1007/s00382-017-3993-2

Prein, A. F., Liu, C., Ikeda, K., Trier, S. B., Rasmussen, R. M., Holland, G. J., & Clark, M. P. (2017b). Increased rainfall volume from future convective storms in the US. *Nature Climate Change*, *7*, 880–884. https://doi.org/10.1038/s41558-017-0007-7

Pritchard, M. S., & Somerville, R. C. J. (2009). Assessing the diurnal cycle of precipitation in a multi-scale climate model. *Journal of Advances in Modeling Earth Systems*, *1*, 12. doi:10.3894/JAMES.2009.1.12.

Pritchard, M. S., Moncrieff, M. W., & Somerville, R. C. J. (2011). Orogenic propagating precipitation systems over the United States in a global climate model with embedded explicit convection. *Journal of the Atmospheric Sciences*, *68*, 1821–1840. https://doi.org/10.1175/2011JAS3699.1

Qian, Y., Yang, Z., Feng, Z., Liu, Y., Gustafson, W. I., Berg, L. K., et al. (2020). Neglecting irrigation contributes to the simulated summertime warm-and-dry bias in the central United States. *NPJ Climate and Atmospheric Science*, *3*, 31. https://doi.org/10.1038/s41612-020-00135-w

Rädel, G., Mauritsen, T., Stevens, B., Dommenget, D., Matei, D., Bellomo, K., & Clement, A. (2016). Amplification of El Niño by cloud longwave coupling to atmospheric circulation. *Nature Geoscience*, *9*, 106–110. https://doi.org/10.1038/ngeo2630

Randall, D., Khairoutdinov, M., Arakawa, A., & Grabowski, W. (2003). Breaking the cloud parameterization deadlock. *Bulletin of the American Meteorological Society, 84*, 1547–1564. https://doi.org/10.1175/BAMS-84-11-1547

Randall, D. A. (2013). Beyond deadlock. *Geophysical Research Letters*, *40*, 5970– 5976. doi:10.1002/2013GL057998

Rasmussen, K. L., & Houze, R. A. (2016). Convective initiation near the Andes in subtropical South America. *Monthly Weather Review*, *144*, 2351–2374. https://doi.org/10.1175/MWR-D-15-0058.1

Rasmussen, K.L., Hill, A.J., Toma, V.E., Zuluaga, M.D., Webster, P.J., & Houze, R.A., Jr. (2015). Multiscale analysis of three consecutive years of anomalous flooding in Pakistan. *Quarterly Journal of the Royal Meteorological Society*, *141*, 1259–1276. https://doi.org/10.1002/qj.2433

Rasmussen, K. L., Chaplin, M. M., Zuluaga, M. D., & Houze, R. A. (2016). Contribution of extreme convective storms to rainfall in South America. *Journal of Hydrometeorology*, *17*, 353–367. https://doi.org/10.1175/JHM-D-15-0067.1

Rickenbach, T.M., Nieto-Ferreira, R., Zarzar, C., & Nelson, B. (2015). A seasonal and diurnal climatology of precipitation organization in the southeastern United States. *Quarterly Journal of the Royal Meteorological Society*, *141*, 1938–1956. https://doi.org/10.1002/qj.2500

Roberts, M. J., Vidale, P. L., Senior, C., Hewitt, H. T., Bates, C., Berthou, S., et al. (2018). The benefits of global high resolution for climate simulation process understanding and the enabling of stakeholder decisions at the regional scale. *Bulletin of the American Meteorological Society*, *99*, 2341–2359. https://doi.org/10.1175/BAMS-D-15-00320.1

Roca, R., Aublanc, J., Chambon, P., Fiolleau, T., & Viltard, N. (2014). Robust observational quantification of the contribution of mesoscale convective systems to rainfall in the tropics. *Journal of Climate*, *27*, 4952–4958. https://doi.org/10.1175/JCLI-D-13-00628.1

Roca, R., Fiolleau, T., & Bouniol, D. (2017). A simple model of the lifecycle of mesoscale convective systems cloud shield in the tropics. *Journal of Climate*, *30*, 4283–4298. https://doi.org/10.1175/JCLI-D-16-0556.1

Roca, R., & Fiolleau, T. (2020). Extreme precipitation in the tropics is closely associated with long-lived convective systems. *Communications Earth & Environment*, *1*, 18. https://doi.org/10.1038/s43247-020-00015-4

Romatschke, U., & Houze, R. A. (2010). Extreme summer convection in South America. *Journal of Climate*, *23*, 3761–3791. https://doi.org/10.1175/2010JCLI3465.1

Rosenfeld, D., Lohmann, U., Raga, G. B., Dowd, C. D., Kulmala, M., Fuzzi, S., et al. (2008). Flood or drought:

How do aerosols affect precipitation? *Science*, *321*(5894), 1309LP–LP1313. https://doi.org/10.1126/science.1160606

Rosenfeld, D., Wood, R., Donner, L. J., & Sherwood, S. C. (2013). Aerosol cloud-mediated radiative forcing: Highly uncertain and opposite effects from shallow and deep clouds. In G. R. Asrar & J. W. Hurrell (Eds.), *Climate science for serving society: Research, modeling and prediction priorities* (pp. 105–149). Dordrecht: Springer Netherlands. https://doi.org/10.1007/978-94-007-6692-1_5

Rosenfeld, D., Sherwood, S., Wood, R., & Donner, L. (2014). Climate effects of aerosol-cloud interactions. *Science*, *343*, 379–380. https://doi.org/10.1126/science.1247490

Rosenfeld, D., Andreae, M. O., Asmi, A., Chin, M., de Leeuw, G., Donovan, D. P., et al. (2014). Global observations of aerosol-cloud-precipitation-climate interactions. *Reviews of Geophysics*, *52*(4), 750–808. https://doi.org/10.1002/2013RG000441

Rosenfeld, D., Zheng, Y., Hashimshoni, E., Pöhlker, M. L., Jefferson, A., Pöhlker, C., et al. (2016). Satellite retrieval of cloud condensation nuclei concentrations by using clouds as CCN chambers. *Proceedings of the National Academy of Sciences of the United States of America*, *113*(21), 5828–5834. https://doi.org/10.1073/pnas.1514044113

Rotunno, R., Klemp, J. B., & Weisman, M. L. (1988). A theory for strong, long-lived squall lines. *Journal of the Atmospheric Sciences*, *45*, 463–485. https://doi.org/10.1175/1520-0469(1988)045<0463:ATFSLL>2.0.CO;2

Rowe, A. K., & Houze, R. A. (2014). Microphysical characteristics of MJO convection over the Indian Ocean during DYNAMO. *Journal of Geophysical Research: Atmospheres*, *119*, 2543–2554. doi:10.1002/2013JD020799.

Ruppert, J. H., & Johnson, R. H. (2015). Diurnally modulated cumulus moistening in the preonset stage of the Madden? Julian Oscillation during DYNAMO. *Journal of the Atmospheric Sciences*, *72*(4), 1622–1647. https://doi.org/10.1175/JAS-D-14-0218.1

Saint-Lu, M., Bony, S., & Dufresne, J.-L. (2020). Observational evidence for a stability Iris effect in the tropics. *Geophysical Research Letters*, *47*, e2020GL089059. https://doi.org/10.1029/2020GL089059

Sakaeda, N., Powell, S. W., Dias, J., & Kiladis, G. N. (2018). The diurnal variability of precipitating cloud populations during DYNAMO. *Journal of the Atmospheric Sciences*, *75*(4), 1307–1326. https://doi.org/10.1175/JAS-D-17-0312.1

Sakaguchi, K., Leung, L. R., Zhao, C., Yang, Q., Lu, J., Hagos, S., et al. (2015). Exploring a multiresolution approach using AMIP simulations. *Journal of Climate*, *28*, 5549–5574. https://doi.org/10.1175/JCLI-D-14-00729.1

Salio, P., Nicolini, M., & Zipser, E. J. (2007). Mesoscale convective systems over southeastern South America and their relationship with the South American low-level jet. *Monthly Weather Review*, *135*, 1290–1309. https://doi.org/10.1175/MWR3305.1

Satoh, M., Stevens, B., Judt, F., Khairoutdinov, M., Lin, S. J., Putman, W. M., & Düben, P. (2019). Global cloud-resolving models. *Current Climate Change Reports*, *5*, 172–184. https://doi.org/10.1007/s40641-019-00131-0

Schalkwijk, J., Jonker, H. J. J., Siebesma, A. P., & Van Meijgaard, E. (2015). Weather forecasting using GPU-based large-eddy simulations. *Bulletin of the American Meteorological Society*, *96*, 715–723. https://doi.org/10.1175/BAMS-D-14-00114.1

Schumacher, C., & Houze, R. A. (2003). Stratiform rain in the tropics as seen by the TRMM precipitation radar. *Journal of Climate*, *16*, 1739–1756. https://doi.org/10.1175/1520-0442(2003)016<1739:SRITTA>2.0.CO;2

Schumacher, C., Houze, R. A., & Kraucunas, I. (2004). The tropical dynamical response to latent heating estimates derived from the TRMM precipitation radar. *Journal of the Atmospheric Sciences*, *61*, 1341–1358. https://doi.org/10.1175/1520-0469(2004)061<1341:TTDRTL>2.0.CO;2

Schumacher, R. S., & Johnson, R. H. (2006). Characteristics of U.S. extreme rain events during 1999–2003. *Weather and Forecasting*, *21*, 69–85. https://doi.org/10.1175/WAF900.1

Schumacher, R. S., & Rasmussen, K. L. (2020). The formation, character and changing nature of mesoscale convective systems. *Nature Reviews Earth and Environment*, *1*, 300–314. https://doi.org/10.1038/s43017-020-0057-7

Seeley, J. T., & Romps, D. M. (2015). Why does tropical convective available potential energy (CAPE) increase with warming? *Geophysical Research Letters*, *42*, 10,429–10,437. doi:10.1002/2015GL066199.

Seinfeld, J. H., Bretherton, C., Carslaw, K. S., Coe, H., DeMott, P. J., Dunlea, E. J., et al. (2016). Improving our fundamental understanding of the role of aerosol–cloud interactions in the climate system. *Proceedings of the National Academy of Sciences*, *113*(21), 5781–5790. https://doi.org/10.1073/pnas.1514043113

Semie, A. G., & Bony, S. (2020). Relationship between precipitation extremes and convective organization inferred from satellite observations. *Geophysical Research Letters*, *47*, e2019GL086927. https://doi.org/10.1029/2019GL086927

Shapiro, M. A., Shukla, J., Brunet, G., & Coauthors. (2010). An Earth-system prediction initiative for the twenty-first century. *Bulletin of the American Meteorological Society, 91*(10), 1377–1388.

Shepherd, J. M., & Burian, S. J. (2003). Detection of urban-induced rainfall anomalies in a major coastal city. *Earth Interactions*, *7*(4), 1–17. https://doi.org/10.1175/1087-3562(2003)007<0001:douira>2.0.co;2

Shepherd, M., Carter, M., Manyin, M., Messen, D., & Burian, S. (2010). The impact of urbanization on current and future coastal precipitation: A case study for Houston. *Environment and Planning B: Planning and Design*, *37*(2), 284–304. https://doi.org/10.1068/b34102t

Shi, M., Liu, J., Worden, J. R., Bloom, A. A., Wong, S., & Fu, R. (2019). The 2005 Amazon drought legacy effect delayed the 2006 wet season onset. *Geophysical Research Letters*, *46*(15), 9082–9090. https://doi.org/10.1029/2019GL083776

Silva Dias, M. A. F., Petersen, W., Silva Dias, P. L., Cifelli, R., Betts, A. K., Longo, M., et al. (2002). A case study of convective organization into precipitating lines in the Southwest Amazon during the WETAMC and TRMM-LBA. *Journal of Geophysical Research*, *107*(D20), 8078. doi:10.1029/2001JD000375

Siqueira, M., Katul, G., & Porporato, A. (2009). Soil moisture feedbacks on convection triggers: The role of soil-plant hydrodynamics. *Journal of Hydrometeorology*, *10*(1), 96–112. https://doi.org/10.1175/2008JHM1027.1

Slingo, J. M., & Annamalai, H. (2000). 1997: The El Niño of the century and the response of the Indian summer monsoon. *Monthly Weather Review*, *128*, 1778–1797. https://doi.org/10.1175/1520-0493(2000)128<1778:TENOOT>2.0.CO;2.

Smith, B. T., Thompson, R. L., Grams, J. S., Broyles, C., & Brooks, H. E. (2012). Convective modes for significant severe thunderstorms in the contiguous United States. Part I: Storm classification and climatology. *Weather and Forecasting*, *27*(5), 1114–1135. https://doi.org/10.1175/WAF-D-11-00115.1

Song, F., Feng, Z., Ruby, L. L., Houze, R. A., Wang, J., Hardin, J., & Homeyer, C. R. (2019). Contrasting spring and summer large-scale environments associated with mesoscale convective systems over the U.S. Great Plains. *Journal of Climate*, *32*, 6749–6767. https://doi.org/10.1175/JCLI-D-18-0839.1

Song, F., Feng, Z., Leung, L. R., Pokharel, B., Wang, S.-Y. S, Chen, X., et al. (2021). Crucial roles of eastward propagating environments in the summer MCS initiation over the U.S. Great Plains. *Journal of Geophysical Research: Atmospheres*, *126*, e2021JD034991. https://doi.org/10.1029/2021JD034991

Stephens, E., Day, J. J., Pappenberger, F., & Cloke, H. (2015). Precipitation and floodiness. *Geophysical Research Letters*, *42*, 10,316–10,323. doi:10.1002/2015GL066779

Stevens, B., & Feingold, G. (2009). Untangling aerosol effects on clouds and precipitation in a buffered system. *Nature*, *461*, 607–613. https://doi.org/10.1038/nature08281

Stevens, B., Satoh, M., Auger, L., Biercamp, J., Bretherton, C. S., Chen, X., et al. (2019). DYAMOND: The Dynamics of the Atmospheric general circulation Modeled On Non-hydrostatic Domains. *Progress in Earth and Planetary Science*, *6*, 61. https://doi.org/10.1186/s40645-019-0304-z

Stevenson, S. N., & Schumacher, R. S. (2014). A -10-Year survey of extreme rainfall events in the central and eastern United States using gridded multisensor precipitation analyses. *Monthly Weather Review*, *53*(12), 2848–2862. https://doi.org/10.1175/MWR-D-13-00345.1

Su, H., Jiang, J. H., Neelin, J. D., Shen, T. J., Zhai, C., Yue, Q., et al. (2017). Tightening of tropical ascent and high clouds key to precipitation change in a warmer climate. *Nature Communications*, *8*, 15771. https://doi.org/10.1038/ncomms15771

Sullivan, S. C., Schiro, K. A., Stubenrauch, C., & Gentine, P. (2019). The response of tropical organized convection to El Niño warming. *Journal of Geophysical Research: Atmospheres*, *124*, 8481–8500. https://doi.org/10.1029/2019JD031026

Sullivan, S. C., Schiro, K. A., Yin, J., & Gentine, P. (2020). Changes in tropical precipitation intensity with El Niño warming. *Geophysical Research Letters*, *47*, e2020GL087663. https://doi.org/10.1029/2020GL087663

Tackett, J. L., & Di Girolamo, L. (2009). Enhanced aerosol backscatter adjacent to tropical trade wind clouds revealed by satellite-based lidar. *Geophysical Research Letters*, *36*, L14804. doi:10.1029/2009GL039264

Takayabu, Y.N., & Tao, W. K. (2020). Latent heating retrievals from satellite observations. In: Levizzani, V., Kidd, C., Kirschbaum, D., Kummerow, C., Nakamura, K., & Turk, F. (Eds.), *Satellite precipitation measurement. Advances in global change research* (vol. *69*). Cham: Springer. https://doi.org/10.1007/978-3-030-35798-6_22

Taschetto, A. S., & England, M. H. (2009). El Niño Modoki impacts on Australian rainfall. *Journal of Climate*, *22*(11), 3167–3174. https://doi.org/10.1175/2008JCLI2589.1

Taylor, C.M., Harris, P.P., & Parker, D.J. (2010). Impact of soil moisture on the development of a Sahelian mesoscale convective system: A case-study from the AMMA Special Observing Period. *Quarterly Journal of the Royal Meteorological Society*, *136*, 456–470. https://doi.org/10.1002/qj.465

Toms, B. A., van Den Heever, S. C., Riley Dellaripa, E. M., Saleeby, S. M., & Maloney, E. D. (2020). The boreal summer Madden–Julian oscillation and moist convective morphology over the Maritime Continent. *Journal of the Atmospheric Sciences*, *77*(2), 647–667. https://doi.org/10.1175/JAS-D-19-0029.1

Trenberth, K. E. (1997). The definition of El Niño. *Bulletin of the American Meteorological Society*, *78*(12), 2771–2778. https://doi.org/10.1175/1520-0477(1997)078<2771:TDOENO>2.0.CO;2

Vant-Hull, B., Rossow, W., & Pearl, C. (2016). Global comparisons of regional life cycle properties and motion of multiday convective systems: Tropical and midlatitude land and ocean. *Journal of Climate*, *29*(16), 5837–5858. https://doi.org/10.1175/JCLI-D-15-0698.1

Varble, A., Zipser, E. J., Fridlind, A. M., Zhu, P., Ackerman, A. S., Chaboureau, J.-P., et al. (2014a). Evaluation of cloud-resolving and limited area model intercomparison simulations using TWP-ICE observations: 1. Deep convective updraft properties. *Journal of Geophysical Research: Atmospheres*, *119*, 13,891–13,918. doi:10.1002/2013JD021371

Varble, A., Zipser, E. J., Fridlind, A. M., Zhu, P., Ackerman, A. S., Chaboureau, J.-P., et al. (2014b). Evaluation of cloud-resolving and limited area model intercomparison simulations using TWP-ICE observations: 2. Precipitation microphysics. *Journal of Geophysical Research: Atmospheres*, *119*, 13,919–13,945. doi:10.1002/2013JD021372

Varble, A., Morrison, H., & Zipser, E. (2020). Effects of under-resolved convective dynamics on the evolution of a squall line. *Monthly Weather Review*, *148*(1), 289–311. https://doi.org/10.1175/MWR-D-19-0187.1

Várnai, T., & Marshak, A. (2009). MODIS observations of enhanced clear sky reflectance near clouds. *Geophysical Research Letters*, *36*, L06807. doi:10.1029/2008GL037089

Virts, K. S., & Houze, R. A. (2016). Seasonal and intraseasonal variability of mesoscale convective systems over the South Asian monsoon region. *Journal of the Atmospheric Sciences*, *73*(12), 4753–4774. https://doi.org/10.1175/JAS-D-16-0022.1

Vondou, D.A., Nzeukou, A., Lenouo, A., & Mkankam Kamga, F. (2010). Seasonal variations in the diurnal patterns of convection in Cameroon–Nigeria and their neighboring areas. *Atmospheric Science Letters*, *11*, 290–300. https://doi.org/10.1002/asl.297

Wang, B., Jin, C., & Liu, J. (2020a). Understanding future change of global monsoons projected by CMIP6 Models. *Journal of Climate*, *33*(15), 6471–6489. https://doi.org/10.1175/JCLI-D-19-0993.1

Wang, D., Giangrande, S. E., Feng, Z., Hardin, J. C., & Prein, A. F. (2020b). Updraft and downdraft core size and intensity as revealed by radar wind profilers: MCS observations and idealized model comparisons. *Journal of Geophysical Research: Atmospheres*, *125*, e2019JD031774. https://doi.org/10.1029/2019JD031774

Wang, J., Bras, R. L., & Eltahir, E. A. B. (2000). The impact of observed deforestation on the mesoscale distribution of rainfall and clouds in Amazonia. *Journal of Hydrometeorology*, *1*(3), 267–286. https://doi.org/10.1175/1525-7541(2000)001<0267:TIOODO>2.0.CO;2

Wang, J., Chagnon, F. J. F., Williams, E. R., Betts, A. K., Renno, N. O., Machado, L. A. T., et al. (2009). Impact of deforestation in the Amazon basin on cloud climatology. *Proceedings of the National Academy of Sciences of the United States of America*, *106*(10), 3670–3674. https://doi.org/10.1073/pnas.0810156106

Wang, J., Fan, J., Feng, Z., Zhang, K., Roesler, E., Hillman, B., et al. (2021). Impact of a new cloud microphysics parameterization on the simulations of mesoscale convective systems in E3SM. *Journal of Advances in Modeling Earth Systems*, *13*, e2021MS002628. https://doi.org/10.1029/2021MS002628

Wehner, M. F., Reed, K. A., Li, F., Prabhat., Bacmeister, J., Chen, C.-T., et al. (2014). The effect of horizontal resolution on simulation quality in the Community Atmospheric Model, CAM5.1. *Journal of Advances in Modeling Earth Systems*, *6*, 980–997. doi:10.1002/2013MS000276

Weisman, M. L., & Rotunno, R. (2004). "A theory for strong long-lived squall lines" revisited. *Journal of the Atmospheric Sciences*, *61*(4), 361–382. https://doi.org/10.1175/1520-0469(2004)061<0361:ATFSLS>2.0.CO;2

Van Weverberg, K., Morcrette, C. J., Petch, J., Klein, S. A., Ma, H.-Y., Zhang, C., et al. (2018). CAUSES: Attribution of surface radiation biases in NWP and climate models near the U.S. *Southern Great Plains. Journal of Geophysical Research: Atmospheres*, *123*. https://doi.org/10.1002/2017JD027188

Wienert, U., & Kuttler, W. (2005). The dependence of the urban heat Island intensity on latitude - A statistical approach. *Meteorologische Zeitschrift*, *14*(5), 677–686. https://doi.org/10.1127/0941-2948/2005/0069

Wing, A. A., Stauffer, C. L., Becker, T., Reed, K. A., Ahn, M.-S., Arnold, N. P., et al. (2020). Clouds and convective self-aggregation in a multimodel ensemble of radiative-convective equilibrium simulations. *Journal of Advances in Modeling Earth Systems*, *12*, e2020MS002138. https://doi.org/10.1029/2020MS002138

Worden, S., Fu, R., Chakraborty, S., Liu, J., & Worden, J. (2021). Where does moisture come from over the Congo Basin? *Journal of Geophysical Research: Biogeosciences*, *126*, e2020JG006024. https://doi.org/10.1029/2020JG006024

Wright, J. S., Fu, R., Worden, J. R., Chakraborty, S., Clinton, N. E., Risi, C., et al. (2017). Rainforest-initiated wet season onset over the southern Amazon. *Proceedings of the National Academy of Sciences of the United States of America*, *114*(32), 8481–8487. https://doi.org/10.1073/pnas.1621516114

Yang, Q., Houze, R. A., Leung, L. R., & Feng, Z. (2017). Environments of long-lived mesoscale convective systems over the central United States in convection permitting climate simulations. *Journal of Geophysical Research: Atmospheres*, *122*, 13,288–13,307. https://doi.org/10.1002/2017JD027033

Yang, Q., Majda, A. J., & Moncrieff, M. W. (2019). Upscale impact of mesoscale convective systems and its parameterization in an idealized GCM for an MJO analog above the Equator. *Journal of the Atmospheric Sciences*, *76*(3), 865–892. https://doi.org/10.1175/JAS-D-18-0260.1

Yang, X., Fei, J., Huang, X., Cheng, X., Carvalho, L. M. V., & He, H. (2015). Characteristics of mesoscale convective systems over China and its vicinity using geostationary satellite FY2. *Journal of Climate*, *28*(12), 4890–4907. https://doi.org/10.1175/JCLI-D-14-00491.1

Yang, X., Hou, Y., & Chen, B. (2011). Observed surface warming induced by urbanization in east China. *Journal of Geophysical Research*, *116*, D14113. doi:10.1029/2010JD015452

Yuan, J., & Houze, R. A. (2010). Global variability of mesoscale convective system anvil structure from A-Train satellite data. *Journal of Climate*, *23*(21), 5864–5888. https://doi.org/10.1175/2010JCLI3671.1

Zhang, C., & Dong, M. (2004). Seasonality in the Madden-Julian oscillation. *Journal of Climate*, *17*(16), 3169–3180. https://doi.org/10.1175/1520-0442(2004)017<3169:SITMO>2.0.CO;2

Zhang, L., Min, J., Zhuang, X., & Schumacher, R. S. (2019). General features of extreme rainfall events produced by MCSS over East China during 2016–17. *Monthly Weather Review*, *147*(7), 2693–2714. https://doi.org/10.1175/MWR-D-18-0455.1

Zhang, W., Villarini, G., Vecchi, G. A., & Smith, J. A. (2018). Urbanization exacerbated the rainfall and flooding caused by hurricane Harvey in Houston. *Nature*, *563*, 384–388. https://doi.org/10.1038/s41586-018-0676-z

Zhang, Z., Varble, A., Feng, Z., Hardin, J., & Zipser, E. (2021). Growth of mesoscale convective systems in observations and a seasonal convection-permitting simulation over Argentina. *Monthly Weather Review*, *149*(10), 3469–3490. https://doi.org/10.1175/mwr-d-20-0411.1

Zhong, S., Qian, Y., Zhao, C., Leung, R., Wang, H., Yang, B., et al. (2017). Urbanization-induced urban heat island and aerosol effects on climate extremes in the Yangtze River Delta region of China. *Atmospheric Chemistry and Physics*, *17*, 5439–5457. https://doi.org/10.5194/acp-17-5439-2017

Zhou, W., Leung, L. R., Song, F., & Lu, J. (2021). Future changes in the Great Plains Low-Level Jet governed by seasonally dependent pattern changes in the North Atlantic Subtropical High. *Geophysical Research Letters*, *48*(4), e2020GL090356. https://doi.org/10.1029/2020GL090356

Zipser, E. J., Cecil, D. J., Liu, C., Nesbitt, S. W., & Yorty, D. P. (2006). Where are the most: Intense thunderstorms on Earth? *Bulletin of the American Meteorological Society*, *87*(8), 1057–1072. https://doi.org/10.1175/BAMS-87-8-1057

Zolman, J. L., Zipser, E. J., & Mohr, K. I. (2000). A comparison of tropical mesoscale convective systems in El Nino and La Nina. *Journal of Climate*, *13*(18), 3314–3326. https://doi.org/10.1175/1520-0442(2000)013<3314:ACOTMC>2.0.CO;2

Part III
Clouds and Circulation

10

Interactions Between the Tropical Atmospheric Overturning Circulation and Clouds in Present and Future Climates

Kathleen A. Schiro[1] **and Hui Su**[2,3]

ABSTRACT

Uncertainty in the tropical cloud feedback has been a root cause of uncertainty in estimating climate sensitivity for decades. Considering that the distribution of cloudiness is tightly coupled to the tropical overturning circulation, improving our mechanistic understanding of circulation changes and their coupling to clouds is of paramount importance. In the tropics, shallow and deep convective clouds are coupled to zonal and meridional deep overturning circulations known as the Walker and Hadley circulations, respectively, through a multitude of dynamic, thermodynamic, microphysical, and kinematic processes. There is substantial modeling and observational evidence, supported by theoretical arguments, that the tropical overturning circulation is slowing down and the tropical ascent area is reducing. In part due to ascent area reduction, as well as increased tropospheric stability, detrainment from tropical deep convection and high cloudiness are generally reducing in response to greenhouse gas forcing. Tropical ascent area reduction has also been linked to a reduction in low cloudiness through energetic constraints on subsidence weakening. Tropical low cloudiness is generally projected to decrease, which constitutes a strong positive feedback on the climate system. An enhanced rate of lower tropospheric mixing and boundary layer dehydration, as well as reduced equatorial upwelling due to a weakening of the Walker circulation, both likely contribute substantially to the positive low cloud feedback. Though causal pathways have historically been difficult to discern, experiments in which cloud-radiation interactions have been disabled have revealed that high clouds act to narrow the Intertropical Convergence Zone (ITCZ) through their effects on longwave radiation while low clouds enhance the precipitation and atmospheric circulation over tropical oceans through their effects on surface turbulent fluxes. We argue that such studies aimed at identifying and explaining causal linkages between circulation and clouds in present and future climates, in novel ways, should be prioritized in future work.

10.1. INTRODUCTION

Clouds, convection, and the large-scale atmospheric overturning circulation in the tropics interact in a multitude of ways across scales, which have been studied using both observations and models. Vertical motion greatly determines the vertical extent and characteristics of cloudiness throughout the tropics as shown in Figure 10.1 (Bony et al., 2004; Su et al., 2013; Su et al., 2014). High clouds dominate in ascent regimes and low clouds dominate in descent regimes, with a majority of tropical ocean basins covered by low clouds (see Chapter 7: Tropical Marine Low Clouds). Other meteorological controls, such as sea surface temperature (SST), column-integrated water vapor (CWV), and lower tropospheric stability (LTS) are strong controls on the vertical

[1]*Department of Environmental Sciences, University of Virginia, Charlottesville, VA, USA*

[2]*Department of Civil and Environmental Engineering, The Hong Kong University of Science and Technology, Hong Kong SAR*

[3]*Department of Atmospheric and Oceanic Sciences, University of California Los Angeles, Los Angeles, CA, USA*

Clouds and Their Climatic Impacts: Radiation, Circulation, and Precipitation, Geophysical Monograph 281, First Edition.
Edited by Sylvia C. Sullivan and Corinna Hoose.

DOI: 10.1002/9781119700357.ch10

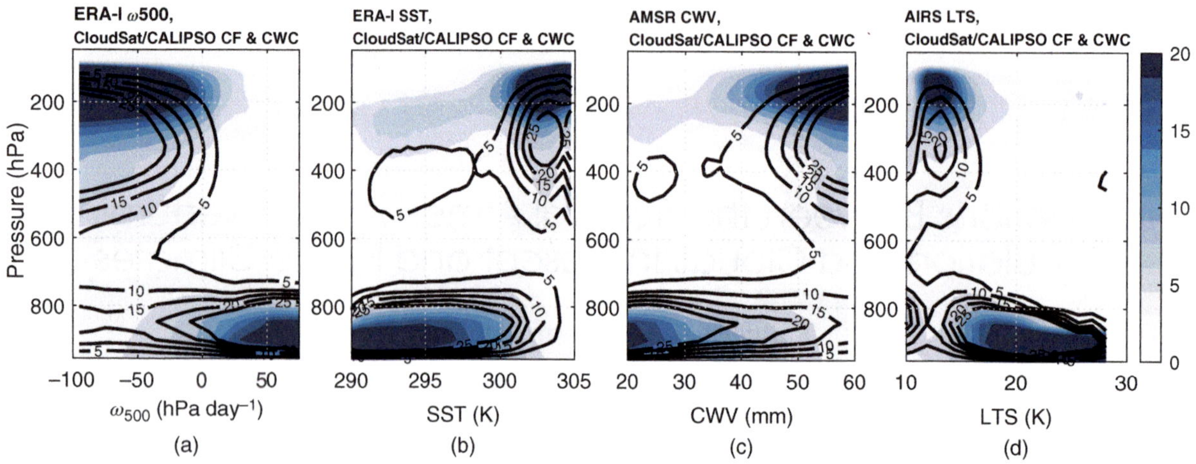

Figure 10.1 (a) The observed mean cloud water content (liquid + ice, black contours; mg m^{-3}) and cloud fraction (color shading; %) from CloudSat/CALIPSO conditionally averaged by (a) pressure velocity at 500 hPa (ω_{500}), (b) sea surface temperature (SST) from the Advanced Microwave Scanning Radiometer (AMSR-E), (c) column-integrated water vapor (CWV) from AMSR-E, and (d) lower tropospheric stability ($\theta_{700} - \theta_{1000}$) from 30S-30N from AIRS. Cloud water content profiles are from version r04 of CloudSat 2B-CWC-RO and cloud fraction is from version r04 of combined radar and lidar retrievals (2B-GEOPROF-LIDAR). The conditional sampling uses monthly cloud water content and cloud fraction regridded at 2.5° (longitude) × 2° (latitude) and onto 40 vertical levels from the surface to 20 km, with an interval of 500 m. The monthly CloudSat/CALIPSO data from August 2006 to July 2010 are used.

distribution of cloudiness (Figure 10.1), as the dynamics and thermodynamics of the tropical atmosphere are tightly coupled. Therefore, as the large-scale overturning circulations in the tropics (the Hadley and Walker circulations) change, either due to internal variability of the climate system or in response to external forcing such as increasing anthropogenic greenhouse gases, the vertical distribution of cloudiness will change accordingly.

Cloud feedbacks are changes to cloud radiative effects in response to changes in radiative forcing, such as the increase in greenhouse gases. Clouds reflect sunlight and thus cool the planet (~−50 Wm^{-2}). They also exert a positive warming effect (~30 Wm^{-2}) due to the trapping of longwave emission from Earth's surface (Stephens et al., 2012). The difference between the mean all-sky and the clear-sky energy balances is called the "cloud radiative effect" (CRE; Ramanathan et al., 1989), which has a shortwave component, an infrared or longwave component, and a total effect (the sum of the longwave and shortwave contributions). The overall magnitude of the cloud feedback in response to anthropogenic forcing is likely positive, increasing the magnitude of the surface temperature change (IPCC AR5, AR6). This consensus is based in part on the theories and evidence suggesting that high-cloud altitude increases with warming and tropical low-cloud amount decreases with warming (Bretherton et al., 2013; Hartmann & Larson 2002; Qu et al., 2015). High cloud amount is also generally projected to decrease, contributing significantly to the intermodel spread in the cloud feedback strength, yet it is unclear whether the decrease will act as a positive or negative feedback.

In this chapter, we discuss the state-of-the-science linking cloud feedbacks to circulation changes through our understanding of changes observed in the satellite record and a hierarchy of modeling experiments, paying special attention to changes in the Intertropical Convergence Zone (ITCZ) and Hadley circulation. While convective, microphysical, and other relevant small-scale processes play a critical role in the cloud-circulation relationships discussed, here we will remain focused on the "large-scale" O(>1,000 km, months to decades) interactions between clouds, convection, and circulation and their responses to internal variability in the climate system and to anthropogenic forcing.

10.2. FUTURE CHANGES TO THE TROPICAL OVERTURNING CIRCULATION

The strength of the atmospheric overturning circulation is projected to decrease as the climate warms (Held & Soden, 2006; Vecchi & Soden, 2007). This includes a weakening of both the Hadley (Frierson et al., 2007; Held & Soden, 2006; Lu et al., 2007; Lau & Kim, 2015; Su et al., 2014; Su et al., 2017; Vecchi & Soden, 2007) and Walker cells (Bayr et al., 2014; DiNezio et al., 2013; Kociuba & Power, 2015; Power & Kociuba, 2011; Vecchi et al., 2006; Vecchi & Soden, 2007; Zhang & Song, 2006).

The large-scale rising motion in the tropics weakens to make up for the fact that a 7% per K increase in saturation water vapor pressure per the Clausius-Clapeyron relationship far exceeds the roughly 2% per K radiatively constrained increase in precipitation (Held & Soden, 2006). In response to this radiative constraint, the tropical circulation slows down, which is shown in Figure 10.2 (Bony et al., 2013).

The Hadley circulation strength is found to be reduced by up to 2.6% K^{-1} for an abrupt quadrupling of carbon dioxide concentration (Feldl & Bordoni, 2016). Moreover, the Hadley cell's meridional extent is projected to widen, with the subsiding branch in the subtropics extending farther poleward (Frierson et al., 2007; Johanson & Fu, 2009; Lu et al., 2007). The poleward shift of the Hadley cell edge is two to three times larger in the Southern Hemisphere (SH) than the Northern Hemisphere in response to increasing greenhouse gases, except during austral spring, in both CMIP5 and CMIP6 models (Grise & Davis, 2020). Hadley circulation widening exhibits large seasonal variations (Hu & Fu, 2007): widening trends in summer and autumn are much more significant than trends in winter and spring. Given the robust widening in the SH summer, the ozone hole has been investigated for its potential role in modifying the width of the Hadley circulation (Polvani et al., 2011). Stratospheric ozone depletion causes widening of the SH cell in austral spring and summer and widening of the Northern Hemispheric cell in boreal spring in CMIP5 models (Tao et al., 2015), while increasing greenhouse gases cause widening of the Hadley circulation in both hemispheres. The role of anthropogenic aerosols on the width of the Hadley circulation is unclear, as Tao et al. (2015) found that they appear to have no significant effects on the width of the Hadley circulation, at least in CMIP5 models while Allen et al. (2012) suggested a strong influence of black carbon. SST changes are also considered to be an important factor contributing to the widening of the Hadley circulation (Hu et al., 2011; Tao et al., 2015). Changes to cloudiness associated with the widening of the Hadley circulation are most pronounced in the extratropics (e.g., Lipat et al., 2017). However, the role of clouds in determining the widening response to warming is non-trivial. For instance, Ceppi and Shepherd (2017) and Voigt et al. (2019) found that most of the poleward circulation expansion in response to increasing surface temperatures can be attributed to changes in clouds and their associated radiative effects. Clouds have also been found to impact the strength of the Hadley circulation, yet the results are model-dependent (Ceppi & Hartmann, 2016; Voigt & Shaw, 2015; Voigt & Albern, 2019). The impact of clouds on the tropical overturning circulation will be discussed in greater detail in Section 10.5 below.

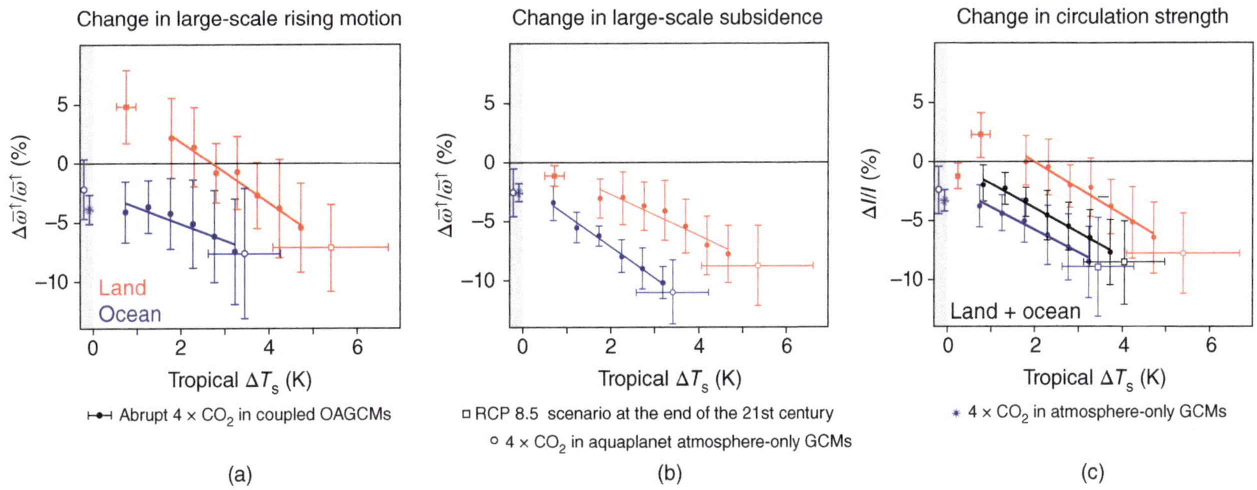

Figure 10.2 (a) CMIP5 multi-model mean fractional change in mean upward motion averaged over tropical grid points that have upward motion in monthly tropics-wide averages. (b) Same as in (a) but for downward vertical pressure velocities. (c) Same as in (a) but for the strength of the overturning circulation (I), which is the difference between upward and downward pressure velocities, predicted by coupled ocean-atmosphere models after an abrupt CO_2 quadrupling (solid circles, shown only when at least 10 model results are available). Results are shown for land (red) and ocean (blue) areas, compared with pre-industrial values. Estimates from the RCP8.5 scenario around 2090 are shown in open squares, and estimates from the 4×CO_2 atmosphere-only experiments with fixed SSTs in realistic (stars) and aquaplanet (open circles) configurations are also shown. The vertical bars show ±1 standard deviation of individual model results around the multi-model mean. Source: Adapted from Bony et al. (2013).

Another key characteristic of the response in the Hadley cell to greenhouse gas forcing is an equatorward narrowing of the tropical ascent zone (Byrne & Schneider, 2016a, 2016b; Byrne et al., 2018; Donohoe et al., 2019; Lau & Kim, 2015; Su et al., 2017). However, the magnitude of the tropical ascent area reduction differs across the latest generation of Coupled Model Intercomparison Project climate models (CMIP3-6; Byrne et al., 2018; Su et al., 2017; Su et al., 2019), owing at least in part to the differences in convective and microphysical process representation (Schiro et al., 2019). Figure 10.3 shows this enhanced equatorial ascent and reduced ascent on its flanks. Subsidence is also reduced in the subtropics. Additionally, the seasonal ITCZs are projected to shift equatorward. These zones widen and weaken, which causes a narrowing of the annual-mean zonal-mean tropical ascent and a widening of the descent regimes (Donohoe et al., 2019; Zhou et al., 2020). The zonal mean ITCZ location shows no robust shift under warming in climate model simulations (Donohoe & Voigt, 2017). However, Mamalakis et al. (2021) pointed out a robust regionally-varying ITCZ shift: a northward shift over eastern Africa and the Indian Ocean and a southward shift in the eastern Pacific and Atlantic oceans.

Projected changes in the ITCZ width are a combined result of various widening and tightening processes (Byrne & Schneider, 2016a,2016b) that need to be more carefully analyzed across a model hierarchy. For instance, Byrne and Schneider (2016b) found that ITCZ narrowing arises because the meridional moisture gradient strengthens with warming. On the other hand, this narrowing is partly compensated by the effect of increased shortwave heating in the ascent region, due to both cloud and clear-sky effects, which acts to widen the ITCZ. Ahmed et al. (2023) developed a simple model based on dry static energy and moisture equations and driven by SST changes to examine controls on ITCZ width. They found that ITCZ narrowing under uniform warming results

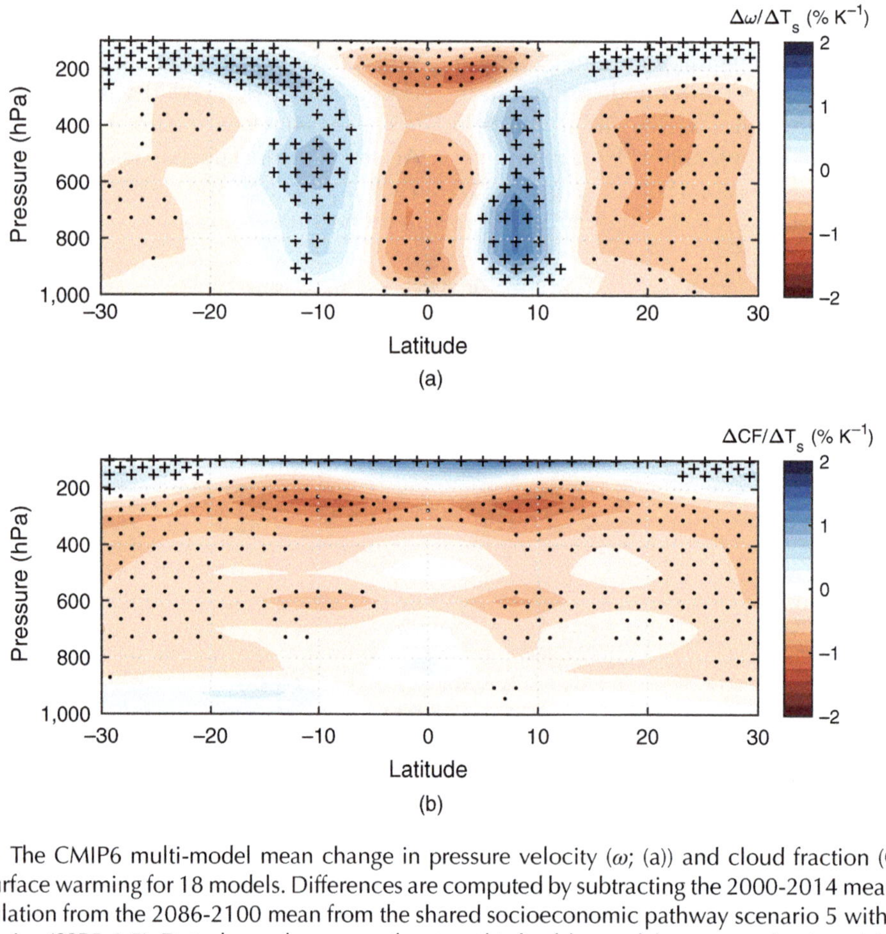

Figure 10.3 The CMIP6 multi-model mean change in pressure velocity (ω; (a)) and cloud fraction (CF; (b)) per degree of surface warming for 18 models. Differences are computed by subtracting the 2000-2014 mean in the historical simulation from the 2086-2100 mean from the shared socioeconomic pathway scenario 5 with 8.5 W m^{-2} radiative forcing (SSP5-8.5). Dots show where more than two thirds of the models agree on the sign of the decrease; pluses show where more than two thirds of the models agree on the sign of the increase. The models included are ACCESS-CM2, CAMS-CSM1-0, CESM2, CNRM-CM6-1, CNRM-ESM2-1, FGOALS-f3-L, FGOALS-g3, GFDL-CM4, GISS-E2-1-G, MIROC6, MIROC-ES2L, MPI-ESM1-2-HR, MPI-ESM1-2-LR, MRI-ESM2-0, NESM3, NorESM2-LM, NorESM2-MM, and UKESM1-0-LL.

from reduced GMS (gross moist stability: the ratio of the vertically integrated horizontal divergence of some intensive quantity conserved in moist adiabatic processes and a measure of the strength of moist convection per unit area; Neelin & Held, 1987; Raymond et al., 2009) and increased shortwave absorption by water vapor in the deep tropics, as well as increased poleward moisture transport. On the other hand, weaker cloud radiative effects with warming – namely, weaker longwave radiative heating in the ascent region – and increased radiative emission in the dry region expand the ITCZ width. Schiro et al. (2019) also inferred from results of a parameter perturbation experiment in the Community Atmosphere Model (CAM) that weaker longwave cloud radiative effects with warming, due to a reduction in high clouds, would counteract the ITCZ narrowing. The role of longwave cloud radiative heating in contracting the ITCZ width has also been noted in experiments where cloud-radiation coupling is disabled (see Section 10.5). In comparing their results from the simple model to atmosphere-only (AMIP) climate model simulations, Ahmed et al. (2023) found that GMS and clear-sky water vapor feedbacks dominate variability in ITCZ width among AMIP models. Lastly, changes to convective clustering across scales may also contribute significantly to changes in ITCZ width (Popp et al., 2020). However, no clear linkages between convective organization and warming have been found in cloud-resolving models (CRMs) or global climate models (GCMs) to-date (Cronin & Wing, 2017; Wing et al., 2020).

Lastly, future Pacific Walker circulation changes are tightly coupled to changes in SSTs due to the Bjerknes feedback, a positive feedback. More specifically, reducing the east-west temperature gradient would reduce the Walker Circulation strength, which would further reduce the equatorial upwelling and surface temperature gradient. Models generally project a weakening of Pacific SST gradients (DiNezio et al., 2013; Power & Kociuba 2015; Ying et al., 2022) and thus a weaker Pacific Walker circulation, though the responses among climate models are varied (Vecchi et al., 2008; Han & Zheng 2023). The Walker circulation slowdown is also closely related to the direct effect of $CO2$ on tropospheric stability (Merlis 2015; Kang et al., 2023), with additional contributions from local net surface heat flux responses (Kang et al., 2023). However, Duffy & O'Gorman (2023) find that the CMIP6 multi-model mean weakening and intermodel spread is mostly related to changes in radiation and GMS.

10.3. FUTURE CHANGES TO CLOUDS

High-cloud fraction in the tropics is generally expected to decrease in response to warming, as shown in both idealized (e.g., Bony et al., 2016; Stauffer & Wing, 2022) and comprehensive coupled modeling experiments (Su et al., 2017) (Fig. 10.3), though there is a large spread among both fully coupled global earth system models (Schiro et al., 2022) and idealized model ensembles (Wing et al., 2020) in anvil cloud area changes. Cloud resolving models have shown both increases (Singh and O'Gorman, 2015) and decreases (Cronin & Wing, 2017; Romps, 2020) in anvil cloud area fraction. Cloud fraction is defined in different ways throughout the literature, but it generally refers to the area of a grid cell (in model output or gridded observational data) that is covered horizontally by clouds, with some assumption of overlap. We will use "cloud fraction" and "cloudiness" interchangeably, as a formal definition of cloud fraction that would be consistent across all studies cannot be stated.

A decrease in high-cloud fraction can be tied to the change in large-scale circulation, namely, the equatorward narrowing of the ITCZ (Byrne et al., 2018; Lau & Kim, 2015; Su et al., 2017; Schiro et al., 2019). A high-cloud fraction decrease is also thought to result from the increase in tropospheric stability that would reduce upper tropospheric radiatively driven divergence (Bony et al., 2016). In CMIP6 experiments, however, only about half of the models project a decrease in ITCZ width, which may partly explain why high-cloud fraction only decreases in about half of the models (Schiro et al., 2022). High-cloud changes may depend strongly on changes to convective organization and tropospheric stability or the precipitation efficiency of deep convection, whereby more organized, efficient systems may detrain less condensate into the upper troposphere (Mauritsen & Stevens, 2015; Zhao et al., 2016), though these relationships need to be explored further. For a discussion of the relationships between deep convective organization and clouds, see Chapter 8: Mechanisms for the Self-Organization of Tropical Deep Convection and Chapter 13: Precipitation Efficiency and Climate Sensitivity. Moreover, the greater lifetime of cloud condensate in the upper troposphere is thought to contribute significantly to anvil cloud formation and thus high cloud fraction (Seeley et al. 2019). Changes to upper tropospheric saturation specific humidity would, therefore, play a key role in high cloud changes in response to greenhouse gas forcing, highlighting the need to include such sensitivity in cloud parameterization. Factors such as resolution dependence of the mean-state anvil cloud amount (Jeevanjee & Zhou, 2022) could also contribute to the intermodel spread of anvil cloud area changes in cloud resolving models, though it is unclear to what extent mean state properties of high clouds influence high cloud changes.

Whether or not this reduction of high cloudiness constitutes a positive or negative feedback on climate is uncertain, which will depend on the competing effects of longwave cooling and a reduction in shortwave reflection due to either the high clouds themselves (Li et al., 2019) or a subsequent reduction in low cloudiness (Schiro et al., 2022). It was hypothesized that a reduction of high cloudiness would lead to a greater increase in outgoing longwave radiation (OLR) and constitute a net negative feedback on surface temperatures (the "Iris Hypothesis"; Lindzen et al., 2001). However, this hypothesis has proven to be highly controversial. Whether or not a reduction of high cloudiness would result in a positive or negative feedback in response to greenhouse gas forcing seems to be model-dependent (Li et al., 2019; Mauritsen & Stevens, 2015; Schiro et al., 2022). Certain observational studies suggest that the feedback may instead be net positive (Chambers et al., 2002; Lin et al., 2002). Li et al. (2019) presents evidence that the feedback due to the Iris effect may be positive due to a disproportionate decrease in incoming shortwave (reduced albedo) in comparison to the increase in OLR.

Lastly, strong positive low-cloud feedbacks arise predominantly in the equatorial Pacific and in subtropical regions (Zelinka et al., 2020) and have been at the heart of cloud feedback uncertainties in climate models for decades (Bony & Dufresne, 2005; Dufresne & Bony, 2008). Low cloudiness is projected to decrease in the tropics in many models, with a greater decrease in low cloudiness seen for models with a higher climate sensitivity. The largest sensitivities of the CREs to forced surface temperature changes have been found in regimes of strong subsidence or deep ascent, but these regimes are thought to contribute less than moderate subsidence regimes containing low clouds, which have a larger statistical weight despite exhibiting a weak sensitivity to surface temperature changes (Bony et al., 2004).

While most work relating to low cloud feedbacks focuses on local meteorological factors controlling cloud changes (see Chapter 7: Tropical Marine Low Clouds), a few recent studies have presented new perspectives on how changing tropical circulations, related to changing characteristics of deep convection, connect to low cloud feedbacks. Sherwood et al. (2014) points to a linkage between circulation changes and low-cloud reduction through a so-called "lower tropospheric mixing" mechanism, which they considered to be acting at small and large scales. Enhanced mixing would leave the free troposphere more humid and the boundary layer drier, which can be accomplished either by larger scale circulations or smaller scale deep convective processes. They found that the rate of mixing increases as the climate warms, and the rate of increase depends on the initial mixing strength. Since there is a link in the mixing rate between current and future climate, and the mixing rate in current climate is linked to the tropical low-cloud feedback, this linkage permits observational constraint of the low-cloud feedback and climate sensitivity. Additionally, Schiro et al. (2022) suggests that a lesser decrease in subsidence weakening resulting from increased OLR due to reduced detrainment of water vapor and condensate in relation to ITCZ narrowing is associated with a greater low-cloud reduction along convective margins. The subsidence rates among models have also been shown to decrease less in models with (a) more stable tropospheres in the mean-state and (b) when the static stabilities increase less in response to greenhouse gas forcing (Schiro et al. 2022). This is generally consistent with the findings of Feldl & Bordoni (2016), who found that the CMIP5 models with the greatest circulation weakening exhibit the most negative tropical cloud feedback, and vice versa for circulation strengthening. On the other hand, Li et al. (2022) presented evidence from the CMIP6 ensemble suggesting that a greater Walker circulation slowdown occurs in models that exhibit increases in precipitation efficiency (PE is broadly defined as the ratio of cloud condensate that falls as precipitation; see Ch. 13 for a discussion of precipitation efficiency and its changes under warming), which leads to enhanced east Pacific warming due to dampened equatorial wind stress, reduced oceanic upwelling, and thus a greater reduction of low clouds. Overall, these works highlight the critical role deep convection plays in the response of the tropical overturning circulation to warming and its relationship to cloud feedbacks.

10.4. OBSERVATIONAL EVIDENCE OF CHANGES TO CLOUDS AND THE TROPICAL OVERTURNING CIRCULATION

Studying cloud-circulation interactions using satellite observations has greatly helped to shape our understanding of future changes and permit more skilled evaluation of Earth's climate sensitivity and cloud feedback strength. While a weakening of the Walker circulation has been found in some studies (Bellomo & Clement, 2015; Deser et al., 2010; Power & Kociuba, 2011; Tokinaga et al., 2012; Vecchi et al., 2006), consistent with expectations from modeling, a number of studies have shown that many metrics related to the strength of tropical circulation – especially related to the Walker circulation – suggest that the circulation has strengthened in the twentieth century (L'Heureux et al., 2013; Mitas & Clement, 2005; Meng et al., 2012; Sohn & Park, 2010; Solomon & Newman, 2012; Sohn et al., 2013; Tanaka et al., 2004). The modeling study of Kang et al. (2023) attributes the strengthening of the Walker circulation to Southern Ocean heat uptake. Moreover, observations have shown that the Hadley circulation has expanded poleward (Choi et al., 2014; Davis & Rosenlof, 2012; Hu & Fu, 2007; Hu et al., 2011; Seidel & Randel, 2007), consistent with expectations from modeling. Some studies have also suggested that the circulation has strengthened in recent decades (Mitas & Clement, 2005). However, the discrepancies among reanalyses raise questions about the robustness of the strengthening (Chemke & Polvani, 2019). Moreover, simulated widening trends are much weaker than those in observations (Johanson & Fu, 2009). Recently, however, a series of working groups concluded that large rates of observed Hadley cell widening reported in these and other earlier studies resulted from metrics that poorly capture changes to the Hadley circulation or from a spurious effect of early-generation reanalysis (Staten et al., 2020). Staten et al. (2018) found using observations that the tropics has widened by about 0.5 degrees per decade since 1979, while Staten et al. (2020) updated this to be 0.25–0.5 degrees per decade.

While the total extent of the Hadley cell is expanding, the observed spatial extent of the deep tropics associated with large-scale ascent has been decreasing at a rate of −1% per decade (−5% per degree K) from 1979 to 2016 (Fig. 10.4; Su et al., 2020), consistent with expectations from modeling. Su et al. (2020) found that this results from the combined effects of anthropogenic forcing and interdecadal variability, with interdecadal variability contributing more than anthropogenic forcing. Wodzicki and Rapp (2016) also found that the ITCZ experienced narrowing and strengthening from 1979 to 2014, and that the narrowing trends vary depending on the location. The root cause of the narrowing of the annual mean ITCZ over the historical record is still unknown, yet Wodzicki and Rapp (2022) present evidence of a link between convective organization at small scales and ITCZ behavior at large scales.

Lastly, examining co-variability among CREs and circulation over interannual timescales have proven useful in considering physical mechanisms linking clouds to circulation and their changes under anthropogenic warming. (For a more general overview of cloud changes detected in observations, see Chapter 4: Cloud-Radiation Interactions and Cloud-Climate Feedbacks). For instance, Bony et al. (1997) examined the linkages between large-scale circulation changes and cloud changes in response to interannual variability in SST and found

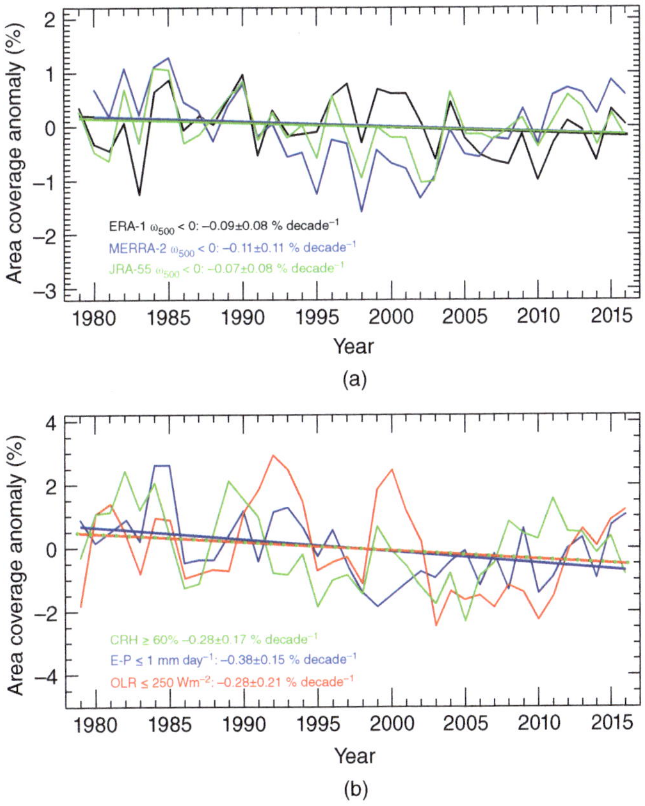

Figure 10.4 Change in annual-mean spatial extent of the tropical ascent region from 1979 to 2016 represented by different metrics: $\omega_{500} < 0$ hPa day^{-1}, column relative humidity (CRH) ≥ 60%, evaporation minus precipitation (E - P) ≤ 1 mm day^{-1}, and outgoing longwave radiation (OLR) ≤ 250 W m^{-2}. The percentage change is based on the tropical ascent area coverage between 30°S and 30°N. Climatological annual means for each variable are removed from the time series. Linear trends with one standard deviation are given in the legend. Source: Adapted from Su et al. (2020).

that for either spatiotemporal, seasonal, or El Niño Southern Oscillation variations, the cloudiness, cloud optical depth, and the shortwave CRE decrease with increasing SST in large-scale subsidence regimes. They also found that over warm SSTs in the ascent regions, both the longwave and shortwave cloud forcing increase with increasing large-scale ascent. Su and Jiang (2013) found, however, that different El Niño events can cause very different changes to the large-scale overturning circulation and CREs. For example, the El Niño event in 2009/2010 showed nearly opposite longwave and shortwave CRE changes in ascent regimes compared with the 2006/2007 El Niño, with differences in the net CRE at the top-of-atmosphere (TOA) between the two events being rather large: on the order of 0.7–1.2 W m^{-2}.

10.5. TOWARD AN IMPROVED MECHANISTIC UNDERSTANDING OF CLOUD-CIRCULATION INTERACTIONS

Thus far, we have mostly considered changes to circulation and clouds in concert, rather than discussing how causality might be assessed. Yet, cloud radiative forcing has been found to contribute about 20% to the magnitude of the tropical overturning circulation (Bergman & Hendon, 2000), emphasizing the role that clouds play in modulating the circulation. Sherwood et al. (1994) found that removing the effects of cloud radiative forcing for high clouds resulted in a near disappearance of the Walker circulation and about a 25% reduction in Hadley circulation strength. Moreover, cloud radiative changes have been found to explain more than half of the total poleward expansion of the Hadley cell, midlatitude jets, and storm tracks under CO_2 quadrupling (Ceppi & Hartmann, 2016; Ceppi & Shepherd, 2017; Voigt et al., 2019). Therefore, studying these interactions in a systematic way using a hierarchy of climate models is critical to our improved understanding of cloud-circulation interactions in present and future climate.

A handful of earlier studies have shown that CREs exert a strong influence on precipitation, convection, and the general circulation of the atmosphere (Ramanathan et al., 1983; Ramaswamy & Ramanathan, 1989; Randall et al., 1989; Slingo & Slingo, 1988; Stuhlmann & Smith, 1988; Sherwood et al., 1994; Raymond 2000). Many recent studies have revisited the question of how CREs modify tropical circulation using atmospheric general circulation models (Fermepin & Bony, 2014; Harrop & Hartmann, 2016; Li et al., 2015; Popp & Silvers, 2017; Watt-Meyer & Frierson, 2017). One commonly used approach to study causality in cloud-circulation interactions is "cloud-locking," where the simulated clouds are replaced with a control (e.g., clouds from year 1 of a model integration) repeatedly throughout the simulation for the purpose of radiation. In another approach, the "COOKIE" method (Clouds On Off Klima Intercomparison Experiment; Stevens et al., 2012), the clouds are transparent to radiation and the cloud fields are allowed to evolve in time. The cloud-locking approach has been shown to more reliably diagnose the cloud-radiative impacts on climate and is also preferred for climate change simulations (Voigt & Albern, 2019), in comparison with the COOKIE method. Cloud-locking is preferred in the context of climate change because by making clouds transparent to radiation, COOKIE drastically alters the control circulation, which can affect the climate change response and lead to misleading conclusions.

A key finding from these modeling experiments is that the ITCZ narrows when CREs are included (Harrop & Hartmann, 2016; Lau et al., 2020; Popp & Silvers, 2017; Randall et al., 1989; Silvers & Robinson, 2021), though this effect can differ greatly across different models (Voigt & Shaw, 2015; Harrop & Hartmann, 2016; Fig. 10.5). This may result from the decrease in mean convective available potential energy (CAPE) due to cloud radiative heating, which ups the ante (Neelin et al., 2003) for convection onset (Harrop & Hartmann, 2016). Dixit et al. (2018), on the other hand, demonstrated that it is the effect of high clouds on lower tropospheric temperatures, rather than upper tropospheric temperatures, that affects ITCZ width the most: associated changes in a shallow meridional overturning circulation drive a feedback via net import of warm, moist air, further destabilizing the ITCZ.

The longwave and shortwave CREs can also be assessed separately within the COOKIE or cloud-locking frameworks. For instance, Popp and Silvers (2017) found that the longwave CREs increase the meridional heating gradient in the tropics and, in response, the ITCZ narrows and the Hadley circulation strengthens. The effect of CREs for high, middle, and low clouds on the circulation can also be assessed. For instance, Stuhlmann and Smith (1988) studied the effects of low and mid-level CREs on available potential energy and found that low clouds cool the lowest layers of the atmosphere while mid-level clouds cool the middle atmosphere and heat the lower atmosphere. Fermepin and Bony (2014) found that low-cloud radiative effects strengthen surface winds over tropical oceans, increase precipitation across the tropics, and amplify the atmospheric overturning circulation. This response results from stronger radiative cooling in the lower troposphere that reduces the near-surface temperature and humidity, enhancing surface turbulent fluxes. These effects highlight the relevance of improved process representation of turbulent boundary-layer processes and understanding of their relationship to low clouds to accurate simulation of climate.

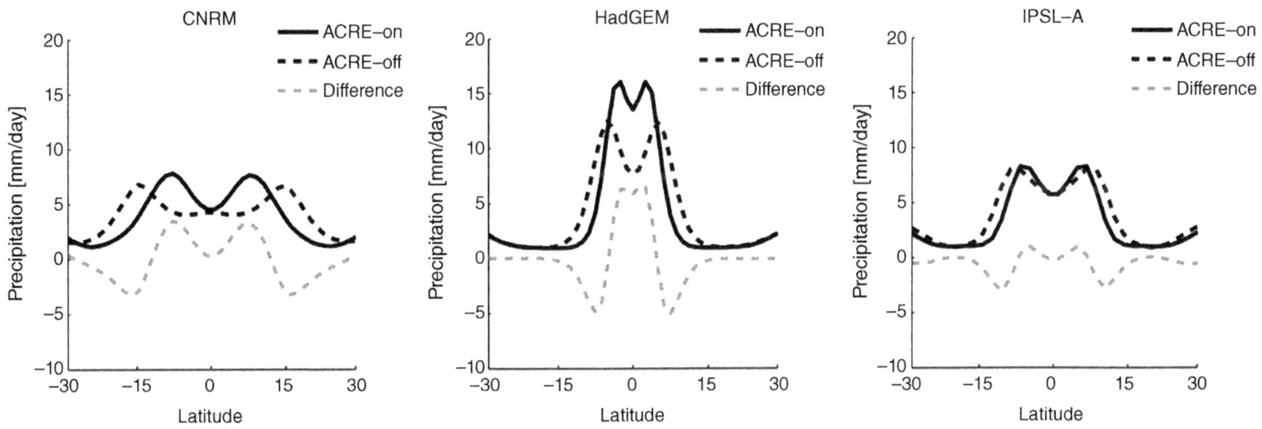

Figure 10.5 The mean precipitation from 30S-30N from six different aquaplanet simulations with three models (CNRM, left; HadGEM, middle; IPSL-A, right) with atmospheric cloud radiative effects (ACRE) turned on (solid) and off (black dashed) as well as the difference (gray dashed). Source: Adapted from Harrop and Hartmann (2016).

In summary, clouds can greatly influence the tropical overturning circulation and climate. (For a more comprehensive review, see Voigt et al., 2021). Therefore, while we often think of circulation as a driver of cloud changes, we must also consider the role of cloud changes in modulating circulation changes. Moreover, while many of these mechanism denial studies do not address the role of cloud changes in response to greenhouse gas forcing in driving corresponding circulation changes, we can infer from the effect of clouds on the circulation in the mean state that such effects could be large. For instance, a key takeaway from many of these mechanism denial experiments synthesized here is that longwave radiative heating by high clouds has been shown to narrow the ITCZ. Thus, one might expect that more longwave radiative heating of high clouds (either due to increases in opacity of high clouds or an increase in high-cloud coverage) would act to narrow the ITCZ. In other words, one might also wonder if a decrease in high cloudiness due to ITCZ narrowing (or other physical changes to the tropical climate, such as stability increases or increases in mesoscale convective organization) may act to counterbalance the ITCZ narrowing itself. This response has been suggested by Ahmed et al. (2023) and Schiro et al. (2019) using a simple model of the tropical atmosphere and AMIP ensembles. Intermodel differences in the magnitudes of these competing effects may contribute to the large intermodel differences in high-cloud feedbacks (Zelinka et al., 2022).

10.6. DISCUSSION

How cloud feedbacks mechanistically relate to changes in the large-scale tropical overturning circulation is one of the biggest open questions in climate science. Our community has thus far made novel use of climate model ensembles and hierarchies to study mechanistic relationships and observationally constrain the large spread in cloud feedback strength in response to increasing greenhouse gases. In summary, the intermodel spread in the Hadley Circulation change among climate models is well correlated with the intermodel spread in the top-of-atmosphere and shortwave CRE changes (Su et al. 2014). In particular, models with the greatest Hadley circulation weakening exhibit the most negative tropical cloud feedback, and vice versa for circulation strengthening (Feldl & Bordoni, 2016). On the other hand, greater weakening of the Walker circulation among models has been linked to a reduction of low cloudiness through a reduction of equatorial trade winds and associated oceanic upwelling (Li et al. 2014). Moreover, enhanced lower tropospheric mixing with warming, which dries the boundary layer and moistens the free troposphere, decreases low cloud cover (Sherwood et al., 2014). Therefore, models with greater increases in lower tropospheric mixing have higher climate sensitivities. Models with a reduced ascent area under warming also have higher climate sensitivities due to associated reductions in both high and low cloudiness, which may be coupled together through energetic constraints on subsidence (Schiro et al. 2022).

Emergent constraints on ECS have been considered based on suggested relationships between cloud and circulation changes. Emergent constraints are defined as "physically explainable empirical relationships between characteristics of the current climate and long-term climate prediction that emerge in collections of climate model simulations" (Klein & Hall, 2015). For instance, Sherwood et al. (2014) established emergent constraints for the rate of lower tropospheric mixing at both small and large scales between the boundary layer and free troposphere. Additionally, Su et al. (2014) presented evidence that ECS may be linked to the intermodel spread in the zonal mean relative humidity structure in the tropical troposphere – which is intimately tied to the intermodel spread in circulation changes – whereby higher ECS models exhibit more realistic zonal mean relative humidity vertical structure compared with observations. Lastly, Lipat et al. (2017) showed that the multi-year average extent of the Hadley cell correlates with ECS in CMIP5 models, which they explained by tying it to the observed correlation of the interannual variability in midlatitude cloudiness. However, these and many other emergent constraints identified among the CMIP5 ensemble have been found to be weaker in the CMIP6 ensemble (Schlund et al., 2020). Moreover, many uncertainties exist surrounding the physical plausibility of emergent constraints, as our knowledge of how the constraints connect back to circulation and cloud changes among the multi-model ensembles is currently limited.

Many opportunities exist for future work exploring cloud-circulation interactions. Examining causal relationships and co-variability between varying deep convective characteristics across scales (e.g., convective clustering, ITCZ width), the overturning circulation strength, and the distribution of high and low cloudiness in observations would be one promising avenue. Linking changes in the overturning circulation to changes in SST patterns and the "pattern effect" – the dependence of outgoing radiation to space on the spatial pattern of surface warming (Andrews et al., 2018; Andrews & Webb, 2018; Ceppi & Gregory, 2017; Dong et al., 2020; Fueglistaler et al., 2019; Stevens et al., 2016; Silvers et al., 2018; Zhou et al., 2016) – would be another promising way forward. Examining the effects of changes in cloud shading on SST gradients and the overturning circulation (Wall et al., 2019) would also tie into this pattern effect exploration. Exploring the sensitivity of convective organization across scales to changing SST and SST patterns will be a critical part of this effort (Coppin & Bony, 2018; Bony et al., 2020).

Future work examining model simulations in novel ways across a model hierarchy will be necessary to identify robust physical linkages between cloud and circulation changes. Examining cloud feedbacks in idealized configurations – such as radiative-convective equilibrium (RCE) simulations with uniform SST and aquaplanet simulations (e.g., Medeiros et al., 2008, 2015; Popp & Silvers, 2017; Reed et al., 2021; Silvers et al., 2016) – would likely be quite instructive. Moreover, simulations of an idealized Walker circulation (a "mock-Walker circulation") contain a host of different cloud types and provide an excellent simplified framework to compare and contrast the interactions between clouds and the large-scale circulation of GCMs and CRMs (Lutsko & Cronin, 2018). Mock-Walker simulations can, therefore, serve as a bridge between RCE and aquaplanet or AMIP simulations. Cloud-locking and COOKIE-type experiments should also continue to be exploited and utilized in the context of understanding cloud-circulation coupling given the wealth of insights gleaned from the recent revival of such studies.

ACKNOWLEDGMENTS

Funding from the University of Virginia (KAS) and NSF (#2225954) supported this review. KAS thanks Clare Singer and Levi Silvers for many insightful discussions on cloud-circulation interactions. The authors also wish to thank three anonymous reviewers for their helpful suggestions for improving the readability and accessibility of this review.

AVAILABILITY STATEMENT

All CMIP6 data used to create Figure 10.4 are publicly available and can be found on the Earth System Federation Grid (ESGF) website: https://esgf-node.llnl.gov/projects/cmip6/.

REFERENCES

Ahmed, F., Neelin, J. D., Hill, S. A., Schiro, K. A., & Su, H. (2023). A process model for ITCZ narrowing under warming highlights clear-sky water vapor feedbacks and gross moist stability changes in AMIP models. *Journal of Climate*, 1–46.

Allen, R., Sherwood, S., Norris, J., & Zender, C. S. (2012). Recent Northern Hemisphere tropical expansion primarily driven by black carbon and tropospheric ozone. *Nature*, *485*, 350–354. https://doi.org/10.1038/nature11097

Andrews, T., & Webb. M. J. (2018). The dependence of global cloud and lapse rate feedbacks on the spatial structure of tropical Pacific warming. *Journal of Climate*, *31*(2), 641–654. https://doi.org/10.1175/JCLI-D-17-0087.1.

Andrews, T., Gregory, J. M., Paynter, D., Silvers, L. G., Zhou, C., Mauritsen, T., et al. (2018). Accounting for changing temperature patterns increases historical estimates of climate sensitivity. *Geophysical Research Letters*, *45*(16), 8490–8499.

Bayr, T., Dommenget, D., Martin, T., & Power, S. B. (2014). The eastward shift of the Walker circulation in response to global warming and its relationship to ENSO variability. *Climate Dynamics, 43*, 2747–2763.

Bellomo, K., & Clement, A. C. (2015). Evidence for weakening of the Walker circulation from cloud observations. *Geophysical Research Letters*, *42*(18), 7758–7766.

Bergman, J. W., & Hendon, H. H. (2000). Cloud radiative forcing of the low-latitude tropospheric circulation: Linear calculations. *Journal of the Atmospheric Sciences*, *57*(14), 2225–2245.

Bony, S., Lau, K. M., & Sud, Y. C. (1997). Sea surface temperature and large-scale circulation influences on tropical greenhouse effect and cloud radiative forcing. *Journal of Climate*, *10*(8), 2055–2077.

Bony, S., Dufresne, J. L., Le Treut, H., Morcrette, J. J., & Senior, C. (2004). On dynamic and thermodynamic components of cloud changes. *Climate Dynamics*, *22*(2–3), 71–86.

Bony, S., & Dufresne, J. L. (2005). Marine boundary layer clouds at the heart of tropical cloud feedback uncertainties in climate models. *Geophysical Research Letters*, *32*(20).

Bony, S., Bellon, G., Klocke, D., Sherwood, S., Fermepin, S., & Denvil, S. (2013). Robust direct effect of carbon dioxide on tropical circulation and regional precipitation. *Nature Geoscience*, *6*(6), 447–451.

Bony, S., Stevens, B., Coppin, D., Becker, T., Reed, K. A., Voigt, A., & Medeiros, B. (2016). Thermodynamic control of anvil cloud amount. *Proceedings of the National Academy of Sciences*, *113*(32), 8927–8932.

Bony, S., Semie, A., Kramer, R. J., Soden, B., Tompkins, A. M., & Emanuel, K. A. (2020). Observed modulation of the tropical radiation budget by deep convective organization and lower-tropospheric stability. *AGU Advances*, *1*(3), e2019AV000155.

Bretherton, C. S., Blossey, P. N., & Jones, C. R. (2013). Mechanisms of marine low cloud sensitivity to idealized climate perturbations: A single-LES exploration extending the CGILS cases. *Journal of Advances in Modeling Earth Systems*, *5*(2), 316–337.

Byrne, M. P., & Schneider, T. (2016a). Narrowing of the ITCZ in a warming climate: Physical mechanisms. *Geophysical Research Letters*, *43*(21), 11–350.

Byrne, M., & Schneider, T. (2016b). Energetic constraints on the width of the Intertropical Convergence Zone. *Journal of Climate*, *29*, 4709–4721.

Byrne, M. P., Pendergrass, A. G., Rapp, A. D., & Wodzicki, K. R. (2018). Response of the intertropical convergence zone to climate change: Location, width, and strength. *Current Climate Change Reports*, *4*(4), 355–370.

Ceppi, P., & Hartmann, D. L. (2016). Clouds and the atmospheric circulation response to warming. *Journal of Climate*, *29*(2), 783–799.

Ceppi, P., & Shepherd, T. G. (2017). Contributions of climate feedbacks to changes in atmospheric circulation. *Journal of Climate*, *30*(22), 9097–9118.

Ceppi, P., & Gregory, J. M. (2017). Relationship of tropospheric stability to climate sensitivity and Earth's observed radiation budget. *Proceedings of the National Academy of Sciences USA*, *114*, 13126–13131. https://doi.org/10.1073/PNAS.1714308114.

Chambers, L. H., Lin, B., & Young, D. F. (2002). Examination of new CERES data for evidence of tropical Iris feedback. *Journal of climate*, *15*(24), 3719–3726.

Chemke, R., & Polvani, L. M. (2019). Opposite tropical circulation trends in climate models and in reanalyses. *Nature Geoscience*, *12*(7), 528–532.

Choi, J., Son, S. W., Lu, J., & Min, S. K. (2014). Further observational evidence of Hadley cell widening in the Southern Hemisphere. *Geophysical Research Letters*, *41*(7), 2590–2597.

Chou, C., & Neelin, J. D. (2004). Mechanisms of global warming impacts on regional tropical precipitation. *Journal of climate*, *17*(13), 2688–2701.

Coppin, D., & Bony, S. (2018). On the interplay between convective aggregation, surface temperature gradients, and climate sensitivity. *Journal of Advances in Modeling Earth Systems*, *10*(12), 3123–3138.

Cronin, T. W., & Wing, A. A. (2017). Clouds, circulation, and climate sensitivity in a radiative-convective equilibrium channel model. *Journal of Advances in Modeling Earth Systems*, *9*(8), 2883–2905.

Davis, S. M., & Rosenlof, K. H. (2012). A multidiagnostic intercomparison of tropical-width time series using reanalyses and satellite observations. *Journal of Climate*, *25*(4), 1061–1078.

Deser, C., Phillips, A. S., & Alexander, M. A. (2010). Twentieth century tropical sea surface temperature trends revisited. *Geophysical Research Letters*, *37*, L10701.

DiNezio, P. N., Vecchi, G. A., & Clement, A. C. (2013). Detectability of changes in the Walker circulation in response to global warming. *Journal of Climate*, *26*(12), 4038–4048.

Dixit, V., Geoffroy, O., & Sherwood, S. C. (2018). Control of ITCZ width by low-level radiative heating from upper-level clouds in aquaplanet simulations. *Geophysical Research Letters*, *45*(11), 5788–5797.

Donohoe, A., & Voigt, A. (2017). Why future shifts in tropical precipitation will likely be small: The location of the tropical rain belt and the hemispheric contrast of energy input to the atmosphere. *Climate Extremes: Patterns and Mechanisms*, *226*, 115.

Donohoe, A., Atwood, A. R., & Byrne, M. P. (2019). Controls on the width of tropical precipitation and its contraction under global warming. *Geophysical Research Letters*, *46*(16), 9958–9967.

Dufresne, J. L., & Bony, S. (2008). An assessment of the primary sources of spread of global warming estimates from coupled atmosphere–ocean models. *Journal of Climate*, *21*(19), 5135–5144.

DiNezio, P. N., Vecchi, G. A., & Clement, A. C. (2013). Detectability of changes in the Walker circulation in response to global warming. *Journal of Climate*, *26*, 4038–4048.

Dong, Y., Armour, K. C., Zelinka, M. D., Proistosescu, C., Battisti, D. S., Zhou, C., & Andrews, T. (2020). Intermodel spread in the pattern effect and its contribution to climate sensitivity in CMIP5 and CMIP6 models. *Journal of Climate*, *33*(18), 7755–7775.

Duffy, M. L., & O'Gorman, P. A. (2023). Intermodel spread in Walker circulation responses linked to spread in moist stability and radiation responses. *Journal of Geophysical Research: Atmospheres*, e2022JD037382.

Feldl, N., & Bordoni, S. (2016). Characterizing the Hadley circulation response through regional climate feedbacks. *Journal of Climate*, *29*(2), 613–622.

Fermepin, S., & Bony, S. (2014). Influence of low-cloud radiative effects on tropical circulation and precipitation. *Journal of Advances in Modeling Earth Systems*, *6*(3), 513–526.

Frierson, D. M., Lu, J., & Chen, G. (2007). Width of the Hadley cell in simple and comprehensive general circulation models. *Geophysical Research Letters*, *34*(18).

Fueglistaler, S. (2019). Observational evidence for two modes of coupling between sea surface temperatures, tropospheric temperature profile, and shortwave cloud radiative effect in the tropics. *Geophysical Research Letters*, *46*, 9890–9898. https://doi.org/10.1029/2019GL083990.

Grise, K. M., & Davis, S. M. (2020). Hadley cell expansion in CMIP6 models. *Atmospheric Chemistry and Physics*, *20*, 5249–5268. https://doi.org/10.5194/acp-20-5249-2020.

Han, Z. W., & Zheng, X. T. (2023). Intermodel uncertainty in response of the Pacific Walker circulation to global warming. *Climate Dynamics*, 1–21.

Harrop, B. E., & Hartmann, D. L. (2016). The role of cloud radiative heating in determining the location of the ITCZ in aquaplanet simulations. *Journal of Climate*, *29*(8), 2741–2763.

Hartmann, D. L., Hendon, H. H., & Houze Jr, R. A. (1984). Some implications of the mesoscale circulations in tropical cloud clusters for large-scale dynamics and climate. *Journal of Atmospheric Sciences*, *41*(1), 113–121.

Hartmann, D. L., & Larson, K. (2002). An important constraint on tropical cloud-climate feedback. *Geophysical Research Letters*, *29*(20), 1951. Doi:10.1029/ 2002GL015835.

Held, I. M., & Soden, B. J. (2006). Robust responses of the hydrological cycle to global warming. *Journal of climate*, *19*(21), 5686–5699.

Hu, Y., & Fu, Q. (2007). Observed poleward expansion of the Hadley circulation since 1979. *Atmospheric Chemistry and Physics*, *7*, 5229–5236.

Hu, Y., Zhou, C., & Liu, J. (2011). Observational evidence for poleward expansion of the Hadley circulation. *Advances in Atmospheric Sciences*, *28*(1), 33–44.

Jeevanjee, N., & Zhou, L. (2022). On the resolution-dependence of anvil cloud fraction and precipitation efficiency in radiative-convective equilibrium. *Journal of Advances in Modeling Earth Systems*, *14*(3), e2021MS002759.

Johanson, C. M., & Fu, Q. (2009). Hadley cell widening: Model simulations versus observations. *Journal of Climate*, *22*, 2713–2725.

Kang, S. M., Shin, Y., Kim, H., Xie, S. P., & Hu, S. (2023). Disentangling the mechanisms of equatorial Pacific climate change. *Science Advances*, *9*(19), eadf5059.

Klein, S. A., & Hall, A. (2015). Emergent constraints for cloud feedbacks. *Current Climate Change Reports*, *1*(4), 276–287.

Klein, S. A., Hall, A., Norris, J. R., & Pincus, R. (2017). Low-cloud feedbacks from cloud-controlling factors: A review. *Surveys in Geophysics*, *38*, 1307–1329. https://doi.org/10.1007/s10712-017-9433-3.

Kociuba, G., & Power, S. B. (2015). Inability of CMIP5 models to simulate recent strengthening of the Walker circulation: Implications for projections. *Journal of Climate*, *28*, 20–35.

L'Heureux, M. L., Lee, S., & Lyon, B. (2013). Recent multidecadal strengthening of the Walker circulation across the tropical Pacific. *Nature Climate Change*, *3*(6), 571–576.

Lau, W. K., & Kim, K. M. (2015). Robust Hadley circulation changes and increasing global dryness due to CO2 warming from CMIP5 model projections. *Proceedings of the National Academy of Sciences*, *112*(12), 3630–3635.

Lau, W. K., Kim, K. M., Chern, J. D., Tao, W. K., & Leung, L. R. (2020). Structural changes and variability of the ITCZ induced by radiation–cloud–convection–circulation interactions: Inferences from the Goddard Multi-scale Modeling Framework (GMMF) experiments. *Climate Dynamics*, *54*(1), 211–229.

Li, Y., Thompson, D. W., & Bony, S. (2015). The influence of atmospheric cloud radiative effects on the large-scale atmospheric circulation. *Journal of Climate*, *28*(18), 7263–7278.

Li, R. L., Studholme, J. H., Fedorov, A. V., & Storelvmo, T. (2023). Increasing precipitation efficiency amplifies climate sensitivity by enhancing tropical circulation slowdown and eastern Pacific warming pattern. *Geophysical Research Letters*, *50*(2), e2022GL100836.

Li, R. L., Storelvmo, T., Fedorov, A. V., & Choi, Y. S. (2019). A positive IRIS feedback: Insights from climate simulations with temperature-sensitive cloud–rain conversion. *Journal of climate*, *32*(16), 5305–5324.

Li, R. L., Studholme, J. H., Fedorov, A. V., & Storelvmo, T. (2022). Precipitation efficiency constraint on climate change. *Nature Climate Change*, *12*(7), 642–648.

Lin, B., Wielicki, B. A., Chambers, L. H., Hu, Y., & Xu, K. M. (2002). The iris hypothesis: A negative or positive cloud feedback?. *Journal of Climate*, *15*(1), 3–7.

Lipat, B. R., Tselioudis, G., Grise, K. M., & Polvani, L. M. (2017). CMIP5 models' shortwave cloud radiative response and climate sensitivity linked to the climatological Hadley cell extent. *Geophysical Research Letters*, *44*(11), 5739–5748.

Lindzen, R. S., Chou, M. D., & Hou, A. Y. (2001). Does the Earth have an adaptive infrared iris? *Bulletin of the American Meteorological Society*, *82*(3), 417–432.

Lu, J., Vecchi, G. A., & Reichler, T. (2007). Expansion of the Hadley cell under global warming. *Geophysical Research Letters*, *34*(6).

Lutsko, N. J., & Cronin, T. W. (2018). Increase in precipitation efficiency with surface warming in radiative-convective equilibrium. *Journal of Advances in Modeling Earth Systems*, *10*(11), 2992–3010.

Mamalakis, A., Randerson, J. T., Yu, J. Y., Pritchard, M. S., Magnusdottir, G., Smyth, P., et al. (2021). Zonally contrasting shifts of the tropical rain belt in response to climate change. *Nature Climate Change*, *11*(2), 143–151.

Mauritsen, T., & Stevens, B. (2015). Missing iris effect as a possible cause of muted hydrological change and high climate sensitivity in models. *Nature Geoscience*, *8*(5), 346–351.

Medeiros, B., Stevens, B., Held, I. M., Zhao, M., Williamson, D. L., Olson, J. G., & Bretherton, C. S. (2008). Aquaplanets, climate sensitivity, and low clouds. *Journal of Climate*, *21*(19), 4974–4991.

Medeiros, B., Stevens, B., & Bony, S. (2015). Using aquaplanets to understand the robust responses of comprehensive climate models to forcing. *Climate Dynamics*, *44*(7), 1957–1977.

Meng, Q., Latif, M., Park, W., Keenlyside, N. S., Semenov, V. A., & Martin, T. (2012). Twentieth century Walker circulation change: Data analysis and model experiments. *Climate Dynamics*, *38*(9–10), 1757–1773.

Merlis, T. M. (2015). Direct weakening of tropical circulations from masked CO2 radiative forcing. *Proceedings of the National Academy of Sciences*, *112*(43), 13167–13171.

Mitas, C. M., & Clement, A. (2005). Has the Hadley cell been strengthening in recent decades? *Geophysical Research Letters*, *32*(3).

Neelin, J. D., & Held, I. M. (1987). Modeling tropical convergence based on the moist static energy budget. *Monthly Weather Review*, *115*(1), 3–12.

Neelin, J. D., Chou, C., & Su, H. (2003). Tropical drought regions in global warming and El Niño teleconnections. *Geophysical Research Letters*, *30*(24).

Polvani, L. M., Waugh, D. W., Correa, G. J. P., & Son, S.-W. (2011). Stratospheric ozone depletion: The main driver of twentieth-century atmospheric circulation changes in the Southern Hemisphere. *Journal of Climate*, *24*, 795–812.

Popp, M., & Silvers, L. G. (2017). Double and single ITCZs with and without clouds. *Journal of Climate*, *30*(22), 9147–9166.

Popp, M., Lutsko, N. J., & Bony, S. (2020). Weaker links between zonal convective clustering and ITCZ width in climate models than in observations. *Geophysical Research Letters*, *47*(22), e2020GL090479.

Power, S. B., & Kociuba, G. (2011). What caused the observed twentieth-century weakening of the Walker circulation? *Journal of Climate*, *24*, 6501–6514.

Qu, X., Hall, A., Klein, S. A., & DeAngelis, A. M. (2015). Positive tropical marine low-cloud cover feedback inferred from cloud-controlling factors. *Geophysical Research Letters*, *42*(18), 7767–7775.

Ramanathan, V., Pitcher, E. J., Malone, R. C., & Blackmon, M. L. (1983). The response of a spectral general circulation model to refinements in radiative processes. *Journal of Atmospheric Sciences*, *40*(3), 605–630.

Ramanathan, V. L. R. D., Cess, R. D., Harrison, E. F., Minnis, P., Barkstrom, B. R., Ahmad, E., & Hartmann, D. (1989). Cloud-radiative forcing and climate: Results from the Earth Radiation Budget Experiment. *Science*, *243*(4887), 57–63.

Ramaswamy, V., & Ramanathan, V. (1989). Solar absorption by cirrus clouds and the maintenance of the tropical upper troposphere thermal structure. *Journal of Atmospheric Sciences*, *46*(14), 2293–2310.

Raymond, D. J. (2000). The Hadley circulation as a radiative–convective instability. *Journal of the Atmospheric Sciences*, *57*(9), 1286–1297.

Raymond, D. J., Sessions, S. L., Sobel, A. H., & Fuchs, Ž. (2009). The mechanics of gross moist stability. *Journal of Advances in Modeling Earth Systems*, *1*(3).

Randall, D. A., Dazlich, D. A., & Corsetti, T. G. (1989). Interactions among radiation, convection, and large-scale dynamics in a general circulation model. *Journal of Atmospheric Sciences*, *46*(13), 1943–1970.

Reed, K. A., Silvers, L. G., Wing, A. A., Hu, I. K., & Medeiros, B. (2021). Using radiative convective equilibrium to explore clouds and climate in the Community Atmosphere Model. *Journal of Advances in Modeling Earth Systems*, *13*(12), e2021MS002539.

Saint-Lu, M., Bony, S., & Dufresne, J. -L. (2020). Observational evidence for a stability Iris effect in the tropics. *Geophysical Research Letters*, *47*, e2020GL089059. https://doi.org/10.1029/2020GL089059.

Schiro, K. A., Su, H., Wang, Y., Langenbrunner, B., Jiang, J. H., & Neelin, J. D. (2019). Relationships between tropical ascent and high cloud fraction changes with warming revealed by perturbation physics experiments in CAM5. *Geophysical Research Letters*, *46*(16), 10112–10121.

Schiro, K. A., Su, H., Ahmed, F., Dai, N., Singer, C. E., Gentine, P., et al. (2022). Model spread in tropical low cloud feedback tied to overturning circulation response to warming. *Nature Communications*, *13*(1), 7119.

Schlund, M., Lauer, A., Gentine, P., Sherwood, S. C., & Eyring, V. (2020). Emergent constraints on equilibrium climate sensitivity in CMIP5: Do they hold for CMIP6? *Earth System Dynamics*, *11*(4), 1233–1258.

Seeley, J. T., Jeevanjee, N., Langhans, W., & Romps, D. M. (2019). Formation of tropical anvil clouds by slow evaporation. *Geophysical Research Letters*, *46*(1), 492–501.

Seidel, D. J., & Randel, W. J. (2007). Recent widening of the tropical belt: Evidence from tropopause observations. *Journal of Geophysical Research: Atmospheres*, *112*(D20).

Sherwood, S. C., Ramanathan, V., Barnett, T. P., Tyree, M. K., & Roeckner, E. (1994). Response of an atmospheric general circulation model to radiative forcing of tropical clouds. *Journal of Geophysical Research: Atmospheres*, *99*(D10), 20829–20845.

Sherwood, S. C., Bony, S., & Dufresne, J. L. (2014). Spread in model climate sensitivity traced to atmospheric convective mixing. *Nature*, *505*(7481), 37–42.

Silvers, L. G., Stevens, B., Mauritsen, T., & Giorgetta, M. (2016). Radiative convective equilibrium as a framework for studying the interaction between convection and its large-scale environment. *Journal of Advances in Modeling Earth Systems*, *8*(3), 1330–1344.

Silvers, L. G., Paynter, D., & Zhao, M. (2018). The diversity of cloud responses to twentieth century sea surface temperatures. *Geophysical Research Letters*, *45*(1), 391–400.

Silvers, L. G., & Robinson, T. (2021). Clouds and radiation in a mock-Walker circulation. *Journal of Advances in Modeling Earth Systems*, *13*(2), e2020MS002196.

Singh, M. S., & O'Gorman, P. A. (2015). Increases in moist-convective updraught velocities with warming in radiative-convective equilibrium. *Quarterly Journal of the Royal Meteorological Society*, *141*(692), 2828–2838.

Slingo, A., & Slingo, J. M. (1988). The response of a general circulation model to cloud longwave radiative forcing. I: Introduction and initial experiments. *Quarterly Journal of the Royal Meteorological Society*, *114*(482), 1027–1062.

Sohn, B. J., & Park, S. C. (2010). Strengthened tropical circulations in past three decades inferred from water vapor transport. *Journal of Geophysical Research: Atmospheres*, *115*(D15).

Sohn, B. J., Yeh, S. W., Schmetz, J., & Song, H. J. (2013). Observational evidences of Walker circulation change over the last 30 years contrasting with GCM results. *Climate Dynamics*, *40*(7), 1721–1732.

Solomon, A., & Newman, M. (2012). Reconciling disparate twentieth-century Indo-Pacific ocean temperature trends in the instrumental record. *Nature Climate Change*, *2*(9), 691–699.

Staten, P. W., Lu, J., Grise, K. M., Davis, S. M., & Birner, T. (2018). Re-examining tropical expansion. *Nature Climate Change*, *8*, 768–775. https://doi.org/10.1038/s41558-018-0246-2.

Staten, P. W., Grise, K. M., Davis, S. M., Karnauskas, K. B., Waugh, D. W., Maycock, A. C., & Son, S. W. (2020). Tropical widening: From global variations to regional impacts. *Bulletin of the American Meteorological Society*, *101*(6), E897–E904.

Stauffer, C. L., & Wing, A. A. (2022). Properties, changes, and controls of deep-convecting clouds in radiative-convective equilibrium. *Journal of Advances in Modeling Earth Systems*, *14*(6), e2021MS002917.

Stephens, G. L., Li, J., Wild, M., Clayson, C. A., Loeb, N., Kato, S., et al. (2012). An update on Earth's energy balance in light of the latest global observations. *Nature Geoscience*, *5*(10), 691–696.

Stevens, B., Bony, S., & Webb, M. (2012). Clouds on-off klimate intercomparison experiment (COOKIE). [online draft].

Stevens, B., Sherwood, S., Bony, S., & Webb, M. (2016). Prospects for narrowing bounds on Earth's equilibrium climate sensitivity. *Earth's Future*, *4*, 512–522.

Stuhlmann, R., & Smith, G. L. (1988). A study of cloud–generated radiative heating and its generation of available potential energy. Part I: Theoretical background. *Journal of Atmospheric Sciences*, *45*(24), 3911–3927.

Su, H., Jiang, J. H., Zhai, C., Perun, V. S., Shen, J. T., Del Genio, A., & Stephens, G. L. (2013). Diagnosis of regime-dependent cloud simulation errors in CMIP5 models using "A-Train" satellite observations and reanalysis data. *Journal of Geophysical Research: Atmospheres*, *118*(7), 2762–2780.

Su, H., & Jiang, J. H. (2013). Tropical clouds and circulation changes during the 2006/07 and 2009/10 El Niños. *Journal of Climate*, *26*(2), 399–413.

Su, H., Jiang, J. H., Zhai, C., Shen, T. J., Neelin, J. D., Stephens, G. L., & Yung, Y. L. (2014). Weakening and strengthening structures in the Hadley Circulation change under global warming and implications for cloud response and climate sensitivity. *Journal of Geophysical Research: Atmospheres*, *119*(10), 5787–5805.

Su, H., Jiang, J. H., Neelin, J. D., Shen, T. J., Zhai, C., Yue, Q., & Yung, Y. L. (2017). Tightening of tropical ascent and high clouds key to precipitation change in a warmer climate. *Nature Communications*, *8*(1), 1–9.

Su, H., Zhai, C., Jiang, J. H., Wu, L., Neelin, J. D., & Yung, Y. L. (2019). A dichotomy between model responses of tropical ascent and descent to surface warming. *NPJ Climate and Atmospheric Science*, *2*(1), 1–8.

Su, H., Wu, L., Zhai, C., Jiang, J. H., Neelin, J. D., & Yung, Y. L. (2020). Observed tightening of tropical ascent in recent decades and linkage to regional precipitation changes. *Geophysical Research Letters*, *47*(3), e2019GL085809.

Tao, L., Hu, Y., & Liu, J. (2015). Anthropogenic forcing on the Hadley circulation in CMIP5 simulations. *Climate Dynamics*, *46*, 3337–3350.

Tanaka, H. L., Ishizaki, N., & Kitoh, A. (2004). Trend and interannual variability of Walker, monsoon and Hadley circulations defined by velocity potential in the upper troposphere. *Tellus A: Dynamic Meteorology and Oceanography*, *56*(3), 250–269.

Tokinaga, H., Xie, S. -P., Timmermann, A., McGregor, S., Ogata, T., Kubota, H., & Okumura, Y. M., (2012). Regional patterns of tropical Indo-Pacific climate change: Evidence of the Walker circulation weakening. *Journal of Climate*, *25*, 1689–1710.

Vecchi, G. A., Clement, A., & Soden, B. J. (2008). Examining the tropical Pacific's response to global warming. *Eos, Transactions American Geophysical Union*, *89*(9), 81–83.

Vecchi, G. A., Soden, B. J., Wittenberg, A. T., Held, I. M., Leetmaa, A., & Harrison, M. J. (2006). Weakening of tropical Pacific atmospheric circulation due to anthropogenic forcing. *Nature, 441*, 73–76.

Vecchi, G. A., & Soden, B. J. (2007). Global warming and the weakening of the tropical circulation. *Journal of Climate*, *20*(17), 4316–4340.

Voigt, A., & Shaw, T. A. (2015). Circulation response to warming shaped by radiative changes of clouds and water vapour. *Nature Geoscience*, *8*(2), 102–106.

Voigt, A., & Albern, N. (2019). No cookie for climate change. *Geophysical Research Letters*, *46*(24), 14751–14761.

Voigt, A., Albern, N., & Papavasileiou, G. (2019). The atmospheric pathway of the cloud-radiative impact on the circulation response to global warming: Important and uncertain. *Journal of Climate*, *32*(10), 3051–3067.

Voigt, A., Albern, N., & Papavasileiou, G. (2019). The atmospheric pathway of the cloud-radiative impact on the circulation response to global warming: Important and uncertain. *Journal of Climate*, *32*(10), 3051–3067.

Voigt, A., Albern, N., Ceppi, P., Grise, K., Li, Y., & Medeiros, B. (2021). Clouds, radiation, and atmospheric circulation in the present-day climate and under climate change. *Wiley Interdisciplinary Reviews: Climate Change*, *12*(2), e694.

Wall, C. J., Hartmann, D. L., & Norris, J. R. (2019). Is the net cloud radiative effect constrained to be uniform over the tropical warm pools? *Geophysical Research Letters*, *46*, 12495–12503.

Watt-Meyer, O., & Frierson, D. M. W. (2017). Local and remote impacts of atmospheric cloud radiative effects onto the eddy-driven jet. *Geophysical Research Letters*, *44*(19), 10–36.

Wing, A. A., Stauffer, C. L., Becker, T., Reed, K. A., Ahn, M. S., Arnold, N. P., & Zhao, M. (2020). Clouds and convective self-aggregation in a multimodel ensemble of radiative-convective equilibrium simulations. *Journal of Advances in Modeling Earth Systems*, *12*(9), e2020MS002138.

Wodzicki, K. R., & Rapp, A. D. (2016). Long-term characterization of the Pacific ITCZ using TRMM, GPCP, and ERA-Interim. *Journal of Geophysical Research: Atmospheres*, *121*(7), 3153–3170.

Wodzicki, K. R., & Rapp, A. D. (2022). More Intense, organized deep convection with shrinking tropical ascent regions. *Geophysical Research Letters*, *49*(15), e2022GL098615.

Ying, J., Collins, M., Cai, W., Timmermann, A., Huang, P., Chen, D., & Stein, K. (2022). Emergence of climate change in the tropical Pacific. *Nature Climate Change*, *12*, 356–364.

Zhang, M., & Song, H. (2006). Evidence of deceleration of atmospheric vertical overturning circulation over the tropical Pacific. *Geophysical Research Letters*, *33*, L12701.

Zhao, M., Golaz, J. C., Held, I. M., Ramaswamy, V., Lin, S. J., Ming, Y., & Guo, H. (2016). Uncertainty in model climate sensitivity traced to representations of cumulus precipitation microphysics. *Journal of Climate*, *29*(2), 543–560.

Zelinka, M. D., Myers, T. A., McCoy, D. T., Po-Chedley, S., Caldwell, P. M., Ceppi, P., & Taylor, K. E. (2020). Causes of higher climate sensitivity in CMIP6 models. *Geophysical Research Letters*, *47*(1), e2019GL085782.

Zelinka, M. D., Klein, S. A., Qin, Y., & Myers, T. A. (2022). Evaluating climate models' cloud feedbacks against expert judgment. *Journal of Geophysical Research: Atmospheres*, *127*(2), e2021JD035198.

Zhou, C., Zelinka, M. D., & Klein, S. A. (2016). Impact of decadal cloud variations on the Earth's energy budget. *Nature Geoscience*, *9*(12), 871–874.

Zhou, W., Leung, L. R., Lu, J., Yang, D., & Song, F. (2020). Contrasting recent and future ITCZ changes from distinct tropical warming patterns. *Geophysical Research Letters*, *47*(22), e2020GL089846.

11

Clouds and Radiatively Induced Circulations

Tra Dinh[1], Blaž Gasparini[2], and Gilles Bellon[1,3]

ABSTRACT

In the atmosphere, there is an intimate relationship between clouds, atmospheric radiative cooling/heating, and radiatively induced circulations at various temporal and spatial scales. This coupling remains not well understood, which contributes to limiting our ability to model and predict clouds and climate accurately. Cloud liquid and ice particles interact with both shortwave (SW) and longwave (LW) radiation, leading to cloud radiative effect (CRE). The CRE includes perturbations of the radiative fluxes at the top of the atmosphere (TOA) and the surface, as well as perturbations of the radiative cooling profile within the atmosphere (see also Chapter 4: Cloud-Radiation Interactions and Cloud-Climate Feedbacks). The effect of clouds that results in atmospheric radiative heating or cooling that is distinct from the clear-sky radiative cooling profile will be termed the CRE on atmospheric heating, or CRE-AH. The CRE-AH can significantly modify the horizontal and vertical gradients of the diabatic heating profile, inducing circulations at various scales in the atmosphere. In turn, circulations govern cloud formation and evolution processes and therefore the properties and distribution of clouds.

11.1. INTRODUCTION

This chapter explores advances in research of the coupling between clouds and radiatively induced circulations and identifies the paths forward. The chapter is organized into three sections. Section 11.2 discusses the relationship between clouds and tropospheric diabatic circulations in which condensate and precipitation are produced. Section 11.3 focuses on low clouds and shallow radiatively induced circulations in the boundary layer and lower free troposphere. Section 11.4 focuses on tropical high clouds and the circulations induced by the CRE-AH in these clouds. All the radiatively induced circulations discussed in the three sections are illustrated in Figure 11.1.

[1]*Department of Physics, The University of Auckland, Auckland, New Zealand*

[2]*Department of Meteorology and Geophysics, University of Vienna, Vienna, Austria*

[3]*Centre National de Recherches Météorologiques, Université de Toulouse, Météo France, CNRS, Toulouse, France*

11.2. CLOUDS AND TROPOSPHERIC DIABATIC CIRCULATIONS

The relationship between clouds and tropospheric diabatic circulations is two-way. On the one hand, clouds are formed in the atmosphere as water vapor condenses within the moist, convective branch of these diabatic circulations. On the other hand, the perturbation in the diabatic heating profile due to the cloud radiative effect (CRE) on atmospheric heating (CRE-AH) can feed back on to weaken or strengthen the circulations in which clouds form and evolve. In this section, we will review progresses in our understanding of this coupling, particularly in the context of the tropospheric diabatic circulations that underlie the atmospheric water cycle.

Following directly from the fact that cloud condensates are produced within the atmospheric water cycle,

Clouds and Their Climatic Impacts: Radiation, Circulation, and Precipitation, Geophysical Monograph 281, First Edition.
Edited by Sylvia C. Sullivan and Corinna Hoose.

DOI: 10.1002/9781119700357.ch11

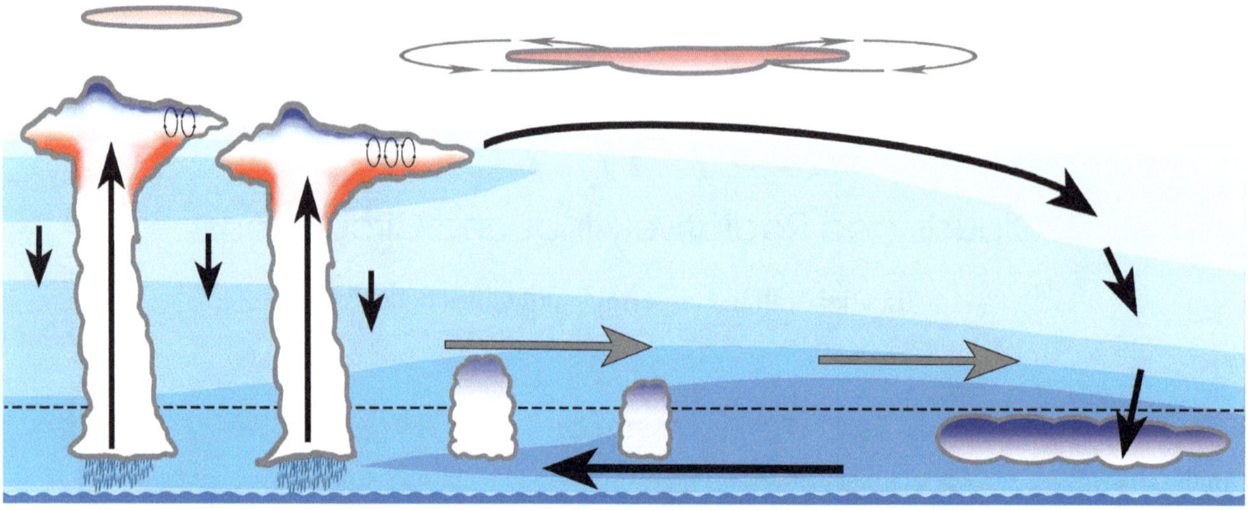

Figure 11.1 Illustration of the radiatively induced circulations discussed in this chapter, including tropospheric diabatic circulations (thick, black arrows, see section 11.2), large-scale shallow circulations (thick, black arrows in the boundary layer and thick, gray arrows in the lower free troposphere, see section 11.3), mesoscale circulations of tropical high clouds (thin, gray arrows, see section 11.4), and small-scale convective motions within the anvils (thin, black arrows, see section 11.4). The thin, dashed line marks the top of the boundary layer, which is located at altitudes of about 1 km to 2 km. The background shading represents the clear-sky radiative cooling. The shading within the clouds indicates the CRE-AH. Blue indicates radiative cooling and red indicates radiative heating.

understanding of cloud behaviors may be obtained by studying the atmospheric water cycle. As a metric of the strength of the atmospheric water cycle, let us use the rate of condensation of water vapor in the atmosphere, that is, the rate (in $kg.kg^{-1}.s^{-1}$) at which water vapor is transformed into condensate at a given location in the atmosphere. The condensation rate by itself does not dictate the percentage of condensates that remain in the atmosphere as clouds rather than falling to the surface as precipitation. But, if we assume that the percentage of cloud condensates (out of the total condensates) does not change under small perturbations of the climate, we will be able to obtain a first-order constraint on clouds based on the constraint on the condensation rate.

To formulate a constraint on the condensation rate – hereafter denoted by m_{cond}, we consider the moisture budget in the atmosphere following Schneider et al. (2010). In the free troposphere, a balanced water vapor budget in steady state is maintained between condensation and vertical advection of water vapor, that is,

$$[m_{\text{cond}}] = -[\omega \partial_p q], \tag{11.1}$$

where ω is the pressure velocity, p is pressure, and q is the specific humidity. The notation $[\cdot]$ indicates that the variable between the brackets is averaged horizontally over a sufficiently large domain for which horizontal advection of water vapor into/out of the domain can be neglected. The entire globe and the tropics are examples of such a domain. Equation (11.1), however, applies only to the free troposphere; in the boundary layer, the source of moisture that results from surface evaporation must be considered. Below, we will first focus on the free troposphere and then touch upon the complications of the boundary layer toward the end of the section.

In a moist convecting atmosphere, vertical advection of water vapor is upward and mostly carried out by moist convection. The updrafts inside moist convective regions are saturated with water vapor above the lifted condensation level (LCL). Thus,

$$[m_{\text{cond}}] \sim -[\omega_{\uparrow} \partial_p q_{\text{sat}}]_{\text{conv}} \sim -[\omega_{\uparrow}]_{\text{conv}} [\partial_p q_{\text{sat}}]_{\text{conv}} \tag{11.2}$$

above the LCL (c.f. equation 5 in Schneider et al., 2010). Here, q_{sat} is the saturation specific humidity, $\omega_{\uparrow}$ is the upward component of the pressure velocity, and $[\cdot]_{\text{conv}}$ indicates that the variable between the square brackets is averaged over moist convective regions within the domain only.

Over this domain, an overturning diabatic circulation can be constructed as consisting of a moist ascending branch in which latent heat is released as water vapor condenses forming condensates and precipitation (see the upward arrows in the deep convective clouds in Fig. 11.1) and a dry descending branch regulated by radiative cooling (see the downward arrows in clear air in Fig. 11.1). A nice illustration of such an overturning diabatic circulation is Figure 11.3 in

Wallace and Hobbs (2006). Following Yano et al. (2002), Zelinka and Hartmann (2010), Bony et al. (2016), and Dinh and Fueglistaler (2017, 2019), we seek to estimate the convective mass flux $[\omega_\uparrow]_{\text{conv}}$ as

$$[\omega_\uparrow]_{\text{conv}} \sim -[\omega_{\text{rad}}], \tag{11.3}$$

where

$$[\omega_{\text{rad}}] \sim \frac{[Q_{\text{rad}}]}{[\sigma]} \tag{11.4}$$

is the radiative subsidence mass flux, Q_{rad} is the radiative cooling rate, σ is the static stability, and T is atmospheric temperature. Furthermore, by assuming that the saturation specific humidity within moist convective regions scales with the domain average profile ($[\partial_p q_{\text{sat}}]_{\text{conv}} \sim [\partial_p q_{\text{sat}}]$), we obtain

$$[m_{\text{cond}}] \sim \frac{[\partial_p q_{\text{sat}}]}{[\sigma]}[Q_{\text{rad}}]. \tag{11.5}$$

Equation (11.5) indicates that the condensation rate depends on the radiative cooling rate as well as the factor

$$[f] = \frac{[\partial_p q_{\text{sat}}]}{[\sigma]}, \tag{11.6}$$

which is a function of atmospheric temperature. The atmospheric water cycle can be considered as driven by atmospheric radiative cooling, but processes that affect the vertical profile of temperature can affect the condensation rate and potentially clouds as well.

The linkage between tropospheric diabatic circulations and clouds as outlined above has been used as the basis to explain how clouds would change following an increase in atmospheric CO_2 concentration and subsequent global warming. The responses of the water cycle and clouds to the radiative forcing from CO_2 increase consist of two components (Dinh & Fueglistaler, 2017): (1) rapid adjustment (Gregory et al., 2004) which occurs in the atmosphere independently of surface temperature changes and (2) a slower response that follows surface warming. Climate models consistently predict that global precipitation and cloud water decrease during rapid atmospheric adjustment (see Dinh and Fueglistaler, 2019, and references therein). Dinh and Fueglistaler (2019) explained that the weakening of the atmospheric water cycle and decrease in cloud water are driven by the decrease in atmospheric radiative cooling, a direct result of the CO_2 increase. Atmospheric radiative cooling decreases because the added CO_2 enhances the total atmospheric absorption of the longwave (LW) radiation emitted by the surface. On the other hand, the slow response following surface warming shows increases in tropical rain rate and high-cloud water content (Larson & Hartmann, 2003). These changes under warming are driven by an increase in atmospheric radiative cooling, which is largely associated with increased water vapor and increased emission of LW radiation by water vapor at higher atmospheric temperatures. Further research is needed to confirm whether the increase in cloud water holds globally and to investigate the differences in the responses of tropical and extratropical clouds to warming.

The radiative cooling constraint on tropospheric diabatic circulations has been shown useful in evaluating the radiative climate feedback of tropical high clouds. Given that the convective mass flux is constrained by the radiative cooling ($[Q_{\text{rad}}]$) following equations (11.3) and (11.4), it can be expected that tropical convection and convective anvil clouds detrain near the level in the upper troposphere where $[Q_{\text{rad}}]$ decreases rapidly with height. For the Earth's atmosphere, $[Q_{\text{rad}}]$ is governed to a large extent by water vapor, whose vertical profile and radiative effect depend strongly on atmospheric temperature. Hartmann and Larson (2002) therefore proposed that the atmospheric temperature at which the LW radiative emission and atmospheric radiative cooling by water vapor become inefficient remains the same regardless of surface temperature. It follows that the temperature at the detrainment level and therefore the emission temperature of tropical anvil clouds remain constant as the surface warms; the clouds absorb more upwelling LW radiation from a warming surface and atmosphere below but emit the same amount of LW radiation. This "fixed anvil temperature" (FAT) hypothesis therefore implies a positive climate feedback of tropical anvil clouds.

The original FAT hypothesis does not account for the increased static stability ($[\sigma]$) at the level where $[Q_{\text{rad}}]$ decreases rapidly with height. In a follow-up work, Zelinka and Hartmann (2010) calculated the convergence profile of tropical tropospheric diabatic circulations from the vertical gradient of $[\omega_{\text{rad}}]$ using the clear-sky $[Q_{\text{rad}}]$ profile (see equation (11.4)). They found that the peak of the convergence profile shifts upward as the surface warms, consistently with the upward shift of the clear-sky $[Q_{\text{rad}}]$. However, the peak convergence level warms slightly and correspondingly the anvil cloud temperature increases slightly as the static stability at this level increases. These findings confirm that the climate feedback of tropical anvil clouds is positive but of a smaller magnitude than that suggested by the original FAT hypothesis.

Using the same argument, Bony et al. (2016) proposed that tropical anvil cloud cover is expected to shrink under warming. As the climate warms, the anvil clouds rise to remain at approximately the same temperature following the FAT hypothesis. As the clouds rise they are embedded in a more stable atmosphere. The increased stability decreases the radiatively driven mass flux (see equation (11.4)) and the convective outflow, leading to a reduction of the anvil cloud fraction. This stability effect has been confirmed based on interannual variations of anvil cloud fraction (Saint-Lu et al., 2020) in data from the Cloud-Aerosol Lidar and Infrared Pathfinder Satellite Observations (CALIPSO).

The original FAT hypothesis was developed for the tropics only, but the same general behavior of atmospheric radiative cooling, which is dominated by water vapor, also holds in the extratropics (Thompson et al., 2017). The pressure level in the upper troposphere at which the clear-sky atmospheric radiative cooling profile decreases rapidly with height thus defines the top of the troposphere and cloud top temperature throughout the globe. This is a strong constraint that can be used to infer cloud top height and how it changes under warming.

Much of the discussions above have focused on the aspect of the circulation-cloud coupling which concerns how the atmospheric water cycle governs large-scale cloudiness. Let us now address the second aspect of this coupling which concerns how the CRE-AH – through its impact on atmospheric radiative cooling – feeds back on diabatic circulations. Referring again to equation (11.5), we see that the CRE-AH can affect the condensation rate and therefore precipitation either by directly modifying the radiative cooling profile or by indirectly modifying the temperature profile through the radiative cooling perturbation. The temperature profile governs both the vertical gradient of the saturation specific humidity and the static stability in the atmosphere, which are the numerator and denominator of the factor $[f]$ defined in equation (11.6).

A number of studies using numerical simulations have shown that the CRE-AH plays an important role in determining the mean tropical circulation (Bergman & Hendon, 2000; Slingo & Slingo, 1988, 1991; Tian & Ramanathan, 2003) as well as the general circulation of the atmosphere and global precipitation (Li et al., 2015; Randall et al., 1989; Voigt et al., 2021). Averaged over the tropics and/or globally, the CRE-AH produces radiative heating in the free troposphere, correspondingly a reduction in the all-sky radiative cooling rate (see, e.g., Fig. 4 in Li et al., 2015). If this were the only effect of the CRE-AH, we would expect that precipitation decreases – particularly in the tropics where the reduction in the radiative cooling rate by the CRE-AH is largest. Indeed, Li et al. (2015) and Voigt et al. (2021) found that the tropical- and global-mean precipitation rates decrease in response to the CRE-AH. However, Randall et al. (1989) found on the contrary that the CRE-AH enhances the globally-averaged precipitation. In Randall et al. (1989)'s simulations, the CRE-AH results in a warmer and deeper troposphere. When integrated over a warmer and deeper troposphere, the $[\partial_p q_{sat}]$ term in equation (11.5) increases significantly. An enhancement in precipitation thus occurs despite reduced radiative cooling.

Given that the role of the CRE-AH on diabatic circulations remains inconclusive, it would be worthwhile to re-examine this topic in future studies. For future analysis, the impact of the CRE-AH on precipitation can be evaluated more accurately by accounting for all of the three terms $[Q_{rad}]$, $[\partial_p q_{sat}]$, and $[\sigma]$ in equation (11.5). Above we have touched upon the impacts of the CRE-AH on $[Q_{rad}]$ and $[\partial_p q_{sat}]$, but the CRE-AH also affects $[\sigma]$ – both locally by modifying the vertical profile of temperature in cloudy air and remotely by modifying the meridional temperature gradient. As shown in Li et al. (2015), the difference in the CRE-AH between the tropics and the extratropics leads to a meridional temperature gradient that affects the meridional transport of energy by the eddies. The transport of energy by the eddies in turn affects the static stability in the extratropics.

The discussions so far have been limited to clouds in the free troposphere above the boundary layer. Strictly speaking, the scaling argument used previously for the free troposphere (see equation (11.1–11.6)) does not apply to the boundary layer because (1) the source of moisture that results from surface evaporation must be added to the moisture budget balance in equation (11.1) for the boundary layer and (2) the updrafts in the boundary layer below the LCL are not saturated. Nevertheless, in the upper boundary layer, the impact of clouds on large-scale diabatic circulations can be qualitatively understood from the radiative cooling constraint. In the upper boundary layer (around 750 hPa to 850 hPa), the CRE-AH is dominated by the LW emission by cloud hydrometeors, leading to enhanced radiative cooling there (Haynes et al., 2013). This effect is apparent in the prevalent marine stratocumulus in the subtropics and higher latitudes (Wood, 2012). The CRE-AH of boundary-layer clouds thus strengthens the overturning diabatic circulations, resulting in an increase in the globally averaged precipitation (Fermepin & Bony, 2014). For other issues regarding the coupling between radiatively induced circulations and low clouds, the readers are referred to section 11.3 below.

In summary, notable progress in our understanding of the coupling between circulations and clouds has been made in the last few decades. In particular, constraints on how clouds change under climate change have been obtained based on basic understanding of tropospheric diabatic circulations. These constraints on cloud properties (vertical and horizontal distributions, water content, and coverage) will enable improved, accurate evaluation of the impacts of the CRE on the Earth's energy budget and the climate feedback of clouds on global warming. However, existing studies have not settled on the roles of the CRE-AH on circulations and precipitation. This leaves significant opportunities for future studies. Progress is anticipated to occur rapidly within the next decade given the advancement in model development and increasing availability of global and long-term observational data of clouds (such as CALIPSO).

11.3. LOW CLOUDS AND SHALLOW CIRCULATIONS

Shallow circulations are characterized by a convergence/divergence in the boundary layer and a compensating flow in the lower free troposphere. The ascending branch of these circulations is generally located in the warm, convective regions and the descending branch in the cooler, less convective regions because of the pressure gradients created by the temperature anomalies. Boundary-layer convergence occurs in the warm, convective regions and divergence in the cool, non-convective regions (see the horizontal black arrow near the bottom of Fig. 11.1), with opposite patterns just above the boundary layer (see the horizontal gray arrows in Fig. 11.1). Shallow circulations appear in numerical model simulations at both mesoscale and large scales and have been observed in the tropics at large scales. In this section, we will first explain the mechanism that gives rise to such circulations. Second, we will review how these circulations influence the spatial organization of convection (see also Chapter 7: Tropical Marine Low Clouds and Chapter 8: Mechanisms for the Self-Organization of Tropical Deep Convection). Lastly, we will review the instances of observed shallow circulations and their influence on the large-scale tropical climatology.

11.3.1. Mechanism of Shallow Circulations

In non-convective regions, the boundary layer experiences radiative cooling at its top, mostly through LW radiation. This cooling results from both low clouds and clear-sky processes. Indeed, the boundary layer is significantly moister, and therefore absorbs and re-emits more LW radiation than the free troposphere above. A large part of the LW radiation emitted upward at the top of the boundary layer is not absorbed by the free troposphere and hence escapes to space, allowing the boundary layer top to cool. Cloud hydrometeors also absorb and emit LW radiation very efficiently, so low clouds experience strong radiative cooling at their tops as well. This cooling is particularly intense in the case of stratocumulus clouds (Bellon & Georoy, 2016; Bellon & Bony, 2020; Wood, 2012). Mixing by turbulence redistributes the cooling from the top of the boundary layer downward. Vertical turbulent mixing is very efficient in a well-mixed boundary layer but a little less so in a shallow convective boundary layer.

The radiative cooling in the boundary layer can give rise to shallow circulations. In the hydrostatic approximation,

$$\frac{\partial \Phi}{\partial p} = -\frac{1}{\rho} \approx -\frac{R_\mathrm{a} T}{p} \tag{11.7}$$

with Φ the geopotential, ρ the density, T the temperature, p the pressure, and R_a the specific gas constant of air. This means that the vertical increase of the geopotential through the boundary layer is smaller in radiatively cooled regions than in others. The baroclinic component of this pattern corresponds to the geopotential near the surface being larger in radiatively cooled regions than in others, and the geopotential at the top of the boundary layer being smaller in these cooled regions than in others. The associated geopotential gradient forces a circulation that includes outflow of air from the cooled regions in the boundary layer and inflow of air toward the cooled regions just above the boundary layer. This mechanism has been illustrated and validated by large-eddy simulation experiments and conceptual models (Naumann et al., 2017, 2019).

11.3.2. Shallow Circulations and Aggregation of Convection

Shallow circulations differ from troposphere-deep circulations particularly in terms of the transport of moist static energy (MSE) $h = c_\mathrm{p} T + \Phi + L_\mathrm{v} q$, with c_p the specific heat capacity of air at constant pressure, L_v the latent heat of vaporization, and q the specific humidity. Shallow circulations transport MSE up-gradient and therefore increase the contrast in MSE between the convective and dry regions, while deep circulations transport MSE down-gradient and reduce such contrast.

The difference in the direction of transport of MSE between troposphere-deep circulations and shallow circulations results from the stratification of MSE in the atmosphere: h is large in the boundary layer where the humidity is high, minimum in the lower free troposphere where the humidity is low, and monotonically increases with altitude in the free troposphere because of the contribution from the geopotential; at the tropopause it reaches values similar to the surface values (see Fig. 4 in Glenn & Krueger, 2017). The vertical integral of h over the middle and upper troposphere is larger than the vertical integral of h over the boundary layer. Therefore, the upper branches of troposphere-deep circulations transport more MSE (from the ascending regions into the subsiding regions) than their lower branches in the boundary layer (from the subsiding regions into the ascending regions). The net effect is that troposphere-deep circulations export MSE from ascending, convective regions of large column-integrated MSE to dry, subsiding regions with small column-integrated MSE. In contrast, shallow circulations import MSE from their subsiding branches (where column-integrated MSE is small) into their ascending branches (where column-integrated MSE is large).

Because shallow circulations increase the horizontal gradient of MSE, they create a positive feedback on the spatial variability: anomalies in MSE are reinforced by the transport associated with these circulations. As a result, shallow circulations can be powerful actors in the spatial organization of convection: warm, convective regions become warmer and moister while dry, cool regions become cooler and drier. This mechanism of spatial organization of convection, named "self-aggregation," is documented both in convection-permitting simulations and in general circulation model (GCM) simulations even without any external circulation nor the effect of rotation. In particular, low-level radiative cooling can cause the development of cooler and drier regions, sometimes called "radiative cold pools," that are instrumental to the development of self-aggregation. Figures 11.2 and 11.3 show examples of such radiative cold pools in a convection-permitting model and a GCM, respectively.

Figure 11.2 shows results of an idealized simulation (see Bellon & Coppin, 2022) using the non-hydrostatic mesoscale model MesoNH (Lafore et al., 1998; Lac et al., 2018) in a convection-permitting configuration. The domain is 90,000 km^2 with a 3 km horizontal resolution and 48 vertical levels. The configuration prescribes a fixed sea surface temperature of 301 K over the entire domain and a horizontally uniform large-scale subsidence ($0.5\ kg.m^{-2}.s^{-1}$ at 400 hPa). Figure 11.2 shows the average variables over one day of the stationary state reached in this simulation.

Figure 11.2a shows the horizontal distribution of precipitable water in the MesoNH simulation described above, with a clear dry patch surrounded by a fairly uniform moist region. Using the bimodality of the distribution of column-integrated MSE, a dry region and a moist region can objectively be identified and the dashed line in Figure 11.2a shows the boundary between these regions. Figures 11.2b and 11.2c show the vertical profiles of the mixing ratios of water vapor and cloud condensates averaged over each of these regions. The figures show that the dry column is much drier than the moist column, both in the boundary layer and in the free troposphere, and with hardly any clouds. The moist region is convective,

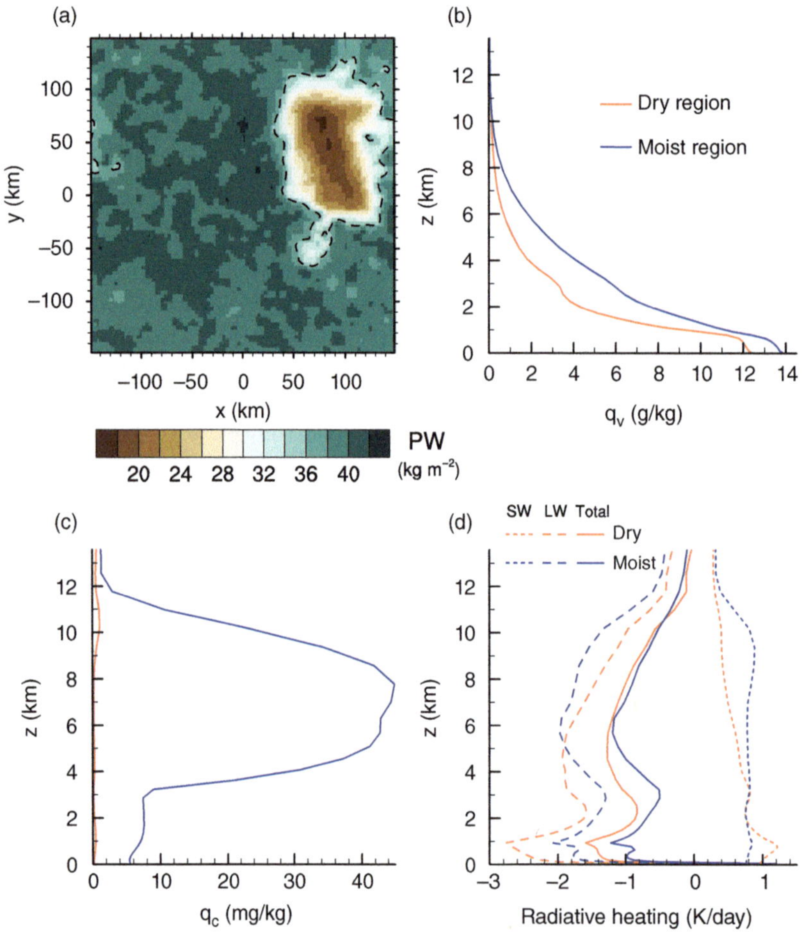

Figure 11.2 One-day average of a convection-permitting simulation: (a) map of precipitable water with a dashed contour showing the boundary between the moist and dry regions; (b) profile of water vapor mixing ratio; (c) profile of water condensate mixing ratio; and (d) profiles of solar (SW, dotted lines), terrestrial (LW, dashes lines), and net (total, solid lines) radiative heating rates. Blue lines refer to the moist region and red lines to the dry region.

with large amount of cloud condensates throughout the troposphere.

Figure 11.2d shows the radiative heating rates (with negative values indicating radiative cooling) in the dry and moist regions. In the lower troposphere, the LW radiative cooling is larger in the dry region because of the smaller greenhouse effect of water vapor and clouds in the free troposphere above. Conversely, in the upper troposphere, the LW radiative cooling is larger in the moist region because of the re-emission of LW radiation by the large amount of water vapor and clouds there. Except in the boundary layer, the shortwave (SW) radiative heating is larger in the moist region where there is more absorption of solar radiation by water vapor and clouds. The difference is opposite in the boundary layer because the downward solar radiative flux reaching the top of the boundary layer is larger in the dry region. In this simulation, the differences in the LW and SW effects compensate each other in the upper troposphere, but not in the lower troposphere where the net radiative cooling is larger (by about 0.5 $K.d^{-1}$) in the dry region than in the moist region. The radiative cooling of the dry region relative to the moist region creates a shallow circulation with winds diverging from the dry region near the surface and a return flow in the lower free troposphere. Consistently with the discussions above, the shallow circulation increases the contrast in MSE between the dry and moist columns (not shown). Note that in this particular simulation, the radiative effect is mostly a clear-sky effect. However, in similar simulations, a strong CRE-AH due to low clouds has been identified, with a similar resulting shallow circulation (Muller & Held, 2012; Muller & Bony, 2015).

Figure 11.3 shows regional snapshots of 925 hPa wind vectors and temperature (left panel) and precipitable water (right panel) at day 151 of a simulation using the GCM LMDz5A (Hourdin et al., 2006). The simulation was in a global radiative-convective-equilibrium configuration (i.e., without rotation and with a prescribed, uniform sea surface temperature) and initialized with horizontally uniform atmospheric conditions. In this case, the sea surface temperature is fairly low (292 K), and the spatial organization of convection results from mechanisms similar to those shown above in convection-permitting simulations. Details on the configuration and the mechanisms of self-aggregation at play in this simulation can be found in Coppin and Bony (2015). Figure 11.3 shows the development of a dry, cool region with near-surface winds diverging from that region. A return flow with converging winds occurs in the lower free troposphere (not shown). The expansion of this radiative cold pool eventually leads to circumscribing convection to a few limited regions of the globe.

As a final point, we note that while shallow circulations appear as a key mechanism of self-aggregation in a large

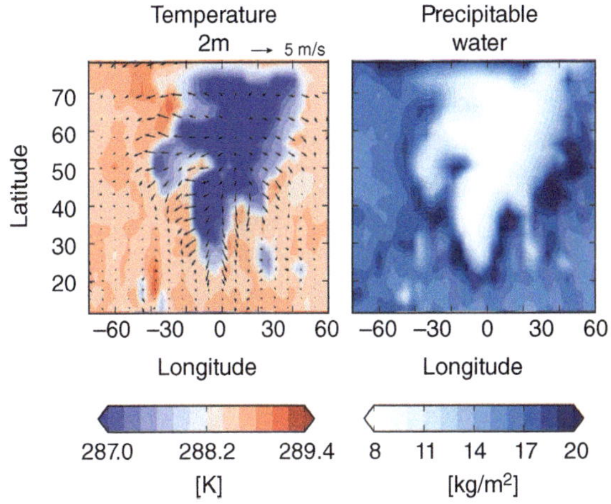

Figure 11.3 Radiative cold pool in a GCM simulation; left panel: temperature (shadings) and 925 hPa winds (vectors) at day 151 of the simulation; right panel: precipitable water. Source: Adapted from Figure 4 in Coppin and Bony (2015).

number of model simulations, other mechanisms can also cause or contribute to self-aggregation (see Wing et al., 2017, for a review). Furthermore, while there are similarities between modeled self-aggregated convection and observed organized convection (see Holloway et al., 2017), it is still unclear whether the mechanisms producing self-aggregation in model simulations are active in the real world.

11.3.3. Observed Shallow Circulations

Observing systems have limited capacity to observe the three-dimensional atmospheric velocity fields and indirect estimates using diabatic heating are not robust for troposphere-deep signal, let alone shallow circulations (Bellon et al., 2017). As a result, mesoscale shallow circulations are yet to be observed.

At large scales, shallow meridional circulations (SMCs) have been identified in the tropics around tropical convergence zones. The first of such observation was in the Intertropical Convergence Zone (ITCZ) of the eastern Pacific (Zhang et al., 2004). Figure 11.4 shows this SMC in the boreal autumn there, with low-level flow convergence around the ITCZ, which is located at 10°–15°N in this season. The northerly cross-equatorial flow in the boundary layer converges in the ITCZ and is compensated by diverging flows both in the lower free troposphere (at 2 km to 6 km in Fig. 11.4) and in the troposphere above 6 km, forming both shallow and deep overturning circulations. Similar SMCs have been observed in the central Atlantic ocean (Zhang et al., 2008) and in the Caribbean (Schulz & Stevens, 2018).

These SMCs in the ITCZ were first thought to result from the cross-equatorial sea surface temperature

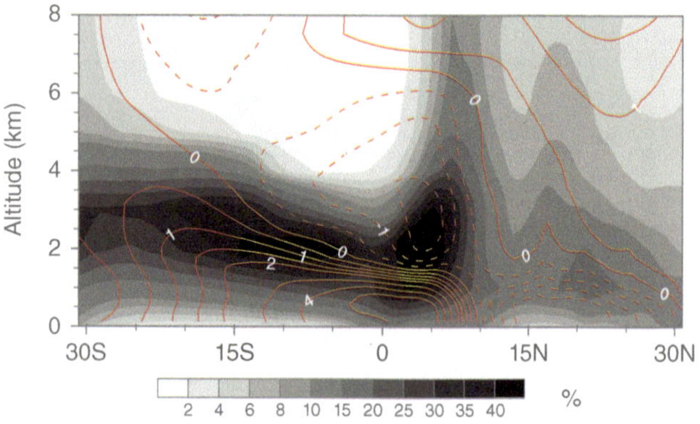

Figure 11.4 Cloud fraction (shadings, dataset GOCCP, Chepfer et al., 2010) and meridional wind (red contours, dataset ERA5, Hersbach et al., 2020) averaged over the eastern Pacific (90°W–120°W) in the boreal autumn (September to November).

gradients that are communicated to the boundary-layer air by turbulent fluxes and create pressure gradients that force the circulations (Nolan et al., 2007, 2010), but more recent work showed the role of the boundary-layer radiative cooling in the descending branch of the SMC. Figure 11.4 shows that the northerly flow in the non-convective region is co-located with the boundary-layer clouds, which lend some credibility to a role of boundary-layer-top cooling, and particularly the CRE-AH. The SMC is also more intense if deep convection is momentarily suppressed and shallow convection is more active (Yokoyama et al., 2014; Zhang et al., 2004), potentially because the boundary-layer temperature follows more closely the sea surface temperature if deep convection is suppressed, but also because it changes the boundary-layer vertical structure and hence its radiative cooling. Additionally to the diabatic forcing, the momentum constraint allows the SMC to develop only if the ITCZ is away from the equator (Dixit & Srinivasan, 2017).

SMCs are also observed in monsoon regions such as West Africa, India, and Australia (Zhang et al., 2008; Zhai & Boos, 2017) with a rising branch over the monsoon heat low. These result from the pressure gradients between the equatorial ocean and the heat low over the continent (Zhang et al., 2008).

SMCs contribute to moistening the convergence zones, although to a lesser extent than troposphere-deep circulations. Recall from section 11.3.2 that SMCs and deep circulations transport MSE in opposite directions because the vertical profile of MSE has a minimum in the lower free troposphere. On the other hand, both types of circulations import moisture into the convergence zones as water vapor decreases monotonically with height in the atmosphere (see, e.g., Fig. 11.2b). However, by exporting air in the lower free troposphere out of the ITCZ, the ITCZ SMCs import less moisture into the ITCZ than deep circulations (Zhang et al., 2008). Monsoon SMCs even include a drying contribution from horizontal advection of moisture, which further decreases the SMCs' moistening effect: as they extend to the heat low poleward of the convergence zones, they export moist air to that region in the boundary layer while importing very dry air into the ITCZ in the lower free troposphere (Zhai & Boos, 2017).

11.4. RESPONSES OF TROPICAL HIGH CLOUDS TO THE CRE-AH

This section describes the motions induced by the CRE-AH in tropical high clouds and the impacts of these motions on the evolution, lifetime, and the CRE at the top-of-the-atmosphere (TOA) of these clouds. The tropical high clouds that are strongly influenced by the motions induced by their own CRE-AH include (1) cirrus in the tropical tropopause layer (TTL); these clouds form either by in situ ice nucleation or from the remnants of anvil clouds, and (2) anvil clouds at various stages of evolution and optical properties, ranging from very reflective fresh anvils with large ice water contents to optically thin anvils that spread far away from deep convective cores.

The radiative heating rates that result from the interaction of ice crystals with radiation are shown in Figures 11.5 and 11.6 for typical high cirrus clouds. The TTL cirrus in Figure 11.5 has a cloud optical depth (COD) of 0.03 and the anvil cirrus in Figure 11.6 have CODs of 50, 5, and 1. The radiative heating rates shown

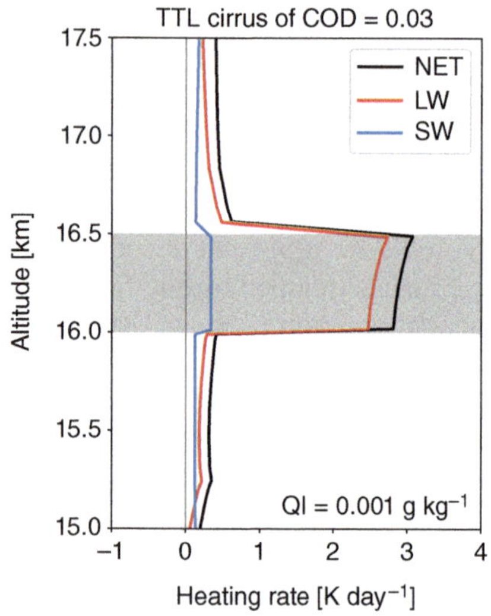

Figure 11.5 Radiative heating rates for a TTL cirrus assuming typical tropical temperature and moisture profiles. QI refers to ice mixing ratio. The ice crystal effective radius is set to values between 10 μm at the cloud top and 16 μm at the cloud base.

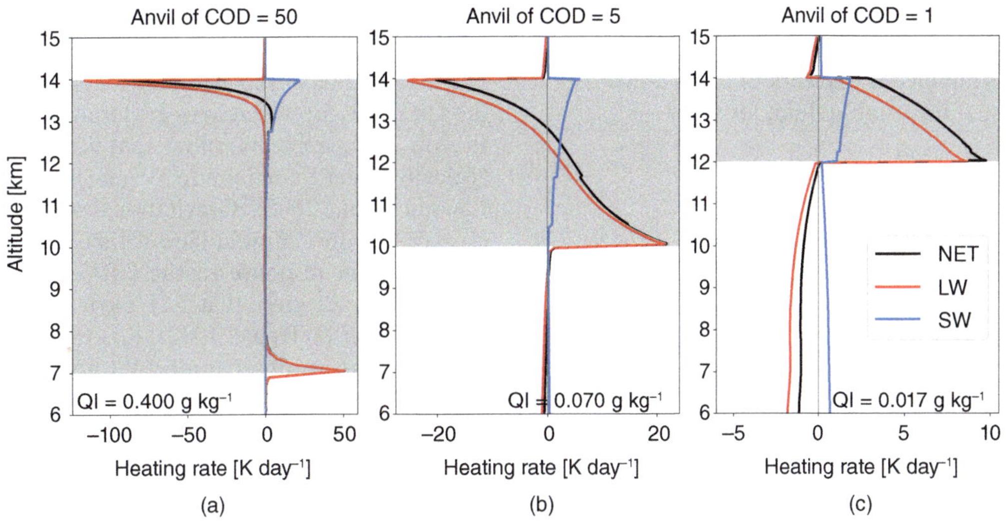

Figure 11.6 Radiative heating rates for (a) thick, (b) intermediately thick, and (c) thin anvil clouds assuming typical tropical temperature and moisture profiles. QI refers to ice mixing ratio. The ice crystal effective radius is set to values between 40 μm at the cloud top and 70 μm at the cloud base in case (a), 30 μm at the cloud top and 60 μm at the cloud base in case (b), and 20 μm at the cloud top and 30 μm at the cloud base in case (c), roughly in line with satellite observations of anvil clouds by Sokol and Hartmann (2020).

in these figures were calculated using the Rapid Radiative Transfer Model (RRTM, Iacono et al., 2008; Mlawer et al., 1997). The absorption of LW radiation results in heating which maximizes either near the cloud top for the TTL cirrus (Fig. 11.5) or near the cloud bases for the anvil clouds (Fig. 11.6). For the thick anvil clouds that are opaque to LW radiation (Fig. 11.6a,b), the large emissivities result in LW cooling near the cloud tops. For all the clouds shown, the absorption of SW radiation results in heating (Figs. 11.5 and 11.6), but the magnitude of the SW heating is significantly smaller than the LW heating and cooling.

The CRE-AH in both TTL cirrus and anvils can drive mesoscale circulations (see Figs. 11.1 and 11.7a). The circulations consist of rising vertical motions within the clouds (or portions of the clouds that are heated diabatically), compensating sinking motions outside the clouds, horizontal outflows near the cloud tops, and horizontal inflows near the cloud bases or below the levels of maximum heating (Ackerman et al., 1988; Durran et al., 2009; Dinh et al., 2010; Gasparini et al., 2019; Lilly, 1988).

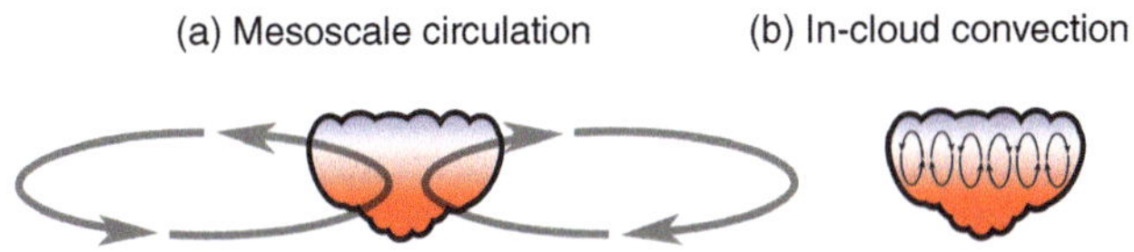

Figure 11.7 Air motions induced by the CRE-AH in tropical high clouds.

In addition, long exposure to a strong CRE-AH can bend the isentropic surfaces, forming unstable, convective layers within the high clouds (see Figs. 11.1 and 11.7b, and also Dobbie & Jonas, 2001; Dinh et al., 2010; Schmidt & Garrett, 2013).

The magnitude and vertical profile of the CRE-AH and the corresponding dynamical responses are highly sensitive to the COD. In the following, we therefore consider separately thin cirrus in the TTL (section 11.4.1) and anvil clouds which are much thicker (section 11.4.2).

11.4.1. TTL Cirrus

Thin cirrus are widespread in the TTL, with reported cloud fractions of 20–50% (Massie et al., 2002; Tseng & Fu, 2017). Such clouds are nearly transparent to the human eye, with CODs on the order of 0.01–0.05 (Dessler & Yang, 2003; Haladay & Stephens, 2009; Lee et al., 2009). They heat radiatively due to the absorption of LW radiation and to a minor extent SW absorption, with typical heating rates of about $5\ K.d^{-1}$ or less (Fig. 11.5).

TTL cirrus modulate the amount of water vapor that enters the stratosphere and therefore the radiative effect of stratospheric water vapor on climate (Dinh & Fueglistaler, 2014a). Furthermore, given that the clear-sky radiative cooling diminishes near zero in the TTL, the CRE-AH of these clouds plays a critical role in driving cross-isentropic transport of air from the TTL into the stratosphere (Corti et al., 2006; Dinh & Fueglistaler, 2014b).

The mesoscale dynamical response to the radiative heating in TTL cirrus can be qualitatively understood based on the simplified framework of a quasi-hydrostatic, two-dimensional Boussinesq fluid, for which

$$\frac{\partial b}{\partial t} + N^2 w = Q_{\text{rad}}, \tag{11.8}$$

$$\frac{\partial u}{\partial x} + \frac{\partial w}{\partial z} = 0 \tag{11.9}$$

(equations 3 and 4 in Durran et al., 2009). Here, u and w are respectively the horizontal and vertical velocities, b is the buoyancy, N is the Brunt–Väisälä frequency, and Q_{rad} is the CRE-AH. As time goes on ($t \rightarrow \infty$),

$$w \rightarrow w_{\text{ss}} = \frac{Q_{\text{rad}}}{N^2}. \tag{11.10}$$

The steady state vertical velocity (w_{ss}) is thus proportional to and has the same shape as the heating. Using equation (11.9) and w_{ss} in equation (11.10), the steady-state horizontal velocity can also be found. The steady state horizontal velocity depends on the vertical gradient of the heating and extends laterally beyond the cloud/heating boundaries (Durran et al., 2009).

Following equations (11.8) and (11.10), rising motions are induced in the clouds where the air is heated radiatively. Outside the lateral edges of the clouds, the buoyancy anomalies result in compensating sinking motions (see Fig. 11.7a and Fig. 3 in Dinh et al., 2010). Mass conservation equation (11.9) leads to horizontal inflows in the lower halves of the cloud layers and horizontal outflows in the upper halves of the cloud layers. The developed circulations narrow the cloud bases, expand the cloud tops, and lead to the spreading and lofting of the clouds (see Figs. 6 and 7 in Dinh et al., 2010).

The mesoscale circulations described above advect environmental air into the TTL cirrus (Dinh et al., 2010, 2012, 2014). Environmental air is brought into the clouds by the inflows in the lower halves of the clouds and then advected upward by the radiatively induced rising motions in the clouds. This leads to water vapor flux convergence (divergence) inside the clouds if the ambient air is moist (dry). Therefore, depending on whether the ambient air is supersaturated or subsaturated (with respect to ice), advection of air by the mesoscale circulations supports or suppresses depositional growth of cloud ice crystals, thereby lengthening or shortening the cloud lifetime (Dinh et al., 2010, 2012, 2014).

The vertical gradient of the radiative heating is negative near the tops of the TTL cirrus (at altitude of 16.5 km in Fig. 11.5). This can lead to the formation of shallow unstable layers at the cloud tops that support small-scale convective motions (see Figs. 4–6 in Dinh et al., 2010). As a result, the cloud tops become heterogeneous with pockets of updrafts with higher ice water contents and areas of downdrafts mixed with environmental air leading to ice sublimation.

The relevance of the mesoscale circulations induced by the CRE-AH for TTL cirrus evolution remains disputed. In situ measurements show that ice crystal radii are typically larger than 5 μm to 10 μm (Krämer et al., 2020; Lawson et al., 2008). Gravitational settling of large ice crystals may not allow sufficient time for the circulations to develop in response to the CRE-AH (Jensen et al., 2011). However, given that TTL cirrus can last for several days (Winker & Trepte, 1998), it is more likely than not that these in situ observations did not sample the cloud formation period (the first 12 hours or so), during which the freshly nucleated ice crystals remain small in size. The presence of wind shear and gravity wave temperature fluctuations (Dinh et al., 2016) was also shown to lead to a faster TTL cirrus decay, therefore diminishing the potential impacts of the circulations (Jensen et al., 2011; Podglajen et al., 2016). These simulations testing the roles of wind shear and temperature fluctuations were carried out for individual cloud cases only. Simulations of large cloud ensembles are needed to further investigate the sensitivity of the strength of the circulations induced by the CRE-AH to initial and environmental conditions.

11.4.2. Anvil Clouds

The deep convective plumes of cloudy air, initially of nearly uniform potential temperature, flatten and stretch after being injected into a stably stratified environment (Lilly, 1988). The convective plumes are self-lofted by the strong latent heating within the convective updrafts until reaching the level of neutral buoyancy, which is constrained by the clear-sky radiative cooling profile (see Hartmann & Larson, 2002 and section 11.2 above). The tops of the convective outflows are colder than the surrounding environment and therefore sink, while the bottoms are warmer and rise, leading to vertical compression and horizontal expansion of the convective outflows (Lilly, 1988) forming thick anvil clouds. From then on, cloud radiative interactions start to play important roles in modulating the morphology and life cycle of anvil clouds.

As a difference from TTL cirrus, freshly detrained anvil clouds of CODs larger than about 4 develop a radiative dipole (Wall et al., 2020) characterized by LW radiative heating near cloud bases and LW cooling near cloud tops (Fig. 10.6a,b). The cloud tops are in addition heated by the absorption of SW radiation by the cloud condensates. The SW heating reaches comparable magnitudes to the LW cooling at the cloud tops in the middle of the day when insolation values exceeds 1,000 W . m^{-2} (Sokol & Hartmann, 2020; Gasparini et al., 2022; Wall et al., 2020).

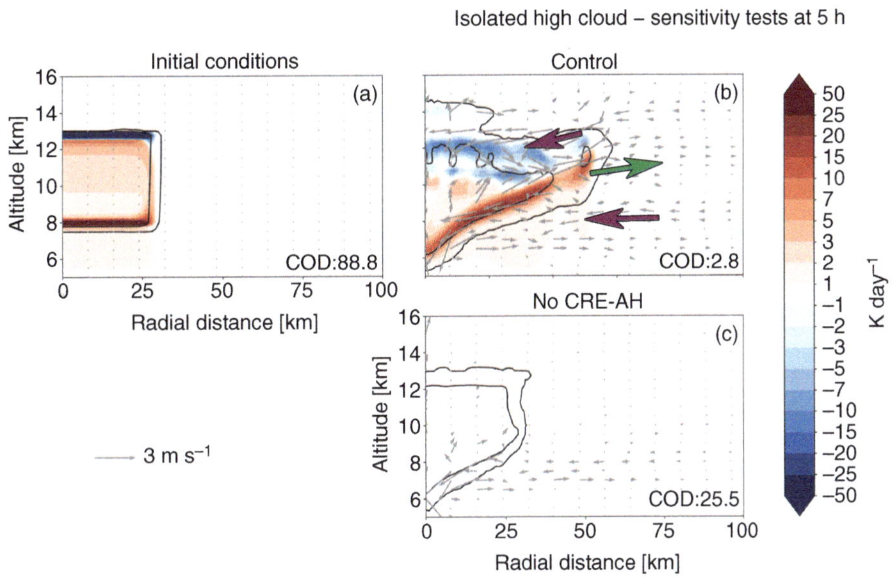

Figure 11.8 CRE-AH and wind vectors at (a) the initialization and (b) and (c) hour 5 after initialization of an optically thick anvil cloud simulated in a cloud-resolving model. Black contours represent ice water mixing ratios of 0.1 and 0.0001 $g \cdot kg^{-1}$. The vertical velocities are multiplied by a factor of 8 for a better visualization. The purple arrows highlight the inflows toward the cloud and the green arrow highlights the outflow.

However, averaged over both day and night, the SW heating is of secondary importance and the net CRE-AH follows closely the LW component.

The resulting CRE-AH drives mesoscale circulations. In the simulations by Gasparini et al. (2019), the primary circulation includes an ascent in the middle of the cloud between 7 km and 11 km and a strong outflow at about 10 km to 12 km (Fig. 11.8b). The primary circulation cell is closed by subsidence at the domain edge and an inflow at about 8 km to 9 km altitude. The circulation spreads the upper half of the cloud and narrows the lower half of the cloud, thereby significantly changing the shape of the cloud relative to the initialization time (Fig. 11.8a).

Also shown in Fig. 11.8b, the cloud top radiative cooling induces downdrafts that sink the cloud top and thin the cloud. The radiative cooling further induces an inflow of drier environmental air that erodes the cloud top. This secondary circulation leads to the formation of an upper thin cirrus cloud (at about 14 km in Fig. 11.8b), which is later detached from the main anvil cloud. Similar clouds were observed in in situ aircraft measurements (Garrett et al., 2004) and satellite remote sensing (Sokol & Hartmann, 2020).

Last but not least, the negative vertical gradient of the radiative heating rate (cooling above 11 km, heating below 11 km) destabilizes the air at the cloud top, resulting in convective motions and cell-like heterogeneous cloud elements there (see the 0.1 $g \cdot kg^{-1}$ contour in Fig. 11.8b).

The CRE-AH strongly affects the CRE at the TOA of anvil clouds. The CRE at the TOA is the difference between the all-sky and clear-sky net (LW plus SW) incoming (downward minus upward) radiative fluxes at the TOA. In the simulations shown in Fig. 11.8, when the CRE-AH is turned off (Fig. 11.8c), the cloud at hour 5 after initialization has a COD higher than 25 and a CRE at the TOA of about $-100\ W \cdot m^{-2}$. In contrast, in the presence of the CRE-AH (Fig. 11.8b), the cloud at hour 5 is optically much thinner with a COD of about 3 and a CRE at the TOA of only about $-6\ W \cdot m^{-2}$. Integrated over the whole cloud life cycle, the net CRE at the TOA is strongly negative in the simulation without the CRE-AH, while the simulation with the CRE-AH is close to being net CRE neutral at the TOA. The latter is consistent with satellite observations that the CRE at the TOA is close to zero for tropical convective clouds over the Indo-Pacific Warm Pool (Hartmann et al., 2001; Hartmann & Berry, 2017; Ramanathan et al., 1989).

Figure 11.9 summarizes the mechanisms by which the CRE-AH of anvil clouds influences their CRE at the TOA. The dominant effect comes from the horizontal spreading of the clouds by the radiatively induced mesoscale circulations. The aging, optically thin, and horizontally extensive anvil clouds have a positive CRE at the TOA (Gasparini et al., 2019; Kubar et al., 2007), counteracting the strongly negative CRE contribution from freshly detrained, thick but much smaller anvil clouds.

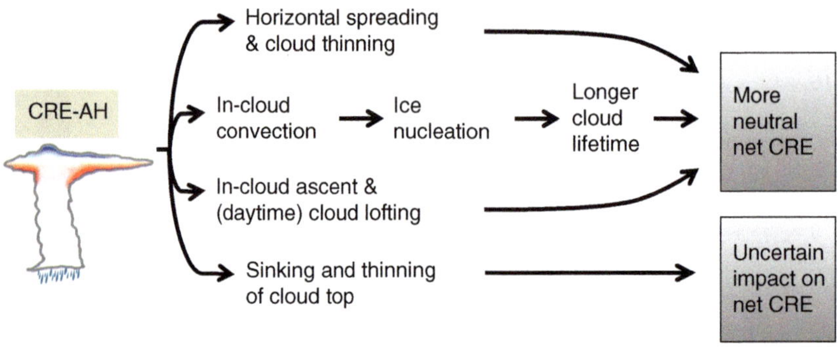

Figure 11.9 Mechanisms of the CRE-AH influencing the morphology and life cycle of tropical anvil clouds and their CRE at the TOA.

The motions induced by the CRE-AH also affect the persistence and lifetime of the clouds. Within the updrafts of the small-scale convective cells at the cloud tops, nucleation of new ice crystals can take place. Freshly nucleated ice crystals are much smaller in size and therefore sediment much more slowly than the pre-existing detrained ice crystals. The matured, optically thin anvil clouds can last a long time as they are replenished with the new ice crystals at their tops. As a result, the net CRE at the TOA integrated over the entire life cycle of anvil clouds becomes more positive (Hartmann et al., 2018).

In addition, the cloud radiative heating drives in-cloud ascents, which counteract ice crystal sedimentation, prolonging the cloud lifetime. On average, the cloud radiative heating comes mainly from the LW absorption, but during the day the SW absorption can significantly increase the net heating rate (Ruppert & Hohenegger, 2018; Wall et al., 2020). Therefore, during the day, the radiatively induced updrafts can be strong enough to overcome ice crystal sedimentation, lifting the clouds to higher levels and colder temperatures, therefore exerting a stronger LW CRE at the TOA (Ruppert & Klocke, 2019; Gasparini et al., 2022; Wall et al., 2020). Finally, the LW cooling at the cloud tops leads to the sinking and erosion of the cloud tops, decreasing the lifetime of the anvil clouds. However, as the sinking and thinning of the clouds decrease both the LW and SW CRE at the TOA, the net effect on the TOA radiative fluxes is uncertain.

In summary, past observational and modeling work have highlighted the importance of the motions induced by the CRE-AH for tropical anvil clouds' evolution and radiative climatic impacts. However, we note that existing modeling studies are limited to individual cloud cases only and may not capture the full range of variation of cloud properties. Both the local CRE-AH and the CRE at the TOA of tropical high clouds are highly sensitive to the microphysical properties of ice crystals, such as their effective radius and asymmetry parameter (Bantges et al., 2020; Järvinen et al., 2018). Simulations of the tropical atmosphere in which the life cycles of a diverse cloud population are captured will be needed to thoroughly evaluate how the CRE-AH influences tropical cloud distribution and radiative effects.

REFERENCES

Ackerman, T. P., Liou, K.-N., Valero, F. P. J., & Pfister, L. (1988). Heating rates in tropical anvils. *Journal of the Atmospheric Sciences*, *45*(10), 1606–1623. doi: 10.1175/1520-0469(1988)045<1606:HRITA>2.0.CO;2.

Bantges, R. J., Brindley, H. E., Murray, J. E., Last, A. E., Russell, J. E., Fox, C., et al. (2020). A test of the ability of current bulk optical models to represent the radiative properties of cirrus cloud across the mid- and far-infrared. *Atmospheric Chemistry and Physics*, *20*(21), 12889–12903. doi: 10.5194/acp-20-12889-2020.

Bellon, G., & Bony, S. (2020). Tropical and subtropical cloud systems. In *Clouds and Climate: Climate Science's Greatest Challenge* (pp. 251–278). Cambridge University Press.

Bellon, G., & Coppin, D. (2022). Sensitivity of convective self-aggregation to subsidence. *Journal of Advances in Modeling Earth Systems*, *14*(12), e2021MS002830.

Bellon, G., & Georoy, O. (2016). Stratocumulus radiative effect, multiple equilibria of the well-mixed boundary layer and transition to shallow convection. *Quarterly Journal of the Royal Meteorological Society*, *142*, 1685–1696. doi: 10.1002/qj.2762.

Bellon, G., Reitebuch, O., & Naumann, A. K. (2017). Shallow circulations: Relevance and strategies for satellite observation. In *Shallow Clouds, Water Vapor, Circulation, and Climate Sensitivity* (pp. 337–356). Springer.

Bergman, J. W., & Hendon, H. H. (2000). Cloud radiative forcing of the low-latitude tropospheric circulation: Linear calculations. *Journal of the Atmospheric Sciences*, *57*(14), 2225–2245. doi: 10.1175/1520-0469(2000)057<2225:CRFOTL>2.0.CO;2.

Bony, S., Stevens, B., Coppin, D., Becker, T., Reed, K. A., Voigt, A., & Medeiros, B. (2016). Thermodynamic control of anvil cloud amount. *Proceedings of the National Academy of Sciences*, *113*(32), 8927–8932. doi: 10.1073/pnas.1601472113.

Chepfer, H., Bony, S., Winker, D., Cesana, G., Dufresne, J. L., Minnis, P., et al. (2010). The GCM-oriented CALIPSO cloud product (CALIPSO-GOCCP). *Journal of Geophysical Research: Atmospheres*, *115*(D4). doi: 10.1029/2009JD012251.

Coppin, D. & Bony, S. (2015). Physical mechanisms controlling the initiation of convective self-aggregation in a general circulation model. *Journal of Advances in Modeling Earth Systems*, *7*(4), 2060–2078. doi: 10.1002/2015MS000571.

Corti, T., Luo, B. P., Fu, Q., VŁomel, H., & Peter, T. (2006). The impact of cirrus clouds on tropical troposphere-to-stratosphere transport. *Atmospheric Chemistry and Physics*, *6*(9), 2539–2547. doi: 10.5194/acp-6-2539-2006.

Dessler, A. E. & Yang, P. (2003). The distribution of tropical thin cirrus clouds inferred from Terra MODIS data. *Journal of Climate*, *16*(8), 1241–1248. doi: 10.1175/1520-0442(2003)16<1241:TDOTTC>2.0.CO;2.

Dinh, T., & Fueglistaler, S. (2014a). Microphysical, radiative and dynamical impacts of thin cirrus clouds on humidity in the tropical tropopause layer and stratosphere. *Geophysical Research Letters*, *41*, 6949–6955. doi: 10.1002/2014GL061289.

Dinh, T., & Fueglistaler, S. (2014b). Cirrus, transport, and mixing in the tropical upper troposphere. *Journal of the Atmospheric Sciences*, *71*, 1339–1352. doi: 10.1175/JAS-D-13-0147.1.

Dinh, T., & Fueglistaler, S. (2017). Mechanism of fast atmospheric energetic equilibration following radiative forcing by CO_2. *Journal of Advances in Modeling Earth Systems*, *9*. doi: 10.1002/2017MS001116.

Dinh, T., & Fueglistaler, S. (2019). On the causal relationship between the moist diabatic circulation and cloud rapid adjustment to increasing CO_2. *Journal of Advances in Modeling Earth Systems*, *11*. doi: 10.1029/2019MS001853.

Dinh, T., Durran, D. R., & Ackerman, T. (2012). Cirrus and water vapor transport in the tropical tropopause layer – Part 1: A specific case modeling study. *Atmospheric Chemistry and Physics*, *12*(20), 9799–9815. doi: 10.5194/acp-12-9799-2012.

Dinh, T., Fueglistaler, S., Durran, D., & Ackerman, T. (2014). Cirrus and water vapour transport in the tropical tropopause layer – Part 2: Roles of ice nucleation and sedimentation, cloud dynamics, and moisture conditions. *Atmospheric Chemistry and Physics*, *14*(22), 12225–12236. doi: 10.5194/acp-14-12225-2014.

Dinh, T., Podglajen, A., Hertzog, A., Legras, B., & Plougonven, R. (2016). Effect of gravity wave temperature fluctuations on homogeneous ice nucleation in the tropical tropopause layer. *Atmospheric Chemistry and Physics*, *16*, 35–46. doi: 10.5194/acp-16-35-2016.

Dinh, T. P., Durran, D. R., & Ackerman, T. P. (2010). Maintenance of tropical tropopause layer cirrus. *Journal of Geophysical Research: Atmospheres*, *115*(D2), 1–15. doi: 10.1029/2009JD012735.

Dixit, V. & Srinivasan, J. (2017). The momentum constraints on the shallow meridional circulation associated with the marine ITCZ. *Meteorology and Atmospheric Physics*, *129*(6), 595–609. doi: 10.1007/s00703-016-0489-2.

Dobbie, S. & Jonas, P. (2001). Radiative influences on the structure and lifetime of cirrus clouds. *Quarterly Journal of the Royal Meteorological Society*, (127), 2663–2682. doi: https://doi.org/10.1002/qj.49712757808.

Durran, D. R., Dinh, T., Ammerman, M., & Ackerman, T. (2009). The mesoscale dynamics of thin tropical tropopause cirrus. *Journal of the Atmospheric Sciences*, *66*(9), 2859–2873. doi: 10.1175/2009jas3046.1.

Fermepin, S., & Bony, S. (2014). Influence of low-cloud radiative effects on tropical circulation and precipitation. *Journal of Advances in Modeling Earth Systems*, *6*(3), 513–526. doi: 10.1002/2013MS000288.

Garrett, T. J., Heymseld, A. J., McGill, M. J., Ridley, B. A., Baumgardner, D. G., Bui, T. P., & Webster, C. R. (2004). Convective generation of cirrus near the tropopause. *Journal of Geophysical Research: Atmospheres*, *109*(21), 1–14. doi: 10.1029/2004JD004952.

Gasparini, B., Blossey, P. N., Hartmann, D. L., Lin, G., & Fan, J. (2019). What drives the life cycle of tropical anvil clouds? *Journal of Advances in Modeling Earth Systems*, *11*(8), 2586 2605. doi: 10.1029/2019MS001736.

Gasparini, G., Sokol, A. B., Wall, C. J., Hartmann, D. L. & Blossey, P. N. (2022). Diurnal fifferences in tropical maritime anvil cloud evolution *Journal of Climate*, *35*(5), 1655–1677.

Glenn, I. B., & Krueger, S. K. (2017). Connections matter: Updraft merging in organized tropical deep convection. *Geophysical Research Letters*, *44*(13), 7087–7094. doi: 10.1002/2017GL074162.

Gregory, J. M., Ingram, W. J., Palmer, M. A., Jones, G. S., Stott, P. A., Thorpe, R. B., et al. (2004). A new method for diagnosing radiative forcing and climate sensitivity. *Geophysical Research Letters*, *31*(3), L03205. doi: 10.1029/2003GL018747.

Haladay, T., & Stephens, G. (2009). Characteristics of tropical thin cirrus clouds deduced from joint CloudSat and CALIPSO observations. *Journal of Geophysical Research: Atmospheres*, *114*(D8). doi: 10.1029/2008JD010675.

Hartmann, D. L., & Berry, S. E. (2017). The balanced radiative effect of tropical anvil clouds. *Journal of Geophysical Research: Atmospheres*, *122*. doi: 10.1002/2017JD026460.

Hartmann, D. L., & Larson, K. (2002). An important constraint on tropical cloud-climate feedback. *Geophysical Research Letters*, *29*(20), 10–13. doi: 10.1029/2002GL015835.

Hartmann, D. L., Moy, L. A., & Fu, Q. (2001). Tropical convection and the energy balance at the top of the atmosphere. *Journal of Climate*, *14*(24), 4495–4511. doi: 10.1175/1520-0442(2001)014<4495:TCATEB>2.0.CO;2.

Hartmann, D. L., Gasparini, B., Berry, S. E., & Blossey, P. N. (2018). The life cycle and net radiative effect of tropical anvil clouds. *Journal of Advances in Modeling Earth Systems*, *10*(12), 3012–3029. doi: 10.1029/2018MS001484.

Haynes, J. M., Vonder Haar, T. H., L'Ecuyer, T., & Henderson, D. (2013). Radiative heating characteristics of Earth's cloudy atmosphere from vertically resolved active sensors. *Geophysical Research Letters*, *40*(3), 624–630. doi: 10.1002/grl.50145.

Hersbach, H., Bell, B., Berrisford, P., Hirahara, S., Horanyi, A., Mu noz-Sabater, J., et al. (2020) The ERA5 global reanalysis. *Quarterly Journal of the Royal Meteorological Society*, *146*(730), 1999–2049. doi: 10.1002/qj.3803.

Holloway, C. E., Wing, A. A., Bony, S., Muller, C., Masunaga, H., L'Ecuyer, T. S., et al. (2017). Observing convective aggregation. In *Shallow Clouds, Water Vapor, Circulation, and Climate Sensitivity* (pp. 26–63). Springer. doi: 10.1007/s10712-017-9419-1.

Hourdin, F., Musat, I., Bony, S., Braconnot, P., Codron, F., Dufresne, J.-L., et al. (2006). The LMDZ4 general circulation model: Climate performance and sensitivity to parametrized physics with emphasis on tropical convection. *Climate Dynamics*, *27* (7–8), 787–813. doi: 10.1007/s00382-006-0158-0.

Iacono, M. J., Delamere, J. S., Mlawer, E. J., Shephard, M. W., Clough, S. A., & Collins, W. D. (2008). Radiative forcing by long-lived greenhouse gases: Calculations with the AER radiative transfer models. *Journal of Geophysical Research: Atmospheres*, *113*(13), 2–9. doi: 10.1029/2008JD009944.

JŁarvinen, E., Jourdan, O., Neubauer, D., Yao, B., Liu, C., Andreae, M. O., et al. (2018). Additional global climate cooling by clouds due to ice crystal complexity. *Atmospheric Chemistry and Physics*, *18*(21), 15767–15781. doi: 10.5194/acp-18-15767-2018.

Jensen, E. J., Pfister, L., & Toon, O. B. (2011). Impact of radiative heating, wind shear, temperature variability, and microphysical processes on the structure and evolution of thin cirrus in the tropical tropopause layer. *Journal of Geophysical Research: Atmospheres*, *116*(12), 1–9. doi: 10.1029/2010JD015417.

KrŁamer, M., Rolf, C., Spelten, N., Afchine, A., Fahey, D., Jensen, E., et al. (2020). A microphysics guide to cirrus – Part 2: Climatologies of clouds and humidity from observations. *Atmospheric Chemistry and Physics*, *20*(21), 12569–12608. doi: 10.5194/acp-20-12569-2020.

Kubar, T. L., Hartmann, D. L., & Wood, R. (2007). Radiative and convective driving of tropical high clouds. *Journal of Climate*, *20*(22), 5510–5526. doi: 10.1175/2007JCLI1628.1.

Lac, C., Chaboureau, J.-P., Masson, V., Pinty, J.-P., Tulet, P., Escobar, J., et al. (2018). Overview of the Meso-NH model version 5.4 and its applications. *Geoscientific Model Development*, *11*(5), 1929–1969. doi: 10.5194/gmd-11-1929-2018.

Lafore, J. P., Stein, J., Asencio, N., Bougeault, P., Ducrocq, V., Duron, J., et al. (1998). The Meso-NH atmospheric simulation system. Part I: Adiabatic formulation and control simulations. *Annales Geophysicae*, *16*(1), 90–109. doi: 10.1007/s00585-997-0090-6.

Larson, K., & Hartmann, D. L. (2003). Interactions among cloud, water vapor, radiation, and large-scale circulation in the tropical climate. Part I: Sensitivity to uniform sea surface temperature changes. *Journal of Climate*, *16*(10), 1425–1440. doi: 10.1175/1520-0442-16.10.1425.

Lawson, R. P., Pilson, B., Baker, B., Mo, Q., Jensen, E., Pfister, L., & Bui, P. (2008). Aircraft measurements of microphysical properties of subvisible cirrus in the tropical tropopause layer. *Atmospheric Chemistry and Physics*, *8*, 1609–1620. doi: 10.5194/acp-8-1609-2008.

Lee, J., Yang, P., Dessler, A. E., Gao, B.-C., & Platnick, S. (2009). Distribution and radiative forcing of tropical thin cirrus clouds. *Journal of the Atmospheric Sciences*, *66*(12), 3721–3731. doi: 10.1175/2009JAS3183.1.

Li, Y., Thompson, D. W. J., & Bony, S. (2015) The influence of atmospheric cloud radiative effects on the large-scale atmospheric circulation. *Journal of Climate*, *28*(18), 7263–7278. doi: 10.1175/JCLI-D-14-00825.1.

Lilly, D. K. (1988). Cirrus outflow dynamics. *Journal of the Atmospheric Sciences*, *45*(10). doi: 10.1175/1520-0469(1988)045<1594:COD>2.0.CO;2.

Massie, S., Gettelman, A., Randel, W., & Baumgardner, D. (2002). Distribution of tropical cirrus in relation to convection. *Journal of Geophysical Research*, *107*(D21), 4591. doi: 10.1029/2001JD001293.

Mlawer, E. J., Taubman, J., Brown, P. D., Iacono, M. J., & Clough, S. A. (1997). Radiative transfer for inhomogeneous atmospheres: RRTM, a validated correlated-k model for the longwave. *Journal of Geophysical Research*, *102*(D14), 16663–16682. doi: doi:10.1029/97JD00237.

Muller, C., & Bony, S. (2015). What favors convective aggregation and why? *Geophysical Research Letters*, *42*(13), 5626–5634. doi: 10.1002/2015GL064260.

Muller, C. J., & Held, I. M. (2012). Detailed investigation of the self-aggregation of convection in cloud-resolving simulations. *Journal of the Atmospheric Sciences*, *69*(8), 2551–2565. doi: 10.1175/JAS-D-11-0257.1.

Naumann, A. K., Stevens, B., Hohenegger, C., & Mellado, J. P. (2017). A conceptual model of a shallow circulation induced by prescribed low-level radiative cooling. *Journal of the Atmospheric Sciences*, *74*(10), 3129–3144. doi: 10.1175/JAS-D-17-0030.1.

Naumann, A. K., Stevens, B., & Hohenegger, C. (2019). A moist conceptual model for the boundary layer structure and radiatively driven shallow circulations in the trades. *Journal of the Atmospheric Sciences*, *76*(5), 1289–1306. doi: 10.1175/JAS-D-18-0226.1.

Nolan, D. S., Zhang, C., & Chen, S.-H. (2007). Dynamics of the shallow meridional circulation around intertropical convergence zones. *Journal of the Atmospheric Sciences*, *64*(7), 2262–2285. doi: 10.1175/JAS3964.1.

Nolan, D. S., Powell, S. W., Zhang, C., & Mapes, B. E. (2010). Idealized simulations of the intertropical convergence zone and its multilevel flows. *Journal of the Atmospheric Sciences*, *67*(12), 4028–4053. doi: 10.1175/2010JAS3417.1.

Podglajen, A., Plougonven, R., Hertzog, A., & Legras, B. (2016). A modelling case study of a large-scale cirrus in the tropical tropopause layer. *Atmospheric Chemistry and Physics*, *16*(6), 3881–3902. doi: 10.5194/acp-16-3881-2016.

Ramanathan, V., Cess, R. D., Harrison, E. F., Minnis, P., Barkstrom, B. R., Ahmad, E., & Hartmann, D. (1989). Cloud-radiative forcing and climate: Results from the Earth radiation budget experiment. *Science*, *243*(4887), 57–63. doi: 10.1126/science.243.4887.57.

Randall, D. A., Harshvardhan, Dazlich, D. A., & Corsetti, T. G. (1989). Interactions among radiation, convection, and large-scale dynamics in a general circulation model. *Journal of the Atmospheric Sciences*, *46*(13), 1943–1970. doi: 10.1175/1520-0469(1989)046<1943:IARCAL>2.0.CO;2.

Ruppert, J. H., & Hohenegger, C. (2018). Diurnal circulation adjustment and organized deep convection. *Journal of Climate*, *31*(12), 4899–4916. doi: 10.1175/JCLI-D-17-0693.1.

Ruppert, J. H., & Klocke, D. (2019). The two diurnal modes of tropical upward motion. *Geophysical Research Letters*, *46*, 2911–2921. doi: 10.1029/2018GL081806.

Saint-Lu, M., Bony, S., & Dufresne, J. L. (2020). Observational evidence for a stability Iris effect in the tropics. *Geophysical Research Letters*, *47*(14). doi: 10.1029/2020GL089059.

Schmidt, C. T., & Garrett, T. J. (2013). A simple framework for the dynamic response of cirrus clouds to local diabatic radiative heating. *Journal of the Atmospheric Sciences*, *70*(5), 1409–1422. doi: 10.1175/JAS-D-12-056.1.

Schneider, T., O'Gorman, P. A., & Levine, X. J. (2010). Water vapor and the dynamics of climate change. *Reviews of Geophysics*, *48*, RG3001. doi: 10.1029/2009RG000302.

Schulz, H., & Stevens, B. (2018). Observing the tropical atmosphere in moisture space. *Journal of the Atmospheric Sciences*, *75*(10), 3313–3330. ISSN 0022-4928. doi: 10.1175/JAS-D-17-0375.1.

Slingo, A., & Slingo, J. M. (1988). The response of a general circulation model to cloud longwave radiative forcing. I: Introduction and initial experiments. *Quarterly Journal of the Royal Meteorological Society*, *114*(482), 1027–1062. doi: 10.1002/qj.49711448209.

Slingo, J. M., & Slingo, A. (1991). The response of a general circulation model to cloud longwave radiative forcing. II: Further studies. *Quarterly Journal of the Royal Meteorological Society*, *117*(498), 333–364. doi: 10.1002/qj.49711749805.

Sokol, A., & Hartmann, D. L. (2020). Tropical anvil clouds: Radiative driving towards a preferred state. *Earth and Space Science Open Archive*, *28*. doi: 10.1002/essoar.10503124.2.

Thompson, D. W. J., Bony, S., & Li, Y. (2017). Thermodynamic constraint on the depth of the global tropospheric circulation. *Proceedings of the National Academy of Sciences*, *114*(31), 8181–8186. doi: 10.1073/pnas.1620493114.

Tian, B., & Ramanathan, V. (2003). A simple moist tropical atmosphere model: The role of cloud radiative forcing. *Journal of Climate*, *16*(12), 2086–2092. doi: 10.1175/1520-0442(2003)016<2086:ASMTAM>2.0.CO;2.

Tseng, H.-H., & Fu, Q. (2017). Tropical tropopause layer cirrus and its relation to tropopause. *Journal of Quantitative Spectroscopy and Radiative Transfer*, *188*, 118–131. doi: 10.1016/j.jqsrt.2016.05.029.

Voigt, A., Albern, N., Ceppi, P., Grise, K., Li, Y., & Medeiros, B. (2021). Clouds, radiation, and atmospheric circulation in the presentday climate and under climate change. *WIREs Climate Change*, 12 (2), e694. doi: 10.1002/wcc.694.

Wall, C. J., Norris, J. R., Gasparini, B., Smith Jr, W. L., Thieman, M. M., & Sourdeval, O. (2020). Observational evidence that radiative heating modifies the life cycle of tropical anvil clouds. *Journal of Climate*, *33*, 8621–8640. doi: 10.1175/JCLI-D-20-0204.1.

Wallace, J. M., & Hobbs, P. V. (2006). *Atmospheric science: An introductory survey*, vol. 92. Elsevier.

Wing, A. A., Emanuel, K., Holloway, C. E., & Muller, C. (2017). Convective self-aggregation in numerical simulations: A review. In *Shallow Clouds, Water Vapor, Circulation, and Climate Sensitivity*, pp. 1–25. Springer. doi: 10.1007/978-3-319-77273-81.

Winker, D. M., & Trepte, C. R. (1998). Laminar cirrus observed near the tropical tropopause by LITE. *Geophysical Research Letters*, *25*(17), 3351–3354. doi: 10.1029/98GL01292.

Wood, R. (2012). Stratocumulus clouds. *Monthly Weather Review*, *140*(8), 2373–2423. doi: 10.1175/mwr-d-11-00121.1.

Yano, J.-I., Grabowski, W. W., & Moncrie, M. W. (2002). Mean-state convective circulations over large-scale tropical SST gradients. *Journal of the Atmospheric Sciences*, *59*(9), 1578–1592. doi: 10.1175/1520-0469(2002)059<1578:MSCCOL>2.0.CO;2.

Yokoyama, C., Zipser, E. J., & Liu, C. (2014). TRMM-observed shallow versus deep convection in the Eastern Pacific related to large-scale circulations in reanalysis datasets. *Journal of Climate*, *27*(14), 5575–5592. doi: 10.1175/JCLI-D-13-00315.1.

Zelinka, M. D., & Hartmann, D. L. (2010). Why is longwave cloud feedback positive? *Journal of Geophysical Research*, *115*(D16), D16117. doi: 10.1029/2010JD013817.

Zhai, J., & Boos, W. R. (2017). The drying tendency of shallow meridional circulations in monsoons. *Quarterly Journal of the Royal Meteorological Society*, *143*(708), 2655–2664. doi: 10.1002/qj.3091.

Zhang, C., McGauley, M., & Bond, N. A. (2004). Shallow meridional circulation in the tropical eastern Pacific. *Journal of Climate*, *17*(1), 133–139. doi: 10.1175/1520-0442(2004)017<0133:SMCITT>2.0.CO;2.

Zhang, C., Nolan, D. S., Thorncroft, C. D., & Nguyen, H. (2008). Shallow meridional circulations in the tropical atmosphere. *Journal of Climate*, *21*(14), 3453–3470. doi: 10.1175/2007JCLI1870.1.

12

The Small-Scale Mixing of Clouds With Their Environment: Impacts on Micro- and Macroscale Cloud Properties

Fabian Hoffmann

ABSTRACT

The small-scale mixing of clouds with their environment is an essential cloud process. Following an entrainment event, turbulent mixing breaks down the entrained air and homogenizes it with the cloud, covering multiple orders of magnitude in lengthscales from the entraining eddies (~100 m) down to the Kolmogorov length (~1 mm). The character of this process, traditionally categorized into homogeneous and inhomogeneous mixing scenarios, can affect the microphysical composition of clouds, with commensurate impacts on large-scale cloud properties such as the cloud albedo and cloud lifetime. Based on the current physical understanding of the small-scale mixing of cloudy and cloud-free air, this chapter will summarize the basic theories describing this process. By considering the wide range of involved scales, we will outline different observational and numerical approaches used to investigate this process in clouds, as well as methods to parameterize it in large-scale numerical models. Finally, we will review the impacts of the small-scale mixing process, focusing on microscale changes in the droplet size distribution as well as macroscale effects relevant to our understanding of clouds in the climate system.

12.1. INTRODUCTION

One major role of clouds in the climate system stems from their ability to reflect incident shortwave radiation back to space, resulting in a major negative forcing on the global radiation budget (e.g., Boucher et al., 2013). This forcing, often quantified as the cloud radiative effect (Betts, 2007), is proportional to the product of cloud albedo and cloud fraction, where the former is primarily determined by the cloud microphysical composition — the number and size of droplets forming the droplet size distribution — and the latter is controlled by the cloud macroscale — the large-scale organization of clouds and their life cycle.

Historically, the influence of cloud microphysics on the cloud radiative effect was attributed to two processes: The *albedo effect* predicts that a higher droplet concentration results in a higher cloud albedo due to the larger integral droplet surface capable of reflecting more shortwave radiation (Twomey, 1974, 1977). The *lifetime effect* indicates that a higher droplet number concentration with commensurately smaller droplets may prevent droplets from colliding, and hence decelerates the decay of clouds by precipitation, increasing cloud lifetime and cloud fraction (Albrecht, 1989). In more recent years, our understanding of these effects and their implications for the global climate has been significantly widened by amending the underlying physical framework by the turbulent nature of clouds (e.g., Bodenschatz et al., 2010), including, inter alia, the entrainment of cloud-free air and its subsequent mixing with the cloud, which is the main topic of this chapter.

The mixing of cloudy and cloud-free air can have important implications for the role of clouds in the climate system, bounded by the canonical scenarios

Meteorological Institute, Faculty for Physics, Ludwig Maximilian University of Munich, Munich, Germany

Clouds and Their Climatic Impacts: Radiation, Circulation, and Precipitation, Geophysical Monograph 281, First Edition.
Edited by Sylvia C. Sullivan and Corinna Hoose.

DOI: 10.1002/9781119700357.ch12

of *homogeneous* and *extreme inhomogeneous mixing* (Baker and Latham, 1979; Warner, 1973). While these terms and their distinct effects on the micro- and macroscale properties of clouds will be defined more carefully in the following, it is important to note that the assumption of one mixing scenario over the other can be of great consequence. For instance, extreme inhomogeneous instead of homogeneous mixing is able to reduce the albedo of a cloud by up to 6 percentage points (Chosson et al., 2007; Slawinska et al., 2008). Furthermore, inhomogeneous mixing might also accelerate the growth of cloud droplets, which can imperil the colloidal stability of clouds by initiating the precipitation process, with commensurate negative effects on cloud lifetime and cover (e.g., Baker et al., 1980).

While turbulence is the main driver for the small-scale mixing of clouds with their environment it has further effects on clouds that are not covered in this chapter. The interested reader is referred to Shaw (2003), Devenish et al. (2012), and Grabowski and Wang (2013) for more information on turbulence-induced supersaturation fluctuations, droplet clustering, including related particle inertia effects, as well as turbulence-enhanced collision rates. Furthermore, this chapter can be seen as an addendum to de Rooy et al. (2013), who reviewed the entrainment process from a large-scale perspective, but did not address the small-scale detail of the subsequent mixing process presented here. Finally, it is necessary to state that this chapter is limited to shallow clouds, free from frozen hydrometeors. While the turbulent mixing of cloudy with cloud-free air is also relevant for mixed-phase and cold clouds, the literature on this topic is still very limited and therefore omitted (e.g., Hoffmann, 2020; Korolev et al., 2017).

The chapter is organized as follows. First, we will summarize the basic processes of entrainment and small-scale mixing in warm clouds (section 12.2). Then, the theoretical framework to describe the small-scale mixing process in clouds is introduced (section 12.3), followed by an overview of observational techniques and numerical models used to investigate it (section 12.4). Based on this, we will explore the effects of small-scale mixing on micro- and macroscale cloud properties (section 12.5), before this chapter is concluded (section 12.6).

12.2. ENTRAINMENT AND MIXING IN WARM CLOUDS

Beginning with Stommel (1947) and Warner (1955), the entrainment of cloud-free air into clouds has been recognized as an essential process to understand the observed liquid water content of clouds, which is lower than the expected adiabatic values due to the mixing with cloud-free air. Today, we generally understand that the cloud-free air is entrained into the cloud by the specific large-scale dynamics of the cloud, that is, the vortical motion of the ascending cumulus cloud top which engulfs cloud-free air laterally (Grabowski & Clark, 1993; Heus et al., 2008; Zhao and Austin, 2005), or the downward branches of the stratocumulus-topped boundary-layer large-eddy circulation which engulf air from above the cloud top (Gerber et al., 2005; Kurowski et al., 2009; Nicholls, 1989; Yamaguchi & Randall, 2012). These entraining motions create regions of negligible liquid water inside the cloud, so-called *cloud holes*, which are continuously deformed (e.g., Krueger, 1993). Initially, this process is driven by the entraining eddy, and later by the developing small-scale turbulence that folds and stretches the cloud hole into increasingly smaller filaments until the Kolmogorov lengthscale is reached. At this lengthscale, molecular diffusion homogenizes the entrained air with the cloud. Throughout this turbulent break-down process, the increasing surface area of the cloud hole exposes more and more droplets to a subsaturated environment where they evaporate. Accordingly, the mixing process combines the turbulent stirring of cloudy and cloud-free air with the commensurate droplet evaporation.

Traditionally, the mixing process in clouds has gained less attention than the entrainment. In fact, mixing is often implicitly included in the broader term entrainment, although only the mixing process causes the evaporation that results in the sub-adiabatic liquid water content initially observed by Stommel (1947) and Warner (1955). Nonetheless, Warner (1969a) hypothesized that mixing is relevant for the development of the broad droplet size distributions typically observed in clouds, but was not able to explain the size distributions theoretically. The reason for this was the common depiction of mixing as an instantaneous process, that is, cloud-free and cloudy air homogenize rapidly after entrainment, which causes all droplets to experience the same subsaturation and to evaporate similarly, with only minuscule effects on the droplet size distribution (Warner, 1973).

Later, this depiction of the mixing process has been challenged by Latham and Reed (1977) and Baker and Latham (1979), who began to consider that the mixing process is not instantaneous. A finite rate mixing process allows some droplets to be located within unblemished cloudy filaments, while only those droplets evaporate that leave the cloud. Since this process allows for a nonidentical development of droplet sizes, a broad droplet size distributions can be produced which is much more in agreement with those observed in clouds (e.g., Warner, 1969a).

Both processes can be idealized as *homogeneous* and *extreme inhomogeneous mixing*, respectively, which are

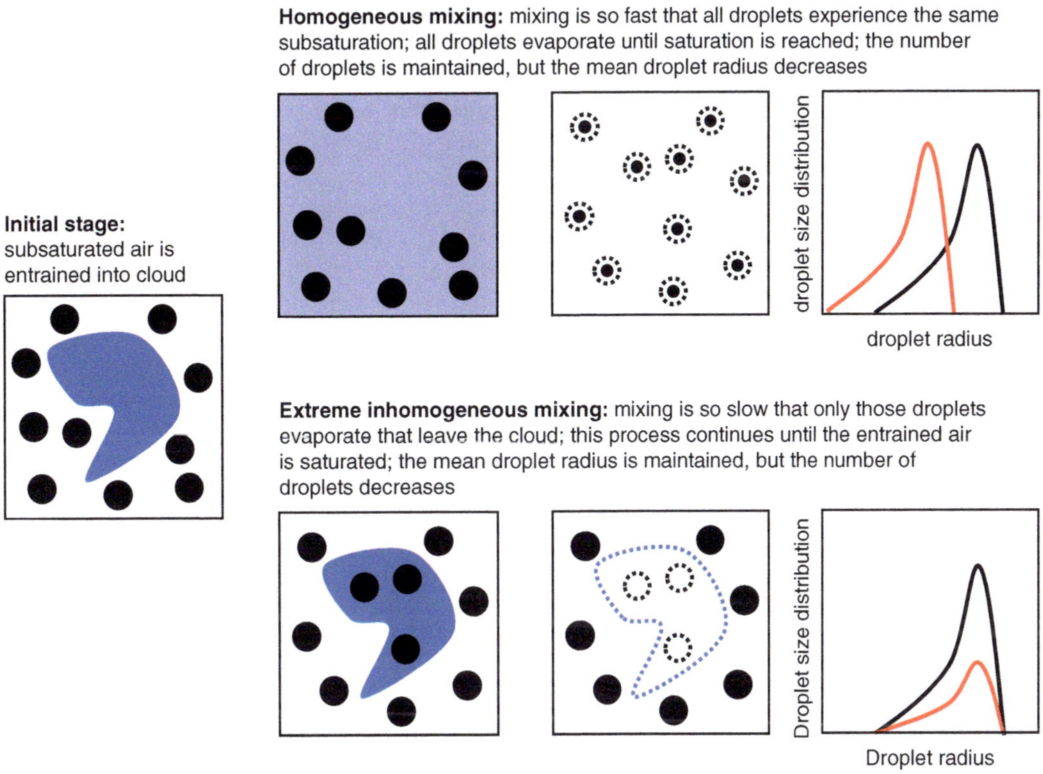

Figure 12.1 This figure illustrates the two limiting scenarios of homogeneous and extreme inhomogeneous mixing following an idealized entrainment event. The entrained air is depicted in shades of blue, indicating the subsaturated cloud-free air. Black dots represent cloud droplets. The rightmost panels show the reaction of the droplet size distribution on the respective mixing scenarios, where the black (red) line indicates the droplet size distribution before (after) mixing.

illustrated in Figure 12.1. During *homogeneous mixing*, which typically occurs under rapid turbulent mixing, all droplets evaporate, but none completely. Accordingly, homogeneous mixing is often identified by a constant droplet number concentration. If, however, the turbulent mixing is comparatively slow, only those droplets evaporate that leave the cloudy filaments. If the droplets that leave the cloud evaporate completely, the mean droplet radius is maintained and the scenario is called *extreme inhomogeneous mixing*. Homogeneous and extreme inhomogeneous mixing limit the range of mixing scenarios that occur in nature. Intermediate mixing scenarios, in which both the mean droplet radius and droplet number change, are generally termed inhomogeneous mixing, and it can be necessary to distinguish them from the case of extreme inhomogeneous mixing.

Finally, it is important to note that real clouds constitute a much more complicated system than this idealized depiction of the mixing process can cover. Real clouds experience several, potentially interacting entrainment events on multiple scales, making it probably impossible to ever reach a final state in which all heterogeneities are homogenized (e.g., Cooper, 1989). Nonetheless, the theoretical framework that builds upon the definitions of homogeneous and extreme inhomogeneous mixing is a starting point to understand the effects of small-scale mixing in clouds, which will be continued in the next section.

12.3. THE THEORY OF SMALL-SCALE MIXING

The last section indicated that small-scale mixing can be understood from two viewpoints, which are either the speed with which the turbulent mixing progresses or the effect of the mixing on the cloud microphysical composition, that is, the droplet number and size. In this section, we will introduce two theoretical frameworks that are based on these viewpoints: the Damköhler number (section 12.3.1) and the microphysical mixing diagram (section 12.3.2).

12.3.1. The Damköhler Number

When does a homogeneous or inhomogeneous mixing scenario occur? While the actual mixing scenario is usually determined by changes in droplet number and size, comparing the respective timescales for the turbulent mixing to the microphysical reaction timescale, that is, the evaporation of droplets, can indicate which mixing scenario is favored. The ratio of these timescales, which will be defined in detail below, is termed the Damköhler number

$$\mathrm{Da} = \frac{\tau_{\mathrm{mixing}}}{\tau_{\mathrm{micro}}}, \tag{12.1}$$

which has been originally proposed in the turbulent combustion literature (e.g., Peters, 2000), and has been adapted to understand the effects of turbulence on cloud microphysics subsequently (e.g., Baker et al., 1980; Shaw, 2003). If $\mathrm{Da} \gg 1$, the turbulent mixing is much slower than the microphysical reaction ($\tau_{\mathrm{mixing}} \gg \tau_{\mathrm{micro}}$). Accordingly, an inhomogeneous mixing scenario is favored since only those droplets evaporate that leave the cloudy filaments. If $\mathrm{Da} \ll 1$, the turbulent mixing is much faster than the microphysical reaction ($\tau_{\mathrm{mixing}} \ll \tau_{\mathrm{micro}}$), and differences between cloudy and cloud-free air vanish rapidly. Thus, a homogeneous mixing scenario is likely since all droplets experience a similar subsaturation and therefore react similarly. Note that distinct filaments of cloudy and cloud-free air may exist on scales as small as the Kolmogorov scale, even in the homogeneous limit of the Damköhler number. However, the microphysical reaction is so slow that these small-scale differences in supersaturation do not affect the evaporation of droplets.

Two timescales are traditionally considered to determine τ_{micro}. First, we can calculate the time a droplet of radius r requires to evaporate completely (e.g., Baker et al., 1980):

$$\tau_{\mathrm{evap}} = r^2 \frac{F_{\mathrm{k}} + F_{\mathrm{D}}}{2|S|}, \tag{12.2}$$

where $S < 0$ is the subsaturation to which the droplets are exposed, and $F_{\mathrm{k}} + F_{\mathrm{D}}$ are parameters depending on heat conduction and molecular diffusion of water vapor, respectively. A second microphysical timescale can be obtained by the time necessary to saturate subsaturated air by the evaporation of droplets (Squires, 1952). The e-folding timescale for this process is called the phase relaxation timescale, and is calculated as

$$\tau_{\mathrm{phase}} = (4\pi D_{\mathrm{v}} r_{\mathrm{m}} N)^{-1}, \tag{12.3}$$

where r_{m} is the arithmetic mean droplet radius and N the droplet number concentration. D_{v} is an effective water vapor diffusion coefficient, which is based on the molecular vapor diffusion coefficient but modified to account for the additional cooling (heating) of droplets during evaporation (condensation), which slows down vapor diffusion (see equations (11) and (14) in Kumar et al. (2013)).

Note, however, there is no consensus on which microphysical timescale is preferable for understanding small-scale mixing in clouds. Feingold and Siebert (2009) argue that under typical conditions, both timescales exhibit similar values between 1 and 10 s. Choosing the minimum of τ_{evap} and τ_{phase} may avoid infinite values occurring in almost saturated conditions or for clouds with vanishing droplet sizes. Lehmann et al. (2009) and Lu et al. (2018) show that τ_{evap} is more appropriate for studies on changes in the droplet size distributions, while τ_{phase} should be applied if changes in the liquid water content are of interest. Furthermore, τ_{micro} can be extended to other microphysical processes than (12.2) and (12.3). The activation of droplets occurs on timescales significantly smaller than 1 s (Arabas & Shima, 2017; Hoffmann et al., 2017) and is therefore highly susceptible to turbulent mixing processes, as also indicated by Abade et al. (2018).

The time to entirely mix the entrained air with the cloud is approximated by the time required to break a blob of entrained air down to the Kolmogorov lengthscale. Below this lengthscale, viscous forces dominate, which naturally terminate turbulent mixing and molecular diffusion concludes homogenization. This so-called *mixing timescale* can be obtained from inertial range scaling as

$$\tau_{\mathrm{mixing}} = \left(\frac{l^2}{\epsilon}\right)^{1/3}, \tag{12.4}$$

where ϵ is the turbulent energy dissipation rate, which is a measure of the turbulence intensity, and l is the characteristic lengthscale of the blob of entrained air (e.g., Baker et al., 1984).

With these definitions, the Damköhler number does not only allow to determine when a mixing process is likely to be homogeneous ($\mathrm{Da} \ll 1$) or inhomogeneous ($\mathrm{Da} \gg 1$), it also allows to determine the scale at which the mixing transitions from inhomogeneous to homogeneous mixing. Following Baker et al. (1980) and Lehmann et al. (2009), this is done by setting $\mathrm{Da} = 1$ and solving equation (12.1) for l. The resulting *transition lengthscale* is

$$l_{\mathrm{trans}} = \epsilon^{1/2} \tau_{\mathrm{micro}}^{3/2}, \tag{12.5}$$

which separates the lengthscales at which spatial differences in the thermodynamic conditions matter and mixing is likely to be inhomogeneous ($l \gg l_{\mathrm{trans}}$) from the lengthscales at which small-scale mixing is rapid, and a homogeneous mixing scenario is favored ($l \ll l_{\mathrm{trans}}$). Accordingly, for typical values of τ_{micro} (2 s) and ϵ ($10^{-3}\,\mathrm{m^2\,s^{-3}}$ for stratocumulus and 10^{-2}–$10^{-1}\,\mathrm{m^2\,s^{-3}}$ for cumulus), the transition from inhomogeneous to homogeneous mixing occurs on a lengthscale between 10 to 100 cm (e.g., Lehmann et al., 2009), which is smaller than the resolution of most numerical models used to simulate clouds. Since most models treat unresolved mixing processes as homogenous by design, the transition lengthscale also defines the minimum resolution that would be required to represent inhomogeneous small-scale mixing and its effects on cloud microphysics successfully. However, resolving l_{trans} imposes significant computational constraints, which limit the application of numerical cloud models to study small-scale mixing processes, as discussed further in section 12.4.

Finally, a refinement on the mixing timescale and the transition lengthscale needs to be made. Turbulence is not the only process that changes the thermodynamical conditions experienced by a droplet. Once a droplet is large enough to sediment significantly, it may fall from a saturated volume of air into a subsaturated, smearing out the boundary conditions for the microphysical reaction (e.g., Grabowski, 1993; Jensen & Baker, 1989; Tölle & Krueger, 2014). The characteristic timescale for this is termed the sedimentation timescale and can be obtained as

$$\tau_{sedi} = \frac{l}{w_{sedi}}, \tag{12.6}$$

where w_{sedi} is the droplet sedimentation velocity and l the aforementioned characteristic lengthscale of the entrained air. The inverse sum of τ_{mix} and τ_{sedi} yields an effective mixing timescale

$$\tau^*_{mix} = \left(\tau^{-1}_{mix} + \tau^{-1}_{sedi}\right)^{-1}, \tag{12.7}$$

which should be used instead of τ_{mix}. Using τ^*_{mix} makes the analytical derivation of an effective transition lengthscale cumbersome. However, by neglecting nonlinear dependencies on w_{sedi}, the effective transition lengthscale can be approximated as

$$l^*_{trans} \approx \epsilon^{1/2}\tau^{3/2}_{micro} + w_{sedi}\tau_{micro}, \tag{12.8}$$

which indicates that droplets starting to be relevant for collision and coalescence ($r > 20\,\mu m$, $w_{sedi} > 5\,cm\,s^{-1}$) may double the transition lengthscale, with commensurate stronger effects for even larger droplets. Accordingly, inhomogeneous mixing and its effects become increasingly irrelevant for larger droplets (e.g., Grabowski & Vaillancourt, 1999).

12.3.2. The Mixing Diagram

The Damköhler number and the equivalent transition lengthscale help understand where to expect inhomogeneous or homogeneous mixing processes. However, they cannot be used to infer a specific mixing scenario. For that, a functional relationship between the droplet size and droplet number is needed, and it is usually presented in the form of a microphysical *mixing diagram* (Burnet & Brenguier, 2007). A typical mixing diagram is depicted in Figure 12.2.

The mixing diagram shows how the (cubed) mean volume radius r_v^3 changes as a function of the droplet number concentration N under a prescribed mixing scenario. The cubed mean volume radius can be calculated from the liquid water content $q_l = 4/3\pi\rho_l r_v^3 N$ once N is known, making it easily available from measurements. ρ_l is the mass density of water. Note that r_v^3 and N represent the microphysical state after the entrainment and mixing process is finished and are normalized by their respective values before entrainment and mixing, $r^3_{v,a}$ and N_a. $r^3_{v,a}$ and N_a are typically unknown in less idealized (i.e., realistic) applications. Therefore, adiabatic values are often used as a proxy.

Changes in $r_v^3/r^3_{v,a}$ for the homogeneous and extreme inhomogeneous limit can be determined from the q_l which is obtained once entrainment and the subsequent mixing are finished. Since the amount of liquid water that needs to be evaporated to saturate the mixture of subsaturated entrained and saturated cloudy air does not depend on the mixing scenario, q_l can be generally expressed as

$$q_l = (1-\mu)q_{l,a} + \mu\delta q. \tag{12.9}$$

Here, the first term on the right-hand side represents the dilution of the considered volume of air by entrainment, where μ is the mass fraction of the entrained air. The second term represents the necessary evaporation to saturate the mixed air, where $\delta q = S/A \leq 0$ is the saturation deficit of the entrained air, which is determined by its saturation ratio S and a parameter A, which is a function of temperature and pressure. (The reader is referred to Korolev et al. (2016) for a detailed derivation of equations (12.9) through (12.14).)

The droplet number concentration after entrainment, but before mixing, can be derived analogously. Since the entrained air is assumed to contain no droplets, the corresponding droplet number concentration decreases as

$$N_h = N_a(1-\mu), \tag{12.10}$$

which is also the droplet number concentration after homogeneous mixing, during which droplets are assumed to not evaporate completely. By introducing (12.10) into (12.9), we determine

$$\frac{r^3_{v,h}}{r^3_{v,a}} = 1 + \frac{1 - N_h/N_a}{N_h/N_a}\frac{\delta q}{q_{l,a}}, \tag{12.11}$$

which describes the reduction in mean volume radius under homogeneous mixing as shown in Figure 12.2 (blue lines). Note that the decrease of $r^3_{v,h}/r^3_{v,a}$ is stronger for drier entrained air. In fact, if the entrained air is very dry (or the fraction of entrained air is sufficiently large), it might not possible to saturate the mixture and a mixing diagram cannot be derived. On the other hand, very humid entrained air requires only minuscule evaporation, making homogeneous and extreme inhomogeneous mixing (red line) almost indistinguishable based on their microphysical reaction. Accordingly, this scenario might be considered a *degenerate* mixing case, in which homogeneous and extreme inhomogeneous mixing result in the same microphysical reaction (Korolev et al., 2016; Pinsky et al., 2016).

During extreme inhomogeneous mixing, the mean volume radius does not change, which results in the constant relationship

$$\frac{r_{\mathrm{v,i}}^3}{r_{\mathrm{v,a}}^3} = 1, \tag{12.12}$$

shown in Figure 12.2 (red line). Inserting (12.12) in (12.9), the droplet number concentration under extreme inhomogeneous mixing can be determined as

$$\frac{N_\mathrm{i}}{N_\mathrm{a}} = \frac{q_\mathrm{l}}{q_{\mathrm{l,a}}}, \tag{12.13}$$

or, equivalently, as

$$N_\mathrm{i} = (1-\mu)N_\mathrm{a} + \frac{\mu\delta q}{\frac{4}{3}\pi\rho_\mathrm{l} r_{\mathrm{v,a}}^3}, \tag{12.14}$$

where the first term on the right-hand side represents the dilution due to entrainment (analogous to homogeneous mixing shown in equation (12.10)), and the second term is the decrease in droplet number concentration due to extreme inhomogeneous mixing, that is, the fraction of droplets that evaporates completely to saturate the entrained air.

But how does microphysics respond to a mixing scenario between extreme inhomogeneous and homogeneous mixing in which both mean volume radius and droplet number concentration change? Since the liquid water content after mixing is independent of the mixing scenario unless all droplets evaporate completely, see equation (12.9), all possible mixing scenarios obey

$$r_\mathrm{v}^3 N = r_{\mathrm{v,h}}^3 N_\mathrm{h} = r_{\mathrm{v,i}}^3 N_\mathrm{i} = const., \tag{12.15}$$

which is marked by the black line in Figure 12.2. This expression can be generalized as

$$\frac{N}{N_\mathrm{h}} = \left(\frac{r_{\mathrm{v,h}}^3}{r_{\mathrm{v,i}}^3}\right)^\alpha, \tag{12.16}$$

which results in homogeneous mixing for $\alpha = 0$, and in extreme inhomogeneous mixing for $\alpha = 1$. Intermediate values of $0 \leq \alpha \leq 1$ are expected for varying degrees of inhomogeneous mixing. Accordingly,

$$\alpha = \frac{\ln(N/N_\mathrm{h})}{\ln(r_{\mathrm{v,h}}^3/r_{\mathrm{v,i}}^3)} \tag{12.17}$$

can be used as a simple metric to characterize observed or simulated mixing scenarios (Andrejczuk et al., 2009; Lu et al., 2013; Morrison & Grabowski, 2008). Furthermore, α can be used to parameterize inhomogeneous mixing in numerical models as explained further in section 12.4.2.3.

Finally, note that even under conditions that favor homogeneous mixing (Da $\ll$ 1), N is not necessarily conserved. If the droplet size distribution exhibits sufficiently small droplets, some of them will evaporate completely, irrespective of the mixing scenario, resulting in an apparently inhomogeneous mixing scenario (Luo et al., 2020; Tölle & Krueger, 2014). Similarly, entrained air might contain aerosol particles that activate to the droplets during the mixing process, which can increase N (e.g., Slawinska et al., 2012). Accordingly, an inhomogeneous mixing scenario can appear more homogeneous than expected (Da $\gg$ 1).

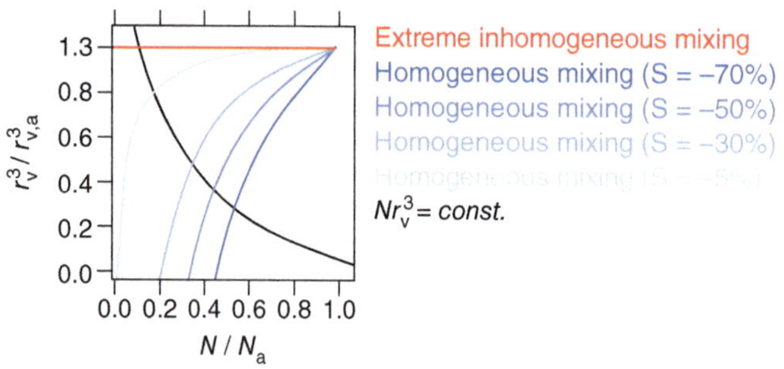

Figure 12.2 This idealized mixing diagram illustrates how the normalized mean volume radius $r_\mathrm{v}^3/r_{\mathrm{v,a}}^3$ depends on the normalized droplet number concentration N/N_a for extreme inhomogeneous mixing (red line) and homogeneous mixing (blue lines) for different saturation ratios of the entrained air (shades of blue). The black line indicates $Nr_\mathrm{v}^3 = const.$, and accordingly all intermediate mixing scenarios between homogeneous and extreme inhomogeneous mixing. Mixing diagrams of this type have been introduced by Burnet et al. (2007). Source: Adapted from Burnet et al. (2007).

12.4. INVESTIGATING SMALL-SCALE MIXING

In this section, we will present approaches to investigate small-scale mixing in observations and numerical modeling. All these approaches have to cope with the multiscale nature of clouds, ranging from entire cloud fields (~100 km) to the smallest scales of turbulent mixing (~1 mm), and cloud microphysics (~10 µm). No approach is able to cover all these scales, and neglecting or idealizing smaller or larger scales is inherent to all methods.

12.4.1. Observations

Observations are a cornerstone of physics, and airborne measurements enabled important insights on entrainment and mixing in clouds (e.g., Blyth et al., 1988; Gerber et al., 2008; Jensen et al., 1985; Paluch & Knight, 1984; Warner, 1969a). Observing the effects of small-scale mixing on clouds is, at first glance, relatively simple, since simultaneous measurements of the droplet number concentration and droplet radii are available for at least 50 years.

However, early instruments, such as sooted glass slides used to collect droplets outside the airplane, exhibited very slow sampling rates on the order of 1 Hz (Warner, 1969a). Due to the comparably high airspeed of planes ($\sim 100\,\mathrm{m\,s^{-1}}$), many airborne measurements of the past did therefore not resolve scales below 100 m, which made it impossible to detect variations on scales as small as the transition lengthscale directly. Thus, by erroneously assuming that the measured droplet concentrations and droplet sizes are spatially uniform, it has been believed that clouds might mix homogeneously. However, the simultaneous presence of small-scale filaments of cloudy and cloud-free air, in agreement with our current depiction of inhomogeneous mixing, can result in the same measurements when averaged over the aforementioned lengthscale imposed by insufficient sampling rates, as first discussed by Paluch (1986) and Paluch and Baumgardner (1989). Today, measurements are commonly provided by instruments such as the forward scattering spectrometer probe (FSSP) (Brenguier, 1993; Brenguier et al., 1998) or the particle volume monitor (PVM) (Gerber et al., 1993) with sampling rates on the order of 1,000 Hz, which enable direct observations of small-scale mixing from airplanes.

A further improvement can be achieved by lowering the airspeed, which can be done by using helicopters instead of airplanes. The Airborne Cloud Turbulence Observation System (ACTOS) (Siebert et al., 2006), for instance, can be suspended from helicopters, providing sufficiently resolved measurements to detect even the finest scales of turbulent processes in clouds (e.g., Siebert et al., 2013). Furthermore, the increasing use of holographic systems, such as the Holographic Detector for Clouds (HOLODEC), allow three-dimensional snapshots of the small-scale distribution of cloud droplets, including sharp gradients between cloudy and non-cloudy filaments as they are expected to be observed during inhomogeneous mixing (Beals et al., 2015). While holographic systems can provide a glimpse into the three-dimensional small-scale distribution of droplets, their sample volume ($\sim 15\,\mathrm{cm^3}$) is too limited to capture all relevant scales of the mixing process. Entire clouds may be sampled in the near future using several remotely piloted aircrafts (RPAs, commonly known as drones), as proposed in a current assessment study (Maury et al., 2021).

Finally, laboratory cloud chamber measurements, as they are currently undertaken in the Π-chamber or similar facilities (e.g., Chang et al., 2016), may also enable deeper observational process-level understanding of entrainment and mixing once appropriate scales can be accommodated (Shaw et al., 2020).

12.4.2. Numerical Modeling

Turbulent mixing processes are ubiquitous and have been extensively investigated in engineering applications such as turbulent combustion (e.g., Peters, 2000). Therefore, many of the following numerical models originate from the engineering literature and have been adapted to explore cloud physics questions. However, to accommodate the large range of relevant scales, multiple modeling approaches are necessary to cover all aspects of the mixing process. Accordingly, we will present models for representing cloud microphysics first (section 12.4.2.1). Then, fluid dynamics models for the investigation of small-scale mixing at the Kolmogorov lengthscale are presented (section 12.4.2.2). Due to the limited computational resources, these models are restricted toward larger scales. Models which represent these larger scales are not able to resolve small-scale mixing explicitly, and parameterizations are applied to represent potentially inhomogeneous mixing scenarios, which are summarized is the last subsection (section 12.4.2.3).

Cloud Microphysical Models

Cloud microphysical models predict parameters describing the microphysical state of a cloud, applying varying degrees of complexity. The reader is referred to Morrison et al. (2020) for a recent review of cloud microphysical modeling approaches.

To a certain degree, all commonly applied cloud microphysical models are able to represent the effects of small-scale mixing on the cloud microphysical composition. Without restrictions, this is possible in detailed models that predict the entire droplet size distribution by solving equations that describe the condensation and evaporation of droplets, their activation from aerosols, as well as collision processes. Thus, changes in the droplets size distribution due to small-scale mixing can be considered directly. Traditionally, these detailed models have been *bin* or *spectral models*, which represent the droplet size distribution on a numerical grid in radius space (e.g., Bott, 1998; Tzivion et al., 1987). Today, an increasing number of similarly detailed *Lagrangian cloud models* is applied, which represent the droplet size distribution by several individually simulated computational particles, each representing just one droplet or a multitude of identical droplets (so-called superdroplets) (e.g., Andrejczuk et al., 2008; Hoffmann et al., 2015; Shima et al., 2009). *Bulk models*, however, represent the droplet size distribution by idealized functions and predict one or several statistical moments of it. To represent the basic impacts of small-scale mixing, bulk models are required to predict at least two moments, N and q_l (e.g., Khairoutdinov & Kogan, 2000; Milbrandt & Yau, 2005; Morrison et al., 2005; Seifert & Beheng, 2006; Thompson & Eidhammer, 2014). Additional moments are necessary to represent the impact of small-scale mixing on the width of the droplet size distribution, as discussed in section 12.5.1.

Direct Modeling of Turbulent Mixing

The most direct approach to simulate small-scale mixing is to employ *direct numerical simulations* (DNSs) that solve the three-dimensional Navier-Stokes equations applying a resolution that matches the Kolmogorov lengthscale (e.g., Moin & Mahesh, 1998). Due to computational constraints, the model domain is restricted to a cube with a typical edge length of a couple of decimeters, even on today's supercomputers (Andrejczuk et al., 2009; Götzfried et al., 2017; Gao et al., 2018; Kumar et al., 2013; Kunishima & Onishi, 2018; Perrin & Jonker, 2015; Vaillancourt et al., 2001, 2002). Accordingly, these models simulate small-scale mixing close to the typical transition lengthscale and provide insights on the transition from inhomogeneous to homogeneous mixing. Note that these models often apply a Lagrangian cloud model in which each computational particle represents one cloud droplet.

DNS models with a coarser resolution and commensurately larger molecular viscosity have also been applied to study small-scale mixing processes in increasingly larger model domains, which enabled deeper insights into inhomogeneous mixing regimes than possible with regular DNS (Dimotakis, 2000; de Lozar & Muessle, 2016; Mellado et al., 2010, 2018; Thomas et al., 2020). However, this approach is limited. The upscaled molecular viscosity might be unable to represent the more complex effects of unresolved turbulent motions at smaller scales adequately, making parameterizations inevitable as further outlined in section 12.4.2.3 below.

An alternative to DNS are one-dimensional models of turbulence and molecular diffusion (Jensen & Baker, 1989; Krueger et al., 1997; Kerstein, 1988; Su et al., 1998). These models are able to resolve the Kolmogorov lengthscale and – due to their reduced dimensionality – can host a model domain of hundreds of meters, even on a simple workstation computer. Of course, turbulence, which naturally requires three dimensions, cannot be represented in these approaches explicitly. However, its effects of compression and folding can be successfully emulated by rearranging or reshaping the model's numerical grid. The specific approach of *linear eddy modeling* (LEM) has been shown to produce results comparable to DNS of the mixing process in clouds (Krueger, 2016).

Parameterizations of Turbulent Mixing

To capture entire clouds and the underlying dynamics that drive entrainment and mixing, today's method of choice are *large eddy simulations* (LESs), which solve the filtered Navier-Stokes equations (e.g., Sagaut, 2006). LESs can represent entire cumulus or stratocumulus fields with a horizontal extent of several tens of kilometers successfully (Ackerman et al., 2009; Siebesma et al., 2003; Stevens et al., 2005; vanZanten et al., 2011). The representation of small-scale processes, however, is limited by their resolution of typically 10–100 m. The impact of turbulent mixing on scales below the resolution is usually modeled by a subgrid-scale (SGS) scheme (e.g., Deardorff, 1980), which represents the unresolved fluxes of momentum, temperature, and water vapor mixing ratio. By design, the effect of unresolved turbulent mixing on cloud microphysics is homogeneous, which corresponds to the notion that resolved LES quantities represent a grid-box average (Schumann, 1975). Introducing a more realistic representation of SGS mixing is therefore necessary to consider the microphysical effects of small-scale mixing in large-scale models.

Morrison and Grabowski (2008) introduced an approach that can be used with two-moment bulk microphysical models to consider different mixing scenarios. In this approach, the order in which SGS turbulent mixing and evaporation occur in the model is used to adjust the droplet number concentration according to the intended mixing scenario:

$$N = N^* \left(\frac{q_l}{q_l^*} \right)^{\alpha}. \tag{12.18}$$

Here, N and q_l represent the *final* droplet number concentration and the liquid water content after SGS mixing and evaporation, respectively, while N^* and q_l^* are the corresponding *intermediate* values after SGS mixing but before evaporation. As in section 12.3.2, α represents the mixing scenario with values between 0 for homogeneous mixing and 1 for extreme inhomogeneous mixing. (It can be shown that (12.18) is equivalent to (12.16).) While Morrison and Grabowski (2008) prescribe fixed values for α, Jarecka et al. (2009, 2013) suggest an approach to predict α based on the Damköhler number. Note that although the parameterization (12.18) is relatively easy to implement, it is not frequently used in LES or other large-scale models.

Lagrangian cloud models offer simpler ways to include SGS processes than (Eulerian) bulk or bin schemes by making use of the individually simulated computational particles (e.g., Grabowski et al., 2019). Probably the first approach has been developed by Grabowski and Abade (2017), who assumed that each computational particle is surrounded by a volume of air. They used a stochastic Wiener process to model unresolved turbulent supersaturation fluctuations in these volumes and considered them in the diffusional growth process of the simulated droplets. Similarly, Hoffmann et al. (2019) used the aforementioned one-dimensional LEM by Kerstein (1988) to represent unresolved turbulent mixing and molecular diffusion among the volumes of air surrounding each computational particle. Accordingly, this approach enables a very detailed representation of cloudy filaments on scales as small as the transition lengthscale, when used as a SGS model in typical LES applications with Lagrangian cloud microphysics, as recently applied in simulations of cumulus (Hoffmann et al., 2019) and stratocumulus (Hoffmann & Feingold, 2019).

12.5. EFFECTS OF SMALL-SCALE MIXING

Observations and numerical modeling enable us to understand the effects of small-scale mixing in frameworks that exceed the simple theoretical considerations presented in section 12.3. These results will be summarized in this section, dividing them into microscale impacts, which predominantly affect the droplet size distribution and its broadening, and macroscale impacts, which are relevant to the role of clouds in the climate system.

12.5.1. Microscale Impacts

The basic effects of small-scale mixing on the droplet size distribution have been outlined in section 12.3: Homogeneous mixing maintains the droplet number and reduces the mean droplet size due to partial evaporation. Extreme inhomogeneous mixing, on the other hand, maintains the mean droplet size, but reduces the droplet number due to the complete evaporation of droplets that leave the cloudy filaments.

An (intermediate) inhomogeneous mixing scenario, however, allows for a large variability in droplet growth histories, including repeated switches between saturated and subsaturated filaments during the mixing process, which leads to a substantial broadening toward smaller droplet sizes by evaporation, as pointed out in numerical studies by Su et al. (1998) and Tölle and Krueger (2014). In fact, the broadening to smaller sizes due to entrainment and mixing has been shown essential for initiating the precipitation process in shallow cumulus clouds (Hoffmann et al., 2017). However, the droplet size distribution does not only experience broadening to smaller sizes. Entrainment generally reduces the droplet concentration in clouds by dilution, irrespective of the mixing scenario (cf. equation (12.10)). Accordingly, the competition for water vapor is reduced and the droplet size distribution might broaden to larger radii by accelerating the diffusional growth of the remaining droplets if the considered air parcel is still being lifted (e.g., Lasher-Trapp et al., 2005). This superadiabatic growth can be enhanced in (intermediate or extreme) inhomogeneous mixing scenarios due to the even lower droplet number concentration, as initially hypothesized by Baker and Latham (1979) and Baker et al. (1980). Detailed modeling studies confirmed that these superadiabatic droplets can reach sizes that may initiate the precipitation process in warm clouds (Hoffmann et al., 2019; Hoffmann & Feingold, 2019; Lasher-Trapp et al., 2005; Su et al., 1998; Tölle & Krueger, 2014).

The activation of aerosols to cloud droplets above the cloud base, so-called secondary activation, is frequently observed in cumulus clouds. It is either caused by a substantial increase in the vertical velocity (Pinsky & Khain, 2002; Warner, 1969b), or by the activation of newly entrained aerosols (Brenguier & Grabowski, 1993; Lasher-Trapp et al., 2005; Paluch & Knight, 1984; Su et al., 1998; Warner, 1969a). The latter pathway, which dominates in shallow cumulus clouds (Chandrakar et al., 2021; Hoffmann et al., 2015), can be affected by the degree of turbulent mixing. Under rapid mixing, the newly entrained aerosols compete directly with the (old) cloud droplets for the available water vapor. Since the (old) cloud droplets absorb the water vapor more efficiently due to their larger surface area, the activation of newly entrained aerosols can be reduced. If the mixing is slower, the newly entrained aerosols may be spatially separated from the (old) cloud droplets, which enables a larger fraction of the entrained aerosols to activate (Lasher-Trapp et al., 2005; Su et al., 1998). While this process broadens the droplet size distribution to smaller radii, it also buffers the negative impact of inhomogeneous mixing on the number of cloud droplets by activating new droplets (Slawinska et al., 2012), complicating the differentiation of homogeneous and inhomogeneous mixing scenarios, as already outlined in section 12.3.2.

Finally, it is important to reiterate that the mixing character can change with time (e.g., Jarecka et al., 2013). During a single entrainment event, an initially inhomogeneous mixing scenario can become homogeneous when the entrained air breaks up into filaments with lengthscales below the transition lengthscale (cf. section 12.3). However, the character of mixing also changes during the cloud life cycle. Schmeissner et al. (2015) derived mixing diagrams from the observation of hundreds of cumulus clouds, showing that actively growing clouds exhibit a more homogeneous mixing character, while clouds in their decaying stage mix more inhomogeneously. Schmeissner et al. (2015) explain this behavior by a humid, almost saturated shell that develops during the cumulus life cycle that prevents the evaporation of cloud droplets in matured cumulus (e.g., Heus & Jonker, 2008). As explained in section 12.3.2, the high humidity of this shell could also favor a degenerate mixing scenario, in which the microphysical reaction to homogeneous and inhomogeneous mixing is almost indistinguishable (Korolev et al., 2016; Pinsky et al., 2016). Schmeissner et al. (2015), however, interpret the humid shell as part of the inhomogeneous mixing process.

This motivates a fundamental question: What are the largest scales to be included in the small-scale mixing process? The actual entrainment process is happening on the scale of the interfacial eddies of cumulus (Grabowski & Clark, 1991) or the holes within a stratocumulus deck (Nicholls, 1989). However, the entrained air on these scales originates from larger scales, namely, the humid shells surrounding cumulus (Heus & Jonker, 2008) or the entrainment interface layer on top of stratocumulus (Caughey et al., 1982; Gerber et al., 2005; Yamaguchi & Randall, 2012). The thermodynamic composition of these regions is dominated by detrained cloudy filaments evaporating and humidifying these areas in the direct vicinity of the cloud, further modified by radiative cooling or heating (e.g., Klinger et al., 2019). Additionally, the isotropic turbulence driving small-scale mixing can be superimposed by larger scale cloud dynamics (e.g., shear instabilities or waves), further changing how the cloud interacts with its environment (e.g., Mellado, 2017). Accordingly, to advance our understanding of small-scale mixing, we need to proceed toward larger scales, requiring us to include processes that have been traditionally neglected.

12.5.2. Macroscale Impacts

Our understanding of small-scale mixing on the microscale allows us to deduce impacts on the macroscale. As indicated in the introduction, the radiative forcing of clouds on the climate can be understood as the product of cloud albedo and cloud fraction (Betts, 2007). The cloud albedo can be approximated as

$$A \approx \frac{\tau}{\tau + \gamma}, \tag{12.19}$$

where τ is the cloud optical thickness, and $\gamma \approx 13.3$ depends on the degree of forward scattering (e.g., Glenn et al., 2020). It can be shown that τ is primarily determined by

$$\tau \propto \mathrm{LWP}^{5/6} N^{1/3}, \tag{12.20}$$

where LWP is the vertically integrated liquid water content, the so-called liquid water path, and N is the droplet number concentration (e.g., Boers & Mitchell, 1994).

Since entrainment determines the amount of non-cloudy air that is engulfed into the cloud, it does not change LWP and N directly. The subsequent small-scale mixing, however, reduces the LWP and N by evaporation and dilution (cf. equations (12.9) and (12.10)). The mixing scenario may lead to a further reduction in N, subject to an increasing degree of inhomogeneous mixing (cf. equation (12.14)), while it does not cause direct changes in LWP. Accordingly, we can already conclude that the mixing scenario affects the cloud albedo by changes in the droplet number concentration, similar to the albedo effect (Twomey, 1974, 1977).

The effect of different mixing scenarios on the albedo of numerically simulated clouds has been studied by Chosson et al. (2007) and Slawinska et al. (2008). They found that the reduction in cloud albedo due to extreme inhomogeneous mixing instead of homogeneous mixing ranges between 2 and 5 percentage points for stratocumulus and between 5 and 6 percentage points for cumulus, respectively. The strongest dependency has been found for very thin clouds, which is expectable since the optical thickness depends much stronger on the liquid water path than on the droplet number concentration, as indicated by the exponents in equation (12.20). Accordingly, the albedo of deeper clouds is less susceptible to changes in the number concentration and hence different mixing scenarios. Based on theoretical arguments, Jeffery (2007) derived a similar dependence of τ on the mixing scenario.

Much more uncertain are the effects of small-scale mixing on cloud lifetime and cover (e.g., Albrecht, 1989). Two pathways are considered for terminating the life cycle of a cloud: entrainment or precipitation, which are both potentially modified by small-scale mixing.

Following an entrainment event, inhomogeneous mixing allows droplets to stay in cloudy filaments for a certain period of time, while droplets are more rapidly exposed to subsaturations under homogeneous mixing (Krueger, 1993). Additionally, inhomogeneous mixing reduces the droplet number concentration, resulting in a stronger increase of the phase relaxation timescale than by homogeneous mixing alone, which can further delay the evaporation of cloud droplets (Hoffmann & Feingold, 2019). Accordingly, both processes decelerate the production of negative buoyancy, which is an important driver for the production of turbulence. Thus, the mixing might slow down even further, which could make the mixing process more inhomogeneous. This positive feedback is analogous to the evaporation-entrainment feedback, in which the faster evaporation of a larger number of cloud droplets is suggested to result in stronger turbulence compared to the same amount of water distributed on a smaller number of droplets (Glassmeier et al., 2021; Wang et al., 2003; Xue & Feingold, 2006). The aforementioned authors even showed that this feedback results in increased entrainment rates in droplet-laden conditions, with a commensurately negative impact on the cloud albedo. Nonetheless, entrainment is driven by large-scale cloud dynamics, as outlined in section 12.2, and a direct impact of small-scale mixing processes must be commensurately weak. Yet, Hoffmann and Feingold (2019) showed slightly lower entrainment rates in stratocumulus LES with SGS inhomogeneous mixing compared to simulations with homogeneous SGS mixing.

A more complex situation arises when droplet sedimentation is considered. As already outlined in section 12.3.1, sedimentation smears out the thermodynamic conditions experienced by a droplet and therefore results in an effectively more rapid mixing scenario (Tölle & Krueger, 2014). Furthermore, sedimentation might also increase the evaporative cooling on the edges of cloudy filaments by exposing droplets to subsaturated air more quickly, as indicated in idealized simulations of Grabowski (1993). As discussed in the last paragraph, a large impact of these small-scale processes on the entrainment rate and hence cloud albedo is unlikely. In fact, LES of stratocumulus (without a special treatment of SGS mixing) indicate generally smaller entrainment rates when sedimentation is considered (Ackerman et al., 2004; Bretherton et al., 2007), which has been confirmed in DNS later (de Lozar & Mellado, 2017; Schulz & Mellado, 2019). This process has been explained by the removal of large droplets from the stratocumulus cloud top by sedimentation, which reduces the potential for evaporative cooling in the entrainment interface layer, with a commensurate effect on the evaporative enhancement of the entrainment rate (Bretherton et al., 2007).

By contributing to the initiation of the precipitation process in warm clouds (Baker & Latham, 1979; Hoffmann et al., 2019; Lasher-Trapp et al., 2005), inhomogeneous mixing may reduce the lifetime of individual clouds and hence affect cloud cover. However, we do not know yet how entire cloud fields are affected. For instance, enhanced precipitation will contribute to the generation of cold pools, which alter the large-scale organization of cloud fields, with commensurate changes in cloud cover (Seifert & Heus, 2013; Zuidema et al., 2017). Accordingly, there are no conclusive answers on how small-scale mixing affects macroscale cloud properties, especially cloud lifetime and cloud cover. We need more studies that investigate the effects of small-scale mixing on entire cloud fields. Using parameterizations (section 12.4.2.3) and improving them with detailed modeling (section 12.4.2.2) seem to be a viable approach.

12.6. CONCLUSIONS

This chapter has been dedicated to our current understanding of the small-scale mixing of clouds with their environment. Although small-scale mixing takes place on scales as small as the Kolmogorov lengthscale, it has fundamental implications for the microphysical composition of clouds and hence their role in the climate system, most significantly by altering the ability of clouds to reflect incident shortwave radiation back to space.

For more than 50 years, small-scale mixing has been investigated in numerous studies employing theory, observations, and numerical modeling. However, our understanding of small-scale mixing is not complete. Limitations primarily consider the interaction of large and small scales. While increasing computational power will solve some problems, especially when larger scales are approached from below, the growing importance of processes that have been traditionally neglected in the discussion of small-scale mixing, for example, radiation and larger scale cloud dynamics, demands careful consideration in the future. Our understanding of small-scale mixing on global scales, however, will still depend on parameterizations in the next decades. Accordingly, developing successful parameterizations of small-scale mixing will be essential to assess the role of small-scale mixing in the climate system. New insights are also expected from the development of new observational approaches such as remotely operated aircrafts (i.e., drones), as well as larger laboratory facilities in which a much wider range of scales relevant to the mixing process can be analyzed.

All in all, small-scale mixing in clouds offers ample opportunities for future research from a large variety of backgrounds and with potentially large impacts on our understanding of clouds in the climate system.

ACKNOWLEDGMENTS

This chapter benefitted from discussions with Graham Feingold and comments by Wojciech Grabowski, Steven Krueger, and one anonymous referee. The author acknowledges support from the Emmy Noether program of the German Research Foundation (DFG) under grant HO 6588/1-1.

REFERENCES

Abade, G. C., Grabowski, W. W., & Pawlowska, H. (2018). Broadening of cloud droplet spectra through eddy hopping: Turbulent entraining parcel simulations. *Journal of the Atmospheric Sciences*, *75*(10), 3365–3379.

Ackerman, A. S., Kirkpatrick, M. P., Stevens, D. E., & Toon, O. B. (2004). The impact of humidity above stratiform clouds on indirect aerosol climate forcing. *Nature*, *432*(7020).

Ackerman, A. S., VanZanten, M. C., Stevens, B., Savic-Jovcic, V., Bretherton, C. S., Chlond, A., et al. (2009). Large-eddy simulations of a drizzling, stratocumulus-topped marine boundary layer. *Monthly Weather Review*, *137*(3).

Albrecht, B. A. (1989). Aerosols, cloud microphysics, and fractional cloudiness. *Science*, *245*(4923), 1227–1230.

Andrejczuk, M., Reisner, J. M., Henson, B., Dubey, M. K., & Jeffery, C. A. (2008). The potential impacts of pollution on a nondrizzling stratus deck: Does aerosol number matter more than type? *Journal of Geophysical Research*, *113*(D19204).

Andrejczuk, M., Grabowski, W. W., Malinowski, S. P., & Smolarkiewicz, P. K. (2009). Numerical simulation of cloud–clear air interfacial mixing: Homogeneous versus inhomogeneous mixing. *Journal of the Atmospheric Sciences*, *66*(8), 2493–2500.

Arabas, S., & Shima, S.-i. (2017). On the CCN (de) activation nonlinearities. *Nonlinear Processes in Geophysics*, *24*(3), 535–542.

Baker, M., & Latham, J. (1979). The evolution of droplet spectra and the rate of production of embryonic raindrops in small cumulus clouds. *Journal of the Atmospheric Sciences*, *36*(8), 1612–1615.

Baker, M., Corbin, R., & Latham, J. (1980). The influence of entrainment on the evolution of cloud droplet spectra: I. A model of inhomogeneous mixing. *Quarterly Journal of the Royal Meteorological Society*, *106*(449), 581–598.

Baker, M. B., Breidenthal, R. E., Choularton, T. W., & Latham, J. (1984). The effects of turbulent mixing in clouds. *Journal of the Atmospheric Sciences*, *41*, 299–304.

Beals, M. J., Fugal, J. P., Shaw, R. A., Lu, J., Spuler, S. M., & Stith, J. L. (2015). Holographic measurements of inhomogeneous cloud mixing at the centimeter scale. *Science*, *350*(6256), 87–90.

Betts, A. K. (2007). Coupling of water vapor convergence, clouds, precipitation, and land-surface processes. *Journal of Geophysical Research*, *112*(D10).

Blyth, A. M., Cooper, W. A., & Jensen, J. B. (1988). A study of the source of entrained air in Montana cumuli. *Journal of the Atmospheric Sciences*, *45*, 3944–3964.

Bodenschatz, E., Malinowski, S. P., Shaw, R. A., & Stratmann, F. (2010). Can we understand clouds without turbulence?. *Science*, *327*(5968), 970–971.

Boers, R., & Mitchell, R. M. (1994). Absorption feedback in stratocumulus clouds influence on cloud top albedo. *Tellus A*, *46*(3), 229–241.

Bott, A. (1998). A flux method for the numerical solution of the stochastic collection equation. *Journal of the Atmospheric Sciences*, *55*(13), 2284–2293.

Boucher, O., Randall, D., Artaxo, P., Bretherton, C., Feingold, G., Forster, P., et al. (2013). Clouds and aerosols. In *Climate change 2013: The physical science basis. Contribution of Working Group I to the Fifth Assessment Report of the Intergovernmental Panel on Climate Change* (pp. 571–657). Cambridge University Press.

Brenguier, J.-L. (1993). Observations of cloud microstructure at the centimeter scale. *Journal of Applied Meteorology and Climatology*, *32*(4), 783–793.

Brenguier, J.-L., & Grabowski, W. W. (1993). Cumulus entrainment and cloud droplet spectra: A numerical model within a two-dimensional dynamical framework. *Journal of the Atmospheric Sciences*, *50*(1), 120–136. doi: 10.1175/1520-0469(1993)050<0120:CEACDS>2.0.CO;2.

Brenguier, J.-L., Bourrianne, T., Coelho, A. A., Isbert, J., Peytavi, R., Trevarin, D., & Weschler, P. (1998). Improvements of droplet size distribution measurements with the Fast-FSSP (Forward Scattering Spectrometer Probe). *Journal of Atmospheric and Oceanic Technology*, *15*(5), 1077–1090.

Bretherton, C., Blossey, P. N., & Uchida, J. (2007). Cloud droplet sedimentation, entrainment efficiency, and subtropical stratocumulus albedo. *Geophysical Research Letters*, *34*(3).

Burnet, F., & Brenguier, J.-L. (2007). Observational study of the entrainment-mixing process in warm convective clouds. *Journal of the Atmospheric Sciences*, *64*(6), 1995–2011.

Caughey, S., Crease, B., & Roach, W. (1982). A field study of nocturnal stratocumulus II turbulence structure and entrainment. *Quarterly Journal of the Royal Meteorological Society*, *108*(455), 125–144.

Chandrakar, K. K., Grabowski, W. W., Morrison, H., & Bryan, G. H. (2021). Impact of entrainment-mixing and turbulent fluctuations on droplet size distributions in a cumulus cloud: An investigation using Lagrangian microphysics with a sub-grid-scale model. *Journal of the Atmospheric Sciences, under review*.

Chang, K., Bench, J., Brege, M., Cantrell, W., Chandrakar, K., Ciochetto, D., et al. (2016). A laboratory facility to study gas–aerosol–cloud interactions in a turbulent environment: The π chamber. *Bulletin of the American Meteorological Society*, *97*(12), 2343–2358.

Chosson, F., Brenguier, J.-L., & Schüller, L. (2007). Entrainment-mixing and radiative transfer simulation in boundary layer clouds. *Journal of the Atmospheric Sciences*, *64*(7), 2670–2682.

Cooper, W. A. (1989). Effects of variable droplet growth histories on droplet size distributions. Part I: Theory. *Journal of the Atmospheric Sciences*, *46*(10), 1301–1311.

de Lozar, A., & Mellado, J. P. (2017). Reduction of the entrainment velocity by cloud droplet sedimentation in stratocumulus. *Journal of the Atmospheric Sciences*, *74*(3), 751–765.

de Lozar, A., & Muessle, L. (2016). Long-resident droplets at the stratocumulus top. *Atmospheric Chemistry and Physics*, *16*(10), 6563–6576.

de Rooy, W. C., Bechtold, P., Fröhlich, K., Hohenegger, C., Jonker, H., Mironov, D., et al. (2013). Entrainment and detrainment in cumulus convection: An overview. *Quarterly Journal of the Royal Meteorological Society*, *139*(670), 1–19. doi: 10.1002/qj.1959.

Deardorff, J. W. (1980). Stratocumulus-capped mixed layers derived from a three-dimensional model. *Boundary-Layer Meteorology*, *18*(4), 495–527.

Devenish, B., Bartello, P., Brenguier, J.-L., Collins, L., Grabowski, W., IJzermans, R., et al. (2012). Droplet growth in warm turbulent clouds. *Quarterly Journal of the Royal Meteorological Society*, *138*(667), 1401–1429.

Dimotakis, P. E. (2000). The mixing transition in turbulent flows. *Journal of Fluid Mechanics*, *409*, 69–98.

Feingold, G., & Siebert, H. (2009). Cloud-aerosol interactions from the micro to the cloud scale. In J. Heintzenberg & R. Charlson (Eds.), *Clouds in the perturbed climate system: Their relationship to energy balance, atmospheric dynamics, and precipitation, Strüngmann Forum Report*. The MIT Press.

Gao, Z., Liu, Y., Li, X., & Lu, C. (2018). Investigation of turbulent entrainment-mixing processes with a new particle-resolved direct numerical simulation model. *Journal of Geophysical Research*, *123*(4), 2194–2214.

Gerber, H., Arends, B. G. & Ackerman, A. S. (1993). New microphysics sensor for aircraft use. *Atmospheric Research*, *31*, 235–252. doi: 10.1016/0169-8095(94)90001-9.

Gerber, H., Frick, G., Malinowski, S., Brenguier, J., & Burnet, F. (2005). Holes and entrainment in stratocumulus. *Journal of the Atmospheric Sciences*, *62*(2), 443–459.

Gerber, H. E., Frick, G. M., Jensen, J. B., & Hudson, J. G. (2008). Entrainment, mixing, and microphysics in trade-wind cumulus. *Journal of the Meteorological Society of Japan*, *86*, 87–106.

Glassmeier, F., Hoffmann, F., Johnson, J. S., Yamaguchi, T., Carslaw, K. S., & Feingold, G. (2021). Aerosol-cloud-climate cooling overestimated by ship-track data. *Science*, *371*(6528), 485–489.

Glenn, I. B., Feingold, G., Gristey, J. J., & Yamaguchi, T. (2020). Quantification of the radiative effect of aerosol–cloud interactions in shallow continental cumulus clouds. *Journal of the Atmospheric Sciences*, *77*(8), 2905–2920. doi: 10.1175/JAS-D-19-0269.1.

Götzfried, P., Kumar, B., Shaw, R. A., & Schumacher, J. (2017). Droplet dynamics and fine-scale structure in a shearless turbulent mixing layer with phase changes. *Journal of Fluid Mechanics*, *814*, 452–483.

Grabowski, W. W. (1993). Cumulus entrainment, fine-scale mixing, and buoyancy reversal. *Quarterly Journal of the Royal Meteorological Society*, *119*(513), 935–956.

Grabowski, W. W., & Abade, G. C. (2017). Broadening of cloud droplet spectra through eddy hopping: Turbulent adiabatic parcel simulations. *Journal of the Atmospheric Sciences*, *74*(5), doi: 10.1175/JAS-D-17-0043.1.

Grabowski, W. W., & Clark, T. L. (1991). Cloud–environment interface instability: Rising thermal calculations in two spatial dimensions. *Journal of the Atmospheric Sciences*, *48*(4), 527–546.

Grabowski, W. W., & Clark, T. L. (1993). Cloud-environment interface instability. Part II: Extension to three spatial dimensions. *Journal of the Atmospheric Sciences*, *50*(4), 555–573. doi: 10.1175/1520-0469(1993)050<0555:CEIIPI>2.0.CO;2.

Grabowski, W. W., & Vaillancourt, P. (1999). Comments on "preferential concentration of cloud droplets by turbulence: Effects on the early evolution of cumulus cloud droplet spectra". *Journal of the Atmospheric Sciences*, *56*(10), 1433–1436.

Grabowski, W. W., & Wang, L.-P. (2013). Growth of cloud droplets in a turbulent environment. *Annual Review of Fluid Mechanics*, *45*, 293–324.

Grabowski, W. W., Morrison, H., Shima, S.-I., Abade, G. C., Dziekan, P., & Pawlowska, H. (2019). Modeling of cloud microphysics: Can we do better? *Bulletin of the American Meteorological Society*, *100*(4), 655–672. doi: 10.1175/BAMS-D-18-0005.1.

Heus, T., & Jonker, H. J. J. (2008). Subsiding shells around shallow cumulus clouds. *Journal of the Atmospheric Sciences*, *65*(3), 1003–1018.

Heus, T., Van Dijk, G., Jonker, H. J., & Van den Akker, H. E. (2008). Mixing in shallow cumulus clouds studied by Lagrangian particle tracking. *Journal of the Atmospheric Sciences*, *65*(8), 2581–2597.

Hoffmann, F. (2020). Effects of entrainment and mixing on the Wegener–Bergeron–Findeisen process. *Journal of the Atmospheric Sciences*, *77*(6), 2279–2296.

Hoffmann, F., & Feingold, G. (2019). Entrainment and mixing in stratocumulus: Effects of a new explicit subgrid-scale scheme for large-eddy simulations with particle-based microphysics. *Journal of the Atmospheric Sciences*, *76*(7), 1955–1973.

Hoffmann, F., Raasch, S., & Noh, Y. (2015). Entrainment of aerosols and their activation in a shallow cumulus cloud studied with a coupled LCM-LES approach. *Atmospheric Research*, *156*, 43–57.

Hoffmann, F., Noh, Y., & Raasch, S. (2017). The route to raindrop formation in a shallow cumulus cloud simulated by a Lagrangian cloud model. *Journal of the Atmospheric Sciences*, *74*(7), 2125–2142.

Hoffmann, F., Yamaguchi, T., & Feingold, G. (2019). Inhomogeneous mixing in Lagrangian cloud models: Effects on the production of precipitation embryos. *Journal of the Atmospheric Sciences*, *76*(1), 113–133. doi: 10.1175/JAS-D-18-0087.1.

Jarecka, D., Grabowski, W. W., & Pawlowska, H. (2009). Modeling of subgrid-scale mixing in large-eddy simulation of shallow convection. *Journal of the Atmospheric Sciences*, *66*(7), 2125–2133. doi: 10.1175/2009JAS2929.1.

Jarecka, D., Grabowski, W. W., Morrison, H., & Pawlowska, H. (2013). Homogeneity of the subgrid-scale turbulent mixing in large-eddy simulation of shallow convection. *Journal of the Atmospheric Sciences*, *70*(9), 2751–2767.

Jeffery, C. A. (2007). Inhomogeneous cloud evaporation, invariance, and Damköhler number. *Journal of Geophysical Research*, *112*(D24).

Jensen, J., Austin, P., Baker, M., & Blyth, A. (1985). Turbulent mixing, spectral evolution and dynamics in a warm cumulus cloud. *Journal of the Atmospheric Sciences*, *42*(2), 173–192.

Jensen, J. B., & Baker, M. B. (1989). A simple model of droplet spectral evolution during turbulent mixing. *Journal of the Atmospheric Sciences*, *46*(18), 2812–2829.

Kerstein, A. R. (1988). A linear-eddy model of turbulent scalar transport and mixing. *Combustion Science and Technology*, *60*(4–6), 391–421.

Khairoutdinov, M., & Kogan, Y. (2000). A new cloud physics parameterization in a large-eddy simulation model of marine stratocumulus. *Monthly Weather Review*, *128*(1), 229–243.

Klinger, C., Feingold, G., & Yamaguchi, T. (2019). Cloud droplet growth in shallow cumulus clouds considering 1-D and 3-D thermal radiative effects. *Atmospheric Chemistry and Physics Discuss.*, *19*(9). https://doi.org/10.5194/acp-2018-1204.

Korolev, A., Khain, A., Pinsky, M., & French, J. (2016). Theoretical study of mixing in liquid clouds–Part 1: Classical concepts. *Atmospheric Chemistry and Physics*, *16*(14).

Korolev, A., McFarquhar, G., Field, P. R., Franklin, C., Lawson, P., Wang, Z., et al. (2017). Mixed-phase clouds: Progress and challenges. *Meteorological Monographs*, *58*, 5–1.

Krueger, S. K. (1993). Linear eddy modeling of entrainment and mixing in stratus clouds. *Journal of the Atmospheric Sciences*, *50*(18), 3078–3090.

Krueger, S. K. (2016). Representing the effects of entrainment and mixing on warm-cloud microphysics in large-eddy simulations. In *AGU Fall Meeting Abstracts*.

Krueger, S. K., Su, C.-W., & McMurtry, P. A. (1997). Modeling entrainment and finescale mixing in cumulus clouds. *Journal of the Atmospheric Sciences*, *54*(23), 2697–2712.

Kumar, B., Schumacher, J., & Shaw, R. A. (2013). Cloud microphysical effects of turbulent mixing and entrainment. *Theoretical and Computational Fluid Dynamics*, *27*(3–4), 361–376.

Kunishima, Y., & Onishi, R. (2018). Direct lagrangian tracking simulation of droplet growth in vertically developing cloud. *Atmospheric Chemistry and Physics*, *18*(22), 16,619–16,630. doi: 10.5194/acp-18-16619-2018.

Kurowski, M. J., Malinowski, S. P., & Grabowski, W. W. (2009). A numerical investigation of entrainment and transport within a stratocumulus-topped boundary layer. *Quarterly Journal of the Royal Meteorological Society*, *135*(638), 77–92. doi: https://doi.org/10.1002/qj.354.

Lasher-Trapp, S. G., Cooper, W. A., & Blyth, A. M. (2005). Broadening of droplet size distributions from entrainment and mixing in a cumulus cloud. *Quarterly Journal of the Royal Meteorological Society*, *131*(605), 195–220.

Latham, J., & Reed, R. (1977). Laboratory studies of the effects of mixing on the evolution of cloud droplet spectra. *Quarterly Journal of the Royal Meteorological Society*, *103*(436), 297–306.

Lehmann, K., Siebert, H., & Shaw, R. A. (2009). Homogeneous and inhomogeneous mixing in cumulus clouds. Dependence on local turbulence structure. *Journal of the Atmospheric Sciences*, *66*(12), 3641–3659.

Lu, C., Liu, Y., Niu, S., Krueger, S., & Wagner, T. (2013). Exploring parameterization for turbulent entrainment-mixing processes in clouds. *Journal of Geophysical Research*, *118*(1), 185–194.

Lu, C., Liu, Y., Zhu, B., Yum, S. S., Krueger, S. K., Qiu, Y., et al. (2018). On which microphysical time scales to use in studies of entrainment-mixing mechanisms in clouds. *Journal of Geophysical Research*, *123*(7), 3740–3756.

Luo, S., Lu, C., Liu, Y., Bian, J., Gao, W., Li, J., et al. (2020). Parameterizations of entrainment-mixing mechanisms and their effects on cloud droplet spectral width based on numerical simulations. *Journal of Geophysical Research, accepted*.

Maury, N., Roberts, G. C., Couvreux, F., Verdu, T., Narvor, P., Villefranque, N., et al. (2021). Use of large-eddy simulations to design an adaptive sampling strategy to assess cumulus cloud heterogeneities by remotely piloted aircraft. *Atmospheric Measurement Techniques Discussions*, 1–26. doi: 10.5194/amt-2021-20.

Mellado, J. P. (2017). Cloud-top entrainment in stratocumulus clouds. *Annual Review of Fluid Mechanics*, *49*, 145–169.

Mellado, J. P., Stevens, B., Schmidt, H., & Peters, N. (2010). Two-fluid formulation of the cloud-top mixing layer for direct numerical simulation. *Theoretical and Computational Fluid Dynamics*, *24*(6), 511–536.

Mellado, J.-P., Bretherton, C., Stevens, B., & Wyant, M. (2018). DNS and LES for simulating stratocumulus: Better together. *Journal of Advances in Modeling Earth Systems* doi: 10.1029/2018MS001312

Milbrandt, J., & Yau, M. (2005). A multimoment bulk microphysics parameterization. Part I: Analysis of the role of the spectral shape parameter. *Journal of the Atmospheric Sciences*, *62*(9).

Moin, P., & Mahesh, K. (1998). Direct numerical simulation: A tool in turbulence research. *Annual Review of Fluid Mechanics*, *30*(1), 539–578.

Morrison, H., & Grabowski, W. W. (2008). Modeling supersaturation and subgrid-scale mixing with two-moment bulk warm microphysics. *Journal of the Atmospheric Sciences*, *65*(3), 792–812.

Morrison, H., Curry, J., & Khvorostyanov, V. (2005). A new double-moment microphysics parameterization for application in cloud and climate models. Part I: Description. *Journal of the Atmospheric Sciences*, *62*(6), 1665–1677.

Morrison, H., van Lier-Walqui, M., Fridlind, A. M., Grabowski, W. W., Harrington, J. Y., Hoose, C., et al. (2020). Confronting the challenge of modeling cloud and precipitation microphysics. *Journal of Advances in Modeling Earth Systems*, *12*(8), e2019MS001,689.

Nicholls, S. (1989). The structure of radiatively driven convection in stratocumulus. *Quarterly Journal of the Royal Meteorological Society*, *115*(487), 487–511.

Paluch, I. R. (1986). Mixing and the cloud droplet size spectrum: Generalizations from the CCOPE data. *Journal of the Atmospheric Sciences*, *43*(18), 1984–1993.

Paluch, I. R., & Baumgardner, D. G. (1989). Entrainment and fine-scale mixing in a continental convective cloud. *Journal of the Atmospheric Sciences*, *46*(2), 261–278.

Paluch, I. R., & Knight, C. A. (1984). Mixing and the evolution of cloud droplet size spectra in a vigorous continental cumulus. *Journal of the Atmospheric Sciences*, *41*(11), 1801–1815. doi: 10.1175/1520-0469(1984)041<1801:MATEOC>2.0.CO;2.

Perrin, V. E., & Jonker, H. J. J. (2015). Lagrangian droplet dynamics in the subsiding shell of a cloud using direct numerical simulations. *Journal of the Atmospheric Sciences*, *72*(10), 4015–4028. doi: 10.1175/JAS-D-15-0045.1.

Peters, N. (2000). *Turbulent combustion*: Cambridge University Press.

Pinsky, M., & Khain, A. (2002). Effects of in-cloud nucleation and turbulence on droplet spectrum formation in cumulus clouds. *Quarterly Journal of the Royal Meteorological Society*, *128*(580), 501–533.

Pinsky, M., Khain, A., & Korolev, A. (2016). Theoretical analysis of mixing in liquid clouds–Part 3: Inhomogeneous mixing. *Atmospheric Chemistry and Physics*, *16*(14), 9273–9297.

Sagaut, P. (2006). *Large eddy simulation for incompressible flows*. Berlin: Springer.

Schmeissner, T., Shaw, R., Ditas, J., Stratmann, F., Wendisch, M., & Siebert, H. (2015). Turbulent mixing in shallow trade wind cumuli: Dependence on cloud life cycle. *Journal of the Atmospheric Sciences*, *72*(4), 1447–1465.

Schulz, B., & Mellado, J. P. (2019). Competing effects of droplet sedimentation and wind shear on entrainment in stratocumulus. *Journal of Advances in Modeling Earth Systems*, *11*(6), 1830–1846. doi: https://doi.org/10.1029/2019MS001617.

Schumann, U. (1975). Subgrid scale model for finite difference simulations of turbulent flows in plane channels and annuli. *Journal of Computational Physics*, *18*(4), 376–404.

Seifert, A., & Beheng, K. D. (2006). A two-moment cloud microphysics parameterization for mixed-phase clouds. Part 1: Model description. *Meteorology and Atmospheric Physics*, *92*, 45–66.

Seifert, A., & Heus, T. (2013). Large-eddy simulation of organized precipitating trade wind cumulus clouds. *Atmospheric Chemistry and Physics*, *13*, 5631–5645.

Shaw, R. A. (2003). Particle-turbulence interactions in atmospheric clouds. *Annual Review of Fluid Mechanics*, *35*(1), 183–227. doi: 10.1146/annurev.fluid.35.101101.161125.

Shaw, R. A., Cantrell, W., Chen, S., Chuang, P., Donahue, N., Feingold, G., et al. (2020). Cloud–aerosol–turbulence interactions: Science priorities and concepts for a large-scale laboratory facility. *Bulletin of the American Meteorological Society*, *101*(7), E1026–E1035.

Shima, S.-I., Kusano, K., Kawano, A., Sugiyama, T., & Kawahara, S. (2009). The super-droplet method for the numerical simulation of clouds and precipitation: A particle-based and probabilistic microphysics model coupled with a non-hydrostatic model. *Quarterly Journal of the Royal Meteorological Society*, *135*(642), 1307–1320.

Siebert, H., Lehmann, K., & Wendisch, M. (2006). Observations of small-scale turbulence and energy dissipation rates in the cloudy boundary layer. *Journal of the Atmospheric Sciences*, *63*(5), 1451–1466.

Siebert, H., Beals, M., Bethke, J., Bierwirth, E., Conrath, T., Dieckmann, K., et al. (2013). The fine-scale structure of the trade wind cumuli over Barbados – an introduction to the CARRIBA project. *Atmospheric Chemistry and Physics*, *13*(19), 10,061–10,077. doi: 10.5194/acp-13-10061-2013.

Siebesma, A. P., Bretherton, C. S., Brown, A. R., Chlond, A., Cuxart, J., Duynkerke, P. G., et al. (2003). A large eddy simulation intercomparison study of shallow cumulus convection. *Journal of the Atmospheric Sciences*, *60*, 1201–1219.

Slawinska, J., Grabowski, W. W., Pawlowska, H., & Wyszogrodzki, A. A. (2008). Optical properties of shallow convective clouds diagnosed from a bulk-microphysics large-eddy simulation. *Journal of Climate*, *21*(7), 1639–1647.

Slawinska, J., Grabowski, W. W., Pawlowska, H., & Morrison, H. (2012). Droplet activation and mixing in large-eddy simulation of a shallow cumulus field. *Journal of the Atmospheric Sciences*, *69*, 444–462.

Squires, P. (1952). The growth of cloud drops by condensation: I. general characteristics. *Australian Journal of Scientific Research*, *5*(1), 59–86.

Stevens, B., Moeng, C.-H., Ackerman, A. S., Bretherton, C. S., Chlond, A., de Roode, S., et al. (2005). Evaluation of large-eddy simulations via observations of nocturnal marine stratocumulus. *Monthly Weather Review*, *133*(6), 1443–1462.

Stommel, H. (1947). Entrainment of air into a cumulus cloud: Paper presented 27 December 1946 at the annual meeting, AMS, Cambridge, Massachusetts. *Journal of Meteorological*, *4*(3), 91–94.

Su, C.-W., Krueger, S. K., McMurtry, P. A., & Austin, P. H. (1998). Linear eddy modeling of droplet spectral evolution during entrainment and mixing in cumulus clouds. *Atmospheric Research*, *47*, 41–58.

Thomas, L., Grabowski, W. W., & Kumar, B. (2020). Diffusional growth of cloud droplets in homogeneous isotropic turbulence: DNS, scaled-up DNS, and stochastic model. *Atmospheric Chemistry and Physics*, *20*(14), 9087–9100.

Thompson, G., & Eidhammer, T. (2014). A study of aerosol impacts on clouds and precipitation development in a large winter cyclone. *Journal of the Atmospheric Sciences*, *71*(10), 3636–3658.

Tölle, M. H., & Krueger, S. K. (2014). Effects of entrainment and mixing on droplet size distributions in warm cumulus clouds. *Journal of Advances in Modeling Earth Systems*, *6*(2), 281–299.

Twomey, S. (1974). Pollution and the planetary albedo. *Atmospheric Environment*, *8*(12), 1251–1256.

Twomey, S. (1977). The influence of pollution on the shortwave albedo of clouds. *Journal of the Atmospheric Sciences*, *34*(7), 1149–1152.

Tzivion, S., Feingold, G., & Levin, Z. (1987), An efficient numerical solution to the stochastic collection equation. *Journal of the Atmospheric Sciences*, *44*(21), 3139–3149.

Vaillancourt, P., Yau, M., Bartello, P., & Grabowski, W. W. (2002). Microscopic approach to cloud droplet growth by condensation. Part II: Turbulence, clustering, and condensational growth. *Journal of the Atmospheric Sciences*, *59*(24), 3421–3435.

Vaillancourt, P. A., Yau, M. K., & Grabowski, W. W. (2001). Microscopic approach to cloud droplet growth by condensation. Part I: Model description and results without turbulence. *Journal of the Atmospheric Sciences*, *58*(14), 1945–1964. doi: 10.1175/1520-0469(2001)058<1945:MATCDG>2.0.CO;2.

vanZanten, M. C., Stevens, B., Nuijens, L., Siebesma, A. P., Ackerman, A. S., Burnet, F., et al. (2011). Controls on precipitation and cloudiness in simulations of trade-wind cumulus as observed during RICO. *Journal of Advances in Modeling Earth Systems*, *3*(2). doi: 10.1029/2011MS000056.

Wang, S., Wang, Q., & Feingold, G. (2003). Turbulence, condensation, and liquid water transport in numerically simulated nonprecipitating stratocumulus clouds. *Journal of the Atmospheric Sciences*, *60*(2), 262–278.

Warner, J. (1955). The water content of cumuliform cloud. *Tellus*, *7*(4), 449–457. doi: https://doi.org/10.1111/j.2153-3490.1955.tb01183.x.

Warner, J. (1969a), The microstructure of cumulus cloud. Part I: General features of the droplet spectrum. *Journal of the Atmospheric Sciences*, *26*(5), 1049–1059. doi: 10.1175/1520-0469(1969)026<1049:TMOCCP>2.0.CO;2.

Warner, J. (1969b). The microstructure of cumulus cloud. Part II: The effect on droplet size distribution of the cloud nucleus spectrum and updraft velocity. *Journal of the Atmospheric Sciences*, *26*(6), 1272–1282. doi: 10.1175/1520-0469(1969)026<1272:TMOCCP>2.0.CO;2.

Warner, J. (1973). The microstructure of cumulus cloud. Part IV: The effect on the droplet spectrum of mixing between cloud and environment. *Journal of the Atmospheric Sciences*, *30*(2), 256–261.

Xue, H., & Feingold, G. (2006). Large-eddy simulations of trade wind cumuli: Investigation of aerosol indirect effects. *Journal of the Atmospheric Sciences*, *63*(6), 1605–1622.

Yamaguchi, T., & Randall, D. A. (2012). Cooling of entrained parcels in a large-eddy simulation. *Journal of the Atmospheric Sciences*, *69*(3), 1118–1136.

Zhao, M., & Austin, P. H. (2005). Life cycle of numerically simulated shallow cumulus clouds. Part II: Mixing dynamics. *Journal of the Atmospheric Sciences*, *62*, 1291–1310.

Zuidema, P., Torri, G., Muller, C., & Chandra, A. (2017). A survey of precipitation-induced atmospheric cold pools over oceans and their interactions with the larger-scale environment. *Surveys in Geophysics*, *38*(6), 1283–1305.

Part IV
Clouds and Precipitation

13

Precipitation Efficiency and Climate Sensitivity

Nicholas J. Lutsko[1], Steven C. Sherwood[2], and Ming Zhao[3]

ABSTRACT

A number of studies have demonstrated strong relationships between precipitation efficiency, particularly its changes under warming, and climate sensitivity. In this chapter, we review the evidence for these relationships, including how they depend on the definition of precipitation efficiency. We identify six mechanisms by which changes in precipitation efficiency may affect Earth's net climate feedback and also discuss evidence for an inverse relationship between present-day precipitation efficiency and climate sensitivity based on several perturbed physics ensembles. This inverse relationship hints at the possibility of developing emergent constraints on climate sensitivity using precipitation efficiency, though it is put in doubt by studies varying convective entrainment rates, which have found the opposite relationship. More work is required to refine our understanding of the mechanisms linking changes in precipitation efficiency to climate sensitivity and more observational data are needed to validate model results. In particular, the precipitation efficiency of midlatitude clouds has been relatively understudied, but deserves more attention in light of the importance of extratropical cloud feedbacks for the high climate sensitivities of CMIP6 models.

13.1. INTRODUCTION

Clouds remain the largest source of uncertainty in Earth's climate sensitivity. While substantial progress has been made in recent years on constraining their response to warming, with a number of lines of evidence pointing to a positive cloud feedback [We define a positive climate feedback as a feedback which enhances greenhouse gas-induced warming, and viceversa for a negative feedback.] (Sherwood et al., 2020), considerable uncertainty remains as to the cloud feedback's magnitude. Uncertainty in the net cloud feedback reflects the immense complexity of clouds: the variety of cloud types, the multiscale turbulent interactions within clouds, the microphysics of phase changes and water droplets, the interactions between clouds and large-scale circulations, the difficulties of observing clouds, etc. Recent progress on many of these aspects of cloud physics is reviewed in other chapters of this monograph.

In this chapter, we focus on the relationship between the precipitation efficiency of clouds and climate sensitivity. A number of studies have demonstrated strong relationships between climate sensitivity and either present-day precipitation efficiency or the response of precipitation efficiency to warming [Note that these studies, reviewed below, have used various methods for evaluating climate sensitivity. For example, some studies used the Cess and Potter (1988) method for diagnosing climate sensitivity, while others performed coupled model simulations. But these ways of estimating climate sensitivity are well correlated

[1]*Scripps Institution of Oceanography, University of California San Diego, La Jolla, CA, USA*

[2]*Climate Change Research Centre and ARC Centre of Excellence for Climate Extremes, University of New South Wales, Sydney, Australia*

[3]*NOAA Geophysical Fluid Dynamics Laboratory, Princeton, NJ, USA*

Clouds and Their Climatic Impacts: Radiation, Circulation, and Precipitation, Geophysical Monograph 281, First Edition.
Edited by Sylvia C. Sullivan and Corinna Hoose.

DOI: 10.1002/9781119700357.ch13

with each other, and with the net cloud feedback, so we will generally ignore the method used to evaluate climate sensitivity in the discussion.]. In both cases, the link reflects the variety of processes which determine precipitation efficiency, including the microphysics within clouds, small-scale mixing between clouds and their environment, and the large-scale organization of clouds and convecting systems (see also Chapter 8: Mechanisms for the Self-Organization of Tropical Deep Convection and Chapter 12: The Small-Scale Mixing of Clouds with their Environment). All of these contribute to the total cloud feedback, making precipitation efficiency a useful bulk metric for many uncertain cloud processes. Constraining precipitation efficiency and its behavior under warming would not eliminate the uncertainty in climate sensitivity, but it would be an important step.

We begin the chapter by reviewing the concept of precipitation efficiency, including its various definitions, which in turn imply different relationships with climate sensitivity. We then review the connections between the response of precipitation efficiency to warming and climate sensitivity (section 13.3), including a discussion of the controversial "Iris hypothesis," followed by a discussion of the potential connections between present-day precipitation efficiency and climate sensitivity (section 13.4). Sections 13.3 and 13.4 are primarily concerned with tropical clouds, so section 13.5 addresses the relationship between the precipitation efficiency of extratropical clouds and climate sensitivity, before we end with conclusions in section 13.6.

We do not attempt to summarize all of the existing literature on precipitation efficiency; the reader is referred to Sui et al. (2020) for a recent overview. Instead, we focus more narrowly on work which has attempted to relate precipitation efficiency and its changes to climate sensitivity. By necessity, this mostly limits our scope to modeling papers, though we touch on observational work where appropriate.

13.2. DEFINING PRECIPITATION EFFICIENCY

Precipitation efficiency can be simply defined as the ratio of surface precipitation to the rate at which cloud particles condense (see Fig. 13.1):

$$\epsilon = P/C, \tag{13.1}$$

where P is surface precipitation in kg m^{-2} s^{-1} and C is the rate of cloud condensation or, equivalently, the sink of atmospheric water vapor, also with units kg m^{-2} s^{-1}. ϵ is unitless and can be thought of in a Lagrangian sense as the probability that water which condenses as a cloud droplet reaches the surface as precipitation at some point (Langhans et al., 2015).

The process of cloud condensate reaching the surface can be divided into two stages. First, the condensate must form (or accrete onto) droplets of precipitation – rain, snow, or graupel. Next, the precipitation must fall through the atmosphere without re-evaporating. This picture leads to the decomposition of surface precipitation into the condensation rate C multiplied by a "conversion efficiency" and a "sedimentation efficiency" (Langhans et al., 2015; Lutsko & Cronin, 2018):

$$P = C\alpha(1-\beta), \tag{13.2}$$

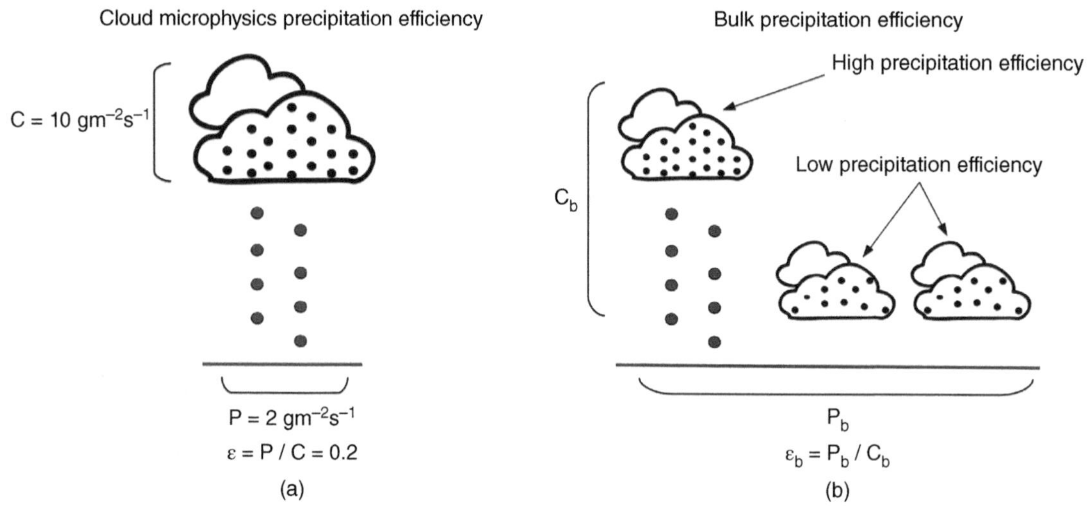

Figure 13.1 (a) Cloud microphysics precipitation efficiency is defined as precipitation P divided by the net condensation in a column C, calculated from microphysical tendencies (vapor condensation, vapor deposition, etc.) only. (b) Bulk precipitation efficiency is the average of cloud microphysics precipitation efficiency across a collection of different cloud types, assuming steady state and averaging over multiple convecting systems.

where α represents the efficiency with which cloud droplets are converted into precipitation, of which a fraction β is re-evaporated, so that $1 - \beta$ is the sedimentation efficiency. Precipitation efficiency can then be written as

$$\epsilon = \alpha(1 - \beta). \tag{13.3}$$

Idealized models of the tropical atmosphere often assume a conversion efficiency $\alpha = 1$, so that precipitation efficiency is set entirely by re-evaporation (e.g., Emanuel, 1987, Yano & Emanuel, 1991), but simulations and observations both suggest that α is substantially less than 1.

These simple definitions are difficult to apply in practice. One has to choose the area and time interval over which precipitation efficiency is defined, and how to calculate C. For example, when considering a limited domain, ϵ can be affected by horizontal advection of cloud condensate and precipitation into or out of the domain, changing the value of α and β for reasons not directly related to microphysics. Another question concerns the re-evaporation or sublimation of cloud water, which can be difficult to distinguish from reduced cloud condensation. For simplicity, we follow Zhao (2014) in conceptually separating out two paths for atmospheric water vapor which undergoes phase changes: a rare/fast process in which water vapor condenses/deposits and precipitates out quickly, and a frequent/slow process in which water vapor condenses/deposits and forms clouds, that are eventually recycled back into the atmosphere through evaporation and sublimation. Precipitation efficiency is then a measure of the efficiency of the fast/rare process in the atmosphere's hydrological cycle: if $\epsilon = 0$ no precipitation falls to the surface and if $\epsilon = 1$ any cloud droplets which form fall out instantaneously.

As for the condensation C, Sui et al. (2007) distinguish between "cloud microphysics" precipitation efficiency, ϵ_{cm}, for which C is only calculated from microphysical tendencies (vapor condensation, vapor deposition, etc.), versus "large-scale" precipitation efficiency, ϵ_{ls}, for which C is calculated as the sum of the large-scale horizontal convergence of water vapor, surface evaporation, and the change in local water vapor storage. The latter definition represents the total water available for precipitation in a region, instead of the water that condenses as cloud droplets only, and is similar to the early definitions of precipitation efficiency in Emanuel (1987) and Yano & Emanuel (1991), as well as to the "drying ratio (e.g., Eidhammer et al. (2018)), for which C is set equal to the horizontal convergence of water vapor.

Because of our focus on climatological precipitation efficiency, we assume here that all quantities are in steady state, in contrast to Sui et al. (2007), and thus neglect changes in vapor loading with time. Furthermore, while we follow the Sui et al. (2007) definition of cloud microphysics precipitation efficiency ϵ_{cm}, we define a "bulk" precipitation efficiency ϵ_b, rather than using ϵ_{ls} for discussing precipitation efficiency at larger scales. ϵ_b is calculated solely from microphysical tendencies, instead of using the total water budget, and integrates P and C over all clouds and updrafts in a given domain. Thus, ϵ_b depends on the particular mix of cloud types in the domain, each having a different typical ϵ_{cm}, and takes into account the higher precipitation efficiency of deep convection versus the lower precipitation efficiency of shallow clouds (see Fig. 13.1b; note that because precipitation efficiency is a ratio ϵ_b is not equal to the average of ϵ_{cm} across different cloud types).

13.3. CHANGES IN PRECIPITATION EFFICIENCY AND CLIMATE SENSITIVITY

Here, we discuss the physical basis for changes in precipitation efficiency to play a role in climate sensitivity. There are two conditions for this. First, changes in global temperature must lead to a systematic change in precipitation efficiency or some related microphysical process. Second, the microphysical change must affect cloud or water vapor properties so as to alter the top-of-atmosphere energy budget. We discuss these two conditions in sections 13.3.1 and 13.3.2, respectively.

13.3.1. Changes in Cloud Microphysics Precipitation Efficiency Under Warming

Observational evidence and detailed modeling suggest that autoconversion of cloud droplets into rain becomes significant only when liquid water amount and/or droplet radii reach a critical threshold (Freud & Rosenfeld, 2012). A liquid amount threshold is represented by the simple and very popular Kessler autoconversion scheme, where cloud condensate is converted into precipitation via an equation of the form:

$$\frac{\partial q_p}{\partial t} = max\left[0, \eta(q_c - q_0)\right], \tag{13.4}$$

where q_p is the mixing ratio of precipitation, η is a rate constant, q_c is the cloud condensate mixing ratio, and q_0 is a threshold cloud condensate mixing ratio. Thus, denser clouds convert condensate into precipitation more efficiently and a minimum liquid mass is required for the process to occur at all. There is strong evidence that aerosol-induced reductions in droplet size can delay the onset of precipitation (e.g., Andrae et al., 2004; Rosenfeld, 2000), consistent with an increase in droplet collision efficiency when radii exceed 15–20 μ (Rosenfeld & Lensky, 1998). Nevertheless, given an initial droplet number (i.e., with cloud condensation nucleus (CCN) number held fixed), the droplet radius and total liquid mass will grow jointly as a cloud rises and vapor condenses, such that either a minimum radius or mass requirement can produce the same behavior. Thus, a warmer climate with higher boundary-layer water vapor concentrations (but no difference in CCN) will produce cloudy updrafts that attain required radii or condensed mass sooner, thereby plausibly increasing the precipitation efficiency, as first suggested by Lindzen et al. (2001, see below). Whether this actually happens may however depend on factors such as cloud mixing and ice processes not considered in the above simple argument.

Relatively few studies have directly examined the climate dependence of tropical precipitation efficiency (studies of midlatitude clouds are covered in section 13.5). Observationally, precipitation efficiency is both difficult to measure and difficult to relate to the large-scale state of the atmosphere. Lau and Wu (2003) suggested that the precipitation efficiency of warm rain increases as the underlying sea surface temperature (SST) increases, but required a microphysical parameterization to convert satellite data into precipitation efficiencies. Conversely, in an analysis of more than 10 years of high-quality radar and sounding data from the tropical Pacific, Narsey et al. (2019) found that ϵ_{cm} is inversely related to surface temperature and to convective available potential energy (which is expected to increase with warming), though they noted that these relationships are modulated by co-varying factors such as relative humidity, which may limit the applicability of their findings to warmer climates.

Idealized cloud-permitting simulations in small domains, with several different microphysics schemes, indicate that cloud microphysics precipitation efficiency should increase with warming in the absence of major changes in circulation or convective mixing (Lutsko & Cronin, 2018). These increases are seen across a wide range of surface temperatures (290–310 K, see Fig. 13.2) and are primarily driven by increases in the condensation efficiency α, with the sedimentation efficiency $1-\beta$ generally showing small decreases with warming. The increases in condensation efficiency reflect increases in cloud condensate density of ~2–6%/ °C of surface warming at almost all heights, consistent with theoretical expectations for entraining plumes rising along a moist adiabat (Betts and Harshvardhan (1987); Lutsko & Cronin (2018)).

In terms of the sedimentation efficiency, Lutsko and Cronin (2018) suggested that β scales as

$$\beta \sim (1 - RH)h/w, \qquad (13.5)$$

where RH is relative humidity, h is the average height at which precipitation forms, and w is the average fall speed of the precipitation. Analysis of cloud-permitting simulations demonstrated that warming-driven changes

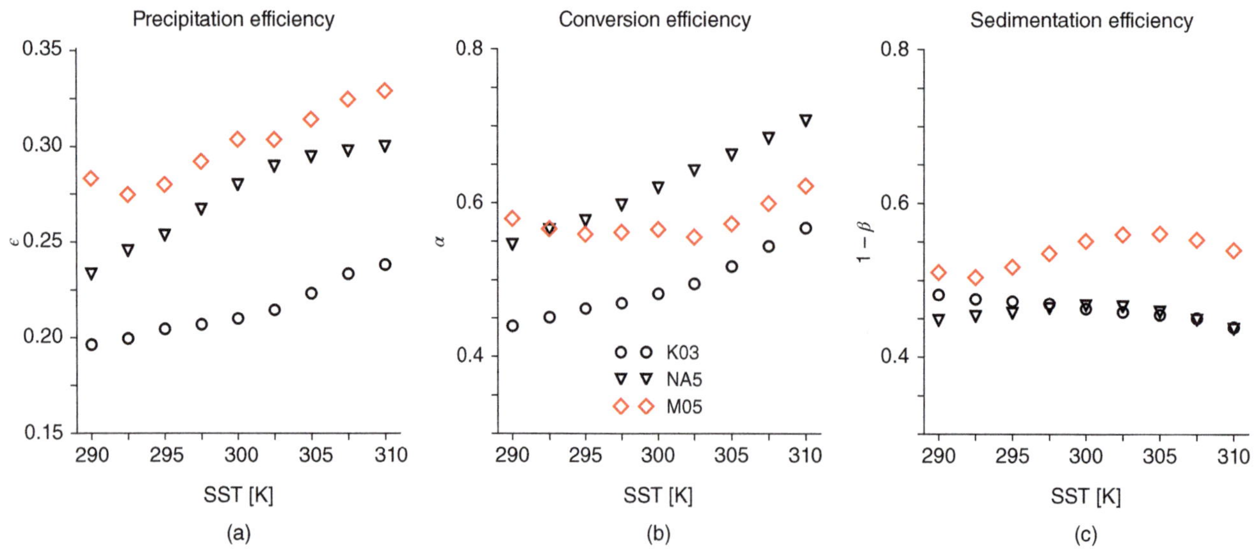

Figure 13.2 (a) Precipitation efficiency ϵ as a function of SST for the three sets of small domain simulations in Lutsko and Cronin (2018) (their Fig. 1). Black circles show results with the single-moment microphysics scheme and the parameter settings used by Khairoutdinov and Randall (2003); black triangles show results with the NOSEDAALIQ5 parameter settings used by Lopez et al. (2009); and red diamonds show results with the Morrison et al. (2005) microphysics scheme. (b) Conversion efficiency α as a function of SST for the same simulations. (c) Sedimentation efficiency $1-\beta$ as a function of SST for the same simulations.

in sedimentation efficiency are primarily caused by increases in the height at which clouds form, so that falling precipitation travels a greater distance through the atmosphere and has a greater chance of being re-evaporated, increasing β and decreasing ϵ. However, these changes are small compared to the changes in condensation efficiency α for temperatures in the range 290–310 K (compare panels b and c of Fig. 13.2), so that increases in cloud density lead to increases in the precipitation efficiency of tropical clouds with warming.

A caveat to these results is phase changes. Clouds are increasingly composed of liquid water droplets rather than cloud ice as the atmosphere warms. In turn, the precipitation in a column transitions from being primarily snow and/or graupel to primarily rain. Ice clouds have a higher conversion efficiency than liquid water clouds, because larger ice crystals grow more rapidly than liquid cloud droplets through collisions and collection of hydrometeors, but snow also re-evaporates much more readily than rain because of its slower fall speeds. Hence, a change from a predominantly snow to a predominantly rain regime results in a decrease in conversion efficiency but an increase in sedimentation efficiency. For relatively warm surface temperatures (cf. 290 K), the increased sedimentation efficiency generally wins out in cloud-permitting simulations and precipitation efficiency increases with warming (Lutsko & Cronin, 2018), though studies of mixed-phase clouds at higher latitudes have found that the reduction in conversion efficiency wins out at colder temperatures, leading to a reduction in precipitation efficiency with warming (see section 13.5 for more).

These arguments ignore changes in convective aggregation, which can affect the thermodynamic environment in which precipitation forms as well as the bulk precipitation efficiency (see below), but suggest that, with or without phase changes, the cloud-microphysics precipitation efficiency of tropical clouds should be expected to increase with warming. These arguments apply to the mean precipitation efficiency, rather than the precipitation efficiency of extreme precipitation, but given our focus on the relationship with cloud feedbacks, we speculate that mean precipitation efficiency is most relevant for the discussion (see Muller et al. (2011) and Singh & O'Gorman (2014) for more on the response of extreme precipitation efficiency to warming).

13.3.2. Precipitation Efficiency-Related Feedbacks on Climate

We now consider various ideas for how climate-driven changes in precipitation efficiency could provide feedbacks on climate. Studies investigating this possibility have often been motivated by the Iris hypothesis. First proposed by Lindzen et al. (2001), the Iris hypothesis is a potential negative high cloud feedback, in which a reduction in cirrus cloudiness under warming leads to an increase in outgoing longwave radiation (OLR). The name comes from an analogy to the iris in the human eye, which contracts in the presence of bright light. Based on an analysis of observed changes in cloud cover coincident with higher surface temperatures in a region of the central Pacific, Lindzen et al. (2001) claimed a 22% reduction in cloud cover per degree warming, implying a strongly negative cloud feedback and a climate sensitivity of $\sim 1^{\circ}$C. However, these results were reported for localized regions, which typically involve confounding dynamical changes or atmospheric influences on the ocean (see Sherwood et al., 2010), and relationships can nearly disappear when averages are taken over the whole tropics (Williams & Pierrehumbert, 2017).

One reason for the Iris hypothesis' controversy is the lack of a clear description of what is contracting and why. Lindzen et al. (2001) suggested that increases in conversion efficiency cause decreases in anvil cloud area, but anvil clouds have only a weak net impact on the global energy budget, so a substantial net radiative effect presumably requires a more general contraction of cloudy conditions in favor of expanded clear sky, with implied reductions in other cloud types and/or relative humidity. This links to the concept of convective aggregation, whereby convection can spontaneously cluster into smaller regions; in idealized radiative convective equilibrium (RCE) models aggregation consistently leads to greater OLR (Wing et al., 2020). Early studies suggested that such aggregation should increase at warmer surface temperatures in some global climate model (GCM) simulations (e.g., Bony, Stevens et al., 2016) and in small-scale simulations, but due to radiative-dynamical mechanisms rather than anything involving precipitation efficiency (Wing & Emanuel, 2014). Recent model intercomparisons are more ambiguous as to whether the process is climate-sensitive (Wing et al., 2020), but observational studies suggest that such contractions are generally associated with larger areas of low relative humidity and/or higher OLR (Bony, Semie, et al., 2016; Hohenegger & Jakob, 2020). Thus, while the tropics appear capable of "Iris"-like fluctuations, it is not clear whether these will deliver a negative feedback on climate. For more discussion of convective aggregation, see Chapter 8: Mechanisms for the Self-Organization of Tropical Deep Convection.

We also emphasize that any substantial Iris-like effect on climate sensitivity would likely have to come from the cloud field, rather than from reductions in relative humidity. While stronger convective clustering could lead to a reduction in re-evaporation and hence to a reduction in tropospheric relative humidity, theory and GCM experiments suggest that the sensitivity of tropospheric relative humidity to re-evaporation – to a change in β – is relatively weak (Romps, 2014; Sherwood & Meyer, 2006). Even in the case of a strong Iris effect, the changes in re-evaporation with warming are likely to be modest, and so are the changes in relative humidity.

Model Investigations of Changing the Response of Precipitation Efficiency to Warming

Since the convective processes potentially responsible for an Iris-like effect are crudely represented in climate models, possible Iris effects on climate sensitivity have been explored by varying the response of ϵ_{cm} with warming. In particular, Mauritsen and Stevens (2015) and Li et al. (2019) modified the autoconversion rate (η) in ECHAM6 and in CESM, respectively, so that

$$\eta(T_s) = \eta_0(1 + I_e)^{T_s - T_0}, \tag{13.6}$$

where T_0 is a reference temperature, set to 25°C in both studies, η_0 is a reference autoconversion rate, and I_e is a parameter which controls the strength of the Iris effect (larger I_e results in a higher ϵ_{cm}). Mauritsen and Stevens (2015) were originally motivated to add this modification to ECHAM6 by their finding that most climate models underestimate the increase in OLR per °C of tropical surface temperature warming on monthly timescales, suggestive of a missing Iris-like effect.

Making the autoconversion rate temperature-dependent successfully increased the OLR per degree warming, primarily through negative longwave cloud feedbacks – more negative than any of the CMIP5 models analyzed by Mauritsen and Stevens, even for the smallest value of I_e tested (0.2). But the inclusion of the temperature-dependent autoconversion also resulted in a more positive shortwave cloud feedback, leading to a partial compensation of the longwave cloud feedback and a modest reduction in ECHAM6's climate sensitivity, from 2.8°C to 2.2°C. Another factor compensating for the negative longwave cloud feedback was a weakened lapse-rate feedback due to reduced warming of the tropical upper troposphere.

Mauritsen and Stevens (2015) noted that the positive shortwave cloud feedback overcompensated for the negative longwave cloud feedback in unpublished experiments with Community Earth System Model (CESM), and this was confirmed by Li et al. (2019). Using the same modifications as Mauritsen and Stevens, Li et al. found that CESM's equilibrium climate sensitivity increased from 3.79°C to 4.59°C as I_e was increased from 0 to 1. The sign of the sensitivity change was robust across different cloud microphysics schemes and was mostly caused by increasingly large reductions in the optical depth of cirrus clouds, rather than by changes in cloud amount, which were small (Fig. 13.3). The reductions in cloud optical depth were caused by thinning of anvil clouds as a stronger Iris feedback was imposed, though the cloud thinning was itself partially compensated by a negative cloud-phase feedback: the liquid water clouds which replace ice clouds in warmer climates are longer lived and more reflective, because they are made up of a large number of small water droplets, rather than a relatively small number of large ice crystals. We return to the negative cloud-phase feedback in section 13.5.

Finally, in a study not directly motivated by the Iris hypothesis, Zhao et al. (2016) found a strong relationship between cloud microphysics precipitation efficiency and climate sensitivity in their experiments with the GFDL-AM4 model. These experiments involved modifying the scheme by which cumulus precipitation is formed to produce either an increase in ϵ_{cm}, a negligible change in ϵ_{cm}, or a decrease in ϵ_{cm} with warming. The configurations each produced good representations of the present-day climate, but had very different climate sensitivities: the version in which ϵ_{cm} increased had a high climate sensitivity, the version in which ϵ_{cm} stayed the same had a moderate climate sensitivity and the version in which ϵ_{cm} decreased had a low climate sensitivity. These differences in climate sensitivity were caused mainly by cloud changes between 800 and 300 hPa, with the response of high clouds being weak in each configuration.

13.4. PRESENT-DAY PRECIPITATION EFFICIENCY AND CLIMATE SENSITIVITY

While changes in precipitation efficiency are likely responsible for the relationship with climate sensitivity, studies with several different GCMs have shown strong links between present-day precipitation efficiency and climate sensitivity. These results demonstrate that microphysical parameters related to precipitation efficiency can be used as tuning parameters for controlling models' climate sensitivity and also imply correlations between baseline ϵ and $\Delta\epsilon$, which could allow observations of present-day precipitation efficiency to be used to constrain its changes.

13.4.1. Mid to High Clouds and Microphysical Mechanisms

The most focused study of the relationship between present-day precipitation efficiency and climate sensitivity is Zhao (2014) who, with the deliberate aim of altering the model's precipitation efficiency, varied two parameters in GFDL's C48HIRAM model: the warm-cloud autoconversion threshold q_0, and c_0, a parameter which controls

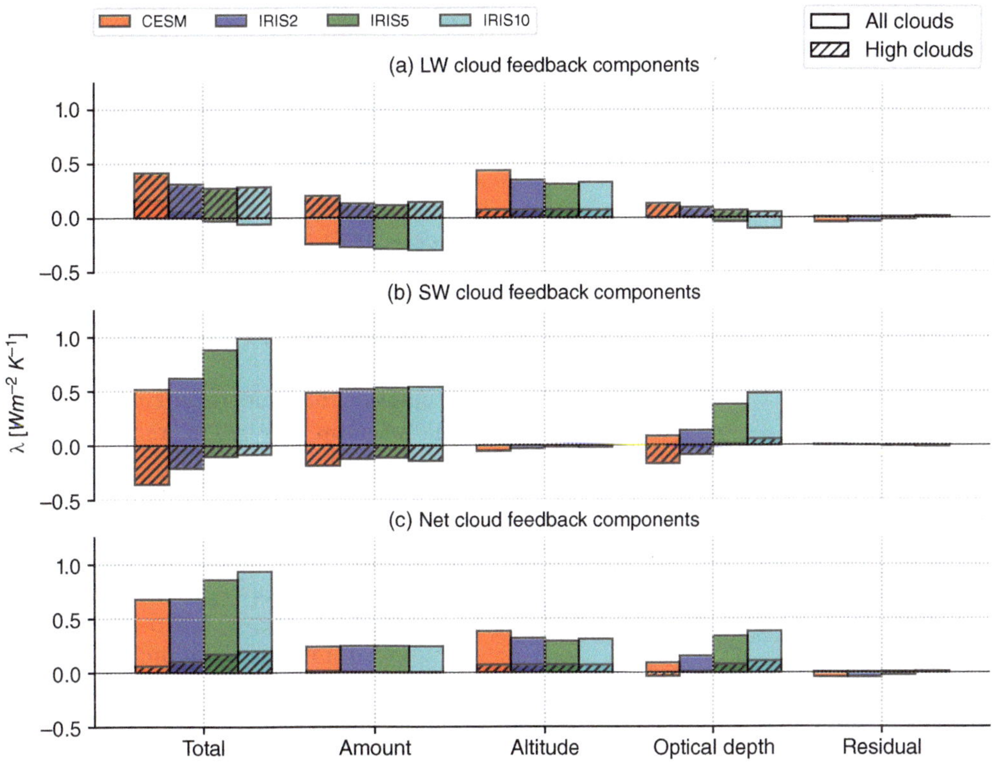

Figure 13.3 Contributions to the global-mean cloud feedbacks in the Iris simulations of Li et al. (2019), separated into cloud-amount, cloud-altitude, and cloud-optical-depth components. Control CESM results are shown in red, results from simulation with $I_e = 0.2$ are shown in blue, results from simulation with $I_e = 0.5$ are shown, in green, and results from simulation with $I_e = 1.0$ are shown in light blue. A separate decomposition is performed for high clouds (<440 hPa), shown with hatched bars. Note that these high-cloud contributions are not additive with the low-cloud (i.e., >440 hPa, not shown) contributions to give back the all-cloud components (all pressure bins). Source: Reproduced, with permission, from Li et al. (2019) / American Meteorological Society (their Fig. 9).

the cumulus entrainment rate (parameterized as c_0/H). Increasing q_0 means that higher cloud water densities are required to form precipitation, reducing the conversion efficiency and thus decreasing the cloud microphysics precipitation efficiency. Large values of q_0 also decrease the vertical velocities of convective plumes by increasing condensate loading, favoring shallow plumes, rather than the deep convective plumes in which the majority of precipitation forms. Increasing c_0 has a similar effect of decreasing plume vertical velocities because of increases in lateral mixing and entrainment. In both cases, the bulk precipitation efficiency decreases as the relative fraction of shallow plumes increases, and the greater low-cloud fraction decreases the net cloud radiative effect (CRE).

The bulk precipitation efficiency consistently increased when the various model configurations were subject to uniform SST warming, with the largest increases for the configurations with the smallest base-state precipitation efficiencies (Fig. 13.4). Zhao (2014) explained this inverse relationship using a conceptual model of tropical convection in which an ascending parcel rises through the atmosphere, exchanging air with the environment such that the parcel conserves its total mass (although this picture is based on a parcel model, it should be thought of as representing the bulk tropical atmosphere). The parcel is assumed to produce total cloud condensate $q_c = aq_b$, where q_b is the boundary-layer specific humidity and a is the fraction of q_b which condenses as the parcel ascends. Of this total condensate, $aq_b - q_0$ reaches the surface as precipitation, where q_0 is again a threshold specific humidity above which precipitation forms (note that the sedimentation efficiency is assumed to be 1, i.e., $\beta = 0$). The bulk precipitation efficiency is then $\epsilon_b = 1 - q_0/(aq_b)$ and the response of ϵ_b to a small perturbation is

$$\Delta\epsilon_b = \frac{q_0}{aq_b}\left(\frac{\Delta a}{a} + \frac{\Delta q_b}{q_b} - \frac{\Delta q_0}{q_0}\right), \qquad (13.7)$$

which can also be written as

$$\Delta\epsilon_b = (1 - \epsilon_b)\left(\frac{\Delta a}{a} + \frac{\Delta q_b}{q_b} - \frac{\Delta q_0}{q_0}\right). \qquad (13.8)$$

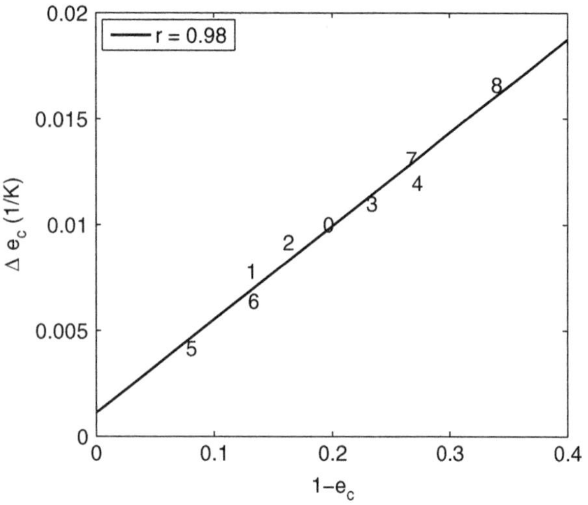

Figure 13.4 Scatterplot of the change in precipitation efficiency with warming $\Delta\epsilon$ versus the present-day precipitation efficiency (i.e., $1-\epsilon$) from the perturbed-physics ensemble in Zhao (2014) (their Fig. 10). 0 denotes the control simulation while numbers 1–8 denote the perturbed-physics simulations. The line shows a linear regression with the correlation coefficient shown in the legend. Source: Zhao et al. (2014) / American Meteorological Society.

Zhao (2014) found that $\Delta\epsilon_b$ was strongly correlated with $1-\epsilon_b$, suggesting that Δa and Δq_b make small contributions to the variations in $\Delta\epsilon_b$ across the model configurations (q_0 is unaffected by warming in this microphysics scheme) [The relationship between $\Delta\epsilon_b$ and $1-\epsilon_b$ holds within different configurations of a single model, but does not necessarily hold when comparing across models. Differences in convective mixing schemes or in microphysics could cause large variations in $\Delta a/a$, $\Delta q_b/q_b$, and $\Delta q_0/q_0$ across models.], and providing the basis for a possible connection between present-day precipitation efficiency and $\Delta\epsilon_b$.

Increases in bulk precipitation efficiency resulted in decreases in liquid and ice water paths with warming in Zhao's experiments, with low- and mid-cloud fractions diminishing at faster rates than high-cloud fractions because higher precipitation efficiencies favor deep convection, rather than shallow plumes. Hence, model configurations with larger $\Delta\epsilon_b$ experienced larger reductions of low- and mid-level clouds, which produced larger decreases in shortwave CRE and more positive cloud feedbacks. In this way, configurations with smaller present-day precipitation efficiencies had higher climate sensitivities.

A potential inverse relationship between present-day precipitation efficiency and climate sensitivity across different configurations of a single model is supported by several other studies, though none were directly investigating precipitation efficiency. Tomassini et al. (2015) found that increasing the autoconversion rate, which increases ϵ_{cm} by increasing α, decreases the climate sensitivity of a coarse-resolution version of the MPI-ESM GCM. Similarly, Mauritsen et al. (2012) found that doubling the autoconversion rate in MPI-ESM-LR decreases the model's climate sensitivity from 3.09 to 2.96 K, while Bender (2008) found that increasing the autoconversion rate by a factor of 5 decreases CAM3.1's climate sensitivity from 2.50 to 2.26 K. These changes are modest, though both sets of experiments involved other changes to the models.

Studies of the connection between the rate of entrainment for deep convection and climate sensitivity have put the inverse relationship into question, however. Tomassini et al. (2015), Mauritsen et al. (2012), and, earlier, Stainforth et al. (2005) found that increasing the rate of entrainment in deep convective plumes led to large decreases in climate sensitivity. This is in contrast to the a priori expectation that higher entrainment rates should produce lower precipitation efficiencies and higher climate sensitivities according to the relationship between ϵ and climate sensitivity established above. However, these studies did not report how ϵ changed in their experiments. We also note that varying the entrainment rate for deep convection is a different modification to Zhao's experiments, in which c_0 governs the entrainment rate throughout the column as part of a bulk plume representation of both shallow and deep convection. Nevertheless, the surprising relationship between deep convective entrainment rates and climate sensitivity indicates that while $\Delta\epsilon$ is often correlated with present-day ϵ, this is not always the case.

The mechanisms connecting deep entrainment rates and climate sensitivity have not been investigated, except for Mauritsen et al. (2012), who suggested that the reductions in climate sensitivity for higher deep entrainment rates are primarily caused by changes in the rapid adjustments to CO_2 forcing, rather than by changes in temperature-dependent feedbacks. In the absence of follow-up studies, further work is needed to verify this claim.

13.4.2. Low Clouds and Convective Mixing

A separate set of studies has proposed a connection between the strength of mixing in shallow convection and climate sensitivity. Air lifted out of the boundary layer can either continue ascending, rain out most of its water vapor and eventually return to lower altitudes, or it can be detrained directly at lower altitudes and retain more of its initial water vapor.

The latter corresponds to a situation with stronger mixing between the boundary layer and the lower troposphere, resulting in a smaller bulk precipitation efficiency, less boundary-layer cloudiness, and a greater net transport of moisture out of the boundary layer for the same mean precipitation rate. Lower tropospheric mixing is expected to strengthen in warmer climates, intensifying the dehydration of the boundary layer and reducing low-cloud cover, with the rate of increase in mixing thought to be proportional to the initial mixing strength (Bretherton, 2015; Brient et al., 2016; Rieck et al., 2012; Stevens, 2007; Vial et al., 2016; Zhang et al., 2013). Hence, a model with a lower initial bulk precipitation efficiency for shallow clouds should experience larger low-cloud reductions with warming and have a higher climate sensitivity – another inverse relationship between present-day precipitation efficiency and climate sensitivity.

In support of this argument, Sherwood et al. (2014) demonstrated that metrics for the strength of lower tropospheric convective mixing could explain about half the variance in CMIP5 models' climate sensitivity, with mixing rates inferred from observations implying a climate sensitivity greater than 3°C. They also confirmed that convective drying of the boundary layer increased with warming in models with stronger metrics of present-day shallow mixing. However, modeling studies in which mixing into shallow convection is directly altered via the parametrized entrainment rate have found a seemingly different result: in Mauritsen and Roeckner (2020), a developmental version of ECHAM6 exhibited a large decrease in climate sensitivity when the shallow convection entrainment rate was increased by a factor of 10, while in an earlier study, MPI-ESM-LR's climate sensitivity declined from 3.09 to 2.86°C when the shallow entrainment rate was increased by a factor of 8/3 (Mauritsen et al., 2012).

These results are not necessarily inconsistent with Sherwood et al. (2014) as, for example, increasing the entrainment rate of shallow convection could favor the development of deep convection and thereby increase ϵ_b for reasons unrelated to changes in shallow mixing. More direct tests of the shallow-mixing idea come from Zhao et al. (2016), who found that the indices defined by Sherwood et al. (2014) were not well correlated with climate sensitivity in their perturbed physics ensemble (PPE) simulations, and that clouds above the boundary layer were more important for differences in cloud feedback between the ensemble members (see section 13.3.2). Finally, Kamae et al. (2016) suggested that the relationship between lower tropospheric mixing and climate sensitivity is model-dependent, as strong correlations between the Sherwood metrics and climate sensitivity were seen in about half the PPEs in their larger multiphysics ensemble. However, PPE ensemble members often exhibit unrealistic feedbacks (e.g., Joshi et al., 2010), reflecting mean-state errors that are typically absent in Coupled Model Intercomparison Project (CMIP)-model runs because of model tuning. Thus, we feel these PPE results should be interpreted with caution, and that the shallow-mixing mechanism described above cannot be definitely ruled out.

There are other factors that might alter the relationship between shallow mixing and the low-cloud feedback. For example, the effects of mixing near the top of the boundary layer on low clouds can be modified by changes in latent heat fluxes and in radiative cooling (Bretherton, 2015; Schneider et al., 2019; Vial et al., 2016). Enhanced dehydration of the boundary layer strengthens the surface latent heat flux, which damps the reduction in low clouds. At the same time, low-cloud reductions stabilize the lower troposphere by decreasing the cloud-top radiative cooling, which in turn decreases the surface latent heat flux and induces further low-cloud reductions. The relative importance of low-cloud mixing versus radiative cooling, and the resulting sign of the latent heat flux response, depends on the convective parameterization (Vial et al., 2016). A convection scheme in which the surface latent heat flux is strongly coupled to low-cloud radiative cooling will have a higher sensitivity of low-cloud fractions to convective mixing parameters, and a stronger low-cloud feedback in response to surface warming. But if radiative cooling is less dominant the latent heat flux may increase with warming, weakening the low-cloud response and reducing the model's climate sensitivity.

Thus, a link between convective mixing and climate sensitivity (or between ϵ_b for low clouds and climate sensitivity) is plausible, but the modeling evidence is inconclusive and the strength of the link is uncertain. The relationship between shallow entrainment rates and climate sensitivity is evidently model-dependent, as is the relative importance of entrainment rates versus other parameters in models' convection and microphysics schemes. Klocke et al. (2011) found that the shallow entrainment rate was the strongest control on climate sensitivity in their PPE and that the rate of autoconversion had a negligible effect, while Tomassini et al. (2015) found that the autoconversion rate exerted a strong control on climate sensitivity and the shallow entrainment rate had a weak effect in their PPE. Putting these model results together, and given the complex mechanisms which govern how shallow entrainment rates affect low-cloud cover, the concept of precipitation efficiency as defined here may not be the best way of framing the link between shallow mixing and climate sensitivity.

13.5. PRECIPITATION EFFICIENCY OF EXTRATROPICAL CLOUDS

Investigations of the relationship between precipitation efficiency and climate sensitivity have mostly focused on tropical clouds, which are often identified as the leading source of uncertainty in Earth's climate sensitivity. Yet, the high climate sensitivities of many of the newest CMIP6 models have been attributed to a reduction of strongly negative extratropical cloud feedbacks compared to CMIP5, particularly over the Southern Ocean (Zelinka et al., 2020), suggesting that the precipitation efficiency of extratropical clouds merits more study. Moreover, while the connection between the precipitation efficiency of extratropical clouds and climate sensitivity has not been investigated explicitly, an inverse relationship between the present-day ϵ_{cm} of extratropical clouds and climate sensitivity has been proposed, mediated by the cloud-phase feedback mentioned in section 13.3.2 (see also Chapter 6: Extratropical Cloud Feedbacks).

The cloud-microphysics precipitation efficiency of extratropical clouds is expected to decrease with warming because of decreases in conversion efficiency associated with the transition from mostly ice clouds to mostly liquid water clouds. The changes in conversion efficiency win out over increases in sedimentation efficiency due to the transition from snow to rain as the dominant form of precipitation (Kirshbaum & Smith, 2008; Storelvmo et al., 2015) [We hypothesize that changes in sedimentation efficiency likely become more important at warmer temperatures, when precipitation forms higher in the atmosphere and differences in fall speed lead to larger changes in re-evaporation. We also note that Eidhammer et al. (2018) found that the precipitation efficiency of orographic precipitation decreases with surface temperature in high-resolution simulations of a region in the Colorado Rockies, and attributed this to decreases in vertical velocities. However, they defined precipitation efficiency as the drying ratio, which is different from our definitions based on microphysical properties.]. Cloud-phase changes also lead to a negative cloud optical depth feedback as liquid clouds with higher optical depths replace ice clouds (see Chapter 6: Extratropical Cloud Feedbacks), which may be a substantial effect: Mitchell et al. (1989) found that the cloud-phase feedback can alter a model's climate sensitivity by a factor of two (see also Ceppi et al., 2017; Li & Le Treut, 1992; McCoy et al., 2015; Mauritsen & Roeckner, 2020; Storelvmo et al., 2015). Hence, models with higher initial ice cloud fractions may experience larger reductions in precipitation efficiency driven by phase changes, and stronger negative cloud optical depth feedbacks under warming. This suggests that models with higher present-day extratropical precipitation efficiencies could have lower climate sensitivities.

Zelinka et al. (2020) noted that one explanation for the weaker extratropical cloud feedbacks in CMIP6 is the increased presence of mean-state supercooled liquid water in mixed-phase clouds in many models. More supercooled water means smaller increases in extratropical cloud optical depth and cloud lifetime compared to models which start with more cloud ice initially. Although Zelinka et al. did not calculate ϵ for extratropical clouds explicitly, it is plausible that the high sensitivity CMIP6 models exhibit smaller decreases in extratropical cloud microphysics precipitation efficiency with warming.

13.6. CONCLUDING REMARKS

The general notion of precipitation efficiency is broad, and it can be defined in several different ways, each of which can be related to various cloud feedback mechanisms. Moreover, the relationships between precipitation efficiency and cloud feedbacks are highly model-dependent, and parameters which are important in one model may be less important in another. This is perhaps best illustrated by the multiphysics, multiparameter ensembles of Kamae et al. (2016), in which some model configurations showed strong relationships between lower tropospheric mixing metrics and climate sensitivity, while others did not. Similarly, Klocke et al. (2011), Zhao (2014), and Tomassini et al. (2015) all demonstrated strong relationships between present-day precipitation efficiency and climate sensitivity, but used different parameter variations to establish these relationships. This model-dependence reflects the fact that precipitation efficiency is a bulk metric for the combined effects of the convective and microphysical processes it represents, and comparing the precipitation efficiencies of models can be deceptive if different processes control ϵ.

Nevertheless, a few robust results do emerge from the literature. For example, there is strong evidence that the cloud microphysics precipitation efficiency of tropical clouds increases with warming, though bulk precipitation efficiency and the cloud microphysics precipitation efficiency of extratropical clouds could decrease. After tracing the various ways in which precipitation efficiency has been connected to climate sensitivity, we also propose that there are six mechanisms by which changes in ϵ could alter the planetary radiative balance (see Fig. 13.5):

1. Increased precipitation efficiency could lead to a reduction in anvil cloud cover. This would increase OLR but also lower albedo, such that the sign of the net feedback is unclear.
2. Increased precipitation efficiency could cause thinning of cloud anvils, producing a positive cloud optical depth feedback.
3. Increased convective organization could produce decreases in cloud cover generally, as well as decreases in

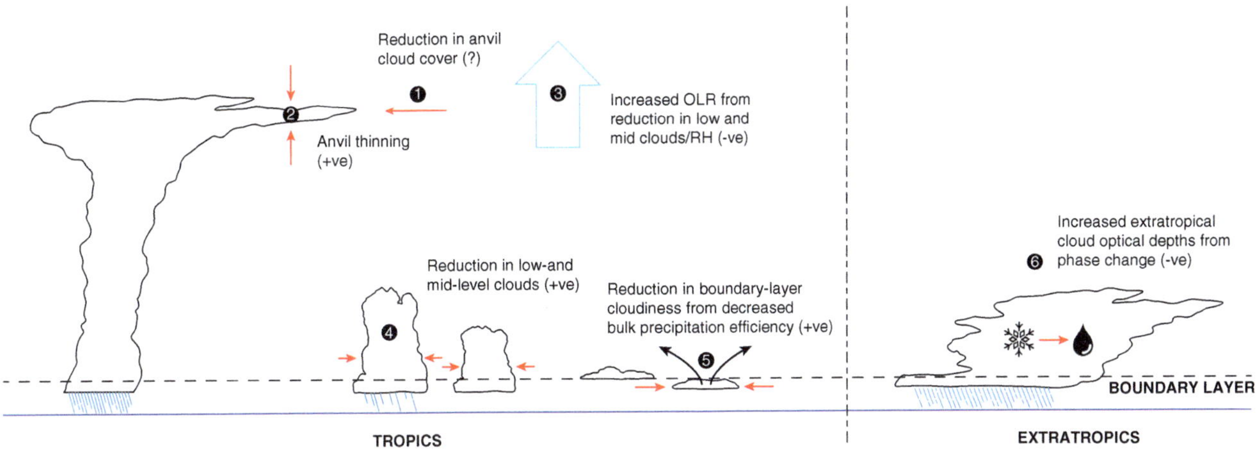

Figure 13.5 Schematic illustration of the six mechanisms by which changes in precipitation efficiency can alter climate sensitivity, as well as the hypothesized signs of their effect on the net climate feedback.

relative humidity, such that OLR increases with warming. This could constitute a negative climate feedback, depending on the response of shallow clouds outside the convective area.

4. Increased microphysical precipitation efficiency could lead to reductions in low- and mid-level cloud cover (independent of changes in convective organization), a positive feedback on warming.

5. Increased demand on boundary-layer moisture due to decreasing bulk precipitation efficiency via shallow mixing could reduce boundary-layer cloudiness, producing a positive feedback.

6. Decreased midlatitude precipitation efficiency could be associated with increased extratropical cloud optical depths, a negative midlatitude cloud feedback.

Mechanisms (1) and (3) correspond most closely to the original "Iris" idea of Lindzen et al. (2001), but (1) is unlikely to produce a strong net radiative feedback while (3) appears to be more of a dynamical than a microphysical mechanism. Changes in anvil cloud amount or tropospheric relative humidity driven by changes in precipitation efficiency are unlikely to be capable of large changes to the planetary radiative balance, leaving changes in other cloud types as the source of changes in sensitivity. Mechanisms (2), (4), (5), and (6) have all been reported in GCMs, and merit further investigation into how they vary across models and to better understand the magnitude of their effects. For example, evidence that increases in precipitation efficiency could result in thinning of cloud anvils (mechanism (2)) comes from a single modeling study (Li et al., 2019). Improving understanding of the precipitation efficiency of extratropical clouds also seems particularly urgent in light of the role of extratropical clouds in the high climate sensitivities of CMIP6 models.

Another important question is clarifying whether there is a relationship between present-day precipitation efficiency and climate sensitivity. A number of model perturbation studies have found inverse relations between present-day precipitation efficiency and climate sensitivity, with the possible exception of modeling studies varying shallow and deep entrainment rates (though these have not actually documented how ϵ changes as parameters are varied). Such a relationship would presumably work through changes in efficiency with warming following one or more of the six mechanisms listed above. A robust relationship in models between present-day precipitation efficiency and climate sensitivity would suggest the possibility of emergent constraints on climate sensitivity, but the link between present-day ϵ and $\Delta\epsilon$ would need to be strengthened before precipitation efficiency could be used in this way, and measuring precipitation efficiency, particularly over large scales and at the required accuracy, presents many challenges.

Despite the issues around model dependence, the various notions of precipitation efficiency clearly play a central role in many types of likely cloud feedback. The concept of precipitation efficiency is also a useful framework for connecting detailed process studies to emergent properties of climate models. This can be helpful in model development and evaluation, while also suggesting new ways of engineering specific climates to explore hypotheses and to investigate observational constraints. Thus, we believe that further study of the impacts of precipitation efficiency on clouds and feedbacks is called for, particularly to refine our understanding of the mechanisms linking changes in precipitation efficiency to climate sensitivity and to provide more observational data with which to validate model results.

ACKNOWLEDGMENTS

We thank two reviewers for close readings and helpful comments which improved this review. Nicholas J. Lutsko was supported by the NOAA Climate Program Office's Modeling, Analysis, Predictions, and Projections program through grant NA20OAR4310387. We thank Gillian Shaffer for help with Figure 13.5. Steven C. Sherwood was supported by the Australian Research Council FL150100035.

REFERENCES

Andrae, M. O., Rosenfeld, D., Artaxo, P., Costa, A. A., Frank, G. P., Longo, K. M., & Silva-Dias, M. A. F. (2004). Smoking rain clouds over the Amazon. *Science*, *303*, 1337–1342.

Bender, F. A.-M. (2008). A note on the effect of GCM tuning on climate sensitivity. *Environmental Research Letters*, *3*(1), 014001.

Betts, A. K., & Harshvardhan. (1987). Thermodynamic constraint on the cloud liquid water feedback in climate models. *Journal of Geophysical Research: Atmospheres*, *92*(D7), 8483–8485.

Bony, S., Semie, A., Kramer, R. J., Soden, B., Tompkins, A. M., & Emanuel, K. A. (2016). Observed modulation of the tropical radiation budget by deep convective organization and lower- tropospheric stability. *AGU Advances*, *1*(3), 1–15.

Bony, S., Stevens, B., Coppin, D., Becker, T., Reed, K. A., Voigt, A., & Medeiros, B. (2016). Thermodynamic control of anvil cloud amount. *Proceedings of the National Academy of Sciences*, *113*(32), 8927–8932.

Bretherton, C. S. (2015). Insights into low-latitude cloud feedbacks from high-resolution models. *Philosophical Transactions of the Royal Society A: Mathematical, Physical and Engineering Sciences*, *373*(2054), 20140415.

Brient, F., Schneider, T., Tan, Z., Bony, S., Qu, X., & Hall, A. (2016). Shallowness of tropical low clouds as a predictor of climate models' response to warming. *Climate Dynamics*, *47*(1–2), 433–449.

Ceppi, P., Brient, F., Zelinka, M. D., & Hartmann, D. L. (2017). Cloud feedback mechanisms and their representation in global climate models: Cloud feedback mechanisms and their representation in GCMs. *Wiley Interdisciplinary Reviews: Climate Change*, *8*(4), e465.

Cess, R. D., & Potter, G. L. (1988). A methodology for understanding and intercomparing atmospheric climate feedback processes in general circulation models. *Journal of Geophysical Research*, *93*, 8305–8314.

Eidhammer, T., Grubišić, V., Rasmussen, R., & Ikdea, K. (2018). Winter precipitation efficiency of mountain ranges in the Colorado Rockies under climate change. *Journal of Geophysical Research: Atmospheres*, *123*(5), 2573–2590.

Emanuel, K. A. (1987). An air-sea interaction model of intraseasonal oscillations in the tropics. *Journal of the Atmospheric Sciences*, *44*(23), 2324–2340.

Freud, E., & Rosenfeld, D. (2012). Linear relation between convective cloud drop number concentration and depth for rain initiation. *Journal of Geophysical Research-Atmospheres*, *117*(D02207).

Hohenegger, C., & Jakob, C. (2020). A relationship between ITCZ organization and subtropical humidity. *Geophysical Research Letters*, *47*(16), e2020GL088515.

Joshi, M. M., Webb, M. J., Maycock, A. C., & Collins, M. (2010). Stratospheric water vapour and high climate sensitivity in a version of the HadSM3 climate model. *Atmospheric Chemistry and Physics*, *10*(15), 7161–7167.

Kamae, Y., Shiogama, H., Watanabe, M., Ogura, T., Yokohata, T., & Kimoto, M. (2016). Lower-tropospheric mixing as a constraint on cloud feedback in a multiparameter multiphysics ensemble. *Journal of Climate*, *29*(17), 6259–6275.

Khairoutdinov, M. F., & Randall, D. A. (2003). Cloud resolving modeling of the ARM summer 1997 IOP: Model formulation, results, uncertainties, and sensitivities. *Journal of the Atmospheric Sciences*, *60*(23), 607–625.

Kirshbaum, D. J., & Smith, R. B. (2008). Temperature and moist-stability effects on midlatitude orographic precipitation. *Quarterly Journal of the Royal Meteorological Society*, *134*(634), 1183–1199.

Klocke, D., Pincus, R., & Quaas, J. (2011). On constraining estimates of climate sensitivity with present-day observations through model weighting. *Journal of Climate*, *24*(23), 6092–6099.

Langhans, W., Yeo, K., & Romps, D. M. (2015). Lagrangian investigation of the precipitation efficiency of convective clouds. *Journal of the Atmospheric Sciences*, *72*(3), 1045–1062.

Lau, K. M., & Wu, H. T. (2003). Warm rain processes over tropical oceans and climate implications. *Geophysical Research Letters*, *30*(3), 2290.

Li, R. L., Storelvmo, T., Fedorov, A. V., & Choi, Y.-S. (2019). A positive iris feedback: Insights from climate simulations with temperature-sensitive cloud–rain conversion. *Journal of Climate*, *32*(16), 5305–5324.

Li, Z.-X., & Le Treut, H. (1992). Cloud-radiation feedbacks in a general circulation model and their dependence on cloud modelling assumptions. *Climate Dynamics*, *7*(3), 133–139.

Lindzen, R. S., Chou, M. D., & Hou, A. U. (2001). Does the Earth have an adaptive infrared iris? *Bulletin of the American Mereorological Society*, *82*(23), 417–432.

Lopez, M. A., Hartmann, D. L., Blossey, P. N., Wood, R., Bretherton, C. S., & Kubar, T. L. (2009). A test of the simulation of tropical convective cloudiness by a cloud resolving model. *Journal of Climate*, *22*(11), 2834–2849.

Lutsko, N. J., & Cronin, T. W. (2018). Increase in precipitation efficiency with surface warming in radiative-convective equilibrium. *Journal of Advances in Modeling Earth Systems*, *10*(11), 2992–3010. doi: 10.1029/2018MS001482

Mauritsen, T., & Roeckner, E. (2020). Tuning the MPI-ESM1.2 global climate model to improve the match with instrumental record warming by lowering its climate sensitivity. *Journal of Advances in Modeling Earth Systems*, *12*(5), e2019MS002037.

Mauritsen, T., & Stevens, B. (2015). Missing iris effect as a possible cause of muted hydrological change and high climate sensitivity in models. *Nature Geoscience*, *8*(346), 346–351.

Mauritsen, T., Stevens, B., Roeckner, E., Crueger, T., Esch, M., Giorgetta, M. et al. (2012). Tuning the climate of a global model. *Journal of Advances in Modeling Earth Systems*, *4*(3), 1–18.

McCoy, D. T., Hartmann, D. L., Zelinka, M. D., Ceppi, P., & Grosvenor, D. P. (2015). Mixed-phase cloud physics and Southern Ocean cloud feedback in climate models. *Journal of Geophysical Research: Atmospheres*, *120*(18), 9539–9554.

Mitchell, J. F. B., Senior, C. A., & Ingram, W. J. (1989). C02 and climate: A missing feedback? *Nature*, *341*(6238), 132–134.

Morrison, H., Curry, J. A., & Khvorostyanov, V. I. (2005). A new double-moment micro-physics parameterization for application in cloud and climate models. Part I: Description. *Journal of the Atmospheric Sciences*, *62*(23), 1665–1677.

Muller, C. J., O'Gorman, P. A., & Back, L. E. (2011). Intensification of precipitation extremes with warming in a cloud resolving model. *Journal of Climate*, *24*(23), 2784–2800.

Narsey, S., Jakob, C., Singh, M. S., Bergemann, M., Louf, V., Protat, A., & Williams, C. (2019). Convective precipitation efficiency observed in the tropics. *Geophysical Research Letters*, *46*(22), 13574–13583.

Rieck, M., Nuijens, L., & Stevens, B. (2012). Marine boundary layer cloud feedbacks in a constant relative humidity atmosphere. *Journal of the Atmospheric Sciences*, *69*(8), 2538–2550.

Romps, D. M. (2014). An analytical model for tropical relative humidity. *Journal of Climate*, *27*(23), 7432–7449.

Rosenfeld, D. (2000). Suppression of rain and snow by urban and industrial air pollution. *Science*, *287*, 1793–1796.

Rosenfeld, D., & Lensky, I. M. (1998). Satellite-based insights into precipitation formation processes in continental and maritime convective clouds. *Bulletin of the American Meteorological Society*, *79*, 2457–2476.

Schneider, T., Kaul, C. M., & Pressel, K. G. (2019). Possible climate transitions from breakup of stratocumulus decks under global warming. *Nature Geoscience*, *12*, 163–167.

Sherwood, S. C., Bony, S., & Dufresne, J.-L. (2014). Spread in model climate sensitivity traced to atmospheric convective mixing. *Nature*, *505*, 37–42.

Sherwood, S. C., & Meyer, C. L. (2006). The general circulation and robust relative humidity. *Journal of Climate*, *19*(24), 6278–6290.

Sherwood, S. C., Roca, R., Weckwerth, T. M., & Andronova, N. G. (2010). Tropospheric water vapor, convection, and climate. *Reviews of Geophysics*, *48*(2).

Sherwood, S. C., Webb, M. J., Annan, J. D., Armour, K. C., Forster, P. M., Hargreaves, J. C. et al. (2020). An assessment of Earth's climate sensitivity using multiple lines of evidence. *Reviews of Geophysics*, e2019RG000678.

Singh, M. S., & O'Gorman, P. A. (2014). Influence of microphysics on the scaling of precipitation extremes with temperature. *Geophysical Research Letters*, *41*(16), 6037–6044.

Stainforth, D. A., Aina, T., Christensen, C., Collins, M., Faull, N., Frame, D. J. et al. (2005). Uncertainty in predictions of the climate response to rising levels of greenhouse gases. *Nature*, *433*(7024), 403–406.

Stevens, B. (2007). On the growth of layers of nonprecipitating cumulus convection. *Journal of the Atmospheric Sciences*, *64*(8), 2916–2931.

Storelvmo, T., Tan, I., & Korolev, A. V. (2015). Cloud phase changes induced by CO2 warming—a powerful yet poorly constrained cloud-climate feedback. *Current Climate Change Reports*, *1*(4), 288–296.

Sui, C. H., Li, X., Yang, M. J., & Huang, H. L. (2007). On the definition of precipitation efficiency. *Journal of the Atmospheric Sciences*, *64*(23), 4506–4513.

Sui, C.-H., Satoh, M., & Suzuki, K. (2020). Precipitation efficiency and its role in cloudradiative feedbacks to climate variability. *Journal of the Meteorological Society of Japan. Ser. II*, *98*(2), 261–282.

Tomassini, L., Voigt, A., & Stevens, B. (2015). On the connection between tropical circulation, convective mixing, and climate sensitivity. *Quarterly Journal of the Royal Meteorological Society*, *9*(141), 1404–1416.

Vial, J., Bony, S., Dufresne, J.-L., & Roehrig, R. (2016). Coupling between lower-tropospheric convective mixing and low-level clouds: Physical mechanisms and dependence on convection scheme. *Journal of Advances in Modeling Earth Systems*, *8*(4), 1892–1911.

Williams, I. N., & Pierrehumbert, R. T. (2017). Observational evidence against strongly stabilizing tropical cloud feedbacks. *Geophysical Research Letters*, *44*(3).

Wing, A. A., & Emanuel, K. A. (2014). Physical mechanisms controlling self-aggregation of convection in idealized numerical modeling simulations. *Journal of Advances in Modeling Earth Systems*, *6*(1), 59–74.

Wing, A. A., Stauffer, C. L., Becker, T., Reed, K. A., Ahn, M.-S., Arnold, N. P. et al. (2020). Clouds and convective self-aggregation in a multimodel ensemble of radiative-convective equilibrium simulations. *Journal of Advances in Modeling Earth Systems*, *12*(9), e2020MS002138.

Yano, J. I., & Emanuel, K. A. (1991). An improved model of the equatorial troposphere and its coupling with the stratosphere. *Journal of the Atmospheric Sciences*, *48*(23), 377–389.

Zelinka, M. D., Myers, T. A., McCoy, D. T., Po-Chedley, S., Caldwell, P. M., Ceppi, P. et al. (2020). Causes of higher climate sensitivity in CMIP6 models. *Geophysical Research Letters*, *47*(1), e2019GL085782.

Zhang, M., Bretherton, C. S., Blossey, P. N., Austin, P. H., Bacmeister, J. T., Bony, S. et al. (2013). CGILS: Results from the first phase of an international project to understand the physical mechanisms of low cloud feedbacks in single column models. *Journal of Advances in Modeling Earth Systems*, *5*(4).

Zhao, M. (2014). An investigation of the connections among convection, clouds, and climate sensitivity in a global climate model. *Journal of Climate*, *68*(27), 1845–1862.

Zhao, M., Golaz, J.-C., Held, I. M., Lin, S.-J., Ming, Y., Ginoux, P. et al. (2016). Uncertainty in model climate sensitivity traced to representations of cumulus precipitation microphysics. *Journal of Climate*, *68*(29), 543–560.

14

Observed Scaling of Precipitation Extremes With Surface Temperature and Convective Available Potential Energy

Wenhao Dong[1,2,3] **and Yanluan Lin**[1]

ABSTRACT

Changes in precipitation extremes remain a key uncertainty as the climate warms. Improved understanding of their evolution is crucial for effective water management. A number of studies have demonstrated various scaling relationships between precipitation extremes and several different environmental variables. In this chapter, we review recent important advances in two of these relationships primarily based on observations: The scaling of precipitation extremes with surface temperature (both air temperature and dew point temperature) and convective available potential energy (CAPE). Two up-to-date global daily datasets are also used to provide a further check on the generality of earlier findings. By applying known scaling relationships, we quantify the impacts of these two factors on precipitation extremes. Results show that both of them play important roles in precipitation extremes, but their impacts vary across regions and timescales, highlighting the challenges of constructing global relationships to explain the changing nature of precipitation extremes.

14.1. INTRODUCTION

Precipitation extremes, referred to in particular as large precipitation events in terms of intensity or accumulation in this study, have substantial impacts on human activity, agriculture, and water resources. These extreme events also play a major role in many hydrological catastrophes and are associated with high economic costs (Ahern et al., 2005; Easterling et al., 2000; Hallegatte et al., 2013; Knapp et al., 2008; Wasko & Sharma, 2017; Westra et al., 2014; Yin et al., 2018). For instance, precipitation-induced flooding is among the top-ranking destructive natural hazards in terms of economic losses worldwide, resulting in substantial damages to infrastructure, ecosystems, and loss of life. Given the practical significance of these extreme events, a great deal of effort has been made by multiple scientific communities to assess whether and to what extent they are changing in frequency and intensity (O'Gorman & Schneider, 2009a; Trenberth, 2011; Westra et al., 2013). Many regions around the world have experienced an increase in precipitation extremes on various timescales in the past several decades (Donat et al., 2016; Min et al., 2011). The increasing tendency, although of varying magnitudes in different geographical locations, is expected to continue as temperature increases (Alexander et al., 2006; Allan & Soden, 2008; Bao et al., 2017; Berg et al., 2013; Fischer & Knutti, 2014; Groisman et al., 2005; Kao & Ganguly, 2011; Kharin et al., 2007; Meehl et al., 2000; O'Gorman, 2014; O'Gorman & Schneider, 2009a; Pall et al., 2007). Therefore, understanding the evolution of these extreme events is crucial for effective water management and mitigating the impacts of hydrological catastrophes.

[1] *Department of Earth System Science, Ministry of Education Key Laboratory for Earth System Modeling, Institute for Global Change Studies, Tsinghua University, Beijing, China*

[2] *Now at NOAA Geophysical Fluid Dynamics Laboratory, Princeton, NJ, USA*

[3] *Cooperative Programs for the Advancement of Earth System Science, University Corporation for Atmospheric Research, Boulder, CO, USA*

Clouds and Their Climatic Impacts: Radiation, Circulation, and Precipitation, Geophysical Monograph 281, First Edition.
Edited by Sylvia C. Sullivan and Corinna Hoose.

DOI: 10.1002/9781119700357.ch14

Benefitting from extensive researches using theoretical studies, station observations, and model simulations (including idealized cloud-resolving models [CRMs] and global/regional climate models [GCMs/RCMs]), significant advances in our understanding of the response of precipitation extremes to a warming climate have been achieved (Muller & Takayabu, 2020, and references therein). Several scaling relationships have been established between precipitation extremes and various measures of temperature, including surface air temperature, surface dew point temperature, tropospheric air temperature, and atmospheric dew point temperature (Ali & Mishra, 2017; Allen & Ingram, 2002; Bui et al., 2019; Lenderink & Van Meijgaard, 2010; Martinkova & Kysely, 2020; Mishra et al., 2012; Roderick et al., 2020), and other environmental properties, such as precipitation efficiency, convective available potential energy (CAPE), vertical wind velocity, and wind shear (Muller et al., 2011; O'Gorman, 2015; Pfahl et al., 2017) (see also Chapter 13: Precipitation Efficiency and Climate Sensitivity). CAPE is an energy-based measure of atmospheric potential instability, which estimates the theoretical maximum velocity that a positive buoyant air parcel could acquire through adiabatic ascent (DeMott & Randall, 2004; North & Erukhimova, 2009).

In this chapter, we focus on the scaling relationship of precipitation extremes with surface temperature, including air temperature (*P_e–Ta*) and dew point temperature (*P_e–Td*), as well as convective available potential energy (*P_e–CAPE*) based on observational records. We do not attempt to summarize all the existing literature on these topics since similar works have been carried out over different regions. Instead, we focus more on works which have come up with different results and reasonable interpretations. A large number of studies have demonstrated a close relationship between precipitation extremes and surface temperature linked to moisture availability (Allen & Ingram, 2002; Held & Soden, 2006; Kharin et al., 2007; Lenderink et al., 2011; Lenderink & Van Meijgaard, 2008, 2010; O'Gorman & Schneider, 2009b; Pall et al., 2007). CAPE is often used to estimate the vertical velocity of convective updrafts. An increase in CAPE indicates enhanced atmospheric instability or the positive buoyancy that would be experienced by a lifted air parcel, which could complicate the moisture-driven intensification of precipitation extremes (DeMott & Randall, 2004; Lepore et al., 2015; Muller & Takayabu, 2020; O'Gorman, 2015; Seeley & Romps, 2015; Singh & O'Gorman, 2013). Exploring these relationships would be an important step to understand the evolution of precipitation extremes and to constrain its future changes. For each relationship, results from pioneering studies are compared and summarized. Considering that most of the previous observational studies are carried out on a regional or local scale, we complement it by analyzing two up-to-date global sets of daily station records (see section 14.3) to provide a further check on the universality of earlier findings. The chapter is structured as follows: we commence by the definition of precipitation extremes and scaling methodology in section 14.2, followed by a description of the data used in this study. We then review the scaling relationship of *P_e–Ta* and *P_e–Td* in section 14.4 and *P_e–CAPE* in section 14.5, respectively. Discussion and summary are presented in section 14.6.

14.2. DEFINITION OF PRECIPITATION EXTREMES AND SCALING METHODOLOGY

How to define precipitation extremes is not a trivial thing. It can introduce considerable complexity and influence the results significantly (Pendergrass et al., 2017). Usually, a definition of precipitation extremes needs to take three aspects into consideration: a metric (e.g. an absolute threshold, a percentile, a return period, etc.), a timescale (the length of precipitation accumulation), and a spatial scale (e.g. point observations, domain-average, contiguous rain area, etc.). Different combinations of these aspects correspond to different types of weather systems across a wide range of spatial and temporal scales, characterized by different impacts. For instance, flash floods are typically associated with short but intense rainstorms at small spatial scales, while riverine floods are generally caused by a long-lasting rainfall system over a larger area. As a result, different extreme precipitation definitions have been proposed, leading to a lack of consistency and comparability across studies. Most commonly, a higher percentile, such as 99th or 99.9th percentile (i.e., the top 1% or 0.1% events), is chosen at a specific location to emphasize the extreme characteristics of the precipitation events (see examples in Table 14.1). In addition, as suggested by previous studies, including non-precipitating events or snowfall records could affect the results significantly (Schleiss, 2018). Large uncertainties were found in snowfall observation, and the response of snowfall to warming is different from that of rainfall (Lute et al., 2015; O'Gorman, 2014, 2015). On account of these impacts, for the dataset used in this study, a precipitation event is defined conventionally with daily accumulated precipitation larger than 0.1 mm. Precipitation extremes are identified using the 99th percentile (hereafter called P_{99}) of all the precipitation events after removing snowfall records.

A binning method based on a single predictor variable is often used to calculate the *P_e*-related scaling. Take *P_e–Ta* scaling as an example, the precipitation events are binned into either equal width or equal sample size according to surface air temperature (Guo

Table 14.1 Summary of selected previous studies of the *P_e–Ta* scaling and the *P_e–Td* scaling in a chronological order. The references are included in the first column. T_{peak} is defined as the temperature at which the fitted scaling curve maximizes.

Region (References)	Temporal resolution	*P_e* definition (percentile)	Super-CC	Peak-like structure	T_{Peak}
Netherlands (Lenderink & Van Meijgaard, 2008)	Hourly/daily *P* & Daily *T*	99 & 99.9th	YES	NO	NA
Europe (Berg et al., 2009)	Daily *P* & *T*	99 & 99.9th	NO	YES	15 ~ 20 °C
Netherlands, Belgium, and Switzerland (Lenderink & Van Meijgaard, 2010)	Hourly/daily *P* & Daily *T*	99 & 99.9th	YES	NO	NA
Australia (Hardwick Jones et al., 2010)	Hourly *P* & Daily *T*	99th	YES	YES	20 ~ 26 °C
Hong Kong and the Netherland (Lenderink et al., 2011)	Hourly *P* & *T*	99 & 99.9th	YES	YES	24 °C
United States (Shaw et al., 2011)	Hourly *T* & *P*	99 & 99.9th	YES	NO	NA
Japan (Utsumi et al., 2011)	10-min *P* & *T*	99th	NO	YES	24 °C
United States (Mishra et al., 2012)	Hourly *P* & Daily *T*	99th	YES	YES	28 °C
Germany (Berg et al., 2013)	5-min *P* & Daily *T*	99th	YES	YES	22 °C
Canada (Panthou et al., 2014)	Hourly *P* & Daily *T*	95th	YES	YES	20 °C
United Kingdom (Blenkinsop et al., 2015)	Hourly *P* & Daily *T*	99th	YES	NO	NA
Mediterranean (Drobinski et al., 2016)	3-hour P & T	99th	NO	YES	15 ~ 18 °C
India (Ali & Mishra, 2018)	Daily *P* & T/T_d	95th	YES	NA	NA
Australia (Bao et al., 2017)	Daily *P* & *T*	99th	YES	NA	NA
South Korea (Park & Min, 2017)	Hourly *P* & Daily *T*	99th	YES	YES	24 °C
Global (Ali et al., 2018)	Daily *P* & T/T_d	95th	YES	NA	NA
Europe (Barbero et al., 2018)	Hourly *P* & Daily *T*	99th	YES	NA	NA
United States (Schleiss, 2018)	Sub-hourly *P* & T/T_d	99th	YES	YES	20 ~ 25 °C
Mediterranean (Pumo et al., 2019)	Hourly *P* & Daily *T*	99th	YES	YES	22 °C
Global (Zhang et al., 2019)	Daily *P* & T/T_d	99th	YES	NA	NA
China (Guo et al., 2020)	Hourly *P* & *T*	99th	YES	YES	24 °C
Six macro-regions (Moustakis et al., 2020)	Hourly *P* & Daily *T*	99th	YES	YES	Wide range
Tibetan Plateau (Yong et al., 2021)	Daily *P* & *T*	90–99.9th	YES	NA	NA
Six macro-regions (Ali et al., 2021)	Hourly *P* & Daily T_d	99th	YES	NA	NA
Global (Fowler et al., 2021)	Sub-daily *P* & *T*	99 & 99.9th	YES	YES	NA
United States (Wu & Wang, 2021)	Hourly *P* & Daily T_d	99th	YES	NA	NA

Note: P = precipitation; T = surface air temperature; T_d = surface dew point temperature.

et al., 2020; Hardwick Jones et al., 2010; Lenderink & Van Meijgaard, 2008; Lepore et al., 2015). Extreme precipitation percentiles are calculated for each bin to determine the scaling rate (α). Other methods, like the quantile regression approach, have also been used in the literature (Ali & Mishra, 2017; Pumo & Noto, 2021; Wasko & Sharma, 2014). Unlike regular linear regression which uses the least squares to calculate the conditional mean of a response variable, quantile regression can calculate any quantile of the response variable. It is an extension of linear regression and is found to be unbiased with sample size. A comparison of different scaling methodology based on the Global Summary of the Day (GSOD) dataset can be found in Ali et al. (2018). Scaling relationships of precipitation extremes with multiple variables have also been explored before (e.g., Lepore et al., 2015; Lepore et al., 2016), but they are not considered here. In this chapter, the *P_e*-related scaling is determined using a least squares linear relationship based on the binning approach with 20 equal sample sizes according to the predictor variables. A minimum of 100 samples is required in each bin (i.e., at least 2,000 effective pairs of data in total) to optimize statistical robustness by balancing the competing needs to reduce noise and to retain enough samples. The main conclusions presented here are qualitatively insensitive to the number of bins (i.e., 10 or 30 bins).

To conclude, for each station analyzed in this study, the precipitation extreme (*P_e*) is identified using the 99th percentile of all the precipitating days (daily precipitation larger than 0.1 mm) after removing snowfall records. While the *P_e–Ta* scaling is determined using a least squares linear relationship based on the binning approach with 20 equal sample sizes of surface air temperature. Only stations with more than 2,000 pairs of observations are analyzed. Similar analyses are performed for the dew point temperature (*P_e–Td* scaling) and the CAPE (*P_e–CAPE* scaling).

14.3. OBSERVATIONAL DATASET

Many efforts have been made to compile precipitation products on various timescales from global weather stations, which are the principal source providing direct measures of precipitation. Products like the Global Sub-Daily Rainfall (GSDR) dataset (Lewis et al., 2019), the GSOD dataset, and the Global Historical Climatology Network (GHCN) datasets (Menne et al., 2012) have been widely used. In this chapter, we utilize daily precipitation and surface air temperature from the GHCN-Daily dataset, which is collected at over 100,000 stations in 180 countries and territories (Menne et al., 2012). The record length and period of record vary largely but, on average, have a span of decades or more. The daily surface temperature in the GHCN dataset is represented by the average of daily maximum and minimum values. In addition, we also utilize daily precipitation, surface air temperature, and dew point temperature from the GSOD dataset, which is available for 13,110 stations global wide. The daily GSOD dataset is derived from the hourly observations archived in the Integrated Surface Hourly (ISH) dataset. The availability of dew point temperature in the GSOD dataset enables us to explore the *P_e–Td* scaling relationship. CAPE values are taken from version 2 of the Integrated Global Radiosonde Archive (IGRA), which contains over 2,700 globally distributed radiosonde stations (Durre et al., 2006; Durre et al., 2009). CAPE is calculated from radiosonde profiles as the vertical integral of the positive portion of the parcel buoyancy between the level of free convection and the level of neutral buoyancy. Bi-daily CAPE values (00 and 12 UTC), when available, are averaged to daily means for the subsequent analysis to reduce diurnal sampling biases.

Data from the GHCN, GSOD, and IGRA during 1950–2019 are first used to identify stations reporting data for more than 10 years. Since our aim is to establish the statistical relationship between precipitation extremes and surface temperature (*Ta* or *Td*) or CAPE, the continuity of these variables is not required. For the *P_e–Ta* and *P_e–Td* scaling relationships, we make use of the GSOD dataset considering the consistency in source data among the three variables and the fact that daily temperature variables calculated from hourly records could be more representative compared to the GHCN dataset, which is merely the average of two records, that is, the maximum and minimum daily temperature. Following the quality check procedure in Ali et al. (2018), for each GSOD station, we remove days with the daily precipitation record covered less than 24 hours. There are 7,491 stations available for the subsequent analysis. A similar *P_e–Ta* scaling relationship is obtained based on the GHCN dataset (figures not shown). While for the *P_e–CAPE* scaling relationship, we pair each radiosonde station with its nearest neighbor among GHCN stations. Though similar results are achieved if we pair the radiosonde station with the GSOD dataset (figures not shown), the GHCN dataset is chosen here because it shows more effective pairs of stations compared to the GSOD dataset. Any station pair with a distance larger than 1 km is excluded from the analysis. The 1 km criterion is justified as one can associate the recorded precipitation amount and temperature with the reported CAPE from IGRA sounding. The pairing procedure translates to 712 pairs of stations global wide. As shown in Figure 14.1, these stations, either from the GSOD dataset or the GHCN dataset, are reasonably well distributed over different continents, enabling a global analysis. On average, larger precipitation amounts and CAPE values are distributed over the deep tropical and coastal regions, while surface air temperature distribution shows a strong dependence on latitude regulated by the solar insolation (Figure 14.1). The dew point temperature has a similar distribution as surface air temperature, but it shows relatively larger values over the coastal region, which is indicative of the impacts of moisture from the surrounding oceans. Note the available number of stations is slightly different among these variables.

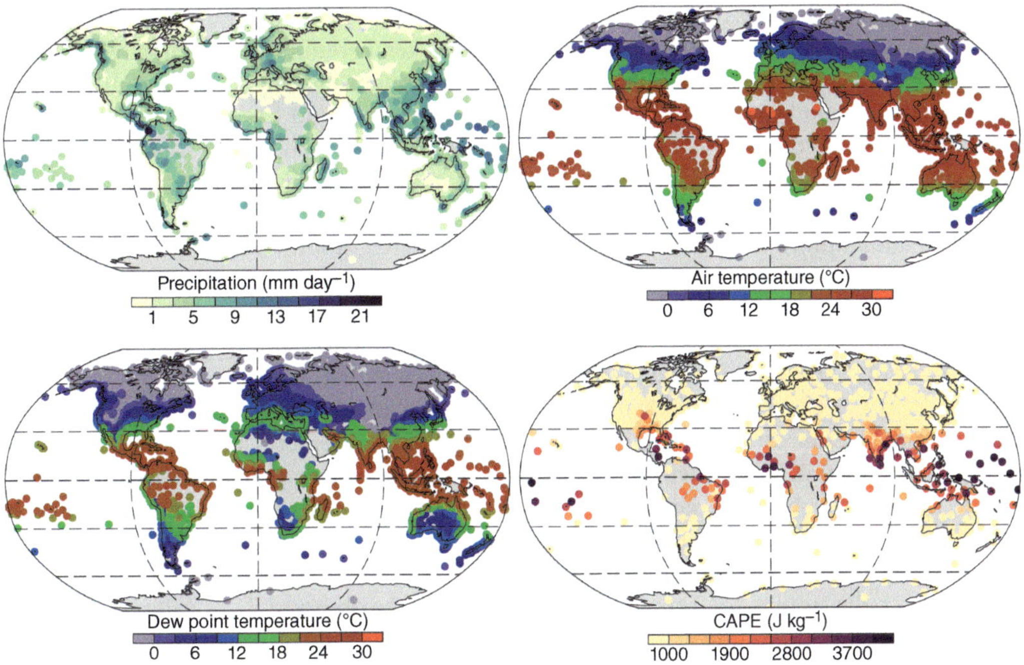

Figure 14.1 Climatological distributions of precipitation (*upper left*), surface air temperature (*upper right*), dew point temperature (*lower left*), and CAPE (*lower right*). The first three variables are based on the GSOD dataset, while the last one is based on the IGRA dataset after matching with the GHCN dataset.

14.4. SCALING OF PRECIPITATION EXTREMES WITH SURFACE TEMPERATURE

We focus on the scaling relationship between precipitation extremes and surface temperature (*P_e–Ta* and *P_e–Td*) in this section. According to the Clausius-Clapeyron (CC) equation, the response of precipitation to temperature is linked to changes in the atmosphere's moisture-holding capacity (Allen & Ingram, 2002; Lenderink & Van Meijgaard, 2008, 2010; Lenderink et al., 2011; Moustakis et al., 2020; O'Gorman & Schneider, 2009a). Under the assumption of constant relative humidity, CC relationship implies an increase in specific humidity at the rate of ~7% per degree of warming, a prediction roughly borne out by plenty of observational and modeling studies (Allen & Ingram, 2002; Held & Soden, 2006; Kharin et al., 2007; Muller et al., 2011; Pall et al., 2007; Romps, 2011; Sherwood et al., 2010). One may expect precipitation to increase at a similar rate as the water vapor holding capacity of overlying air. However, global mean precipitation is found to increase with warming at a rate of 2–3% °C^{-1} across multiple observations and climate model simulations, much slower than the CC scaling (Allen & Ingram, 2002; Attema et al., 2014; Emori & Brown, 2005; Held & Soden, 2006; Norris et al., 2019; Panthou et al., 2014; Pfahl et al., 2017; Utsumi et al., 2011). Unlike the mean precipitation, precipitation extremes show a much larger increasing rate with warming, albeit with significant variations across different geographic locations (Barbero et al., 2017; Chan et al., 2016; Drobinski et al., 2018; Guerreiro et al., 2018; Lenderink & Attema, 2015; Lenderink et al., 2011; Prein et al., 2017). Table 14.1 lists part of the previous studies investigating the *P_e–Ta* scaling rate (α) over different regions. Most of these studies are evaluated in Europe and Australia, with a few recent studies in North America, East Asia, and South Asia. Generally, the *P_e–Ta* scaling can be classified into three categories: a monotonic increasing scaling, a monotonic decreasing scaling, and a peak-like structure where P_{99} first increases with *Ta* until reaching a maximum at a threshold temperature and then declines with a further increase in *Ta* (Berg et al., 2009; Hardwick Jones et al., 2010). The third category is also known as a hook shape or a parabolic shape of the scaling, in which two regimes emerge: one at low temperatures with the scaling rate approach CC or super-CC (i.e., scaling rate larger than 7% °C^{-1}) rates while the other one at higher temperatures with sub-CC rates (i.e., scaling rate smaller than 7% °C^{-1}) or even negative rates. The temperature threshold, usually called peak-point temperature (T_{Peak}), is defined as the temperature at which the fitted curve maximizes. We calculate T_{Peak} by applying a locally weighted regression (LOWESS) smoothing method (Chan et al., 2016; Utsumi et al., 2011).

In the following analysis, we re-examine the *P_e–Ta* scaling for each station using the up-to-date GSOD dataset. The requirement of an effective sample size of 2,000 reduces the available number of stations to 5,227. As shown in Figure 14.2, more than 96% of the available stations report a value of α less than 7% °C^{-1}. Only a few stations (<5%) have α greater than the CC rate. Positive α is observed within midlatitudes to high latitudes over 81% of the total stations, while negative values (19%) are mainly observed in the tropics. This finding is consistent with previous studies on regional or local scales over tropical regions (Ali et al., 2018; Hardwick Jones et al., 2010; Maeda et al., 2012; Utsumi et al., 2011) as well as midlatitudes (Berg & Haerter, 2013; Berg et al., 2013; Lenderink et al., 2011; Lenderink & Van Meijgaard, 2008, 2010; Shaw et al., 2011). But obviously, latitude cannot solely explain the nature of this dependence, small-scale differences are noted over many regions. The negative scaling relationship over the tropics has been attributed to different processes. Most precipitation extremes in the tropics result from synoptic systems, which are usually accompanied by drops in temperature caused by colder air masses intrusion or evaporative cooling. While for higher temperature bins, moisture availability becomes the dominant limiting factor, the drier condition usually leads to fewer and less intense precipitation (Bao et al., 2017; Barbero et al., 2018; Lenderink et al., 2011, 2017; Lenderink & Van Meijgaard, 2011; Park & Min, 2017). Taken together, these two factors contribute to a negative scaling. Additionally, Ali et al. (2018) pointed out that the smaller seasonality of surface air temperature in the tropics also plays a role in the observed negative scaling. After accounting for the temperature seasonality, they found that most locations show a positive scaling relationship with surface air temperature.

In addition, there are 1,713 out of 5,227 stations (~33%) showing a peak-like structure. Among these stations, high-latitude stations or stations over mountainous regions generally have lower T_{Peak}, supporting the idea that T_{Peak} might be related to the local temperatures. On average, the observed T_{Peak} values fall within the range of 18–26 °C for North America, East Asia, and Australia, while a relatively smaller T_{Peak} of about 10–18 °C is noted over Europe, which is in agreement with previous studies

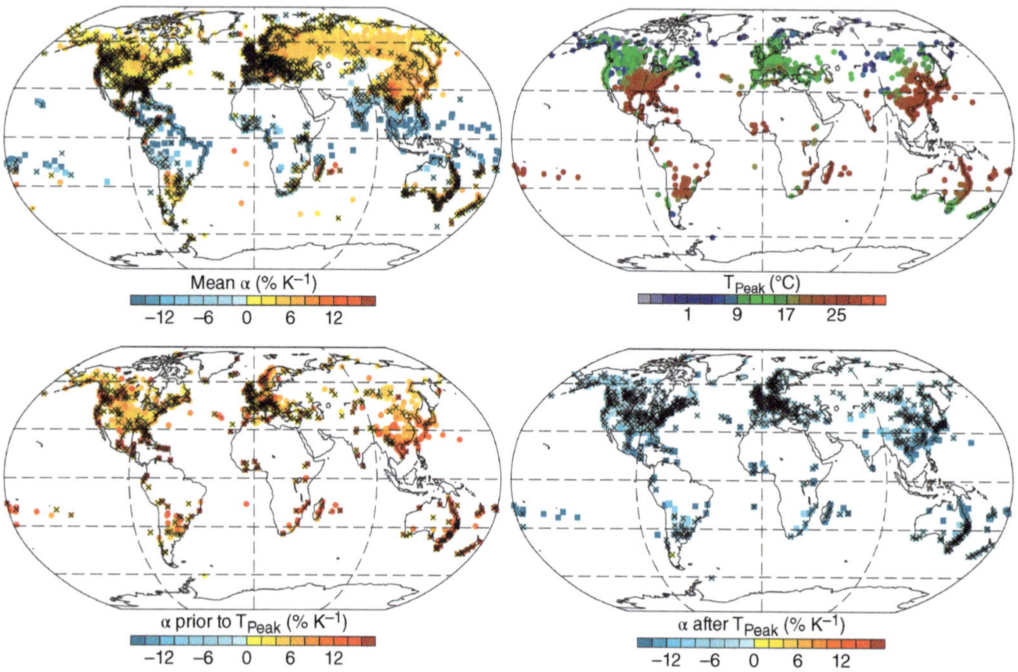

Figure 14.2 Global distribution of mean *P_e–Ta* scaling rate (*upper left*), peak-point temperatures (*upper right*), *P_e–Ta* scaling rate prior to the peak-point temperature (*lower left*), and *P_e–Ta* scaling rate after the peak-point temperature (*lower right*). For each station, the *P_e–Ta* scaling rate is determined using a least squares linear relationship based on the binning approach with 20 equal sample sizes of air temperature. The peak-like temperature is calculated by applying a locally weighted regression (LOWESS) smoothing method. The scaling rate prior to and after the peak-like temperature is calculated similarly as the *P_e–Ta* scaling rate but over different temperature ranges. Dots (squares) denote the positive (negative) scaling rates. Crosses in the *P_e–Ta* scaling rate distribution maps denote the least squares linear relationships that are not statistically significant ($P > 0.05$).

(Drobinski et al., 2016; Guo et al., 2020; Lenderink et al., 2011; Lepore et al., 2015; Mishra et al., 2012; Panthou et al., 2014; Park & Min, 2017; Wasko & Sharma, 2014). A number of arguments have been proposed to explain this peak-like structure, including a decrease in precipitation intensity (Berg et al., 2009; Hardwick Jones et al., 2010) and duration (Haerter & Berg, 2009; O'Gorman, 2012, 2015; Utsumi et al., 2011; Wasko et al., 2015; Wei et al., 2013) at higher temperatures, which is realized through the impact of humidity as discussed above. Several studies have found that relative humidity either flattens or decreases for surface temperature above ~25 °C (Hardwick Jones et al., 2010; Lenderink et al., 2011; Singh et al., 2019; Utsumi et al., 2011; Wasko et al., 2018).

To improve the accuracy of rainfall-temperature sensitivity calculations at higher temperatures, researchers have suggested using the dew point temperature instead of surface temperature. (Ali et al., 2018; Barbero et al., 2017; Lenderink & Van Meijgaard, 2010; Wasko et al., 2018; Zhang et al., 2019). Using dew point temperature instead of surface temperature in the assessment has been found to increase the consistency in scaling relationship across different temperature regimes (Ali et al., 2018; Barbero et al., 2018; Lenderink & Attema, 2015; Panthou et al., 2014; Wasko et al., 2015). This is evident in Figure 14.3 as one can find that there is a significant difference in scaling relationship with *Ta* and *Td*, particularly in the tropics. Positive scaling rates closer to the Clausius-Clapeyron (CC) relationship are obtained for the majority of stations when using dew point temperature, with the number of stations with negative scaling rates reduced from 19% to 9%. These contrasting results of scaling rate for *Ta* and *Td* are consistent with the previous studies analyzed at the regional and global scale (Ali & Mishra, 2017; Ali et al., 2018; Barbero et al., 2018; Lenderink & Van Meijigaard, 2010; Lenderink et al., 2011; Wasko et al., 2018). But similar to Ali et al. (2018), negative scaling rates of *P_e–Td* can still be observed in some deep tropical stations (Figure 14.3). And the peak-like structure is also evident in the *P_e–Td* scaling but with fewer stations (~17%) when compared to the *P_e–Ta* scaling. The value of T_{peak} based on the dew point temperature is close to that of surface air temperature in some of the tropical stations, suggesting a limited role of moisture in the formation of this peak-like structure in these stations. Instead, it could be partly related to the argument that the temperature measured during the occurrence of a rainfall event may not be the most relevant to calculate the scaling rate, as it may be affected by evaporative cooling or cold air advection associated with the synoptic systems (Bao et al., 2017). Some studies found negative scaling rates could be turned to positive ones in some tropical stations when temperatures 3 days prior to rainfall events are used (Ali & Mishra, 2017; Visser et al., 2020; Zhang et al., 2017). Other factors like larger scale dynamics and orography could also affect the *P_e–Ta* scaling (Moustakis et al., 2020).

The *P_e–Ta*/*P_e–Td* relationship also exhibits a unique features of super-CC scaling, which has been reported in several studies conducted across Europe, North America, Australia, and Hong Kong. This phenomenon is characterized by super-CC scaling at low to moderate temperature, followed by negative rates at moderate to high temperatures (Berg & Haerter, 2013; Berg et al., 2013; Blenkinsop et al., 2015; Drobinski et al., 2016; Lenderink et al., 2017; Lenderink & Van Meijgaard, 2008, 2010; Panthou et al., 2014; Westra et al., 2014; Wasko et al., 2015). Similar results can be seen over many stations with the temperature prior to the T_{Peak} based on the daily precipitation extremes as shown in Figures 14.2 and 14.3. This type of super-CC scaling is part of the peak-like structure as discussed above, which is commonly valid over certain temperature ranges. In addition, the existence of super-CC scaling was also found in higher frequency precipitation dataset (see Table 14.1). An example can be found in Fowler et al. (2021; their figure 2), based on a same dataset over the Netherlands, the scaling relationship can change from CC rates for daily precipitation extremes to a 2CC rate (~14% $°C^{-1}$) for 10-min precipitation extremes.

There has been debate about the reason for the reported super-CC relationship. Some studies suggest that there could be a positive feedback in which excessive latent heat release increases upward motions in clouds and enhances moisture convergences (Berg et al., 2013; Fowler et al., 2021; Lenderink et al., 2017; Lenderink & Van Meijgaard, 2008, 2010; Trenberth et al., 2003). But their exact role is uncertain because of the very small cloud-scale dynamics involved. A shift in the circulation may lead to either enhancement or suppression of rainfall over different regions, making the local assessment of changes in extremes challenging. In addition, the super-CC relationship was also found to be linked to changes in weather regimes with temperature, such as a change in rainfall type from lower intensity stratiform (large-scale frontal precipitation) to higher intensity convective precipitation (Fowler et al., 2021; Haerter & Berg, 2009). This argument suggests that super-CC scaling may be most prevalent in regions where both convective and large-scale rainfall events coexist. Other viewpoints, like the mixing of large-scale flow conditions that vary substantially between seasons (Berg et al., 2009; Zhang et al., 2017), may also influence the scaling. Readers are recommended to Fowler et al. (2021), O'Gorman (2015), and Westra et al. (2014) for detailed reviews about these arguments.

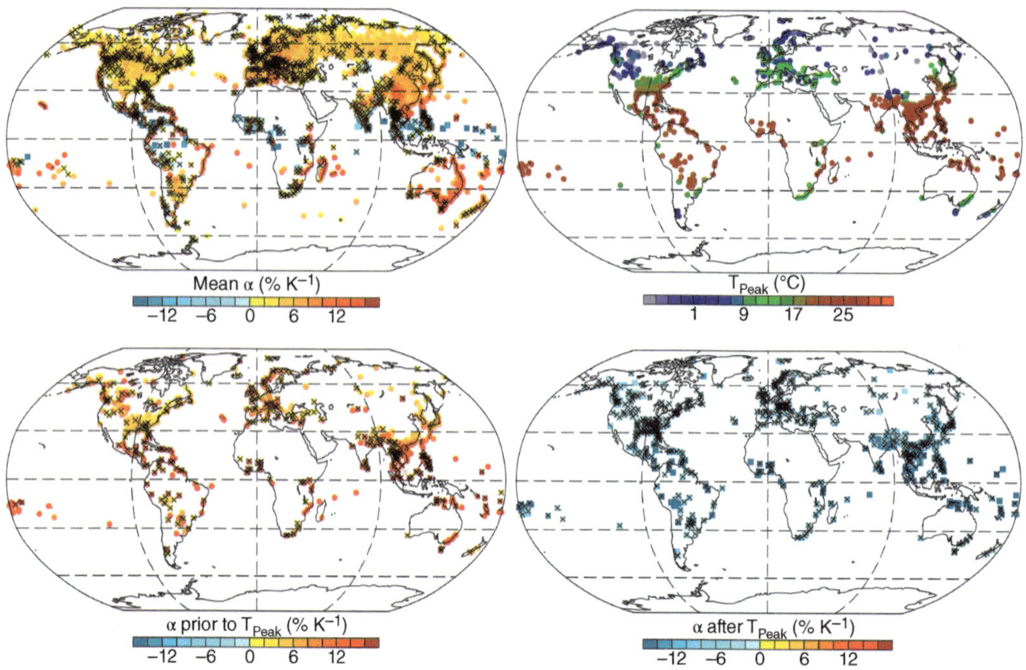

Figure 14.3 Similar as Figure 14.2, but for the *P_e–Td* scaling relationship.

14.5. SCALING OF PRECIPITATION EXTREMES WITH CAPE

This section focuses on the scaling relationship between precipitation extremes and CAPE (*P_e–CAPE*), which has received relatively less attention than *P_e–Ta* scaling. Observational studies on P_e–CAPE scaling are sparse due to the requirement of reliable temperature profiles, which are more difficult to obtain than those of surface variables. Early studies made use of reanalysis products to probe the *P_e–CAPE* scaling. Lepore et al. (2015) interpolated the 12-hour predicted CAPE field from ERA-Interim (Dee et al., 2011) to investigate its relationship with precipitation extremes. Some follow-up studies used radiosonde observation to calculate CAPE (Dong et al., 2019; Guo et al., 2020) or retrieve it from satellite observation (Gartzke et al., 2017; Murali Krishna et al., 2019). The ERA-Interim reanalysis product was found to substantially underestimate the sounding-based measurements of CAPE across Europe (Taszarek et al., 2018). But some studies found that rainfall intensity was better correlated with CAPE from ERA-Interim reanalysis than those computed from atmospheric soundings (Barbero et al., 2018; Lepore et al., 2015). Despite this, they have found ERA-Interim reanalysis to be more closely correlated with rainfall intensity than sounding-based measurements, arguing that it is more proximal to the rainfall measurements given its simultaneity with these station observations. Nevertheless, results based on these studies have been inconsistent.

Given that precipitation is linked to the cloud-scale vertical velocity (W) in updrafts and CAPE provides an upper bound for $\frac{1}{2} W^2$ (Lepore et al., 2015; Muller et al., 2011; Muller & Takayabu, 2020), one may expect precipitation extremes scale with the square root of the CAPE ($CAPE^{\frac{1}{2}}$). This relationship is based on the assumption that the conversion of CAPE to kinetic energy is strong in the lower part of the clouds since there is not much condensation left in the upper part of the clouds. However, the real fraction of CAPE converted to upward velocity varies largely by locations and seasons (Dong et al., 2019; Guo et al., 2020; Lepore et al., 2015; Lepore et al., 2016). Following these studies, we first calculate the natural logarithm of the precipitation and CAPE and then use the following equation $ln\ P{\sim}\beta ln\ CAPE$ to examine their relationship. The advantage of this form is that β could directly reflect the fraction of CAPE converted to upward velocity, with a larger β indicating more CAPE is converted to the kinetic energy of a rising air parcel. According to idealized parcel theory, β has an upper bound of 0.5 when all the CAPE is acquired by the air parcel (Lepore et al., 2015; North & Erukhimova, 2009).

We re-examine the relationship between precipitation extremes and CAPE using the up-to-date global dataset. The number of available stations is further reduced to 263, which is mainly caused by the intermittent instrument failures of radiosonde. Different from the *P_e–Ta*/*P_e–Td* relationship, there are no clear regional-scale differences in the *P_e–CAPE* relationship (Figure 14.4). But it shares the similarity of the three major patterns of the *P_e–Ta*

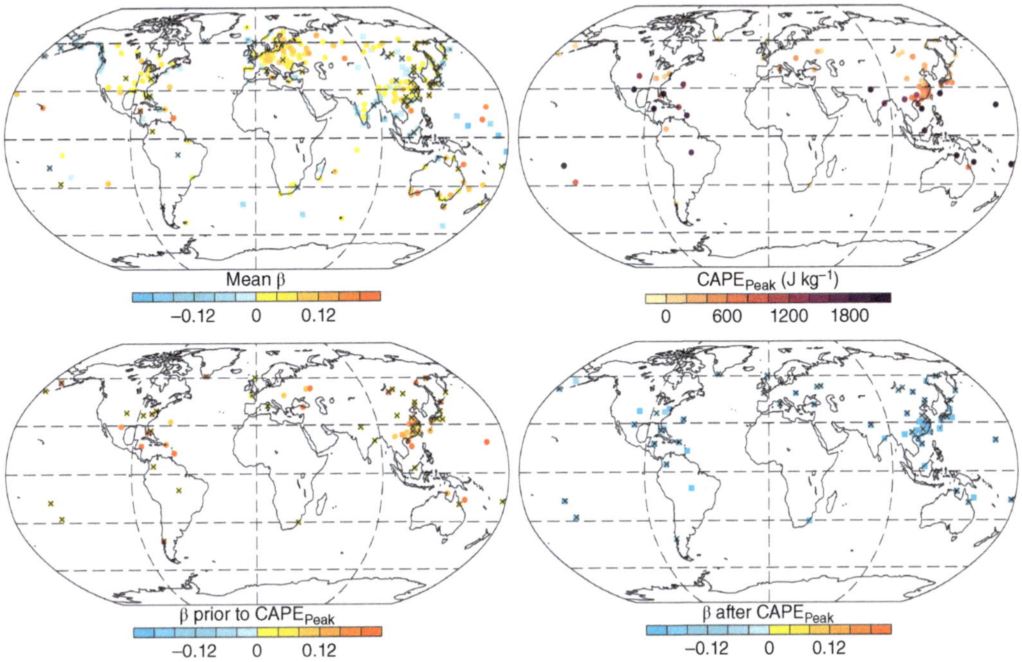

Figure 14.4 Similar as Figure 14.2, but for the dependence of natural logarithms of precipitation on natural logarithms of CAPE. The upper right panel shows the CAPE values when a peak-like structure is detected. β is the regression coefficient between the natural logarithm of precipitation and CAPE.

scaling relationship, with most values falling between –0.2t and 0.2. Larger positive values were observed over eastern North America, Europe, East Asia, and Australia, which is consistent with previous studies analyzed over North America (Lepore et al., 2015; Lepore et al., 2016) and East Asia (Dong et al., 2019; Guo et al., 2020). Negative values were observed over coastal stations or stations over the oceanic island. The deviation from the theoretical value of 0.5 (assume all the CAPE is converted to kinetic energy) may arise from a combination of different influential factors. The value of β was found to depend on the environmental humidity of the entrained air, vertical wind shear, and other factors (Derbyshire et al., 2004; Lepore et al., 2015), which varies substantially between different geographical locations. In terms of the negative *P_e–CAPE* scaling, some studies argued that convective inhibition (CIN) should be taken into consideration (Meredith et al., 2019). CIN is a measure of the energy barrier inhibiting an air parcel from rising from the surface to the level of free convection, which can undermine the relationship between precipitation extremes and CAPE (Barbero et al., 2018; Chen et al., 2020; Dong et al., 2019; Kirkpatrick et al., 2011). If CIN is small, even a modest amount of CAPE can produce updrafts strong enough for precipitation particles to coalesce effectively. Conversely, large values of CIN can suppress the occurrence of updrafts even in the presence of large values of CAPE. Rasmussen et al. (2020) pointed out that over North America if the CIN < -200 J kg^{-1}, the strength of the capping inversion is often too much to overcome and convection is suppressed. Additionally, higher CAPE with strong vertical wind shear or strong steering flow can lead to fast-moving deep convective systems, which may result in less local precipitation accumulation (Barbero et al., 2019). Moreover, the daily data used here may not be able to represent the phase relationship between CAPE and precipitation (Subrahmanyam et al., 2015). The most relevant values of CAPE for a given precipitation event are those prior to the event. The daily averaged data may weaken the implied relationship between precipitation and CAPE (Berg et al., 2009; Haerter & Berg, 2009). This has been found by Guo et al. (2020) over East Asia, in which higher values of β were found when radiosonde-derived CAPE values that are more relevant to the occurrence of precipitation extremes are used.

In addition to a simple *P_e–CAPE* scaling observed over North America (Lepore et al., 2015), a non-monotonic dependence of precipitation extremes on CAPE was noted over East Asia (Dong et al., 2019; Guo et al., 2020), and several sub-regions around the globe (Barbero et al., 2019). In total, 67 out of 263 stations show a parabolic shape with the peak CAPE value ranging from a few hundred J kg^{-1} in midlatitudes to high latitudes to more than 2,000 J kg^{-1} in the tropics. But note that values of β over many stations prior to or after the peak-like CAPE threshold are not significant, calling the robustness of this structure into question. And little is known about the tendency that the *P_e–CAPE* scaling levels off beyond a certain CAPE threshold. CAPE is affected by many physical processes that alter the vertical thermodynamic structure. Williams and Renno (1993) have disclosed a linear relationship between CAPE and surface wet-bulb potential temperature using tropical radiosonde data, indicating that the dependence of CAPE on surface temperature and/or moisture availability may be influential factors for this peak-like structure.

14.6. CONCLUDING REMARKS

This chapter surveys observational studies that investigates the changing nature of precipitation extremes from the perspective of surface temperature (both air temperature and dew point temperature) and CAPE. The contributions from both factors are found to be important for changes in precipitation extremes. Given that previous studies are analyzed separately over different regions, we re-evaluate these relationships by utilizing two up-to-date global datasets to provide a complementary test on the generality of earlier findings. Following the known scaling equations, precipitation extremes are found to respond differently to these two factors, some general patterns are apparent:

1. The *P_e–Ta*/*P_e–Td* scaling is much more complex than suggested by the CC relationship. Different scaling relationships are observed across the globe, which can be generally classified into three categories: a monotonic increasing scaling, a monotonic decreasing scaling, and a peak-like structure where precipitation extremes first increase with surface temperature until reaching a maximum at a threshold temperature and then decline with a further increase in surface temperature. A schematic plot illustrating several common relationships is provided in Figure 14.5. Most of the stations examined here show an intensification of daily precipitation extremes with increasing temperatures, especially over the midlatitude to high-latitude regions, while tropical stations show a negative scaling relationship even when dew point temperatures are used. But the number of stations characterized by negative scaling rates has been reduced from 19% based on the surface air temperature to 9% when the dew point temperature is used to calculate the scaling rates. Several factors are found to contribute to this negative scaling. Limited moisture availability at higher temperatures is one of the major reasons. Another important factor would be that precipitation extremes associated with synoptic systems are usually accompanied by drops in temperature caused by colder air mass intrusion or evaporative cooling. Besides, super-CC scaling is also noted over some stations, but this scaling is particularly prominent for high-frequency data and when analyzed over a certain temperature range. As shown in Figure 14.5, super-CC scaling is often found over a temperature range prior to the T_{peak} as part of the peak-like structure. While in other cases, it is also noted over a temperature range between two different precipitation regimes with small stratiform precipitation at low temperatures and large convective precipitation at high temperatures. Many factors are found to contribute to these departures from the CC scaling and direct comparison between previous works is difficult due to the adoption of different definitions of precipitation extreme, different time periods, various temporal and spatial resolutions, the mixture of different precipitation types, local versus remote moisture availability, etc. Consequently, there is still debate in the literature as to which type of scaling is expected to be dominant in a warming climate.

2. Though the *P_e–CAPE* scaling is comparatively weaker than the *P_e–Ta*/*P_e–Td* scaling, similar scaling patterns are identified based on the daily dataset. The schematic plot in Figure 14.5 may also apply to *P_e–CAPE* scaling. Positive *P_e–CAPE* scaling is observed over North America, Europe, Australia, and East Asia. Negative values are observed over some coastal regions, suggesting that larger CAPE does not necessarily lead to larger precipitation. However, there are many unknowns and uncertainties associated with this scaling relationship. The CAPE prior to the precipitation extremes should be considered in the future works, as they may be more relevant and representative . Moreover, the dependence of CAPE on surface temperature and/or moisture availability is worth further investigation.

We are aware that the depth of our understanding of the changes in precipitation extremes relies on the quality of available observations. Gauge-based datasets, like the GHCN, GSOD, and IGRA analyzed here, often suffer from systematic measurement errors and sampling issues associated with limited spatial coverage and time-varying number of weather stations. These variables, such as precipitation, CAPE, air temperature, and dew point temperature, are observed merely over the land gauge, which only reflect the statistics at the sampled location. But precipitation extremes do not necessarily scale with the local land temperature considering moisture can be advected from surrounding or even remote water bodies. Alternatively, satellite-based products offer more complete spatial coverage than station records despite the overall short period of availability, their ability to reproduce the extreme precipitation is sensitive to the retrieval algorithms (Timmermans et al., 2019) (see also Chapter 15: Satellite Precipitation Measurements). While climate models may be useful to characterize precipitation globally on a longer timescale, they have limited ability to

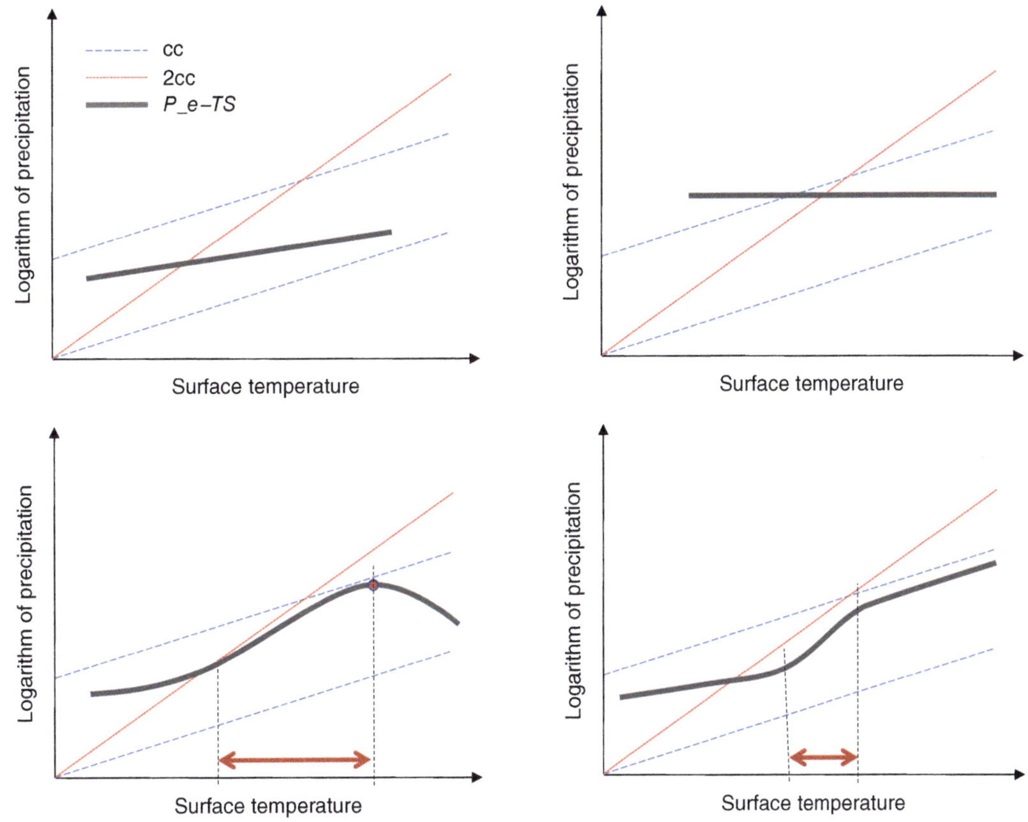

Figure 14.5 Schematic plot for the different types of *P_e–TS* scaling (*TS* represents both the surface air temperature and the dew point temperature) reported in pioneer studies based on daily records, including positive scaling rate (*upper left*), negative scaling rate (*upper right*), peak-like structure (*lower left*), and super-CC (*lower right*) based on two precipitation regimes. Black dashed (dotted) line indicates CC (2CC) scaling relationship. Gray dot in *lower left* panel denotes T_{peak} while gray arrows in *lower left* and *lower right* panels denote the temperature ranges during which super-CC scaling are observed.

simulate small-scale convective processes associated with short-duration precipitation extremes, especially in the tropics where they largely rely on convective parameterizations to represent deep convection dynamics (Davies et al., 2013; Kharin et al., 2013; Kharin et al., 2007; Meehl et al., 2000). Even when convective dynamics is resolved at convection-permitting resolutions, precipitation extremes at short durations are sensitive to the parameterization of cloud and precipitation microphysics (Bryan et al., 2003; Dwyer & O'Gorman, 2017; Lin et al., 2012). Further research is needed to quantify the relevance of these processes to the simulation of precipitation extremes. In addition, growing evidence has indicated that the intensity of sub-daily (hourly or even higher frequency) extreme rainfall is more sensitive to climate change (Fowler et al., 2021 and references therein). These higher frequency precipitation extremes are particularly relevant for flash floods and risk assessment, and offer a better understanding of the intermittent nature of precipitation. Therefore, we encourage further observational and modeling studies to investigate the underlying mechanisms driving changes in precipitation extremes, especially those occurring over shorter time scales. This will help improve our understanding of the impacts of climate change on extreme precipitation events, and inform more effective risk management strategies.

In addition to the two scaling relationships summarized here, there are several other contributors playing important roles in the evolution of precipitation extremes. We want to underline another two important factors: precipitation efficiency and large-scale circulation. The impact of precipitation efficiency on precipitation extreme involves several key microphysical processes, among which the convective organization and its sensitivity to warming are actively researched topics. Convective aggregation can affect the thermodynamic environment for precipitation formation (Chapter 8: Mechanisms for the Self-Organization of Tropical Deep Convection). Observational studies have shown that changes in tropical precipitation were largely associated with changes in the frequency of organized convection (Dong et al., 2020a; Dong et al., 2021; Muller, 2013; Pendergrass, 2020; Tan et al., 2015). The capability of the climate model in representing these systems would provide an opportunity to achieve accurate rainfall projections in a warming climate. A detailed review of precipitation efficiency and climate sensitivity can be found in Chapter 13: Precipitation Efficiency and Climate Sensitivity.

In terms of large-scale circulation impacts, considerable studies have shown that precipitation extremes are often associated with distinct synoptic and large-scale circulation patterns over different regions (Ali et al., 2021; Catto & Pfahl, 2013; Dong et al., 2018; Dong et al., 2020b; Field & Wood, 2007; Lau et al., 2008; Liu et al., 2019; Loriaux et al., 2017; Pfahl & Wernli, 2012; Sudharsan et al., 2020; Utsumi et al., 2017). Precipitation extremes are often forced by local and mesoscale factors, with the larger environment providing a favorable condition for triggering or maintaining these smaller scale processes. The observed relationships between the strength of large-scale circulation patterns and the occurrence of precipitation extremes are diverse. Various types of systems (tropical cyclones, extratropical cyclones, mesoscale convective systems, trough-ridge couplets, low-pressure systems, etc.) and mechanisms (fronts, atmospheric rivers, orographic ascent, etc.) have been shown to be important for precipitation extremes (see also Chapter 9: An Overview of Mesoscale Convective Systems). Sometimes, two or more systems (e.g., extratropical cyclones and fronts) may be associated with a single precipitation extreme, with one of them acting as a trigger while the other one providing a favorable environment to sustain its development. Even the same category of a certain type of system can have multiple sub-types, which might have important regional differences. For instance, different types of mesoscale convective systems are found to be characterized with various structures, dynamical processes, and large-scale circulation patterns (Houze, 2004), leading to different impacts on the associated precipitation. Therefore, it is clear that large-scale circulation patterns are crucial for the occurrence of precipitation extremes, but detailed and process-oriented analyses are limited to regional or case studies. More analysis on these links could be a focus area for future studies. A better understanding of how different large-scale circulation patterns relate to precipitation extremes, and how these relationships vary by season and region would help to interpret any possible future changes.

Despite many advances have been made in understanding changes of precipitation extremes, new challenges are emerging for both observational efforts and model development. To obtain a coherent picture of changes in precipitation extremes over larger areas, more accurate and extensive observations are needed. Future studies should adopt consistent methodologies to enable direct comparison. Continued efforts are recommended for model development to resolve mesoscale and small-scale convective dynamics with advanced cloud and precipitation microphysics and to better simulate short-duration precipitation extremes. Intermodel comparison, including those CRM simulations with the same resolution and forcing, would be very helpful to examine whether the changes in scaling rate depend on the underlying model dynamics and physics or the imposed forcing. We recommend continued efforts in these areas to improve our understanding of the evolution of precipitation extremes in a warming climate.

REFERENCES

Ahern, M., Kovats, R. S., Wilkinson, P., Few, R., & Matthies, F. (2005). Global health impacts of floods: Epidemiologic evidence. *Epidemiologic Reviews*, *27*(1), 36–46.

Alexander, L. V., Zhang, X., Peterson, T. C., Caesar, J., Gleason, B., Klein Tank, A. M. G., et al. (2006). Global observed changes in daily climate extremes of temperature and precipitation. *Journal of Geophysical Research: Atmospheres*, *111*(D5).

Ali, H., & Mishra, V. (2017). Contrasting response of rainfall extremes to increase in surface air and dew point temperatures at urban locations in India. *Scientific Reports*, *7*(1), 1–15.

Ali, H., & Mishra, V. (2018). Contributions of dynamic and thermodynamic scaling in subdaily precipitation extremes in India. *Geophysical Research Letters*, *45*(5), 2352–2361.

Ali, H., Fowler, H. J., & Mishra, V. (2018). Global observational evidence of strong linkage between dew point temperature and precipitation extremes. *Geophysical Research Letters*, *45*(22), 12–320.

Ali, H., Fowler, H. J., Lenderink, G., Lewis, E., & Pritchard, D. (2021). Consistent large-scale response of hourly extreme precipitation to temperature variation over land. *Geophysical Research Letters*, *48*, e2020GL090317.

Allan, R. P., & Soden, B. J. (2008). Atmospheric warming and the amplification of precipitation extremes. *Science*, *321*(5895), 1481–1484.

Allen, M. R., & Ingram, W. J. (2002). Constraints on future changes in climate and the hydrologic cycle. *Nature*, *419*(6903), 228–232.

Attema, J. J., Loriaux, J. M., & Lenderink, G. (2014). Extreme precipitation response to climate perturbations in an atmospheric mesoscale model. *Environmental Research Letters*, *9*(1), 014003.

Bao, J., Sherwood, S. C., Alexander, L. V., & Evans, J. P. (2017). Future increases in extreme precipitation exceed observed scaling rates. *Nature Climate Change*, *7*(2), 128–132.

Barbero, R., Fowler, H. J., Lenderink, G., & Blenkinsop, S. (2017). Is the intensification of precipitation extremes with global warming better detected at hourly than daily resolutions? *Geophysical Research Letters*, *44*(2), 974–983.

Barbero, R., Westra, S., Lenderink, G., & Fowler, H. (2018). Temperature-extreme precipitation scaling: A two-way causality? *International Journal of Climatology*, *38*, e1274–e1279.

Barbero, R., Fowler, H. J., Blenkinsop, S., Westra, S., Moron, V., Lewis, E., et al. (2019). A synthesis of hourly and daily precipitation extremes in different climatic regions. *Weather and Climate Extremes*, *26*, 100219.

Berg, P., & Haerter, J. O. (2013). Unexpected increase in precipitation intensity with temperature — A result of mixing of precipitation types? *Atmospheric Research*, *119*, 56–61.

Berg, P., Haerter, J. O., Thejll, P., Piani, C., Hagemann, S., & Christensen, J. H. (2009). Seasonal characteristics of the relationship between daily precipitation intensity and surface temperature. *Journal of Geophysical Research: Atmospheres*, *114*(D18102).

Berg, P., Moseley, C., & Haerter, J. O. (2013). Strong increase in convective precipitation in response to higher temperatures. *Nature Geoscience*, *6*(3), 181–185.

Blenkinsop, S., Chan, S. C., Kendon, E. J., Roberts, N. M., & Fowler, H. J. (2015). Temperature influences on intense UK hourly precipitation and dependency on large-scale circulation. *Environmental Research Letters*, *10*(5), 054021.

Bryan, G. H., Wyngaard, J. C., & Fritsch, J. M. (2003). Resolution requirements for the simulation of deep moist convection. *Monthly Weather Review*, *131*(10), 2394–2416.

Bui, A., Johnson, F., & Wasko, C. (2019). The relationship of atmospheric air temperature and dew point temperature to extreme rainfall. *Environmental Research Letters*, *14*(7), 074025.

Catto, J. L., & Pfahl, S. (2013). The importance of fronts for extreme precipitation. *Journal of Geophysical Research: Atmospheres*, *118*(19), 10,791–10,801.

Chan, S. C., Kendon, E. J., Roberts, N. M., Fowler, H. J., & Blenkinsop, S. (2016). Downturn in scaling of UK extreme rainfall with temperature for future hottest days. *Nature Geoscience*, *9*(1), 24–28.

Chen, J., Dai, A., Zhang, Y., & Rasmussen, K. L. (2020). Changes in convective available potential energy and convective inhibition under global warming. *Journal of Climate*, *33*(6), 2025–2050.

Davies, L., Jakob, C., May, P., Kumar, V. V., & Xie, S. (2013). Relationships between the large-scale atmosphere and the small-scale convective state for Darwin, *Australia. Journal of Geophysical Research: Atmospheres*, *118*(20), 11–534.

Dee, D. P., Uppala, S. M., Simmons, A. J., Berrisford, P., Poli, P., Kobayashi, S., et al. (2011). The ERA-Interim reanalysis: Configuration and performance of the data assimilation system. *Quarterly Journal of the Royal Meteorological Society*, *137*(656), 553–597.

DeMott, C. A., & Randall, D. A. (2004). Observed variations of tropical convective available potential energy. *Journal of Geophysical Research: Atmospheres*, *109*(D02102).

Derbyshire, S. H., Beau, I., Bechtold, P., Grandpeix, J.-Y., Piriou, J.-M., Redelsperger, J.-L., & Soares, P. M. M. (2004). Sensitivity of moist convection to environmental humidity. *Quarterly Journal of the Royal Meteorological Society*, *130*(604), 3055–3079.

Donat, M. G., Lowry, A. L., Alexander, L. V., O'Gorman, P. A., & Maher, N. (2016). More extreme precipitation in the world's dry and wet regions. *Nature Climate Change*, *6*(5), 508–513.

Dong, W., Lin, Y., Wright, J. S., Xie, Y., Yin, X., & Guo, J. (2019). Precipitable water and CAPE dependence of rainfall intensities in China. *Climate Dynamics*, *52*(5), 3357–3368.

Dong, W. H., Lin, Y. L., Wright, J. S., Xie, Y. Y., Xu, F. H., Yang, K., et al. (2018). Connections between a late summer snowstorm over the southwestern Tibetan Plateau and a concurrent Indian monsoon low pressure system. *Journal of Geophysical Research: Atmospheres*, *123*(24), 13,676–13,691.

Dong, W. H., Lin, Y. L., Zhang, M. H., & Huang, X. M. (2020a). Footprint of tropical mesoscale convective system variability on stratospheric water vapor. *Geophysical Research Letters*, *47*(5), e2019GL086320.

Dong W.H., Ming, Y., & Ramaswamy, V. (2020b). Projected changes in Indian monsoon low-pressure system. *Journal of Climate, 33*(17), 7275–7287.

Dong, W. H., Zhao, M., Ming, Y., & Ramaswamy, V. (2021). Representation of tropical mesoscale convective systems in a general circulation model: Climatology and response to global warming. *Journal of Climate*, *34*(14), 5657–5671.

Drobinski, P., Alonzo, B., Bastin, S., Silva, N. D., & Muller, C. (2016). Scaling of precipitation extremes with temperature in the French Mediterranean region: What explains the hook shape? *Journal of Geophysical Research: Atmospheres*, *121*(7), 3100–3119.

Drobinski, P., Da Silva, N., Panthou, G., Bastin, S., Muller, C., Ahrens, B., et al. (2018). Scaling precipitation extremes with temperature in the Mediterranean: Past climate assessment and projection in anthropogenic scenarios. *Climate Dynamics*, *51*(3), 1237–1257.

Durre, I., Vose, R. S., & Wuertz, D. B. (2006). Overview of the integrated global radiosonde archive. *Journal of Climate*, *19*(1), 53–68.

Durre, I., Williams Jr., C. N., Yin, X., & Vose, R. S. (2009). Radiosonde-based trends in precipitable water over the Northern Hemisphere: An update. *Journal of Geophysical Research: Atmospheres*, *114*(D05112).

Dwyer, J., & O'Gorman, P. (2017). Changing duration and spatial extent of midlatitude precipitation extremes across different climates. *Geophysical Research Letters*, *44*(11), 5863–5871.

Easterling, D. R., Meehl, G. A., Parmesan, C., Changnon, S. A., Karl, T. R., & Mearns, L. O. (2000). Climate extremes: Observations, modeling, and impacts. *Science*, *289*(5487), 2068–2074.

Emori, S., & Brown, S. (2005). Dynamic and thermodynamic changes in mean and extreme precipitation under changed climate. *Geophysical Research Letters*, *32*(L17706).

Field, P. R., & Wood, R. (2007). Precipitation and cloud structure in midlatitude cyclones. *Journal of Climate*, *20*(2), 233–254.

Fowler, H. J., et al. (2021). Anthropogenic intensification of short-duration rainfall extremes. *Nature Reviews Earth & Environment*, 1–16.

Fischer, E. M., & Knutti, R. (2014). Detection of spatially aggregated changes in temperature and precipitation extremes. *Geophysical Research Letters*, *41*(2), 547–554.

Gartzke, J., Knuteson, R., Przybyl, G., Ackerman, S., & Revercomb, H. (2017). Comparison of satellite-, model-, and radiosonde-derived convective available potential energy in the Southern Great Plains region. *Journal of Applied Meteorology and Climatology*, *56*(5), 1499–1513.

Groisman, P. Y., Knight, R. W., Easterling, D. R., Karl, T. R., Hegerl, G. C., & Razuvaev, V. N. (2005). Trends in intense precipitation in the climate record. *Journal of Climate*, *18*(9), 1326–1350.

Guerreiro, S. B., Fowler, H. J., Barbero, R., Westra, S., Lenderink, G., Blenkinsop, S., et al. (2018). Detection of continental-scale intensification of hourly rainfall extremes. *Nature Climate Change*, *8*(9), 803–807.

Guo, J., Yan, Y., Chen, D., Lv, Y., Han, Y., Guo, X., et al. (2020). The response of warm-season precipitation extremes in China to global warming: An observational perspective from radiosonde measurements. *Climate Dynamics*, *54*(9), 3977–3989.

Haerter, J. O., & Berg, P. (2009). Unexpected rise in extreme precipitation caused by a shift in rain type? *Nature Geoscience*, *2*(6), 372–373.

Hallegatte, S., Green, C., Nicholls, R. J., & Corfee-Morlot, J. (2013). Future flood losses in major coastal cities. *Nature Climate Change*, *3*(9), 802–806.

Hardwick Jones, R., Westra, S., & Sharma, A. (2010). Observed relationships between extreme sub-daily precipitation, surface temperature, and relative humidity. *Geophysical Research Letters*, *37*(L22805).

Held, I. M., & Soden, B. J. (2006). Robust responses of the hydrological cycle to global warming. *Journal of Climate*, *19*(21), 5686–5699.

Houze, R. A. (2004). Mesoscale convective systems. *Reviews of Geophysics*, *42*, RG4003.

Kao, S. C., & Ganguly, A. R. (2011). Intensity, duration, and frequency of precipitation extremes under 21st-century warming scenarios. *Journal of Geophysical Research: Atmospheres*, *116*(D16119).

Kharin, V. V., Zwiers, F., Zhang, X., & Wehner, M. (2013). Changes in temperature and precipitation extremes in the CMIP5 ensemble. *Climatic Change*, *119*(2), 345–357.

Kharin, V. V., Zwiers, F. W., Zhang, X., & Hegerl, G. C. (2007). Changes in temperature and precipitation extremes in the IPCC ensemble of global coupled model simulations. *Journal of Climate*, *20*(8), 1419–1444.

Kirkpatrick, C., McCaul, E. W., & Cohen, C. (2011). Sensitivities of simulated convective storms to environmental CAPE. *Monthly Weather Review*, *139*(11), 3514–3532.

Knapp, A. K., Beier, C., Briske, D. D., Classen, A. T., Luo, Y., Reichstein, M., et al. (2008). Consequences of more extreme precipitation regimes for terrestrial ecosystems. *BioScience*, *58*(9), 811–821.

Lau, K. M., Zhou, Y., & Wu, H. T. (2008). Have tropical cyclones been feeding more extreme rainfall? *Journal of Geophysical Research: Atmospheres*, *113*(D23113).

Lenderink, G., & Attema, J. (2015). A simple scaling approach to produce climate scenarios of local precipitation extremes for the Netherlands. *Environmental Research Letters*, *10*(8), 085001.

Lenderink, G., Barbero, R., Loriaux, J. M., & Fowler, H. J. (2017). Super-Clausius–Clapeyron scaling of extreme hourly convective precipitation and its relation to large-scale atmospheric conditions. *Journal of Climate*, *30*(15), 6037–6052.

Lenderink, G., Mok, H. Y., Lee, T. C., & van Oldenborgh, G. J. (2011). Scaling and trends of hourly precipitation extremes in two different climate zones – Hong Kong and the Netherlands. *Hydrology and Earth System Sciences*, *15*(9), 3033–3041.

Lenderink, G., & Van Meijgaard, E. (2008). Increase in hourly precipitation extremes beyond expectations from temperature changes. *Nature Geoscience*, *1*(8), 511–514.

Lenderink, G., & van Meijgaard, E. (2010). Linking increases in hourly precipitation extremes to atmospheric temperature and moisture changes. *Environmental Research Letters*, *5*(2), 025208.

Lepore, C., Allen, J. T., & Tippett, M. K. (2016). Relationships between hourly rainfall intensity and atmospheric variables over the contiguous United States. *Journal of Climate*, *29*(9), 3181–3197.

Lepore, C., Veneziano, D., & Molini, A. (2015). Temperature and CAPE dependence of rainfall extremes in the eastern United States. *Geophysical Research Letters*, *42*(1), 74–83.

Liu, M., Vecchi, G. A., Smith, J. A., & Knutson, T. R. (2019). Causes of large projected increases in hurricane precipitation rates with global warming. *NPJ Climate and Atmospheric Science*, *2*(1), 1–5.

Lewis, E., Fowler, H., Alexander, L., Dunn, R., Mcclean, F., Barbero, R., et al. (2019). GSDR: A global sub-daily rainfall dataset. *Journal of Climate*, *32*(15), 4715–4729.

Lin, Y., Donner, L. J., Petch, J., Bechtold, P., Boyle, J., Klein, S. A., et al. (2012). TWP-ICE global atmospheric model intercomparison: Convection responsiveness and resolution impact. *Journal of Geophysical Research: Atmospheres*, *117*(D09111).

Loriaux, J. M., Lenderink, G., & Siebesma, A. P. (2017). Large-scale controls on extreme precipitation. *Journal of Climate*, *30*(3), 955–968.

Lute, A. C., Abatzoglou, J. T., & Hegewisch, K. C. (2015). Projected changes in snowfall extremes and interannual variability of snowfall in the western United States. *Water Resources Research*, *51*(2), 960–972.

Maeda, E. E., Utsumi, N., & Oki, T. (2012). Decreasing precipitation extremes at higher temperatures in tropical regions. *Natural Hazards*, *64*(1), 935–941.

Martinkova, M., & Kysely, J. (2020). Overview of observed Clausius-Clapeyron scaling of extreme precipitation in midlatitudes. *Atmosphere*, *11*(8), 786.

Meehl, G. A., Zwiers, F., Evans, J., Knutson, T., Mearns, L., & Whetton, P. (2000). Trends in extreme weather and climate events: Issues related to modeling extremes in projections of future climate change, edited, pp. 427–436.

Menne, M. J., Durre, I., Vose, R. S., Gleason, B. E., & Houston, T. G. (2012). An overview of the global historical climatology network-daily database. *Journal of Atmospheric and Oceanic Technology*, *29*(7), 897–910.

Meredith, E. P., Ulbrich, U., & Rust, H. W. (2019). The diurnal nature of future extreme precipitation intensification. *Geophysical Research Letters*, *46*(13), 7680–7689.

Min, S.-K., Zhang, X., Zwiers, F. W., & Hegerl, G. C. (2011). Human contribution to more-intense precipitation extremes. *Nature*, *470*(7334), 378–381.

Mishra, V., Wallace, J. M., & Lettenmaier, D. P. (2012). Relationship between hourly extreme precipitation and local air temperature in the United States. *Geophysical Research Letters*, *39*(L16403).

Moustakis, Y., Onof, C. J., & Paschalis, A. (2020). Atmospheric convection, dynamics and topography shape the scaling pattern of hourly rainfall extremes with temperature globally. *Communications Earth & Environment*, *1*(1), 1–9.

Muller, C. J. (2013). Impact of convective organization on the response of tropical precipitation extremes to warming. *Journal of Climate*, *26*(14), 5028–5043.

Muller, C. J., & Takayabu, Y. (2020). Response of precipitation extremes to warming: What have we learned from theory and idealized cloud-resolving simulations, and what remains to be learned? *Environmental Research Letters*, *15*(3), 035001.

Muller, C. J., & O'Gorman, P. A. (2011). An energetic perspective on the regional response of precipitation to climate change. *Nature Climate Change*, *1*(5), 266–271.

Muller, C. J., O'Gorman, P. A., & Back, L. E. (2011). Intensification of precipitation extremes with warming in a cloud-resolving model. *Journal of Climate*, *24*(11), 2784–2800.

Murali Krishna, U. V., Das, S. K., Uma, K. N., & Pandithurai, G. (2019). Retrieval of convective available potential energy from INSAT-3D measurements: Comparison with radiosonde data and their spatial-temporal variations. *Atmospheric Measurement Techniques*, *12*, 777.

Norris, J., Chen, G., & Neelin, J. D. (2019). Thermodynamic versus dynamic controls on extreme precipitation in a warming climate from the Community Earth System Model Large Ensemble. *Journal of Climate*, *32*(4), 1025–1045.

North, G. R., & Erukhimova, T. L. (2009). *Atmospheric thermodynamics: Elementary physics and chemistry*. Cambridge: Cambridge University Press.

O'Gorman, P. A., & Schneider, T. (2009a). The physical basis for increases in precipitation extremes in simulations of 21st-century climate change. *Proceedings of the National Academy of Sciences*, *106*(35), 14773–14777.

O'Gorman, P. A. (2012). Sensitivity of tropical precipitation extremes to climate change. *Nature Geoscience*, *5*(10), 697–700.

O'Gorman, P. A. (2014). Contrasting responses of mean and extreme snowfall to climate change. *Nature*, *512*(7515), 416–418.

O'Gorman, P. A. (2015). Precipitation extremes under climate change. *Current Climate Change Reports*, *1*(2), 49–59.

O'Gorman, P. A., & Schneider, T. (2009b). Scaling of precipitation extremes over a wide range of climates simulated with an idealized GCM. *Journal of Climate*, *22*(21), 5676–5685.

Pall, P., Allen, M. R., & Stone, D. A. (2007). Testing the Clausius–Clapeyron constraint on changes in extreme precipitation under CO2 warming. *Climate Dynamics*, *28*(4), 351–363.

Panthou, G., Mailhot, A., Laurence, E., & Talbot, G. (2014). Relationship between surface temperature and extreme rainfalls: A multi-time-scale and event-based analysis. *Journal of Hydrometeorology*, *15*(5), 1999–2011.

Park, I.-H., & Min, S.-K. (2017). Role of convective precipitation in the relationship between subdaily extreme precipitation and temperature. *Journal of Climate*, *30*(23), 9527–9537.

Pendergrass, A. G. (2020). Changing degree of convective organization as a mechanism for dynamic changes in extreme precipitation. *Current Climate Change Reports*, 1–8.

Pendergrass, A. G., Knutti, R., Lehner, F., Deser, C., & Sanderson, B. M. (2017). Precipitation variability increases in a warmer climate. *Scientific Reports*, *7*(1), 17966.

Pfahl, S., O'Gorman, P. A., & Fischer, E. M. (2017). Understanding the regional pattern of projected future changes in extreme precipitation. *Nature Climate Change*, *7*(6), 423–427.

Pfahl, S., & Wernli, H. (2012). Quantifying the relevance of cyclones for precipitation extremes. *Journal of Climate*, *25*(19), 6770–6780.

Prein, A. F., Rasmussen, R. M., Ikeda, K., Liu, C., Clark, M. P., & Holland, G. J. (2017). The future intensification of hourly precipitation extremes. *Nature Climate Change*, *7*(1), 48–52.

Pumo, D., Carlino, G., Blenkinsop, S., Arnone, E., Fowler, H., & Noto, L. (2019). Sensitivity of extreme rainfall to temperature in semi-arid Mediterranean regions. *Atmospheric Research*, *225*, 30–44.

Pumo, D., & Noto, L. (2021). Exploring the linkage between dew point temperature and precipitation extremes: A multi-time-scale analysis on a semi-arid Mediterranean region. *Atmospheric Research*, *254*, 105508.

Rasmussen, K. L., Prein, A. F., Rasmussen, R. M., & Liu, C. (2020). Changes in the convective population and thermodynamic environments in convection-permitting regional climate simulations over the United States. *Climate Dynamics*, *55*, 383–408.

Roderick, T. P., Wasko, C., & Sharma, A. (2020). An improved covariate for projecting future rainfall extremes? *Water Resources Research*, *56*(8), e2019WR026924.

Romps, D. M. (2011). Response of tropical precipitation to global warming. *Journal of the Atmospheric Sciences*, *68*(1), 123–138.

Schleiss, M. (2018). How intermittency affects the rate at which rainfall extremes respond to changes in temperature. *Earth System Dynamics*, *9*(3), 955–968.

Seeley, J. T., & Romps, D. M. (2015). Why does tropical convective available potential energy (CAPE) increase with warming? *Geophysical Research Letters*, *42*(23), 10,429–10,437.

Shaw, S. B., Royem, A. A., & Riha, S. J. (2011). The relationship between extreme hourly precipitation and surface temperature in different hydroclimatic regions of the United States. *Journal of Hydrometeorology*, *12*(2), 319–325.

Sherwood, S. C., Roca, R., Weckwerth, T. M., & Andronova, N. G. (2010). Tropospheric water vapor, convection, and climate. *Reviews of Geophysics*, *48*(2).

Singh, M. S., & O'Gorman, P. A. (2013). Influence of entrainment on the thermal stratification in simulations of radiative-convective equilibrium. *Geophysical Research Letters*, *40*(16), 4398–4403.

Singh, M. S., Warren, R. A., & Jakob, C. (2019). A steady-state model for the relationship between humidity, instability, and precipitation in the tropics. *Journal of Advances in Modeling Earth Systems*, *11*(12), 3973–3994.

Subrahmanyam, K. V., Kumar, K. K., & Narendra Babu, A. (2015). Phase relation between CAPE and precipitation at diurnal scales over the Indian summer monsoon region. *Atmospheric Science Letters*, *16*(3), 346–354.

Sudharsan, N., Karmakar, S., Fowler, H. J., & Vittal, H. (2020). Large-scale dynamics have greater role than thermodynamics in driving precipitation extremes over India. *Climate Dynamics*, *55*, 2603–2614.

Tan, J., Jakob, C., Rossow, W. B., & Tselioudis, G. (2015). Increases in tropical rainfall driven by changes in frequency of organized deep convection. *Nature*, *519*(7544), 451–454.

Taszarek, M., Brooks, H. E., Czernecki, B., Szuster, P., & Fortuniak, K. (2018). Climatological aspects of convective parameters over Europe: A comparison of ERA-Interim and sounding data. *Journal of Climate*, *31*(11), 4281–4308.

Timmermans, B., Wehner, M., Cooley, D., O'Brien, T., & Krishnan, H. (2019). An evaluation of the consistency of extremes in gridded precipitation data sets. *Climate Dynamics*, *52*(11), 6651–6670.

Trenberth, K. E. (2011). Changes in precipitation with climate change. *Climate Research*, *47*(1–2), 123–138.

Trenberth, K. E., Dai, A., Rasmussen, R. M., & Parsons, D. B. (2003). The changing character of precipitation. *Bulletin of the American Meteorological Society*, *84*(9), 1205–1218.

Utsumi, N., Seto, S., Kanae, S., Maeda, E. E., & Oki, T. (2011). Does higher surface temperature intensify extreme precipitation? *Geophysical Research Letters*, *38*(L16708).

Utsumi, N., Kim, H., Kanae, S., & Oki, T. (2017). Relative contributions of weather systems to mean and extreme global precipitation. *Journal of Geophysical Research: Atmospheres*, *122*(1), 152–167.

Visser, J. B., Wasko, C., Sharma, A., & Nathan, R. (2020). Resolving inconsistencies in extreme precipitation-temperature sensitivities. *Geophysical Research Letters*, *47*(18), e2020GL089723.

Wasko, C., & Sharma, A. (2014). Quantile regression for investigating scaling of extreme precipitation with temperature. *Water Resources Research*, *50*(4), 3608–3614.

Wasko, C., & Sharma, A. (2017). Global assessment of flood and storm extremes with increased temperatures. *Scientific Reports*, *7*(1), 7945.

Wasko, C., Sharma, A., & Johnson, F. (2015). Does storm duration modulate the extreme precipitation-temperature scaling relationship? *Geophysical Research Letters*, *42*(20), 8783–8790.

Wasko, C., Lu, W., & Mehrotra, R. (2018). Relationship of extreme precipitation, dry-bulb temperature, and dew point temperature across Australia. *Environmental Research Letters*, *13*(7), 074031.

Wei, S., Jian, L., & Ru-Cong, Y. (2013). Corresponding relation between warm season precipitation extremes and surface air temperature in South China. *Advances in Climate Change Research*, *4*(3), 160–165.

Westra S., Alexander, L. V., & Zwiers, F. W. (2013). Global increasing trends in annual maximum daily precipitation. *Journal of Climate*, *26*(11), 3904–3918.

Westra, S., Fowler, H. J., Evans, J. P., Alexander, L. V., Berg, P., Johnson, F., et al. (2014). Future changes to the intensity and frequency of short-duration extreme rainfall. *Reviews of Geophysics*, *52*(3), 522–555.

Williams, E., & Renno, N. (1993). An analysis of the conditional instability of the tropical atmosphere. *Monthly Weather Review*, *121*(1), 21–36.

Wu, G., & Wang, K. (2021). Observed response of precipitation intensity to dew point temperature over the contiguous US. *Theoretical and Applied Climatology*, *144*(3), 1349–1362.

Yin, J., Gentine, P., Zhou, S., Sullivan, S. C., Wang, R., Zhang, Y., & Guo, S. (2018). Large increase in global storm runoff extremes driven by climate and anthropogenic changes. *Nature Communications*, *9*(1), 4389.

Yong, Z., Xiong, J., Wang, Z., Cheng, W., Yang, J., & Pang, Q. (2021). Relationship of extreme precipitation, surface air temperature, and dew point temperature across the Tibetan Plateau. *Climatic Change*, *165*(1), 1–22.

Zhang, W., Villarini, G., & Wehner, M. (2019). Contrasting the responses of extreme precipitation to changes in surface air and dew point temperatures. *Climatic Change*, *154*(1), 257–271.

Zhang, X., Zwiers, F. W., Li, G., Wan, H., & Cannon, A. J. (2017). Complexity in estimating past and future extreme short-duration rainfall. *Nature Geoscience*, *10*(4), 255–259.

15

Satellite Precipitation Measurements: What Have We Learnt About Cloud-Precipitation Processes From Space?

Maki Kikuchi[1], Scott A. Braun[2], Kentaroh Suzuki[3], Guosheng Liu[4], and Alessandro Battaglia[5,6]

ABSTRACT

Precipitation is one of the fundamental elements that define global and regional climatology. Precipitation systems consist of a broad spectrum of three-dimensional structures in which microphysical processes interact with macroscale processes in the cloud system and the ambient environment that prescribe the evolution of the system. Since 1970s, satellite observations of precipitation have been a fundamental tool in quantifying this complex interaction. They have first quantified the frequency and intensity of global precipitation, including remote areas over open oceans and polar regions, thus providing today's precipitation climatology. More recently, satellite observations of cloud and precipitation have been exploited for understanding the physical mechanisms governing precipitation systems. A subset of these studies also provided observation-based metrics to probe physical processes operating in cloud-precipitation systems and to apply them as diagnostic measures for evaluating the representation of the processes in numerical models for better projections of future climate. In this chapter, we review the theoretical basis of precipitation remote sensing from space and describe how it is practically applied in satellite missions. In the first part of the chapter, a historical overview of the satellite missions is described, summarizing the instruments and retrieval algorithms developed in the missions. In the second part, we introduce a set of studies discussing the fundamental mechanisms behind precipitation formation, highlighting what we have learnt to date on cloud-precipitation processes from satellite observations.

15.1. INTRODUCTION

Precipitation is condensed water (liquid and ice particles or their mixtures) that forms in clouds when cloud particles grow large enough to fall. Precipitating particles are distributed vertically in the atmosphere and fall typically as rain, snow, graupel, or hail. As cloud growth and precipitation formation progress, they supply water to the surface, constituting a major component in the hydrological cycle. Precipitation also plays a fundamental role in the Earth's energy budget. When solar energy is consumed to evaporate the surface water, energy is transported through the atmosphere as latent heat associated with water vapor. The latent heat is then released to warm the atmosphere via condensation and deposition processes, thus moving the energy from the surface to atmosphere.

In precipitation formation, there are several paths for cloud particles to grow into precipitation. In warm clouds where clouds are entirely composed of liquid droplets, precipitation is formed through a microphysical process

[1]*Earth Observation Research Center, Japan Aerospace Exploration Agency, Ibaraki, Japan*

[2]*NASA Goddard Space Flight Center, Greenbelt, MD, USA*

[3]*Atmosphere and Ocean Research Institute, The University of Tokyo, Chiba, Japan*

[4]*Department of Earth, Ocean and Atmospheric Science, Florida State University, Tallahassee, FL, USA*

[5]*Department of Environment, Land and Infrastructure Engineering, Polytechnic University of Turin, Turin, Italy*

[6]*Department of Physics and Astronomy, University of Leicester, Leicester, UK*

Clouds and Their Climatic Impacts: Radiation, Circulation, and Precipitation, Geophysical Monograph 281, First Edition.
Edited by Sylvia C. Sullivan and Corinna Hoose.

DOI: 10.1002/9781119700357.ch15

called *warm-rain processes*. During this process, the cloud droplets are formed by condensation of water vapor and become larger by swelling water vapor further until they reach a certain size (typically around 15–20 μm). Beyond this size, the process is replaced by *collision and coalescence*, where larger droplets grow by collecting surrounding smaller droplets. In cold clouds under sub-freezing temperatures, *cold-rain processes* become the dominant precipitation mechanism. Whenever ice and liquid coexist, ice crystals grow rapidly at the expense of liquid droplets by the *Wegener-Bergeron-Findeisen (WBF) process*, where the liquid droplets evaporate and serve as a continuous source of water vapor to ice crystals due to the lower saturation vapor pressure of ice compared to that of liquid (Bergeron, 1935; Findeisen, 1938; Wegener, 1911). Other mechanisms for ice particle growth involve liquid-ice and ice-ice interactions, including *riming* where supercooled droplets freeze directly onto the crystal surface and *aggregation* where crystal arms of multiple particles stick together to form one larger particle. Latent heat is released during most of these processes (e.g., vapor deposition and freezing of liquid to ice) and heats the surrounding atmosphere which invigorates convection and precipitation formation. Ultimately, these ice crystals become large enough to fall as snow, or otherwise melt to form rain drops.

What makes the precipitating cloud systems complex is the fact that these microphysical processes, typically occurring at the scale of micrometers to millimeters, underpin individual convective storms (of few kilometers in size) that interact with large-scale dynamics to generate larger systems such as mesoscale convective systems (MCS) (see also Chapter 9: An Overview of Mesoscale Convective Systems). Such a wide spectrum of scales with high temporal and spatial variabilities makes it challenging to quantify precipitation over the globe. For over a century, rain gauges have been providing detailed observations essential to characterizing the surface precipitation. Later, satellite mission concepts were considered with the aim to understand global precipitation and its impact on climate given that the area covered by gauges is rather limited, equivalent to "less than half a football field" (Kidd et al., 2017). It was in the 1970s when Earth observation from satellites started. Thanks to subsequent measurement advances, our understanding of global precipitation has improved drastically, particularly for its frequency and intensity across the globe, enabling better characterization of precipitation climatology. Building upon these advances, a number of studies have been devoted to analyzing multiple satellite measurements to examine not only precipitation itself but also its interactions with cloud, radiation, and the ambient environment, in an attempt to reveal signals of precipitation-related processes behind the snap-shot observations.

Generally speaking, the term "process" refers to a mechanism responsible for a temporal change in physical states. With regard to precipitation, the relevant processes include microphysical and thermodynamical changes of hydrometeors themselves (e.g., particle evolutions, phase changes, and the resultant radiative changes), as well as dynamical changes in the cloud system (e.g., latent heat impact on storm invigoration). These "internal" processes interact with "external" factors including aerosols (e.g., aerosol seeding of cloud droplets or crystals, precipitation removal of aerosols, aerosol impacts on precipitation, and radiative forcing associated with them) and further interact on broader factors related to land-surface and cryospheric processes, the water cycle and the radiation budget of the Earth. The search for a deeper understanding of these physical processes governing precipitation has leveraged studies linking individual measurement data to diagnostics of the related processes underpinning the observed precipitation beyond scrutinizing precipitation climatology, the evolution of these processes in a warming climate, and better representation of the processes in numerical models (Stephens et al., 2018).

In this chapter, an overview of the progress in satellite precipitation measurement is addressed, with a particular focus on outlining the insights into precipitation-related processes obtained from recent studies. Section 15.2 reviews the fundamental principles of precipitation measurement in satellite remote sensing. The historical overview of satellite precipitation measurement and their retrieval algorithms are described in section 15.3. Selected examples of observation-based analysis of cloud-precipitation processes are then highlighted in section 15.4. Future satellite missions planned for precipitation measurement to further advance cloud-precipitation science are discussed in section 15.5. Note that the chapter highlights recent advances in cloud-precipitation processes rather than on water vapor processes, although some aspects of water vapor in convective processes are touched upon. Finally, the chapter is summarized in section 15.6.

15.2. SATELLITE REMOTE SENSING OF PRECIPITATION – PHYSICAL BASIS

Hydrometeors emit, absorb, and scatter radiative energy and precipitation remote sensing seeks to convert the radiative energy received by a satellite-borne radar or radiometer to water flux at the surface. Therefore, the physical basis of retrieving precipitation involves two fundamental relations, that is, (1) how hydrometeors interact with radiation, which is often determined by their size, shape, and single-scattering properties and (2) how microphysical properties distributed in the vertical column relate to surface water flux, which depends on the vertical variability of natural precipitating clouds. For radars, the second relation is less of an issue since radar reflectivity can be determined in individual atmospheric layers. However, in the case of passive remote sensing by radiometers, the vertical distribution of hydrometeors plays a critical role in determining satellite-observed radiative energy. Thus, early SSM/I-era retrieval methods relied on statistical relations between hydrometers aloft and precipitation rate at the surface (Grody, 1991).

For passive remote sensing, satellite received radiances come from either reflected sunlight or thermal emission by targets in the sensor's field of view. Precipitating clouds are opaque in the visible and infrared spectrum and precipitation estimations in these spectral ranges are chiefly related to the cloud top structure, relying on the second relation mentioned above. The general assumption is that the more reflective or the taller the cloud top is, the more intense the precipitation will be at the surface. In the microwave (including millimeter wave) spectrum, however, the satellite received radiation is a result of scattering/emission of the surface emission as it interacts with all hydrometeors and constituent gases throughout the entire atmospheric column, and the absorption and scattering by hydrometeors dictate the upwelling radiative energy. Using the frequency of 166 GHz, which is used in Global Precipitation Measurement mission microwave imager, as an example, the absorption and scattering properties of several types of hydrometeors in this wave spectrum are shown in Figure 15.1. Because radiation at higher frequencies is more sensitive to ice scattering, the 166 GHz is used here to make a clear contrast among the scattering properties by different types of hydrometeors. In this figure, raindrops and hail are assumed to be spheres with density of 1.0 and 0.8 g/cm^3, respectively, while dendrite crystals and aggregates are taken from ice scattering database of Liu (2008) and Nowell et al. (2013). First of all, absorption by ice particles is about 2 orders of magnitude weaker than that by liquid drops, which gives rise to little emission by ice particles in the microwave spectrum. Second, given the same liquid-equivalent radius (or mass), the scattering intensity depends strongly on particle shape, which dramatically complicates the retrieval problem because of the large variety of ice particle shapes in natural clouds.

Due to the relative transparency of the atmosphere in the microwave spectrum, surface emissivity is another impacting factor on upwelling radiation. Emissivity over ocean is polarized with a magnitude lower than unity, and fairly well modeled except at high wind conditions (Kilic et al., 2019), while over land it is high (close to 1) and highly variable. As explained above, passive microwave precipitation remote sensing relies on emission and scattering signatures from hydrometeors. Generally speaking, the emission signature is generated from liquid drops that enhances upwelling radiances over ocean at low frequencies (e.g., <80 GHz) and is less of an influence over most land surfaces because of the near 0.9 surface emissivity. On the other hand, the scattering signature is generated by ice particles and only strong enough at high frequencies, although it can be observed over both land and ocean. Therefore, precipitation retrieval

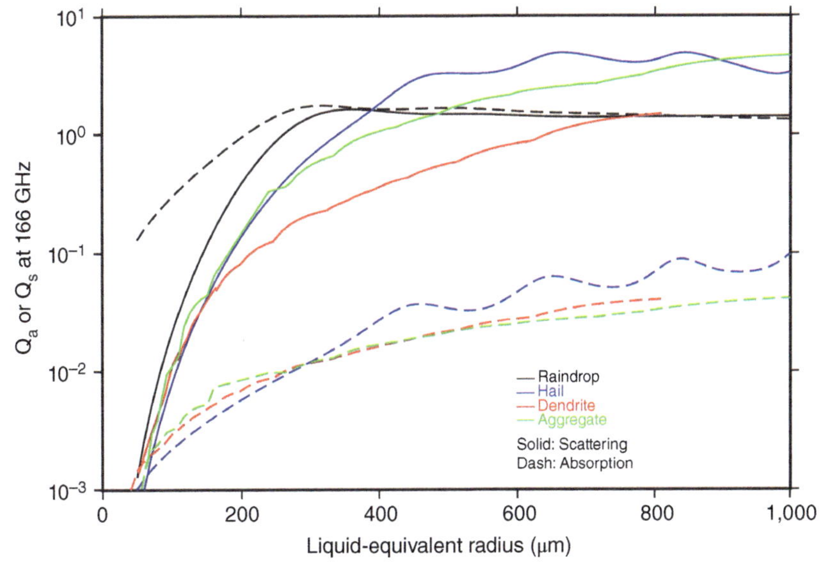

Figure 15.1 Scattering and absorption at 166 GHz for raindrop, hail, dendrite crystal, and ice aggregate. Q_s and Q_a are defined by the scattering and absorption cross-sections divided by cross-section area of a liquid-equivalent drop.

by passive microwave observations is based on the assumption that a greater amount of liquid or ice water in the atmospheric column leads to heavier precipitation at surface. Although this relation is more direct than the cloud top-precipitation relation assumed for visible and infrared remote sensing, it is still a large source of error, especially for the ice-precipitation relation assumed for scattering signatures.

For active remote sensing by radars, the observed signature is the backscattered energy by hydrometeors in a finite volume (namely, radar reflectivity). The relation between radar reflectivity and precipitation rate depends strongly on the particle size distribution. The satellite dual-frequency radar measurements (explained in section 15.3) are introduced to minimize this uncertainty, in which one frequency operates in the Rayleigh and the other operates in the non-Rayleigh regime. Assuming that the particle size distribution can be expressed by a two-parameter function – for example, mean diameter and total particle number – the precipitation rate can then be uniquely determined by the two radar reflectivity values. As in passive remote sensing, additional complication arises for retrieving solid precipitation (i.e., snowfall), in which ice particles are non-spherical with a large variety of shapes, and often have preferred orientations.

In the next section, we introduce how these theoretical characteristics of precipitation remote sensing are practically applied in the precipitation measurement from space.

15.3. HISTORICAL OVERVIEW OF SATELLITE PRECIPITATION REMOTE SENSING

15.3.1. Passive Remote Sensing

Building upon the principles explained above, satellite precipitation remote sensing has essentially two pillars of observation techniques: active remote sensing by microwave radars and passive remote sensing by microwave radiometers. Passive microwave instrumentation had undergone multiple generational upgrades since the earliest single-frequency imager was used for precipitation remote sensing in the 1970s (Wilheit, 1972); current satellite microwave imagers in operation cover frequencies from 6 to 183 GHz and make measurements at both horizontal and vertical polarizations (Aonashi & Ferraro, 2020), among which the following sensors are worth mentioning. First, the Special Sensor Microwave Imager (SSM/I) was a conically scanning radiometer operating at frequencies near 19, 22, 37, and 85.5 GHz with dual polarizations at all but the 22 GHz channel (Hollinger et al., 1987). With its multispectral coverage and stable calibration, the SSM/I is often considered to mark the start of the modern era for microwave radiometry for precipitation (Kummerow, 2020); most currently operational radiometers can find their roots from SSM/Is. For example, the Tropical Rainfall Measuring Mission (TRMM) Microwave Imager (TMI) operated at channels similar to SSM/I, with added channels at a lower frequency of 10.7 GHz, and the Advanced Microwave Scanning Radiometer (AMSR) series further adds channels near 7 GHz, primarily for ocean surface parameter retrievals. The Special Sensor Microwave Imager/Sounder (SSMIS) continues the legacy of SSM/I, but consolidates channels previously covered by two other sounding sensors on the Defense Meteorological Satellite Program (DMSP) satellites to include channels near 60 GHz oxygen and 183 GHz water vapor absorption frequencies and at a 150 GHz window frequency. Likewise, the Global Precipitation Measurement (GPM) mission microwave imager (GMI) continues the legacy of TMI but includes observations at higher frequencies of 166 and 183 GHz mainly for snowfall estimation. Most modern era microwave radiometer precipitation algorithms have their origins in concepts developed earlier with SSM/I observations.

Satellite microwave radiometer received energy (as expressed by equivalent blackbody brightness temperature, or TB) is an integral of the emission, absorption, and scattering from the surface, atmospheric gases, and hydrometeors. Because of the complexity involved in the combinations of these geophysical variables, any given set of observed brightness temperatures could correspond to a number of precipitation rates. In other words, the retrieval problem is ill-conditioned with multiple solutions from a single input. To obtain a reasonable solution, retrieval algorithms must set some constraints to reduce the range within which the solution is to be searched. In the following, we briefly describe how satellite algorithms retrieve precipitation rates, by taking the GPM's Goddard Profiling (GPROF) algorithm (Kummerow et al., 2015) as an example. The GPROF algorithm retrieves precipitation using the Bayes' theorem with an *a priori* database linking hydrometeor profiles and brightness temperatures. Estimation is computed by a weighted average of precipitation rates for those profiles that are radiometrically consistent with observations, and the weights are determined by the distance between observed and database brightness temperature vectors. The key component of the algorithm is the *a priori* database in which the hydrometeor profiles need to be representative to those occurring in nature. The most current version uses hydrometeor profiles retrieved from the GPM radar-radiometer combined algorithm (Grecu et al., 2016) and their associated brightness temperatures are computed by radiative transfer simulations. To constrain the algorithm solution by the associated environmental conditions, ancillary data of surface class,

2-m temperature, and precipitable water are introduced to subset the database, so that solutions can only be found in a specified portion within the database.

With the advantage of frequent sampling, visible and infrared measurements, particularly from geostationary satellites, have also played an important role in precipitation retrieval. The imaging spectroradiometers onboard these platforms have gradually evolved in spectral, spatial, and temporal resolution. For example, the Advanced Baseline Imager (ABI) on the GOES-R Series and the Advanced Himawari Imager (AHI) on the Himawari-8 satellites can produce a full disk image every 10–15 minutes with 0.5–2 km spatial resolution; both instruments have 16 channels covering 0.455–13.3 μm wavelength. Due to the physical indirectness to surface precipitation, visible/infrared retrieval algorithms are prone to large errors. To compensate for this shortcoming, techniques have been introduced to estimate precipitation total over a long time and a large area, for example, the GOES Precipitation Index (GPI, Arkin & Meisner, 1987). On the other hand, efforts for instantaneous precipitation retrieval have been made by classifying clouds using horizontal texture, cloud-patch identification, and multi-spectral information (Adler & Negri, 1988; Ba & Gruber, 2001; Bellerby, 2004; Behrangi et al., 2009; Hong et al., 2004). Taking the advantage of frequent sampling by infrared measurements and the physical directness of microwave measurements, some great successes have been reported recently by combining the two, in which microwave retrievals are used to calibrate the infrared algorithm while infrared measurements are used to enhance spatial and temporal coverages of precipitation estimates (Huffman et al., 2007; Hsu & Sorooshian, 2009; Joyce & Xie, 2011; Ushio et al., 2009).

15.3.2. Active Remote Sensing

Radars represent the other pillar of precipitating cloud remote sensing, as demonstrated by the continuous operation record of the three major atmospheric radar missions in the past 22 years: the TRMM, CloudSat, and the GPM mission. Figure 15.2 illustrates the timeline of the launch and operational period of the missions. The launch (November 1997) of the TRMM Ku-band (13.6 GHz) Precipitation Radar (PR), part of a joint mission between the National Aeronautics and Space Administration (NASA) and the Japanese Aerospace Exploration Agency (JAXA), spearheaded a new era of space-borne atmospheric radar missions (Kummerow et al., 2000). The TRMM PR was the first weather radar in space providing unprecedented three-dimensional (3D) structure of storm systems (Houze Jr. et al., 2015; Okamoto, 2003), mapping precipitation over a swath of more than 215 km with a 250 m vertical resolution. The TRMM PR was followed (in April 2006) by the NASA CloudSat W-band (94 GHz) Cloud Profiling Radar (CPR) (Tanelli et al., 2008) as part of the A-Train, a constellation of active and passive instruments in a sun-synchronous orbit with ascending equatorial crossing around 1:30 p.m. local solar time. Thanks to its superior sensitivity and its synergy with other instruments in the A-Train (Stephens et al., 2018), the CloudSat CPR has provided the first ever collection of global vertical profile observations not only of clouds but also of light rain and snow (Lebsock et al., 2020). The global precipitation record begun by TRMM was continued with the launch (February 2014) of the GPM-core observatory (Skofronick-Jackson et al., 2018). The GPM Dual-frequency Precipitation Radar (DPR), the first of its kind to be flown in space with the Ku-band (13.6 GHz) and Ka-band (35.5 GHz) radars (Iguchi, 2020), expands the capability of the TRMM PR in terms of coverage of high latitudes and improved sensitivity, thus enabling detection of lighter precipitation and snow (Kidd et al., 2020).

The diverse capabilities of these different radar systems, spanning frequencies from Ku to W band, are demonstrated in the GPM and CloudSat coincident observations of a precipitation system over the Gulf of Thailand (top panels of Fig. 15.2). The CloudSat CPR depicts a cloud system extending up to 15 km, including three precipitating clusters with a melting layer at about 5 km and an upper cirrus cloud anvil. The CPR signal is attenuated in the rain region below the melting layer and decreases below the noise level in correspondence to the first rain cell at around 80 km along-track distance. As an indication of the attenuation of the signal the path integrated attenuation (PIA) derived from the surface reference technique (Meneghini et al., 2015) is overplotted as a black line on all reflectivity panels in Figure 15.2. The GPM DPR misses most of the ice cloud but can easily penetrate the rain layer, with different levels of attenuation for the two frequencies (rain attenuation at Ka is approximately from two to six times larger than at Ku, compare the values corresponding to the black lines in the top two panels on the left). Note also the better sensitivity of the Ku against the matched scan Ka. Despite the lower sensitivity, it must be kept in mind that the DPR, with its 245 km swath, does provide a full 3D view of the storm and not simply a curtain through it like the CPR. This case study epitomizes the complementarity of these systems (Battaglia et al., 2020a). Luo et al. (2017) exploited the TRMM and CloudSat matchups (with the co-located measurements of Cloud-Aerosol Lidar and Infrared Pathfinder Observation, CALIPSO) to study tropical cloud and precipitation regimes and their internal vertical structures. Berg et al. (2010) combined the TRMM-PR and the CloudSat CPR, best positioned to detect and quantify the high/medium and drizzle/light

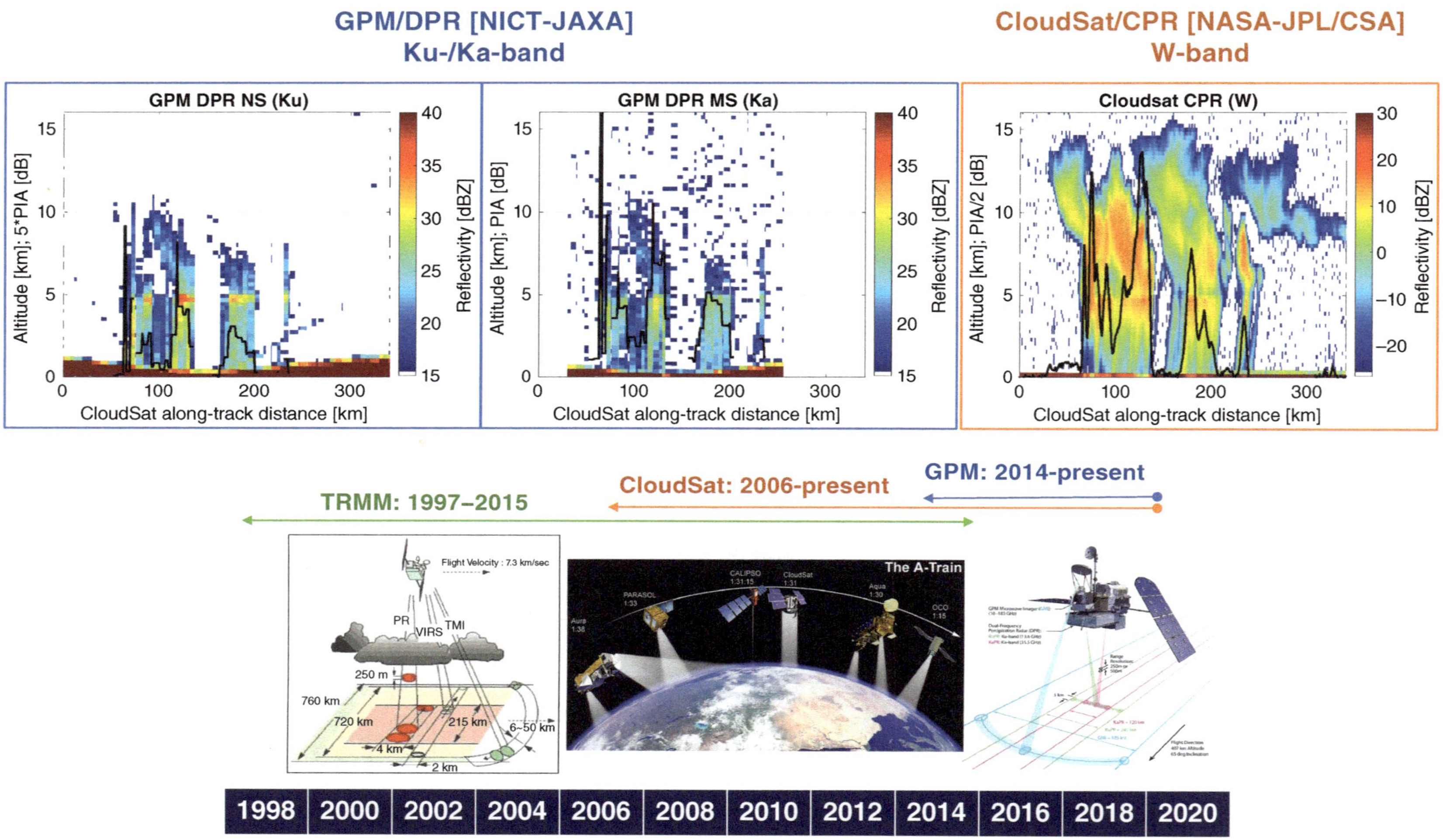

Figure 15.2 Timeline of active cloud and precipitation radar missions with, above, drawings of the TRMM PR scanning pattern, the CloudSat CPR A-Train constellation formation, and the GPM core satellite. The top three panels provide an example of the current space-borne cloud and precipitation radar capabilities with almost coincident vertical profile reflectivities of the GPM Ku (left panel), the GPM Ka (center panel), and the CloudSat W-band (right panel) corresponding to a precipitating systems observed on 4 June 2014. The over plotted black lines (with scaling factors as indicated in the y-labels) indicate the path integrated attenuation as derived from surface reference techniques. The case study is from the GPM product 2B-CSATGPM. Source: NASA / Public Domain.

rain, respectively, to provide the best estimates of the total rain volume in the tropical oceans.

Reflectivity profiles like those shown in Figure 15.2 are used to derive microphysical profiles of hydrometeors. A list and details of TRMM, CloudSat, and GPM products can be found at https://gpm.nasa.gov/data/directory, www.cloudsat.cira.colostate.edu/data-products, and https://gportal.jaxa.jp/gpr/, respectively. There are three fundamental issues when converting the measured radar range-resolved quantities into precipitation profiles:

1. ambiguities in the effective radar reflectivity factors versus precipitation rates conversion which are related to the natural variability of the particle size distribution and, for solid precipitation, of the particle shapes and densities;
2. the attenuation correction required to derive the effective radar reflectivity factors from the measured reflectivity factors. Rain attenuation is increasingly important with increasing frequency and precipitation rate; graupel and snow may produce non-negligible attenuation already at Ka and W bands (Lhermitte, 1990);
3. non-uniform beam filling (Nakamura, 1991) and multiple-scattering effects (Battaglia et al., 2010) that are more pronounced in the presence of large footprints and strongly attenuating media.

The first issue is generally addressed by adopting *a priori* assumptions on the particle size distribution shape (e.g., for rain in TRMM and GPM algorithms (Iguchi & Haddad, 2020) or for snow in the CloudSat 2C-snow algorithm (Wood & L'Ecuyer, 2013)); caution must be taken because such assumptions depend on precipitation regimes and may show regional biases (e.g., Protat et al., 2019). In fact, dual- and multi-frequency observations have been proposed for mitigating such uncertainties (Battaglia et al., 2020b; Iguchi & Haddad, 2020). The second issue, attenuation correction, is an essential step in all retrieval algorithms and is performed either via analytical solutions (Seto & Iguchi, 2015) or via optimal estimation-based iterative processes (L'Ecuyer & Stephens, 2002). Because errors in all of these techniques are accumulated recursively with increasing range (Hitschfeld & Bordan, 1954), additional constraints, such as the PIA estimated from the surface reference technique (Meneghini et al., 2015), are usually exploited in order to stabilize the retrieval. Finally, non-uniform beam filling corrections have been accounted for in TRMM and DPR algorithms (Seto et al., 2015), whereas multiple scattering is intrinsically accounted for when adopting appropriate forward models (e.g., Hogan & Battaglia, 2008) as in CloudSat rain products. However, there are instances associated with extreme events where the two effects occur simultaneously (Mroz et al., 2018); tackling both remains challenging and is topic of current research.

Despite the above issues, these radar measurements have mapped for the first time the global vertical distributions of clouds and precipitation. What is noteworthy is that the precipitation radars not only identified precipitation itself, but also gave observational insights into latent heating in the tropical atmosphere, including its diurnal, intraseasonal, seasonal, and annual variability, as well as its role in tropical and extratropical circulations (Simpson et al., 1988) (see also Chapter 10: Interactions Between the Tropical Atmospheric Overturning Circulation and Clouds in Present and Future Climates). One of the science drivers of the missions was the improvement of climate modeling and prediction, and the provision of precipitation microphysics and latent heat release were viewed as being needed to improve Earth system models (Hou et al., 2014). Studies have used the latent heating products to examine the response of atmospheric circulation to heating (Huaman & Schumacher, 2018; Mathew & Kumar, 2019; Schumacher et al., 2004), to describe the Madden-Julian Oscillation (MJO) (Barnes et al., 2015; Chang & L'Ecuyer, 2019; Vincent & Lane, 2018), and to characterize the contributions to latent heating of precipitation systems with different sizes, depths, and intensities (Liu et al., 2015).

Further, a number of recent studies employed the combinations of radar/radiometers to characterize the properties of hailstorms in an attempt to develop hail climatologies. For example, Marra et al. (2017) identified DPR and GMI properties for a severe hailstorm over Italy and placed it in the context of 2 years of GPM data. Mroz et al. (2018) examined the vertical structure of more than 800 hailstorms using DPR and noted particular features of Ka-band reflectivity profiles associated with hail, including multiple scattering and non-uniform beam filling. Bang and Cecil (2019) used 16 years of TRMM data to pair TMI brightness temperatures with surface hail reports and apply those relationships to GMI data to construct a near-global climatology of hail.

The precipitation measurements by these active radars and passive radiometers better shaped the distribution and properties of global precipitation. In light of this, more efforts were made in later years to understand coupled cloud-precipitation processes from the measurements and propose observation-based metrics in evaluating the representation of those processes in numerical models. The next section introduces examples of such process studies, attempting to reveal the mechanisms behind the precipitation phenomena identified.

15.4. UNDERSTANDING CLOUD-PRECIPITATION PROCESSES

Cloud-precipitation processes are central to the fundamental question with regard to atmospheric water:

How does precipitation form from clouds? As should be obvious from this question, precipitation formation accompanies consecutive changes in physical properties of cloud and precipitating particles, and as such, there has been emerging recognition in the importance of analyzing both cloud and precipitation measurement simultaneously, rather than investigating cloud and precipitation formed separately. The following sections introduce selected example studies that analyzed active and passive measurements synergistically in an attempt to address this fundamental science question.

15.4.1. Warm Rain Processes

The unique capability of CloudSat CPR to vertically profile the cloud and light precipitation offered a novel opportunity to study rain formation processes, particularly when combined with passive measurements of clouds included in A-Train. Early studies that exploited the simultaneous measurement of cloud and light precipitation (e.g., Kubar et al., 2009; Leon et al., 2008; Wood et al., 2009) investigated how CloudSat-observed radar reflectivity relates to MODIS-derived cloud properties such as cloud top particle size, cloud optical depth, and liquid water path (LWP) to find that the radar reflectivity systematically varies with these cloud properties. The light precipitation measurement by CloudSat was also combined with aerosol information obtained from MODIS to assess how rain formation tends to be suppressed by increasing aerosols (Lebsock et al., 2008; L'Ecuyer et al., 2009; Wang et al., 2012). Specifically, these studies investigated probability of precipitation (POP) quantified by light precipitation measurement of CloudSat as a function of LWP from MODIS or AMSR-E for differing conditions of aerosol turbidity also characterized by MODIS. Aerosol impacts on precipitation were then assessed by analyzing how POP tends to increase with increasing LWP at a rate that differs depending on aerosol turbidity (measured by aerosol optical depth or aerosol index). The global statistics thus constructed depicted the aerosol impacts on the warm rain formation process. Wang et al. (2012) further extended this analysis to introduce the sensitivity or "susceptibility" of POP to perturbed aerosols, denoted by S_{pop} as an analog to the precipitation susceptibility S_p (Sorooshian et al., 2009), and showed that global models tend to overestimate S_{pop} relative to satellite observations, implying a too pronounced response of the cloud-precipitation system to perturbed aerosols in climate models.

The vertical profiling capability of CloudSat CPR was further exploited by some studies (e.g., Nakajima et al., 2010; Suzuki et al., 2010) to explore cloud-to-precipitation processes. These studies constructed particular statistics of CloudSat radar profiles rescaled by cloud optical depth and classified according to cloud top particle size obtained from MODIS to observationally "fingerprint" the cloud-to-precipitation processes. The statistics illustrated how vertical microphysical structures vary systematically from non-precipitating to precipitating characteristics as a fairly monotonic function of the cloud top particle size, and such a microphysical signature found in the statistics was interpreted in the context of drop collection process.

In light of such fingerprints from the analysis, subsequent studies have used these results to assess the credibility of cloud-precipitation processes in numerical models (such as climate models and cloud-resolving models). As in other satellite-based model evaluations, these studies employed satellite simulators that translate model geophysical parameters into satellite-observed radiances/backscatters (Bodas-Salcedo et al., 2011; Haynes et al., 2007; Hashino et al., 2013; Masunaga et al., 2010; Voors et al., 2007). As seen in section 15.3, each observation sensor has its own sensitivity/attenuation characteristics, as well as spatial and temporal resolutions. Consequently, comparing satellite measurements to model simulations requires these satellite characteristics to be taken into account, rather than comparing satellite measurements directly to native model outputs. These satellite simulator frameworks provide what a satellite would have measured if it flew over the modeled atmosphere. This reduces the ambiguities attributed to satellite measurement characteristics, enabling a more consistent "scale-aware and definition-aware" evaluations (Kay et al., 2018).

Model evaluations in this manner exposed that state-of-the-art global models, including cloud-resolving models and climate models, share a common propensity to produce rain even when the clouds consist of small droplets (Jing et al., 2017; Suzuki et al., 2011, 2015), as a key cause of the "too-light, too-frequent" precipitation bias in global models identified against satellite observations (Dai, 2006; Stephens et al., 2010). The model-satellite comparisons also offered a process-based constraint on key uncertainties of climate models that have typically been regarded as "tunable knobs". A threshold cloud particle size that triggers rain formation is one particular example for such a tuning parameter in climate models (Golaz et al., 2013). It is shown that the model assumption on this parameter that *best* represents the satellite-observed rain statistics leads to a historical temperature trend that provides the *worst* match to the observed trend and vice versa (Suzuki et al., 2013). This "dichotomy" between the process-based constraint and the top-down energy balance requirement is caused by the radiative forcing due to aerosol-cloud interaction that is too large and negative such that it cancels much of the global warming in twentieth century, as a result of pronounced cloud responses to aerosols enhanced by wet-scavenging of aerosols (Jing & Suzuki, 2018; Jing et al., 2019). This underscores a critical need for better modeling the aerosol-cloud-precipitation coupling.

These studies imply the presence of error compensations in modern climate models, which need to be mitigated for enhancing model reliability of climate projections. The process-based analysis of satellite observations such as those described above also can serve as a guide to improve the models at a fundamental process level. Recently, Michibata and Suzuki (2020) demonstrated that a sophisticated prognostic precipitation parameterization in a climate model can reconcile the error compensation described above to simultaneously represent both rain processes consistent with satellite observations and reasonable magnitudes of radiative forcing due to aerosol-cloud interaction (Fig. 15.3). This highlights how satellite measurement of warm-rain process can be exploited to advance climate modeling (Stephens et al., 2019b) and underscores that such a process-oriented effort in combining models and satellite observations should be extended to understanding of the cold-rain processes, which is another source of uncertainty in climate projections.

15.4.2. Mixed-phase Cloud and Snowfall Processes

Mixed-phase clouds are ubiquitous globally in temperature conditions between 0°C and –40°C. Commonly found in the Southern Ocean and Arctic are ice clouds with thin supercooled liquids at the cloud top and snow falling out near the ground (Morrison et al., 2012; Shupe, 2011). This three-layered structure has been observed using CloudSat CPR with the CALIPSO's scattering lidar, Cloud-Aerosol Lidar with Orthogonal Polarization (CALIOP). The constellation formation of the CloudSat and CALIPSO captures the cloud-precipitation structure almost simultaneously, where the CALIPSO lidar observes the liquid layer at the top and the CloudSat radar observes the ice cloud and snowfall underneath (Forbes & Ahlgrimm, 2014; Huang et al., 2012). According to ground-based observations, the structure often persists for several days (Shupe, 2011), which is relatively unexpected given that the coexistence of liquid and ice is microphysically unstable. Whenever ice and liquid water exist in subfreezing temperatures, the WBF process enhances the evaporation of liquid droplets into water vapor, which are then condensed onto ice, meaning the process supports ice growth at the expense of neighboring liquid droplets (c.f. section 15.1). As such, the microphysical instability can transit mixed-phase clouds into all-iced clouds in a short period of time and the persistence of these clouds for several days is quite surprising. Still, the supercooled liquid layer lasts, suggesting a strong liquid condensation at the cloud top with continuous supply of water vapor that balance continuous liquid mass-loss by ice growth and precipitation. Although not fully understood, the preserving processes involves cloud-scale upward air motion (acting to increase relative humidity), radiative cooling at the cloud top (acting to decrease the static stability and thus driving turbulence), and large-scale advection of water vapor near the cloud top (Morrison et al., 2012).

The complexity of mixed-phase clouds in association with local processes as exemplified above gives rise to fundamental challenges for representing them in global climate and weather prediction models. In particular, it has been recognized as a longstanding mixed-phase cloud problem within the climate modeling community that models tend to underestimate supercooled liquid clouds over the Southern Ocean and (to a lesser extent) in the Arctic (Bodas-Salcedo et al., 2012, 2014, 2016; Kay et al., 2016; Kawai et al., 2019; Trenberth & Fasullo, 2010). In general, for a given cloud water content, cloud liquid droplets are much smaller than cloud ice crystals, which leads liquid clouds to become optically thicker and reflect more solar radiation than ice clouds (Cesana & Storelvmo, 2017). As such, the lack of supercooled water over the regions induces the models to absorb excessive solar radiation and overestimate sea surface temperature (SST). For example, McIlhattan et al. (2017) found the insufficient liquid in Arctic clouds in the climate model of the Community Earth System Model Large Ensemble (CESM-LE), and, in turn, generates too much snowfall than what CloudSat precipitation measurements suggested. They speculated that the WBF process in the model was progressing too fast, contributing in part to scavenging the liquid clouds too efficiently. Recognizing the substantial impacts on cloud-precipitation-ocean-radiation processes (Kay et al., 2016), a number of studies in recent years worked on reducing the supercooled liquid deficit by adjusting phase partitioning of clouds in the model, essentially slowing cloud freezing processes and sustaining more liquids in clouds (e.g., Forbes et al., 2016).

The changes in the phase partitioning subsequently exert profound impact on future climate projections since the melting of cloud ice to liquid in a warming climate increases cloud albedo and thereby induces a negative cloud phase feedback onto the climate (Storelvmo et al., 2015). Tan et al. (2016) illustrated this using the National Center for Atmospheric Research's Community Atmosphere Model (CAM) model by adjusting cloud phase partitioning in the models to what satellite observations implied. They found that the adjustment decreased cloud glaciation rate in a warmer climate that weakened the negative feedback and induced higher climate sensitivity. The follow-up work by Tan and Storelvmo (2019)

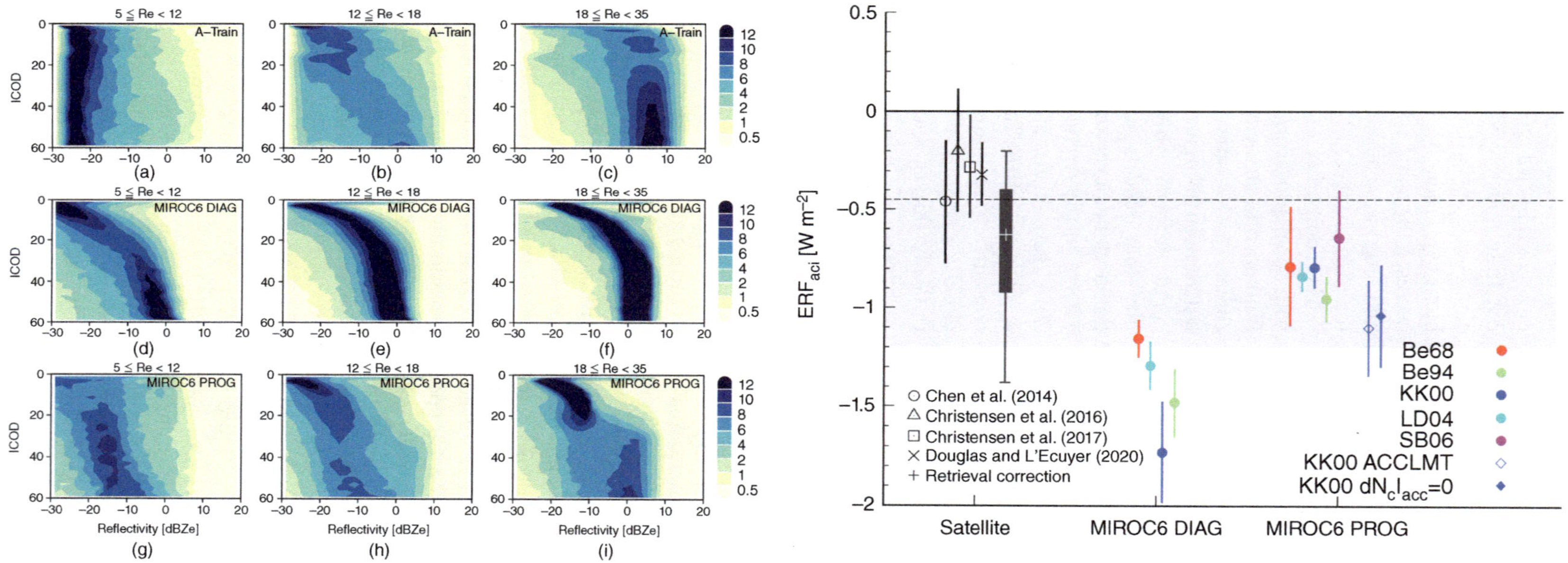

Figure 15.3 (Left) Statistical representations of radar reflectivity profile as a function of in-cloud optical depth (ICOD) classified according to cloud top particle size (Re) from (top) A-Train, (middle) MIROC6 old version, and (bottom) MIROC6 new version. (Right) Effective radiative forcing due to aerosol-cloud interaction estimated by satellite and two versions of MIROC6. Source: Michibata et al. (2020) / John Wiley & Sons/ Public Domain.

interestingly explained the contradicting effects of the cloud phase change on the faster warming of the Arctic compared to the rest of the globe – commonly referred to as the "Arctic amplification." They showed that the cloud phase change during a warming atmosphere initially *reduces* the warming by the negative cloud phase feedback, but at the same time could ultimately *enhance* the Arctic amplification by increase in downward longwave radiation trapped by the prolonged clouds as a result of the smaller liquid droplets (than ice clouds) that produce less precipitation.

The studies above underscored the importance of realistic representation of mixed-phase cloud-to-precipitation processes in the models. Some studies compared the models against CloudSat measurements to find that the models generally produce snowfall too frequently (Kay et al., 2018; Palerme et al., 2017). In an attempt to mitigate this bias, Reitter et al. (2011) introduced a prognostic precipitation scheme that interacts with cloud and ambient environment during sedimentation, instead of the conventional diagnostic scheme with no such interaction. Since snow falls out instantaneously in the diagnostic scheme, the authors found considerable underestimates of the large ice water content (IWC) compared to CloudSat, while the prognostic scheme resulted in overestimates of IWC with the excess ice clouds remained in the atmosphere due to the insufficient snow fall speed, as also suggested by comparison with CloudSat. More recently, Michibata et al. (2019) also introduced a prognostic precipitation scheme in another climate model to achieve better agreement with CloudSat/CALIPSO in total cloud ice content and showed a substantial radiative impact by falling snow of 6.4 W/m^2 in longwave and 5.1 W/m^2 in shortwave.

The mixed-phase clouds and ice precipitation also have profound influences on the polar ice sheet through radiation and snowfall. This is particularly important in the context of evolving mass budget of the Arctic ice sheet with its substantial impact on sea level rise. Boening et al. (2012) used the CloudSat snowfall observations (with the ERA Interim reanalysis) and showed the significant mass increase during 2009–2011 observed by satellite gravitational measurements of Gravity Recovery and Climate Experiment (GRACE) was caused by extreme precipitation events during the austral winter. Palerme et al. (2017) showed that the models in the Fifth Climate Model Intercomparison Project (CMIP5) archive generally predict increases in the Antarctic precipitation (from 5.5% to 24.5%) between the periods 1986–2005 and 2080–2099, with the subset models in agreement with CloudSat snowfall (within 20% of error) estimating larger increases (from 7.4% to 29.3%) implying larger impacts on sea level rise. This underscores a critical role of satellite precipitation measurement in evaluating the processes controlling sensitive balance between the snowfall accumulation and glacial discharge, and their impact on sea level rise in a warming climate. Despite these attempts, process-level understandings of mixed-phase precipitation have generally not yet advanced as much as of warm-rain processes due to the microphysical complexity of ice and snow hydrometeors. This called for ice and mixed-phase processes to be one of the major scientific questions in future missions as highlighted in section 15.5.

15.4.3. Latent Heat and Convective Processes

The latent heat released during precipitation formation is a fundamental factor in atmospheric circulation and convective storm evolution. Stephens et al. (2019a) emphasizes analyzing simultaneous measurements of cloud and precipitation with a focus on the radiative and latent heating characteristics related to different cloud-precipitation vertical structures. They employed CloudSat cloud and precipitation observations combined with GPM precipitation, CALIPSO lidar, and MODIS radiance data to investigate the cloud properties of precipitation. They considered two cloud top height (CTH) parameters, the maximum or highest CTH (designated HCTH), typical of what might be estimated from radiometric observations, and the lowest CTH of the raining column (designated RCTH) derived from an active instrument capable of detecting overlapping cloud layers. Not too surprisingly, they found that HCTH, indicative of the deepest raining clouds in the tropics, was a poor gauge of rainfall intensity and led to misclassification of rainfall associated with shallower cloud systems overtopped by higher clouds. For HCTH, precipitation was associated with a bimodal distribution of CTHs, with rain appearing to occur primarily with deeper convective cloud systems and shallow, primarily subtropical clouds. For RCTH, the proportion of rainfall from deeper systems was reduced, while rainfall from shallow and congestus systems increased, with the CTH distribution exhibiting more of a trimodal nature of shallow, midlevel, and deep systems.

Using a clustering methodology, they identified three main modes of organization: a deep, optically thick mode associated with deep convection; a deep, optically thin (designated "mixed") mode; and a shallow mode representing shallow convection and stratiform precipitation. The mixed mode consisted of shallow to deep precipitating clouds underneath high clouds for the HCTH distribution and deep, but optically thinner clouds for RCTH. Stephens et al. (2019a) showed that for the RCTH distribution, the shallow/mixed/deep modes made the largest/middle/least contributions to light rainfall (<1 mm hr^{-1}), with the reverse true for heavy (>10 mm h^{-1}) rainfall. They also related these cloud

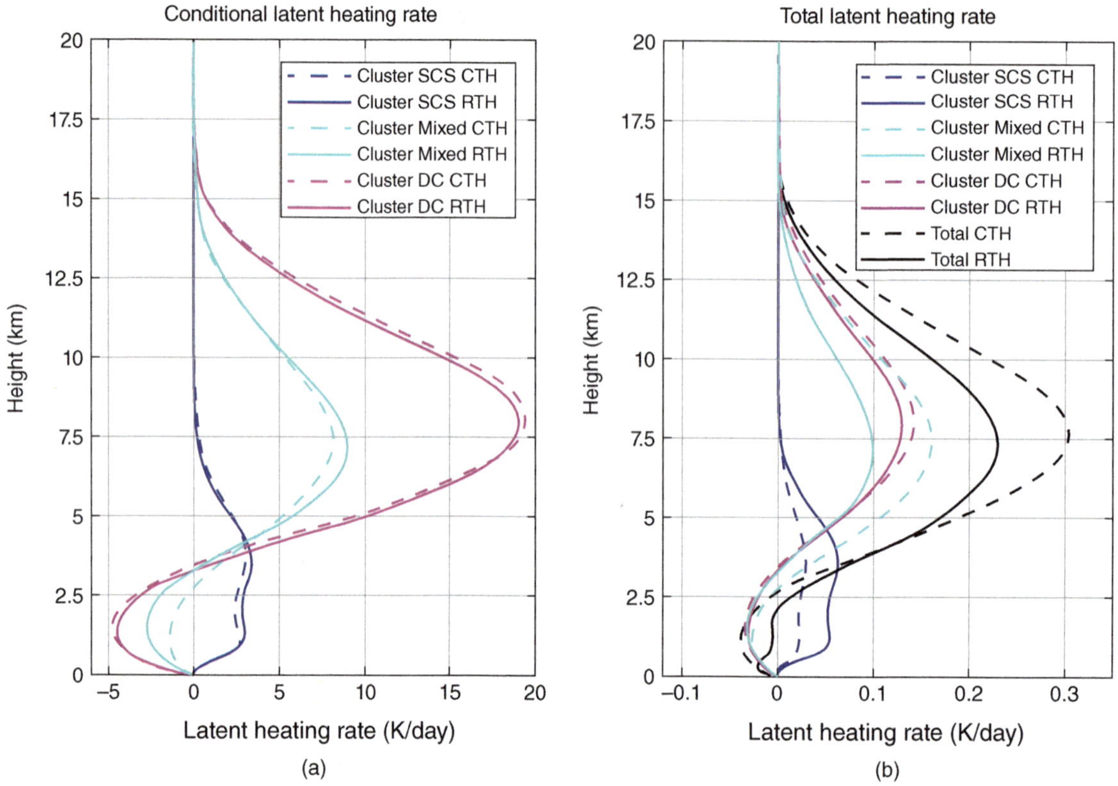

Figure 15.4 (a) The conditional profile latent heating averaged over the tropics when precipitation occurs. (b) The equivalent profiles weighted by the occurrence of each k-means cloud regime, interpreted as the total latent heat contribution from oceanic raining clouds to the total tropical atmosphere. Source: Stephens et al. (2019) / John Wiley & Sons.

modes to latent heating rate profiles (Fig. 15.4) derived from the approach of Schumacher et al. (2004). Averaged heating rates (Fig. 15.4a, conditioned on when precipitation occurs) were similar for the HCTH and RCTH datasets. Latent heating rates for deep systems exhibited a top-heavy profile typical of deep convective systems (convective and stratiform regions). The mixed mode had a similar shape, but with smaller heating/cooling rates, while the shallow mode was characterized by heating up to 7.5 km altitude. When weighted by the frequency of occurrence of the different modes (Fig. 15.4b), the heating rates exhibited more sensitivity to the measure of CTH, with slightly reduced deep-mode heating and substantially reduced mixed-mode heating aloft and increased shallow heating at lower-to-middle levels. Thus, the shallow mode plays a greater role when the more realistic treatment of CTH is used, with the shallow heating largely cancelling the cooling from the deep and mixed modes at low levels.

Huaman and Schumacher (2018) examined the relationship between the distribution of latent heating in the eastern Pacific intertropical convergence zone (ITCZ) and the meridional circulation that it drives. They calculated latent heating rates using an approach similar to Schumacher et al. (2004) in which heating from convective and stratiform precipitation regions was derived from TRMM rainfall rates and heating from lighter shallow precipitation was estimated from CloudSat data. While Huaman and Schumacher (2018) discussed all seasons, only results for June-July-August (JJA) are shown here (Fig. 15.5a). They found distinct seasonal variability, with the northern ITCZ strongest in JJA, with a low- and high-level heating peak. The first six months of the year, in contrast, were characterized by a low-level peak near 800–700 hPa. Other than March-April-May, the southern ITCZ was associated with shallow heating only, with a peak just below 800 hPa. Huaman and Schumacher (2018) also noted that the northern ITCZ sloped northward with height due to a predominance of shallow (deep) heating on its southern (northern) side. Average latent heating rates in the ITCZ (Fig. 15.5b) showed considerable variability between the different estimates. The two TRMM/GPM standard products (CSH and SLH) were reasonably similar, with a top-heavy heating profile peaking near 400 hPa. While the Huaman and Schumacher PR-CPR latent heating profile was similar to CSH and SLH above 500 hPa, it showed a distinct low-level heating peak near 700 hPa associated with the CPR estimates of shallow-cloud heating.

Huaman and Schumacher (2018) compared the satellite-derived heating rates to vertical motions and circulation vectors from two reanalysis datasets of the Modern-Era Retrospective analysis for Research and Applications, version 2 (MERRA2; Fig. 15.5c), and the National Centers for Environmental Prediction–the National Center for Atmospheric Research reanalysis (NCEP-NCAR reanalysis; Fig. 15.5d). The patterns of vertical velocity were consistent with the heating rates, with the largest differences between reanalyses occurring in the JJA and September-October-November (SON) timeframes. The deep, top-heavy heating profile produced a deep overturning circulation, while the low-level heating peak gave rise to a shallow meridional overturning circulation. While not as apparent during JJA, the shallow circulation was apparent in the MERRA2 data (southerly flow in the 800–700 hPa layer, Fig. 15.5c), but was absent in the NCEP-NCAR reanalysis (Fig. 15.5d), which had only weak shallow heating. Their results suggest potentially significant issues with the representation of organized convection and shallow clouds in global circulation models.

The latent heat released during cloud condensation and precipitation formation warms the air and enhances instability inside clouds in adiabatic conditions that invigorates the storms. Deep convective storms are a key source of intense precipitation. The linkage between convection and extreme precipitation has been investigated using the precipitation radars of TRMM and DPR (Gingrey et al., 2018: Kuo et al., 2020; Sohn et al., 2013; Wang & Tang, 2020). Extreme rainfall events were studied by Hamada et al. (2015) using long-term TRMM PR data record. Interestingly, they showed that extreme rainfall events associated with deep convective clouds tend to *not* be associated with extremely tall convective systems. This is illustrated in Figure 15.6 as a statistical difference in vertical profiles of PR-observed radar reflectivity between extreme rainfall and convection cases. A remarkable finding is a downward increase of reflectivity below 8 km in the extreme rainfall ("R-only") case, indicative of a substantial contribution of the warm rain process to formation of the extreme rainfall. The extremely tall ("H-only") case, in contrast, is characterized by more vertically aligned structures even with a slight downward decrease of reflectivity below the melting level. This difference in occurrences between the extreme rainfall and convection cases highlights a critical need for understanding

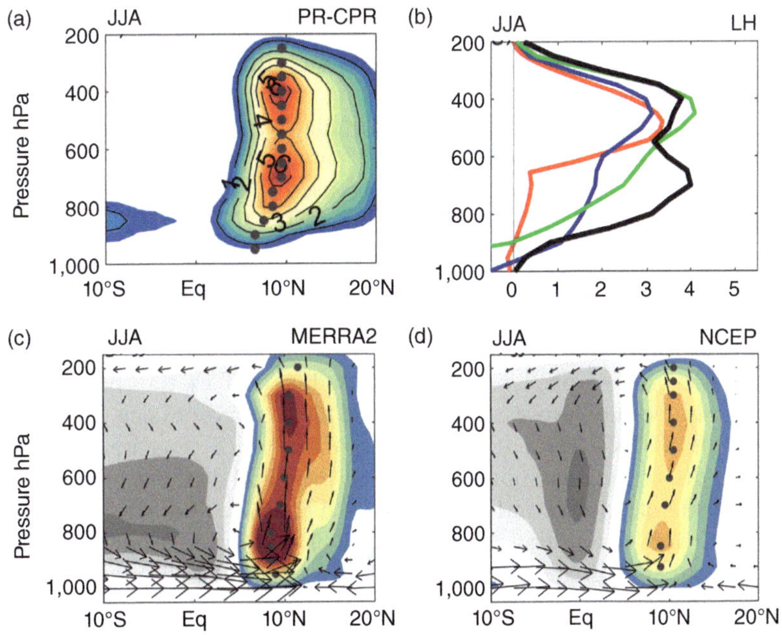

Figure 15.5 (a) Mean latitude-pressure cross-section of latent heating (K day-1) for June-July-August for 130°–90°W from the CloudSat CPR and TRMM PR. (b) Latent heating profiles in the ITCZ (region of precipitation >3 mm day^{-1}) derived from TRMM/CloudSat (black) and TRMM PR-based retrievals: CSH (green), SLH (blue), and PRH (red). (c)–(d) Mean meridional mass flux vectors (kg m^{-2}s^{-1}) and vertical velocity (Pa s^{-1}; shaded) from (c) MERRA-2 and (d) and NCEP-NCAR. The vertical flux component has been amplified 100 times to account for the aspect ratio of the plot. The maximum heating position at each level in (a, c, and d) is indicated by gray dots. Source: Huaman et al. (2018) / with permission of American Meteorological Society.

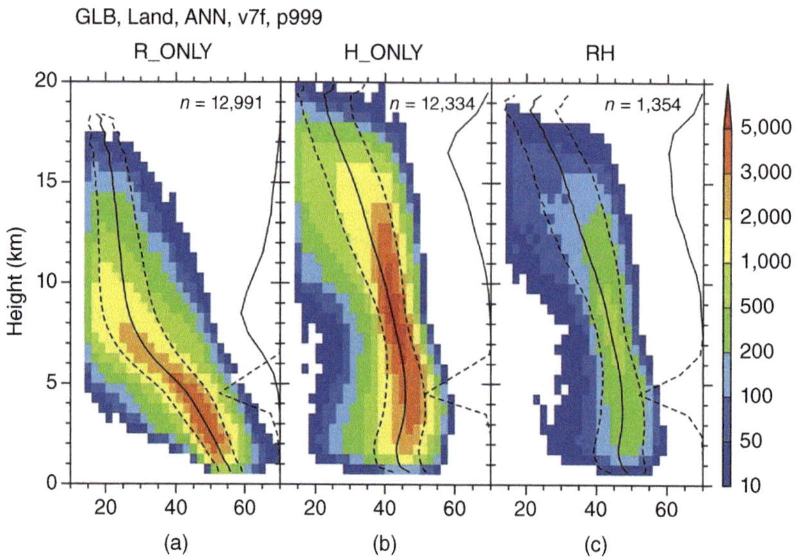

Figure 15.6 Vertical profiles of radar reflectivity statistically represented in the form of joint histograms of radar reflectivity and height over tropical ocean for extreme rainfall (R-only), extreme vertical extent (H-only), and extreme relative humidity (RH) cases. Source: Hamada et al. (2015) / Springer Nature / CC BY 4.0.

relative roles of microphysical and dynamical processes in cloud-precipitation systems. Although the two different aspects (i.e., microphysics and dynamics) have started to be linked with some combinations of active and passive sensors for a spectrum of shallow to deep convective systems (e.g., Kikuchi & Suzuki, 2019), their process-level understanding is still generally missing. Chapter 14: Observed Scaling of Precipitation Extremes discusses dependence of these extreme events on both surface temperature and atmospheric instability in more detail.

15.4.4. Tracking Temporal Evolution

Strictly speaking, the term "process" means the mechanisms responsible for *temporal* evolution of physical states (as discussed in section 15.1). In this sense, the analysis discussed above in this section just "probe" processes statistically, rather than readily "seeing" processes, with a lack of *time* dimension in the analysis. These statistical probing processes are mainly due to the limitation of low Earth orbiting satellites that sample clouds and precipitation only intermittently at a given location and the fact that temporally continuous measurements enabled by recent geostationary satellites (e.g., Himawari-8 and GOES-R) have neither cloud nor precipitation radars onboard. Despite this limitation, however, there have been several attempts to extract "temporal" information of cloud-precipitation system from existing satellite measurements (Fiolleau & Roca, 2013; Letu et al., 2019; Wardah et al., 2008). One such approach is proposed by Luo et al. (2009, 2010) who combined cloud-precipitation states measured by A-Train multi-sensor satellites with information on the ambient environment to infer dynamical characteristics of deep convective clouds, such as whether a given convective system is "transient" or "terminal" in terms of its buoyancy (Luo et al., 2009) and an estimate of the cloud top buoyancy and entrainment rate (Luo et al., 2010). Along with these "indirect" approaches, a more "direct" approach to infer cloud top dynamics is proposed by Luo et al. (2014), who suggested the use of A-Train multiple satellite measurements (i.e., Aqua/MODIS and CALIPSO/IIR) with a slight time lag (~1–2 minutes). They derived convective vertical velocities by taking a difference in cloud top temperatures (measured by the IR channels on both satellites) and related them to cloud internal structures inferred from the CloudSat profiles, thus bringing together cloud dynamics and cloud microphysics. These A-Train analyses are also combined with geostationary satellite measurements to put the cloud dynamic characteristics into the context of cloud life cycle (Takahashi & Luo, 2014).

Such time differences in multiple satellite measurements are also utilized in different and novel ways by Masunaga (2012), who analyzed data from multiple satellite platforms that overpass a given location at *different* local times to construct temporal composites of the satellite observables, often referred to as "*delta-t*" measurements. Specifically, Masunaga (2012) exploited the temporal gap between TRMM and Aqua satellite overpasses to project their measurements onto a composite temporal

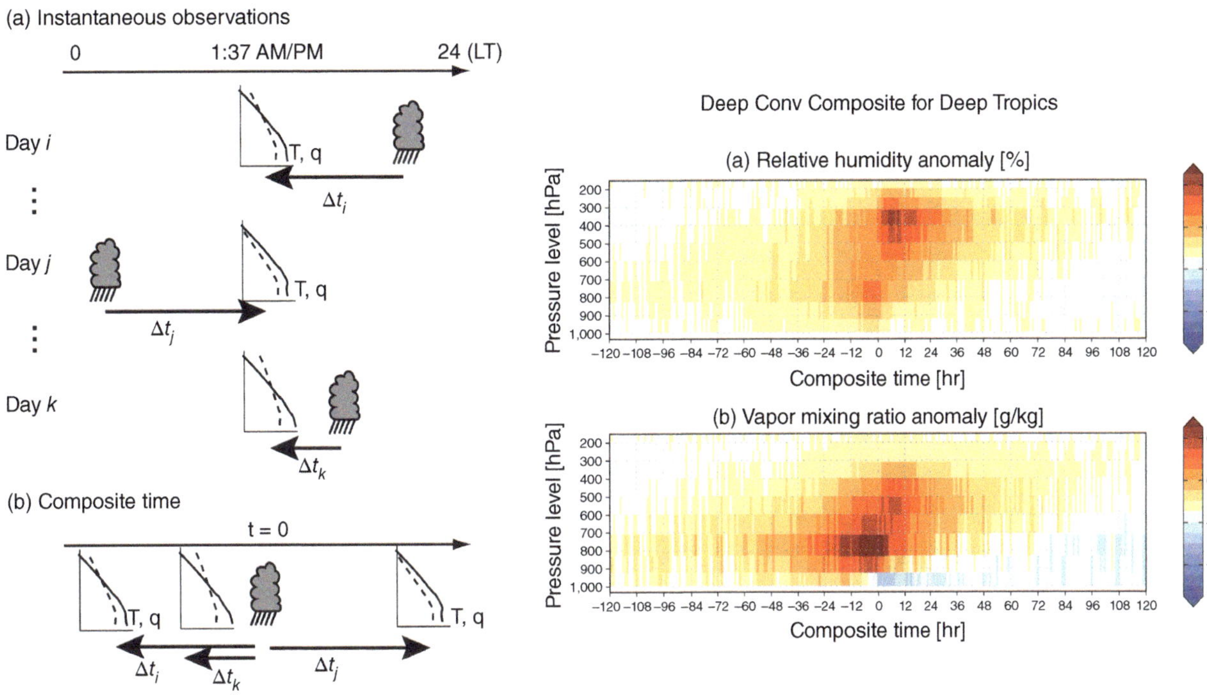

Figure 15.7 (Left) Schematic illustrations of the temporal composite analysis with (a) TRMM and Aqua measurements on individual days and (b) their projections onto the composite space. (Right) Temporal sequence of relative humidity anomaly (upper) and vapor mixing ratio anomaly (lower) composited with respect to deep convection. Source: Masunaga et al. (2012) / with permission of American Meteorological Society.

sequence of, for instance, water vapor mixing ratio as a function of relative time prior and subsequent to precipitation occurrence (Fig. 15.7, left panels), depicting temporal evolutions of water vapor profile associated with precipitating convection (Fig. 15.7, right panels). This analysis methodology was then applied to a variety of satellite platforms and sensors to investigate thermodynamic evolution of atmospheric phenomena relevant to convection and precipitation in the tropics and subtropics over relatively short (hourly to daily) timescales (e.g., Masunaga, 2013; Masunaga & Bony, 2018). This work exemplifies a promising approach to further advance our understanding of "dynamical" characteristics of cloud-precipitation processes, when combined with independent approaches of diagnosing cloud top dynamics described above (Luo et al., 2010, 2014) to infer convective and large-scale mass fluxes separately (Masunaga & Luo, 2016) and also through being applied to investigation of aerosol scavenging by convection (Sauter & L'Ecuyer, 2017; Sauter et al., 2019). These analysis approaches readily add the *time* dimension to satellite observations of cloud-precipitation systems and their interactions with aerosols and convection, particularly expected by the enhanced capabilities in future missions.

15.5. FUTURE SATELLITE MISSIONS FOR CLOUD-PRECIPITATION PROCESS STUDIES

The next radar mission, the Earth, Clouds, Aerosols and Radiation Explorer (EarthCARE) satellite, is developed jointly by the European Space Agency (ESA), JAXA, and National Institute of Information and Communications (NICT). EarthCARE is regarded as a historical successor of the vertical cloud and aerosol observations by the A-Train satellites, loading four sensors, (1) Cloud Profiling Radar (CPR), (2) Atmospheric Lidar (ATLID), (3) Multispectral Imager (MSI), and (4) Broadband Radiometer (BBR), all of which are specifically designed to investigate global profiles of cloud, aerosol, and precipitation and associated radiative properties (Illingworth et al., 2015). In terms of the precipitating cloud measurement, the EarthCARE CPR is expected to be the first space-borne weather radar with a Doppler measurement capability, enabling the measurement of vertical motions of hydrometeors globally (Kollias et al., 2014; Hagihara et al., 2021). Together with the High Spectral Resolution Lidar (HSRL) of ATLID, the measurements will offer novel information, that is, cloud dynamics and better particle habit identifications, to enhance our understanding on the precipitation processes described in previous sections particularly for mixed-phase clouds with cold rain processes, which had been challenging for previous measurements. In particular, the dynamical information offered by the CPR Doppler measurement is expected to add a temporal context to the cloud-precipitation vertical measurement achieved by the CloudSat.

With a particular focus in observing storm life cycles, the Time-Resolved Observations of Precipitation structure and storm Intensity with a Constellation of Smallsats (TROPICS) was selected by NASA as part of the Earth Venture-Instrument (EVI-3) program (Blackwell et al., 2018). The mission, led by the Massachusetts Institute of Technology/Lincoln Lab, was designed to obtain rapid-revisit passive microwave observations of tropical cyclones, with median revisit rates of less than 50 minutes. TROPICS comprises a constellation of four Cubesats, two satellites in each of two 30°-inclination orbits, each satellite carrying a passive microwave sensor with seven channels near 118 GHz for temperature profiling, three channels near 183 GHz for moisture profiling, and imaging channels at 92 and 205 GHz (Note that it originally planned for six satellites across three orbits, but the launch of the first two satellites failed to achieve orbit, thereby reducing the mean revisit rate to about 60 minutes). Its science objectives are focused on relating the evolution of precipitation structure, including convective bursts, to storm intensity change and warm-core development, relating environmental moisture to storm structure (including size) and intensity, and assessing the impact of the observations on numerical and statistical tropical cyclone prediction models.

Another promising observing system currently under formulation at NASA is the Atmosphere Observing System (AOS, https://aos.gsfc.nasa.gov), representing two designated observables identified in the 2017 Earth Science Decadal Survey (National Academies of Science, Engineering, and Medicine, 2018). The Aerosol designated observable focuses on vertical profiles of aerosol and clouds, with a candidate payload including a backscatter lidar and multi-channel, multi-angle, polarization imaging radiometer. For Clouds, Convection, and Precipitation (CCP), the science focuses on coupled cloud-precipitation properties and dynamics, with a candidate measurement approach that includes a possible dual-frequency Doppler radar with a multi-frequency passive microwave and sub-millimeter radiometer. The overall science objectives of AOS are defined to address diverse research objectives, including low- and high-cloud feedbacks; the dynamics of convective storm systems; cold cloud and precipitation processes (including snowfall and mixed-phase properties); aerosol attribution and air quality; aerosol processing, wet removal, and vertical redistribution; aerosol direct effects and absorption; and aerosol indirect effects. The architecture under formulation includes the candidate sensors identified above as well as important international contributions focused on convection and precipitation, the properties of ice clouds, and profiles of aerosols and moisture in the upper troposphere and lower stratosphere.

15.6. SUMMARY

Over the course of 50 years, satellite passive microwave instruments have provided unprecedented measurements for characterizing global precipitation. The measurement was essentially initiated by the microwave radiometer of SSM/I, which estimated precipitation using the microwave emission and scattering nature of hydrometeors. The successful continuity of microwave radiometry – including TRMM TMI, the AMSR series, SSMIS, and GPM GMI – established them as the backbone of the global precipitation satellite observing system. The first microwave radar measurements, acquired late in the twentieth century by the TRMM PR, added a vertical dimension to the 2D picture provided by the radiometers. Given the intrinsic 3D nature of cloud and precipitation systems, the measurements from the TRMM, CloudSat, and GPM radars accelerated process studies in the satellite community, looking deeper into physical mechanisms behind the precipitation observed. Such studies have shed light on key processes such as latent heat release, warm rain, and convective storm processes, and have even explored extracting temporal information from the snapshot observations. The "golden era" of satellite observations will be further nurtured by the planned EarthCARE mission and AOS observing system currently under study, with their enhanced capabilities that will further add a temporal dimension to global precipitation measurements and thus deepen our understanding of cloud-precipitation dynamics and their feedback mechanisms.

REFERENCES

Adler, R. F., & Negri, A. J. (1988). A satellite infrared technique to estimate tropical convective and stratiform rainfall. *Journal of Applied Meteorology and Climatology*, *27*, 30–51. https://doi.org/10.1175/1520-0450(1988)027<0030:ASITTE>2.0CO;2

Aonashi, K., & Ferraro, R. R. (2020). Microwave sensors, imagers and sounders. In V. Levizzani, C. Kidd, D. B. Kirschbaum, C. D. Kummerow, K. Nakamura, & F. J. Turk (Eds.), *Satellite precipitation measurement, advances in global change research* (Vol. 67, pp. 63–81). Cham: Springer. https://doi.org/10.1007/978-3-030-24568-9_4

Arkin, P. A., & Meisner, B. N. (1987). The relationship between large-scale convective rainfall and cold cloud over the Western Hemisphere during 1982–84. *Monthly Weather Review*, *115*, 51–74. https://doi.org/10.1175/1520-0493(1987)115<0051:TRBLSC>2.0.CO;2

Ba, M. B., & Gruber, A. (2001). GOES multispectral rainfall algorithm (GMSRA). *Journal of Applied Meteorology and Climatology*, *40*, 1500–1514. https://doi.org/10.1175/1520-0450(2001)040<1500:GMRAG>2.0.CO;2

Bang, S. D., & Cecil, D. J. (2019). Constructing a multifrequency passive microwave hail retrieval and climatology in the GPM domain. *Journal of Applied Meteorology and Climatology*, *58*, 1889–1904. https://doi.org/10.1175/JAMC-D-19-0042.1

Barnes, H. C., Zuluaga, M. D., & Houze Jr., R. A. (2015). Latent heating characteristics of the MJO computed from TRMM Observations. *Journal of Geophysical Research: Atmospheres*, *120*, 1322–1334. https://doi.org/10.1002/2014JD022530

Battaglia, A., Tanelli, S., Kobayashi, S., Zrnic, D., Hogan, R. J., & Simmer, C. (2010). Multiple-scattering in radar systems: A review. *Journal of Quantitative Spectroscopy & Radiative Transfer*, *111*(6), 917–947. https://doi.org/10.1016/j.jqsrt.2009.11.024

Battaglia, A., Kollias, P., Dhillon, R., Roy, R., Tanelli, S., Lebsock, M., et al. (2020a). Spaceborne cloud and precipitation radars: Status, challenges, and ways forward. *Reviews of Geophysics*, *58*(3), e2019RG000686. https://doi.org/10.1029/2019RG000686

Battaglia, A., Tanelli, S., Tridon, F., Kneifel, S., Leinonen, J., & Kollias, P. (2020b). Triple-frequency radar retrievals. In V. Levizzani, C. Kidd, D. B. Kirschbaum, C. D. Kummerow, K. Nakamura, & F. J. Turk (Eds.), *Satellite precipitation measurement, advances in global change research* (Vol. 67, pp. 211–229). Cham: Springer. https://doi.org/10.1007/978-3-030-24568-9_13

Behrangi, A., Hsu, K., Imam, B., Sorooshian, S., Huffman, G. J., & Kuligowski, R. J. (2009). PERSIANN-MSA: A precipitation estimation method from satellite-based multispectral analysis. *Journal of Hydrometeorology*, *10*(6), 1414–1429. https://doi.org/10.1175/2009JHM1139.1

Bellerby, T. J. (2004). A feature-based approach to satellite precipitation monitoring using geostationary IR imagery. *Journal of Hydrometeorology*, *5*, 910–921. https://doi.org/10.1175/1525-7541(2004)005<0910:AFATSP>2.0.CO;2

Berg, W., LEcuyer, T., & Haynes, J. M. (2010). The distribution of rainfall over oceans from spaceborne radars. *Journal of Applied Meteorology and Climatology*, *49*(3), 535–543. https://doi.org/10.1175/2009JAMC2330.1

Bergeron, T. (1935). On the physics of cloud and precipitation. In *Proceedings of the 5th General Assembly of the UGGI* (Vol. 2, pp. 156–175) UGGI, Lisbon.

Blackwell, W. J., Braun, S., Bennartz, R., Velden, C., DeMaria M., Atlas, R., et al. (2018). An overview of the TROPICS NASA Earth venture mission. *Quarterly Journal of the Royal Meteorological Society*, *144*(S1), 16–26. https://doi.org/10.1002/qj.3290

Bodas-Salcedo, A., Webb, M. J., Bony, S., Chepfer, H., Dufresne, J., Klein, S. A., et al. (2011). COSP: Satellite simulation software for model assessment. *Bulletin of the American Meteorological Society*, *92*(8), 1023–1043. https://doi.org/10.1175/2011BAMS2856.1

Bodas-Salcedo, A., Williams, K. D., Field, P. R., & Lock, A. P. (2012). Downwelling solar radiation surplus over the Southern Ocean in the Met Office Model: The role of midlatitude cyclone clouds. *Journal of Climate*, *25*, 7467–7486. https://doi.org/10.1175/JCLI-D-11-00702.1

Bodas-Salcedo, A., Williams, K. D., Ringer, M. A., Beau, I., Cole, J. N. S., Dufresne, J.-L., et al. (2014). Origins of the solar radiation biases over the Southern Ocean in CFMIP2 models. *Journal of Climate*, *27*(1), 41–56. https://doi.org/10.1175/JCLI-D-13-00169.1

Bodas-Salcedo, A., Hill, P. G., Furtado, K., Williams, K. D., Field, P. R., Manners, J. C., et al. (2016). Large contribution of supercooled liquid clouds to the solar radiation budget of the Southern Ocean. *Journal of Climate*, *29*(11), 4213–4228. https://doi.org/10.1175/JCLI-D-15-0564.1

Boening, C., Lebsock, M., Landerer, F., & Stephens, G. (2012). Snowfall-driven mass change on the East Antarctic ice sheet. *Geophysical Research Letters*, *39*, L21501. https://doi.org/10.1029/2012GL053316

Chang, K.-W., & L'Ecuyer, T. (2019). Role of latent heating vertical distribution in the formation of the tropical cold trap. *Journal of Geophysical Research: Atmospheres*, *124*, 7836–7851. https://doi.org/10.1029/2018JD030194

Cesana, G., & Storelvmo, T. (2017). Improving climate projections by understanding how cloud phase affects radiation. *Journal of Geophysical Research: Atmospheres*, *122*, 4594–4599. https://doi.org/10.1002/2017JD026927

Dai, A. (2006). Precipitation characteristics in eighteen coupled climate models. *Journal of Climate*, *19*(18), 4605–4630. https://doi.org/10.1175/JCLI3884.1

Findeisen, W. (1938). Kolloid-meteorologische Vorgänge bei Niederschlagsbildung. *Meteorologische Zeitschrift*, *55*, 121–133. (trans. and ed. by E. Volken, A.M. Giesche, & S. Brönnimann, (2015). *Meteorologische Zeitschrift*, *24*(4), 443–454. https://doi.org/10.1127/metz/2015/0675).

Fiolleau, T. & Roca, R. (2013). Composite life cycle of tropical mesoscale convective systems from geostationary and low Earth orbit satellite observations: Method and sampling considerations. *Quarterly Journal of the Royal Meteorological Society*, *139*, 941–953. https://doi.org/10.1002/qj.2174

Forbes, R. M., & Ahlgrimm, M. (2014). On the representation of high-latitude boundary layer mixed-phase cloud in the ECMWF global model. *Monthly Weather Review*, *142*(9), 3425–3445. https://doi.org/10.1175/MWR-D-13-00325.1

Forbes, R., Geer, A., Lonitz, K., & Ahlgrimmm M. (2016). Reducing systematic errors in cold-air outbreaks. *ECMWF Newsletter*, *146*, 17–22. https://doi.org/10.21957/s41h7q7l

Gingrey, A., Varble, A., & Zipser, E. (2018). Relationships between extreme rain rates and convective intensities from the perspectives of TRMM and WSR-88D radars. *Journal of Applied Meteorology and Climatology*, *57*(6), 1353–1369. https://doi.org/10.1175/JAMC-D-17-0240.1

Golaz, J.-C., Horowitz, L. W., & Levy II, H. (2013). Cloud tuning in a coupled climate model: Impact on 20th century warming. *Geophysical Research Letters*, *40*, 2246–2251. https://doi.org/10.1002/grl.50232

Grecu, M., Olson, W. S., Munchak, S. J., Ringerud, S., Liao, L., Haddad, Z. S., et al. (2016). The GPM combined algorithm. *Journal of Atmospheric and Oceanic Technology*, *33*(10), 2225–2245. https://doi.org/10.1175/JTECH-D-16-0019.1

Grody, N. C. (1991). Classification of snow cover and precipitation using the special sensor microwave imager. *Journal of Geophysical Research: Atmospheres*, *96*(D4), 7423–7435. https://doi.org/10.1029/91JD00045

Hagihara, Y., Ohno, Y., Horie, H., Roh, W., Satoh, M., Kubota, T., & Oki, R. (2021). Assessments of Doppler velocity errors of EarthCARE cloud profiling radar using global cloud system resolving simulations: Effects of Doppler broadening and folding. *IEEE Transactions on Geoscience and Remote Sensing, accepted.* https://doi.org/10.1109/TGRS.2021.3060828

Hamada, A., Takayabu, Y. N., Liu, C., & Zipser, E. J. (2015). Weak linkage between the heaviest rainfall and tallest storms. *Nature Communications*, *6*, 6213. https://doi.org/10.1038/ncomms7213

Hashino, T., Satoh, M., Hagihara, Y., Kubota, T., Matsui, T., Nasuno, T., & Okamoto, H. (2013). Evaluating cloud microphysics from NICAM against CloudSat and CALIPSO. *Journal of Geophysical Research: Atmospheres*, *118*, 7273–7292. https://doi.org/10.1002/jgrd.50564

Haynes, J. M., Marchand, R. T., Luo, Z., Bodas-Salcedo, A., & Stephens, G. L. (2007). A multipurpose radar simulation package: QuickBeam. *Bulletin of the American Meteorological Society*, *88*(11), 1723–1727. https://doi.org/10.1175/BAMS-88-11-1723

Hitschfeld, W., & Bordan, J. (1954). Errors inherent in the radar measurement of rainfall at attenuating wavelengths. *Journal of Meteorology*, *11*, 58–67. https://doi.org/10.1175/1520-0469(1954)011<0058:EIITRM>2.0.CO;2

Hollinger, J., Lo, R., & Poe, G. (1987). *Special sensor microwave/imager user's guide*. Washington, DC: Naval Research Laboratory.

Hong, Y., Hsu, K.-L., Sorooshian, S., & Gao, X. (2004). Precipitation estimation from remotely sensed imagery using an acritical neural network cloud classification system. *Journal of Applied Meteorology and Climatology*, *43*, 1834–1853. https://doi.org/10.1175/JAM2173.1

Hogan, R., & Battaglia, A. (2008). Fast lidar and radar multiple-scattering models. Part 2: Wide-angle scattering using the time-dependent two-stream approximation. *Journal of the Atmospheric Sciences*, *65*(12), 3636–3651. https://doi.org/10.1175/2008JAS2643.1

Hou, A. Y., Kakar, R. K., Neeck, S., Azarbarzin, A. A., Kummerow, C. D., Kojima, M., et al. (2014). The global precipitation measurement mission. *Bulletin of the American Meteorological Society*, *95*, 701–722. https://doi.org/10.1175/BAMS-D-13-00164.1

Houze Jr., R. A., Rasmussen, K. L., Zuluaga, M. D., & Brodzik, S. R. (2015). The variable nature of convection in the tropics and subtropics: A legacy of 16 years of the Tropical Rainfall Measuring Mission satellite. *Reviews of Geophysics*, *53*(3), 994–1021. https://doi.org/10.1002/2015RG000488

Huffman, G. J., Bolvin, D. T., Nelkin, E. J., Wolff, D. B., Adler, R. E., Gu, G., et al. (2007). The TRMM multi-satellite precipitation analysis: Quasi-global, multi-year, combined-sensor precipitation estimates at fine scale. *Journal of Hydrometeorology*, *8*, 38–55. https://doi.org/10.1175/JHM560.1

Hsu, K.-L., & Sorooshian, S. (2009). Satellite-based precipitation measurement using PERSIANN system. In S. Sorooshian, K.-L. Hsu, E. Coppola, B. tomasserti, M. Verdecchia & G. Visconti (Eds.), *Hydrological modeling and the water cycle* (pp. 27–48). Berlin/Heidelberg: Springer. ISBN: 978-3-540-77843-1.

Huaman, L., & Schumacher, C. (2018). Assessing the vertical latent heating structure of the east Pacific ITCZ using the CloudSat CPR and TRMM PR. *Journal of Climate*, *31*, 2563–2577. https://doi.org/10.1175/JCLI-D-17-0590.1

Huang, Y., Siems, S. T., Manton, M. J., Protat, A., & Delanoë, J. (2012). A study on the low-altitude clouds over the Southern Ocean using the DARDAR-MASK. *Journal of Geophysical Research: Atmospheres*, *117*, D18204. https://doi.org/10.1029/2012JD017800

Iguchi, T. (2020). Dual-frequency Precipitation Radar (DPR) on the Global Precipitation Measurement (GPM) mission's core observatory. In V. Levizzani, C. Kidd, D. B. Kirschbaum, C. D. Kummerow, K. Nakamura, & F. J. Turk (Eds.), *Satellite precipitation measurement, advances in global change research* (Vol. 67, pp. 183–192). Cham: Springer. https://doi.org/10.1007/978-3-030-24568-9_11

Iguchi, T., & Haddad, Z., (2020). Introduction to radar rain retrieval methods. In V. Levizzani, C. Kidd, D. B. Kirschbaum, C. D. Kummerow, K. Nakamura, & F. J. Turk (Eds.), *Satellite precipitation measurement, advances in global change research* (Vol. 67, pp 169–182). Cham: Springer. https://doi.org/10.1007/978-3-030-24568-9_10

Illingworth, A. J., Barker, H. W., Beljaars, A., Ceccaldi, M., Chepfer, H., Clerbaux, N., et al. (2015). The EarthCARE satellite: The next step forward in global measurements of clouds, aerosols, precipitation, and radiation. *The Bulletin of the American Meteorological Society*, *96*. https://doi.org/10.1175/BAMS-D-12-00227.1

Jing, X., & Suzuki, K., (2018). The impact of process-based warm rain constraints on the aerosol indirect effect. *Geophysical Research Letters*, *45*, 10729–10737. https://doi.org/10.1029/2018GL079956

Jing, X., Suzuki, K., Guo, H., Goto, D., Ogura, T., Koshiro, T., & Mulmenstadt, J. (2017). A multi-model study on warm precipitation biases in global models compared to satellite observations. *Journal of Geophysical Research: Atmospheres*, *122*, 11806–11824. https://doi.org/10.1002/2017JD027310

Jing, X., Suzuki, K., & Michibata, T. (2019). The key role of warm rain parameterization in determining the aerosol indirect effect in a global climate model. *Journal of Climate*, *32*, 4409–4430. https://doi.org/10.1175/JCLI-D-18-0789.1

Joyce, R. J., & Xie, P. (2011). Kalman filter-based CMORPH. *Journal of Hydrometeorology*, *12*, 1547–1563. https://doi.org/10.1175/JHM-D-11-022.1

Kawai, H., Yukimoto, S., Koshiro, T., Oshima, N., Tanaka, T., Yoshimura, H., & Nagasawa, R. (2019). Significant improvement of cloud representation in the global climate model MRI-ESM2. *Geoscientific Model Development*, *12*, 2875–2897. https://doi.org/10.5194/gmd-12-2875-2019

Kay, J. E., Bourdages, L., Miller, N. B., Morrison, A., Yettella, V., Chepfer, H., & Eaton, B. (2016). Evaluating and improving cloud phase in the Community Atmosphere Model version 5 using spaceborne lidar observations. *Journal of Geophysical Research: Atmospheres*, *121*, 4162– 4176. https://doi.org/10.1002/2015JD024699

Kay, J. E., L'Ecuyer, T., Pendergrass, A., Chepfer, H., Guzman, R., & Yettella, V. (2018). Scale-aware and definition-aware evaluation of modeled near-surface precipitation frequency using CloudSat observations. *Journal of Geophysical Research: Atmospheres*, *123*, 4294–4309. https://doi.org/10.1002/2017JD028213

Kidd, C., Becker, A., Huffman, G. J., Muller, C.L., Joe, P., Skofronick-Jackson, G., & Kirschbaum, D. B. (2017). So, how much of the Earth's surface is covered by rain gauges? *Bulletin of the American Meteorological Society*, *98*, 69–78. https://doi.org/10.1175/BAMS-D-14-00283.1

Kidd, C., Takayabu, Y. N., Skofronick-Jackson, G. M., Huffman, G. J., Braun, S. A., Kubota, T., & Turk, F. J. (2020). The Global Precipitation Measurement (GPM) mission. In V. Levizzani, C. Kidd, D. B. Kirschbaum, C. D. Kummerow, K. Nakamura, & F. J. Turk (Eds.), *Satellite precipitation measurement, advances in global change research* (Vol. 67, pp. 3–23). Cham: Springer Nature. https://doi.org/10.1007/978-3-030-24568-9_1

Kikuchi, M., & Suzuki, K. (2019). Characterizing vertical particle structure of precipitating cloud system from multiplatform measurements of A-Train constellation. *Geophysical Research Letters*, *46*, 1040–1048. https://doi.org/10.1029/2018GL081244

Kilic, L., Prigent, C., Boutin, J., Meissner, T., English, S., & Yueh, S. (2019). Comparisons of ocean radiative transfer models with SMAP and AMSR2 observations. *Journal of Geophysical Research: Oceans*, *124*, 7683–7699. https://doi.org/10.1029/2019JC015493

Kollias, P., Tanelli, S., Battaglia, A., & Tatarevic, A. (2014). Evaluation of EarthCARE cloud profiling radar Doppler velocity measurements in particle sedimentation regimes. *Journal of Atmospheric and Oceanic Technology*, *31*(2), 366–386. https://doi.org/10.1175/JTECH-D-11-00202.1

Kubar, T. L., Hartmann, D. L., & Wood, R. (2009). Understanding the importance of microphysics and macrophysics for warm rain in marine low clouds. Part I: Satellite observations. *Journal of the Atmospheric Sciences*, *66*, 2953–2972. https://doi.org/10.1175/2009JAS3071.1

Kummerow, C. D. (2020). Introduction to passive microwave retrieval methods. In V. Levizzani, C. Kidd, D. B. Kirschbaum, C. D. Kummerow, K. Nakamura, & F. J. Turk (Eds.), *Satellite precipitation measurement, advances in global change research* (Vol. 67, pp. 123–140). Cham: Springer Nature. https://doi.org/10.1007/978-3-030-24568-9_7.

Kummerow, C. D., Randel, D. L., Kulie, M., Wang, N., Ferraro, R., Munchak, S. J., & Petkovic, V. (2015). The evolution of the Goddard profiling algorithm to a fully parametric scheme. *Journal of Atmospheric and Oceanic Technology*, *32*, 2265–2280. https://doi.org/:10.1175/jtech-d-15-0039.1.

Kummerow, C., Simpson, J., Thiele, O., Barnes, W., Chang, A. T. C., Stocker, E., et al. (2000). The status of the Tropical Rainfall Measuring Mission (TRMM) after two years in orbit. *Journal of Applied Meteorology and Climatology*, *39*, 1965–1982. https://doi.org/10.1175/1520-0450(2001)040%3C1965:TSOTTR%3E2.0.CO;2

Kuo, Y., Neelin, J. D., Chen, C., Chen, W., Donner, L. J., Gettelman, A., et al. (2020). Convective transition statistics over tropical oceans for climate model diagnostics: GCM evaluation. *Journal of the Atmospheric Sciences*, *77*(1), 379–403. https://doi.org/10.1175/JAS-D-19-0132.1

Lebsock, M., Stephens, G. L., & Kummerow, C. (2008). Multisensor satellite observations of aerosol effects on warm clouds. *Journal of Geophysical Research: Atmospheres*, *113*, D15205. https://doi.org/10.1029/2008JD009876

Lebsock, M. D., L'Ecuyer, T. S., Wood, N. B., Haynes, J. M. & Smalley, M. A. (2020). Status of the CloudSat mission. In V. Levizzani, C. Kidd, D. B. Kirschbaum, C. D. Kummerow, K. Nakamura, & F. J. Turk (Eds.), *Satellite precipitation measurement, advances in global change research* (Vol. 67, pp. 211–229). Cham: Springer Nature. https://doi.org/10.1007/978-3-030-24568-9_2

Leon, D. C., Wang, Z., & Liu, D. (2008). Climatology of drizzle in marine boundary layer clouds based on 1 year of data from CloudSat and Cloud-Aerosol Lidar and Infrared Pathfinder Satellite Observations (CALIPSO). *Journal of Geophysical Research: Atmospheres*, *113*, D00A14. https://doi.org/10.1029/2008JD009835

L'Ecuyer, T. S., & Stephens, G. L. (2002). An estimation-based precipitation retrieval algorithm for attenuating radars. *Journal of Applied Meteorology and Climatology*, *41*(3), 272–285. https://doi.org/10.1175/1520-0450(2002)041%3C0272:AEBPRA%3E2.0.CO;2

L'Ecuyer, T. S., Berg, W., Haynes, J., Lebsock, M., & Takemura, T. (2009). Global observations of aerosol impacts on precipitation occurrence in warm maritime clouds. *Journal of Geophysical Research: Atmospheres*, *114*, D09211. https://doi.org/10.1029/2008JD011273

Letu, H., Nagao, T. M., Nakajima, T. Y., Riedi, J., Ishimoto, H., Baran, A. J., et al. (2019). Ice cloud properties from Himawari-8/AHI next-generation geostationary satellite: Capability of the AHI to monitor the DC cloud generation process. *IEEE Transactions on Geoscience and Remote Sensing*, *57*. https://doi.org/10.1109/TGRS.2018.2882803

Lhermitte, R. (1990). Attenuation and scattering of millimeter wavelength radiation by clouds and precipitation. *Journal of Atmospheric and Oceanic Technology*, *7*, 464–479. https://doi.org/10.1175/1520-0426(1990)007%3C0464:AASOMW%3E2.0.CO;2

Liu, G. (2008). A database of microwave single-scattering properties for nonspherical ice particles. *Bulletin of the American Meteorological Society*, *89*, 1563–1570. https://doi.org/10.1175/2008bams2486.1

Liu, C., Shige, S., Takayabu, Y., & Zipser, E. (2015). Latent heating contribution from precipitation systems with different sizes, depths, and intensities in the tropics. *Journal of Climate*, *28*, 186–203. https://doi.org/10.1175/JCLI-D-14-00370.1

Luo, Z. J., Liu, G. Y., Stephens, G. L., & Johnson, R. H. (2009). Terminal versus transient cumulus congestus: A CloudSat perspective. *Geophysical Research Letters*, *36*, L05808. https://doi.org/10.1029/2008GL036927

Luo, Z. J., Liu, G. Y., & Stephens, G. L. (2010). Use of A-Train data to estimate convective buoyancy and entrainment rate.

Geophysical Research Letters, *37*, L09804. https://doi.org/10.1029/2010GL042904

Luo, Z. J., Jeyaratnam, J., Iwasaki, S., Takahashi, H., & Anderson, R. (2014). Convective vertical velocity and cloud internal vertical structure: An A-Train perspective. *Geophysical Research Letters*, *41*. https://doi.org/10.1002/2013GL058922

Luo, Z. J., Anderson, R., Rossow, W. B., & Takahashi, H. (2017). Tropical cloud and precipitation regimes as seen from near-simultaneous TRMM, CloudSat, and CALIPSO observations and comparison with ISCCP. *Journal of Geophysical Research: Atmospheres, 122*. https://doi.org/10.1002/2017JD026569

Marra, A. C., Porcu, F., Baldini, L., Petracca, M., Casella, D., Dietrich, S., et al. (2017). Observational analysis of an exceptionally intense hailstorm over the Mediterranean area: Role of the GPM core observatory. *Atmospheric Research*, *192*, 72–90. https://doi.org/10.1016/j.atmosres.2017.03.019

Masunaga, H., Matsui, T., Tao, W., Hou, A. Y., Kummerow, C. D., Nakajima, T., et al. (2010). Satellite data simulator unit: A multisensor, multispectral satellite simulator package. *Bulletin of the American Meteorological Society*, *91*, 1625–1632. https://doi.org/10.1175/2010BAMS2809.1

Masunaga, H. (2012). A satellite study of the atmospheric forcing and response to moist convection over tropical and subtropical oceans. *Journal of the Atmospheric Sciences*, 150–167. https://doi.org/10.1175/JAS-D-11-016.1

Masunaga, H. (2013). A satellite study of tropical moist convection and environmental variability: A moisture and thermal budget analysis. *Journal of the Atmospheric Sciences*, *70*, 2443–2465. https://doi.org/10.1175/JAS-D-12-0273.1

Masunaga, H., & Bony, S. (2018). Radiative invigoration of tropical convection by preceding cirrus clouds. *Journal of the Atmospheric Sciences*, *75*, 1327–1342. https://doi.org/10.1175/JAS-D-17-0355.1

Masunaga, H., & Luo, Z. J. (2016). Convective and large-scale mass flux profiles over tropical oceans determined from synergistic analysis of a suite of satellite observations. *Journal of Geophysical Research: Atmospheres*, *121*, 7958–7974. https://doi.org/10.1002/2016JD024753

Mathew, S. S., & Kumar, K. K. (2019). On the role of precipitation latent heating in modulating the strength and width of the Hadley circulation. *Theoretical and Applied Climatology*, *136*, 661–673. https://doi.org/10.1007/s00704-018-2515-4

McIlhattan, E. A., L'Ecuyer, T. S., & Miller, N. B. (2017). Observational evidence linking Arctic supercooled liquid cloud biases in CESM to snowfall processes. *Journal of Climate*, *30*(12), 4477–4495. https://doi.org/10.1175/JCLI-D-16-0666.1

Meneghini, R., Kim, H., Liao, L., Jones, J. A., & Kwiatkowski, J. M. (2015). An initial assessment of the surface reference technique applied to data from the Dual-Frequency Precipitation Radar (DPR) on the GPM satellite. *Journal of Atmospheric and Oceanic Technology*, *32*, 2281–2296. https://doi.org/10.1175/JTECH-D-15-0044.1

Michibata, T., Suzuki, K., Sekiguchi, M., & Takemura, T. (2019). Prognostic precipitation in the MIROC6-SPRINTARS GCM: Description and evaluation against satellite observations. *Journal of Advances in Modeling Earth Systems*, *11*, 839–860. https://doi.org/10.1029/2018MS001596

Michibata, T., & Suzuki, K. (2020). Reconciling compensating errors between precipitation constraints and the energy budget in a climate model. *Geophysical Research Letters*, *47*, e2020GL088340. https://doi.org/10.1029/2020GL088340

Morrison, H., de Boer, G., Feingold, G., Harrington, J., Shupe, M. D., & Sulia, K. (2012). Resilience of persistent Arctic mixed-phase clouds. *Nature Geoscience*, *5*, 11–17. https://doi.org/10.1038/ngeo1332

Mroz, K., Battaglia, A., Lang, T. J., Tanelli, S., & Sacco, G. F. (2018). Global precipitation measuring Dual-Frequency Precipitation Radar observations of hailstorm vertical structure: Current capabilities and drawbacks. *Journal of Applied Meteorology and Climatology*, *57*, 2161–2178. https://doi.org/10.1175/JAMC-D-18-0020.1

Nakamura, K. (1991). Biases of rain retrieval algorithms for spaceborne radar caused by nonuniformity of rain. *Journal of Atmospheric and Oceanic Technology*, *8*, 363–373. https://doi.org/10.1175/1520-0426(1991)008<0363:BORRAF>2.0.CO;2

Nakajima, T. Y., Suzuki, K., & Stephens, G. L. (2010). Droplet growth in warm water clouds observed by the A-Train. Part II: A multisensor view. *Journal of the Atmospheric Sciences*, *67*, 1897–1907. https://doi.org/10.1175/2010JAS3276.1

National Academies of Sciences, Engineering, and Medicine (2018). *Thriving on our changing planet: A decadal strategy for Earth observation from space*. Washington, DC: The National Academies Press. https://doi.org/10.17226/24938

Nowell, H., Liu, G., & Honeyager, R. (2013). Modeling the microwave single-scattering properties of aggregate snowflakes. *Journal of Geophysical Research: Atmospheres*, *118*, 7873–7885. https://doi.org/10.1002/jgrd.50620

Okamoto K. (2003). A short history of the TRMM precipitation radar. In W. K. Tao, & R. Adler (Eds.), *Cloud systems, hurricanes, and the Tropical Rainfall Measuring Mission (TRMM), meteorological monographs*. Boston, MA: American Meteorological Society. https://doi.org/10.1007/978-1-878220-63-9_16

Palerme, C., Genthon, C., Claud, C., Kay, J. E., Wood, N. B., & L'Ecuyer, T. (2017). Evaluation of current and projected Antarctic precipitation in CMIP5 models. *Climate Dynamics*, *48*, 225–239. https://doi.org/10.1007/s00382-016-3071-1

Protat, A., Klepp, C., Louf, V., Petersen, W. A., Alexander, S. P., Barros, A., et al. (2019). The latitudinal variability of oceanic rainfall properties and its implication for satellite retrievals: 1. Drop size distribution properties. *Journal of Geophysical Research: Atmospheres*, *124*(23), 13291–13311. https://doi.org/10.1029/2019JD031010

Reitter, S., Fröhlich, K., Seifert, A., Crewell, S., & Mech, M. (2011). Evaluation of ice and snow content in the global numerical weather prediction model GME with CloudSat. *Geoscientific Model Development*, *4*(3), 579–589. https://doi.org/10.5194/gmd-4-579-2011

Sauter, K., & L'Ecuyer, T. S. (2017). Observational evidence for the vertical redistribution and scavenging of Saharan dust by tropical cyclones. *Geophysical Research Letters*, *44*, 6421–6430. https://doi.org/10.1002/2017GL074166

Sauter, K., L'Ecuyer, T. S., van den Heever, S. C., Twohy, C., Heidinger, A., Wanzong, S., & Wood, N. (2019). The observed influence of tropical convection on the Saharan dust layer. *Journal of Geophysical Research: Atmospheres*, *124*, 10896–10912. https://doi.org/10.1029/2019JD031365

Schumacher, C., Houze, R. A., & Kraucunas, I. (2004). The tropical dynamical response to latent heating estimates derived from the TRMM precipitation radar. *Journal of the Atmospheric Sciences*, *61*, 1341–1358. https://doi.org/10.1175/1520-0469(2004)061<1341:TTDRTL>2.0.CO;2

Seto, S., & Iguchi, T. (2015). Intercomparison of attenuation correction methods for the GPM Dual-Frequency Precipitation Radar. *Journal of Atmospheric and Oceanic Technology*, *32*, 915–926. https://doi.org/10.1175/JTECH-D-14-00065.1

Seto, S., Iguchi, T., Shimozuma, T., & Hayashi, S. (2015). NUBF correction methods for the GPM/DPR level-2 algorithms. *IEEE International Geoscience Remote Sensing Symposium (IGARSS)*, 2612–2614. https://doi.org/10.1109/IGARSS.2015.7326347

Shupe, M. D. (2011). Clouds at Arctic atmospheric observatories. Part II: Thermodynamic phase characteristics. *Journal of Applied Meteorology and Climatology*, *50*(3), 645–661. https://doi.org/10.1175/2010JAMC2468.1

Simpson, J., Adler, R. F., & North, G. R. (1988). A proposed Tropical Rainfall Measuring Mission (TRMM) satellite. *Bulletin of the American Meteorological Society*, *69*, 278–295. https://doi.org/10.1175/1520-0477(1988)069<0278:APTRMM>2.0.CO;2

Skofronick-Jackson, G., Kirschbaum, D., Petersen, W., Huffman, G., Kidd, C., Stocker, E., & Kakar, R. (2018). The Global Precipitation Measurement (GPM) mission's scientific achievements and societal contributions: Reviewing four years of advanced rain and snow observations. *Quarterly Journal of the Royal Meteorological Society*, *144*(S1), 27–48. https://doi.org/2789 10.1002/qj.3313

Sohn, B. J., Ryu, G., Song, H., & Ou, M. (2013). Characteristic features of warm-type rain producing heavy rainfall over the Korean Peninsula inferred from TRMM measurements. *Monthly Weather Review*, *141*(11), 3873–3888. https://doi.org/10.1175/MWR-D-13-00075.1

Sorooshian, A., Feingold, G., Lebsock, M. D., Jiang, H., & Stephens, G. L. (2009). On the precipitation susceptibility of clouds to aerosol perturbations. *Geophysical Research Letters*, *36*, L13803. https://doi.org/10.1029/2009GL038993

Stephens, G. L., L'Ecuyer, T., Forbes, R., Gettelman, A., Golaz, J.-C., Bodas-Salcedo, A., et al. (2010). Dreary state of precipitation in global models. *Journal of Geophysical Research: Atmospheres*, *115*, D24211. https://doi.org/10.1029/2010JD014532

Stephens, G., Winker, D., Pelon, J., Trepte, C., Vane, D., Yuhas, C., et al. (2018). CloudSat and CALIPSO within the A-Train: Ten years of actively observing the Earth system. *Bulletin of the American Meteorological Society*, *99*, 569–581. https://doi.org/10.1175/BAMS-D-16-0324.1

Stephens, G. L., Smalley, M. A., & Lebsock, M. D. (2019a). The cloudy nature of tropical rains. *Journal of Geophysical Research: Atmospheres*, *124*, 171–188. https://doi.org/10.1029/2018JD029394

Stephens, G. L., Christensen, M., Andrews, T., Haywood, J., Malavelle, F. F., Suzuki, K., et al. (2019b). Cloud physics from space. *Quarterly Journal of the Royal Meteorological Society*, *145*, 2854–2875. https://doi.org/10.1002/qj.3589

Storelvmo, T., Tan, I., & Korolev, A.V. (2015). Cloud phase changes induced by CO2 warming—a powerful yet poorly constrained cloud-climate feedback. *Current Climate Change Reports*, *1*, 288–296. https://doi.org/10.1007/s40641-015-0026-2

Suzuki, K., Nakajima, T. Y., & Stephens, G. L. (2010). Particle growth and drop collection efficiency of warm clouds as inferred from joint CloudSat and MODIS observations. *Journal of Geophysical Research: Atmospheres*, *67*, 3019–3032. https://doi.org/10.1175/2010JAS3463.1

Suzuki, K., Stephens, G. L., van den Heever, S. C., & Nakajima, T. Y. (2011). Diagnosis of the warm rain process in cloud-resolving models using joint CloudSat and MODIS observations. *Journal of the Atmospheric Sciences*, *68*, 2655–2670. https://doi.org/10.1175/JAS-D-10-05026.1

Suzuki, K., Golaz, J.-C., & Stephens, G. L. (2013). Evaluating cloud tuning in a climate model with satellite observations. *Geophysical Research Letters*, *40*, 4464–4468. https://doi.org/10.1002/grl.50874

Suzuki, K., Stephens, G., Bodas-Salcedo, A., Wang, M., Golaz, J.-C., Yokohata, T., & Koshiro, T. (2015). Evaluation of the warm rain formation process in global models with satellite observations. *Journal of the Atmospheric Sciences*, *72*, 3996–4014. https://doi.org/10.1175/JAS-D-14-0265.1

Takahashi, H., & Luo, Z. J. (2014). Characterizing tropical overshooting deep convection from joint analysis of CloudSat and geostationary satellite observations. *Journal of Geophysical Research: Atmospheres*, *119*, 112–121. https://doi.org/10.1002/2013JD020972

Tan, I., Storelvmo, T., & Zelinka, M. D. (2016). Observational constraints on mixed-phase clouds imply higher climate sensitivity. *Science*, *352*(6282), 224–227. https://doi.org/10.1126/science.aad5300

Tan, I., & Storelvmo, T. (2019). Evidence of strong contributions from mixed-phase clouds to Arctic climate change. *Geophysical Research Letters*, *46*, 2894–2902. https://doi.org/10.1029/2018GL081871

Tanelli, S., Durden, S. L., Im, E., Pak, K. S., Reinke, D. G., Partain, P., et al. (2008). CloudSat's cloud profiling radar after two years in orbit: Performance, calibration, and processing. *IEEE Transactions on Geoscience and Remote Sensing*, *46*(11), 3560–3573. https://doi.org/10.1109/TGRS.2008.2002030

Trenberth, K. E., & Fasullo, J. T. (2010). Simulation of present-day and twenty-first-century energy budgets of the Southern Oceans. *Journal of Climate*, *23*, 440–454. https://doi.org/10.1175/2009JCLI3152.1

Ushio, T., Kubota, T., Shige, S., Okamoto, K., Aonashi, K., Inoue, T., et al. (2009). A Kalman filter approach to the Global Satellite Mapping of Precipitation (GSMaP) from combined passive microwave and infrared radiometric data. *Journal of the Meteorological Society of Japan*, *87A*, 137–151. https://doi.org/10.2151/jmsj.87A.137

Vincent, C. L., & Lane, T. P. (2018). Mesoscale variation in diabatic heating around Sumatra, and its modulation with the

Madden–Julian Oscillation. *Monthly Weather Review*, *146*, 2599–2614. https://doi.org/10.1175/MWR-D-17-0392.1

Voors, R., Donovan, D., Acarreta, J., Eisinger, M., Franco, R., Lajas, D., et al. (2007). ECSIM: The simulator framework for EarthCARE. *Proceedings SPIE 6744, Sensors, Systems, and Next-Generation Satellites XI*, 67441Y. https://doi.org/10.1117/12.737738

Wardah, T., Abu Bakar, S. H., Bardossy, A., & Maznorizan, M. (2008). Use of geostationary meteorological satellite images in convective rain estimation for flash-flood forecasting. *Journal of Hydrology*, *356*, 3–4, 283–298. https://doi.org/10.1016/j.jhydrol.2008.04.015

Wang, T., & Tang, G. (2020). Spatial variability and linkage between extreme convections and extreme precipitation revealed by 22-year space-borne precipitation radar data. *Geophysical Research Letters*, *46*, e2020GL088437. https://doi.org/10.1029/2020GL088437

Wang, M., Ghan, S., Liu, X., L'Ecuyer, T. S., Zhang, K., Morrison, H., et al. (2012). Constraining cloud lifetime effects of aerosols using A-Train satellite observations. *Geophysical Research Letters*, *39*, L15709. https://doi.org/10.1029/2012GL052204

Wegener, A. (1911). *Thermodynamik der Atmosphäre* (p. 331). – Leipzig, Germany: Barth.

Wilheit, T. T. (1972). The Electrically Scanning Microwave Radiometer (ESMR) experiment. In R. R. Sabatini (Eds.), *The Nimbus 5 user's guide* (pp. 87–108). Greenbelt, MD: NASA/Goddard Space Flight Center.

Wood, N., &. L'Ecuyer, T. S. (2013). Level 2C snow profile process description and interface control document, version 0. *CloudSat Project: A NASA Earth System Science Pathfinder Mission*, 23. http://www.cloudsat.cira.colostate.edu/sites/

Wood, R., Kubar, T. L., & Hartmann, D. L. (2009). Understanding the importance of microphysics and macrophysics for warm rain in marine low clouds. Part II: Heuristic models of rain formation. *Journal of the Atmospheric Sciences*, *66*, 2973–2990. https://doi.org/10.1175/2009JAS3072.1

Part V
Outlook

16

Machine Learning for Clouds and Climate

Tom Beucler[1,2,3], Imme Ebert-Uphoff[4,5], Stephan Rasp[6,7,8], Michael Pritchard[1], and Pierre Gentine[2]

ABSTRACT

Machine learning (ML) algorithms are powerful tools to build models of clouds and climate that are more faithful to the rapidly increasing volumes of Earth system data than commonly used semiempirical models. Here, we review ML tools, including interpretable and physics-guided ML, and outline how they can be applied to cloud-related processes in the climate system, including radiation, microphysics, convection, and cloud detection, classification, emulation, and uncertainty quantification. We additionally provide a short guide to get started with ML and survey the frontiers of ML for clouds and climate.

16.1. INTRODUCTION

Machine learning (ML) describes algorithms that learn to perform a task from data without being explicitly programmed for that task. This is in contrast to traditional algorithms, such as existing cloud parameterizations, that are explicitly programmed based on human expertise. Because ML can extract knowledge from large data volumes, it has revolutionized computer vision, natural language processing, and recommender systems. As we get an unprecedented amount of Earth system data from diverse observations (remote sensing, in situ measurements, citizen science) and models (high-resolution simulations, large ensembles of simulations), ML is quickly permeating geosciences (Bergen et al. 2019; Irrgang et al. 2021, Chapter 1; Karpatne et al. 2018) while ML practitioners are increasingly interested in tackling climate change-related problems (Rolnick et al., 2019). Taking the example of numerical weather prediction, ML has already improved post-processing (Grönquist et al., 2020), statistical forecasting (McGovern et al., 2017), and nowcasting (Sønderby et al., 2020), along with promising attempts at purely data-driven weather forecasting (Rasp & Thuerey, 2020).

In contrast to numerical weather prediction, climate science exhibits challenges that have limited direct ML applications. First, many more of the variables used in climate models cannot be directly observed, such as cloud condensation rates and radiative effects, confining most ML attempts to the emulation of numerical models. Second, as understanding is often more important than accuracy for key challenges in clouds and climate science (Bony et al., 2015), climate scientists may avoid methods they deem uninterpretable. Third, it is notoriously hard to benchmark ML models of climate because perfectly labeled climate datasets are rare; two recent attempts had

[1]Department of Earth System Science, University of California Irvine, Irvine, CA, USA
[2]Department of Earth and Environmental Engineering, Columbia University, New York, NY, USA
[3]Institute of Earth Surface Dynamics, University of Lausanne, Lausanne, Switzerland
[4]Cooperative Institute for Research in the Atmosphere, Colorado State University, Fort Collins, CO, USA
[5]Department of Electrical and Computer Engineering, Colorado State University, Fort Collins, CO, USA
[6]Google Research, San Francisco, CA, USA
[7]ClimateAi, Inc, San Francisco, CA, USA
[8]Meteorological Institute, Faculty for Physics, Ludwig Maximilian University of Munich, Munich, Germany

Clouds and Their Climatic Impacts: Radiation, Circulation, and Precipitation, Geophysical Monograph 281, First Edition.
Edited by Sylvia C. Sullivan and Corinna Hoose.

DOI: 10.1002/9781119700357.ch16

Table 16.1 Acronyms used in this chapter.

Acronym	Description	Cross-reference
cGAN	Conditional generative adversarial network	16.3.4
CNN	Convolutional neural network	16.3.4, 16.3.7, 16.4.1
LRP	Layer-wise relevance propagation	16.2.2
ML	Machine learning	All sections
MRMS	Multi-Radar Multi-Sensor	16.2.2
NN	Neural network	All sections
RF	Random forest	16.2.1, 16.2.3, 16.3.1, 16.3.3, 16.4.2
SR	Super-resolution	16.3.4
XAI	eXplainable Artificial Intelligence	16.2.2, 16.3.5, 16.5

to rely on human labeling to benchmark shallow cloud classification (Rasp, Schulz et al., 2020) and extreme weather events (Kashinath et al., 2021), while the most recent ML benchmark for data-driven weather forecasting exclusively relied on meteorological reanalysis data (Rasp, Dueben et al., 2020). Finally, making long-term predictions in a changing climate is an extrapolation problem, meaning that ML algorithms solely trained on present-day climate data might fail to make predictions in the future (unobserved) climate (Beucler et al., 2020; Schneider, Teixeira et al., 2017) with current methods.

In this final chapter, we argue that despite these challenges, ML algorithms are promising tools to consistently capture climate statistics from large datasets in their full complexity (Watson-Parris, 2020), which purely physical models struggle to do (Table 16.1). After an overview of ML in section 16.2, we present promising applications of ML to clouds and climate in section 16.3, and give advice on how to get started with scientific ML in section 16.4 before concluding in section 16.5.

16.2. MACHINE LEARNING

To address the challenges specific to clouds and climate, we survey both physics-guided (section 16.2.3) and explainable ML (section 16.2.2) after introducing common ML tools (section 16.2.1).

16.2.1. Overview of Machine Learning Tools

To identify the ML algorithm most appropriate for the task at hand, it is useful to introduce two hierarchies. First, we can classify algorithms based on the availability of external supervisory information from a human expert or other sources. In *supervised* learning, by far the most common ML approach, the algorithm learns to map an input (or features) to an output (or target) from a training dataset consisting of input-output pairs. The algorithm is then trained to make output predictions as close as possible to the training targets. This is usually done via the minimization of a cost or loss function (e.g., mean square error, cross-entropy loss), which maps the algorithm's output to a single number that can be minimized. In contrast, *unsupervised* learning extracts features from data and without the need for labeled input-output pairs. Examples include dimensionality reduction (via, e.g., principal component analysis, autoencoders, self-organizing maps, see Chapter 8 of Géron (2019)), clustering algorithms (e.g., k-means, DBSCAN, Gaussian mixture models, see Chapter 9 of Géron (2019)), and generative learning, which can be broadly defined as a probabilistic model for how a dataset is generated (e.g., variational autoencoders, generative adversarial networks, see Foster (2019)).

Second, we can classify algorithms based on the number of parameters that are fitted as part of the training process, as illustrated in Figure 16.1 (orange axis). It is always preferable to start with simple models because they are easier to interpret (green axis) and less computationally expensive to train. For *regression* tasks that seek to predict continuous outputs, linear regressions are arguably the simplest models, while logistic regressions are simple models for *classification* tasks that seek to predict categorical outputs. Decision trees, which use simple conditional statements (the "branches") to match each input's characteristics to corresponding outputs (the "leaves"), can be trained and averaged to form random forests (RFs) that exhibit lower variance in their predictions than single decision trees. Finally, neural networks (NNs), which have computational architectures loosely based on biological networks of neurons, are powerful nonlinear regression and classification tools that are typically expensive to train if deep, that is, composed of many neuron layers.

While in theory complex models such as NNs can fit any nonlinear mapping given enough data (Scarselli & Tsoi, 1998), this high representation power comes at a cost: Simple models such as linear regressions optimized using least squares have a unique set of optimal parameters, but the parameters of more complex models often need to be optimized stochastically with no guarantee of an optimal solution. As the number of required training samples increases with the number of model parameters, complex models tend to easily *overfit* the training dataset, that is, they memorize the dataset "by heart," leading to poor performance when the model is applied to data samples it has not been trained on. To mitigate this issue,

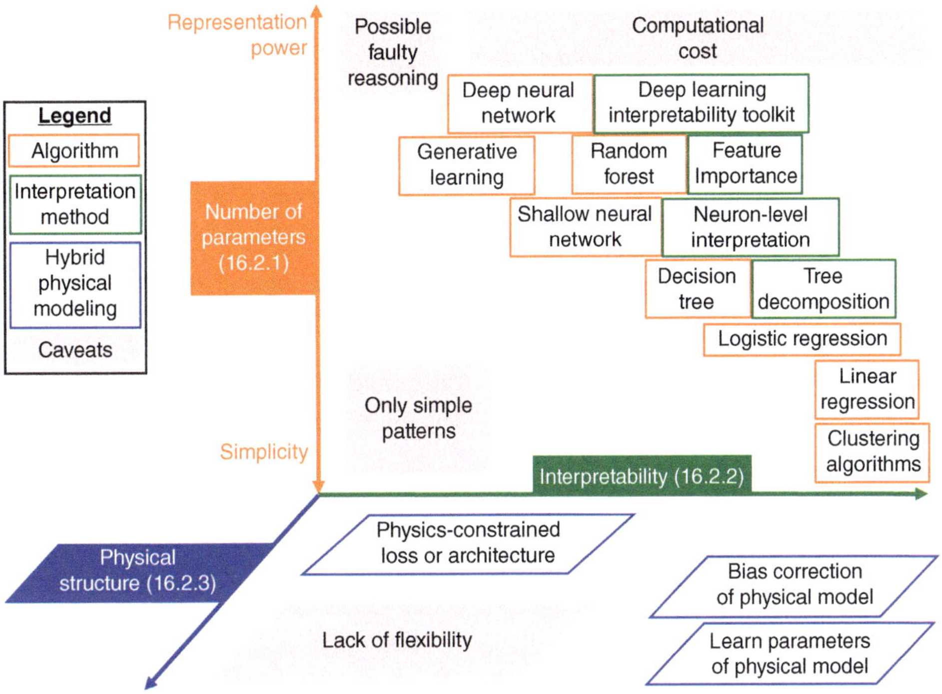

Figure 16.1 ML algorithms (orange boxes), corresponding interpretation methods (green boxes), and hybrid ML-physical modeling methods (blue boxes) organized in the three-dimensional space formed by the number of parameters (section 16.2.1), interpretability (section 16.2.2), and physical structure (section 16.2.3). Caveats of different configurations are indicated in light red boxes.

best ML practices require to split the available data into three datasets: A *training* dataset to optimize the model's parameters, a *validation* dataset to detect overfitting to the training dataset and to optimize the model's hyperparameters (e.g., number of layers of an NN), and a *test* dataset to evaluate the final model on data samples it has never seen before. In Earth science, splitting the data into independent training/validation/test datasets can be particularly challenging as samples tend to exhibit high spatiotemporal correlation (Karpatne et al., 2018), meaning that, for example, splitting the data by long time interval (such as years) may be preferable to a random split. Ying (2019) surveys common ML methods to avoid overfitting, including regularization of the model's parameters (e.g., L1, L2, Dropout) and early stopping of the model's training.

If simple models fail to capture the data and a complex model is required, another caveat is that complex models such as NNs require large amounts of (usually labeled) data samples to train their many free parameters, often more than is available from, for example, observations. An elegant workaround, referred to as *transfer learning*, is to build on an NN trained for a different but closely related task. Examples include leveraging NNs classifying the 50M images from the famous "ImageNet" dataset (Deng et al., 2009) to make predictions from satellite observations (Marmanis et al., 2015), and leveraging NNs trained on climate model data to make predictions from meteorological reanalysis data (Ham et al., 2019; Rasp & Thuerey, 2020). In these cases, building on existing NNs reduces not only the required sample size, but also development time and cost. That being said, the complex resulting NN may still be hard to interpret, motivating methods to explain the strategies the NN uses to make accurate predictions.

16.2.2. Interpretable/Explainable Machine Learning

The literature often distinguishes *interpretable* ML, which generally refers to models that are designed to be a priori understandable, and *explainable* ML (known as eXplainable Artificial Intelligence or XAI in the literature), which refers to methods that try to explain a posteriori the prediction of a trained model, often for specific samples (Rudin, 2019). Both frameworks help design more transparent and hence trustworthy ML models. Two books (Molnar, 2019; Samek et al., 2019) provide in-depth discussion of both approaches. In this section, we focus on XAI to help interpret trained ML models for climate science.

In Figure 16.1, methods to interpret ML algorithms are indicated with green boxes (simple methods such as linear regressions or decision trees are interpretable by construction): For example, tree decomposition follows the path of a decision tree to decompose a given prediction as the sum of each input's contribution. While methods that seek to interpret individual neurons of an NN have been successful in computer vision (Bau et al., 2017; Carter et al., 2019; Olah et al., 2017, 2018; Yosinski et al., 2015), they are difficult to apply in climate science where objects have fuzzy boundaries (e.g., heat waves and atmospheric rivers). Exceptions involve cases where we can focus on only a few neurons, including McGovern et al. (2020) who ranked individual neurons in an NN trained for tornado prediction using the neurons' ability to discriminate output classes.

Therefore, XAI for climate science often focuses on understanding the prediction of overall ML models for specific samples, most notably via *attribution* methods. Attribution methods, most common in NNs for image classification, ask: Given a sample and its label, which pixels in the input image are most important to correctly predict the label? The result is a *heatmap* (or attribution map) indicating the most important areas in the input images. Attribution methods include saliency maps (Simonyan et al., 2013), SmoothGrad (Smilkov et al., 2017), Integrated Gradients (Sundararajan et al., 2017), Layer-wise relevance propagation (Montavon et al., 2017, 2018) (see Fig. 16.2), DeepLIFT (Shrikumar et al., 2017), SHAP (Lundberg & Lee, 2017), GradCAM (Selvaraju et al., 2017), and occlusion methods (Fong & Vedaldi, 2019). Attribution methods have been successfully applied in atmospheric science (Barnes et al., 2020; Brenowitz, Beucler et al., 2020; Gagne II et al., 2019; Lagerquist et al., 2020; McGovern & Lagerquist, 2020), and we refer the curious reader to recent reviews by McGovern et al. (2019) and Ebert-Uphoff and Hilburn (2020) on the use of XAI methods in meteorological applications.

Beyond attribution methods, other methods seeking to interpret ML models as a whole include *backwards optimization*, which calculates the input maximizing confidence in a given output (McGovern et al., 2019), and *ablation* studies (Raghu & Schmidt, 2020), which remove certain capabilities from the model's architecture and retrain the resulting model to test how important these capabilities are (see Sønderby et al. (2020) and Ebert-Uphoff and Hilburn (2020) for meteorological applications). Finally, when using XAI methods one should keep in mind (1) potential limitations of the attribution methods (Adebayo et al., 2018; Bansal et al., 2020; Kindermans et al., 2017), and (2) the importance of providing explanations that are tailored to the user's concerns and needs (Ras et al., 2018; Rutjes et al., 2019).

16.2.3. Physics-Guided Machine Learning

Even highly interpretable ML models of physical processes such as clouds are often physically inconsistent in two major ways, limiting their impact on climate science. First, they may violate physical laws we know should hold, such as mass and energy conservation. Second, they may fail to generalize to unseen conditions such as extreme weather events and shifts in the distribution of variables (e.g., geographical shift and climate change).

To address these issues, we can restrict the ML output space to physically plausible solutions by adding physical

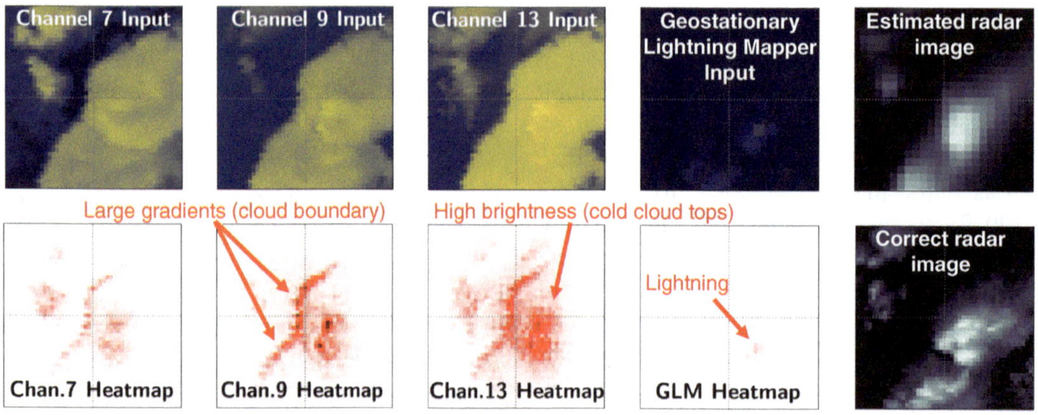

Figure 16.2 Use of an attribution method, namely, Layer-Wise Relevance Propagation (LRP), to analyze a neural network model that seeks to translate imagery from the Geostationary Operational Environmental Satellite into Multi-Radar Multi-Sensor (MRMS) imagery. LRP results (bottom row) show *where* in the input channels the neural network is focusing when predicting the MRMS value for a single (central) pixel. Using LRP indicated for various samples three strategies for the neural network to trigger large MRMS values, namely, (i) cloud boundaries, (ii) cold cloud tops, and (iii) presence of lightning in the Geostationary Lightning Mapper channel. Source: Ebert-Uphoff and Hilburn (2020) / American Meteorological Society.

structure to the ML framework. This is referred to as *physics-guided ML* (Reichstein et al., 2019; Willard et al., 2020) or hybrid ML-physical modeling (blue axis in Fig. 16.1). We can group physics-guided ML approaches into three categories (three blue boxes in Fig. 16.1), noting that the optimal approach depends on the task and data at hand.

First, we can integrate physical knowledge without changing the ML algorithm's architecture by adding a penalizing term to its loss function, akin to a Lagrange multiplier (de Bezenac et al., 2019; Karpatne et al., 2017; Wu et al., 2020; Willard et al., 2020). However, these soft constraints are often insufficient in climate science, where, for example, conservation laws need to hold exactly since climatic trends are driven by small energy imbalances in the Earth system. This motivates changing the model's architecture to enforce conservation laws to within machine precision, for example, by using RFs that enforce conservation laws by construction, as long as these laws are a linear combination of the outputs (O'Gorman & Dwyer, 2018), or by augmenting the NN's architecture with physical constraint layers so as to exactly enforce conservation laws (Beucler, Pritchard et al., 2019). However, enforcing physical laws in this way is often not enough to properly generalize outside of the training set (Beucler et al., 2020; O'Gorman & Dwyer, 2018; Rasp et al., 2018; Reichstein et al., 2019). To address this, the penalizing term in the loss function may enforce assumed (and often approximate) dynamics based on a physical model (Camps-Valls et al., 2018; Gao et al., 2020; Mao et al., 2020; Pang et al., 2019; Raissi et al., 2017; Raissi et al., 2019; Sun et al., 2020; Wang et al., 2017; Wu et al., 2020; Zhang et al., 2020). Even if imperfect, such constraints improve predictions in unseen conditions by reducing the range of possible outputs, unlike pure ML approaches that have to make predictions in much larger phase spaces (Karpatne et al., 2017; Yang, Sun, et al., 2019).

If loss-based constraints do not provide enough physical structure, we can use ML to bias correct a physical prior or calibrate a physical model's free parameters. Examples include ML the parameters of Earth System Model parameterizations within the structure of the known governing equations (Schneider, Lan et al., 2017), learning an effective diffusion coefficient to represent turbulent processes in the boundary layer rather than learning the full turbulent fluxes (Camps-Valls et al., 2018; Reichstein et al., 2019), or only learning certain coefficients of the Reynolds stress tensor to preserve Galilean invariance (Ling et al., 2016). A caveat of imposing too much physical structure is the resulting lack of flexibility to model the data; for instance, at standard climate model resolutions, even the best ML fit of an eddy-diffusion model (Siebesma et al., 2007) cannot properly capture the non-local transport associated with shallow convection in the boundary layer.

16.3. APPLICATION TO CLOUDS AND CLIMATE

We now give concrete examples of how ML can be applied to clouds and climate. Sections are ordered by spatial scales whenever possible, from radiative transfer operating at the atomic level (section 16.3.1) to cloud classification at the planetary scale (section 16.3.7).

16.3.1. Radiative Transfer

Radiative transfer is defined as the energy transfer in the form of electromagnetic radiation. As solar radiation is the atmosphere's largest energy source, while thermal radiation to space is its largest energy sink, atmospheric models cannot forego calculating the heat transfer resulting from solar (shortwave) radiation and thermal (longwave) radiation. While we have excellent empirical knowledge of how molecules in the atmosphere absorb and emit radiation at each electromagnetic spectral line, it is computationally intractable to calculate radiative transfer line-by-line and atmospheric models rely on different levels of approximation, including (1) integrating radiative fluxes over spectral bands of predetermined width, (2) neglecting the three-dimensional nature of radiation by assuming strictly vertical fluxes (plane-parallel approximation), and (3) calculating radiation using a resolution in time and/or space that is coarser than the model's standard resolution.

Aside from recent work using RFs and NNs to reproduce the variability in surface solar irradiance resulting from the three-dimensional interaction between radiation and shallow cumuli (Gristey et al. (2020), addressing approximation 2), ML has mostly been used to improve the *temporal* resolution of radiative transfer (approximation 3). By replacing the original, computationally expensive radiative scheme with its ML-emulated counterpart (computationally inexpensive once trained), atmospheric models can be accelerated, with the possibility of calling the radiative scheme every time step to improve the quality of cloud-radiation interactions and numerical weather predictions. To keep using physical equations when they are available while still accelerating atmospheric models, it is also possible to only replace expensive computations within radiative schemes, such as gas optics (Ukkonen et al., 2020; Veerman et al., 2020). ML emulation of radiative transfer has been tested in meteorological (Chevallier et al., 2000; Krasnopolsky & Fox-Rabinovitz, 2006), global climate (Belochitski et al., 2011; Krasnopolsky et al., 2008; Krasnopolsky et al., 2010; Pal et al., 2019), and cloud-resolving

(Roh & Song, 2020) models. Recent efforts using ML to emulate subgrid-scale thermodynamics (Rasp et al. (2018),
see section 16.3.3) usually include subgrid-scale radiative cooling, making it an example where ML helps improve the *spatial* resolution of radiative transfer, for example, by capturing the effect of subgrid clouds on grid-scale radiative cooling.

Other notable applications of ML to atmospheric radiative transfer are the retrieval of cloud properties from satellite images (Min et al., 2020) and statistical predictions of surface radiative fluxes. This includes solar forecasting, where ML algorithms have been successful at post-processing physical forecasts, nowcasting, 6-hour forecasting (see review by Voyant et al. (2017) and Gala et al. (2016)), and land-atmosphere modeling, where ML can help emulate surface radiative properties that are distorted by coarse climate models, for example, sun-induced chlorophyll fluorescence spectra (Rivera et al., 2015).

16.3.2. Microphysics

Microphysics refers to small-scale (sub-μm to cm) processes that affect cloud and precipitation particles (Chapters 3, 4, and 12). Microphysical parameterization schemes, which model the effect of cloud and precipitation particle populations on weather and climate, currently face two major challenges (Morrison, van Lier-Walqui, Fridlind et al., 2020): (1) how to represent this effect despite the impossibility of simulating all particles individually; and (2) uncertainties in microphysical process rates owing to critical gaps in cloud physics knowledge, especially for ice-phase processes. As microphysics links various components of Earth's atmospheric water and energy cycles, its overly simplistic representation, typically limited to one or two moments from the particle distribution of each hydrometeor species, remains a large source of uncertainty in numerical weather forecast and climate simulations (Zelinka et al., 2020).

Thankfully, simulations describing the hydrometeor size distribution more faithfully via a bin approach improve accuracy. Although such simulations are too computationally expensive to be run for long-term climate predictions, they can provide high-quality training data for ML emulation (which should typically be much faster), with potential progress toward addressing challenge (1). Gettelman et al. (2020) trained multiple NNs to emulate the formation of warm rain from a bin scheme, and the NN-powered climate simulation was able to match the bin scheme's accuracy at a significantly reduced computational cost. Seifert and Rasp (2020) trained an NN to emulate microphysical conversion rates in a two-moment scheme from a Monte Carlo super-droplet simulation, but while process rates were reproduced with greater accuracy, the resulting solutions to the collision-coalescence ordinary differential equations did not match the reference simulation as well as the heuristically designed parameterization. Analysis hints at the ill-posed nature of two-moment schemes and highlights the importance of evaluating ML models in long-term simulations. To address structural uncertainty in existing parameterizations, Morrison, van Lier-Walqui, Kumjian et al. (2020) propose a Bayesian framework to flexibly relate microphysical process rates to moments of the hydrometeor size distribution via generalized power series (Loeb, 1991), while ML-based short-term forecasts tend to forego microphysical schemes by directly fitting parameters of the size distribution to observational data, for example, for hail prediction (Gagne et al., 2017).

16.3.3. Convection

Atmospheric convection, defined as the vertical motion driven by air density differences, is notoriously hard to simulate because of its multiscale nature, as it leads to the formation of stratocumulus decks ($\sim 10-100$ km, Chapters 6 and 10) while interacting with planetary-scale dynamics ($\sim 10^4$ km, Chapters 9 and 11). Since clouds are radiatively active at all scales while convection vertically transports heat and water from the surface to the atmosphere, misrepresenting convection and clouds in atmospheric models leads to large errors in the energy balance and hence remains the largest source of uncertainty in long-term climate predictions (Bony et al., 2015). In standard global climate models, convection and clouds are not explicitly represented and their effect on the climate is approximated by *subgrid closures*. Traditionally, designing subgrid closures involved a heuristic process and manual tuning to observations. Recent storm-resolving simulations can explicitly represent tropical deep convection (Stevens et al., 2019) but are too expensive for climate change simulations. Therefore, emulating the effect of convection and clouds in storm-resolving simulations using statistical algorithms, including ML, could provide a shortcut toward data-driven subgrid closure in global climate models.

The first step in designing a data-driven subgrid closure is to create the training data; this can be done by coarse-graining a storm-resolving simulation to derive the correction term from coarse to high resolution (Brenowitz & Bretherton, 2019; Yuval & O'Gorman, 2020), by running a super-parameterized model that explicitly separates the storm and the coarse scales and directly provides that correction term (Gentine et al., 2018; Rasp et al., 2018), or by nudging a standard climate model to observations or meteorological reanalysis (McGibbon & Bretherton, 2019; Watt-Meyer et al., 2020). The second step is to choose the ML algorithm: While NNs are commonly used to emulate subgrid-scale parameterization because they usually yield the best fits (Brenowitz & Bretherton, 2018; Gentine et al., 2018; Krasnopolsky, 2013; Krasnopolsky et al., 2013), RF-based parameterizations have the advantage of making bounded predictions that respect linear physical constraints (O'Gorman & Dwyer, 2018). This could explain why RFs tend to be more stable once coupled back to a climate model (Brenowitz, Henn et al., 2020), although NN parameterizations can be designed to be unconditionally stable (Brenowitz, Beucler et al., 2020; Yuval et al., 2020) and conserve mass/energy (Beucler, Rasp et al., 2019). Recent success in emulating subgrid thermodynamics with Earth-like boundary conditions (Han et al. (2020), Mooers, Pritchard et al. (2020), see Fig. 16.3) is a promising step toward ML-powered long-term climate predictions, while simple proxys of subgrid parameterization such as the Lorenz 96 system (Lorenz, 1996) can help quickly test new ML algorithms (Crommelin & Vanden-Eijnden, 2008; Gagne et al., 2020) and frameworks (Mouatadid et al., 2019; Rasp, 2020).

16.3.4. Downscaling

Operational meteorological forecasts and climate predictions usually require variables at the local (e.g., 1 km) scale, but global atmospheric models usually output variables at a coarser (e.g., 50–200 km) scale, are not tuned for regional-scale predictions, and rely on physical variables that may not directly be societally relevant (e.g., momentum fluxes instead of wind bursts and heating fluxes instead of surface solar irradiance). Ground-based observations provide large datasets of such variables, which allow ML algorithms to be trained to predict societally relevant variables at the local scale (Sharifi et al., 2019) from coarse-scale model output. We note that this is a particular case of statistical downscaling (reviewed in Wilby et al. (1998)), which is a data-driven and computationally inexpensive alternative to dynamical downscaling where a regional-scale physical model is run with boundary conditions derived from coarse-scale model output (Feser et al., 2011; Xue et al., 2014).

From an ML perspective, downscaling is analogous to super-resolution (SR), which aims to obtain a high-resolution output from its low-resolution version (see review by Yang, Zang, et al. (2019)). We note that SR is an ill-posed problem, as a single low-resolution image corresponds to an infinite number of high-resolution images from an unknown probability distribution. Convolutional neural networks (CNNs), which hierarchically extract data from image patches, have been successfully leveraged to super-resolve precipitation predictions (Vandal et al., 2017, 2019), satellite images (Pouliot et al., 2018), and idealized turbulent flows (Fukami et al., 2018). Despite outperforming simple interpolation baselines (Baño-Medina et al., 2020), CNNs typically underestimate extremes (Sachindra et al., 2018) as they tend to predict the average of all possible solutions to minimize the error at each pixel. While Sachindra et al. (2018) found that Relevance Vector Machine (a Bayesian approach to learning probabilistic sparse generalized linear models, Tipping (2000)) improved the SR of precipitation extremes, recent work has focused on generative modeling for SR (Singh et al., 2019), in particular conditional

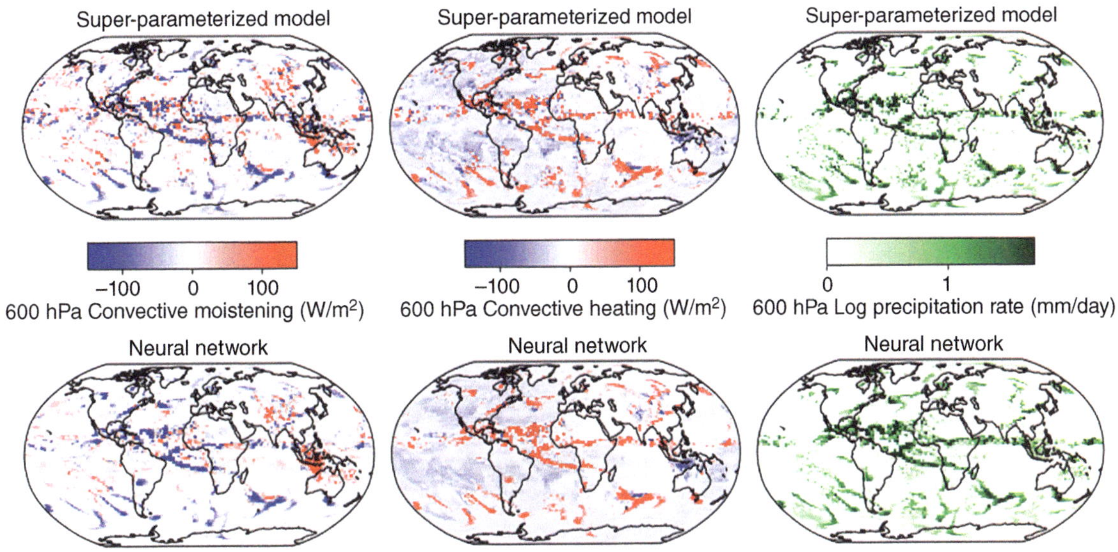

Figure 16.3 Lower tropospheric subgrid moistening and heating tendencies (left) and subgrid precipitation (right) from a super-parameterized climate model with realistic boundary conditions (top) and as emulated by a deep NN (bottom). Source: Adapted from Mooers, Pritchard et al. (2020).

generative adversarial networks (cGANs, Mirza & Osindero (2014)). For instance, Stengel et al. (2020) used a sequence of two cGANs to super-resolve wind and solar power at the 2 km-scale using 100 km-scale climate model output, and showed that the resulting turbulent kinetic energy spectra and solar irradiance semi-variograms were more consistent with high-resolution climate model output than those generated by interpolation and simple CNNs (Fig. 16.4). Finally, in an effort to represent the full distribution of possible super-resolved temperature and precipitation fields from coarse model output, Groenke et al. (2020) adapted recent work in normalizing flows for variational inference (Grover et al., 2020) to develop an unsupervised NN approach that generates the joint distribution of high- and low-resolution climate maps.

16.3.5. Climate Analysis and Understanding

The successful ML attempts described above at modeling atmospheric processes already improve our understanding of climate dynamics. Taking the example of subgrid-scale thermodynamics parameterization (Brenowitz & Bretherton, 2018; Rasp et al., 2018; Yuval & O'Gorman, 2020), the ability of ML models that are local in time and space to capture the effect of clouds and fine-scale turbulence on large-scale thermodynamics demonstrates that it is possible to approximately close large-scale moist thermodynamics equations without knowledge of small-scale stochasticity, convective-scale organization, and convective memory.

In addition, modern ML algorithms can be used to find relevant patterns and detect climate signals. Ebert-Uphoff and Deng (2012) used Bayesian networks, a type of graphical model, to detect causal relationships between low-frequency patterns in the atmosphere (see Runge et al. (2019) for a broad overview of causal discovery methods in climate science). Wills et al. (2020) show that signal-to-noise maximal pattern filtering extracts forced climate signals with up to 10 times fewer climate models, even on regional scales. Barnes, Hurrell et al. (2019) and Barnes et al. (2020) show that XAI applied to an NN tasked with categorizing the year in a forced climate data record, based on inputs of global temperature maps, provides a simple new way to identify time-varying regional indicators of the external forcing, with the possibility of distinguishing, for example, greenhouse gas from aerosol forcing (Labe & Barnes, 2021). Toms et al. (2020) shows how XAI applied to NNs can also update views of El-Nino intrinsic predictability – compared to linear regression, NNs de-emphasize the roles of the Atlantic and Indian Ocean and boost the relevance of the northwest tropical Pacific. Applied to West Coast temperature anomalies, the same approach reveals that NNs correctly identify near-coastal surface temperatures as the strongest source of predictability for 60-day forecasts. In contrast, linear regressions overemphasize El-Nino-related surface temperature signals in the tropical ocean that should only dominate at longer lead times. Beyond NNs, Barnes, Samarasinghe et al. (2019) used probabilistic graph models to hypothesize a two-branch interaction between the Madden-Julian Oscillation and the North Atlantic Oscillation, while Di Capua et al. (2020) combined maximum covariance analysis with causal discovery networks to understand the causal influence of tropical convection on midlatitude dynamics.

16.3.6. Uncertainty Quantification

Uncertainties can be divided into four main categories:

1. Observational: due to measurement and representation errors (Janjić et al., 2018);

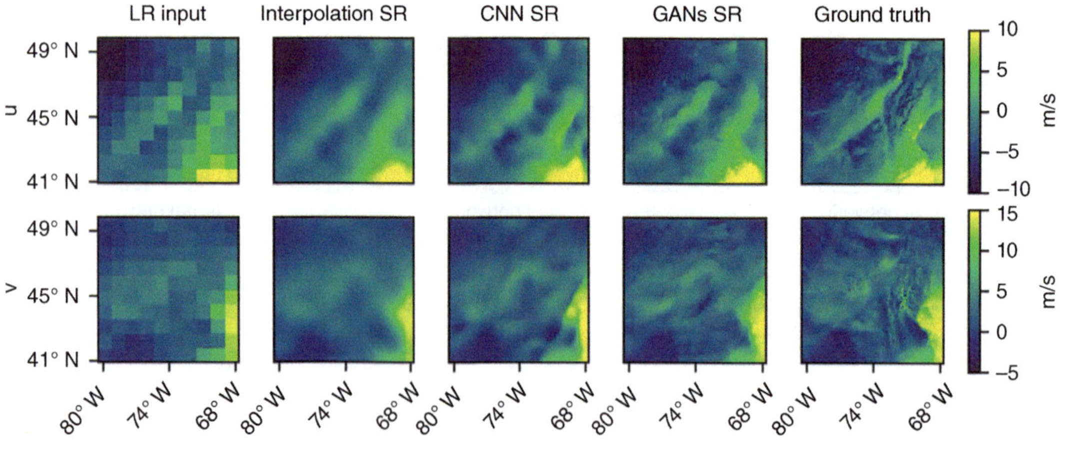

Figure 16.4 SR of east-west (U) and north-south (V) wind velocity from low-resolution (LR) input using bi-linear interpolation, a CNN, and GANs. Source: Fig. 3 from Stengel et al. (2020) / National Academy of Science.

2. Structural: due to incorrect model structure;
3. Parametric: due to incorrect model parameters;
4. Stochastic: due to internal climate variability or the chaotic nature of the flow (Deser et al., 2012; Lorenz, 1963).

These categories do not intend to be mutually exclusive but rather emphasize separate philosophical issues. For instance, consider the uncertain parameterization of cloud microphysics wherein multiple equivalently plausible equations to represent autoconversion rates are in use despite structural differences (structural uncertainty). Meanwhile, even within a given set of equations, the coefficients and parameters are left to be empirically constrained (parametric uncertainty). This requires confronting observational uncertainty (e.g., indirect radiometric measurement of precipitating cloud droplets from radar backscatter with associated sampling and inverse model uncertainty) as well as stochastic uncertainty (e.g., a perfect model can make counterfactual chaotic trajectories that will eventually diverge from nature necessitating multiple runs per comparison).

While observational, structural, and parametric errors should be reduced as much as possible, stochasticity should be reproduced as well as possible to increase the fidelity of the simulations (Berner et al., 2017). Many simple ML algorithms, such as individual feed-forward NNs, are primarily deterministic such that once trained their predictions do not characterize uncertainties by construction. Ensembles of such NNs, constructed by, for example, using different initial random weights or shuffling the training set, may capture *stochastic* uncertainty. But systematically characterizing *parameteric* uncertainties is at the heart of *data assimilation* (Eknes & Evensen, 1997; Evensen et al., 1998; Evensen, 2009), which adopts Bayesian approaches to infer the posterior distribution of either the state or parameters given some initial prior and observational uncertainty. However, despite characterizing the parameteric uncertainty (Dunbar et al., 2020), these Bayesian approaches do not always address structural uncertainty.

This motivates methods that inherently characterize uncertainties, such as Gaussian processes (Camps-Valls et al., 2016) that, for example, were recently trained on perturbed climate simulation ensembles to better characterize aerosol forcing uncertainty from observed aerosol optical depth (Watson-Parris, Bellouin et al., 2020) and cloud droplet number (McCoy et al., 2020). A strategy that has not yet seen many applications in Earth sciences is the use of generative models such as variational autoencoders (Pu et al., 2016) or generative adversarial networks (Radford et al., 2015; Yang, Wu, et al., 2019) to build intrinsically probabilistic models. These approaches aim at reproducing the underlying distribution as a function of the state and thus can be promising approaches for uncertainty quantification. Finally, dropout layers in NNs (Gal & Ghahramani, 2016) and Bayesian NNs (Bate, 1998; Khan & Coulibaly, 2006) extend deterministic NNs to represent structural uncertainty, making them promising tools to quantify uncertainty.

16.3.7. Cloud Detection and Classification

The ability to detect and classify clouds from satellite or ground-based observations has a wide range of important applications, ranging from short-term forecasting of hazardous weather to gaining insights into the climate system by detecting the occurrence of different cloud types. Cloud detection and classification is a problem especially suited to modern ML as it mostly relies on image data for which CNNs are particularly well suited. Some of the earliest work predates the surge of CNNs, namely, Lee et al. (1990) and Tian et al. (1999) use fully connected NNs with engineered features to detect different cloud types (stratus, cirrus, and cumulus). Wood & Hartmann (2006) use fully connected NNs to detect different types of cloud organization (closed/open cell shallow convection). Muhlbauer et al. (2014) create a climatology of these cloud classes.

Detecting different cloud types and patterns is of particular interest to the climate community since different cloud structures have widely different impacts on the Earth's energy balance (Bony et al. (2015), Chapter 14). Mahajan & Fataniya (2019) provide an overview of methods to detect cloud properties from satellite imagery. Marais et al. (2020) and Rasp, Schulz et al. (2020) try to detect larger cloud patterns from satellite imagery. Marais et al. (2020) use a CNN to classify satellite imagery (MODIS and VIIRS) into categories such as clear-air, closed-stratiform, and high-altitude clouds. Rasp, Schulz et al. (2020) focus on modes of organization in subtropical shallow cumulus clouds, based on cloud pattern classes (sugar, flower, fish, and gravel) defined by Stevens et al. (2020). Watson-Parris, Sutherland et al. (2020) used a CNN to detect pockets of open cells, trained on a small dataset of hand-labeled features, and used this algorithm to investigate their radiative impact.

A common challenge for these applications is the need for a large number of labeled data samples which are often hard to obtain. Solutions include (1) generating labels manually, (2) generating labels from other modalities that may only be available intermittently or in certain locations, (3) using transfer learning to reduce the number of samples needed (see section 16.2.1), and (4) using unsupervised learning. For example, both Marais et al. (2020) and Rasp, Schulz et al. (2020) developed a labeling interface that allowed scientists to label many thousands of images. Yet, even that amount of training labels is barely enough to train large modern CNNs. Thus, both groups used CNNs pretrained on a huge number of natural images, fine-tuned on the cloud patterns, thus using both Strategies 1 and 3. An example of Strategy 2 is to use ground-based radar (only available in some locations) to generate convection labels for Geostationary Operational Environmental Satellite (GOES) imagery (Lee Y. et al., 2021), or to use CloudSat data (only available intermittently) to generate cloud-type labels for Himawari-8 satellite imagery (Zhang, Zhuge et al., 2019). Similarly, Zantedeschi et al. (2019) used a semi-supervised learning approach to leverage a small number of classified satellite images for creating a much larger labeled dataset to cloud types. Approaches for Strategy 4, unsupervised learning for cloud classification, range from very intuitive to quite abstract. Haynes et al. (2011) use k-means clustering on physical features to categorize satellite imagery into cloud types. Since the clustering is performed in a space with clear physical meaning the resulting cloud categories are easily interpretable. In contrast, recent methods propose to use deep NNs for this purpose. Denby (2020) uses an unsupervised NN (Tile2Vec architecture by Jean et al. (2019)), then performs clustering in the abstract space of an NN layer. Visualization of samples shows some obvious similarity within each cluster (see Fig. 16.5); however, it is not yet clear what exactly the obtained cloud categories represent, or whether a network that is trained slightly differently would yield a similar categorization. Kurihana et al. (2019) also use an unsupervised NN approach for cloud classification, namely, classifying satellite (MODIS) imagery using a convolutional autoencoder. In summary, the motivation for NN-based approaches is that NNs might discover yet unknown patterns, but these patterns remain underexplored and the corresponding cloud categories may not be robust.

Other important applications include detecting convection from satellite imagery (Cintineo et al., 2020; Lee et al., 2021; Zhang, Wang, et al., 2019), detecting convective storms from ground-based radar observations (Gooch & Chandrasekar, 2020), detecting cloud types from pictures taken from the surface (Zhang et al., 2018), and detecting fronts from numerical weather model output (Lagerquist et al., 2019, 2020). Finally, Liu et al. (2016), Racah et al. (2017), and Kashinath et al. (2021) describe an effort to build a database of expert-labeled tropical cyclones and atmospheric rivers in climate model output, using a custom labeling interface. This is especially relevant to estimate how the frequency of these extreme weather events will change under global warming.

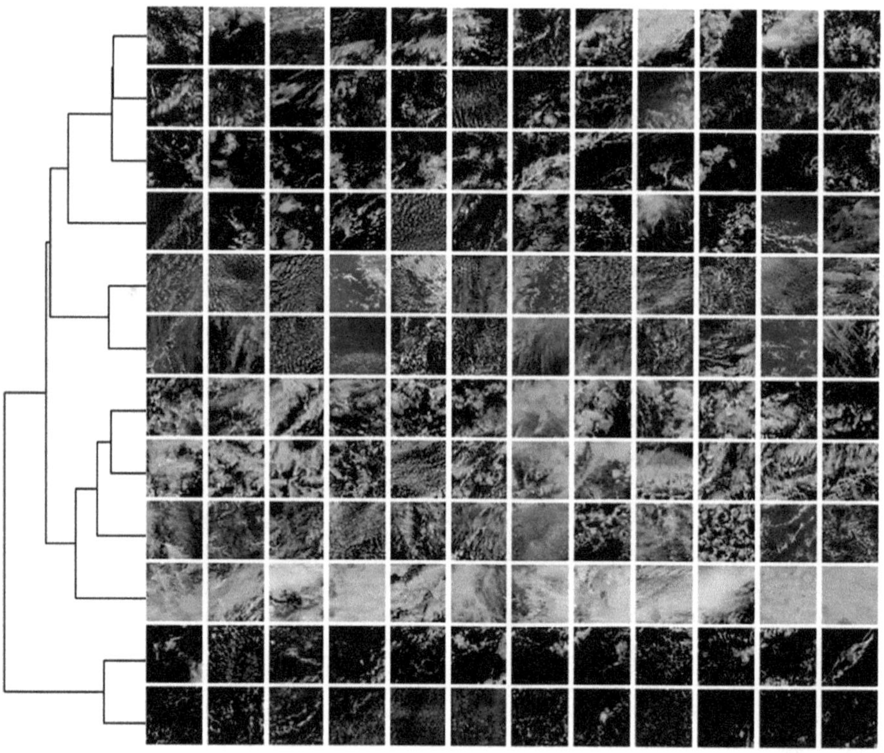

Figure 16.5 Cloud classification by applying hierarchical clustering to the latent space representation of satellite images in an NN trained with the Tile2Vec method. Each row shows samples from one distinct cloud category identified by the NN. Source: Reproduced from Fig. 2 of Denby (2020) / John Wiley & Sons.

16.4. GETTING STARTED WITH MACHINE LEARNING

16.4.1. A Workflow for Scientific Machine Learning

The first step in a scientific ML workflow (Fig. 16.6) is clearly defining a reproducible task to be solved by the ML algorithm and the data to be used for training. In some cases, the data might already be available but in many other cases creating the training data is the first major challenge. For the use case of training an ML subgrid parameterization discussed in section 16.3.3, reference model simulations need to be run, often with different output requirements than traditional model simulations in terms of variables and spatiotemporal frequency. If the parameterization is to be trained on high-resolution simulations, an additional coarse-graining step has to be performed to provide the training dataset. For cloud detection and classification, as described in section 16.3.7, high-quality labels of the objects to be classified are often not readily available. Here, traditional algorithms can be used to create labels (Racah et al., 2017). However, as the motivation for using ML is to outperform traditional hand-crafted algorithms, crowd-sourcing has turned out to be a powerful tool to collect labels from domain experts (Kashinath et al., 2021; Rasp, Schulz et al., 2020). The final task in creating a good dataset is to set aside representative and independent validation/test sets that can be used to monitor the performance of various ML algorithms on previously unseen data.

With the data available, the next step is training an ML model that is appropriate for the task. Before doing so, however, it is paramount to have solid baselines to compare to. These baselines can be traditional non-ML techniques or very simple linear ML algorithms such as linear or logistic regressions. Only with baselines and comparable evaluation is it possible to judge how well more complex algorithms are performing. This makes benchmark problems such as in WeatherBench (Rasp, Dueben et al., 2020) crucial for the advancement of ML. With baselines set, the question turns to which ML algorithm to use. If possible, it is helpful to search for similar problems previously solved with ML (Reichstein et al., 2019). Generally, NNs and RFs are the most common modern ML techniques. For spatially structured data, CNNs are often the algorithms of choice. For time series problems with memory, recurrent NNs are commonly used. Another important factor to consider when choosing an algorithm is the number of samples, as it will determine the complexity of the algorithm one can use as discussed in section 16.2.

Finally, it is important to visualize and analyze the models' predictions thoroughly and check for unexpected behavior that might not be visible from a simple validation score. Feature importance and other interpretability methods can help understand the inner workings of the algorithms and point toward potential shortcomings and ways to improve (section 16.2.2). If one detects physically unrealistic behavior, physics-guided ML methods to incorporate various degrees of physical structure (section 16.2.3) can help, especially when data are limited or when the algorithm will be applied in unfamiliar climate regimes.

Combined, all of these steps in an ML project can take several months with lots of trial and error. For this reason,

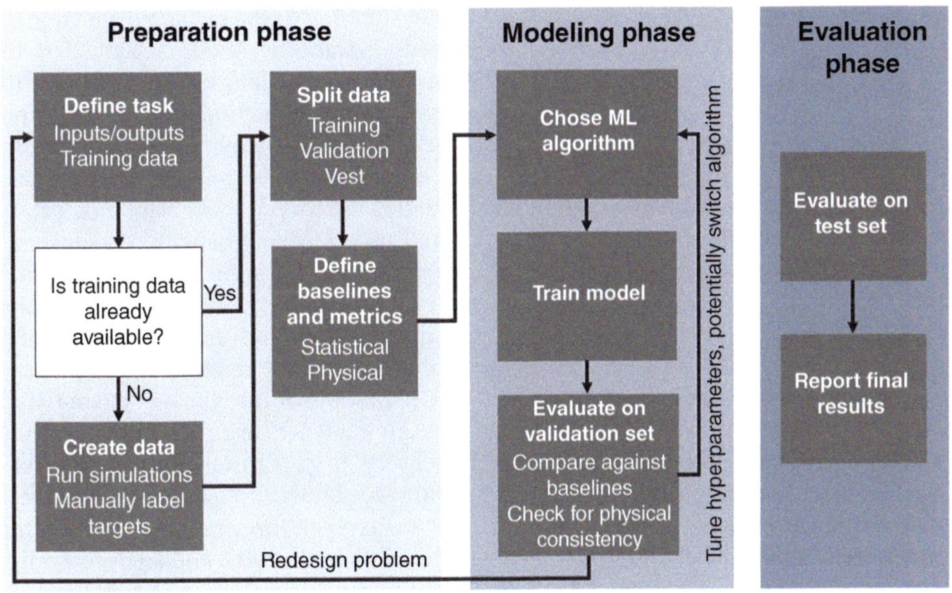

Figure 16.6 Workflow for a machine learning project in physical sciences.

it is paramount to make this workflow reproducible by, for example, using a version control system like Git and creating frequent checkpoints.

16.4.2. Machine Learning Software Ecosystem and Resources for Learning

The ML ecosystem changes significantly more rapidly than many other fields. For this reason, the software and learning resources listed below are also only a snapshot taken at the time of writing and might change in the future.

`Python` is by far the most popular programming language for modern ML and we will focus on Python for the remainder of this section. However, there are some alternatives: `R` still has a strong standing for statistical applications, for example, post-processing, and is still being updated with the latest ML developments. `Matlab`, an old favorite, has also recently added support for deep learning. Finally, `Julia` is a newcomer with strong core support in the atmospheric science community and growing ML capabilities. In Python, `scikit-learn` is a well-established library that has implemented a huge number of supervised and unsupervised learning algorithms, such as linear regression, RFs, various clustering methods all the way up to simple NNs. For deep learning the two most popular choices in Python are `Tensorflow` with its easy-to-use wrapper `Keras` and `Pytorch`. Most deep learning algorithms found in literature will have code in one of these two `Python` libraries available.

As for learning resources, there are a great number of books and courses available. For books, Géron (2019) is a great starting choice. So is Chollet (2017), which specifically focuses on deep learning. For courses, `deeplearning.ai` hosts several well-produced courses on `Coursera`, and we additionally recommend the deep learning courses of `fast.ai`.

16.5. OUTLOOK

Exploration of ML for clouds and climate is still in its infancy with many underexplored research frontiers spanning observations, modeling, and understanding.

For observations, the main challenge remains the lack of samples in the observational record: While the *volume* of observed data increases drastically from year to year, the number of observed events can still be very small for many applications (e.g., extreme events such as cyclones and atmospheric rivers) and often there are no labels. Traditional ML techniques to mitigate these issues, such as data augmentation (image translation, rotation, and mirroring to create more samples, Yu et al. (2017)), may only be valid for some two-dimensional data so long as the Earth's spherical geometry can be respected (Weyn et al., 2020). For that reason, it is promising to use transfer learning to fine-tune an ML model on the (potentially) sparse observational data of interest after training it on model data that can be generated at will (section 16.2.1), or meta-learning to efficiently adjust the ML algorithm's hyperparameters with only a few samples (Finn et al., 2017; Rußwurm et al., 2020). Finally, the mathematical simplicity of (fitted) ML models compared to dynamical models makes them easier to integrate within data assimilation frameworks (Brajard et al., 2020), opening the door to ML-powered bias correction of operational atmospheric models (Bonavita & Laloyaux, 2020).

For modeling, ML emulations of individual processes such as clouds are flourishing in climate models and can replace known equations that are hard to discretize (Bar-Sinai et al., 2019) or fit process-resolving simulations when the equations are unknown as illustrated by the microphysics (section 16.3.2) and subgrid parameterization (section 16.3.3) cases. This could readily be extended to parameterize subgrid processes at progressively smaller scales as global climate model resolution improves. Despite recent progress, it should be noted that ML emulation for climate modeling faces challenges that are not fully addressed, such as stability when coupled back to the climate model and generalization to different climates (section 16.2.3). A related issue is the quantification of uncertainties with probabilistic or stochastic ML techniques (section 16.3.6). Attempts to date have also not explored the potential of multi-node Graphics Processing Unit/Tensor Processing Unit (GPU/TPU) based high-performance computing for ML at scale due to software infrastructure challenges (Ben-Nun & Hoefler, 2019).

Finally, leveraging modern ML tools to improve our understanding of the climate from large datasets remains mostly untapped. While recent data-driven equation discovery tools (Long et al., 2019; Zhang & Lin, 2018) show promising preliminary results in physics (Brunton et al., 2016; Rudy et al., 2017) and oceanography (Zanna & Bolton, 2020) in simple settings, partial differential equation discovery tools have not yet been applied to cloud processes and climate modeling in Earth-like settings. XAI helps translate successes in emulation into improved climate understanding (section 16.2.2), but it does not extract causes of particular phenomena, motivating causal research to improve our theoretical understanding of the climate system (Runge et al., 2019). Causal inference (e.g., Granger causality) and causal discovery (e.g., causal effect networks) methods have been successfully applied to establish a feedback of ocean surface temperatures on the North Atlantic Oscillation (Mosedale et al., 2006) and analyze Arctic drivers of mid-latitude winter circulation (Kretschmer et al., 2016), but they remain underexplored in the analysis of convection and cloud-related climactic processes. A recent exception is Hirt et al. (2020) who used linear causal graph analysis to show that low model resolution decreased the frequency of convective initiation by reducing upward mass flux at gust fronts. Finally, as generative modeling can combine dimensionality reduction with prediction, further exploring generated latent spaces may reveal new sources of predictability in observational and simulation data (Behrens et al., 2022; Mooers, Tuyls et al., 2020).

ACKNOWLEDGMENTS

Tom Beucler and Michael Pritchard acknowledge funding from NSF grants OAC1835863 and AGS-1734164. Michael Pritchard acknowledges partial funding from the DOE "Enabling Aerosol-cloud interactions at GLobal convection-permitting scalES (EAGLES)" project (74358), funded by the U.S. Department of Energy, Office of Science, Office of Biological and Environmental Research, Earth System Model Development (ESMD) program. Additionally, this material is based upon work supported by the National Science Foundation under grants ICER-2019758 (IE) and OAC-1934668 (IE). Pierre Gentine thanks NSF grant AGS-1734164 as well as USMILE European Research Council grant. We thank Griffin Mooers for Figure 16.3 and Elizabeth Barnes, Alexei Belochitski, Derek Chang, Vladimir Krasnopolsky, Jeremy McGibbon, Sylvia Sullivan, Duncan Watson-Parris, and two anonymous reviewers for helpful advice that greatly improved the content and clarity of the manuscript.

REFERENCES

Adebayo, J., Gilmer, J., Muelly, M., Goodfellow, I., Hardt, M., & Kim, B. (2018). Sanity checks for saliency maps (pp. 9505–9515). *32nd Conference on Neural Information Processing Systems (NeurIPS 2018)*. Montral, Canada.

Baño-Medina, J. L., García Manzanas, R., & Gutiérrez Llorente, J. M. (2020). Configuration and intercomparison of deep learning neural models for statistical downscaling. *Geoscientific Model Development*, *13*, 2109–2124.

Bansal, N., Agarwal, C., & Nguyen, A. (2020). Sam: The sensitivity of attribution methods to hyperparameters (pp. 8673–8683). *Proceedings of the IEEE/CVF Conference on Computer Vision and Pattern Recognition*. Virtual (COVID).

Barnes, E. A., Hurrell, J. W., Ebert-Uphoff, I., Anderson, C., & Anderson, D. (2019). Viewing forced climate patterns through an AI lens. *Geophysical Research Letters*, *46*(22), 13389–13398.

Barnes, E. A., Samarasinghe, S. M., Ebert-Uphoff, I., & Furtado, J. C. (2019, Aug). Tropospheric and stratospheric causal pathways between the MJO and NAO. *Journal of Geophysical Research D: Atmospheres*, *124*(16), 9356–9371.

Barnes, E. A., Toms, B., Hurrell, J. W., Ebert-Uphoff, I., Anderson, C., & Anderson, D. (2020). Indicator patterns of forced change learned by an artificial neural network. *Journal of Advances in Modeling Earth Systems (JAMES)*, *12*(9), e2020MS002195. doi: 10.1029/2020MS002195

Bar-Sinai, Y., Hoyer, S., Hickey, J., & Brenner, M. P. (2019). Learning data-driven discretizations for partial differential equations. *Proceedings of the National Academy of Sciences*, *116*(31), 15344–15349.

Bate, A., Lindquist, M., Edwards, I. R., Olsson, S., Orre, R., Lansner, A., & De Freitas, R. M. (1998). A Bayesian neural network method for adverse drug reaction signal generation. *European Journal of Clinical Pharmacology*, *54*(4), 315–321.

Bau, D., Zhou, B., Khosla, A., Oliva, A., & Torralba, A. (2017). Network dissection: Quantifying interpretability of deep visual representations. In *Proceedings of the IEEE Conference on Computer Vision and Pattern Recognition* (pp. 6541–6549). Honolulu, HI, USA.

Behrens, G., Beucler, T., Gentine, P., Iglesias-Suarez, F., Pritchard, M., & Eyring, V. (2022). Nonlinear dimensionality reduction with a variational encoder decoder to understand convective Processes in climate models. *Journal of Advances in Modeling Earth Systems*, *14*(8), e2022MS003130.

Belochitski, A., Binev, P., DeVore, R., Fox-Rabinovitz, M., Krasnopolsky, V., & Lamby, P. (2011). Tree approximation of the long wave radiation parameterization in the NCAR CAM global climate model. *Journal of Computational and Applied Mathematics*, *236*(4), 447–460.

Ben-Nun, T., & Hoefler, T. (2019, Aug). Demystifying parallel and distributed deep learning: An in-depth concurrency analysis. *ACM Computing Surveys*, *52*(4), 1–43.

Bergen, K. J., Johnson, P. A., Maarten, V., & Beroza, G. C. (2019). Machine learning for data-driven discovery in solid Earth geoscience. *Science*, *363*(6433), eaau0323.

Berner, J., Achatz, U., Batte, L., Bengtsson, L., Camara, A. d. l., Christensen, H. M., et al. (2017). Stochastic parameterization: Toward a new view of weather and climate models. *Bulletin of the American Meteorological Society*, *98*(3), 565–588.

Beucler, T., Pritchard, M., Gentine, P., & Rasp, S. (2020). Towards physically-consistent, data-driven models of convection. *In IGARSS 2020–2020 IEEE international geoscience and remote sensing symposium*, 3987–3990 *arXiv preprint arXiv:2002.08525*.

Beucler, T., Pritchard, M., Rasp, S., Gentine, P., Ott, J., & Baldi, P. (2019). Enforcing analytic constraints in neural-networks emulating physical systems. *Physical Review Letters*, 126(9), 098302.

Beucler, T., Rasp, S., Pritchard, M., & Gentine, P. (2019). Achieving conservation of energy in neural network emulators for climate modeling. *arXiv preprint arXiv:1906.06622*.

Bonavita, M., & Laloyaux, P. (2020). Machine learning for model error inference and correction. *Earth and Space Science Open Archive*, *36*, e2020MS002232. doi: 10.1002/essoar.10503695.1

Bony, S., Stevens, B., Frierson, D. M., Jakob, C., Kageyama, M., Pincus, R., et al. (2015). Clouds, circulation and climate sensitivity. *Nature Geoscience*, *8*(4), 261–268.

Brajard, J., Carassi, A., Bocquet, M., & Bertino, L. (2020). Combining data as-similation and machine learning to emulate a dynamical model from sparse and noisy observations: A case study with the Lorenz 96 model. *arXiv preprint arXiv:2001.01520*.

Brenowitz, N. D., Beucler, T., Pritchard, M., & Bretherton, C. S. (2020). Interpreting and stabilizing machine-learning parametrizations of convection. *arXiv preprint arXiv:2003.06549*.

Brenowitz, N. D., & Bretherton, C. S. (2018). Prognostic validation of a neural network unified physics parameterization. *Geophysical Research Letters*, *45*(12), 6289–6298.

Brenowitz, N. D., & Bretherton, C. S. (2019). Spatially extended tests of a neural network parametrization trained by coarse-graining. *Journal of Advances in Modeling Earth Systems*, *11*(8), 2728–2744.

Brenowitz, N. D., Henn, B., McGibbon, J., Clark, S. K., Kwa, A., Perkins, W. A., et al. (2020). Machine learning climate model dynamics: Offline versus online performance. *arXiv preprint arXiv:2011.03081*.

Brunton, S. L., Proctor, J. L., & Kutz, J. N. (2016). Discovering governing equations from data by sparse identification of nonlinear dynamical systems. *Proceedings of the National Academy of Sciences*, *113*(15), 3932–3937.

Camps-Valls, G., Martino, L., Svendsen, D. H., Campos-Taberner, M., Muñoz-Marí, J., Laparra, V., et al. (2018). Physics-aware Gaussian processes in remote sensing. *Applied Soft Computing*, *68*, 69–82.

Camps-Valls, G., Verrelst, J., Muñoz-Marí, J., Laparra, V., Mateo-Jiménez, F., & Gomez-Dans, J. (2016). A survey on Gaussian processes for Earth-observation data analysis: A comprehensive investigation. *IEEE Geoscience and Remote Sensing Magazine*, *4*(2), 58–78.

Carter, S., Armstrong, Z., Schubert, L., Johnson, I., & Olah, C. (2019). *Exploring neural networks with activation atlases.*Distill.

Chevallier, F., Morcrette, J.-J., Chéruy, F., & Scott, N. (2000). Use of a neural- network-based long-wave radiative-transfer scheme in the ECMWF atmospheric model. *Quarterly Journal of the Royal Meteorological Society*, *126*(563), 761–776.

Chollet, F. (2017). *Deep learning with python.* New York, NY: Manning Publications Company.

Cintineo, J. L., Pavolonis, M. J., Sieglaff, J. M., Wimmers, A., Brunner, J., & Bellon, W. (2020). A deep-learning model for automated detection of intense mid-latitude convection using geostationary satellite images. *Weather and Forecasting*, *35*(6), 2567–2588.

Crommelin, D., & Vanden-Eijnden, E. (2008). Subgrid-scale parameterization with conditional Markov chains. *Journal of the Atmospheric Sciences*, *65*(8), 2661–2675.

de Bezenac, E., Pajot, A., & Gallinari, P. (2019). Deep learning for physical processes: Incorporating prior scientific knowledge. *Journal of Statistical Mechanics: Theory and Experiment*, *2019*(12), 124009.

Denby, L. (2020). Discovering the importance of mesoscale cloud organization through unsupervised classification. *Geophysical Research Letters*, *47*(1), e2019GL085190.

Deng, J., Dong, W., Socher, R., Li, L.-J., Li, K., & Fei-Fei, L. (2009). ImageNet: A large-scale hierarchical image database. In *2009 IEEE Conference on Computer Vision and Pattern Recognition* (pp. 248–255). Miami, FL.

Deser, C., Phillips, A., Bourdette, V., & Teng, H. (2012). Uncertainty in climate change projections: The role of internal variability. *Climate Dynamics*, *38*(3–4), 527–546.

Di Capua, G., Runge, J., Donner, R. V., van den Hurk, B., Turner, A. G., Vellore, R., et al. (2020). Dominant patterns of interaction between the tropics and mid-latitudes in boreal summer: Causal relationships and the role of timescales. *Weather and Climate Dynamics*, *1*(2), 519–539.

Dunbar, O. R., Garbuno-Inigo, A., Schneider, T., & Stuart, A. M. (2020). Calibration and uncertainty quantification of convective parameters in an idealized gcm. *arXiv preprint arXiv:2012.13262*.

Ebert-Uphoff, I., & Deng, Y. (2012). Causal discovery for climate research using graphical models. *Journal of Climate*, *25*(17), 5648–5665.

Ebert-Uphoff, I., & Hilburn, K. A. (2020). Evaluation, tuning and interpretation of neural networks for working with images in meteorological applications. *Bulletin of the American Meteorological Society (BAMS)*, *101*(12), E2149–E2170.

Eknes, M., & Evensen, G. (1997). Parameter estimation solving a weak constraint variational formulation for an Ekman model. *Journal of Geophysical Research: Oceans*, *102*(C6), 12479–12491.

Evensen, G. (2009). The ensemble Kalman filter for combined state and parameter estimation. *IEEE Control Systems Magazine*, *29*(3), 83–104.

Evensen, G., Dee, D. P., & Schröter, J. (1998). Parameter estimation in dynamical models. In E. P. Chassignet & J. Verron (Eds.), *Ocean Modeling and Parameterization* (pp. 373–398). NATO Science Series, Vol 516. Dordrecht: Springer.

Feser, F., Rockel, B., von Storch, H., Winterfeldt, J., & Zahn, M. (2011). Regional climate models add value to global model data: A review and selected examples. *Bulletin of the American Meteorological Society*, *92*(9), 1181–1192.

Finn, C., Abbeel, P., & Levine, S. (2017). Model-agnostic meta-learning for fast adaptation of deep networks. *arXiv preprint arXiv:1703.03400*.

Fong, R., & Vedaldi, A. (2019). Explanations for attributing deep neural network predictions. In W. Samek, G. Montavon, A. Vedaldi, L. K. Hansen, & K.-R. Müller (Eds.), *Explainable AI: Interpreting, Explaining and Visualizing Deep Learning* (pp. 149–167). Springer.

Foster, D. (2019). *Generative deep learning: Teaching machines to paint, write, compose, and play*. O'Reilly Media.

Fukami, K., Fukagata, K., & Taira, K. (2018). Super-resolution reconstruction of turbulent flows with machine learning. *arXiv preprint arXiv:1811.11328*.

Gagne, D. J., Christensen, H. M., Subramanian, A. C., & Monahan, A. H. (2020). Machine learning for stochastic parameterization: Generative adversarial networks in the Lorenz'96 model. *Journal of Advances in Modeling Earth Systems*, *12*(3), e2019MS001896.

Gagne, D. J., McGovern, A., Haupt, S. E., Sobash, R. A., Williams, J. K., & Xue, M. (2017). Storm-based probabilistic hail forecasting with machine learning applied to convection-allowing ensembles. *Weather and Forecasting*, *32*(5), 1819–1840.

Gagne II, D. J., Haupt, S. E., Nychka, D. W., & Thompson, G. (2019). Interpretable deep learning for spatial analysis of severe hailstorms. *Monthly Weather Review*, *147*(8), 2827–2845.

Gal, Y., & Ghahramani, Z. (2016). Dropout as a Bayesian approximation: Representing model uncertainty in deep learning. In *International Conference on Machine Learning* (pp. 1050–1059). New York City, NY, USA.

Gala, Y., Fernández, Á., Díaz, J., & Dorronsoro, J. R. (2016). Hybrid machine learning forecasting of solar radiation values. *Neurocomputing*, *176*, 48–59.

Gao, H., Sun, L., & Wang, J.-X. (2020). PhyGeoNet: Physics-informed geometry-adaptive convolutional neural networks for solving parametric PDEs on irregular domain. *arXiv preprint arXiv:2004.13145*.

Gentine, P., Pritchard, M., Rasp, S., Reinaudi, G., & Yacalis, G. (2018). Could machine learning break the convection parameterization deadlock? *Geophysical Research Letters*, *45*(11), 5742–5751.

Géron, A. (2019). *Hands-on machine learning with Scikit-learn, Keras, and Tensor-flow: Concepts, tools, and techniques to build intelligent systems*. CA, USA: O'Reilly Media. CA, USA.

Gettelman, A., Gagne, D. J., Chen, C.-C., Christensen, M., Lebo, Z., Morrison, H., & Gantos, G. (2020). Machine learning the warm rain process. *Journal of Advances in Modeling Earth Systems*, *13*(2), e2020MS002268.

Gooch, S. R., & Chandrasekar, V. (2020). Improving historical data discovery in weather radar image data sets using transfer learning. *IEEE Transactions on Geoscience and Remote Sensing*, *59*(7), 5619–5629.

Gristey, J. J., Feingold, G., Glenn, I. B., Schmidt, K. S., & Chen, H. (2020). On the relationship between shallow cumulus cloud field properties and surface solar irradiance. *Geophysical Research Letters*, *47*(22), e2020GL090152.

Groenke, B., Madaus, L., & Monteleoni, C. (2020). ClimAlign: Unsupervised statistical downscaling of climate variables via normalizing flows. In *Proceedings of the 10th International Conference on Climate Informatics (CI2020)*. Oxford University on 23-25 September 2020

Grönquist, P., Yao, C., Ben-Nun, T., Dryden, N., Dueben, P., Li, S., & Hoefler, T. (2020). Deep learning for post-processing ensemble weather forecasts. *arXiv preprint arXiv:2005.08748*.

Grover, A., Chute, C., Shu, R., Cao, Z., & Ermon, S. (2020). AlignFlow: Cycle consistent learning from multiple domains via normalizing flows. *Proceedings of the AAAI Conference on Artificial Intelligence*, *34*(04), 4028–4035.

Ham, Y.-G., Kim, J.-H., & Luo, J.-J. (2019). Deep learning for multi-year ENSO fore-casts. *Nature*, *573*(7775), 568–572.

Han, Y., Zhang, G. J., Huang, X., & Wang, Y. (2020). A moist physics parameterization based on deep learning. *Journal of Advances in Modeling Earth Systems*, *12*(9), e2020MS002076.

Haynes, J. M., Jakob, C., Rossow, W. B., Tselioudis, G., & Brown, J. (2011). Major characteristics of southern ocean cloud regimes and their effects on the energy budget. *Journal of Climate*, *24*(19), 5061–5080.

Hirt, M., Craig, G. C., Schäfer, S. A., Savre, J., & Heinze, R. (2020). Cold-pool-driven convective initiation: Using causal graph analysis to determine what convection-permitting models are missing. *Quarterly Journal of the Royal Meteorological Society*, *146*(730), 2205–2227.

Irrgang, C., Boers, N., Sonnewald, M., Barnes, E. A., Kadow, C., Staneva, J., & Saynisch-Wagner, J. (2021). Will artificial intelligence supersede Earth system and climate models? *arXiv preprint arXiv:2101.09126*.

Janjić, T., Bormann, N., Bocquet, M., Carton, J., Cohn, S., Dance, S., et al. (2018). On the representation error in data assimilation. *Quarterly Journal of the Royal Meteorological Society*, *144*(713), 1257–1278.

Jean, N., Wang, S., Samar, A., Azzari, G., Lobell, D., & Ermon, S. (2019). Tile2Vec: Unsupervised representation learning for spatially distributed data. In *Proceedings of the AAAI Conference on Artificial Intelligence* (Vol. 33, pp. 3967–3974). Honolulu, Hawaii, USA.

Karpatne, A., Ebert-Uphoff, I., Ravela, S., Babaie, H. A., & Kumar, V. (2018). Machine learning for the geosciences: Challenges and opportunities. *IEEE Transactions on Knowledge and Data Engineering*, *31*(8), 1544–1554.

Karpatne, A., Watkins, W., Read, J., & Kumar, V. (2017). Physics-guided neural networks (PGNN): An application in lake temperature modeling. *arXiv preprint arXiv:1710.11431*.

Kashinath, K., Mudigonda, M., Kim, S., Kapp-Schwoerer, L., Graubner, A., Karais-mailoglu, E., et al. (2021). ClimateNet: An expert-labeled open dataset and deep learning architecture for enabling high-precision analyses of extreme weather. *Geoscientific Model Development*, *14*(1), 107–124.

Khan, M. S., & Coulibaly, P. (2006). Bayesian neural network for rainfall-runoff modeling. *Water Resources Research*, *42*(7). https://agupubs.onlinelibrary.wiley.com/doi/full/10.1029/2005WR003971

Kindermans, P.-J., Hooker, S., Adebayo, J., Alber, M., Schutt, K. T., Dahne, S., et al. (2017). The (un) reliability of saliency methods. *arXiv preprint arXiv:1711.00867*.

Krasnopolsky, V., Fox-Rabinovitz, M., Hou, Y., Lord, S., & Belochitski, A. (2010). Accurate and fast neural network emulations of model radiation for the NCEP coupled climate forecast system: Climate simulations and seasonal predictions. *Monthly Weather Review*, *138*(5), 1822–1842.

Krasnopolsky, V. M. (2013). The application of neural networks in the earth system sciences. Neural network emulations for complex multidimensional mappings. *Atmospheric and Oceanic Science Library*, *46*, 81–180.

Krasnopolsky, V. M., & Fox-Rabinovitz, M. S. (2006). Complex hybrid models combining deterministic and machine learning components for numerical climate modeling and weather prediction. *Neural Networks*, *19*(2), 122–134.

Krasnopolsky, V. M., Fox-Rabinovitz, M. S., & Belochitski, A. A. (2008). Decadal climate simulations using accurate and fast neural network emulation of full, longwave and shortwave, radiation. *Monthly Weather Review*, *136*(10), 3683–3695.

Krasnopolsky, V. M., Fox-Rabinovitz, M. S., & Belochitski, A. A. (2013). Using ensemble of neural networks to learn stochastic convection parameterizations for climate and numerical weather prediction models from data simulated by a cloud resolving model. *Advances in Artificial Neural Systems* *2013*, 5–5.

Kretschmer, M., Coumou, D., Donges, J. F., & Runge, J. (2016). Using causal effect networks to analyze different Arctic drivers of midlatitude winter circulation. *Journal of Climate*, *29*(11), 4069–4081.

Kurihana, T., Foster, I., Willett, R., Jenkins, S., Koenig, K., Werman, R., et al. (2019). Cloud classification with unsupervised learning. In *9th International Workshop on Climate Informatics* (CI2019). Paris, France, October 2–4, 2019.

Labe, Z. M., & Barnes, E. A. (2021). Detecting climate signals using explainable AI with single-forcing large ensembles. *Journal of Advances in Modeling Earth Systems*, *13*(6), e2021MS002464.

Lagerquist, R., McGovern, A., & Gagne II, D. J. (2019). Deep learning for spatially explicit prediction of synoptic-scale fronts. *Weather and Forecasting*, *34*(4), 1137–1160.

Lagerquist, R., McGovern, A., Homeyer, C. R., Gagne, D. J., & Smith, T. (2020). Deep learning on three-dimensional multiscale data for next-hour tornado prediction. *Monthly Weather Review*. *148*(7), 2837–2861.

Lee, J., Weger, R. C., Sengupta, S. K., & Welch, R. M. (1990). A neural network approach to cloud classification. *IEEE Transactions on Geoscience and Remote Sensing*, *28*(5), 846–855.

Lee, Y., Kummerow, C. D., & Ebert-Uphoff, I. (2021). Applying machine learning methods to detect convection using GOES-16 ABI data. *Submitted to Atmospheric Measurement Techniques (in review)*.

Ling, J., Kurzawski, A., & Templeton, J. (2016). Reynolds averaged turbulence modelling using deep neural networks with embedded invariance. *Journal of Fluid Mechanics*, *807*, 155–166.

Liu, Y., Racah, E., Correa, J., Khosrowshahi, A., Lavers, D., Kunkel, K., et al. (2016). Application of deep convolutional neural networks for detecting extreme weather in climate datasets. *arXiv preprint arXiv:1605.01156*.

Loeb, D. E. (1991). Series with general exponents. *Journal of Mathematical Analysis and Applications*, *156*(1), 184–208.

Long, Z., Lu, Y., & Dong, B. (2019). PDE-Net 2.0: Learning PDEs from data with a numeric-symbolic hybrid deep network. *Journal of Computational Physics*, *399*, 108925.

Lorenz, E. N. (1963). Deterministic nonperiodic flow. *Journal of the Atmospheric Sciences*, *20*(2), 130–141.

Lorenz, E. N. (1996). Predictability: A problem partly solved. In *Proceedings Seminar on Predictability* (Vol. 1). ECMWF, Reading, Berkshire, UK, 1–18.

Lundberg, S. M., & Lee, S.-I. (2017). A unified approach to interpreting model predictions. In I. Guyon, U. Von Luxburg, S. Bengio, H. Wallach, R. Fergus, S. Vishwanathan, & R. Garnett (Eds.), *Advances in Neural Information Processing Systems* (pp. 4765–4774). NeurIPS Proceedings.

Mahajan, S., & Fataniya, B. (2019). Cloud detection methodologies: Variants and development—A review. *Complex and Intelligent Systems*, 1–11.

Mao, Z., Jagtap, A. D., & Karniadakis, G. E. (2020). Physics-informed neural networks for high-speed flows. *Computer Methods in Applied Mechanics and Engineering*, *360*, 112789.

Marais, W. J., Holz, R. E., Reid, J. S., & Willett, R. M. (2020). Leveraging spatial textures, through machine learning, to identify aerosols and distinct cloud types from multispectral observations. *Atmospheric Measurement Techniques*, *13*(10), 5459–5480.

Marmanis, D., Datcu, M., Esch, T., & Stilla, U. (2015). Deep learning Earth observation classification using imagenet pretrained networks. *IEEE Geoscience and Remote Sensing Letters*, *13*(1), 105–109.

McCoy, I. L., McCoy, D. T., Wood, R., Regayre, L., Watson-Parris, D., Grosvenor, D. P., et al. (2020). The hemispheric contrast in cloud microphysical properties constrains aerosol forcing. *Proceedings of the National Academy of Sciences*, *117*(32), 18998–19006.

McGibbon, J., & Bretherton, C. S. (2019). Single-column emulation of reanalysis of the northeast Pacific marine boundary layer. *Geophysical Research Letters*, *46*(16), 10053–10060.

McGovern, A., Elmore, K. L., Gagne, D. J., Haupt, S. E., Karstens, C. D., Lagerquist, R., et al. (2017). Using artificial intelligence to improve real-time decision-making for high-impact weather. *Bulletin of the American Meteorological Society*, *98*(10), 2073–2090.

McGovern, A., Lagerquist, R., John Gagne, D., Jergensen, G. E., Elmore, K. L., Homeyer, C. R., & Smith, T. (2019). Making the black box more transparent: Understanding the physical implications of machine learning. *Bulletin of the American Meteorological Society*, *100*(11), 2175–2199.

McGovern, A., & Lagerquist, R. A. (2020). Using machine learning and model interpretation and visualization techniques to gain physical insights in atmospheric science. In *International Conference on Learning Representations (ICLR 2020), AI for Earth Sciences Workshop*. Virtual (COVID).

McGovern, A., Lagerquist, R. A., & Gagne, D. J. I. (2020). Using machine learning and model interpretation and visualization techniques to gain physical insights in atmospheric science. In *International Conference on Learning Representations (ICLR2020)*. Virtual (COVID).

Min, M., Li, J., Wang, F., Liu, Z., & Menzel, W. P. (2020). Retrieval of cloud top properties from advanced geostationary satellite imager measurements based on machine learning algorithms. *Remote Sensing of Environment*, *239*, 111616.

Mirza, M., & Osindero, S. (2014). Conditional generative adversarial nets. *arXiv preprint arXiv:1411.1784*.

Molnar, C. (2019). Interpretable machine learning. *Lulu. com*.

Montavon, G., Lapuschkin, S., Binder, A., Samek, W., & Müller, K.-R. (2017). Explaining nonlinear classification decisions with deep Taylor decomposition. *Pattern Recognition*, *65*, 211–222.

Montavon, G., Samek, W., & Müller, K.-R. (2018). Methods for interpreting and understanding deep neural networks. *Digital Signal Processing*, *73*, 1–15.

Mooers, G., Pritchard, M., Beucler, T., Ott, J., Yacalis, G., Baldi, P., & Gentine, P. (2020). Assessing the potential of deep learning for emulating cloud super-parameterization in climate models with real-geography boundary conditions. *arXiv preprint arXiv:2010.12996*.

Mooers, G., Tuyls, J., Mandt, S., Pritchard, M., & Beucler, T. (2020). Generative modeling for atmospheric convection. *arXiv preprint arXiv:2007.01444*.

Morrison, H., van Lier-Walqui, M., Fridlind, A. M., Grabowski, W. W., Harrington, J. Y., Hoose, C., et al.

(2020). Confronting the challenge of modeling cloud and precipitation microphysics. *Journal of Advances in Modeling Earth Systems*, e2019MS001689.

Morrison, H., van Lier-Walqui, M., Kumjian, M. R., & Prat, O. P. (2020). A Bayesian approach for statistical–physical bulk parameterization of rain microphysics. Part I: Scheme description. *Journal of the Atmospheric Sciences*, *77*(3), 1019–1041.

Mosedale, T. J., Stephenson, D. B., Collins, M., & Mills, T. C. (2006). Granger causality of coupled climate processes: Ocean feedback on the north Atlantic oscillation. *Journal of Climate*, *19*(7), 1182–1194.

Mouatadid, S., Gentine, P., Yu, W., & Easterbrook, S. (2019). Recovering the parameters underlying the Lorenz-96 chaotic dynamics. *arXiv preprint arXiv:1906.06786*.

Muhlbauer, A., McCoy, I. L., & Wood, R. (2014). Climatology of stratocumulus cloud morphologies: Microphysical properties and radiative effects. *Atmospheric Chemistry and Physics*, *14*(13), 6695–6716. doi: 10.5194/acp-14-6695-2014.

O'Gorman, P. A., & Dwyer, J. G. (2018). Using machine learning to parameterize moist convection: Potential for modeling of climate, climate change, and extreme events. *Journal of Advances in Modeling Earth Systems*, *10*(10), 2548–2563.

Olah, C., Mordvintsev, A., & Schubert, L. (2017). Feature visualization. *Distill*, *2*(11), e7.

Olah, C., Satyanarayan, A., Johnson, I., Carter, S., Schubert, L., Ye, K., & Mord-vintsev, A. (2018). The building blocks of interpretability. *Distill*, *3*(3), e10.

Pal, A., Mahajan, S., & Norman, M. R. (2019). Using deep neural networks as cost-effective surrogate models for super-parameterized E3SM radiative transfer. *Geophysical Research Letters*, *46*(11), 6069–6079.

Pang, G., Lu, L., & Karniadakis, G. E. (2019). fPINNS: Fractional physics-informed neural networks. *SIAM Journal on Scientific Computing*, *41*(4), A2603–A2626.

Pouliot, D., Latifovic, R., Pasher, J., & Duffe, J. (2018). Landsat super-resolution enhancement using convolution neural networks and Sentinel-2 for training. *Remote Sensing*, *10*(3), 394.

Pu, Y., Gan, Z., Henao, R., Yuan, X., Li, C., Stevens, A., & Carin, L. (2016). Variational autoencoder for deep learning of images, labels and captions. *In Proceedings of the 30th International Conference on Neural Information Processing Systems (NIPS'16)*. (pp. 2352–2360). Curran Associates Inc., Red Hook, NY, USA.

Racah, E., Beckham, C., Maharaj, T., Kahou, S. E., Prabhat, M., & Pal, C. (2017). Extreme weather: A large-scale climate dataset for semi-supervised detection, localization, and understanding of extreme weather events. *In Proceedings of the 31st International Conference on Neural Information Processing Systems (NIPS'17)*. (pp. 3402–3413). Curran Associates Inc., Red Hook, NY, USA.

Radford, A., Metz, L., & Chintala, S. (2015). Unsupervised representation learning with deep convolutional generative adversarial networks. *arXiv preprint arXiv:1511.06434*.

Raghu, M., & Schmidt, E. (2020). A survey of deep learning for scientific discovery. *arXiv preprint arXiv:2003.11755*.

Raissi, M., Perdikaris, P., & Karniadakis, G. E. (2017). Physics informed deep learning. Part I: Data-driven solutions of nonlinear partial differential equations. *arXiv preprint arXiv:1711.10561*.

Raissi, M., Perdikaris, P., & Karniadakis, G. E. (2019). Physics-informed neural networks: A deep learning framework for solving forward and inverse problems involving nonlinear partial differential equations. *Journal of Computational Physics*, *378*, 686–707.

Ras, G., van Gerven, M., & Haselager, P. (2018). Explanation methods in deep learning: Users, values, concerns and challenges. In *Explainable and Interpretable Models in Computer Vision and Machine Learning* (pp. 19–36). Springer.

Rasp, S. (2020). Coupled online learning as a way to tackle instabilities and biases in neural network parameterizations: General algorithms and Lorenz 96 case study (v1. 0). *Geoscientific Model Development*, *13*(5), 2185–2196.

Rasp, S., Dueben, P. D., Scher, S., Weyn, J. A., Mouatadid, S., & Thuerey, N. (2020). Weatherbench: A benchmark dataset for data-driven weather fore-casting. *Journal of Advances in Modeling Earth Systems*, e2020MS002203. doi: https://doi.org/10.1029/2020MS002203

Rasp, S., Pritchard, M. S., & Gentine, P. (2018). Deep learning to represent subgrid processes in climate models. *Proceedings of the National Academy of Sciences*, *115*(39), 9684–9689.

Rasp, S., Schulz, H., Bony, S., & Stevens, B. (2020). Combining crowd-sourcing and deep learning to explore the meso-scale organization of shallow convection. *Bulletin of the American Meteorological Society*, *101*(11), E1980–E1995.

Rasp, S., & Thuerey, N. (2020). Data-driven medium-range weather prediction with a resnet pretrained on climate simulations: A new model for weatherbench. *arXiv preprint arXiv: 2008.08626*.

Reichstein, M., Camps-Valls, G., Stevens, B., Jung, M., Denzler, J., Carvalhais, N., et al. (2019). Deep learning and process understanding for data-driven Earth system science. *Nature*, *566*(7743), 195–204.

Rivera, J. P., Verrelst, J., Gómez-Dans, J., Muñoz-Marí, J., Moreno, J., & Camps-Valls, G. (2015). An emulator toolbox to approximate radiative transfer models with statistical learning. *Remote Sensing*, *7*(7), 9347–9370.

Roh, S., & Song, H.-J. (2020). Evaluation of neural network emulations for radiation parameterization in cloud resolving model. *Geophysical Research Letters*, *47*(21), e2020GL089444.

Rolnick, D., Donti, P. L., Kaack, L. H., Kochanski, K., Lacoste, A., Sankaran, K., et al. (2019). Tackling climate change with machine learning. *arXiv preprint arXiv:1906.05433*.

Rudin, C. (2019). Stop explaining black box machine learning models for high stakes decisions and use interpretable models instead. *Nature Machine Intelligence*, *1*(5), 206–215.

Rudy, S. H., Brunton, S. L., Proctor, J. L., & Kutz, J. N. (2017). Data-driven discovery of partial differential equations. *Science Advances*, *3*(4), e1602614.

Runge, J., Bathiany, S., Bollt, E., Camps-Valls, G., Coumou, D., Deyle, E., et al. (2019). Inferring causation from time series in Earth system sciences. *Nature Communications*, *10*(1), 1–13.

Rußwurm, M., Wang, S., Korner, M., & Lobell, D. (2020). Meta-learning for few-shot land cover classification. In *Proceedings of the IEEE/CVF Conference on Computer Vision and Pattern Recognition Workshops* (pp. 200–201). Virtual (COVID).

Rutjes, H., Willemsen, M., & IJsselsteijn, W. (2019). Considerations on explainable AI and users' mental models. In *Chi 2019*

Workshop: Where Is the Human? Bridging the Gap Between AI and HCI. Glasgow, Scotland.

Sachindra, D., Ahmed, K., Rashid, M. M., Shahid, S., & Perera, B. (2018). Statistical downscaling of precipitation using machine learning techniques. *Atmospheric Research*, *212*, 240–258.

Samek, W., Montavon, G., Vedaldi, A., Hansen, L. K., & Müller, K.-R. (2019). *Explainable AI: Interpreting, explaining and visualizing deep learning* (Vol. *11700*). Springer Nature.

Scarselli, F., & Tsoi, A. C. (1998). Universal approximation using feedforward neural networks: A survey of some existing methods, and some new results. *Neural Networks*, *11*(1), 15–37.

Schneider, T., Lan, S., Stuart, A., & Teixeira, J. (2017). Earth system modeling 2.0: A blueprint for models that learn from observations and targeted high-resolution simulations. *Geophysical Research Letters*, *44*(24), 12–396.

Schneider, T., Teixeira, J., Bretherton, C. S., Brient, F., Pressel, K. G., Schär, C., & Siebesma, A. P. (2017). Climate goals and computing the future of clouds. *Nature Climate Change*, *7*(1), 3–5.

Seifert, A., & Rasp, S. (2020). Potential and limitations of machine learning for modeling warm-rain cloud microphysical processes. *Journal of Advances in Modeling Earth Systems*, *12*(12), e2020MS002301.

Selvaraju, R. R., Cogswell, M., Das, A., Vedantam, R., Parikh, D., & Batra, D. (2017). Grad-CAM: Visual explanations from deep networks via gradient-based localization. In *Proceedings of the IEEE International Conference on Computer Vision* (pp. 618–626). Honolulu, HI, USA.

Sharifi, E., Saghafian, B., & Steinacker, R. (2019). Downscaling satellite precipitation estimates with multiple linear regression, artificial neural networks, and spline interpolation techniques. *Journal of Geophysical Research: Atmospheres*, *124*(2), 789–805.

Shrikumar, A., Greenside, P., & Kundaje, A. (2017). Learning important features through propagating activation differences. *arXiv preprint arXiv:1704.02685*.

Siebesma, A. P., Soares, P. M., & Teixeira, J. (2007). A combined Eddy-diffusivity mass-flux approach for the convective boundary layer. *Journal of the Atmospheric Sciences*, *64*(4), 1230–1248.

Simonyan, K., Vedaldi, A., & Zisserman, A. (2013). Deep inside convolutional networks: Visualising image classification models and saliency maps. *arXiv preprint arXiv:1312.6034*.

Singh, A., White, B. L., & Albert, A. (2019). Downscaling numerical weather models with GANs. In *AGU Fall Meeting 2019*. San Francisco, CA, USA.

Smilkov, D., Thorat, N., Kim, B., Viégas, F., & Wattenberg, M. (2017). Smooth-Grad: Removing noise by adding noise. *arXiv preprint arXiv:1706.03825*.

Sønderby, C. K., Espeholt, L., Heek, J., Dehghani, M., Oliver, A., Salimans, T., et al. (2020). MetNet: A neural weather model for precipitation forecasting. *arXiv preprint arXiv:2003.12140*.

Stengel, K., Glaws, A., Hettinger, D., & King, R. N. (2020). Adversarial super-resolution of climatological wind and solar data. *Proceedings of the National Academy of Sciences*, *117*(29), 16805–16815.

Stevens, B., Bony, S., Brogniez, H., Hentgen, L., Hohenegger, C., Kiemle, C., et al. (2020). Sugar, gravel, fish and flowers: Mesoscale cloud patterns in the trade winds. *Quarterly Journal of the Royal Meteorological Society*, *146*(726), 141–152. doi: 10.1002/qj.3662.

Stevens, B., Satoh, M., Auger, L., Biercamp, J., Bretherton, C. S., Chen, X., et al. (2019). DYAMOND: The dynamics of the atmospheric general circulation modeled on non-hydrostatic domains. *Progress in Earth and Planetary Science*, *6*(1), 1–17.

Sun, L., Gao, H., Pan, S., & Wang, J.-X. (2020). Surrogate modeling for fluid flows based on physics-constrained deep learning without simulation data. *Computer Methods in Applied Mechanics and Engineering*, *361*, 112732.

Sundararajan, M., Taly, A., & Yan, Q. (2017). Axiomatic attribution for deep networks. *arXiv preprint arXiv:1703.01365*.

Tian, B., Shaikh, M. A., Azimi-Sadjadi, M. R., Haar, T. H. V., & Reinke, D. L. (1999). A study of cloud classification with neural networks using spectral and textural features. *IEEE Transactions on Neural Networks*, *10*(1), 138–151.

Tipping, M. E. (2000). The relevance vector machine. In S. Solla and T. Leen and K. Müller (Eds.), *Advances in Neural Information Processing Systems* (pp. 652–658). NeurIPS Proceedings.

Toms, B. A., Kashinath, K., & Yang, D. (2020). Testing the reliability of interpretable neural networks in geoscience using the Madden–Julian oscillation. *Geoscientific Model Development Discussions*, *14*(7), 4495–4508.

Ukkonen, P., Pincus, R., Hogan, R. J., Pagh Nielsen, K., & Kaas, E. (2020). Accelerating radiation computations for dynamical models with targeted machine learning and code optimization. *Journal of Advances in Modeling Earth Systems*, *12*(12), e2020MS002226.

Vandal, T., Kodra, E., & Ganguly, A. R. (2019). Intercomparison of machine learning methods for statistical downscaling: The case of daily and extreme precipitation. *Theoretical and Applied Climatology*, *137*(1–2), 557–570.

Vandal, T., Kodra, E., Ganguly, S., Michaelis, A., Nemani, R., & Ganguly, A. R. (2017). DeepSD: Generating high resolution climate change projections through single image super-resolution. In *Proceedings of the 23rd ACM SIGKDD International Conference on Knowledge Discovery and Data Mining* (pp. 1663–1672). Long Beach, CA, USA.

Veerman, M. A., Pincus, R., Stoffer, R., van Leeuwen, C., Podareanu, D., & van Heerwaarden, C. C. (2020). Predicting atmospheric optical properties for radiative transfer computations using neural networks. *arXiv preprint arXiv:2005.02265*.

Voyant, C., Notton, G., Kalogirou, S., Nivet, M.-L., Paoli, C., Motte, F., & Fouilloy, A. (2017). Machine learning methods for solar radiation forecasting: A review. *Renewable Energy*, *105*, 569–582.

Wang, J.-X., Wu, J.-L., & Xiao, H. (2017). Physics-informed machine learning approach for reconstructing Reynolds stress modeling discrepancies based on DNS data. *Physical Review Fluids*, *2*(3), 034603.

Watson-Parris, D. (2020). Machine learning for weather and climate are worlds apart. *arXiv preprint arXiv:2008.10679*.

Watson-Parris, D., Bellouin, N., Deaconu, L., Schutgens, N. A., Yoshioka, M., Regayre, L. A., et al. (2020). Constraining uncertainty in aerosol direct forcing. *Geophysical Research Letters*, *47*(9), e2020GL087141.

Watson-Parris, D., Sutherland, S., Christensen, M., & Stier, P. (2020). A large-scale analysis of pockets of open cells and their radiative impact. *Geophysical Research Letters*, *48*(6), e2020GL092213.

Watt-Meyer, O., Brenowitz, N., Bretherton, C. S., Clark, S., Henn, B. M., Kwa, A., et al. (2020). Correcting weather models by learning nudging tendencies from hindcast simulations. In *AGU Fall Meeting 2020*. Virtual (COVID).

Weyn, J. A., Durran, D. R., & Caruana, R. (2020). Improving data-driven global weather prediction using deep convolutional neural networks on a cubed sphere. *arXiv preprint arXiv:2003.11927*.

Wilby, R. L., Wigley, T., Conway, D., Jones, P., Hewitson, B., Main, J., & Wilks, D. (1998). Statistical downscaling of general circulation model output: A comparison of methods. *Water Resources Research*, *34*(11), 2995–3008.

Willard, J., Jia, X., Xu, S., Steinbach, M., & Kumar, V. (2020). Integrating physics-based modeling with machine learning: A survey. *arXiv preprint arXiv:2003.04919*.

Wills, R. C. J., Battisti, D. S., Armour, K. C., Schneider, T., & Deser, C. (2020, Oct). Pattern recognition methods to separate forced responses from internal variability in climate model ensembles and observations. *Journal of Climate*, *33*(20), 8693–8719.

Wood, R., & Hartmann, D. L. (2006). Spatial variability of liquid water path in marine low cloud: The importance of mesoscale cellular convection. *Journal of Climate*, *19*(9), 1748–1764. doi: 10.1175/JCLI3702.1

Wu, J.-L., Kashinath, K., Albert, A., Chirila, D., & Xiao, H. (2020). Enforcing statistical constraints in generative adversarial networks for modeling chaotic dynamical systems. *Journal of Computational Physics*, *406*, 109209.

Xue, Y., Janjic, Z., Dudhia, J., Vasic, R., & De Sales, F. (2014). A review on regional dynamical downscaling in intraseasonal to seasonal simulation/prediction and major factors that affect downscaling ability. *Atmospheric Research*, *147*, 68–85.

Yang, T., Sun, F., Gentine, P., Liu, W., Wang, H., Yin, J., et al. (2019). Evaluation and machine learning improvement of global hydrological model-based flood simulations. *Environmental Research Letters*, *14*(11), 114027.

Yang, W., Zhang, X., Tian, Y., Wang, W., Xue, J.-H., & Liao, Q. (2019). Deep learning for single image super-resolution: A brief review. *IEEE Transactions on Multimedia*, *21*(12), 3106–3121.

Yang, Z., Wu, J.-L., & Xiao, H. (2019). Enforcing deterministic constraints on generative adversarial networks for emulating physical systems. *arXiv preprint arXiv:1911.06671*.

Ying, X. (2019). An overview of overfitting and its solutions. *Journal of Physics: Conference Series*, *1168*, 022022.

Yosinski, J., Clune, J., Nguyen, A., Fuchs, T., & Lipson, H. (2015). Understanding neural networks through deep visualization. *arXiv preprint arXiv:1506.06579*.

Yu, X., Wu, X., Luo, C., & Ren, P. (2017). Deep learning in remote sensing scene classification: A data augmentation enhanced convolutional neural network framework. *GIScience and Remote Sensing*, *54*(5), 741–758.

Yuval, J., Hill, C. N., & O'Gorman, P. A. (2020). Use of neural networks for stable, accurate and physically consistent parameterization of subgrid atmospheric processes with good performance at reduced precision. *arXiv preprint arXiv:2010.09947*.

Yuval, J., & O'Gorman, P. A. (2020). Stable machine-learning parameterization of subgrid processes for climate modeling at a range of resolutions. *Nature Communications*, *11*(1), 1–10.

Zanna, L., & Bolton, T. (2020). Data-driven equation discovery of ocean mesoscale closures. *Geophysical Research Letters*. *47*(17), e2020GL088376.

Zantedeschi, V., Falasca, F., Douglas, A., Strange, R., Kusner, M. J., & Watson- Parris, D. (2019). CUMULO: A dataset for learning cloud classes. *arXiv preprint arXiv:1911.04227*.

Zelinka, M. D., Myers, T. A., McCoy, D. T., Po-Chedley, S., Caldwell, P. M., Ceppi, P., et al. (2020). Causes of higher climate sensitivity in CMIP6 models. *Geophysical Research Letters*, *47*(1), e2019GL085782.

Zhang, C., Zhuge, X., & Yu, F. (2019). Development of a high spatiotemporal resolution cloud-type classification approach using Himawari-8 and CloudSat. *International Journal of Remote Sensing*, *40*(16), 6464–6481.

Zhang, D., Guo, L., & Karniadakis, G. E. (2020). Learning in model space: Solving time-dependent stochastic PDES using physics-informed neural networks. *SIAM Journal on Scientific Computing*, *42*(2), A639–A665.

Zhang, J., Liu, P., Zhang, F., & Song, Q. (2018). Cloud-Net: Ground-based cloud classification with deep convolutional neural network. *Geophysical Research Letters*, *45*(16), 8665–8672.

Zhang, S., & Lin, G. (2018). Robust data-driven discovery of governing physical laws with error bars. *Proceedings of the Royal Society A: Mathematical, Physical and Engineering Sciences*, *474*(2217), 20180305.

Zhang, X., Wang, T., Chen, G., Tan, X., & Zhu, K. (2019). Convective clouds extraction from Himawari–8 satellite images based on double-stream fully convolutional networks. *IEEE Geoscience and Remote Sensing Letters*, *17*(4), 553–557.

INDEX

Note: Page numbers followed by *f* or *t* indicate material in figures or tables respectively

Clouds and Their Climatic Impacts: Radiation, Circulation, and Precipitation, Geophysical Monograph 281, First Edition.
Edited by Sylvia C. Sullivan and Corinna Hoose.

DOI: 10.1002/9781119700357.bindex

Printed in the USA/Agawam, MA
September 16, 2026

910661.212